COMPARATIVE MAGNETIC MINIMA: CHARACTERIZING QUIET TIMES IN THE SUN AND STARS

IAU SYMPOSIUM No. 286

COVER ILLUSTRATION:

Mendocinean landscape showing a typical vineyard plantation with the Andes mountains in the background. Surrounded by a mixture of arid and semiarid landscapes, the city of Mendoza and its rural outskirts have been turned into a fertile oasis, sustained by the melting of glaciers and snow and manmade dams, channels, and drains. Also called "The land of Sun and good wine", its diaphanous skies and wine-producing fields attract over a million tourists every year.

Our Mendoza IAU Symposium on "Comparative Magnetic Minima" brought together scientists who studied the Sun, stars, and effects of magnetic activity on planetary space environments. One such "space weather" effect is that of beautiful aurorae, as illustrated here on a star field background courtesy of NASA and The Hubble Heritage Team (STScI/AURA). The solar disc image is courtesy of SDO (NASA) and the AIA consortium, while the solar corona is courtesy of Williams College Eclipse Expedition (Jay M. Pasachoff, Muzhou Lu, and Craig Malamut), captured on July 11, 2010.

INTERNATIONAL ASTRONOMICAL UNION

UNION ASTRONOMIQUE INTERNATIONALE

COMPARATIVE MAGNETIC MINIMA: CHARACTERIZING QUIET TIMES IN THE SUN AND STARS

PROCEEDINGS OF THE 286th SYMPOSIUM OF THE INTERNATIONAL ASTRONOMICAL UNION HELD IN THE CITY OF MENDOZA, MENDOZA, ARGENTINA
OCTOBER 3–7, 2011

Edited by

Cristina H. Mandrini
Instituto de Astronomía y Física del Espacio, IAFE, CC. 67, Suc. 28, 1428 Buenos Aires, Argentina

and

David F. Webb
ISR, Boston College, Kenny Cottle, 106A, 885 Centre St., Newton, MA 02459, USA

CAMBRIDGE UNIVERSITY PRESS
The Edinburgh Building, Cambridge CB2 2RU, UnitedKingdom
40 West 20th Street, New York, NY 10011–4211, USA
10 Stamford Road, Oakleigh, Melbourne 3166, Australia

First published 2012

Printed in the United Kingdom at the University Press, Cambridge

Typeset in System LaTeX 2_ε

A catalogue record for this book is available from the British Library

Library of Congress Cataloguing in Publication data

This journal issue has been printed on FSC-certified paper and cover board. FSC is an independent, non-governmental, not-for-profit organization established to promote the responsible management of the worlds forests. Please see www.fsc.org for information.

ISBN 9781107019867 hardback
ISSN 1743-9213

Table of Contents

Preface

IAU Symposium 286, "Comparative Magnetic Minima: Characterizing Quiet Times in the Sun and Stars", was coordinated through Division II, with the strong support of Division IV, including several of their associated commissions. It was held in Mendoza, Argentina, from 3 to 7 October 2011, and attracted nearly 100 scientists expert on various pertinent topics from 23 countries. The goal of the symposium was to consider solar and stellar minima, from generative dynamo mechanisms to in-depth analyses from Sun to Earth for recent well-observed and modeled minima, to a range of stellar cyclic activity, to outlier "grand minima". Solar, heliospheric, geospace, atmospheric, stellar, and planetary sciences were included in the meeting's scope.

Solar and stellar minima represent times of low magnetic activity and simple helio/asterospheres. They are, thus, excellent targets for interdisciplinary, system-wide studies of the origins of stellar variability and consequent impacts on planetary systems. The recent solar minimum extended longer and was "quieter" than any we have observed in the Space Age, inspiring both scientific and public interest. A rich variety of satellite and ground-based observations, in conjunction with theoretical and numerical modeling advances, have allowed us to probe the peculiarities of this minimum as never before. The implications are far-reaching, connecting Earth to Sun to stars, radio to X-rays to cosmic rays, and the plethora of observations of recent minima to the Sun's past behavior as preserved in cosmogenic isotopes and historical sunspot and auroral records.

At the meeting, the keynote talk on "The nature and significance of solar minima" was given by Eric Priest. This was followed by 28 invited, 6 solicited talks and 28 contributed presentations spread over five sessions: Solar and Stellar Minima, Dynamos and Cycle Variability, Comparative Solar Minima from Sun to Earth, Stellar Cycles and Grand Minima, and Historical Records. A closing discussion on whether we are entering a grand minimum was led by Karel Schrijver. Thirty one poster presentations were put up and remained during the entire meeting. A public outreach talk on global warming and solar activity was given by Pablo Mauas at the end of the symposium.

The presentations described how magnetic fields can be cyclically generated in solar and stellar interiors via various dynamo processes. Numerical models have increased in complexity to the point where many observed aspects of the cycles in the Sun and stars are captured, although mysteries remain such as the origins of extended, or "grand" minima. Both stellar observations and historical and cosmogenic records at the Earth were presented, forming a basis of understanding of such intervals, and of solar/stellar long-term variability in general. A simple method to reconcile the Zürich Sunspot Number and the Group Sunspot Number was presented, with important and wide ranging implications towards an agreed-upon and vetted single sunspot series for use in the future.

The recent extended minimum was the lowest and longest minimum in about a century, having weak polar magnetic fields, a complex corona and heliosphere, and recurrent high-speed streams. Simultaneously, it was found that solar minima do not all look alike, given that the Sun can have different magnetic flux configurations even during very quiet times, yielding distinct 3D magnetic flux distributions and, therefore, diverse structure of the corona and heliosphere. During this recent minimum, the solar magnetic field achieved a solar maximum-like corona and solar wind source situation, but with weak magnetic fields and associated weak heating. The discussed results point out the need for textbooks and solar physics educators to revise the way they describe the solar wind and its sources.

In addition, the recent minimum provoked discussions on the possibility of a trend in the Sun's current magnetic cycles towards a grand minimum and the potential implications for the Earth's climate. For instance, there is evidence that a strong decrease of solar activity can lead to a delay of ozone recovery, partially compensating greenhouse warming, and that irradiance variability is the most important forcing for global problems. A combination of the bottom-up and top-down models seems appropriate for radiative solar forcing of the atmosphere. Although the forcing due to anthropogenic influences is about seven times larger than the radiative solar forcing, solar activity certainly does affect climate, and all relevant observations need to be maintained or extended.

The question of the origins and implications of cyclic behavior, for the Sun-Earth system and also for other stellar-planetary systems, was the subject of several presentations. For instance, it was shown that induced magnetospheres directly interact with the solar wind and, therefore, are more prone to atmospheric evolution than intrinsic magnetospheres.

This symposium was undoubtedly unique in the sense that it brought together a diverse group of scientists that were able to take part in discussions, appreciate the scientific disciplines of others, and discover the common aspects of the physical processes involved in the different studied environments from Sun to Earth, and stars to planets. The editors take this opportunity to thank Germán Cristiani and Marcelo López-Fuentes for their valuable assistance in preparing this volume. We also are grateful to the following reviewers who assisted us in improving the papers: Drs. Thomas Ayres, Alisson Dal Lago, Sergio Dasso, Marcelo López-Fuentes, Daniel Gómez, Manuel Güdel, Gustavo Guerrero, Jeffrey Hall, Margit Haberreiter, Kanya Kusano, Georgeta Maris, Leif Svalgaard, Andrey Tlatov, Ilya Usoskin, Adriana Valio, and Alberto Vásquez. Please note that many of the papers contain color figures, which are printed here in black and white but which can be viewed online in color.

Sarah Gibson and Hebe Cremades, co-chairs SOC
Cristina H. Mandrini, chair LOC
Cristina H. Mandrini and David F. Webb, Proceedings Editors
Buenos Aires, Argentina, 29 March 2012

THE ORGANIZING COMMITTEES

Scientific

Hebe Cremades (co-chair, Argentina)
Sarah Gibson (co-chair, USA)
Thomas Ayres (USA)
Alisson Dal Lago (Brazil)
Daniel Gómez (Argentina)
Manuel Güdel (Austria)
Gustavo Guerrero (Sweden)
Margit Haberreiter (Switzerland)
Johanna Haigh (UK)
Jeffrey Hall (USA)
Kanya Kusano (Japan)
Cristina Mandrini (Argentina)
Georgeta Maris (Romania)
Valentín Martínez-Pillet (Spain)
Andrey Tlatov (Russia)
Ilya Usoskin (Finland)
Adriana Valio (Brazil)

Local

Cristina Mandrini (chair)
Laura Balmaceda
Hebe Cremades
Germán Cristiani
Sergio Dasso
Marcelo López-Fuentes
María Luisa Luoni

Acknowledgements

The symposium was coordinated through IAU Division II (Sun and Heliosphere) and sponsored and supported by IAU Divisions III (Planetary System Sciences) and IV (Stars), including several of their associated Commissions: 10 (Solar Activity), 12 (Solar Radiation and Structure), 49 (Interplanetary Plasma and Heliosphere), and 36 (Theory of Stellar Atmospheres).

The Local Organizing Committee operated under the auspices of the Instituto de Astronomía y Física del Espacio (IAFE) and the Universidad Tecnológica Nacional - Facultad Regional Mendoza (UTN-FRM).

Funding support by the
International Astronomical Union (IAU),
Agencia Nacional de Promoción Científica y Tecnológica (ANPCyT),
Consejo Nacional de Investigaciones Científicas y Técnicas (CONICET),
Air Force Office of Scientific Research - Southern Office of Aerospace Research and Development (AFOSR-SOARD),
Scientific Committee on Solar-Terrestrial Physics (SCOSTEP),
Committee on Space Research (COSPAR),
Comisión Nacional de Actividades Espaciales (CONAE),
and
Centro Latinoamericano de Física (CLAF),
is gratefully acknowledged.

CONFERENCE PHOTOGRAPH

1. Lela Taliashvili
2. Blanca Mendoza
3. Inez Batista
4. Werner Schmutz
5. Madhulika Guhathakurta
6. Caius Selhorst
7. Eric Priest
8. Adriana Valio
9. Cristina Mandrini
10. Hebe Cremades
11. Cesar Bertucci
12. María Luisa Luoni
13. Lidia van Driel-Gesztelyi
14. Ximena Abrevaya
15. Mariela Vieytes
16. Bernardo Vargas Cárdenas
17. Sarah Gibson
18. Romina García
19. Nadezhda Zolotova
20. Maximiliano Crescitelli
21. Deysi Cornejo Espinoza
22. Lurdes Martínez Meneses
23. Lois Linsky
24. Accompanying person (Raphael Steinitz)
25. Raphael Steinitz
26. Jaime Osorio Rosales
27. Jeffrey Linsky
28. Mariano Poisson
29. Marlos Rockenbach da Silva
30. Carolina Salas Matamoros
31. Andrey Tlatov
32. Ilya Usoskin
33. Francesco Zuccarello
34. Sasha Brun
35. Michael Thompson
36. Kate Thompson
37. Ivonne Elsworth
38. Hiroko Miyahara
39. David Webb
40. Heidi Korhonen
41. Marcelo López-Fuentes
42. Katja Poppenhaeger
43. Dibyendu Nandi
44. Volker Bothmer
45. Romina Petrucci
46. Eugene Rozanov
47. Alisson Dal Lago
48. Pablo Mauas
49. Ezequiel Echer
50. Margit Haberreiter
51. Carlene Skeffington
52. Ed Cliver
53. Joern Warnecke
54. John Linker
55. Dmitry Sokoloff
56. Matthias Rheinhardt
57. Marc DeRosa
58. Andres Muñoz Jaramillo
59. Leif Svalgaard
60. Laura Balmaceda
61. Pedro Corona Romero
62. Matthew Browning
63. Karel Schrijver
64. Bidya B. Karak
65. Raul Terrazas Ramos
66. Giuliana de Toma
67. Cristian Ferradas Alva
68. Emre Isik
69. Alberto Vásquez
70. Germán Cristiani
71. Steven Saar
72. Jurgen Schmitt
73. Abraham Chian
74. Mark Giampappa
75. Arnab Choudhuri
76. Axel Brandenburg
77. Federico Nuevo
78. Simon Candelaresi
79. José Vaquero
80. Gustavo Guerrero
81. Jenny Rodríguez Gómez
82. Marialejandra Luna Cardozo
83. Fabio del Sordo
84. Fernando López
85. Daniel Gómez
86. María Emilia Ruiz
87. Ramon Lopez
88. Paulo Batista
89. Sergio Dasso
90. Francisco Iglesias
91. Janet Luhman
92. Vera Hurlburt
93. Neal Hurlburt
94. Kazuoki Munakata

Participants

Ximena **Abrevaya**, Instituto de Astronomía y Física del Espacio, IAFE, Buenos Aires, Argentina — abrevaya@iafe.uba.ar

Laura **Balmaceda**, Inst. de Cs. Astronómicas de la Tierra y del Espacio, ICATE, San Juan, Argentina — lbalmaceda@icate-conicet.gob.ar

Inez **Batista**, Instituto Nacional de Pesquisas Espaciais, INPE, São José dos Campos, Brazil — inez@dae.inpe.br

César **Bertucci**, Instituto de Astronomía y Física del Espacio, IAFE, Buenos Aires, Argentina, — cbertucci@iafe.uba.ar

Volker **Bothmer**, Institute for Astrophysics, University of Göttingen, Göttingen, Germany — bothmer@astro.physik.uni-goettingen.de

Axel **Brandenburg**, Nordic Institute for Theoretical Physics, NORDITA, Sweden — brandenb@nordita.org

Matthew **Browning**, CITA, University of Exeter, UK — browning@cita.utoronto.ca

Allan S. **Brun**, CEA, Saclay, France — sacha.brun@cea.fr

Simon **Candelaresi**, Nordic Institute for Theoretical Physics, NORDITA, Sweden — iomsn@physto.se

Abraham **Chian**, Paris Observatory, Meudon, Paris, France — abraham.chian@gmail.com

Arnab **Choudhuri**, Department of Physics, Indian Institute of Science, Bangalore, India — arnab@physics.iisc.ernet.in

Rodolfo **Cionco**, Facultad Regional San Nicolás, Universidad Tecnológica Nacional, Argentina — gcionco@frsn.utn.edu.ar

Edward **Cliver**, Air Force Research Laboratory, USA — edcliver@gmail.com

Deysi **Cornejo Espinoza**, Comisión Nacional de Investigación y Desarrollo Aeroespacial, CONIDA, Perú — veronicadce@gmail.com.pe

Pedro **Corona Romero**, Instituto de Geofísica, UNAM, México — piter.cr@gmail.com

Hebe **Cremades**, Facultad Regional Mendoza, Universidad Tecnológica Nacional, Argentina — hebe.cremades@frm.utn.edu.ar

Maximiliano **Crescitelli**, Facultad Regional Mendoza, Universidad Tecnológica Nacional, Argentina — albertut@hotmail.com

Germán **Cristiani**, Instituto de Astronomía y Física del Espacio, IAFE, Buenos Aires, Argentina — gcristiani@iafe.uba.ar

Alisson **Dal Lago**, Instituto Nacional de Pesquisas Espaciais, INPE, São José dos Campos, Brazil — dallago@dge.inpe.br

Sergio **Dasso**, Instituto de Astronomía y Física del Espacio, IAFE, Buenos Aires, Argentina — sdasso@iafe.uba.ar

Giuliana **De Toma**, High Altitude Observatory, National Center for Atmospheric Research, USA — detoma@ucar.edu

Fabio **Del Sordo**, Nordic Institute for Theoretical Physics, NORDITA, Sweden — brandenb@nordita.org

Ezequiel **Echer**, Instituto Nacional de Pesquisas Espaciais, INPE, São José dos Campos, Brazil — ezequiel.echer@gmail.com

Yvonne **Elsworth**, School of Physics and Astronomy University of Birmingham, UK — y.p.elsworth@bham.ac.uk

Cristian **Ferradas Alva**, Comisión Nacional de Investigación y Desarrollo Aeroespacial, CONIDA, Perú — cristian.ferradas@pucp.edu.pe

Romina **García**, Universidad Nacional de San Juan, San Juan, Argentina — rominita dance@hotmail.com

Mark **Giampapa**, National Solar Observatory, NOAO, USA — giampapa@noao.edu

Sarah **Gibson**, High Altitude Observatory, National Center for Atmospheric Research, USA — sgibson@ucar.edu

Daniel **Gómez**, Departamento de Física, FCEN, UBA, Buenos Aires, Argentina — dgomez@df.uba.ar

Walter **González**, Instituto Nacional de Pesquisas Espaciais, INPE, São José dos Campos, Brazil — gonzalez@dge.inpe.br

Gustavo **Guerrero**, Nordic Institute for Theoretical Physics, NORDITA, Sweden — guerrero@nordita.org

Madhulika **Guhathakurta**, Heliophysics Division, NASA Headquarters, USA — madhulika.guhathakurta@nasa.gov

Heidy **Gutiérrez**, Centro de Investigaciones Espaciales, Universidad de Costa Rica, Costa Rica — heidy.gutierrez@ucr.ac.cr

Margit **Haberreiter**, Physikalisch-Meteorologisches Observatorium Davos, WRC, Switzerland — margit.haberreiter@pmodwrc.ch

Neal **Hurlburt**, Lockheed Martin Advanced Technology Center, USA — hurlburt@Lmsal.com

Francisco **Iglesias**, Facultad Regional Mendoza, Universidad Tecnológica Nacional, Argentina — franciscoaiglesias@hotmail.com

Emre **Isik**, Istanbul Kultur University, Istanbul, Turkey — e.isik@iku.edu.tr

Bidya **Karak**, Department of Physics, Indian Institute of Science, Bangalore, India — bidya karak@physics.iisc.ernet.in

Heidi **Korhonen**, Niels Bohr Institute University of Copenhagen, Copenhagen, Denmark — heidi.h.korhonen@utu.fi

Jon **Linker**, Predictive Science Inc., USA — linkerj@predsci.com

Jeffrey **Linsky**, Joint Institute for Laboratory Astrophysics, University of Colorado, — jlinsky@jila.colorado.edu

Fernando **López**, Universidad Nacional de San Juan, San Juan — ferl1983@hotmail.com

Ramón **Lopez**, Department of Physics, University of Texas at Arlington, USA — relopez@uta.edu

Marcelo **López Fuentes**, Instituto de Astronomía y Física del Espacio, IAFE, Buenos Aires, Argentina — lopezf@iafe.uba.ar

Janet **Luhmann**, Space Sciences Laboratory, University of California, Berkeley, USA — jgluhman@ssl.berkeley.edu

Marialejandra **Luna Cardozo**, Instituto de Astronomía y Física del Espacio, IAFE, Buenos Aires, Argentina — mluna@iafe.uba.ar

María L. **Luoni**, Instituto de Astronomía y Física del Espacio, IAFE, Buenos Aires, Argentina — mluoni@iafe.uba.ar

Cristina H. **Mandrini**, Instituto de Astronomía y Física del Espacio, IAFE, Buenos Aires, Argentina — mandrini@iafe.uba.ar

Lurdes **Martínez Meneses**, Universidad Nacional San Luis Gonzaga de Ica, Perú — lurdesmartinez5@yahoo.es

Pablo **Mauas**, Instituto de Astronomía y Física del Espacio, IAFE, Buenos Aires, Argentina — pablo@iafe.uba.ar

Blanca **Mendoza Ortega**, Instituto de Geofísica, UNAM, México blanca@geofisica.unam.mx
Hiroko **Miyahara**, The University of TokyoLeiden Observatory, Tokio, Japan hmiya@icrr.u-tokyo.ac.jp
Kazuoki **Munakata**, Physics Department, Shinshu University, Japan kmuna00@shinshu-u.ac.jp
Andrés **Muñoz Jaramillo**, Harvard-Smithsonian Center for Astrophysics, USA amunoz@cfa.harvard.edu
Dibyendu **Nandi**, Indian Institute of Science Education and Research, Kolkata, India dnandi@iiserkol.ac.in
Federico **Nuevo**, Instituto de Astronomía y Física del Espacio, IAFE, Buenos Aires, Argentina federico@iafe.uba.ar
Constantin **Oprea**, Institute of Geodynamics, Romanian Academy, Romania const_oprea@yahoo.com
Jaime **Osorio Rosales**, Instituto de Geofísica, UNAM, México jaime@geofisica.unam.mx
Romina **Petrucci**, Instituto de Astronomía y Física del Espacio, IAFE, Buenos Aires, Argentina romina@iafe.uba.ar
Mariano **Poisson**, Instituto de Astronomía y Física del Espacio, IAFE, Buenos Aires, Argentina mpoisson@iafe.uba.ar
Katja **Poppenhaeger**, Hamburg Observatory, Hamburg, Germany katja.poppenhaeger@hs.uni-hamburg.de
Eric **Priest**, Saint Andrews University, UK eric@mcs.st-and.ac.uk
Matthias **Rheinhardt**, Nordic Institute for Theoretical Physics, NORDITA, Sweden mreinhardt@nordita.org
Marlos **Rockenbach da Silva**, Universidade do Vale do Paraíba, UNIVAP, Brazil marlosrs@gmail.com
Jenny **Rodríguez Gómez**, Observatorio Astronómico Nacional, Universidad Nacional de Colombia, Colombia jemfisi@hotmail.com
Eugene **Rozanov**, Physikalisch-Meteorologisches Observatorium Davos, WRC, Switzerland e.rozanov@pmodwrc.ch
María E. **Ruiz**, Instituto de Astronomía y Física del Espacio, IAFE, Buenos Aires, Argentina meruiz@iafe.uba.ar
Steven **Saar**, Smithsonian Astrophysical Observatory, USA saar@cfa.harvard.edu
Carolina **Salas Matamoros**, Centro de Investigaciones Espaciales, Univ. de Costa Rica, Costa Rica carolina.salas@planetario.ucr.ac.cr
Jürgen **Schmitt**, Hamburger Sternwarte, Germany jschmitt@hs.uni-hamburg.de
Werner **Schmutz**, Physikalisch-Meteorologisches Observatorium Davos, WRC, Switzerland werner.schmutz@pmodwrc.ch
Karel **Schrijver**, Lockheed Martin Advanced Technology Center, USA schrijver@lmsal.com
Caius **Selhorst**, Universidade do Vale do Paraíba, UNIVAP, Brazil caiuslucius@gmail.com
Dmitry **Sokoloff**, Moscow State University, Moscow, Russia sokoloff@dds.srcc.msu.su
Raphael **Steinitz**, Ben Gurion University, Israel raphael@bgu.ac.il
Leif **Svalgaard**, Stanford University, USA leif@leif.org
Natalia **Szajko**, Instituto de Astronomía y Física del Espacio, IAFE, Buenos Aires, Argentina pajarin@gmail.com
Lela **Taliashvili**, Centro de Investigaciones Espaciales, Universidad de Costa Rica, Costa Rica lela.taliashvili@cinespa.ucr.ac.cr
Raúl **Terrazas Ramos**, Universidad Nacional San Luis Gonzaga de Ica, Perú raulterrazas81@gmail.com
Michael **Thompson**, High Altitude Observatory, National Center for Atmospheric Research, USA mjt@ucar.edu
Andrey **Tlatov**, Kislovodsk Mountain Astronomical Station, Pulkovo Observatory, Russia tlatov@mail.ru
Ilya **Usoskin**, Department of Physics, University of Oulu, Finland ilya.usoskin@oulu.fi
Adriana **Válio**, CRAAM, Mackenzie University, São Paulo, Brazil avalio@craam.mackenzie.br
Lidia **van Driel-Gesztelyi**, Konkoly Observatory, Hungary Lidia.vanDriel@obspm.fr
José **Vaquero**, Universidad de Extremadura, España jvaquero@unex.es
Bernardo **Vargas Cárdenas**, Instituto de Geofísica, UNAM, México bernardo@geofisica.unam.mx
Alberto **Vásquez**, Instituto de Astronomía y Física del Espacio, IAFE, Buenos Aires, Argentina albert@iafe.uba.ar
Mariela **Vieytes**, Instituto de Astronomía y Física del Espacio, IAFE, Buenos Aires, Argentina mariela@iafe.uba.ar
Joern **Wernecke**, Nordic Institute for Theoretical Physics, NORDITA, Sweden joern@nordita.org
David **Webb**, Institute of Scientific Research, Boston College, USA david.webb@bc.edu
Nadezhda **Zolotova**, Saint Petersburg State University, Saint Petersburg, Russia myagkalapka@gmail.com
Francesco **Zuccarello**, Centrum voor Plasma-Astrofysic, KU Leuven, Belgium francesco.zuccarello@wis.kuleuven.be

Session 1

Solar and Stellar Minima

Comparative Magnetic Minima:
Characterizing Quiet Times in the Sun and Stars
Proceedings IAU Symposium No. 286, 2011
C. H. Mandrini & D. F. Webb, eds.

doi:10.1017/S1743921312004577

The nature and significance of solar minima

Eric Priest

Mathematics Institute, University of St Andrews,
St Andrews, KY16 9SS, UK
email: eric@mcs.st-andrews.ac.uk

Abstract. As an introduction to the theme of this symposium, I give a simple review of the photospheric magnetic field, the properties of the solar cycle, the way in which the magnetic field is thought to be generated by dynamo action, and finally the unusual properties of the recent solar minimum. This has awakened an interest in improving predictions of the solar cycle and in the nature of solar minima not just as gaps between maxima but as phenomena of intrinsic interest in their own right.

Keywords. magnetic fields, MHD, plasmas, Sun: general, Sun: magnetic fields, Sun: photosphere, sunspots

1. Introduction

Many colourful headlines have appeared in newspapers over the past 2 years, such as: the mystery of the missing sunspots; this minimum is weird; is the Sun dead? Sun shows signs of life; Earth may head into a mini-ice-age within a decade; the next ice age - now; 10 reasons to be cheerful about the coming new ice age. So what lies behind the headlines?

As we shall discover, the solar minimum of the Sun is much more interesting than it looks at first sight. Before discussing the recent solar minimum, we need to lay the ground by describing the photospheric magnetic field, the solar cycle and dynamo activity.

2. The Photosphere

The Sun's radius is 700 Mm and the outer 30% of the interior is a turbulent convection zone. The atmosphere consists of the photosphere (the top of the convection zone), the chromosphere and corona.

The photosphere itself is covered with turbulent convection cells, namely, granulation on scales of 1 Mm and supergranulation on scales of 15–30 Mm. Global photospheric magnetographs reveal active regions forming a bipolar pattern of one sign in a band in the northern hemisphere and one of opposite sign in the southern hemisphere. These represent the effect of two large flux tubes below the surface, segments of which occasionally emerge through the surface to give the active regions. One flux tube is directed to the right and the other to the left.

In the core of complex active regions are to be found sunspots, but the sunspots represent only a fraction of the photospheric flux. In addition, the whole surface is covered with tiny intense flux tubes that are carried to the edges of supergranule cells and accumulated there.

In the 1980s and 90s the general picture outside active regions was of the photospheric magnetic field being mainly vertical and mainly located in supergranulation boundaries (Fig. 1) and with a hint of unresolved flux inside supergranules (Livingston & Harvey, 1971). However, this paradigm has now been changed. High-resolution photospheric observations in white light show tiny bright points and lines located between granules.

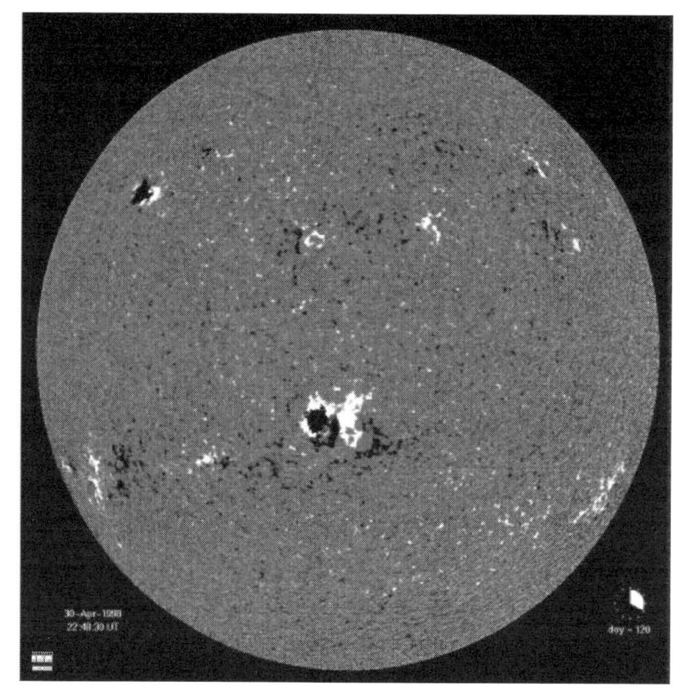

Figure 1. A global magnetogram of the Sun with white and back showing regions where the magnetic field is pointing towards and away from you (from the MDI instrument on the ESA/NASA SoHO spacecraft).

Figure 2. Local magnetograms of part of the quiet Sun, showing the line-of-sight magnetic field when the threshold for magnetic flux is (left) 100 Gauss and (right) 25 gauss (from the Hinode spacecraft), courtesy of Bruce Lites.

Furthermore, with Hinode if you reduce the threshold magnetic flux you see more and more line-of-sight flux, with a huge amount in the interiors of supergranules (Fig. 2). An even greater surprise was the discovery of transverse (horizontal) magnetic fields located at the edges of granules (Lites et al. 2008).

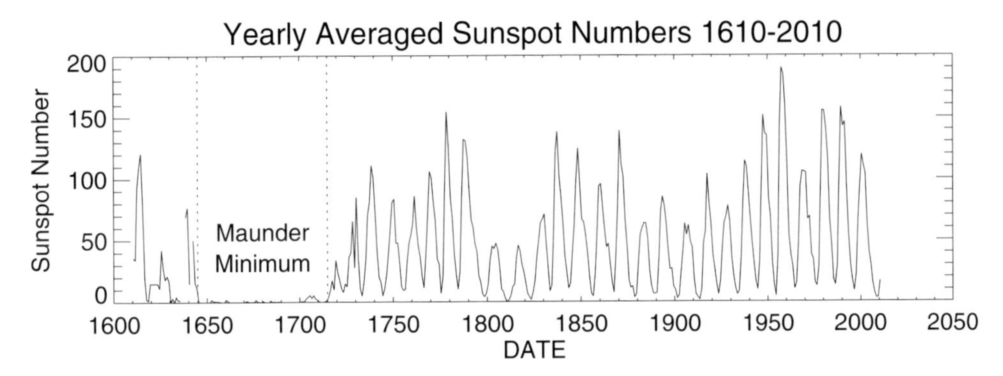

Figure 3. The yearly averaged sunspot number from 1610 to 2010, showing the 11-year cycle with a long-term modulation. The Maunder minimum of 1645–1715 had virtually no sunspots and the Dalton minimum of 1800–1820 had very few (courtesy David Hathaway).

The Sunrise balloon mission has for the first time resolved these kilogauss vertical fields and has detected quiet-Sun magnetic fields with a flux that is lower by a factor of 10 than Hinode (i.e., down to 2×10^{15} Mx) (Solanki *et al.* 2010). Consequently, it sees ten times as many features as Hinode.

Images of sunspots from space and from the Swedish Solar Telescope show much fine structure and puzzling behaviour in the penumbra. However, impressive computational models by Rempel (2011) have recently led to a breakthrough in understanding with amazingly realistic-looking images and a realisation that all the observed features are a natural consequence of convection in an inclined magnetic field.

In the convection zone, there are two important gobal effects of rotation on compressible turbulence. The first is the appearance of strong *differential rotation* in which the equator rotates much faster than the polar regions, with periods of 25.4 days rather than 36 days. The main drivers are the Reynolds stresses ($< v_r v_\phi >$ and $< v_\theta v_\phi >$) which produce angular momentum fluxes. Of particular importance is the tilting of convection cells by Coriolis forces, especially in the downflowing plumes.

The second effect of rotation on turbulence is the creation of a weak *meridional flow* towards the poles at the photosphere of strength 20 m s^{-1}. This is due to small departures from magnetogeostrophic balance between large terms (namely, buoyancy, Reynolds stresses, pressure gradients and Coriolis forces) as well as small temperature differences between pole and equator.

3. The Solar Cycle

The number of sunspots oscillates with an 11.1-year cycle, with the period varying between 8 and 15 years and the maximum varying substantially. When the rise phase is faster then the cycle tends to be larger and longer. Also, there is a long-term modulation known as the Gleissberg cycle. Around 1910, 1810 and 1710 the maxima were smaller than average and the minima deeper (see talks by Svalgaard and Miyahara).

Jack Eddy realised that there were hardly any sunspots at all in most of the 17th century (Fig. 3), a period of 70 years from 1645 to 1715 known as the *Maunder minimum.* This period was also known as the Little Ice Age, since the climate of Europe was considerably cooler than normal with the river Thames occasionally freezing over.

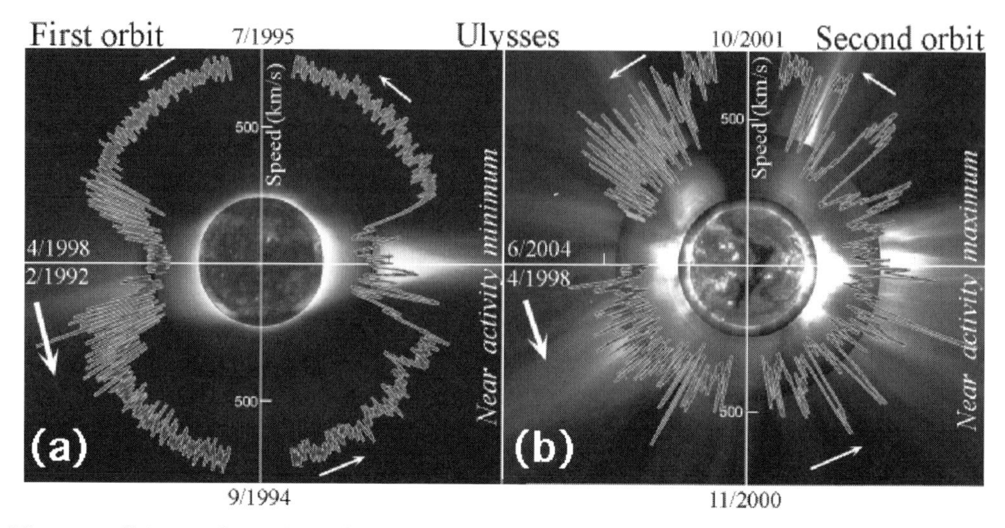

Figure 4. Solar wind speed as a function of latitude in a polar plot from the ULYSSES spacecraft during (a) solar minimum and (b) solar maximum. Superimposed are coronal images from the Mauna Loa K coronameter, plus SoHO EIT and LASCO (from Meyer 2007).

Detailed sunspot numbers started in 1750, but solar cycle variations can also be seen much further back for 30,000 years in ^{10}Be ice cores and ^{14}C tree rings (see the talks by Saar and Usoskin).

So, how typical is the current solar cycle behaviour? Well it depends how far you look – looking back 400 years the current modern maximum is unusual (but see L Svalgaard's presentation here). However, looking back say 10,000 years, the recent maximum and minimum are very common and in fact 9000 years ago the maxima were considerably greater than the recent maxima.

It is interesting to note that the ^{10}Be oscillation continues through the Maunder minimum, so that the magnetic cycle did not switch off but just reduced in strength so that sunspots could not form readily (Beer *et al.* 1998). Furthermore, it has recently been shown that the decline into the Maunder minimum was gradual rather than sudden (Vaquero 2011). The cause of the Maunder minimum may be *intermittency*, when there is a random change from one state of behaviour to another (e.g., Gomez & Mininni, 2006). This may be due to stochastic noise, nonlinearities, threshold effects or time delays (e.g., Charbonneau 2010).

During the solar cycle the whole solar atmosphere varies, not just the sunspot number. For example, the chromosphere has quite a different appearance at solar minimum and solar maximum, as does the global magnetic field revealed in white-light eclipse images: thus, at solar minimum the corona has a dipole shape with prominent open plumes at the poles and helmet streamers at the equator, whereas at solar maximum the corona is much more isotropic with streamers stretching out from all latitudes (see the talks by Tlatov and Vasquez). Furthermore, the coronal intensity in soft x-rays increases by a factor of a hundred from minimum to maximum. In addition, the corona is much more highly structured and varied at solar maximum.

The solar wind velocity varies with the solar cycle. At a normal solar minimum, there are long-lasting fast solar wind streams of 700 km s^{-1} spreading over a large angle from both poles and sporadic slow solar wind at 300 km s^{-1} from large equatorial streamers (Fig. 4a). At solar maximum, the corona is much more isotropic with mixed fast and

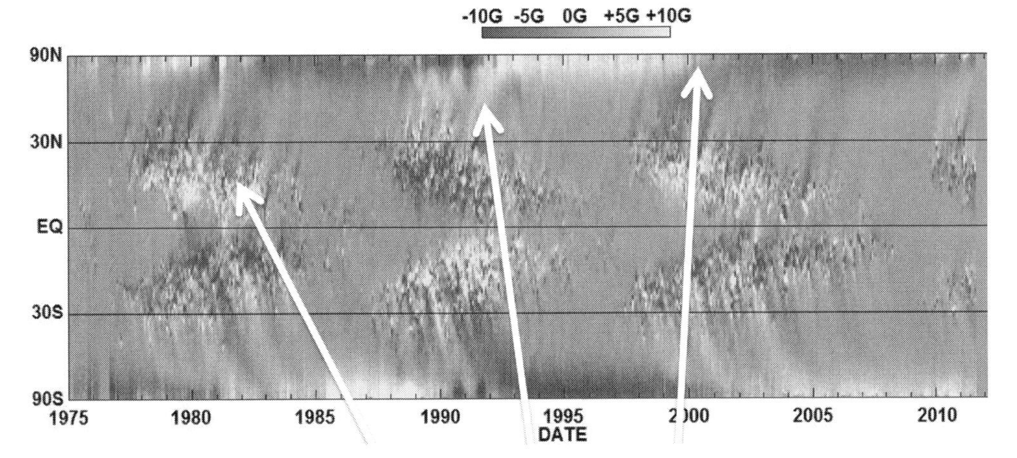

Figure 5. A magnetic butterfly diagram showing the variation in sunspots of positive (light) and negative polarity (dark) as a function of time from 1975. A series of arrows (from left to right) indicate: the migration of sunspots towards the equator; the poleward migration of trailing flux by a meridional flow; and the reversal of the polar field about 2 years after sunspot maximum (courtesy David Hathaway).

slow streams at all latitudes (Fig. 4b). As we shall see in this conference, the present solar minimum has a very different appearance from normal (see talk by de Toma). The interplanetary open magnetic flux also varies with the cycle and was much lower than normal in the recent solar minimum. Again, along with the variations in active regions, the solar cycle produces an oscillation in the locations and frequency of solar flares, prominences and coronal mass ejections (see the talks by Cliver, Cremades, Webb, Bothmer, Gibson).

Magnetic butterfly diagrams such as Fig. 5 are highly revealing. They show how the sunspots migrate equatorward during the solar cycle, and indicate the polewards migration of trailing flux, especially near sunspot maximum. This leads to a reversal of the polar field about a couple of years after maximum, which is also clearly visible.

Finally, there are many effects of the solar cycle at Earth, including aspects of space weather, geomagnetic activity, cosmic rays, the Earth's atmosphere and climate (see talks by Luhmann, Echer, Kazuoki, Munakata, Rozanov, Batista, Guhathakurta, Mendoza and Bertucci). Indeed, Lockwood (2010) has shown that the Sun cannot be the main cause of the present increase in the global temperature of the Earth, since currently the effect of the Sun is declining rather than increasing. Also, Feulner & Rahmstorf (2010) has estimated that, even if a Maunder minimum were taking place just now, its effect would be to decrease the global temperature by only 0.3 ^{o}C by 2100, which is far smaller than the temperature increase expected from greenhouse gases emitted by humans.

4. Generating the Magnetic Field by a Dynamo

The dynamo problem is an interesting nonlinear example of regular behaviour with turbulent or chaotic aspects. After Cowling (1934) had shown that generating an axisymmetric magnetic field is impossible, Parker (1955a, 1955b) made conceptual breakthroughs by showing how flux tubes rise by *magnetic buoyancy* and by suggesting how

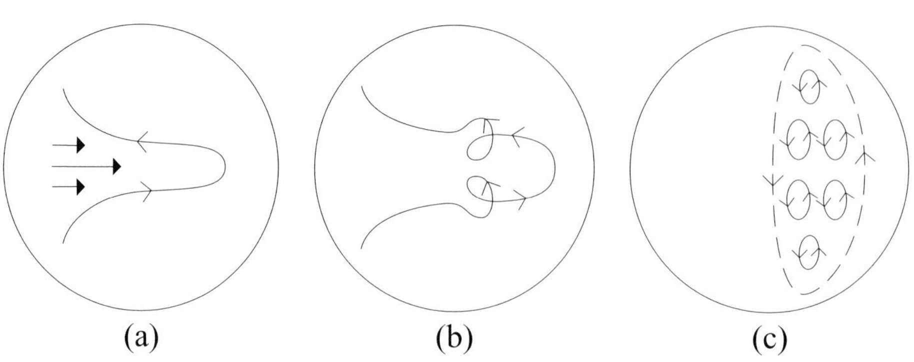

Figure 6. Parker's dynamo model in which: (a) toroidal flux is generated from poloidal flux by differential rotation (the ω-effect); (b) the effect of two helically rising blobs on the toroidal field; (c) the merging of the resulting closed loops of many cyclonic eddies to give new large-scale (dashed) poloidal flux (the α-effect).

both the toroidal and poloidal field components (B_{tor}, B_{pol}) could be generated. His idea was that differential rotation generates a toroidal field from a poloidal field (the ω-effect) and turbulent cyclonic convection in turn generates poloidal field from a toroidal one (the *α-effect*) (Fig. 6). He modelled the latter process by a term of the form $\nabla \times (\alpha \mathbf{B})$, although he used the notation Γ in place of α. This physical idea was formalised as *mean-field theory* by Steenbeck *et al.* (1966) and Moffatt (1978), who wrote the magnetic field as the sum of a large-scale mean field and a small-scale turbulent field in the MHD equations. A key result of the theory is that an angular velocity that increases with depth ($d\Omega/dr < 0$) is required to give migration of dynamo activity towards the equator.

In the 1980's, cracks started appearing in the above framework for producing B_{pol} by the α-effect, although the ω-effect remains accepted to this day as the mechanism for generating B_{tor}. It was realised that the properties of emerging fields, such as the latitudes of emergence and the tilts of bipoles, require fields of 10^5 Gauss, but these would rise through the convection zone very quickly and would be resistant to turbulence, so that the α-effect would stall. Also, global simulations failed to give solar-like dynamos and doubts appeared about the validity of mean-field theory and the derivation of the α-effect, in particular the assumption that the fluctuating fields are much smaller than the mean field.

However, the final nail in the coffin of the previous theory appeared when helioseismology showed that the angular velocity is constant with radius ($d\Omega/dr = 0$) in the convection zone rather than increasing outwards as required by the theory. It had long been known that the solar rotation at the surface increases from poles to equator and had been expected that in the solar interior the rotation would be constant on cylinders and the magnetic field would be generated throughout the convection zone. Surprises from helioseismology were that the angular velocity is instead constant on cones and that the rotation below the convection zone is uniform, so that there is a strong shear layer, called the *tachocline*, at the base of the convection zone. This is now thought to be the site of the main dynamo that produces active regions and sunspots. However, there may well be another dynamo just below the photosphere that generates the small-scale magnetic field seen in ephemeral regions and intense flux tubes at the edges of granules and supergranules.

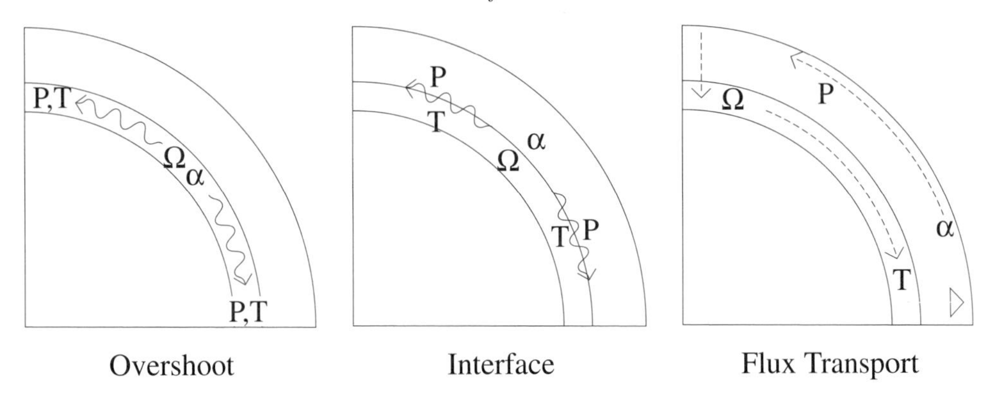

Figure 7. Magnetic field generation by overshooting, interface and flux-transport dynamos, indicating where the α- and ω-effects are located and where poloidal (P) and toroidal (T) components are generated. Curly, dashed and double arrows represent transport by a dynamo wave, meridional flow and buoyant flux emergence, respectively.

In the 1990's, two new ideas were proposed for generating a poloidal field from a toroidal one (Fig. 7). The first is by some kind of *tachocline dynamo* at the tachocline, namely, either locating both the α-effect and ω-effect in the overshoot region just below the base of the tachocline or in an interface dynamo separating these effects spatially and placing the ω-effect below the interface and the α-effect above it (Parker 1993, Charbonneau 1978). The second idea was to propose a *flux-transfer dynamo* that develops the earlier Babcock (1961)-Leighton (1969) dynamo by solving the axisymmetric kinematic dynamo equations with an imposed meridional flow and an ω-effect focussed near the tachocline together with an α-effect at the solar surface (Choudhuri 1995, Dikpati 1994, Charbonneau 1997, Nandi 2002); Mininni & Gomez 2002).

Many other effects are potentially important in dynamo theories, such as: shear instabilities, magnetic buoyancy instabilities, flux tube instabilities in the tachocline or the overshoot layer; the back-reaction of the Lorentz force on the flow and the efficiency of the α-effect; time delays or stochastic forcing to modulate the dynamo; and a proper treatment of sub-grid physics in numerical experiments (see the talk by Brun and the living review by Charbonneau (2010)).

Full MHD global computations have been conducted by a number of authors, including Brun (2004) and Ghizaru (2010). They are now able to resolve supergranulation and generate reasonable behaviour for differential rotation and meridional circulation, as well as a turbulent α-effect and reversals of the magnetic fields (see talks by Brandenburg and Browning).

Predicting the solar cycle is, however, a tough endeavour. Many methods have been employed, including climatalogical effects, dynamo theory, neural networks, polar fields and geomagnetic indices. The maximum sunspot numbers predicted for cycle 24 have ranged between 40 and 170 among the 75 or so attempts. For example, flux transport dynamo theory has been used by Dikpati *et al.* (2006) and Choudhuri *et al.* (2007). They adopted different values for the magnetic diffusivity, differential rotation, meridional circulation, poloidal flux source and alpha quenching. The former predicted a strong cycle with a maximum of about 140, whereas the latter predicted a weak cycle with a maximum of about 80. David Hathaway is one of the experts at predictions and the way in which his predictions have varied in time is illustrated in Fig. 8.

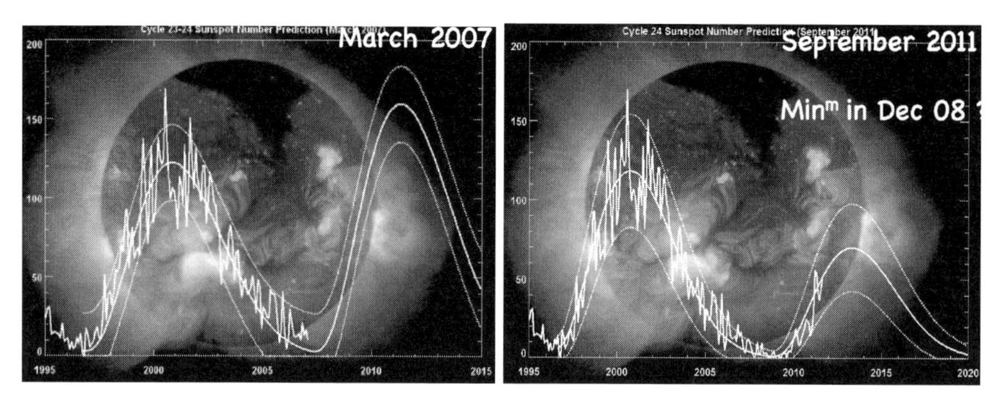

Figure 8. Predictions of the future sunspot number in (a) March 2007 and (b) September 2011 (courtesy David Hathaway).

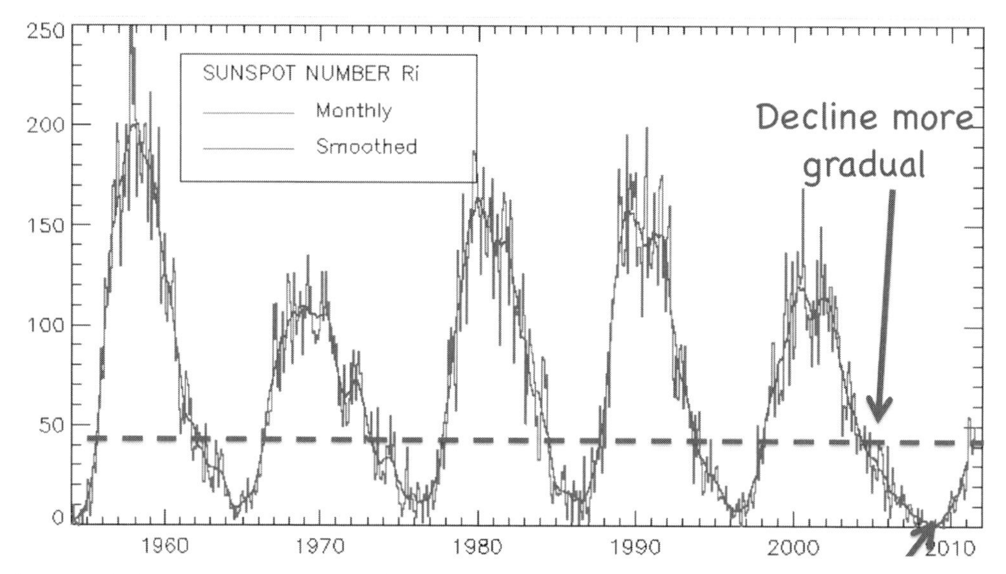

Figure 9. The recent sunspot number, indicating the slow decline into the mimimum and the depth and length of the mimimum (from Solar Influences Data Centre, Brussels).

5. How has this Solar Minimum been Different?

The present solar minimum has been unusual in many ways. The sunspot number (Fig. 9) illustrates how the decline in sunspot number into the minimum was much more gradual and how the minimum itself was much deeper and longer than in the previous few cycles. In 2008, 75% of the days were spotless, whereas in 2009 this figure rose to 90%. The duration of the last cycle was 12.6 years, the longest for 100 years and the next maximum could well be the lowest for 200 years.

The butterfly diagram shows that usually the cycles overlap, with the new sunspots appearing at high latitudes at the start of a new cycle at the same time as spots from the old cycle are still appearing at low latitudes. However, this was not the case this time, since a clear gap between the two cycles has been present. The same was true back in 1900. The magnetic butterfly diagram also shows how the sunspots have approached

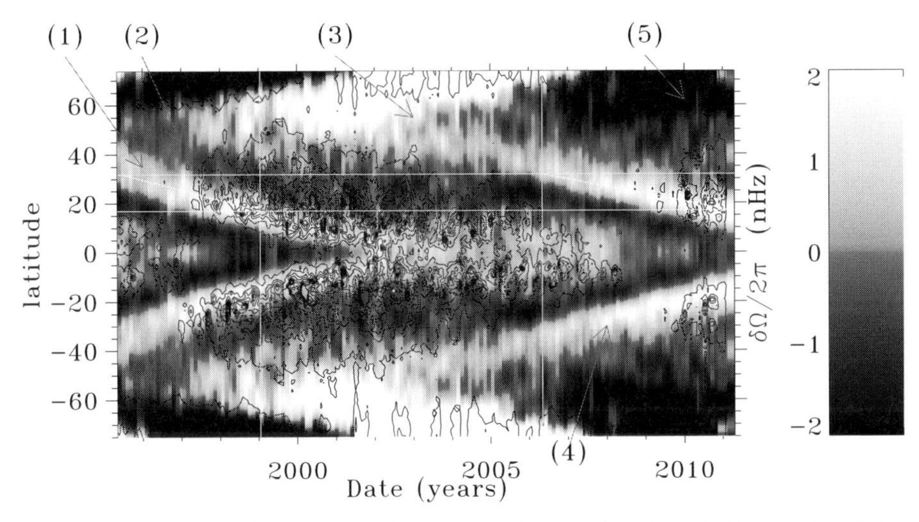

Figure 10. Torsional oscillations showing alternating bands (1) and (2) rotating faster and slower than normal, the migration of the bands (4) from mid-latitudes to equator and the fact that the polar branch is late in starting (compare (3) and (5)) (courtesy Rachel Howe).

the equator much more slowly in the last cycle and the polar fields have been much weaker.

The mean interplanetary open magnetic flux varies with the cycle, and its value at solar minimum fell from 3.82×10^{16} Wb in 1987 to 1.98×10^{14} Wb in 2007 (Lockwood *et al.* 2009). (See the talks by de Toma and Dasso for the way the global and interplanetary fields have varied.)

For sunspots, their brightness varies in phase with the solar cycle, while their radius is independent of the cycle. Recently, Livingston & Penn (2009) have suggested that the magnetic field strength in sunspots is weakening by 50 Gauss per year. If the variation is a straight line and continues in future, they point out that it will fall below the value of 1500 Gauss needed for sunspots to form in 2020. However, there is a large scatter in the data, which may also be fit by a curve that reaches a minimum in future and then increases.

Furthermore, the total solar irradiance has been lower during the recent minimum than in the previous two (see talk by Schmutz), although the mechanism is unclear.

An intriguing recent discovery is that differential rotation varies with the solar cycle, as shown in Fig.10 (Howe *et al.* 2009, 2011). Alternating bands of rotation that are faster and slower than normal are located polewards of the active-region belts and migrate from mid-latitudes towards the equator. By comparison with the previous cycle, the polar branch has been late starting at this minimum (see talk by Thompson). The meridional flow varies too: Mount Wilson observations for the last two cycles show that during cycle 22 there was a counter-flow away from the poles as well as the normal flow towards the poles, whereas in cycle 23 there was no counter-flow. Although meridional flow can so far be measured only in the top 15 Mm of the convection zone, this suggests that perhaps the meridional flow formed a double-cell pattern in cycle 22 but a single-cell pattern in cycle 23 (Dikpati 2010).

The effect of a variable meridional flow on flux transport dynamos has been evaluated by Nandi, Muñoz-Jaramillo & Martens (2010). Their simulated butterfly diagram (Fig.11) indicates how a fast meridional flow produces no overlap in cycles, whereas a

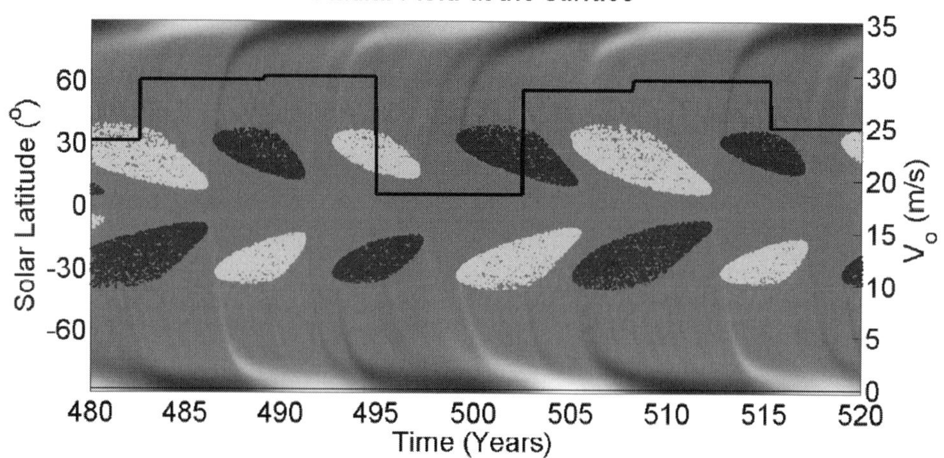

Figure 11. A butterfly diagram produced by a flux-transport model in the which the meridional flow is imposed to vary in a manner indicated by the dark black line (courtesy Dibyendu Nandi and Andres Muñoz).

slow flow makes the cycles overlap. What happens in their scenario is that the value of the flow in the rise phase affects the cycle overlap and the polar field at solar minimum. Thus, a fast meridional flow sweeps the poloidal field more rapidly along the base of the tachocline, so that a weaker toroidal field is built up, leading to a smaller sunspot cycle.

6. Conclusion

For the solar dynamo, great progress has been made over the past few years with many new ideas. There is a healthy tension between the pure dynamo theorists, who for example wonder about the validity of mean-field theory, and the applied dynamo theorists who are more motivated by an attempt to explain observed features of the solar cycle. Clearly, both approaches are needed for a full understanding. However, many aspects are unclear: where and by what kind of alpha-effect is the toroidal field converted into poloidal flux? what is the relation between sunspots and the strength of a magnetic cycle? what value should be put on the strength of the turbulent magnetic diffusivity?

For the solar cycle, it is not clear what is the best way of defining solar minimum. Also, it is a challenge to be able to predict the values and dates of the next solar maximum and solar minimum. When will the next Maunder minimum take place? Also, what is the precise effect of the Sun on the Earth's climate?

Furthermore, the present solar minimum has revealed many unusual features that highlight fundamental problems in understanding about the nature of the dynamo and the solar cycle. I look forward with great anticipation to hearing the latest results and ideas during this symposium and hopefully to Karel Schrijver's answers to some of our questions. In the meantime, let's have fun showing just how interesting the solar minimum is.

References

Babcock, H. W. 1961, *Astrophys. J.*, 133, 572–587.

Beer, J., Tobias, S., & Weiss, N. O. 1998, *Solar Phys.*, 181, 237–249.

Brun, A. S., Miesch, M. S., & Toomre, J. 2004, *Astrophys. J.*, 614, 1073–1098.

Charbonneau, P. & MacGregor, K. B. 1997, *Astrophys. J.*, 486, 502–520.

Charbonneau, P., Beaubien, G., & St-Jean, C. 2007, *Astrophys. J.*, 658, 657–662.

Charbonneau, P. 2010, *Living Reviews in Solar Physics*, 7, 3.

Choudhuri, A. R., Schüssler, M., & Dikpati, M. 1995, *Astron. Astrophys.*, 303, L29–L32.

Choudhuri, A. R., Chatterjee, P., & Jiang, J. 2007, *Phys. Rev. Letts.*, 98, I. 13, id. 131103.

Cowling, T. G. 1934, *Mon. Not. Roy. Astron. Soc.*, 94, 39–48.

Dikpati, M. & Choudhuri, A. R. 1994, *Astron. Astrophys.*, 291, 975–989.

Dikpati, M. & Gilman, P. A. 2006, *Astrophys. J.*, 649, 498–514.

Dikpati, M., Gilman, P. A., & Ulrich, R. K. 2010, *Astrophys. J.*, 722, 774–778.

Feulner, G. & Rahmstorf, S. 2010, *Geophys. Res. Lett.*, 370, L05707.

Ghizaru, M., Charbonneau, P., & Smolarkiewicz, P. K. 2010, *Astrophys. J. Letts.*, 715, L133–L137.

Gómez, D. O. & Mininni, P. D. 2006, *Advances in Space Research*, 38, 856–861.

Howe, R., Christensen-Dalsgaard, J., Hill, F., Komm, R., Schou, J., & Thompson, M. J. 2009, *Astrophys. J. Letts.*, 701, L87–L90.

Howe, R., Hill, F., Komm, R., Christensen-Dalsgaard, J., Larson, T. P., Schou, J., Thompson, M. J., & Ulrich, R. 2011, *Journal of Physics Conference Series 271,* 012074.

Leighton, R. B. 1969, *Astrophys. J.*, 156, 1–26.

Lites, B. W., Kubo, M., Socas-Navarro, H., Berger, T., Frank, Z., Shine, R., Tarbell, T., Title, A., Ichimoto, K., Katsukawa, Y., Tsuneta, S., Suematsu, Y., Shimizu, T. & Nagata, S. 2008, *Astrophys J.*, 672, 1237.

Livingston, W. & Harvey, J. 1971, IAU Symposium 43: *Solar Magnetic Fields*, 51.

Livingston, W. & Penn, M. 2009, *EOS Transactions*, 90, 257–258.

Lockwood, M., Owens, M., & Rouillard, A. P. 2009, *J. Geophys. Res.*, 114, A111014.

Lockwood, M., Harrison, R. G., Woollings, T., & Solanki, S. K. 2010, *Environmental Research Letters 5,* 2 (Apr.), 024001.

Meyer-Vernet, N. 2007, *Basics of the Solar Wind*, Cambridge University Press (Cambridge UK)

Mininni, P. D. & Gómez, D. O. 2002, *Astrophys. J.*, 573, 454–463.

Moffatt, H. K. 1978, *Magnetic Field Generation in Electrically Conducting Fluids*, (Cambridge University Press, Cambridge, England).

Muñoz-Jaramillo, A., Nandy, D., & Martens, P. C. H. 2009, *Astrophys. J.*, 698, 461–478.

Muñoz-Jaramillo, A., Nandy, D., Martens, P. C. H., & Yeates, A. R. 2010, *Astrophys. J. Letts.*, 720, L20–L25.

Nandy, D. 2006, *J. Geophys. Res.*, 111, A12S01.

Nandy, D. & Choudhuri, A. R. 2002, *Science*, 296, 1671–1673.

Nandy, D., Muñoz-Jaramillo, A., & Martens, P. C. H. 2011, *Nature*, 471, 80–82.

Parker, E. N. 1955a, *Astrophys. J.*, 121, 491–507.

Parker, E. N. 1955b, *Astrophys. J.*, 122, 293–314.

Parker, E. N. 1993, *Astrophys. J.*, 408, 707–719.

Rempel, M. 2011, *Astrophys. J.*, 729, 5.

Solanki, S. K., Barthol, P., Danilovic, S., Feller, A., Gandorfer, A., Hirzberger, J., Riethmüller, T. L., Schüssler, M., Bonet, J. A., Martínez Pillet, V., del Toro Iniesta, J. C., Domingo, V., Palacios, J., Knölker, M., Bello González, N., Berkefeld, T., Franz, M., Schmidt, W., & Title, A. M. 2010, *Astrophys. J. Letts.*, 723, L127–L133.

Steenbeck, M., Krause, F., & Rädler, K. H. 1966, *Z. Naturforsch*, 21a, 369–376.

Vaquero, J. M., Gallego, M. C., Usoskin, I. G., & Kovaltsov, G. A. 2011, *Astrophys. J. Letts.*, 731, L24.

Discussion

MICHAEL THOMPSON: I understand small scale flux closes in the lower atmosphere. As a rule of thumb, how high in the atmosphere would you say small-scale flux extends?

ERIC PRIEST: In 2003, Rob Close, Clare Parnell, Duncan MacKay and myself calculated potential field lines in the quiet Sun for MDI magnetograms and found that 50% of the flux closed within 2.5 Mm of the photosphere and 90% within 25 Mm. This needs to be redone with the greater resolution of HMI on SDO. One consequence, as Schrijver has emphasised is that the canopy over a supergranular cell is punctured randomly by field lines that extend up into the corona from the cell interior.

Comparative Magnetic Minima:
Characterizing quiet times in the Sun and Stars
Proceedings IAU Symposium No. 286, 2011
C. H. Mandrini & D. F. Webb, eds.

doi:10.1017/S1743921312004589

Solar and stellar activity: diagnostics and indices

Philip G. Judge and Michael J. Thompson

High Altitude Observatory, National Center for Atmospheric Research
PO Box 3000, Boulder CO 80307-3000, USA
email: judge@ucar.edu, mjt@ucar.edu

Abstract. We summarize the fifty-year concerted effort to place the "activity" of the Sun in the context of the stars. As a working definition of solar activity in the context of stars, we adopt those *globally–observable* variations on time scales below thermal time scales, of $\sim 10^5$ yr for the convection zone. So defined, activity is dominated by magnetic–field evolution, including the 22–year Hale cycle, the typical time it takes for the quasi-periodic reversal in which the global magnetic–field takes place. This is accompanied by sunspot variations with 11 year periods, known since the time of Schwabe, as well as faster variations due to rotation of active regions and flaring. "Diagnostics and indices" are terms given to the *indirect* signatures of varying magnetic–fields, including the photometric (broad-band) variations associated with the sunspot cycle, and variations of the accompanying heated plasma in higher layers of stellar atmospheres seen at special optical wavelengths, and UV and X-ray wavelengths. Our attention is also focussed on the theme of the Symposium by examining evidence for deep and extended minima of stars, and placing the 70–year long solar Maunder Minimum into a stellar context.

Keywords. techniques: photometric, techniques: spectroscopic, Sun: magnetic fields, Sun: UV radiation, Sun: chromosphere, Sun: helioseismology, stars: activity, stars: evolution

---

1. Motivations: Do we remotely understand the Sun?

Globally, the Sun appears to a remarkably constant star. RMS variations in irradiance (observed bolometric flux), as measured from Earth orbit since 1978, are $\approx 0.04\%$. The theory of stellar structure and atmospheres has reached a remarkable level of agreement with critical measurements, now including the remarkable tool of astero-seismology of solar-like stars. While there remain important debates, such as the apparent disagreement between solar interior abundances derived from helioseismology and surface abundances derived from spectroscopy, we have nevertheless gained confidence that our basic theory and understanding has withstood many onslaughts and experimental challenges. A particular success is the resolution of the "solar neutrino problem" in terms of particle, not solar physics.

The above statements might by some be considered a reasonable summary of solar physics. The global and long time scale solar behavior might even be considered a "dead" subject for most astrophysicists, were it not for one inconvenient fact. The 0.04% RMS variations occur on time scales of decades and less, orders of magnitude smaller than the thermal relaxation (Kelvin-Helmholz) timescale of 10^5 years for the Sun's convection zone. This is a further 4 orders of magnitude smaller than the "diffusion time" for global magnetic–fields. Remarkably, overwhelming evidence indicates that the solar magnetic–field, evolving *globally on decadal time scales*, is the culprit.

The global Sun is well described by magneto-hydrodynamics (MHD), in which equations of hydrodynamics are coupled to an equation for magnetic–field evolution (the "induction equation") because the solar plasma is highly conducting. The system of MHD equations is highly non-linear; fluid motions generate electric currents which generate magnetic–fields; magnetic–fields with the electric currents act through the Lorentz force on the fluid, and so forth. Because of enormous physical scales, inductance effects dominate the electromagnetic–fields. Because the plasma is highly conducting, steady electric fields are essentially zero; all the EM energy lies in the magnetic–field. By radius, the outer 30% of the Sun is fully turbulent, as radiation is unable to carry the energy flux and thermal convection takes over. Just beneath the observed photosphere, turbulent fluid motions carry all of this energy flux. Under these conditions we would expect that the Sun would exhibit the rich landscape of non-linear phenomena, including chaotic behavior. Such phenomena are of course observed in the form of small scale, dynamic granulation. However, when we view the behavior of the global solar magnetic–field in this fashion, several questions come to mind. For example, why should the Sun's magnetic–field appear so prominently in the form of intense concentrations- sunspots? Why does the solar cycle have so much *order* (quasi-cycling behavior; Hale's polarity law, Joy's law of tilt of sunspot bipoles, active longitudes)? Why does the global field reverse every 22–years? These are profound, unanswered questions of solar physics, related in some way to the order imposed by the (differential) solar rotation (Parker 1955). When viewed in terms of first principles, such ordered behavior is surely *unexpected.* As we will see below, many solar-like stars do not exhibit this level of order. In this sense, some components of solar magnetic variability, such as extended and Maunder-like minima studied at this Symposium, are just some examples of stochastic behavior in our limited historical record of our non-linearly varying star.

To add insult to injury, the extent of our ignorance the Sun's variable magnetism is highlighted by recent sobering results. Brown *et al.* (2010) made numerical experiments of "rapidly rotating Suns". General consensus was that long-lived, ordered fields in stellar interiors, needed to explain the order in sunspot behavior, should exist should exist mostly outside convection zones. Yet Brown and colleagues found coherent "wreaths" of magnetic–field living entirely *within* highly turbulent convection zones, for many convective turnover times. Further, in the abstract of Brown *et al.* (2011), we read

> "Striking magnetic wreaths span the convection zone and coexist with the turbulent convection. A surprising feature of this wreath-building dynamo is its rich time dependence. The dynamo exhibits cyclic activity and undergoes quasi-periodic polarity reversals where both the global-scale poloidal and toroidal fields change in sense on a roughly 1500 day timescale. These magnetic activity patterns emerge spontaneously from the turbulent flow.."

Given this state of affairs, we review solar-stellar research to shed light on these basic issues, in special relation to states of minimum magnetic activity, such as the Sun's recent extended minimum and the Maunder Minimum. The stars offer the opportunity to "run the solar experiment again", with the caveats that (1) no two stars are identical, and (2) that we only observe the Sun from our special viewpoint in the ecliptic plane, only 7° from the solar equatorial plane.

2. The need for diagnostics and indices

"Diagnostics and indices" – proxies for magnetic activity on solar-like stars – are required because "direct" measurement of magnetism of the Sun-as-a-star is difficult. Measurements of stellar magnetic–fields are based almost exclusively upon the polarization of spectral lines induced through the Zeeman effect, or the increased width of certain Zeeman-sensitive lines, because most lines are not fully split (Zeeman splitting is less than linewidths). Hemisphere-integrated Zeeman signals for stars of *solar type* (mass, age, rotation rate) are particularly difficult to measure. Considering the Sun itself, the peak-to-peak, disk-integrated variation of polarized light from the kilo-Gauss sunspot fields varying with the solar cycle is estimated to be equivalent to a mere 2 Mx cm^{-2} average flux density (Plachinda & Tarasova 2000). For typical spectral lines at visible wavelengths, this leads to a tiny polarization (a few times $\sim 10^{-4}$)†.

It is important to remind oneself that remotely sensed "magnetic–field strengths" (in units of Gauss or Tesla) through spectral lines, even in the Sun, are direct measurements only when spectral lines are fully split (Zeeman splitting > line widths, usually dominated by Doppler broadening). So direct measurements are possible for very intense fields and at longer (infrared) wavelengths, but generally this is not the case in practice. Solar "magnetograms" generated by ground- and spaced- based instruments exclusively work in the unsplit regime, where the first order polarization signature is proportional to $\int \mathbf{B} \cdot \mathbf{dS}$, $\mathbf{S}$ being a vector along the line of sight, observed with pixels of projected area S. Thus if one has a field- (and polarization-) free surface of area S and a single magnetic structure with line-of-sight strength $B_{\|}$ occupying an area $s < S$, one measures an average "flux density" $\frac{s}{S} B_{\|}$ Mx cm^{-2}, where the actual field strength is B G. This difference between the kG field strengths of sunspots and the average surface flux density in Mx cm^{-2} in unresolved solar-like stars makes the disk-integrated net polarization small.

The magnetic polarization of Sun-as-a-star spectra is also limited for other physical reasons. Sunspot groups are small compared with stellar hemispheres: the absence of magnetic monopoles means that opposite polarities appear together in spot pairs, in this case the dominant (first-order) Zeeman induced polarization almost cancels in the integrated light (the same flux emerges in one polarity as returns through the other polarity in a given active region). Force balance in the photosphere limits the field strengths to near-equipartition values where $B^2/8\pi < \frac{3}{2}nkT$, a few thousand G. Thus, stellar Stokes V/I measurements are intrinsically very weak. Few solar-like stars have been targeted using polarimetry, and those which have are not really of solar type: they are younger, more rapid rotators.

This boils down to the *necessity*, for all but a few special (rapidly rotating and active) targets, to look for other signatures of magnetic activity. Hence those "diagnostics and indices" of our title: the variable radiation in well-known chromospheric (optical, UV lines), transition region (UV) and coronal (EUV/X-ray) features. These are important because they correlate with spatially resolved magnetic structures measured on the Sun (e.g., Schrijver & Zwaan 2000), they vary considerably with the solar sunspot cycle, and, being radiators of dissipated magnetic energy ($B^2/8\pi$), they are not subject to the cancellation of signals arising from opposite polarity fields on the visible hemisphere. Recently, helio- and astro- *seismology* have become important additions to the toolkit for activity indicators.

The correlation between Zeeman signals of magnetic–field and the classical "Ca II" (chromospheric brightness) index has been demonstrated clearly for at least one solar-like star, other than the Sun itself (Fig. 1). The star, ξ Boo A of spectral type G8 V, with rotation period $\sim$ 6 days, is significantly younger and more active than the Sun. The figure also shows a correlation between low-ℓ oscillation frequencies and the F10.7 radio flux, both measured for the Sun-as-a-star, from Salabert *et al.* (2003).

† If the Sun were observed nearly pole-on, the solar cycle might be seen with a higher amplitude as the poles are dominated by large areas of magnetic–field of the *same* polarity whose flux density varies with an average amplitude of $\sim 10 - 20$ Mx cm^{-2} Schrijver & Harvey (1994).

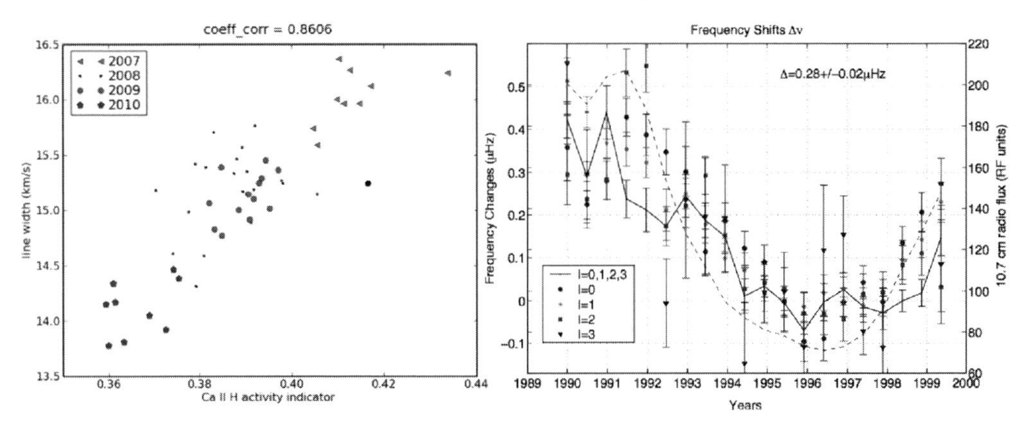

Figure 1. Left panel: variations in a Zeeman-sensitive line width (ordinate) versus chromospheric emission index (abscissa), for the young and active solar-like G 8 V star ξ Boo A (Morgenthaler *et al.* 2010). Right panel: solar measurements acquired as if the Sun were a star, showing relative mode oscillation frequency differences versus time. The 10.7 cm radio flux, a well-known magnetic activity index forming through free-free emission near the coronal base, is plotted as a dashed line (Salabert *et al.* 2003).

In summary, *for Sun-like stars* we must rely on the fact that stellar magnetic–fields are correlated with variations in stellar optical photometric measurements (both in global helioseismic techniques and in photometry revealing the passage of spots and plages across stellar disks), in UV fluxes and in X-ray fluxes, through their presumed connection to the magnetic features - spots, plages and network - which are directly seen on the Sun. There is a large literature on these relationships for the Sun, both seen-as-a star and for features across the solar disk. Much material is nicely reviewed by Schrijver & Zwaan (2000).

3. Comparing magnetic minima in the Sun and stars

3.1. *Practical considerations*

Systematic observations of stellar magnetic activity began in 1966, when Wilson (1968) began the ground-based "Mt. Wilson survey". Table 1 lists significant programs that have contributed to the database of measurements of magnetic activity on decade and longer time scales. The table includes the important Fraunhofer Ca II "H and K" lines at 397 and 393 nm, whose line cores form in chromospheric plasmas.

The Ca II lines are important since they represent the longest continuously observed diagnostic of stellar surface magnetism, and they contain data both on rotation and dynamo action on stars. A traditional "index" is the "S-index", which measures the line core (= chromospheric component) integrated over a triangular filter, relative to the neighboring "continuum", thereby giving a normalized measure of the chromospheric to broad-band flux. The S index will suffice for our use below, but the reader should be aware that others are in use, some of which are more closely related to basic stellar parameters. This issue is discussed, for example, by Hall (2008).

A fundamental difficulty in comparing stellar and solar activity indices is that it is a "bandwidth–limited" exercise. At best, any star has been observed for 45 years on a daily basis, subject to seasonal observability constraints. If in an ensemble of such stars we truly had *identical* suns with their rotation axes, like the Sun, just a few degrees from the plane-of-the-sky, then we could invoke ergodicity and compare the Sun's statistical behavior in time with the variations seen among the stellar sample. Unfortunately, this

Table 1. Significant synoptic observational programs of solar and stellar magnetic variability

Target	Program	dates	Observable	notes
The Sun				
	Various[a]	1608-	Sunspot counts	∼ daily
	Ottawa[a]	1947-	F10.7 (10.7 cm radio flux)	daily
	Sacramento Peak[b]	1974-	Solar disk Ca II	∼ daily
	Kitt Peak[b]	1974	Solar disk Ca II	∼ daily
	Various spacecraft[c]	1978-	total irradiance	daily
	Various spacecraft[d]	1981-	IR-X-ray spectral irradiances	daily
Stars				
	Mt. Wilson[e]	1966-1996	Ca II field G&K stars	daily/seasonal
	Lowell/Fairborn[f]	1984-	Stromgren b, y colors	daily/seasonal
	HAO SMARTS[g]	2007-	S. hemisphere field G&Kstars	weekly/seasonal
Sun and stars				
	SSS/Lowell[h]	1994-	Ca II field G&K stars, integrated sunlight	weekly/seasonal

Notes: (a) See, for example, Hufbauer (1991). (b) Livingston et al. (2010). (c) Frohlich (2011). (d) Rottman (2006). (e) Baliunas et al. (1995). (f) Radick et al. (1998). (g) Metcalfe et al. (2010). (h) Hall (2008).

exercise is made complicated by significant dependences of the magnetic indices on stellar mass, age, metallicity, and orientation of the rotation axes. There is no solar "identical twin", so we cannot yet exactly "rerun the solar experiment" under controlled conditions, and some care is needed†.

Now if we had observed the Sun daily for 45 years, we would have captured 2 complete Hale magnetic 22–year cycles, unless we had been observing during the Maunder Minimum (see Fig. 2). Importantly, if we place a 45 year window across any particular

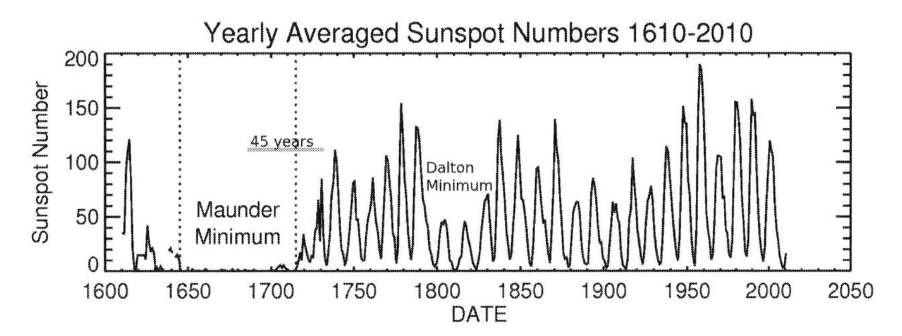

Figure 2. Yearly averaged sunspot numbers. The bar marked "45 years", placed arbitrarily near the end of the Maunder Minimum, shows the span of time for which we have continuous observations of magnetic activity in stars.

period in the sunspot record, it is clear that at most we would observe four minima in the sunspot cycle. Further, if we were to observe the Sun somewhere near the Maunder Minimum, we might conclude that the Sun is in not a regular cycling state, but instead is perhaps a "flat activity" or even "irregularly varying" star. The one thing that we would expect, though, is that the *mean value* of the chromospheric activity index (the Ca II "S" index for example) is near the low end of the stellar distribution. This is because solar "S-index" data were obtained during the last 40 years or so of high sunspot numbers (Fig. 2) and yet they are relatively low in a stellar context (Fig. 3).

Henceforth we will therefore discuss *relatively inactive stars, which are also slow rotators* e.g. Noyes *et al.* (1984a). Techniques requiring strong magnetic–fields in which different spot polarities are spectrally separated by differential Doppler shifts in rapidly

† This is an area we expect future progress from astroseismology, see section 4.

rotating stars, such as Zeeman Doppler Imaging Semel (1989), cannot therefore be applied. Lastly, even to examine "comparative minima", the stars must be

(*a*) Single (weak star-star/ star-exoplanet interactions)
(*b*) Cycling, or
(*c*) In a GM state (flat)

Mostly we will be restricted to single stars similar in mass, age and (hence) activity to the Sun.

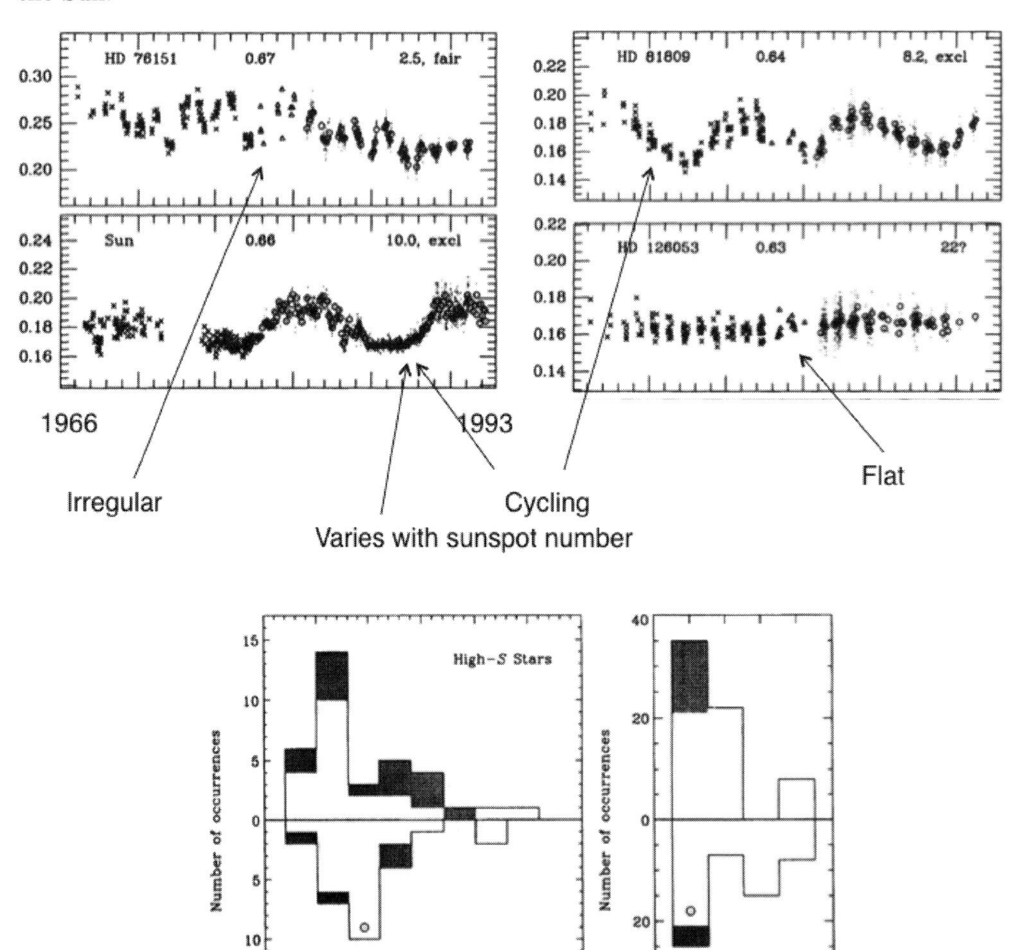

Figure 3. A selection of different stellar behavior measured in the Mt. Wilson survey up to 1995, reported by Baliunas *et al.* (1995). The upper four panels show representative time series of the "S-index", the lower panels the number of stars as a function of cycle period, and the number of stars which were classified as cyclic (C, like the current Sun), irregularly variable (V), flat (F, the Maunder Minimum Sun?) and with a long-term trend (L).

3.2. *The stellar records – "settled" issues*

Analysis of histograms such as those shown in Fig. 3 suggests that the Sun is a typical "low-S" star. In the stellar Ca II data, almost as many such stars don't cycle as do

Baliunas *et al.* (1995). Such results are broadly confirmed by the Lowell and other (non-synoptic, but otherwise relevant) observing programs and campaigns (such as a campaign on solar-age stars in M67 by Giampapa *et al.* 2006). So, in spite of practical difficulties, the present Sun does appear broadly to behave like a significant fraction of G and K main sequence stars, at least according to the Ca II and broad-band photometric records.

Several contentious issues in the comparison of the Sun and stars are now largely, if not completely, resolved as we have gained more data and a better understanding of existing data. Most points of debate arose because of "small number statistics", but there are interesting biases of a physical origin that have also led to confusion.

A debate concerning the statistical occurrence of stars in a state perhaps equivalent to the Maunder Minimum arose soon after a publication of Baliunas+Jastrow (1990), which showed a bi-model distribution of chromospheric S-indices which seemed to correlate with whether low activity stars were in a cycling or flat state. It was suggested that the two distributions corresponded to the Sun in its cycling versus non-cycling states. But the published correlation has not survived scrutiny from two perspectives: first, Hall & Lockwood (2004) showed that when S-index data are analyzed according to seasonal averages (i.e. equal weights given to equal spans of time) the bimodal distribution disappears. Secondly, Wright (2004) used accurate parallaxes from the *HIPPARCOS* mission to show that many flat activity stars appear to be significantly evolved above the main sequence.

Another question arose concerning stellar X-rays and activity cycles. It has been known for decades that the variances in solar activity indices increase with the temperature of the plasma from which they originate. Thus, the Ca II index varies more than radiation from the cool, dense photosphere; vacuum UV radiation from the mid-upper chromosphere varies more than Ca II, EUV radiation from well-known coronal lines (with no change in principle quantum number) observed routinely by missions such as *SOHO*, *TRACE* and *SDO* varies more still, and then X- and γ- rays vary the most. If we can see stellar cycles in Ca II so clearly (Fig. 3), we should see them with enormous amplitudes in X- and γ-rays. Yet, for a decade or so, we did not.

An important feature of X-rays from solar type stars was highlighted by Schmitt (1997). Based on a volume limited sample of stars observed by *ROSAT*, he showed that *the soft X-ray flux in the 0.1-2.4 keV bandpass has a lower limit of* 10^4 *erg cm*$^{-2}$ *s*$^{-1}$. To have such a lower limit seemed to contrast with solar data from the Soft X-ray Telescope on *Yohkoh*, which exhibited enormous variations in count rates over the solar cycle†. Asking the question, "where are the stellar cycles in X-rays?" Stern, Alexander & Acton (2003) computed the soft X-ray irradiance variations from the *Yohkoh* data showing a maximum to minimum ratio, effectively smoothed with a 4th order polynomial fit, of 30. Inspection of their figure 3, in which the lowest count rates are systematically overestimated, suggests this to be a lower limit, more likely ratios appear closer to 100 and can approach 1000. They compared these data with soft X-ray data for solar-like stars in the Hyades group from *ROSAT* IPC data. Discounting unlikely fortuitous phases linking stellar cycles with the epochs of stellar observations, they concluded that "Hyades F-G dwarfs have either very long X-ray cycles, weak cycles or no cycle at all". This puzzle has since been resolved noting that inter-instrumental calibrations must be done taking particular care to define the precise response of the different detectors used for solar and stellar work Judge, Solomon & Ayres (2003). In a result anticipated by Ayres (1997), Judge and colleagues simulated ROSAT soft X-ray count rates of the Sun using solar soft X-ray data from

† See, for example, the "solar cycle in X-rays" images which were widely distributed among the community, at `http://solar.physics.montana.edu/sxt/`

the SNOE experiment. They found that through the 0.1-2.4 keV, ROSAT channel, the Sun would would have observed factors of between 5 and 10, maximum / minimum flux. Cyclic soft X-ray variations of a factor of several (max/min) have in fact since been seen in this channel in the K5 V star 61 Cyg A (Robrade,Schmitt & Hempelmann 2007). The X-ray variations were in phase with Ca II emission over a period of 12 years.

A related problem was the reported "disappearance" of the corona of α Cen A, a G2 V star similar to but some 20% larger than the Sun itself (Robrade, Schmitt & Favata 2005). They reported a factor 25 reduction in the "X ray luminosity" of this star over two years of observations with the XMM-Newton satellite, ostensibly between energies of 0.2-2.0 keV. This seemed to suggest that the Sun's corona had the possibility of almost disappearing in a span of 2 years, something unprecedented since X-ray data were first acquired some 6 decades ago. The dilemma was resolved when Ayres *et al.* (2008) obtained LETGS spectra with the *Chandera* satellite. The results, highlighted in Fig. 4, show clearly that while the higher energy soft X-rays decrease enormously, the EUV transitions now so familiar to us in solar images from SOHO, TRACE and SDO, remain strong. A slight drop in the average coronal temperature serves to remove soft X-rays from the spectrum, originating from plasma near 2MK, while at the same time keeping the EUV coronal transitions (17.1nm, 19.5nm) strong.

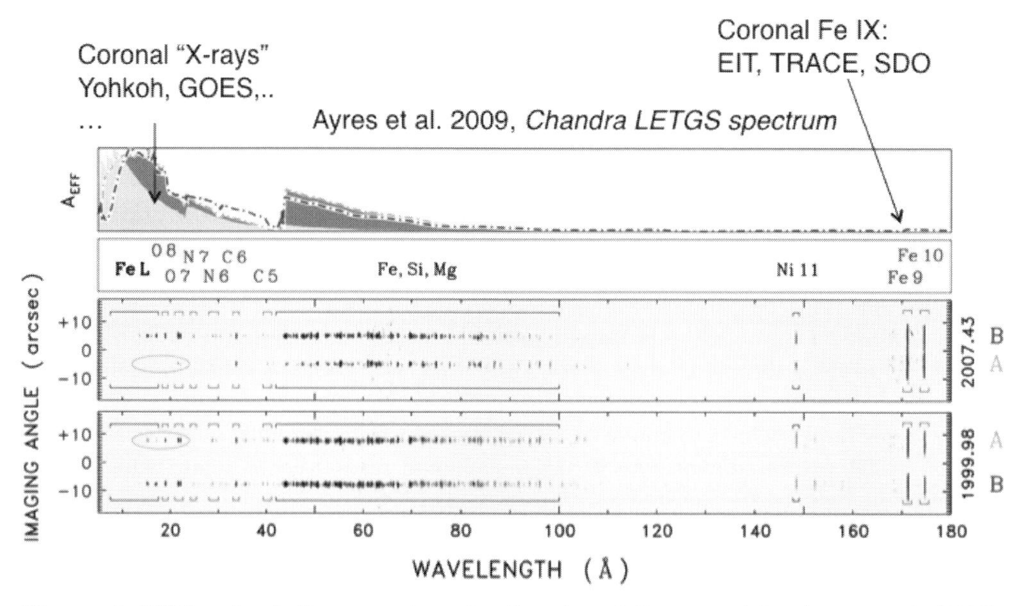

Figure 4. EUV and soft X-ray spectra of α Cen A and B reported by Ayres *et al.* (2008). Ellipses highlight the spectral comings and goings of soft X-rays in α Cen A at two different epochs, while the EUV data at energies some 10× lower remain similar.

In summary, the higher energy behavior of the Sun in the context of sun-like stars is no longer generally believed by most to be anomalous. Returning to lower energies, it was earlier believed that perhaps the Sun's broad-band ($\propto$ irradiance) variations were lower than other stars for a given variation in S-index (Radick *et al.* 1998, Lockwood *et al.* 2007). But the evidence that the Sun is anomalous in this sense is not so clear (Hall *et al.* 2009). More sensitive stellar photometry has revealed stars with lower photometric variability (the Sun's variability measured since 1978 is close to the detection limit for ground-based telescopes), and the nearest "solar twin" (18 Sco) has a photometric behavior much like that of the Sun (Hall 2008).

3.3. *The stellar records- unsettled issues*

It may not be possible to invoke "small number statistics" to explain the apparently anomalous behavior of the Sun shown in Fig. 5. Using carefully vetted data from Saar & Brandenburg (1999), in which stellar S indices were used to derive both rotation periods and, for those cycling stars, cycle periods, Böhm-Vitense (2007) plotted the derived cycle periods against rotation period. Her motivation was to examine the role of deep–seated and near–surface shear layers as potential sources for the re-generation of magnetic–fields (dynamos) on stars, discussed, for example, by Durney, Mihalas & Robinson (1981). Böhm-Vitense proposed that the appearance of the two branches (I= "inactive" and A='"active") in the figure (with some stars plotted twice if they showed two cycle periods) may correspond to the actions of shear-generated magnetic–fields beneath the convection zone (I) and above it (A). The Sun lies squarely between the two branches. This kind of result, although based on a limited stellar sample, is exciting because it may represent a clear departure of the Sun from two relatively simple proposed sources for dynamo action in stars. It is precisely this kind of disagreement which can lead to advances in our understanding.

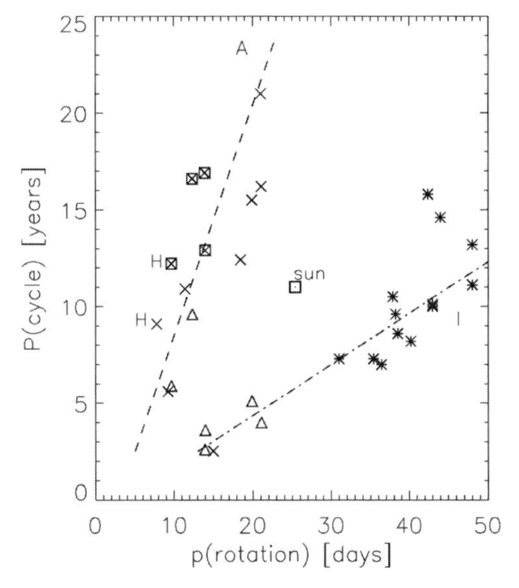

Figure 5. Periodic behavior derived from the Ca II S index data (Böhm-Vitense 2007). H indicates Hyades group stars, A and I active and inactive sequences. Squares show stars with B -V < 0.62. Triangles indicate secondary periods for stars on the A sequence.

Another question often asked is, "where is a star entering/ exiting a Maunder Minimum-like phase?" If the sunspot record of the last 400 years is typical we should see perhaps 1 in 10 cycling stars evolving into/ out of such grand minima. This is where the 45 year span of synoptic measurements is of prime importance- the probability of not finding such a star dramatically decreases with increasing observing time spans, as one can become more and more confident that a star has resumed or has stopped cycling, as the Sun appeared to to in 1745 and 1645 respectively. The ensemble of solar-like stars cannot address this point, since we do not really know what fraction of "flat activity" stars truly correspond to Sun-like stars, and not sub-giants for example Wright (2004). One clue might be the observed anti-correlation of photometric ($b + y$ Stromgren colors) and Ca II S indices for inactive stars as well as the Sun. In the Sun, the brightening due to plage emission seems to dominate over the sunspot darkening, at least when viewed from near the solar equator. Hall (2008) discusses one candidate star, HD 140538, in which the correlation appears to have changed sign from negative to positive as the star has become more active, appearing to begin cycling. However, more data are really needed to see if this is indeed a star emerging from a "grand minimum".

The sub-convection zone shear layer ("tachocline") in the Sun is commonly assumed to be an essential component of the solar dynamo, although this has been questioned solely on the basis of solar observations (Spruit 2011). The question then arises, can cycles exist in fully convective stars? Well, if the proposal of Böhm-Vitense (2007) survives further scrutiny, then a tachocline is not a pre-requisite for a dynamo that can produce solar-like sunspot cycles. In 2007, Cincunegui and colleagues tentatively identified a "cycle" with period 442 days in the fully convective star Proxima Cen dM5.5e.

4. Asteroseismology through Kepler

Asteroseismic signals complement the chromospheric and coronal signatures discussed above. The high degree of correlation between the solar magnetic cycle as seen in the varying 10.7 cm radio emission, and the shifts in acoustic oscillation frequencies, shown the right hand panel of Fig. 1, indicates that asteroseismology has the capability to reveal solar-like signatures of the solar sunspot cycle in global oscillations of stars. Both were derived from solar integrated light data such as might be obtained from more distant stars. Indeed, asteroseismic observing missions, notably the *Kepler* mission, are beginning to make this capability a reality.

Launched in March 2009, Kepler is a NASA mission which is staring at a star field in the constellation of Cygnus to look for planets orbiting around other stars. But Kepler is also permitting high-precision asteroseismology of the stars within its view. The broad sweep of asteroseismology achieved in the early phase of the mission has been summarized by Chistensen-Dalsgaard & Thompson (2011).

Stellar activity affects not only the frequencies of the oscillation modes. In the Sun, the activity cycle and the p-mode *amplitudes* are observed to be anti-correlated. The same anti-correlation is found in the CoRoT star HD49933. Kepler promises to enable fundamental contributions to the understanding of stellar activity by exploiting two complementary aspects of the precision photometry. The first aspect is the modulation by star spots, which reveals surface activity and surface rotation rate (including differential rotation). The second aspect is asteroseismology, which reveals internal rotation, differential rotation, and internal structure. Observed over an extended period, both can reveal stellar cycles. These ideas have been explored in a preliminary analysis of Kepler target stars by Garcia *et al.* (2011). The initial results are very promising, with more than 100 stars already observed with rotation periods below the 10-day period being revealed by these methods. Moreover, the asteroseismic analysis has enabled the masses and radii of these stars to be determined. However, there are also challenges for this approach, since on the one hand the stars have to be active enough for starspots to be present and to create a robust modulation signal of the integrated stellar light, and yet on the other hand the modes have still to be of sufficiently large magnitude that the asteroseismology remains feasible. A further complication revealed already by CoRoT is that in many solar-like oscillators the mode lifetimes are rather shorter than in the Sun: shorter lifetimes means broader p-mode peaks in the power spectrum, which makes the measurement of internal rotation more difficult. But the continued study of stars with fast surface rotation is very promising.

5. Comparative minima: prospects

In our short narrative we have suggested that we are only beginning to examine the magnetic minima of stars for comparison with the recent extended solar minimum, and episodes like the Maunder Minimum. We can say that the Sun lies at an overall low level of magnetic activity for a star of its spectral type, but its activity appears normal for a star of its age. We have insufficiently long stellar time series to understand if these unusual episodes of minimal solar activity have counterparts in stars (do stars have Maunder Minima?), and so we cannot really tell yet if the Sun's documented variations are in any way unusual. There are speculations, based upon stars similar to the Sun, and based on the recent extended solar minimum, that the Maunder Minimum was a time of significant small-scale magnetic activity (Judge & Saar 2007, Schrijver *et al.* 2011). It does seem clear, however, from cosmogenic isotope records, that the Maunder Minimum was a period when the global solar field reversal continued, even in the absence of a strong sunspot count (Beer, Tobias & Weiss 1998).

How then are we to make progress in this area? At least three lines of attack seem worthwhile: (1) We must continue to get much more "boring" data, monitoring the photometric and Ca II emission for decades into the future: we will test if stars can be found, for example, entering/exiting grand minima. (2) Asteroseismology with Kepler and other experiments will clarify the evolutionary states of large numbers of stars. The evolution of magnetism in stellar samples will be set more quantitatively than is at present possible. (3) We must observe in detail those stars of special interest to our Sun. The star 18 Sco is the closest "solar twin", and the two G2 V stars of the 16 Cygni system have recently been studied asteroseismically (Metcalfe *et al.* 2012), being two of the brightest targets in the Kepler field of view. The two stars are 6.8 Gyr old stars just slightly more massive than the Sun itself. The combination of asteroseismic determinations of mass, age etc., with continued careful monitoring of Ca II and other indices presents us with a powerful tool for probing the magnetic–field evolution of the Sun and stars.

References

Ayres, T. R. 1997, *J. Geophys. Res.*, **102**, 1641
Ayres, T. R., Judge, P. G., Saar, S. H., & Schmitt, J. H. M. M. 2008, *ApJL*, **678**, L121
Baliunas, S. & Jastrow, R. 1990, *Nature*, **348**, 520
Baliunas, S. L., Donahue, R. A., Soon *et al.*, 1995, *ApJ*, **438**, 269
Beer, J., Tobias, S., & Weiss, N. 1998, *Solar Phys.*, **181**, 237
Böhm-Vitense, E. 2007, *ApJ*, **657**, 486
Brown, B. P., Browning, M. K., Brun, A. S. *et al.*, 2010, *ApJ*, **711**, 424
Brown, B. P., Miesch, M. S., Browning, M. K., Brun, A. S., & Toomre, J. 2011, *ApJ*, **731**, 69
Christensen-Dalsgaard, J. & Thompson, M. J. 2011, in N. H. Brummell, A. S. Brun, M. S. Miesch, & Y. Ponty (Ed.), *IAU Symposium*, Vol. 271 of *IAU Symposium*, p. 32
Cincunegui, C., Díaz, R. F., & Mauas, P. J. D. 2007, *A&A*, **461**, 1107
Durney, B. R., Mihalas, D., & Robinson, R. D. 1981, *PASP*, **93**, 537
Fröhlich, C. 2011, *Space Sci. Rev.*, http://dx.doi.org/10.1007/s11214-011-9780-1
García, R. A., Ceillier, T., Campante *et al.*, 2011, *ArXiv e-prints*, number 1109.6488
Giampapa, M. S., Hall, J. C., Radick, R. R., & Baliunas, S. L. 2006, *ApJ*, **651**, 444
Hall, J. C. 2008, *Living Reviews in Solar Physics*, **5**, 2
Hall, J. C., Henry, G. W., Lockwood, G. W. *et al.*, 2009, *Astron. J.*, **138**, 312

Hall, J. C. & Lockwood, G. W. 2004, *ApJL*, **614**, 942
Hufbauer, K. 1991, *Exploring the sun: solar science since Galileo*, Baltimore: JHU press
Judge, P. G. & Saar, S. H. 2007, *ApJ*, **663**, 643
Judge, P. G., Solomon, S., & Ayres, T. R. 2003, *ApJ*, **593**, 534
Livingston, W., White, O. R., Wallace, L., & Harvey, J. 2010, *Mem. Ast. Sco. It.*, **81**, 643
Lockwood, G. W., Skiff, B. A., Henry, G. W., Henry *et al.*, 2007, *ApJS*, **171**, 260
Metcalfe, T. & 33 co-authors 2012, *ApJL*, in press
Metcalfe, T. S., Basu, S., Henry, T. J. *et al.*, 2010, *ApJL*, **723**, L213
Morgenthaler, A., Petit, P., Aurière, M. *et al.*, 2010, in S. Boissier, M. Heydari-Malayeri, R. Samadi, & D. Valls-Gabaud (Ed.), *SF2A-2010*, 269
Noyes, R. W., Hartmann, L. W., Baliunas, S. L. *et al.*, 1984, *ApJ*, **279**, 763
Parker, E. N. 1955, *ApJ*, **122**, 293
Plachinda, S. I. & Tarasova, T. N. 2000, *ApJ*, **533**, 1016
Radick, R. R., Lockwood, G. W., Skiff, B. A., & Baliunas, S. L. 1998, *ApJS*, **118**, 239
Robrade, J., Schmitt, J. H. M. M., & Favata, F. 2005, *A&A*, **442**, 315
Robrade, J., Schmitt, J. H. M. M., & Hempelmann, A. 2007, *Mem. Ast. Sco. It.*, **78**, 311
Rottman, G. 2006, *Space Sci. Rev.*, **125**, 39
Saar, S. H. & Brandenburg, A. 1999, *ApJ*, **524**, 295
Salabert, D., Jiménez-Reyes, S. J., & Tomczyk, S. 2003, *A&A*, **408**, 729
Schmitt, J. H. M. M. 1997, *A&A*, **318**, 215
Schrijver, C. J. & Harvey, K. L. 1994, *Solar Phys.*, **150**, 1
Schrijver, C. J., Livingston, W. C., Woods, T. N., & Mewaldt, R. A. 2011, *GRL*, **38**, 6701
Schrijver, C. J. & Zwaan, C. 2000, *Solar and Stellar Magnetic Activity*, Cambridge Univ. Press, Cambridge, UK
Semel, M. 1989, *Astron. Astrophys.*, **225**, 456
Spruit, H. C. 2011, in Miralles, M. P. & Sánchez Almeida, J. (Ed.), *The Sun, the Solar Wind, and the Heliosphere*, IAGA Special Sopron Book Series, Vol. 4. Berlin: Springer, 39
Stern, R. A., Alexander, D., & Acton, L. W. 2003, in A. Brown, G. M. Harper, & T. R. Ayres (Ed.), *12th Cambridge Workshop on Cool Stars, Stellar Systems, and the Sun*, **12**, 906
Wilson, O. C. 1968, *ApJ*, **153**, 221
Wright, J. T. 2004, *Astron. J.*, **128**, 1273

Discussion

Jeff Linsky: There is a new diagnostic, the far ultraviolet continuum emission observed by Hubble COS. We find that active solar-mass dwarf stars have FUV continuum fluxes very similar to bright solar faculae and inactive solar-mass stars have FUV fluxes similar to centers of solar granules. There is a paper by Linsky *et al.* (2011) now available in Astroph.

Michael Thompson: Thank you for pointing this out.

Mark Giampapa: You mentioned that several hundred solar-type stars have detected p-mode oscillations. Why haven't all solar-type stars in the Kepler sample shown p-mode oscillations?

Michael Thompson: First I would point out that "solar-type" doesn't mean "solar-twin", for a solar-twin star I would certainly expect an oscillation spectrum essentially like the Sun. Secondly, the solar-type stars without detected oscillations may be oscillating but at an amplitude below that which even Kepler can detect.

Yvonne Elsworth: Also, as you pointed out in your talk, activity suppresses p-mode amplitudes. So, active stars in the sample may have unobserved p-modes.

Comparative Magnetic Minima: characterizing quiet times in the Sun and stars
Proceedings IAU Symposium No. 286, 2011
C. H. Mandrini & D. F. Webb, eds.

doi:10.1017/S1743921312004590

How well do we know the sunspot number?

Leif Svalgaard
Hansen Experimental Physics Laboratory, Stanford University
650 Via Ortega, Stanford, CA 94304, USA
email: leif@leif.org

Abstract. We show that only two adjustments are necessary to harmonize the Group Sunspot Number with the Zürich Sunspot Number. The latter has been increased from the 1940s on to the present by 20% due to weighting of sunspot counts according to size of the spots and can be corrected by increasing the earlier values as well. The Group Sunspot Number before ~1885 is too low by ~50%. With these adjustments a single sunspot number series results. Of note is that there is no longer a distinct Modern Grand Maximum.

Keywords. Sun: activity, sunspots

1. Motivation

A hundred years after Rudolf Wolf's death, Hoyt *et al.* (1994) asked "Do we have the correct reconstruction of solar activity?" After a heroic effort to find and tabulate many more early sunspot reports than were available to Wolf, Hoyt *et al.* thought to answer that question in the negative and to provide a revised measure of solar activity, the Group Sunspot Number (GSN) based solely on the number of sunspot *groups*, normalized by a factor of 12 to match the Wolf numbers 1874–1991. Implicit in that normalization is the assumption or stipulation that the 'Wolf' number is 'correct' over that period. In this paper we shall show that that assumption is likely false and that the Wolf number (WSN) must be corrected. With this correction, the difference between the GSN and WSN (Fig. 1) becomes disturbing: The GSN shows either a 'plateau' until the 1940s followed by a Modern Grand Maximum, or alternatively a steady rise over the past three hundred years, while the (corrected) WSN shows no significant secular trend. As the sunspot number is often used as the basic input to models of the future evolution of the Earth's environment (e.g. Emmert & Beig (2011)) and of the climate (e.g. Lean & Rind (2009)), having the correct reconstruction becomes of utmost importance, and the difference illustrated in Fig. 1 becomes unacceptable.

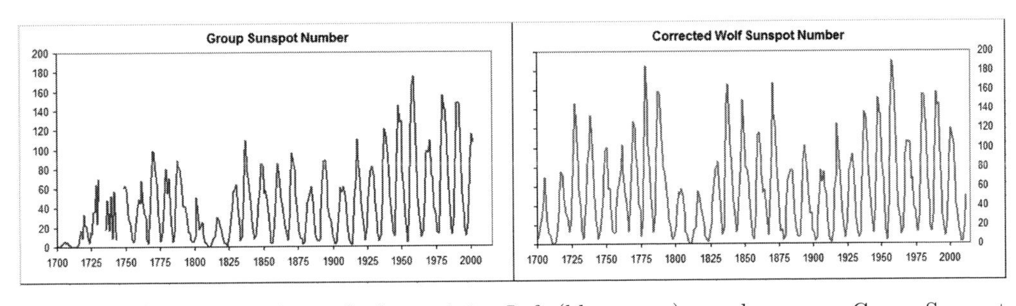

Figure 1. Contrasting views of solar activity. Left (blue curve): yearly average Group Sunspot Number. Right (red curve): The yearly average Wolf Sunspot Number corrected as described in the text. We use the designation 'Wolf" number to include the values derived by Wolf's successors in Zürich and after 1981 in Brussels.

2. The Wolf Number

Sunspots occur in groups with a coherent magnetic configuration (although this detail was not known to Wolf) and a definite evolution over several days – emergence and coalescence of smaller spots, spreading in longitude, and decay. Today we call such a group an active region and give each a number for tracking purposes. Recognizing that the birth of a new group was a much more significant event than the addition of a single new spot to the several spots already there, Wolf fashioned his Relative Sunspot Number, R, as the sum of a weighted number of groups, G, and the total number, S, of individual spots regardless of their size: $R = 10G + S$. The weight factor of 10 was chosen based both on experience (average number of spots per group) and on convenience for calculation. As telescopes of different apertures allow a different population of sunspots to be seen, Wolf introduced a scale factor, k, in the definition: $R = k(10G + S)$ to be able to reduce the observations to a common standard. For Wolf himself when using the standard (80 mm aperture) Fraunhofer refractor at magnification 64 the k-factor is unity. The k-factor depends on many things: telescope, seeing, observer visual acuity and experience, counting method, etc. and although these factors do not impact the number of groups and of spots the same way, the practice is to stick with the simplest scheme with a single k-factor for each observer at a given time. It is now clear that the homogeneity of the series and a correct measure of long-term trends depend critically on the k-factor, determined or adopted, for each observer.

Wolf published several versions and (importantly) revisions of this sunspot series. For observations before Wolf's own began in 1849, he relied primarily on Staudach's drawings 1749-1799 (Arlt (2009)) and on Schwabe's observations 1825-1848 (Arlt & Abdolvand (2011)). Up to the mid 1860s Wolf used two superb refractors built by renowned optician Joseph von Fraunhofer. One of these still exists and is used to this day to continue the Wolf number. But increasing travel (he became president of the Swiss Geodetic Commission) forced Wolf to use, eventually exclusively, smaller portable refractors (aperture 37 mm at magnification 40), which also still exist in use to this day (Fig. 2). With the smaller telescopes fewer spots (and a third fewer groups) could be seen, and Wolf determined a k-factor of 1.5 for those to match the count from the larger telescope. Observing with the larger telescope, Wolf decided (following Schwabe) not to count the smallest spots and (as he called them –"grayish") pores, visible only during exceptional seeing. With the smaller, portable telescopes these spots could not be seen anyway.

Wolf hired several assistants from the 1870s on. One of those, Alfred Wolfer, eventually became Wolf's successor. Wolfer found that Wolf's practice of omitting the smallest spots was not reproducible and advocated counting all spots that could be seen. During a

Figure 2. Left: the 80 mm Fraunhofer refractor used since 1855 by Wolf and successors until 1996. At the eye piece can be seen the Merz polarizing helioscope employed to reduce light intensity to comfortable levels. Center: the same telescope in use today by Thomas Friedli (person at right). Right: the 37 mm portable telescope used by Wolf since the mid 1860s.

16-year period of simultaneous observations, Wolf determined that the sunspot number using Wolfer's counting method should be multiplied by a k-factor of 0.60 to match the Wolf scale for the 80 mm telescope, although Wolf compared Wolfer's count not to own counts using the 80 mm telescope, but to the count with the 37 mm portable, multiplied by 1.5. Subsequent observers at the Zürich and Brussels centers to this day adopt that same k-factor, which actually cannot be measured as Wolf is not around anymore. For the years 1876-1893, the published sunspot numbers are an average of Wolf's [using the 37 mm] and the assistants' [using the 80 mm], with k-factors applied. In 1980 the responsibility for production of the sunspot number was transferred to the SIDC in Brussels. There, rather than relying on a principal observer supplemented by secondary observers, observations are averaged over a network of ~65 observers, all scaled to the reference station, Locarno in Southern Switzerland for which a k-factor of 0.60 is simply adopted.

3. The Weighting Scheme

In 1945 Max Waldmeier became director of the Zürich Observatory and in charge of production of the Wolf Number. He noted (Waldmeier 1948) Wolfer's different counting procedure, namely also counting the smallest spots that Wolf was omitting. But Waldmeier also claimed that Wolfer started (around 1882) weighting the sunspot count by the size and structure of each spot. In Waldmeier (1968) the weighting scheme is described as follows: "A spot appearing as a fine point is counted as one spot; a larger spot, but still without penumbra, gets the statistical weight 2, a smallish spot with penumbra gets 3, and a larger one gets 5." Presumably there would also be spots with weight 4. However, we have found *no* mention of this scheme in any of the, otherwise meticulous, reports or papers by Wolf, Wolfer, and Brunner (director 1926-1945), as was also noted by Kopecký *et al.* (1980). Nor do other sunspot observers (professional or amateur) not affiliated with the Zürich observers employ such a weighting scheme (or any), which is, indeed, generally unknown.

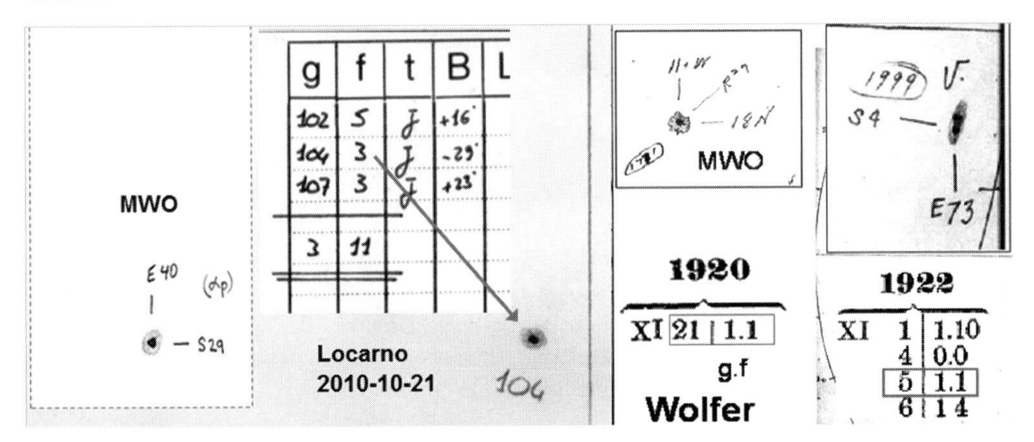

Figure 3. Left: Part of a drawing made at Locarno, showing that the spot with penumbra designated '104' was counted three times (weight 3). The insert shows the spot as observed at Mount Wilson Observatory (MWO). Right: Two spots with the same area (~220 μhemispheres) on drawings from MWO and counted with weight 1 by Wolfer (in each case as one group with one spot (Wolf's notation 'groups.spots' = '1.1'), as the group was the only group on the disk).

It is easy to show (Fig. 3) that Wolfer did not employ any weighting, contrary to the claim by Waldmeier. There are many days (43, in fact) when there was exactly one group on the disk having exactly one spot with an area of $\gtrsim\sim 200$ μhemispheres that should have been counted with weight 3, for a total sunspot number of at least $0.6\cdot(10\cdot 1+3) = 7.8 \equiv 8$, yet were all reported as $0.6 \cdot (10 \cdot 1 + 1) = 6.6 \equiv 7$.

4. The Effect of the Weighting

To determine how much the weighting scheme incrases the sunspot numbers we for each day 'undo' the weighting by counting on Locarno's drawings each spot only once for comparison with the sum of the weighted counts (Fig. 4). Locarno was founded as 'backup' station for Zürich. The station, located on the other side of the Alps has often complementary weather.

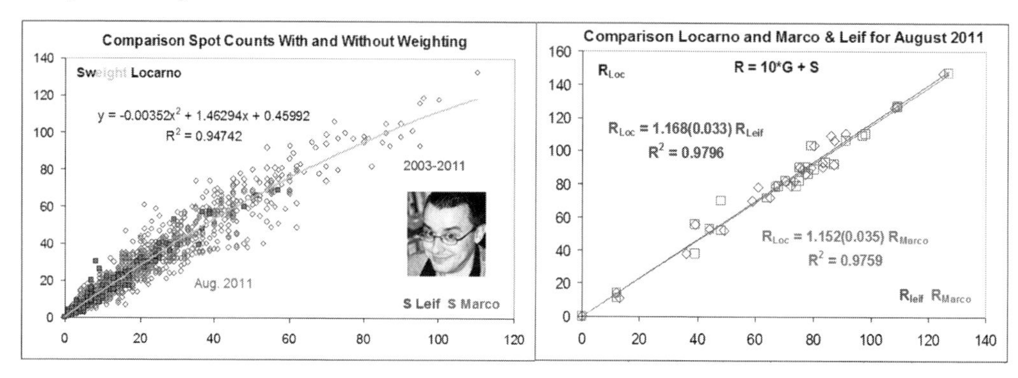

Figure 4. Left: (blue open diamonds) The weighted sunspot count compared to the simple count for 2557 days, 2003-2011. The double-blind control count is shown by pink squares. Right: Comparison of calculated relative sunspot numbers for August, 2011; blue open diamonds: Leif Svalgaard, pink open squares: Marco Cagnotti, Locarno (pictured). The standard errors of the slopes of the regression lines are given in parentheses.

As the drawings from the projected images are at times only indicative (and thus not always a true reflection of the actual, reported observations done visually at the eyepiece) it was felt necessary to perform a double-blind control test. The observer at Locarno would also record the unweighted count for each day in addition to the reported 'official' visual count. This was done for the month of August, 2011 (and is continuing). As is clear from Fig. 4, there is no significant difference between the resident observer and the probing solar physicist, thus preliminarily validating the technique. The result is that the weighting for a typical number of spots increases the spot count by 30-50%. But we are interested in the effect on the relative sunspot number where the group count will dilute the effect by about a factor of two, as shown in the right panel of Fig. 4 where an inflation of 15-17% is evident. The inflation weight factor, W, can be determined for each of the years 2003-2011 and we find that there is a weak dependence on the sunspot number: $W = 1.13 + R/2500$. For a typical sunspot number of $R = 100$, W becomes 1.17.

Waldmeier also introduced a new classification of groups (the Zürich classification) based on understanding of the evolution of the group rather than mere proximity of the spots. This tends to increase the number of groups above what proximity would dictate. We find that, on average, on a fifth of all days an additional group is reported which means that the relative sunspot number increases by about 3% due to this inflation of the group count brought about by the better understanding of what constitutes a group. Kopecký *et al.* (1980) quote the observer Zelenka suggesting a possible influence of the new Zürich classification of groups. The combined effect of the weighting and the classification might thus be of the order of 20%, which can be corrected for by increasing the pre-1945 values by a factor 1.20, as done in Fig. 1.

5. The Waldmeier Discontinuity

The Group Sunspot Number during most of the 20th century is based on the Greenwich Helio Photographs and might be suspected of having a constant calibration over that interval. We can form the ratio WSN/GSN when neither is too small to examine its stability. As shown in upper left of Fig. 5 there is a clear discontinuity corresponding to a jump of a factor 1.18 between 1945 and 1946 as we would expect if the weighting was introduced at that time. The sunspot area, S_A, (Balmaceda *et al.* (2009)) has a strong (albeit slightly non-linear) relationship with the sunspot number: $R = A\,S_A^{0.732}$. The ratio $A = R/\,S_A^{0.732}$ plotted in the right-hand panel of Fig. 5 shows the same discontinuity around 1945, with an increase by a factor of 1.21 (as established by the histograms of A in the lower left of the Figure).

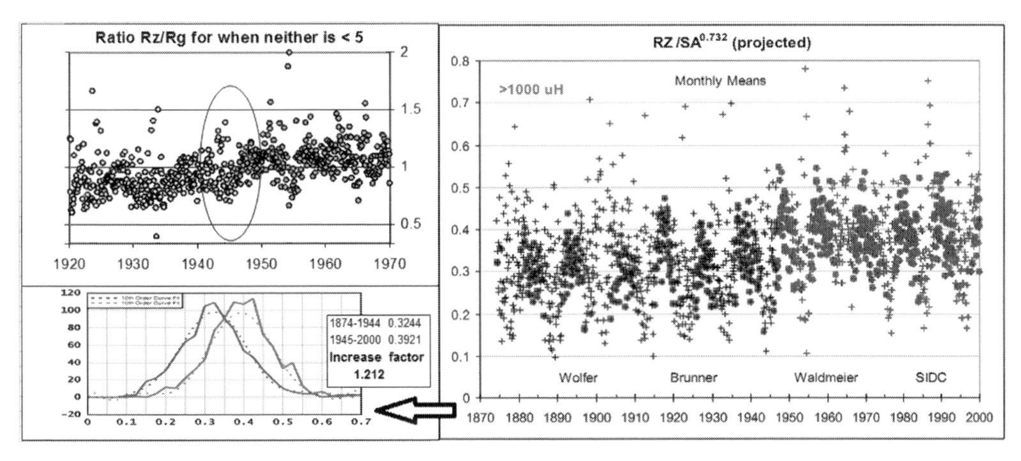

Figure 5. Upper left: Ratio between WSN and GSN, monthly values 1920–1969. Right: Ratio between WSN and linearized sunspot area. Monthly averages greater than 1000 μhemispheres are marked by pink squares. Lower left: Histograms of that ratio before and after 1945.

From ~40,000 Ca II K-line spectroheliograms from the 60-foot tower at Mount Wilson between 1915 and 1985 a daily index of the fractional area of the visible solar disk occupied by plages and active network has been constructed (Bertello *et al.* (2008)). Monthly averages of this index is strongly correlated with the sunspot number, with the expression $R = 27235\,CaK - 67.14$ before 1946. After that year, the observed sunspot number compared to the value calculated from this expression is too high by 19% (Fig. 6). An equivalent conclusion was reached by Foukal (1998).

6. Terrestrial Evidence

The F2 layer critical frequency is the maximum radio frequency that can be reflected by the F2 region of of the ionosphere at vertical incidence and has been found to have a profound solar cycle dependence which showed a difference between cycle 17 and cycle 18 (Ostrow & PoKempner (1952). It was necessary to shift the sunspot number 21% between the two cycles in 1945 to obtain the same relationship. Again we see the same discontinuity.

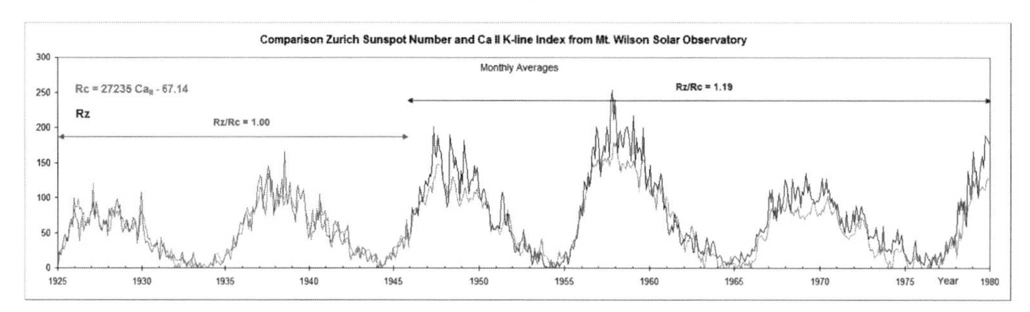

Figure 6. Sunspot number (red curve) calculated from MWO CaII K-line spectroheliograms compared to observed Wolf Numbers (blue curve). From now on reference to colors in figures correspond to the online version of the paper.

Wolf (1859) discovered the linear relationship between the amplitude, v, of the diurnal variation of the Declination of the 'magnetic needle': $v = a + b\,R$, and used it to calibrate the Wolf Number. The variation of the Declination (Fig. 7) was discovered in 1722 by George Graham and reliable measurements stretch back in to the 1780s.

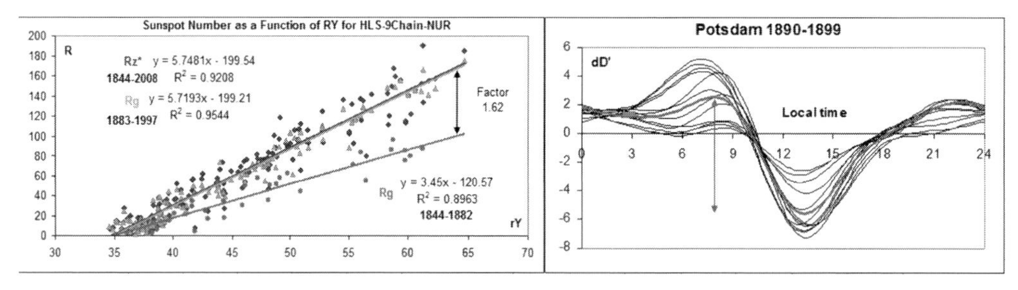

Figure 7. Right: Twelve monthly curves showing the variation during the day of the magnetic Declination. The red curve and arrow show the yearly average variation which is a strong function of the sunspot number (controlled by Far Ultraviolet (FUV) radiation). Left: The relationship between a many-station series of diurnal variation of the East-West component of the geomagnetic field – which is calculated from the variation of the Declination and which is the actual physical parameter controlled by solar FUV – and the Wolf Number (blue diamonds) back to 1844. The Group Sunspot Number has the same relationship (green triangles) after 1882, but is seriously too small before (pink dots).

It is clear that there is a consistent relationship between the diurnal range and the Wolf sunspot number, establishing an objective calibration, but that the Group Sunspot Number is seriously too low before the 1880s.

7. Implications

The possible implications from this reassessment of the sunspot series are so serious that a series of workshops are planned to resolve these questions. The first was held at Sunspot, NM in September, 2011, (`http://ssnworkshop.wikia.com/wiki/Home`).

References

Arlt, R. 2009 *Sol. Phys.*, 255, 143

Arlt, R. & Abdolvand, A. 2011 *Proc. IAU Symposium 273*, 6, 286

Balmaceda, L. A., Solanki, S. K., Krivova, N. A., & Foster, S. 2009, *J. Geophys. Res.*, 114, A07104

Emmert, J. T. & Beig, G. 2011, *J. Geophys. Res.*, 116, A00H01
Foukal, P. 1998, *Geophys. Res. Lett.*, 25, 2909
Hoyt, D. V., Schatten, K. H., & Nesmes-Ribes, E. 1994, *Geophys. Res. Lett.*, 21, 2067
Kopecký, M., Ružičková-Topolová, B., & Kuklin, G. V. 1980, *Bull. Astron. Inst. Czech*, 31, 267
Lean, J. L. & Rind, D. H. 2009, *Geophys. Res. Lett.*, 36, L15708
Ostrow, S. M. & PoKempner, M. 1952, *J. Geophys. Res.*, 57, 473
Waldmeier, M. 1948, *Astron. Mitteil.*, Nr. 152
Waldmeier, M. 1968, *Astron. Mitteil.*, Nr. 285
Wolf, J. R. 1859, *Mittheil. Sonnenfl.*, Nr. 9

Discussion

Michael Thompson: If you were inventing an index today would you use sunspot number?

Leif Svalgaard: Yes. However, since we are aiming for a single series it would make sense to call it "the" sunspot number because we want to only have one agreed-upon and vetted index.

Session 2

Dynamos and Cycle Variability

Comparative Magnetic Minima:
Characterizing Quiet Times in the Sun and Stars
Proceedings IAU Symposium No. 286, 2011
C. H. Mandrini & D. F. Webb, eds.

doi:10.1017/S1743921312004619

Cycles and cycle modulations

Axel Brandenburg[1,2] **and Gustavo Guerrero**[1,3]

[1]Nordita, Roslagstullsbacken 23, SE-10691 Stockholm, Sweden
email: brandenb@nordita.org

[2]Department of Astronomy, Stockholm University, SE-10691 Stockholm, Sweden

[3]Solar Physics, HEPL, Stanford University, Stanford, CA 94305-4085, USA

Abstract. Some selected concepts of the solar activity cycle are reviewed. Cycle modulations through a stochastic α effect are being identified with limited scale separation ratios. Three-dimensional turbulence simulations with helicity and shear are compared at two different scale separation ratios. In both cases the level of fluctuations shows relatively little variation with the dynamo cycle. Prospects for a shallow origin of sunspots are discussed in terms of the negative effective magnetic pressure instability. Tilt angles of bipolar active regions are discussed as a consequence of shear rather than the Coriolis force.

Keywords. MHD, turbulence, Sun: activity, Sun: magnetic fields, sunspots

---

1. Solar cycle

The solar cycle manifests itself through spots at the Sun's surface. To understand activity variations, we have to understand not only their source, but also the detailed connection between variations in the strength of the dynamo and its effect on the number and size of sunspots. In this paper, we address both aspects.

The physics of the solar cycle is not entirely clear. The models that work best are not necessarily those that would emerge from first principles. Even the reason for the equatorward migration of the activity belts is not completely clear. Following Parker's seminal paper of 1955, this migration seemed to be a simple property of an $\alpha\Omega$ dynamo, i.e., a dynamo that works with α effect and shear. What matters for equatorward migration is not the Ω gradient in the latitudinal direction, but that in the radial one, $\partial\Omega/\partial r$. However, in the bulk of the convection zone, $\partial\Omega/\partial r$ is mostly positive. This, together with an α effect of positive sign in the northern hemisphere results in poleward migration (Yoshimura 1975), which is not what is observed. On the other hand, according to the flux transport dynamos, magnetic fields are advected by the meridional circulation. Assuming that there is a coherent circulation with equatorward migration at the bottom of the convection zone, this would then turn the dynamo wave around so as to explain the solar butterfly diagram and that sunspots emerge from progressively lower latitudes (Choudhuri *et al.* 1995; Dikpati & Charbonneau 1999; Guerrero & de Gouveia Dal Pino 2008). This requires that most of the field resides at the bottom of the convection zone. Moreover, the α effect is taken to be non-vanishing only near the very top of the convection zone, i.e., the mean electromotive force has to be written formally as a convolution of the mean magnetic field with an integral kernel to account for this non-locality (see, e.g., Brandenburg & Käpylä 2007). Furthermore, from the observed tilt angles of bipolar regions it is inferred that the Sun's magnetic field at the bottom of the convection zone reaches strengths of the order of 100 kG (D'Silva & Choudhuri 1993), which is nearly 100 times over the equipartition value. Finally, there are assumptions about the turbulent magnetic diffusivity. In all cases, the magnetic diffusivity in the evolution equation for the toroidal field in the bulk of the convection zone is rather small, below $10^{11}\,\mathrm{cm^2\,s^{-1}}$ (see, e.g., Chatterjee & Choudhuri 2006). The magnetic diffusivity for the poloidal field is assumed to be larger and similar to the values expected from mixing length theory (see below).

In any case, these assumptions are hardly in agreement with standard formulae that the magnetic diffusivity is given by $\frac{1}{3}\tau u_{\mathrm{rms}}^2$, where τ is the turnover time and u_{rms} is the rms value of the turbulent velocity. The turnover time is $\tau = (u_{\mathrm{rms}} k_{\mathrm{f}})^{-1}$, where k_{f} is the wavenumber of the energy-carrying eddies. This result for η_t is well confirmed by simulations (Sur *et al.* 2008). For the Sun, mixing length theory appears to be reasonably good and gives $\eta_t \approx (1...3) \times 10^{12}\,\mathrm{cm^2\,s^{-1}}$. Also, in contrast to the assumptions of some flux transport dynamo models, a strong degree of anisotropy of the η tensor is not expected from theory (Brandenburg *et al.* 2012).

An alternate approach is to use turbulent transport coefficients from theory, which give rise to what is called a distributed dynamo, i.e., the induction effects are non-vanishing and distributed over the entire convection zone. In addition, there is the hypothesis that the near-surface shear layer may be important for the equatorward migration (Brandenburg 2005), but this has never been confirmed by simulations either. In any case, based on such models one would not expect there to be a 100 kG magnetic field, but only a much weaker field of around 0.3–1 kG. This calls then for an alternative explanation for the magnetic field concentrations of up to 3 kG seen in sunspots and active regions. Various proposals were already discussed in Brandenburg (2005), and meanwhile there are direct numerical simulations (DNS) confirming the validity of the physics assumed in one of those proposals. This will be addressed in Section 3.

2. Cycle modulation

Early ideas for cycle modulations go back to Tavakol (1978) who argued that the solar cycle may be a chaotic attractor. This explanation became very popular in the following years (Ruzmaikin 1981; Weiss *et al.* 1984). These ideas were elaborated upon in the framework of low-order truncations of mean-field dynamo models, having in mind that the same idea applies also to the underlying fully nonlinear three-dimensional equations of magnetohydrodynamics. Another line of thinking is that in mean-field theory (MFT) the physics of the cycle models can be explained by random fluctuations in the turbulent transport coefficients (Choudhuri *et al.* 1992; Moss *et al.* 1992; Schmitt *et al.* 1996; Brandenburg & Spiegel 2008). For high-dimensional attractors there is hardly any difference between both approaches. A completely different proposal for cycle modulation is related to variations in the meridional circulation (Nandy *et al.* 2011). This proposal still lacks verification from DNS of a dynamo whose cycle period is indeed controlled by meridional circulation. By contrast, fluctuations in the turbulent transport coefficients have indeed been borne out by simulations (see Brandenburg *et al.* 2008).

To illustrate this, let us now consider a physical realization of a simple $\alpha\Omega$ dynamo in a periodic domain. In the language of MFT, this corresponds to solving the following set of mean-field equations,

$$\frac{\partial \overline{\boldsymbol{B}}}{\partial t} = \boldsymbol{\nabla} \times \left(\overline{\boldsymbol{U}} \times \overline{\boldsymbol{B}} + \overline{\boldsymbol{\mathcal{E}}} - \eta\mu_0 \overline{\boldsymbol{J}}\right) \tag{2.1}$$

in a Cartesian domain, (x, y, z), in one dimension, $-\pi < z < \pi$, where $\overline{\boldsymbol{U}} = \overline{\boldsymbol{U}}_S \equiv (0, Sx, 0)$ is a linear shear flow velocity (assuming $S = \mathrm{const}$), $\overline{\boldsymbol{J}} = \boldsymbol{\nabla} \times \overline{\boldsymbol{B}}/\mu_0$ is the mean current density, μ_0 is the vacuum permeability, η is the microphysical (molecular) magnetic diffusivity, and

$$\overline{\boldsymbol{\mathcal{E}}} = \alpha \overline{\boldsymbol{B}} - \eta_t \mu_0 \overline{\boldsymbol{J}} \tag{2.2}$$

is the mean electromotive force. In DNS, on the other hand, one solves directly the equation

$$\frac{\partial \boldsymbol{B}}{\partial t} = \boldsymbol{\nabla} \times (\boldsymbol{U} \times \boldsymbol{B} - \eta \mu_0 \boldsymbol{J}), \tag{2.3}$$

together with corresponding equations governing the evolution of the turbulent velocity $\boldsymbol{U}$. Here, one often make the assumption of an isothermal gas with constant sound speed $c_{\rm s}$. This will also be done in the present work.

In the following we present results of simulations using shearing–periodic boundary conditions. To maintain the solenoidality of the magnetic field, we write $\boldsymbol{B} = \boldsymbol{\nabla} \times \boldsymbol{A}$ and solve for the magnetic vector potential $\boldsymbol{A}$. Using in the following the velocity for the deviations from the shear flow, $\boldsymbol{U}$, our equations are

$$\frac{\partial \boldsymbol{A}}{\partial t} + \boldsymbol{U}_S \cdot \boldsymbol{\nabla} \boldsymbol{A} = -S A_y \boldsymbol{x} + \boldsymbol{U} \times \boldsymbol{B} + \eta \boldsymbol{\nabla}^2 \boldsymbol{A}, \tag{2.4}$$

$$\frac{\mathcal{D}\boldsymbol{U}}{\mathcal{D}t} = -S U_x \boldsymbol{y} - c_{\rm s}^2 \boldsymbol{\nabla} \ln \rho + \boldsymbol{f} + \frac{1}{\rho} (\boldsymbol{J} \times \boldsymbol{B} + \boldsymbol{\nabla} \cdot 2\nu\rho \mathsf{S}), \tag{2.5}$$

$$\frac{\mathcal{D} \ln \rho}{\mathcal{D}t} = -\boldsymbol{\nabla} \cdot \boldsymbol{U}, \tag{2.6}$$

where $\mathrm{D}/\mathrm{D}t = \partial/\partial t + (\boldsymbol{U} + \boldsymbol{U}_S) \cdot \boldsymbol{\nabla}$ is the advective derivative with respect to the total flow, $\boldsymbol{U} + \boldsymbol{U}_S$, ρ is the gas density, ν is the viscosity, $\mathsf{S}_{ij} = \frac{1}{2}(U_{i,j} + U_{j,i}) - \frac{1}{3}\delta_{ij} \boldsymbol{\nabla} \cdot \boldsymbol{U}$ is the trace-less rate of strain matrix, and $\boldsymbol{f}$ is a forcing function that drives both turbulence and a linear shear flow. Alternatively, turbulence can also be the result of some instability (Rayleigh-Bénard instability, magneto-rotational instability, etc). In the following we restrict ourselves to a random forcing function with wavevectors whose modulus is in a narrow interval around an average wavenumber $k_{\rm f}$. This has the additional advantage that we can arrange the forcing function such that it has a part that is fully helical, i.e., $\boldsymbol{\nabla} \times \boldsymbol{f} = k_{\rm f} \boldsymbol{f}$ (the part driving the shear flow is of course non-helical). Because of the presence of helicity, we should expect there to be an α effect operating in the system, but if the number of turbulent eddies in the domain is not very large, there can be significant fluctuations in the resulting α effect.

Important control parameters are the magnetic Reynolds and Prandtl numbers, $\mathrm{Re}_M = u_{\rm rms}/\eta k_{\rm f}$ and $\mathrm{Pr}_M = \nu/\eta$. In addition, there is the non-dimensional shear parameter defined here as $\mathrm{Sh} = S/u_{\rm rms} k_{\rm f}$. The smallest possible wavenumber in a triply-periodic domain of size $L \times L \times L$ is $k_1 = 2\pi/L$. For the purpose of presenting exploratory results, we restrict ourselves here to a resolution of 64^3 meshpoints. We use the fully compressible PENCIL CODE† for all our calculations.

In Figure 1 we present the results of two simulations with scale separation ratios $k_{\rm f}/k_1$ of 1.5 and 2.2. In both cases, $k_{\rm f}/k_1$ is still relatively small, but the difference in the results is already quite dramatic. For $k_{\rm f}/k_1 = 2.2$ the cycle is more regular while for 1.5 it is quite erratic. We show the toroidal field (i.e. the component B_y in the direction of the mean shear flow) at an arbitrarily chosen mesh point as well as its squared value (which could be taken as a proxy of the sunspot number), the mean magnetic energy in the full domain, as well as its contributions from the mean and fluctuating fields. In all cases, the magnetic field is normalized by the equipartition value, $B_{\rm eq} = \sqrt{\mu_0 \rho_0} u_{\rm rms}$, where $\rho_0 = \langle \rho \rangle$ is the mean density, which is conserved for periodic and shearing–periodic boundary conditions.

† http://www.pencil-code.googlecode.com

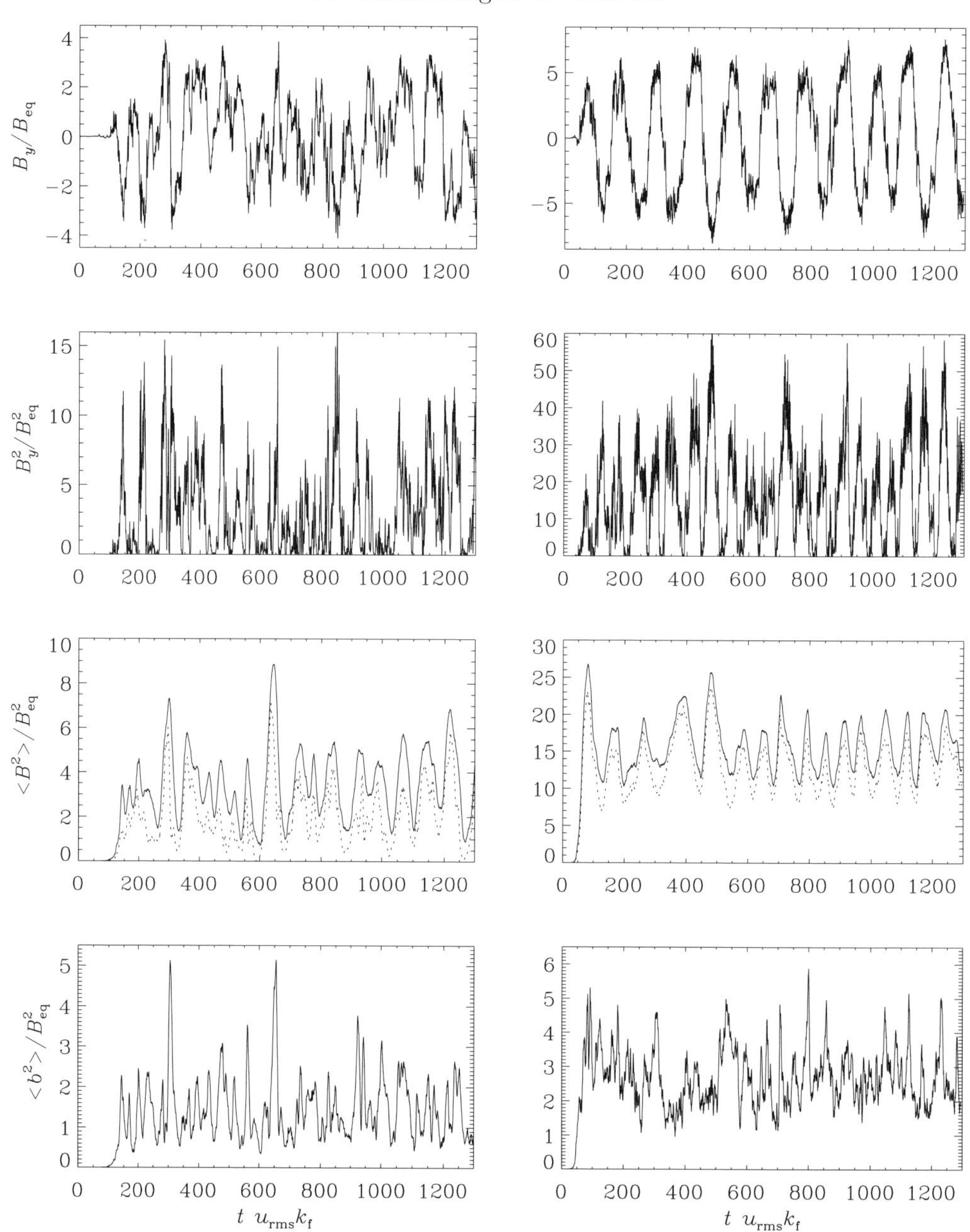

Figure 1. Time sequences of B_y, B_y^2, $\langle \boldsymbol{B}^2 \rangle$ (solid line) together with $\langle \overline{\boldsymbol{B}}^2 \rangle$ (dotted line), and $\langle \boldsymbol{b}^2 \rangle$. The field is always normalized by B_{eq}. Here, $\mathrm{Re}_M = 22$ for $k_f/k_1 = 1.5$ (left column) and $\mathrm{Re}_M = 9$ for $k_f/k_1 = 2.2$ (right column). In both cases, $\mathrm{Pr}_M = 5$ and $\mathrm{Sh} \approx -2$.

Here, angle brackets denote volume averages and overbars are defined as xy averages, so

$$\overline{\boldsymbol{B}}(z,t) = \int \boldsymbol{B}(x,y,z,t)\,\mathrm{d}x\,\mathrm{d}y/L_xL_y, \tag{2.7}$$

which implies that

$$\langle \boldsymbol{B}^2 \rangle = \langle \overline{\boldsymbol{B}}^2 \rangle + \langle \boldsymbol{b}^2 \rangle. \tag{2.8}$$

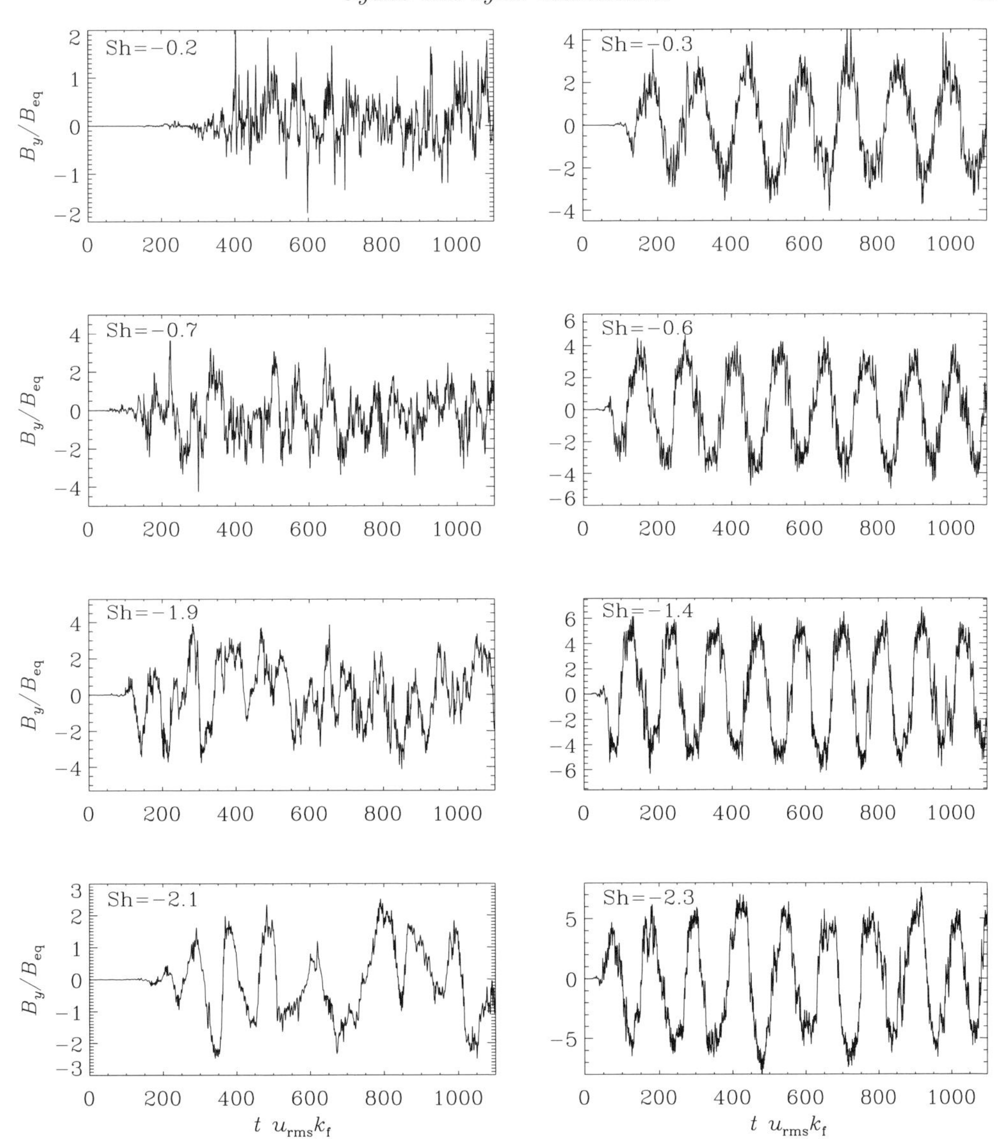

Figure 2. Time series of B_y/B_{eq} for $k_f/k_1 = 1.5$ (left column) and $k_f/k_1 = 2.2$ (right column) for different values of Sh. Again, $\mathrm{Re}_M = 22$ for $k_f/k_1 = 1.5$ (left column) and $\mathrm{Re}_M = 9$ for $k_f/k_1 = 2.2$ (right column), and $\mathrm{Pr}_M = 5$ in both cases. For $k_f/k_1 = 2.2$ the oscillations tend to become less sinusoidal as Sh becomes larger, while for $k_f/k_1 = 1.5$ there are strong fluctuations that tend to become somewhat weaker for larger values of Sh.

One sees that $\langle \boldsymbol{b}^2 \rangle$ shows fluctuations that are not strongly correlated with the variations of the mean field (Figure 1). This is important because the lack of a correlation is sometimes used to argue that the Sun's small-scale magnetic field must be created by a local small-scale dynamo and disconnected with the large-scale dynamo.

There are several other interesting differences between the two cases. The cycle period is given by $\omega_{\mathrm{cyc}} \approx \eta_t k_1^2$ (Käpylä & Brandenburg 2009), and with $\eta_t \approx u_{\mathrm{rms}}/3k_f$ (Sur *et al.* 2008) we have $\omega_{\mathrm{cyc}} \approx \frac{1}{3} u_{\mathrm{rms}} k_f (k_1/k_f)^2$; thus the normalized cycle period is $T_{\mathrm{cyc}} u_{\mathrm{rms}} k_f \approx 2\pi u_{\mathrm{rms}} k_f/\omega_{\mathrm{cyc}} \approx 6\pi (k_f/k_1)^2 \approx 91$ for $k_f/k_1 = 2.2$, which agrees with the result shown in the upper right panel of Figure 1. Next, for $k_{\rm f}/k_1 = 2.2$ the mean magnetic energy and $\langle \boldsymbol{B}^2 \rangle$ are about 4 times larger than for $k_{\rm f}/k_1 = 1.5$. This value is larger than the one expected from the theory where this ratio should be equal to the ratio of the respective values of $k_{\rm f}$ (Blackman & Brandenburg 2002), namely $2.2/1.5 \approx 1.5$.

Next, we ask how the properties of the dynamo change as it becomes more supercritical. This is shown in Figure 2 where we plot time series of $B_y/B_{\rm eq}$ for $k_{\rm f}/k_1 = 1.5$ and $k_{\rm f}/k_1 = 2.2$ for different values of Sh. Note that for $k_{\rm f}/k_1 = 2.2$ the oscillations become less sinusoidal as Sh becomes larger, while for $k_{\rm f}/k_1 = 1.5$ there are strong fluctuations that become somewhat weaker for larger values of Sh.

A more quantitative way of assessing the properties of the large-scale dynamo is by looking at the scaling of the magnetic energy of the mean field and the cycle frequency as a function of the nominal dynamo number, $D = C_\alpha C_S$, where

$$C_\alpha = \alpha/\eta_{\rm T} k_1 = \iota \epsilon_{\rm f} k_{\rm f}/k_1, \quad C_S = S/\eta_{\rm T} k_1 = 3\iota\,{\rm Sh}\,(k_{\rm f}/k_1)^2, \tag{2.9}$$

are non-dimensional numbers measuring the expected value of the α effect (assuming $\alpha \approx \frac{1}{3}\tau\langle \boldsymbol{\omega} \cdot \boldsymbol{u} \rangle$, $\eta_t \approx \frac{1}{3}\tau\langle \boldsymbol{u}^2 \rangle$, with $\tau = (u_{\rm rms} k_{\rm f})^{-1}$ and $\langle \boldsymbol{\omega} \cdot \boldsymbol{u} \rangle \approx k_{\rm f} \langle \boldsymbol{u}^2 \rangle$), as well as the shear or Ω effect. In these approximations, $\eta_{\rm T} = \eta + \eta_t$ is the expected total magnetic diffusivity and $\iota = (1 + 3/{\rm Re}_M)^{-1}$ is a correction factor that takes into account finite conductivity effects resulting from the fact that $\eta_t \neq \eta_{\rm T}$. The results are shown in Figure 3.

For $k_{\rm f}/k_1 = 1.5$, the scaling of $\langle \overline{\boldsymbol{B}}^2 \rangle / B_{\rm eq}^2$ with D suggests that the critical value is between 1 and 2, i.e., somewhat smaller than the theoretical value of 2 (Brandenburg & Subramanian 2005). For $k_{\rm f}/k_1 = 2.2$ the critical value is < 1. The cycle frequencies are approximately independent of D, except that for $k_{\rm f}/k_1 = 1.5$ there is a sharp drop for $D > 10$. Owing to fluctuations, a Fourier spectrum of the time series is not sharp but has a certain width. We determine the quality or width, w_0, by fitting the spectrum to a Gaussian proportional to $P(\omega) \sim \exp[-(\omega - \omega_0)^2/2w_0^2]$. Also the values of w_0, shown in the last panel of Figure 3, are approximately independent of D. For $k_{\rm f}/k_1 = 2.2$, w_0 is substantially smaller than for $k_{\rm f}/k_1 = 1.5$. This indicates that the cycle period is better defined for larger scale separation ratios.

3. Active regions and their inclination angle

One should expect that the sunspot number depends in a complicated way on the magnetic field strength. If sunspots are indeed relatively shallow phenomena, the field must be locally concentrated to field strengths of up to 3 kG. A candidate for a mechanism that can concentrate mean fields of ~ 300 G, which is about 10% of the local equipartition field strength, is the negative effective magnetic pressure instability (NEMPI). This is a remarkable phenomenon resulting from the suppression of turbulent pressure by a moderately strong large-scale magnetic field. This suppression is stronger than the added magnetic pressure from the mean field itself, so the net effect is a negative one.

The fact that this phenomenon can lead to an instability in a stratified layer was first found in mean-field models (Brandenburg *et al.* 2010, 2012; Käpylä *et al.* 2012), and more recently in DNS (Brandenburg *et al.* 2011). However, NEMPI has not yet been able to explain flux concentration in the direction along the mean magnetic field, i.e., the large-scale structures remain essentially axisymmetric. To discuss the theoretical origin of this, we need to look at the underlying mean-field theory. Similar to the effective magnetic diffusivity in the mean electromotive force, the sum of Reynolds and Maxwell stresses from the small-scale field depends on the mean magnetic field in a way that looks like a Maxwell stress from the mean field, but with renormalized coefficients. The concept

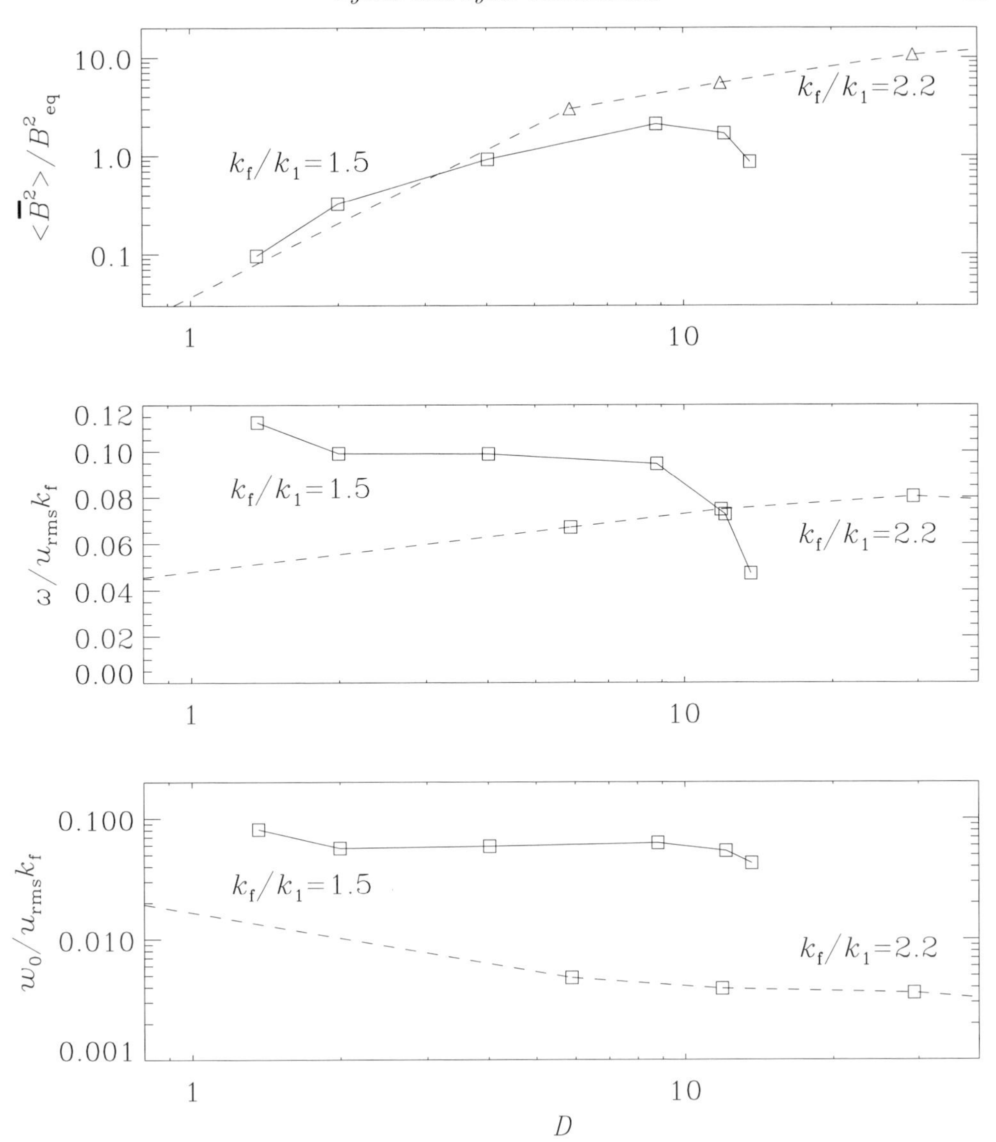

Figure 3. Scaling of the relative magnetic energy of the mean field, $\langle \overline{\boldsymbol{B}}^2 \rangle / B_{eq}^2$, normalized cycle frequency, and normalized quality versus dynamo $D = C_\alpha C_S$.

of expressing the Reynolds stress from the fluctuating velocities, $\overline{u_i u_j}$, by the mean flow $\overline{\boldsymbol{U}}$ is of course familiar and leads to the usual turbulent viscosity term, $-\nu_t(\overline{U}_{i,j} + \overline{U}_{j,i})$. However, in the presence of a mean magnetic field, symmetry arguments allow one to write down additional components, in particular those proportional to $\delta_{ij}\overline{\boldsymbol{B}}^2$ and $\overline{B}_i\overline{B}_j$. The sum of Reynolds and Maxwell stresses from the fluctuating velocity and magnetic fields is given by

$$\overline{\Pi}_{ij}^{\rm f} \equiv \overline{\rho}\,\overline{u_i u_j} - \overline{b_i b_j}/\mu_0 + \tfrac{1}{2}\overline{\boldsymbol{b}^2}/\mu_0, \qquad (3.1)$$

where the superscript f indicates contributions from the fluctuating field. Expressing $\overline{\Pi}_{ij}^{\rm f}$ in terms of the mean field, the leading terms are (Kleeorin *et al.* 1990; Kleeorin &

Rogachevskii 1994; Kleeorin *et al.* 1996; Rogachevskii & Kleeorin 2007)

$$\overline{\Pi}_{ij}^{\mathrm{f}} = q_{\mathrm{s}} \overline{B}_i \overline{B}_j / \mu_0 - \tfrac{1}{2} q_{\mathrm{p}} \delta_{ij} \overline{\boldsymbol{B}}^2 / \mu_0 + ... \tag{3.2}$$

where the dots indicate the presence of additional terms that enter when the effects of stratification affect the anisotropy of the turbulence further. Note in particular the definition of the signs of the terms involving the functions $q_{\mathrm{s}}(\overline{\boldsymbol{B}})$ and $q_{\mathrm{p}}(\overline{\boldsymbol{B}})$. This becomes obvious when writing down the mean Maxwell stress resulting from both mean and fluctuating fields, i.e.,

$$-\overline{B}_i \overline{B}_j / \mu_0 + \tfrac{1}{2} \delta_{ij} \overline{\boldsymbol{B}}^2 / \mu_0 + \overline{\Pi}_{ij}^{\mathrm{f}} = -(1 - q_{\mathrm{s}}) \overline{B}_i \overline{B}_j / \mu_0 + \tfrac{1}{2} (1 - q_{\mathrm{p}}) \delta_{ij} \overline{\boldsymbol{B}}^2 / \mu_0 + ... \tag{3.3}$$

A broad range of different DNS have now confirmed that q_{p} is positive for $\mathrm{Re}_M > 1$, but q_{s} is small and negative. A positive value of q_{s} (but with large error bars) was originally reported for unstratified turbulence (Brandenburg *et al.* 2010). Later, stratified simulations with isothermal stable stratification (Brandenburg *et al.* 2012) and convectively unstable stratification (Käpylä *et al.* 2012) showed that it is small and negative. Nevertheless, $q_{\mathrm{p}}(\overline{\boldsymbol{B}})$ is consistently positive provided $\mathrm{Re}_M > 1$ and $\overline{B}/B_{\mathrm{eq}}$ is below a certain critical value that is around 0.5. This implies that it is probably not possible to produce flux concentrations stronger than half the equipartition field strength. So, making sunspots with this mechanism alone is maybe unlikely.

The significance of a positive q_{s} value comes from mean-field simulations with $q_{\mathrm{s}} > 0$ indicating the formation of three-dimensional (non-axisymmetric) flux concentrations (Brandenburg *et al.* 2010). This result was later identified to be a direct consequence of having $q_{\mathrm{s}} > 0$ (Kemel *et al.* 2012). Before making any further conclusions, it is important to assess the effect of other terms that have been neglected. Two of them are related to the vertical stratification, i.e. terms proportional to $g_i g_j$ and $g_i \overline{B}_j + g_j \overline{B}_i$ with $\boldsymbol{g}$ being gravity. The coefficient of the former term seems to be small (Brandenburg *et al.* 2012; Käpylä *et al.* 2012) and the second only has an effect when there is a vertical imposed field. However, there could be other terms such as $\overline{J}_i \overline{J}_j$ as well as $\overline{J}_i \overline{B}_j$ and $\overline{J}_j \overline{B}_i$ that have not yet been looked at.

Yet another alternative for causing flux concentrations is the suppression of turbulent (convective) heat transport which might even be strong enough to explain the formation of sunspots (Kitchatinov & Mazur 2000). It is conceivable that effects from heat transport become more important near the surface, so that a combination of the negative effective magnetic pressure and the suppression of turbulent heat transport are needed. Another advantage of the latter is that this mechanism works for vertical fields and is isotropic with respect to the horizontal plane, so one should expect the formation of three-dimensional non-axisymmetric structures.

Once a bipolar active region is formed, we must ask ourselves how to explain the observed tilt angle. This question cannot be answered within the framework of NEMPI alone, but it requires a connection with the underlying dynamo. Here, we can refer to the work of Brandenburg (2005) where bipolar regions occur occasionally at the surface of a domain in which shear-driven turbulent dynamo action was found to operate; see Figure 4. The reason for the tilt is here not the Coriolis force, as is usually assumed, but shear; see also Kosovichev & Stenflo (2008). In the simulations of Brandenburg (2005), shear was admittedly rather strong compared with the turbulent velocity, so the effect is exaggerated compared to what we should expect to happen in the Sun. However, even then there are a few other problems. One of them is that the bipolar regions appear usually quite far away from each other (Figure 4). This may not be realistic. On the other hand, it is not clear how to scale this model to the Sun. In this model, the scale

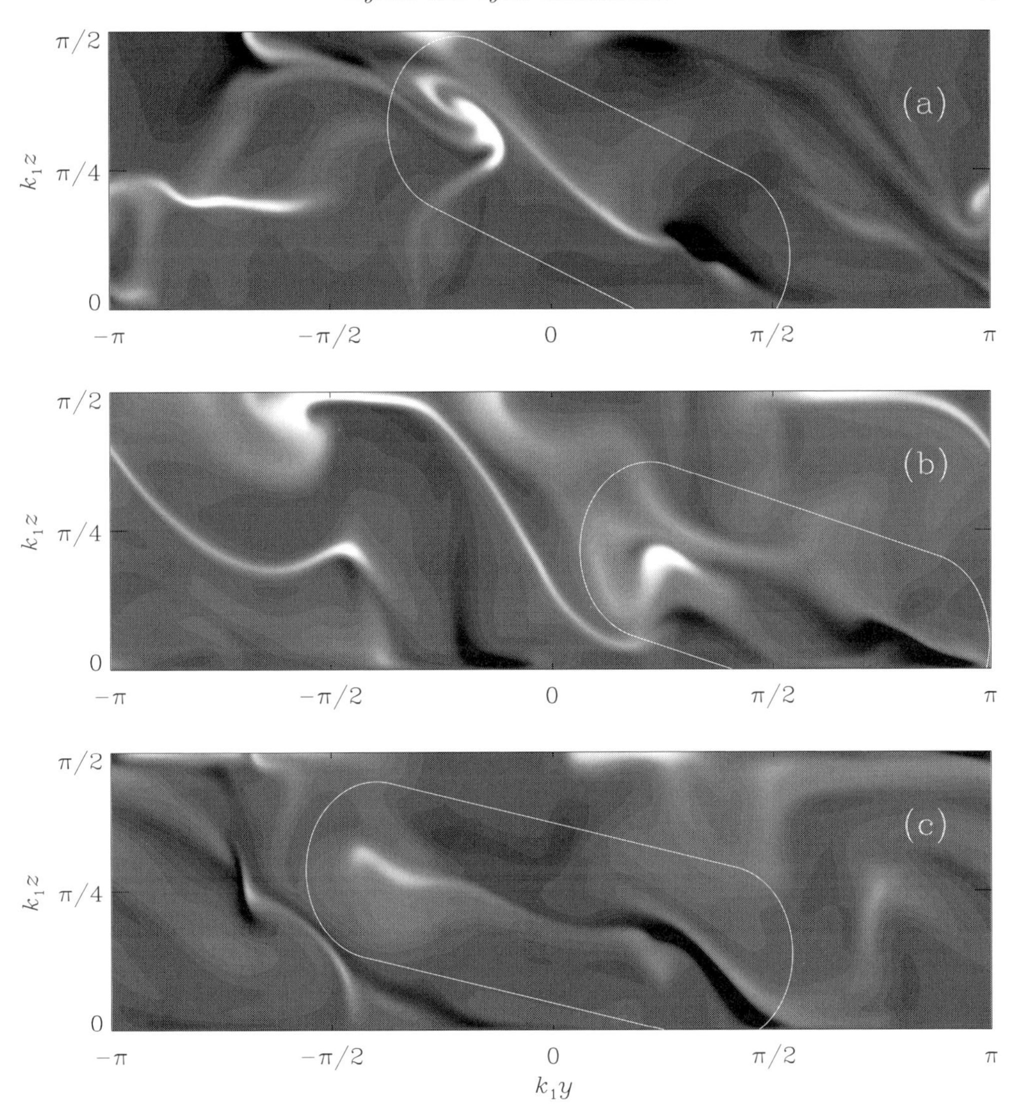

Figure 4. Magnetograms of the radial field at the outer surface on the northern hemisphere at different times for a simulation presented in Brandenburg (2005). Light shades correspond to field vectors pointing out of the domain, and dark shades correspond to vectors pointing into the domain. The elongated rings highlight the positions of bipolar regions. Note the clockwise tilt relative to the y (or toroidal) direction, and the systematic sequence of polarities (white left and dark right) corresponding to $\overline{B}_y > 0$. Here, the z direction corresponds to latitude.

separation ratio is rather small, so the extent of the bipolar regions is comparable to a few times the convective eddy size, which, for the solar surface, is not very big (a few Mm). However, on that small scale one would not expect the effects of differential shear to be very important.

Another issue is that in the model of Brandenburg (2005), bipolar regions occur only occasionally. To illustrate this, we show in Figure 5 the resulting magnetograms for three times that are separated by about 2 turnover times. Clearly, other structures can appear too and the field is not always bipolar.

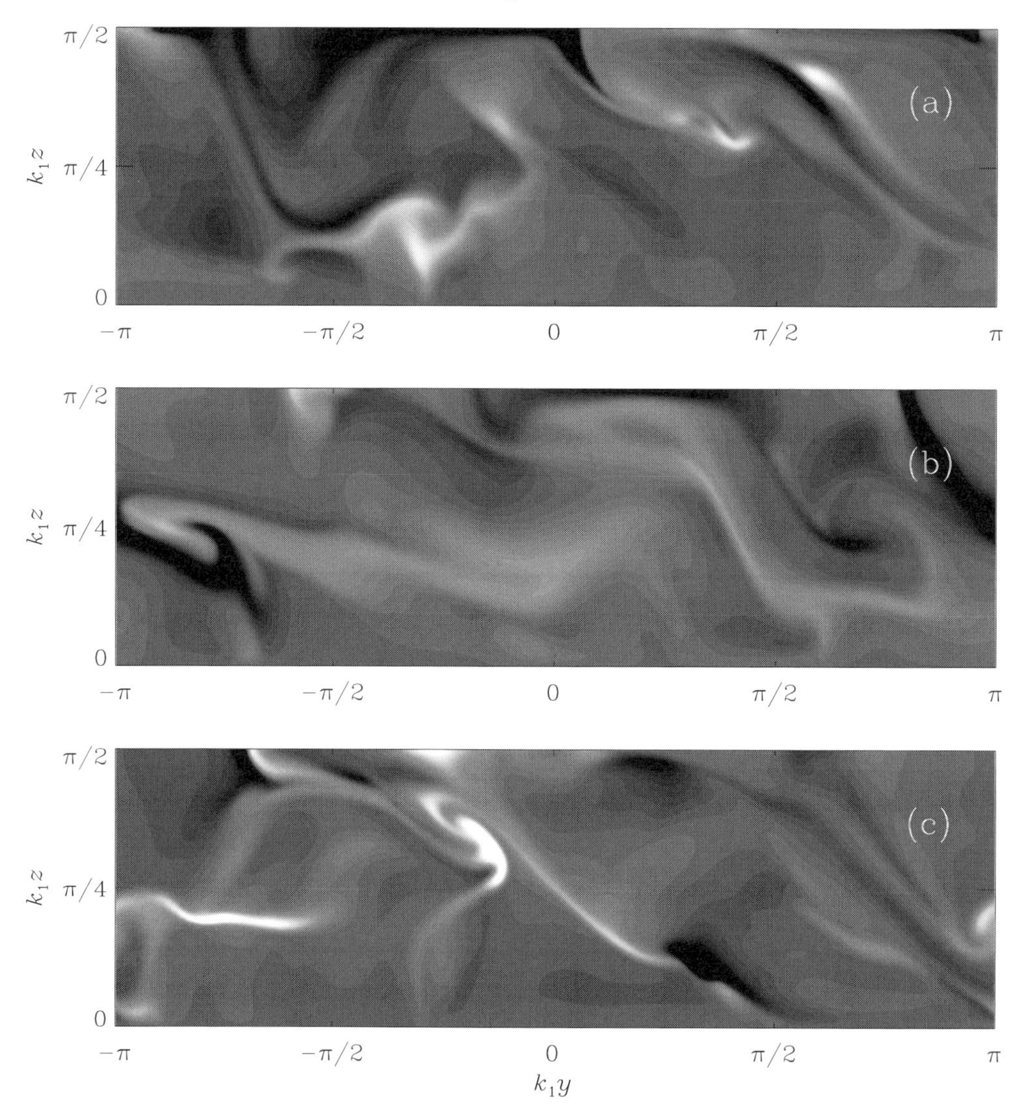

Figure 5. Similar to Figure 4a, showing also the results 4 and 2 turnover times earlier. The last panel is identical to Figure 4a.

4. Conclusions

It is clear that the magnetohydrodynamics of mean magnetic and velocity fields is quite rich and full of important effects. The standard idea that sunspots and bipolar regions form as a result of an instability in the tachocline (Gilman & Dikpati 2000; Cally *et al.* 2003; Parfrey & Menou 2007) may need to be re-examined in view of several new alternative proposals being on the horizon. In addition to comparing models with observations at the solar surface, there are ways of comparison both beneath the surface and above. Particularly exciting are the recent determinations of Ilonidis *et al.* (2011) of some sort of activity at $\approx 60\,\mathrm{Mm}$ depth. It is also of interest to explain magnetic activity in the solar wind, and especially its magnetic helicity which has recently been found to be bi-helical, i.e., of opposite signs at large and small length scales (Brandenburg *et al.* 2011) and positive at small length scales in the north. This is particularly interesting, because such a result has recently been reproduced by distributed dynamo simulations of Warnecke *et al.* (2011) who also find positive magnetic helicity at small length scales in the north. More detailed and varied comparisons between the different approaches are thus required to fully understand the Sun's activity cycles and their long term variations.

Acknowledgments

We acknowledge the allocation of computing resources provided by the Swedish National Allocations Committee at the Center for Parallel Computers at the Royal Institute of Technology in Stockholm and the National Supercomputer Centers in Linköping. This work was supported in part by the European Research Council under the AstroDyn Research Project 227952 and the Swedish Research Council grant 621-2007-4064.

References

Blackman, E. G. & Brandenburg, A. 2002, *ApJ,* 579, 359
Brandenburg, A. 2005, *ApJ,* 625, 539
Brandenburg, A. & Käpylä, P. J. 2007, *New J. Phys.*, 9, 305, 1
Brandenburg, A., Kemel, K., Kleeorin, N., Mitra, D., & Rogachevskii, I. 2011, *ApJ,* 740, L50
Brandenburg, A., Kemel, K., Kleeorin, N., & Rogachevskii, I. 2012, *ApJ*, 749, 179
Brandenburg, A., Kleeorin, N., & Rogachevskii, I. 2010, *Astron. Nachr.,* 331, 5
Brandenburg, A., Rädler, K.-H., & Kemel, K. 2012, *A&A,* 539, A35
Brandenburg, A., Rädler, K.-H., Rheinhardt, M., & Käpylä, P. J. 2008, *ApJ,* 676, 740
Brandenburg, A. & Spiegel, E. A. 2008, *Astron. Nachr.,* 329, 351
Brandenburg, A. & Subramanian, K. 2005, *Phys. Rep.*, 417, 1
Brandenburg, A., Subramanian, K., Balogh, A., & Goldstein, M. L. 2011, *ApJ,* 734, 9
Cally, P. S., Dikpati, M., & Gilman, P. A. 2003, *ApJ,* 582, 1190
Chatterjee, P. & Choudhuri, A. R. 2006, *Solar Phys.,* 239, 29
Choudhuri, A. R. 1992, *A&A,* 253, 277
Choudhuri, A. R., Schüssler, M., & Dikpati, M. 1995, *A&A,* 303, L29
Dikpati, M. & Charbonneau, P. 1999, *ApJ,* 518, 508
D'Silva, S. & Choudhuri, A. R. 1993, *A&A,* 272, 621
Gilman, P. A. & Dikpati, M. 2000, *ApJ,* 528, 552
Guerrero, G. & de Gouveia Dal Pino, E. M. 2008, *A&A,* 485, 267
Ilonidis, S., Zhao, J., & Kosovichev, A. 2011, *Science,* 333, 993
Käpylä, P. J., Brandenburg, A., Kleeorin, N., Mantere, M. J., & Rogachevskii, I. 2012, *MNRAS,* 2465
Käpylä, P. J. & Brandenburg, A. 2009, *ApJ,* 699, 1059
Kemel, K., Brandenburg, A., Kleeorin, N., & Rogachevskii, I. 2012, *Astron. Nachr.,* 333, 95
Kitchatinov, L. L. & Mazur, M. V. 2000, *Solar Phys.,* 191, 325
Kleeorin, N., Mond, M., & Rogachevskii, I. 1996, *A&A,* 307, 293
Kleeorin, N. & Rogachevskii, I. 1994, *Phys. Rev. E,* 50, 2716
Kleeorin, N. I., Rogachevskii, I. V., & Ruzmaikin, A. A. 1990, *Sov. Phys. JETP,* 70, 878
Kosovichev, A. G., & Stenflo, J. O. 2008, *ApJ,* 688, L115
Moss, D., Brandenburg, A., Tavakol, R. K., & Tuominen, I. 1992, *A&A,* 265, 843
Nandy, D., Muñoz-Jaramillo, A., & Martens, P. C. H. 2011, *Nature,* 471, 80
Parfrey, K. P. & Menou, K. 2007, *ApJ,* 667, L207
Rogachevskii, I. & Kleeorin, N. 2007, *Phys. Rev. E,* 76, 056307
Ruzmaikin, A. A. 1981, *Comments Astrophys.*, 9, 85
Schmitt, D., Schüssler, M., & Ferriz-Mas, A. 1996, *A&A,* 311, L1
Sur, S., Brandenburg, A., & Subramanian, K. 2008, *MNRAS,* 385, L15
Tavakol, R. K. 1978, *Nature,* 276, 802
Warnecke, J., Brandenburg, A., & Mitra, D. 2011, *A&A,* 534, A11

Weiss, N. O., Cattaneo, F., Jones, C. A. 1984, *Geophys. Astrophys. Fluid Dyn.*, 30, 305
Yoshimura, H. 1972, *ApJ*, 178, 863

Discussion

Andrés Muñoz-Jaramillo: If sunspots are produced at the surface, which processes would lead to their decay?

Axel Brandenburg: I think it could be the continued submersion of magnetic structures. The system remains time-dependent and new structures will form near the surface, while old ones disappear from view.

Janet Luhmann: Axel, you are one of the few dynamo modelers that include the corona and larger heliosphere in your models and thinking. How important is that to the results of the models, and do models not including that aspect have compromised results?

Axel Brandenburg: The magnetic helicity flux divergence is crucial for alleviating catastrophic alpha quenching; see the next talk by Candaleresi *et al.* (2011). Helicity fluxes through surface carry about 30% of the total; the rest goes through the equator. The observed magnetic helicity spectra support our understanding in terms of the magnetic helicity evolution equation. Models not including helicity fluxes suffer artificially strong catastrophic quenching, but only if their magnetic Reynolds numbers are really large and magnetic helicity evolution is actually included, which is often not the case either.

Arnab Choudhuri: Flux rise simulations based on the idea that the toroidal field forms in the tachocline explained Joy's law other characteristics of sunspot graphs. Can these results be recovered if sunspots form from near-surface fields?

Axel Brandenburg: The tilt angles of near-surface produced flux concentrations is determined by latitudinal shear, as was demonstrated by Brandenburg (2005). Models with the same shear, but different helicity, give still the same tilt angles

Dibyendu Nandy: For the near-surface shallow dynamo to work, you need processes that are slow enough to store, amplify, and transport fields on 11 year timescales. However, in the upper convection zone, eddy turnover timescales are short. Can you comment on how you reconcile this with your shallow dynamo?

Axel Brandenburg: Magnetic helicity conservation is generally responsible for prolonging the time scales. In fact, the partial alleviation of catastrophic quenching by magnetic helicity fluxes means that the timescales are not infinite. The dynamo period is proportional to $(\alpha\partial\Omega/\partial r)^{-1/2}$, so α quenching prolongs the period. As far as active region formation by NEMPI is concerned, the relevant time scale was shown to be longer than the eddy turnover time by a factor that is equal to the square of the scale separation ratio (Brandenburg *et al.* 2011).

Comparative Magnetic Minima:
Characterizing quiet times in the Sun and stars
Proceedings IAU Symposium No. 286, 2011
C. H. Mandrini & D. F. Webb, eds.

doi:10.1017/S1743921312004620

Magnetic helicity fluxes and their effect on stellar dynamos

Simon Candelaresi and Axel Brandenburg

NORDITA, AlbaNova University Center, Roslagstullsbacken 23, SE-10691 Stockholm, Sweden; and Department of Astronomy, Stockholm University, SE-10691 Stockholm, Sweden

Abstract. Magnetic helicity fluxes in turbulently driven α^2 dynamos are studied to demonstrate their ability to alleviate catastrophic quenching. A one-dimensional mean-field formalism is used to achieve magnetic Reynolds numbers of the order of 10^5. We study both diffusive magnetic helicity fluxes through the mid-plane as well as those resulting from the recently proposed alternate dynamic quenching formalism. By adding shear we make a parameter scan for the critical values of the shear and forcing parameters for which dynamo action occurs. For this $\alpha\Omega$ dynamo we find that the preferred mode is antisymmetric about the mid-plane. This is also verified in 3-D direct numerical simulations.

Keywords. Sun: magnetic fields, dynamo, magnetic helicity

---

1. Introduction

The magnetic field of the Sun and other astrophysical objects, like galaxies, show field strengths that are close to equipartition and length scales that are much larger than that of the underlying turbulent eddies. Their magnetic field is assumed to be generated by a turbulent dynamo. Heat is transformed into kinetic energy, which then generates magnetic energy, which reaches values close to the kinetic energy, i.e. they are in equipartition. The central question in dynamo theory is under which circumstances strong large-scale magnetic fields occur and what the mechanisms behind it are.

During the dynamo process, large- and small-scale magnetic helicities of opposite signs are created. The presence of small-scale helicity works against the kinetic α-effect, which drives the dynamo (Pouquet *et al.* 1976; Brandenburg 2001; Field & Blackman 2002). As a consequence, the dynamo saturates on resistive timescales (in the case of a periodic domain) and to magnetic field strengths well below equipartition (in a closed domain). This behavior becomes more pronounced with increasing magnetic Reynolds number $\mathrm{Re_M}$, such that the saturation magnetic energy of the large-scale field decreases with $\mathrm{Re_M}^{-1}$ (Brandenburg & Subramanian 2005), for which it is called catastrophic. Such concerns were first pointed out by Vainshtein & Cattaneo (1992). The quenching is particularly troublesome for astrophysical objects, since for the Sun $\mathrm{Re_M} = 10^9$ and galaxies $\mathrm{Re_M} = 10^{18}$.

2. Magnetic helicity fluxes

The first part of this work addresses if fluxes of small-scale magnetic helicity in an α^2 dynamo can alleviate the catastrophic quenching. We want to reach as high magnetic Reynolds numbers as possible. Consequently we consider the mean-field formalism (Moffatt 1980; Krause & Rädler 1980) in one dimension, where a field $\boldsymbol{B}$ is split into a mean part $\overline{\boldsymbol{B}}$ and a fluctuating part $\boldsymbol{b}$. In mean-field theory the induction equation reads

$$\partial_t \overline{\boldsymbol{B}} = \eta \nabla^2 \overline{\boldsymbol{B}} + \boldsymbol{\nabla} \times (\overline{\boldsymbol{U}} \times \overline{\boldsymbol{B}} + \overline{\boldsymbol{\mathcal{E}}}), \tag{2.1}$$

with the mean magnetic field $\overline{\boldsymbol{B}}$, the mean velocity field $\overline{\boldsymbol{U}}$, the magnetic diffusivity η, and the electromotive force $\overline{\boldsymbol{\mathcal{E}}} = \overline{\boldsymbol{u} \times \boldsymbol{b}}$, where $\boldsymbol{u} = \boldsymbol{U} - \overline{\boldsymbol{U}}$ and $\boldsymbol{b} = \boldsymbol{B} - \overline{\boldsymbol{B}}$ are fluctuations. A common approximation for $\overline{\boldsymbol{\mathcal{E}}}$, which relates small-scale with the large-scale fields, is

$$\overline{\boldsymbol{\mathcal{E}}} = \alpha \overline{\boldsymbol{B}} - \eta_{\rm t} \boldsymbol{\nabla} \times \overline{\boldsymbol{B}}, \tag{2.2}$$

where $\eta_{\rm t} = u_{\rm rms}/(3k_{\rm f})$ is the turbulent magnetic diffusivity in terms of the rms velocity $u_{\rm rms}$ and the wavenumber $k_{\rm f}$ of the energy-carrying eddies, and $\alpha = \alpha_{\rm K} + \alpha_{\rm M}$ is the sum of kinetic and magnetic α, respectively. The kinetic α is the forcing term, i.e. the energy input to the system. In this model $\alpha_{\rm K}$ vanishes at the mid-plane and grows approximately linearly with height until it rapidly falls off to 0 at the boundary. The magnetic α can be approximated by the magnetic helicity in the fluctuating fields: $\alpha_{\rm M} \approx \overline{h}_{\rm f} \times (\mu_0 \rho_0 \eta_{\rm t} k_{\rm f}^2 / B_{\rm eq}^2)$, where μ_0 is the vacuum permeability, ρ_0 is the mean density, $B_{\rm eq} = (\mu_0 \rho_0)^{1/2} u_{\rm rms}$ is the equipartition field strength and $\overline{h}_{\rm f} = \overline{\boldsymbol{a} \cdot \boldsymbol{b}}$ the magnetic helicity in the small-scale fields.

The advantage of this approach is that we can use the time evolution equation for the magnetic helicity to obtain the evolution equation for the magnetic α (Brandenburg *et al.* 2009)

$$\frac{\partial \alpha_{\rm M}}{\partial t} = -2\eta_{\rm t} k_{\rm f}^2 \left(\frac{\overline{\boldsymbol{\mathcal{E}}} \cdot \overline{\boldsymbol{B}}}{B_{\rm eq}^2} + \frac{\alpha_{\rm M}}{{\rm Re}_{\rm M}} \right) - \boldsymbol{\nabla} \cdot \overline{\boldsymbol{\mathcal{F}}}_\alpha, \tag{2.3}$$

where $\overline{\boldsymbol{\mathcal{F}}}_\alpha$ is the magnetic helicity flux term. To distinguish this from the algebraic quenching (Vainshtein & Cattaneo 1992) it is called dynamical α-quenching.

For the flux term on the RHS of equation (2.3) we either choose it to be diffusive, i.e. $\overline{\boldsymbol{\mathcal{F}}}_\alpha = -\kappa_\alpha \boldsymbol{\nabla} \alpha_{\rm M}$, or we take it to be proportional to $\overline{\boldsymbol{\mathcal{E}}} \times \overline{\boldsymbol{A}}$, where $\overline{\boldsymbol{A}}$ is the vector potential of the mean field $\overline{\boldsymbol{B}} = \boldsymbol{\nabla} \times \overline{\boldsymbol{A}}$. The latter expression follows from the recent realization (Hubbard & Brandenburg 2012) that terms involving $\overline{\boldsymbol{\mathcal{E}}}$ should not occur in the expression for the flux of the total magnetic helicity. This will be referred to as the alternate quenching model.

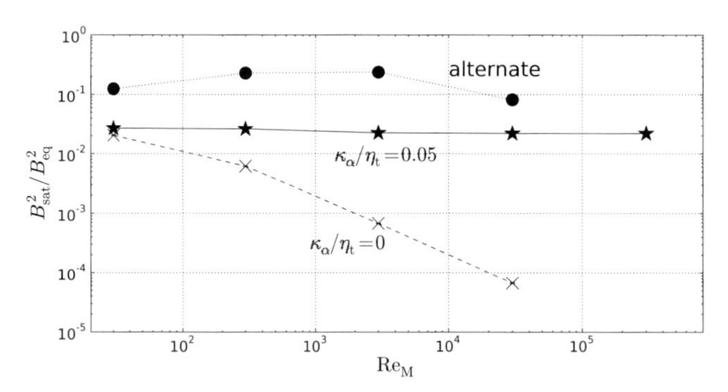

Figure 1. Saturation magnetic energy for different magnetic Reynolds numbers with closed boundaries and diffusive fluxes (solid line) and without (dashed line), as well as the alternate quenching formalism (dotted line).

Without diffusive magnetic helicity fluxes ($\kappa_\alpha = 0$), quenching is not alleviated and the equilibrium magnetic energy decreases as ${\rm Re}_{\rm M}^{-1}$ (Fig. 1). We find that diffusive magnetic helicity fluxes through the mid-plane can alleviate the catastrophic α quenching and allow for magnetic field strengths close to equipartition. The diffusive fluxes ensure that magnetic helicity of the small-scale field is moved from one half of the domain to the other where it has opposite sign. With the alternate quenching formalism we obtain larger values than with the usual dynamical α-quenching–even without the diffusive flux term. The magnetic energies are however higher than expected from simulations (Brandenburg & Subramanian 2005; Hubbard & Brandenburg 2012), which raises questions about the accuracy of the model or its implementation.

3. Behavior of the $\alpha\Omega$ dynamo

In this second part we address the implications arising from adding shear to the system and study the symmetry properties of the magnetic field in a full domain. The large scale velocity field in equation (2.1) is then $\overline{\boldsymbol{U}} = (0, Sz, 0)$, where S is the shearing amplitude and z the spatial coordinate. We normalize the forcing amplitude α_0 and the shearing amplitude S conveniently:

$$C_\alpha = \frac{\alpha_0}{\eta_t k_1} \qquad C_S = \frac{S}{\eta_t k_1^2}, \tag{3.1}$$

with the smallest wave vector k_1.

First we perform runs for the upper half of the domain using closed (perfect conductor or PC) and open (vertical field or VF) boundaries and impose either a symmetric or an antisymmetric mode for the magnetic field by adjusting the boundary condition at the mid-plane. A helical forcing is applied, which increases linearly from the mid-plane. The critical values for the forcing and the shear parameter for which dynamo action occurs are shown in Fig. 2.

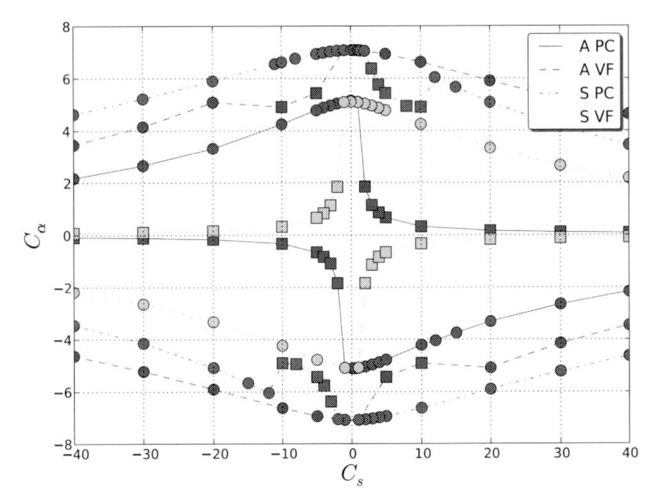

Figure 2. Critical values for the forcing amplitude C_α and the shear amplitude C_s for an $\alpha\Omega$-dynamo in 1-D mean-field to get excited. The circles denote oscillating solutions, while the squares denote stationary solutions.

Imposing the parity of the magnetic field is however unsatisfactory, since it a priori excludes mixed modes. Accordingly we compute the evolution of full domain systems with closed boundaries and follow the evolution of the parity of the magnetic field. The parity is defined such that it is 1 for a symmetric magnetic field and −1 for an antisymmetric one:

$$p = \frac{E_S - E_A}{E_S + E_A}, \quad E_{S/A} = \int_0^H \left[\overline{\boldsymbol{B}}(z) \pm \overline{\boldsymbol{B}}(-z)\right]^2 \, dz, \tag{3.2}$$

with the domain height H. In direct numerical simulations $B_x(z)$ and $B_y(z)$ are horizontal averages. The field reaches an antisymmetric solution after some resistive time $t_{\rm res} = 1/(\eta k_1^2)$ (Fig. 3), which depends on the forcing amplitude C_α. To check whether symmetric modes can be stable, a symmetric initial field is imposed. This however evolves into a symmetric field too (Fig. 4), from which we conclude that it is the stable mode.

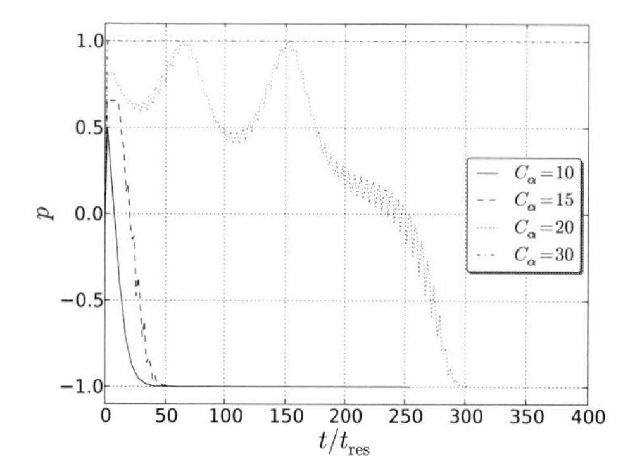

Figure 3. Parity of the magnetic field versus time for a random initial field in 1-D mean–field.

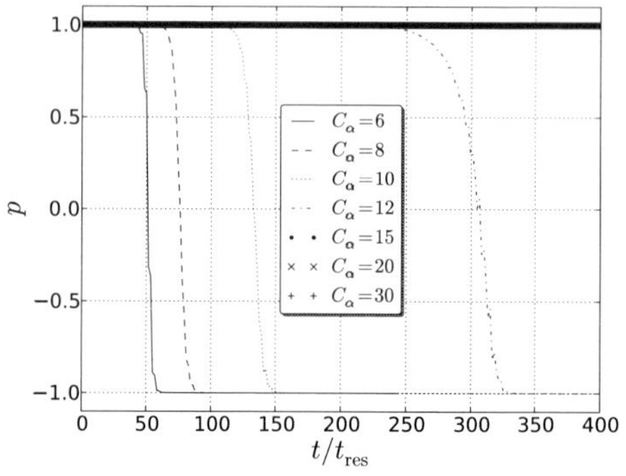

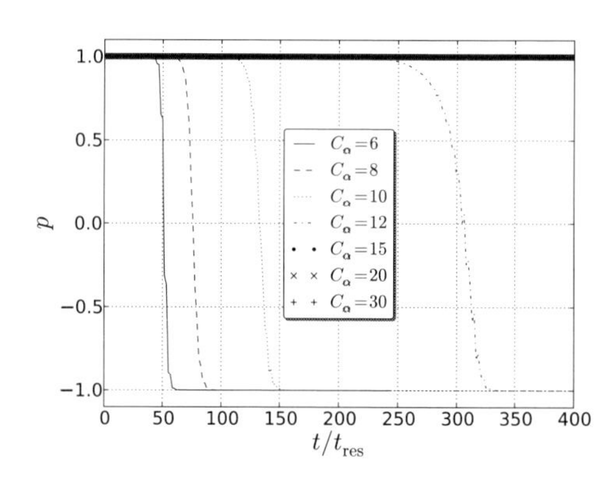

Figure 4. Parity of the magnetic field versus time for a symmetric initial field in 1-D mean–field.

The mean-field results are tested in 3-D direct numerical simulations (DNS); Figs. 5 and 6. The behavior is similar to the mean-field results. The preferred mode is always the antisymmetric one and the time for flipping increases with the forcing amplitude C_α. This is however very preliminary work and has to be studied in more detail.

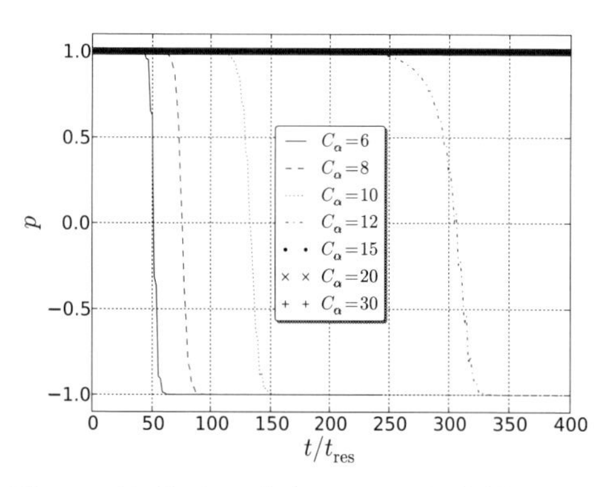

Figure 5. Parity of the magnetic field versus time for a random initial field in 3-D DNS.

Figure 6. Parity of the magnetic field versus time for a symmetric initial field in 3-D DNS.

4. Conclusions

The present work has shown that the magnetic helicity flux divergences within the domain are able to alleviate catastrophic quenching. This is also true for the fluxes implied by the alternate dynamical quenching model of Hubbard & Brandenburg (2012). However, those results deserve further numerical verification. Further, we have shown that, for the model with magnetic helicity fluxes through the mid-plane, the preferred mode is indeed dipolar, i.e. of odd parity. Here, both mean-field models and DNS are found to be in agreement.

References

Brandenburg, A. 2001, *ApJ*, 550, 824
Brandenburg, A., Candelaresi, S., & Chatterjee, P. 2009, *MNRAS*, 398, 1414
Brandenburg, A. & Subramanian, K. 2005, *Astron. Nachr.*, 326, 400
Field, G. B. & Blackman, E. G. 2002, *ApJ*, 572, 685
Hubbard, A. & Brandenburg, A. 2012, *ApJ*, 748, 51.
Krause, F. & Rädler, K.-H., *Mean-field Magnetohydrodynamics and Dynamo Theory.* Oxford: Pergamon Press (1980).
Moffatt, H. K. *Magnetic Field Generation in Electrically Conducting Fluids.* Cambridge: Cambridge Univ. Press (1978).
Pouquet, A., Frisch, U., & Leorat, J. 1976, *J. Fluid Mech.*, 77, 321
Vainshtein, S. I. & Cattaneo, F. 1992, *ApJ*, 393, 165

Discussion

Sacha Brun: Is there a reason that your system prefers antisymmetric solutions? It seems linked to your choice of parameters.

Simon Candelaresi: So far we do not see a reason for that. But we see a parameter dependence of the transition time. We will look at the dependence of the growth rate of the modes on the parameters. This will give us some better clue if also mixed or symmetric modes are preferred.

Gustavo Guerrero: Is there a regime where the advective flux removes all the mean field out of the domain?

Simon Candelaresi: If the advective flux is too high the magnetic field gets shed before it is enhanced, which kills the dynamo. So, there is a window for the advection strength for which it is beneficial for the dynamo.

Comparative Magnetic Minima:
Characterizing quiet times in the Sun and stars
Proceedings IAU Symposium No. 286, 2011
C. H. Mandrini & D. F. Webb

doi:10.1017/S1743921312004632

Modeling the solar cycle: what the future holds

Dibyendu Nandy

Indian Institute of Science Education and Research, Kolkata,
Mohanpur 741252, West Bengal, India
email: dnandi@iiserkol.ac.in

Abstract. Stellar magnetic fields are produced by a magnetohydrodynamic dynamo mechanism working in their interior – which relies on the interaction between plasma flows and magnetic fields. The Sun, being a well-observed star, offers an unique opportunity to test theoretical ideas and models of stellar magnetic field generation. Solar magnetic fields produce sunspots, whose number increases and decreases with a 11 year periodicity – giving rise to what is known as the solar cycle. Dynamo models of the solar cycle seek to understand its origin, variation and evolution with time. In this review, I summarize observations of the solar cycle and describe theoretical ideas and kinematic dynamo modeling efforts to address its origin. I end with a discussion on the future of solar cycle modeling – emphasizing the importance of a close synergy between observational data assimilation, kinematic dynamo models and full magnetohydrodynamic models of the solar interior.

Keywords. Sun: activity, Sun: magnetic fields, Sun: interior, magnetohydrodynamics.

1. Introduction to the Solar Cycle

Sunspots have been observed for many years and systematically so from the early 17th Century since the invention of the telescope. The number of sunspots on the surface of the Sun varies cyclically going through successive maxima and minima with an average period of 11 years (Fig. 1-top). Given that sunspots are strongly magnetized regions (Hale 1908) with typical field strengths on the order of 1000 Gauss (G), this 11 year solar cycle is essentially a magnetic cycle. At the beginning of the cycle, sunspots are observed to appear at high latitudes and with the progress of the cycle, more and more sunspots appear at lower and lower latitudes with the cycle ending close to the equator. This equatorward migration of the sunspot formation belt takes place in both the hemispheres generating the well known solar butterfly diagram of sunspots (Fig. 1-bottom). At the same time, the weaker, radial field outside of sunspots is seen to move poleward with the progress of the cycle – reaching the poles and starting to reverse the old cycle polar field around the time of sunspot maxima. The polar field reaches its strongest intensity at sunspot minimum, thereby having a 90° phase lag with the sunspot cycle. The polar field reversal occurs with an average period of 11 years as well and is therefore coupled to the sunspot cycle.

The solar cycle is not strictly periodic. There are variations from one cycle to another in both the period and more notably in the amplitude of the cycle. Even the length and nature of minima varies from one cycle to another, which was highlighted recently in the unusually long minimum of solar cycle 23 (see Supplementary Information–Fig. 1 from Nandy, Muñoz-Jaramillo & Martens 2011). In the recorded history of sunspot observations there is a period between 1645–1715 AD when hardly any sunspots were seen on the solar surface (Fig. 1-top). This period is termed as the Maunder minimum and it

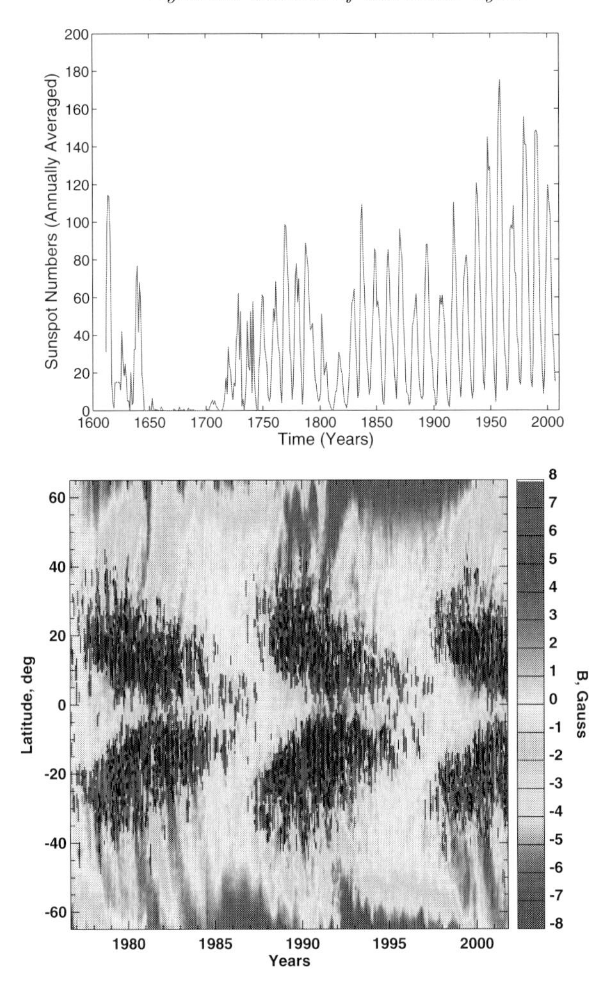

Figure 1. Top: The variation in the number of sunspots observed on the Sun with time over the last fours centuries. The variation is for the most part, cyclic, ranging typically from 9–14 year cycles and with an overall average period of 11 years. The amplitude of the cycle is observed to fluctuate significantly from one cycle to another. A notable period is the Maunder minimum between 1645–1715 AD, when the sunspot cycle pretty much stopped. Solar activity reconstruction based on cosmogenic isotopes indicate that there have been many such episodes of reduced or no activity. Bottom: A butterfly diagram (latitude-time plot) of solar surface magnetic field showing the evolution of latitudes where sunspots emerge (black regions) and the surface radial field (outside of sunspots; scale on right-hand y-axis) with the progress of time. About two and half sunspot cycles are depicted here. The butterfly diagram shows that the sunspot emergence belt moves equatorward, whereas, the weak surface radial field moves poleward with the progress of the solar cycle. The polar field reversal starts at about the time of sunspot maximum, and the polar field is the strongest at sunspot minimum. See colour figures in the online version of the paper.

is thought that the sunspot cycle stopped during this phase but restarted again. Solar activity reconstructions based on cosmogenic isotopes indicate that in the past, several such episodes have occurred (Usoskin, Solanki & Kovaltsov 2007). Such grand minima are therefore an integral part of solar activity over long timescales. Some evidence based on tree-ring data also suggests that there were weak cycles – at least in the solar open flux and polar fields – during the Maunder minima (Miyahara *et al.* 2004).

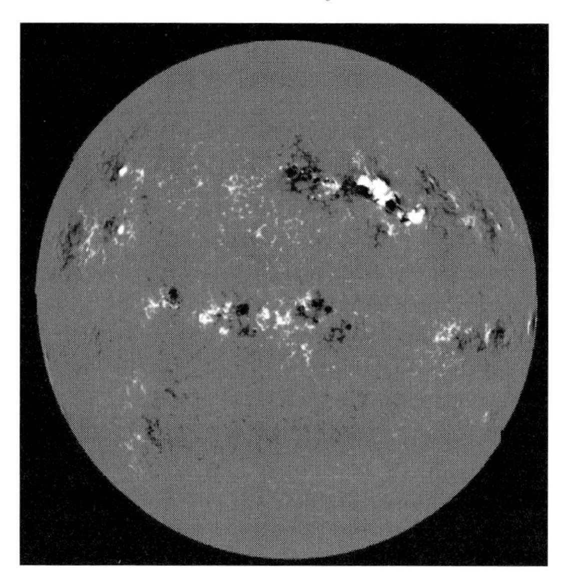

Figure 2. A magnetogram showing the vertical component of the solar magnetic field on the Sun's surface. The typical strength of the magnetic field within sunspots is on the order of 1000 G. White regions denote sunspots which have positive vertical magnetic field and dark regions show sunspots with negative vertical magnetic field. Sunspots occur in bipolar pairs which have opposite polarities. Their polarity orientation is reversed across the equator, which is evident in the above figure. The line joining the two polarities of a sunspot pair are systematically tilted relative to the local parallel of latitude and this tilt increases with increasing latitude away from the equator.

The sunspots themselves are observed to mainly emerge in pairs of opposite polarities (see Fig. 2). The bipolar sunspot pairs have a systematic tilt with pairs at higher latitudes being more tilted than those at low latitudes – following what is known as the Joy's law distribution of tilt angles (Hale *et al.* 1919). It is also seen that the orientation of these bipolar sunspot pairs are always reversed across the equator, i.e., if the leading polarity spots on the Northern-hemisphere are positive, then the leading polarity spots in the Southern-hemisphere are negative. This relative orientation reverses from one 11 year cycle to another and therefore the full magnetic cycle constitutes 22 years.

It is to be noted that similar magnetic cycles – exhibiting a wide range of activity behavior – including Maunder minima like states are also observed in other solar-like stars (Baliunas & Vaughan 1985; Giampapa *et al.* 2006; Poppenhäger *et al.* 2009). This indicates that magnetic cycles are a common and ubiquitous phenomena in stellar physics and plausibly have a common underlying mechanism at their origin (see e.g., Saar Brandenburg 1999, Nandy 2004).

Plasma flows in the Sun are believed to play a crucial role in the solar cycle and are now observed routinely. Tracking sunspots was an early indicator of the differential rotation of the Sun and it is known that the equator of the Sun rotates faster than its poles. Helioseismology makes it possible to probe the solar internal rotation and we know now that in the near-surface layers there is a weak radial gradient in the rotation which changes to a predominantly latitudinal gradient in the main body of the SCZ. At the SCZ base a strong radial differential rotation is concentrated in a layer known as the tachocline (Charbonneau *et al.* 1999). The differential rotation of the Sun constitutes the ϕ-component of the plasma flow in the solar interior. Another component of the plasma flow in the Sun's interior in the meridional plane (r-θ plane) known as the meridional flow of plasma, is observed to be poleward in the surface and near-surface layers (González Hernández *et al.* 2006) and it is believed to pervade the SCZ (Giles *et al.* 1997), with an equatorward counterflow theoretically expected at or below the base of the SCZ (Nandy & Choudhuri 2002). Convective turbulence is also expected to be a feature of the plasma in the Sun's interior and visible manifestations of super-granular cells in the surface are an indicator of the underlying convective turbulent motions.

With these ideas of solar plasma flows and magnetic field dynamics in the background, I will trace the history of development of ideas in solar dynamo theory in the next section.

2. Solar Dynamo Theory: Development of Ideas

The behavior of magnetic fields in plasma systems (such as that in stellar interiors) is governed by the induction equation

$$\frac{\partial \mathbf{B}}{\partial t} = \nabla \times (\mathbf{v} \times \mathbf{B} - \eta \nabla \times \mathbf{B}), \tag{2.1}$$

where $\mathbf{B}$ is the magnetic field, $\mathbf{v}$ the velocity field and η the effective magnetic diffusivity of the system. Astrophysical plasma systems have a high characteristic magnetic Reynolds number (the ratio of the first to the second term on the R.H.S. of the above equation). The concept of flux-freezing holds in such systems (Alfvén 1942), wherein, the magnetic fields remain frozen in the fluid flow – essentially because the diffusion timescale is much larger compared to the flow timescale. This allows the energy of convective flows in the solar convection zone (SCZ) to be drawn into producing and amplifying magnetic fields against dissipation – which is the dynamo mechanism in a nutshell!

Assuming spherical symmetry (which is applicable to stellar interiors), the magnetic and velocity fields can be expressed as

$$\mathbf{B} = B_\phi \hat{\mathbf{e}}_\phi + \nabla \times (A \hat{\mathbf{e}}_\phi) \tag{2.2}$$

$$\mathbf{v} = r \sin(\theta) \Omega \hat{\mathbf{e}}_\phi + \mathbf{v}_p. \tag{2.3}$$

The first term on the R.H.S. of Eqn. 2.2 is the toroidal component (i.e, in the ϕ-direction) and the second term is the poloidal component (i.e., in the r-θ plane) of the magnetic field. In the case of the velocity field (Eqn. 2.3), these two terms are the rotation Ω and meridional circulation v_p, respectively. Due to the differentially rotating Sun, any pre-existing poloidal field would get stretched in the direction of rotation creating a toroidal component. It is believed that these strong toroidal flux tubes are stored and amplified in the tachocline region – which coincides with a region of sub-adiabatic temperature gradient known as the overshoot layer and where magnetic buoyancy is suppressed (Spiegel & Weiss 1980; van Ballegooijen 1982). These horizontal toroidal flux tubes in the solar interior become unstable to magnetic buoyancy (Parker 1955a) when they come out in to the SCZ (through overshooting turbulence or meridional up-flows) and erupt through the solar surface creating bipolar sunspot pairs. During their buoyant rise, these flux tubes acquire tilt due to the Coriolis force generating the Joy's law distribution of tilt angles (D'Silva & Choudhuri 1993).

For the dynamo chain of events to be completed, the toroidal field has to regenerate the poloidal component so that the solar cycle can go on through a recycling of magnetic flux between the two components (in Eqn. 2.2) mediated via the plasma flows. The first mechanism proposed as a regeneration mechanism for the poloidal component of the magnetic field is the action of small-scale helical turbulence on rising magnetic flux tubes (Parker 1955b) – which twists the toroidal field back into the meridional place (thereby

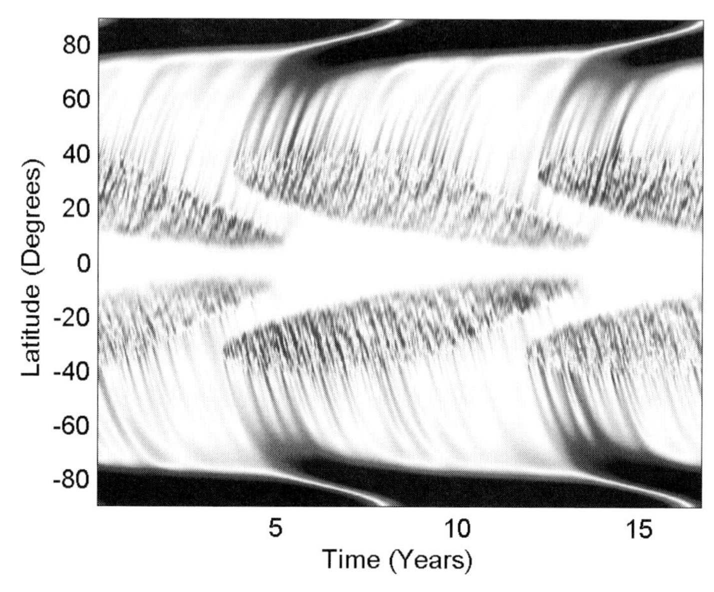

Figure 3. A simulated solar butterfly diagram showing the evolution of the Sun's radial magnetic field at the surface from a recently developed kinematic solar dynamo model. Eruptions of bipolar sunspot pairs are evident at lower latitudes; the redistribution of their flux through surface flux transport processes involving meridional circulation, differential rotation and diffusion generates a large-scale global dipolar field. This cancels the older cycle polar fields at high latitudes and creates the polar fields of the new cycle. The observational features of the solar cycle are reproduced well through such simulations. This particular simulation is from a kinematic dynamo model that includes a realistic algorithm for the emergence of tilted bipolar sunspot pairs (Muñoz-Jaramillo, Nandy & Martens 2010). See colour figure in the online version of the paper.

generating a r-θ or poloidal component). This mechanism relying on helical turbulence is traditionally known as the mean-field α-effect.

Simulations of the buoyant rise of toroidal flux tubes showed that the field strength of these flux tubes have to be close to 10^5 G at the base of the SCZ to match the emergence properties of bipolar sunspot pairs (Choudhuri & Gilman 1987) and their tilt angle distribution (D'Silva & Choudhuri 1993; Fan, Fisher & DeLuca 1993). The required field strength is much stronger than the equipartition magnetic field strength in the SCZ (at which the energy in the fields and convection are in equipartition). This called into question the viability of the mean-field α-effect which is likely to get quenched by such strong fields.

An alternative mechanism, originally proposed by Babcock (1961) and Leighton (1969), is now the leading candidate for the regeneration of the poloidal component of the magnetic field. In this mechanism, the decay of tilted bipolar sunspot pairs, mediated via diffusion and meridional circulation creates a large-scale solar dipole from the emergence of individual bi-poles with a non-zero tilt (which have a net dipole moment). Thus the large-scale solar poloidal field is created and manifests as the global solar dipole configuration – most intense at solar minimum – from which the next cycle toroidal field is generated. This keeps the cycle going. This observed process of surface flux distribution is reproduced by surface flux transport simulations (Mackay & Lockwood 2002; Jiang *et al.* 2011) and dynamo models (Muñoz-Jaramillo, Nandy & Martens 2010) based on the Babcock and Leighton (BL) idea (see also Charbonneau 2010 and references therein). More importantly, recently inferred correlations between observed proxies of the BL source term and the strength of the next sunspot cycle (Dasi-Espuig *et al.* 2010) lend strong support to the BL mechanism for poloidal field creation.

While full magnetohydrodynamic (MHD) simulations of the solar interior are becoming more and more sophisticated due to rapid improvements in our numerical capabilities they are still not capable of reproducing the rich variety of solar cycle observations. Kinematic dynamo models play a useful role here, wherein, their simplicity allows for a transparent underlying physics, and the capability to explore the full breadth of solar cycle properties through numerical investigations. In these models, the plasma flow is prescribed (and is not a dynamical variable) and the magnetic induction equation (Eqn. 2.1) is solved with suitable boundary conditions and parameterizations and algorithms necessary for accounting for the physics of the solar interior. Kinematic flux transport dynamo models (as opposed to models based on dynamo waves) rely on the plasma flows to transport flux between different source layers for the field components and have now become the leading tool for exploring solar activity (Jouve *et al.* 2008; see also the review by Charbonneau 2010) and recently developed models are good enough to reproduce not only the global features of the solar cycle, but also the dynamics of surface flux transport and polar field reversal accurately (Fig. 3).

We do not provide a pedagogical description of the mean field dynamo equations here and instead refer (interested) readers to Nandy (2010a) – where a more complete description is provided, and to Nandy (2010b) – where some issues of current interest are discussed that will complement this review. I now jump ahead and devote the next section to discussing what I deem to be fertile grounds for dynamo theory in the coming years.

3. A Future Outlook

It is interesting times for solar dynamo theory. In the last decade major improvements have happened in the field of kinematic dynamo modeling and the so-called "flux transport" (kinematic) dynamo models have become useful tools for explaining various features of the solar cycle. *I must note here in passing that I do not necessarily endorse the usage of the term "flux-transport" in this context as many dynamo modelers who declare their models to be "flux-transport" dynamo models do not include either magnetic buoyancy or downward pumping of magnetic flux – which are plausibly the most efficient of all flux transport mechanisms in the solar interior.* Simultaneously, in the last several years or so, full MHD models have developed to the point where structured large-scale magnetic fields (Brown *et al.* 2011) and cyclic reversals in solar-like conditions are starting to be seen (Ghizaru, Charbonneau & Smolarkiewicz 2010). With the available long-term sunspot record in conjunction with polar magnetic field measurements stretching over three cycles and helioseismic observations covering at least a full cycle, we also now have the requisite observational data to constrain and indeed, drive dynamo simulations. I have no doubt that the most important advances in dynamo theory are going to come at the confluence of all of these, where kinematic and full MHD simulations of the solar cycle complement each other and assimilate magnetic field and plasma flow data. Below, I elaborate on some specific aspects – which are important and which I believe are going to garner sustained attention in the near future.

3.1. *It takes Two to Tango: Kinematic Dynamos and full MHD Simulations*

The history of solar dynamo theory has been such that kinematic dynamo modeling and full MHD simulations of solar and stellar interiors have often progressed in parallel with very limited feedback from each other. This is changing. A prominent example is explorations of the nature of the meridional flow which is largely unconstrained by observations. Kinematic dynamo models, especially those based on the BL mechanism – wherein the source layers for the poloidal and toroidal field are spatially separated, find that the meridional circulation is an important factor in determining various aspects of the solar cycle, including fluctuations in the period (Charbonneau & Dikpati 2000) and the latitudinal distribution of sunspots (Nandy & Choudhuri 2002); the latter proposed that the flow has to penetrate below the base of the SCZ for reproducing the observed solar butterfly diagram. While this suggestion was controversial at the time, MHD models are now beginning to find overshooting plumes and plasma flows that can indeed penetrate below the base of the SCZ (Rogers, Glatzmaier & Jones 2006; Garaud & Brummell 2008). This penetration of plasma plumes and flows into the radiative interior is important because this region is stable to magnetic buoyancy; this in conjunction with the longer diffusion timescales there, allow for magnetic field storage and slow transport (by the equatorward meridional counterflow) that are relevant for establishing an 11 year solar cycle. Clearly, this convergence of kinematic dynamo models with MHD simulations of the solar interior reiterate the importance of meridional circulation vis-a-vis the solar cycle.

It is to be noted that kinematic dynamo models do not necessarily self-consistently include the full range of physics possible in stellar interiors and therefore must assimilate those physical processes that full MHD simulations point out to be important. A case in point is the downward pumping of magnetic flux in the presence of rotating, stratified convection – a process that is very efficient at pumping magnetic fields down into the stable interior below the base of the SCZ (Tobias *et al.* 2001; Käpylä *et al.* 2006). Only few kinematic dynamo models include downward pumping – those that do show that it has a significant influence on the dynamo (Guerrero & de Gouveia Dal Pino 2008); evidently, given its short timescales, the process of flux pumping competes with meridional flow and turbulent diffusion in transporting flux between the source layers of the dynamo and should be considered in kinematic dynamo models. Note that MHD simulations also point out that the meridional flow could be multi-cellular deeper down in the SCZ – which could impact the flux transport process. I believe useful exchanges between kinematic dynamo modeling and full MHD simulations of the kind highlighted above are going to lead to more meaningful advances in the understanding of the solar cycle.

3.2. *Beyond Normal: Irregularities in the Solar Cycle*

Fig. 1 shows that the amplitude of the solar cycle varies significantly. While magnetic buoyancy presumably sets the order of magnitude field strength (Nandy 2002), smaller fluctuations around these values likely have a different origin. Then there are grand minima phases like the Maunder minimum when the sunspot cycle basically stops. There are also variations in the period of the cycle, the length of the minimum between solar cycles and the strength of the polar field. Given that dynamo models are now reasonably successful in reproducing and explaining the regular features of the solar cycle, the challenge ahead lies in confronting models of magnetic field generation with irregularities in the solar cycle.

What causes these variations – is it due to stochastic fluctuations, e.g., in the meridional flow and the dynamo α-effect (Charbonneau & Dikpati 2000), or is it the non-linear feedback of the fields on the flows (Tobias 1997) that may lead to chaotic modulation or is it due to sudden changes in some dynamo ingredient like the meridional flow (Karak 2010). Some observational analysis seems to rule out the existence of low-dimensional chaos in the solar cycle (Mininni, Gómez & Mindlin 2002), however, the underlying non-linearities of the system can certainly exhibit chaotic modulation when the forcing is strong; the jury is still out on this one.

While it is tempting to surmise that the same process(es) that produce cycle to cycle fluctuations also sometimes bring down the activity levels to produce grand minima like episodes – that would be intellectually satisfying – does it hold in reality? The unusually long minimum of Solar Cycle 23 and a dynamo based explanation of this episode (Nandy, Muñoz-Jaramillo, & Martens 2011) suggests that such variations in the nature of solar minima are caused by different processes than those that cause grand minima episodes. The way the Sun gets into Maunder-minima like episodes and the way it gets out from these quiescent phases also holds keys to understanding many of the subtleties of the dynamo process. We are currently just scratching the surface here and need to know more about the origin and implications of solar cycle fluctuations and extreme events.

3.3. *Data Assimilation and Solar Cycle Forecasting*

The holy grail of solar dynamo theory is to be able to forecast the solar cycle. It is easier said than done, especially given the uncertainties regarding the relative roles of flux transport processes such as diffusion, circulation and downward flux pumping in the solar interior. Dynamo based predictions for the next cycle do not converge (Dikpati, de Toma & Gilman 2006; Choudhuri, Chatterjee & Jiang 2007) and this is due to differing importance laid on meridional circulation and turbulent diffusion (see Yeates, Nandy & Mackay 2008). The memory of the solar cycle is related to the timescale of the dominant flux transport process and since the build up of memory makes prediction possible, accurate predictions will require understanding the subtleties of various flux transport processes in the solar interior. Prediction also requires the assimilation of data. At the least, the data could be polar field (or surface magnetic field) observations – which are precursors of the next cycle (Dasi-Espuig *et al.* 2010) – and which can be fed into kinematic dynamo models to generate the toroidal field of the next cycle.

However, more sophisticated treatment of the prediction problem, including predictions of solar cycle timing would also necessitate the realistic assimilation of observed plasma flow data in solar dynamo models; such efforts are just beginning (see Muñoz-Jaramillo, Nandy & Martens 2009). Efforts are also on to take methods used in meteorology and develop them for solar cycle predictions (see e.g., Katiashvili & Kosovichev 2008 and Jouve, Brun & Talagrand 2011). Given the success of such techniques in weather prediction, this would seem to a pragmatic approach. In this approach, an observed state of the Sun is integrated forward in time using a solar model, and continuous corrections based on real-time observations are made to keep the model on track. Evidently, such an approach requires a good dynamo model to begin with and techniques for assimilating both plasma flow and magnetic field observations within this model. Given the need of reliable space weather forecasts, it would seem that attempts to predict the solar cycle would also drive the motivation to understand it better in the coming years!

3.4. *Solar-Stellar Connections: Stellar Activity and Dynamo Theory*

The Sun is one amongst many stars with similar properties. There is no reason than that a dynamo model for magnetic field generation would be unique to the Sun; in fact, one would think that the Sun just offers one data point for the dynamo, whilst observations of other solar-like stars offer multiple data points – in the parameter space of e.g., differential rotation or convective turn-over time – to constrain the dynamo mechanism with. An extreme form of this argument would imply that an acceptable dynamo model for the Sun should also be able to explain the activity of other stars with suitably adjusted parameters conforming to the interior of these stars. Typically however, not all dynamo models are confronted with stellar observational data. This is because the stellar data does not have well-constrained plasma flow observations (differential rotation and meridional circulation) and in the absence of these, there are too many free parameters in spatially extended dynamo models for the whole exercise to be meaningful. Often, low-order mathematical models have been used instead to explore the wide range of stellar activity observations and they provide useful insight on how the exhibited dynamics is related to the underlying structure of the dynamo equations (see e.g., Wilmot-Smith *et al.* 2005, 2006 and references therein).

Instead of modeling individual, unresolved stars, an alternative and promising approach is to extend solar dynamo models to explore well-known relationships that exist in the stellar data, e.g., the rotation rate, activity amplitude and period relationship (Noyes, Vaughan & Weiss 1984; Brandenburg, Saar & Turpin 1998; Nandy 2004; Jouve, Brown & Brun 2010). Surface flux transport ideas gleaned from the Sun can also be adapted to explain the formation of polar caps and starspot dynamics (Schrijver & Title 2001; Holzwarth, Mackay & Jardine 2006; Isik, Schüssler & Solanki 2007). Such approaches sometimes throw up surprises; it appears that the BL dynamo as envisaged for the Sun runs into some difficulty when confronted with the activity-period relationship from stellar observations (Jouve, Brown & Brun 2010). Given that the BL dynamo is supported both by theoretical models and solar observations are we missing something when we confront such models with stellar data – perhaps a realistic handling of downward flux pumping? A combination of full MHD simulations of stellar interiors and kinematic dynamo modeling would be required to address such vexing issues; certainly more effort should be devoted to using stellar activity relationships to constrain dynamo theory more meaningfully.

3.5. *Space Climate: The Solar Dynamo in Time*

I would like to end by bringing up a topic that has not received much attention but which is going to become important in the years to come. This pertains to uncovering the long-term evolution of the solar dynamo over the life-time of the Sun and using this in conjunction with ideas of solar luminosity variations with radius and age to come up with a complete profile of the radiative history of the Sun. Theoretical and observational studies of solar-like stars at different main-sequence ages are going to play an important role in this regard (Nandy & Martens 2007). With increasing importance of understanding the causes and consequences of global climate change, the need to figure out how our parent star has shaped our atmosphere and climate over time will also become an important quest.

Acknowledgments

I would like to acknowledge the Department of Science and Technology of the Government of India for supporting my research through the Ramanujan Fellowship. I am grateful to the Argentinians – specifically for being wonderful hosts, and in general – for giving tango to the world.

References

Alfvén, H. 1942, *Ark. f. Mat. Astr. o Fysik*, 29B, No. 2
Babcock, H. W. 1961, *ApJ*, 133, 572
Baliunas, S. L. & Vaughan, A. H. 1985, *ARAA*, 23, 379
Brandenburg, A., Saar, S. H., & Turpin, C. R. 1998, *ApJ*, 498, L51

Brown, B. P. *et al.* 2011, *ApJ*, 731, 69
Charbonneau, P. 2010, *Living Reviews in Solar Physics*, 7, 3
Charbonneau, P. *et al.* 1999, *ApJ*, 527, 445
Charbonneau, P. & Dikpati, M. 2000, *ApJ*, 543, 1027
Choudhuri, A. R. & Gilman, P. A. 1987, *ApJ*, 316, 788
Choudhuri, A. R., Chatterjee, P., & Jiang, J. 2007, *Phys. Rev. Lett.*, 98, 131103
Dasi-Espuig, M. *et al.* 2010, *A&A*, 518, A7
Dikpati, M., de Toma, G., & Gilman, P. A. 2006, *Geophys. Res. Lett.*, 33, 5102
D'Silva, S. & Choudhuri, A. R. 1993, *A&A*, 272, 621
Fan, Y., Fisher, G. H., & Deluca, E. E. 1993, *ApJ*, 405, 390
Garaud, P. & Brummell, N. H. 2008, *ApJ*,674, 498
Ghizaru, M., Charbonneau, P., & Smolarkiewicz, P. K. 2010, *ApJ*, 715, L133
Giampapa, M. S., Hall, J. C., Radick, R. R., & Baliunas, S. L. 2006, *ApJ*, 651, 444
Giles *et al.* 1997, *Nature*, 390, 52
González Hernández, I. *et al.* 2006, *ApJ*, 638, 576
Guerrero, G. & de Gouveia Dal Pino, E. M. 2008, *A&A*, 484, 267
Hale, G. E. 1908, *ApJ*, 28, 315
Hale, G. E., Ellerman, F., Nicholson, S. B., & Joy, A. H. 1919, *ApJ*, 49, 153
Holzwarth, V., Mackay, D. H., & Jardine, M. 2006, *MNRAS*, 369, 1703
Isik, E., Schüssler, M., & Solanki, S. K. 2007, *A&A*, 464, 1049
Jiang, J., Cameron, R. H., Schmitt, D., & Schüssler, M. 2011, *A&A*, 528, A83
Jouve, L. *et al.* 2008, *A&A*, 483, 949
Jouve, L., Brown, B.P., & Brun, A.S. 2010. *A&A*, 509, A32
Jouve, L., Brun, A. S., & Talagrand, O. 2011, *ApJ*, 735, 31
Käpylä, P. J., Korpi, M. J., Ossendrijver, M., & Stix, M. 2006, *A&A*, 455, 401
Karak, B. B. 2010, ApJ, 724, 1021
Kitiashvili, I. & Kosovichev, A. G. 2008, *ApJ*, 688, L49
Leighton, R. B. 1969, ApJ, 156, 1
Mackay, D. H. & Lockwood, M. 2002, *Solar Phys.*, 209, 287
Mininni, P. D., Gómez, D. O., & Mindlin, G. B. 2002, *Phys. Rev. Lett.*, 89, 061101
Miyahara, H. *et al.* 2004, *Solar Phys.*, 224, 317
Muñoz-Jaramillo, A., Nandy, D., & Martens, P. C. H. 2009, ApJ, 698, 461
Muñoz-Jaramillo, A., Nandy, D., Martens, P. C. H.., & Yeates, A. R. 2010, ApJL, 720, L20
Nandy, D. 2002, Astrophysics and Space Science, 282, 209
Nandy, D. 2004, *Solar Phys.*, 224, 161
Nandy, D. 2010a, in Heliophysical Processes, ed. N. Gopalswamy, S. Hasan & A. Ambastha, Springer (Berlin), 35
Nandy, D. 2010b, in Magnetic Coupling between the Interior and Atmosphere of the Sun, ed. S. S. Hasan & R. J. Rutten, Springer (Berlin), 86
Nandy, D. & Choudhuri, A. R. 2002, Science, 296, 1671
Nandy, D. & Martens, P. C. H.. 2007, *Adv. Sp. Res.*, 40, 891
Nandy, D., Muñoz-Jaramillo, A., & Martens, P. C. H. 2011, *Nature*, 471, 80
Noyes, R. W., Weiss, N. O., & Vaughan, A. H. 1984, *ApJ.* 287, 769
Parker, E. N. 1955a, ApJ, 121, 491
Parker, E. N. 1955b, ApJ, 122, 293
Poppenhäger, K., Robrade, J., Schmitt, J. H. M.. M., & Hall, J. C. 2009, *A&A*, 508, 1417
Rogers, T. M., Glatzmaier, G. A., & Jones, C. A. 2006, *ApJ*, 653, 765
Saar, S. H. & Brandenburg, A. 1999, *ApJ*, 524, 295
Schrijver, C. J. & Title, A. M. 2001, *ApJ*, 551, 1099
Spiegel, E. A. & Weiss, N. O. 1980, *Nature*, 287, 616
Tobias S. M. 1997, *A&A*, 322, 1007
Tobias, S. M., Brunnell, N. H., Clune, T. L., & Toomre, J. 2001, ApJ, 549, 1183
Usoskin, I. G., Solanki S. K. & Kovaltsov, G. A. 2007, *A&A*, 471, 301
van Ballegooijen, A. A. 1982, *A&A*, 113, 99

Wilmot-Smith, A. L. *et al.* 2005, MNRAS, 363, 1167
Wilmot-Smith, A. L. *et al.* 2006, ApJ, 652, 696
Yeates, A. R., Nandy, D., & Mackay, D. H. 2008, ApJ, 673, 544

Discussion

Michael Thompson: One comment and one question. The comment is that I strongly agree with you that short-term prediction is still possible in a chaotic system. The question is about the flux transport dynamo, in which the meridional circulation regulates the cycle length. I assume that a model of the solar dynamo should also explain dynamos in other stars. The observations show that faster rotators have shorter cycle periods, which should require faster meridional circulation. But Sacha Brun asserted from his models that faster rotators have slower circulation. So, is Sacha wrong or are flux transport dynamos dead?

Dibyendu Nandy: I appreciate your comment on the issue of predictability in chaotic systems. Too often this issues has been trivialized by strong opinions lacking strong physical foundations. I reiterate that the inherent memory of a dynamical system allows for short-term predictions and this is certainly the case for the solar dynamo. In your question you point out an important inconsistency. I believe the answer to this dilemma lies in proper accounting of downward magnetic flux pumping in flux transport dynamo models, which is not being done currently. Downward flux pumping short-circuits meridional circulation and makes the latter relatively less important. If faster rotating stars have stronger downward flux pumping, this would reconcile flux transport dynamos with Sacha Brun's simulations and stellar observations.

Arnab Choudhuri: In your otherwise comprehensive review, you have not covered one topic which I consider very important: hemispheric coupling. Why is it that strong cycles are strong in both hemispheres and weak cycles are weak in both hemispheres (see Chatterjee & Choudhuri 2006, Solar Phys. 239, 299)? Is hemispheric coupling important for driving Grand Minima? Some of these issues should be investigated in greater depth.

Dibyendu Nandy: The kind of hemispheric coupling you are talking about follows naturally from strong turbulent diffusivity in solar convection zones and the consequent maintenance of solar-like parity, which I did discuss. In fact, it is my conviction that the parity issue and hemispheric coupling are causally connected and inter-dependent phenomena.

Comparative Magnetic Minima:
Characterizing quiet times in the Sun and stars
Proceedings IAU Symposium No. 286, 2011
C. H. Mandrini & D. F. Webb, eds.

doi:10.1017/S1743921312004644

Spontaneous chiral symmetry breaking in the Tayler instability

Fabio Del Sordo[1,2], Alfio Bonanno[3], Axel Brandenburg[1,2], and Dhrubaditya Mitra[1]

[1]Nordita, Roslagstullsbacken 23, SE-10691 Stockholm, Sweden, email: fabio@nordita.org
[2]Department of Astronomy, Stockholm University, SE 10691 Stockholm, Sweden
[3]INAF- Catania Astrophysical Observatory, Via S.Sofia 78, 95123 Catania ITALY

Abstract. The chiral symmetry breaking properties of the Tayler instability are discussed. Effective amplitude equations are determined in one case. This model has three free parameters that are determined numerically. Comparison with chiral symmetry breaking in biochemistry is made.

Keywords. Sun: magnetic fields, dynamo, magnetic helicity

---

1. Introduction

An important ingredient to the solar dynamo is the α effect. Mathematically speaking α is a pseudo scalar that can be constructed using gravity $\boldsymbol{g}$ (a polar vector) and angular velocity $\boldsymbol{\Omega}$ (an axial vector): $\boldsymbol{g}\cdot\boldsymbol{\Omega}$ is thus a pseudo scalar and is proportional to $\cos\theta$, where θ is the colatitude. This pseudo scalar changes sign across the equator. This explanation for large-scale astrophysical dynamos works well and therefore one used to think that the existence of the α effect in dynamo theory requires always the existence of a pseudo scalar in the problem. This has indeed been general wisdom, although it has rarely been emphasized in the literature. That this is actually not the case has only recently been emphasized and demonstrated. One example is the magnetic buoyancy instability in the *absence* of rotation, but with a horizontal magnetic field $\boldsymbol{B}$ and vertical gravity $\boldsymbol{g}$ being perpendicular to each other, so the pseudo scalar $\boldsymbol{g}\cdot\boldsymbol{B}$ vanishes (Chatterjee *et al.* 2011). Another example is the Tayler instability of a purely toroidal field in a cylinder (Tayler 1973; Gellert *et al.* 2011). Thus, the magnetic field is again perpendicular to all possible polar vectors that can be constructed, for example the gradient of the magnetic energy density which points in the radial direction. In both cases, kinetic helicity and a finite α, both of either sign, emerge in the nonlinear stage of the instability. In the first example (Chatterjee *et al.* 2011), the α tensor has been computed using the test-field method. In the second (Gellert *et al.* 2011), the components of the α tensor have been computed using the imposed-field method (see Hubbard *et al.* 2009, for a discussion of possible pitfalls in the nonlinear case).

The purpose of the present paper is to examine spontaneous chiral symmetry breaking in the Tayler instability and to estimate numerically the coefficients governing the underlying amplitude equations. This allows us then to make contact with a system of chemical reactions that can give rise to the same type of spontaneous symmetry breaking.

The connection with chemical systems is of interest because the question of spontaneous symmetry breaking has a long history ever since Pasteur (1853) discovered the preferential handedness of certain organic molecules. The preferential handedness of biomolecules is believed to be the result of a bifurcation event that took place at the origin of life itself (Kondepudi & Nelson 1984; Sandars 2003; Brandenburg *et al.* 2005).

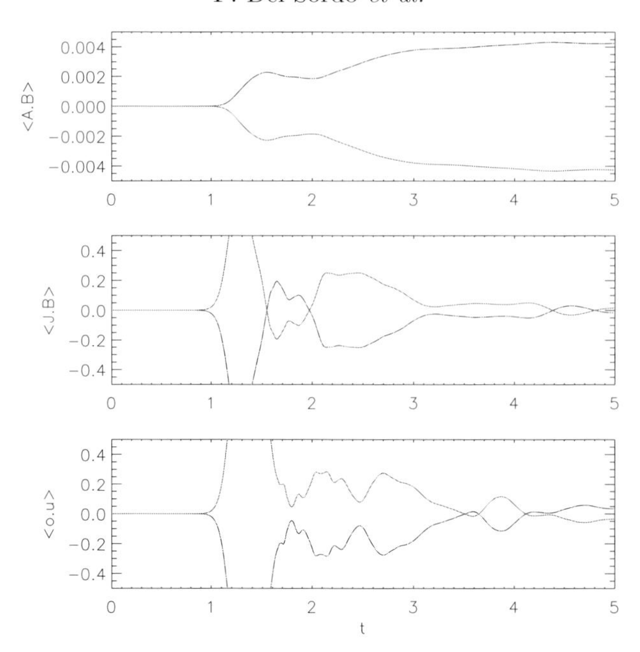

Figure 1. The two lines show the evolution of the volume averaged magnetic helicity $\langle \boldsymbol{A} \cdot \boldsymbol{B} \rangle$ for two initial conditions differing only in the parity of their initial perturbations. After the exponential growth magnetic helicity levels off. The inset shows the same in a semi-logarithmic representation. Here, to normalize magnetic helicity and time, we have used the sound speed c_s, $R \equiv s_{\rm in}$ and B_0, that is a normalization constant defining the initial magnetic field.

2. Numerical simulations

Our setup consists of an isothermal cylinder with a radial extent from $s_{\rm in}$ to $s_{\rm out}$ and vertical size h. We use cylindrical coordinates s, φ, z and we solve the time dependent ideal MHD equations with periodic boundary conditions in z, reflection in s and periodic in φ and a resolution ranging from 64^3 to 128^3 in the three directions.

The azimuthal field in the basic state is taken of the form

$$B_\varphi = B_0\ (s/s_0) \exp[-(s - s_0)^2/\sigma^2]$$

with B_0 being a normalization constant; the axial field B_z is chosen to be zero. In the basic state, the Lorentz force is balanced with a gradient of pressure, and we have checked that our setup was numerically stable if no perturbation was introduced in the system. For the actual calculations we have chosen $h = 2$, $s_{\rm in} = 1$, $s_{\rm out} = 3$, $s_0 = 2$ and $\sigma^2 = 0.2$. For a Sun-like star with an average density $\rho \approx 1 g/cm^3$ the Alfvén travel time is of the order of a year for a 1 kG magnetic field. We therefore expect that everywhere below the surface the sound speed is much greater than the Alfvén speed c_A (Bonanno and Urpin 2011) and we then assume $c_s \geqslant c_A$, in order to have a sub-thermal magnetic field.

At the beginning of the simulation we perturb the magnetic field. We add a perturbation of amplitude 10^{-7} of the background field. The perturbing field has a given helicity that is either positive or negative. During the development of the instability we observe a net increase of the helicity, as shown in Fig. 1 where we plot time series of the normalized magnetic helicity, which exhibits an initial exponential growth, reaches a peak and then levels off.

3. Amplitude equations

The linear stability analysis of this instability shows that there exist helical growing modes. But the left and right handed modes have exactly the same growth rate independently of their helicity. Hence the growth of helical perturbations cannot be described by a linear theory. However a weakly nonlinear theory is able to describe it as we show below. Let us begin by considering two helical modes of right and left handed varieties, respectively, each of which satisfy the Beltrami relation $\boldsymbol{\nabla}\times\boldsymbol{R}=\Lambda\boldsymbol{R}$ and $\boldsymbol{\nabla}\times\boldsymbol{L}=-\Lambda\boldsymbol{L}$, where Λ is a coefficient. We can deal with the Fourier transform of these modes,

$$\boldsymbol{L}(\boldsymbol{x})=\int\hat{\boldsymbol{L}}(\boldsymbol{q})d^dq \quad\text{and}\quad \boldsymbol{R}(\boldsymbol{x})=\int\hat{\boldsymbol{R}}(\boldsymbol{q})d^dq \tag{3.1}$$

For the left helical mode, total helicity and energy are given by

$$E_L=\frac{1}{2}\int\boldsymbol{L}^2(\boldsymbol{x})d^dx=\frac{1}{2}\int\hat{\boldsymbol{L}}\cdot\hat{\boldsymbol{L}}^*d^dq \quad\text{and}\quad \mathcal{H}_L=\int\boldsymbol{L}\cdot\boldsymbol{\nabla}\times\boldsymbol{L}d^dx=-2\Lambda E_L, \tag{3.2}$$

where $*$ denotes complex conjugation. We then have $E=E_L+E_R$ being the total energy and $\mathcal{H}=\mathcal{H}_L+\mathcal{H}_R$ the total helicity. An analogous relation holds also for E_R and $\mathcal{H}_R$.

In the weakly nonlinear regime the evolution of these modes can be described by general equations of the form:

$$\frac{\partial\hat{\boldsymbol{L}}}{\partial t}=\frac{\delta\mathcal{L}}{\delta\hat{\boldsymbol{L}}} \quad\text{and}\quad \frac{\partial\hat{\boldsymbol{R}}}{\partial t}=\frac{\delta\mathcal{L}}{\delta\hat{\boldsymbol{R}}}, \tag{3.3}$$

where the form of the Lagrangian $\mathcal{L}$ can be written down from symmetry considerations. In the present case one has to consider the fact that under parity transformation L and R can interchange into each other. With this additional symmetry the simplest Lagrangian takes the following form (Fauve *et al.* 1991)

$$\mathcal{L}[\hat{\boldsymbol{L}},\hat{\boldsymbol{R}}]=\int\left[\gamma\left(|\hat{\boldsymbol{L}}|^2+|\hat{\boldsymbol{R}}|^2\right)-\mu\left(|\hat{\boldsymbol{L}}|^4+|\hat{\boldsymbol{R}}|^4-\mu_*|\hat{\boldsymbol{L}}|^2|\hat{\boldsymbol{R}}|^2\right)\right]d^dq, \tag{3.4}$$

The coefficients γ, μ and μ_* cannot be found from symmetry considerations. Note that in order to show the simplest form, in writing down the Lagrangian we have ignored dissipation. This gives rise to the following set of amplitude equations,

$$\frac{\partial\hat{\boldsymbol{L}}}{\partial t}=\gamma\hat{\boldsymbol{L}}-\left(\mu|\hat{\boldsymbol{L}}|^2+\mu_*|\hat{\boldsymbol{R}}|^2\right)\hat{\boldsymbol{L}}, \quad \frac{\partial\hat{\boldsymbol{R}}}{\partial t}=\gamma\hat{\boldsymbol{R}}-\left(\mu|\hat{\boldsymbol{R}}|^2+\mu_*|\hat{\boldsymbol{L}}|^2\right)\hat{\boldsymbol{R}}. \tag{3.5}$$

For certain range of parameters these coupled equations allow the growth of one mode at the expense of the other (Fauve *et al.* 1991), a phenomenon known to biologists by the name "mutual antagonism" (Frank 1953).

Using Eqs. (3.2) and (3.5) and defining $H=\mathcal{H}/2\Lambda$ we obtain equations for E and H,

$$\frac{dE}{dt} = 2\gamma E-2(\mu+\mu_*)E^2-2(\mu-\mu_*)H^2, \tag{3.6}$$

$$\frac{dH}{dt} = 2\gamma H-4\mu EH. \tag{3.7}$$

Hence, by calculating the total energy and helicity from direct numerical simulations (DNS) we can determine the unknown coefficients γ,μ and μ_*.

To determine the coefficients γ, μ, and μ_*, we define the instantaneous logarithmic time derivatives of E and H, $\gamma_E=\frac{1}{2}\mathrm{d}\ln E/dt$ and $\gamma_H=\frac{1}{2}\mathrm{d}\ln H/dt$, so we have

$$\gamma=\gamma_H+2\mu E, \quad \mu=(\gamma-\gamma_H)/2E, \quad \mu_*=[(\gamma-\gamma_E)E-\mu(E^2+H^2)]/(E^2-H^2). \tag{3.8}$$

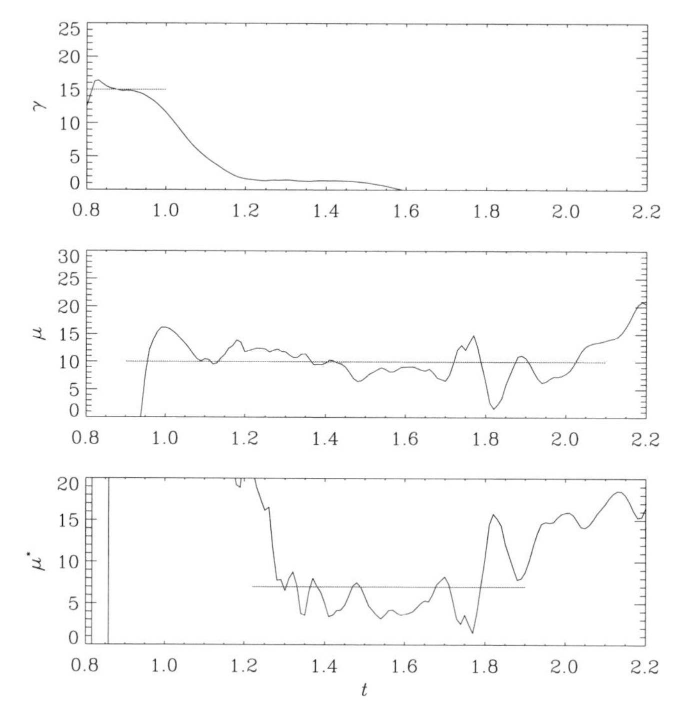

Figure 2. Time dependence of γ, μ, and μ_*, normalized in terms of inner radius and sound speed. The horizontal (red in the online version) lines give the fit results $\gamma R/c_s \approx 14$, $\mu R/c_s^3 \approx 10$, and $\mu^* R/c_s^3 \approx 7$.

The result is shown in Fig. 2, where we can identify first the value of $\gamma R/c_s \approx 14$ during the initial linear growth phase of the instability, and then the values $\mu R/c_s^3 \approx 10$ and $\mu^* R/c_s^3 \approx 7$ during the nonlinear stage. Here R is the inner radius of the cylinder.

4. Conclusions

The present work has demonstrated that the Tayler instability can produce parity-breaking and that it is possible to empirically determine fit parameters that reproduce the nonlinear evolution of energy and helicity. So far, no rigorous derivation of the amplitude equations exists, so this would be an important next step. However, it should be emphasized that chiral symmetry breaking instabilities in biochemistry is described by equations that are identical to those used here; see Bonanno *et al.* (2012) for details.

References

Bonanno, A. & Urpin, V. 2011, *Phys. Rev.*, E 84, 056310

Bonanno, A., Brandenburg, A., Del Sordo, F., & Mitra, D. 2012, Phys. Rev. E, (submitted, arXiv:1204.0081)

Brandenburg, A., Andersen, A. C., Höfner, S., & Nilsson, M. 2005, *Orig. Life Evol. Biosph.*, 35, 225

Chatterjee, P., Mitra, D., Brandenburg, A., & Rheinhardt, M. 2011, *Phys. Rev.* E 84, 025403R

Fauve, S., Douady, S., & Thual, O. 1991, *J. Phys.*, II 1, 311

Frank, F. C. 1953, *Biochim. Biophys. Acta*, 11, 459

Gellert, M., Rüdiger, G., & Hollerbach, R. 2011, *Mon. Not. R. Astron. Soc.*, 414, 2696

Hubbard, A., Del Sordo, F., Käpylä, P. J., & Brandenburg, A. 2009, *Mon. Not. R. Astron. Soc.*, 398, 1891

Kondepudi, D. K., & Nelson, G. W. 1984, *Phys. Lett.*, 106A, 203
Pasteur, L. 1853, *Ann. Phys.*, 166, 504
Sandars, P. G. H. 2003, *Orig. Life Evol. Biosph.*, 33, 575
Tayler, R. J. 1973, *Mon. Not. R. Astron. Soc.*, 161, 365

Discussion

GUSTAVO GUERRERO: Is this instability also happening in the solar radiative zone?

FABIO DEL SORDO: There are no direct observations of the appearance of this instability in the radiative zone of the Sun. Nevertheless, a toroidal field is likely present in that zone and the Tayler instability, in this case, would play an important role.

Comparative Magnetic Minima:
Characterizing quiet times in the Sun and stars
Proceedings IAU Symposium No. 286, 2011
C. H. Mandrini & D. F. Webb, eds.

doi:10.1017/S1743921312004656

Magnetic feature tracking, what determines the speed?

Gustavo Guerrero[1,3], Matthias Rheinhardt[2,3] and Mausumi Dikpati[4]

[1]Solar Physics, HEPL, Stanford University, Stanford, CA, 94305-4085, USA
email: gag@stanford.edu

[2]Department of Physics, FI-00014 University of Helsinki, Finland
email: rheinhar@mappi.helsinki.fi

[3]NORDITA, Roslagstullsbacken 17, Stockholm, Sweden
[4]High Altitude Observatory, NCAR, Boulder, CO, 80301, USA
email: dikpati@ucar.edu

Abstract. Recent observations revealed that small magnetic elements abundant at the solar surface move poleward with a velocity which seems to be lower than the plasma velocity $\boldsymbol{U}$. Guerrero *et al.* (2011) explained this discrepancy as a consequence of diffusive spreading of the magnetic elements due to a positive radial gradient of $|U_\theta|$. As the gradient's sign (inferred by local helioseismology) is still unclear, cases with a negative gradient are studied in this paper. Under this condition, the velocity of the magnetic tracers turns out to be larger than the plasma velocity, in disagreement with the observations. Alternative mechanisms for explaining them independently are proposed. For the turbulent magnetic pumping it is shown that it has to be unrealistically strong to reconcile the model with the observations.

Keywords. Sun: activity, Sun: magnetic fields

1. Introduction

Amongst the axisymmetric constituents of the plasma flow in the Sun's convection zone, the meridional circulation (MC), although being much slower than the differential rotation, is gaining growing interest which arises from its potential importance for the solar dynamo. Certainly any advanced non-linear mean-field model of this dynamo will have to include the MC to cover the full interaction between mean field and motion. In particular however, flux-transport dynamo models depend already on the *kinematic level* crucially on the MC as it is the ingredient which allows to close the dynamo cycle. At the same time it explains naturally both the equatorward migration of active regions and the poleward drift of weak poloidal field elements during the solar cycle.

The MC in the Sun is accessible to a variety of techniques out of which Doppler measurements and helioseismological time-distance or ring diagram analyses tend to agree in the surface speed profiles with a peak of ≈ 20m/s at a latitude $\approx 35°$. However, attempts to measuring the MC velocity using small magnetic structures as tracers of the flow (Komm *et al.* 1993, Hathaway & Rightmire 2010) reveal systematic differences from these results. Compared to Doppler measurements (Ulrich 2010) magnetic feature tracking (MFT) speeds can be as much as $\approx 30\%$ smaller. This indicates that the magnetic field is *not completely frozen* in the plasma, which is reasonable given the assumed high turbulent magnetic diffusivity $\eta_{\rm T}$. Due to diffusive spreading of the field, its advection is affected by the depth dependence of the MC velocity. On top of advection and diffusion, turbulent motions may be contributing by an extra advection term due to turbulent magnetic pumping or by the more complicated interaction with dynamo waves.

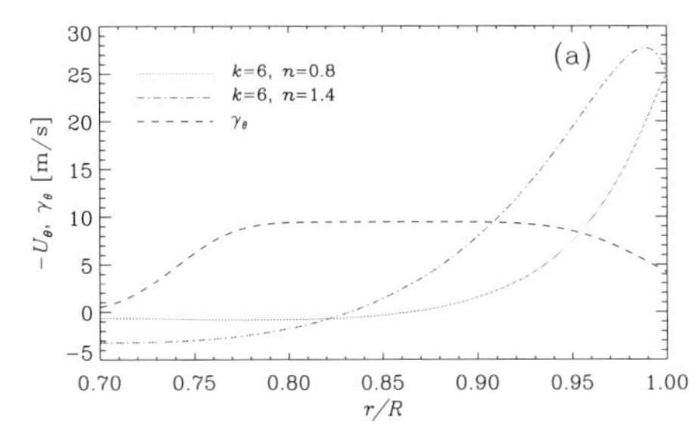

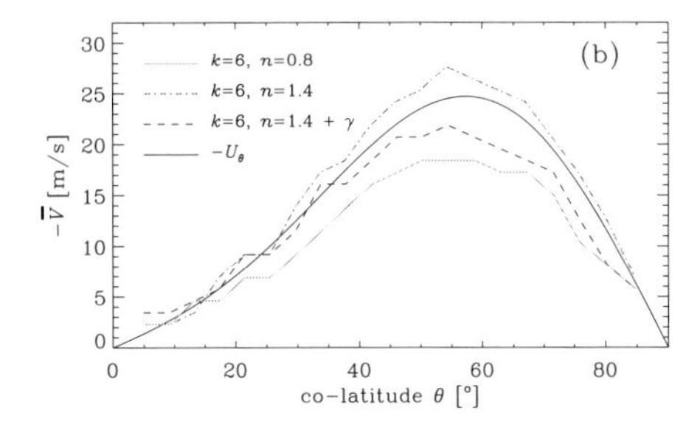

Figure 1. a) Radial profile of $U_\theta(r, 57^\circ)$ for different values of n. Broken line: radial profile of γ_θ, see eq. (3.1). b) Corresponding tracking velocities, $\overline{V}$, for three cases: $\partial_r |U_\theta| > 0$ (yellow/continuous) and $\partial_r |U_\theta| < 0$ with (broken) and without (red/dot-dashed) pumping. $\eta_\mathrm{T} = 10^{12}\mathrm{cm}^2/\mathrm{s}$ throughout. Colors refer to the online version of the paper.

In a previous work (Guerrero *et al.* 2011, hereafter GRBD) we studied numerically the kinematic evolution of small bipolar magnetic elements at different latitudes in the northern hemisphere. For simplicity we considered turbulent diffusion and advection due to a predefined MC as the only magnetic transport mechanisms. In a 1D (i.e., surface) version of the model, the northern spot acquires an effective velocity which is higher than the flow speed, but the southern one travels slower than the fluid. However, on average, or considering the center of the bipolar region, its speed clearly fits the fluid speed. This result is independent of the value of η_T. In the 2D model, on the other hand, we found that the MFT speed differs from the fluid velocity provided that the frozen-in condition is not satisfied. Moreover, we demonstrated that the relevant threshold value of η_T depends on the radial gradient of the latitudinal velocity, $\partial_r |U_\theta|$, assumed positive. The difference between MFT and flow speeds increases with $\partial_r |U_\theta|$. Thus, this simple model is able to explain the discrepancy between the Doppler and MFT observations.

Modern local helioseismological techniques allow also to infer the radial variation of the MC speed. Unfortunately, time-distance and ring-diagram analyses have not given consistent results. The first method yields that at the first few megameters beneath the photosphere the latitudinal velocity decreases inwards, i.e., a positive $\partial_r |U_\theta|$ (Zhao *et al.* 2011). By contrast, the results from the second method indicate a slight inward *increase*, i.e., a negative gradient albeit restricted to middle latitudes. (Gonzáles Hernández *et al.* 2006). Such velocity profiles were not considered in GRBD and the aim of this paper is to fill this gap as well as to discuss other effects that may influence MFT speeds.

2. MFT speed with a negative radial gradient of $|U_\theta|$

The method followed here to simulate the evolution of small bipolar magnetic structures is described in detail in GRBD. We also use the same notation. Briefly, we solve the induction equation in axisymmetry for the azimuthal component of the vector potential, A_ϕ in the variables r and θ. The prescribed meridional flow fed into the model is given by eqs. (4-6) of GRBD and has a single cell per hemisphere. In the present paper η_T is assumed to be constant across the domain. Bipolar regions (with spot separation 3° and depth $0.04R$) at 20 different latitudes from the pole to equator are evolved over a time interval of two weeks. Their velocity is computed as an average over this time span through the differences in final and initial latitudes (see eq. (8) of GRBD). To obtain a negative radial gradient of $|U_\theta|$ we preserve its general mathematical form (eq. (5) of GRBD), but choose the parameters of this ansatz as $k = 6$ and $n = 1.4$ which results in

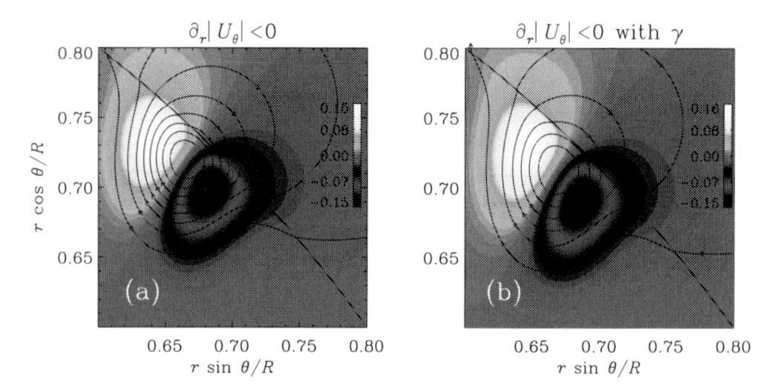

Figure 2. Magnetic field lines and r component (color coded) after two weeks propagation, $\eta_T = 10^{12}\,\mathrm{cm}^2/\mathrm{s}$. Velocity profile as in Fig. 1, red curve. a) without, b) with pumping, $\gamma_{0\theta} = 10\,\mathrm{m/s}$, see eq. (3.1). Colors refer to the online version of the paper

a negative $\partial_r|U_\theta|$ within a shallow surface-near layer of roughly 2% of the thickness of the convection zone (see Fig. 1a). It is worth mentioning that the stress-free boundary condition cannot be satisfied with a negative gradient. With this U_θ profile the velocities of the magnetic tracers turn out to be larger than the flow velocity (red/dot-dashed line in Fig. 1b). This result does not come unexpected since the field lines underneath the surface are pushed northwards by a flow faster than the surface one. By magnetic tension the field lines at the surface are then also accelerated. A snapshot of the magnetic field after two weeks of evolution is shown in Fig. 2a.

3. Additional transport mechanisms

Given that a negative $\partial_r|U_\theta|$ does not allow to reproduce the observations of *decelerated* magnetic tracer motion at the surface, one could be tempted to argue that the results of the ring diagram analysis could be incorrect. This conclusion, however, would be premature as our model is by no means complete. As mentioned in GRBD, studying axisymmetric, i.e., averaged fields requires the inclusion of the full mean electromotive force, from which only the turbulent diffusion term has so far been considered.

What about other constituents? Under anisotropic conditions, turbulence acts like a mean advective motion, called *turbulent magnetic pumping* and is described by a vector $\boldsymbol{\gamma}$. In the northern hemisphere, its θ-component is found to be positive, i.e., equatorwards whereas its radial component is mainly negative (downwards), see Käpylä *et al.* (2006a). Being helical, convective turbulence in the rotating Sun exhibits also an α-effect which together with differential rotation in the uppermost 35Mm of the convection zone (the so called near-surface shear layer) may well enable a turbulent dynamo of the $\alpha\Omega$ type. As $\partial_r\Omega < 0$ there and α is expected to be positive in the northern hemisphere, the dynamo wave should, according to the Parker-Yoshimura rule, propagate equatorwards, contrary to the MC. This scenario is difficult to study as it requires establishing a full nonlinear mean-field dynamo model of the surface shear layer. Turbulent pumping, in contrast, can easily be introduced as an additive contribution to the MC. We do so employing profiles for $\boldsymbol{\gamma}$ adapted from Guerrero & de Gouveia Dal Pino (2008) (hereafter GDP08):

$$\begin{aligned}
\gamma_\theta &= \gamma_{0\theta}\left[1+\mathrm{erf}\left(\frac{r-0.74}{0.035}\right)\right]\left[1-\mathrm{erf}\left(\frac{r-0.995}{0.05}\right)\right]\cos\theta\sin^4\theta, \\
\gamma_r &= -\gamma_{0r}\left[1+\mathrm{erf}\left(\frac{r-0.7}{0.015}\right)\right]\left[1-\mathrm{erf}\left(\frac{r-0.975}{0.1}\right)\right]\left[\exp\left(\frac{r-0.715}{0.25}\right)^2\cos\theta+1\right].
\end{aligned} \tag{3.1}$$

Similar to GDP08, we set γ_{0r} and $\gamma_{0\theta}$ such that the maximum of γ_θ is 2.5 times larger than that of γ_r.† The radial profile of γ_θ, which is the more relevant component for this study, is shown with a broken line in Fig. 1a. From direct numerical simulations (DNS) Käpylä *et al.* (2006a) have estimated that the maximum γ_θ should be ≈ 2.5m/s. According to eq. (3.1) its value at the surface would then be 1.11m/s. Given that the surface amplitude of $|U_\theta| \approx 25$m/s, this small value cannot significantly modify the results found in the previous section. Hence we have increased $\gamma_{0\theta}$ progressively and found that an MFT speed profile that roughly coincides with that of the flow speed is obtained with $\gamma_{0\theta} = 5$m/s (i.e., 2.2m/s at the surface or $\approx 10\%$ of the flow speed). For larger values of $\gamma_{0\theta}$ the MFT speed is smaller than the flow speed. In Fig. 1b we show the result obtained with the extraordinarily high value $\gamma_{0\theta} = 10$m/s (see broken lines). Note that the latitudinal profile of γ_θ (eq. (3.1)) peaks at 63° while the flow profile does at 57°. This is interesting because the resulting profile of the MFT speed peaks at a different latitude than the flow, just as it is obtained in observations (see fig. 10 of Ulrich 2010). Fig. 2 compares the morphology of the magnetic field from the simulation with pumping (b) to that without (a). Note that the magnetic field is transported further to the north in the case without magnetic pumping

4. Discussion

Negative gradients $\partial_r|U_\theta|$ at the solar surface, as suggested by helioseismological ring-diagram analyses, result in MFT speeds higher than the plasma flow speed. By enhancing our flux transport model with turbulent magnetic pumping, we tried to reconcile the helioseismological results with the actually oberserved *lower* MFT speeds. However, high amplitudes of the pumping velocity not supported by DNS are needed. We have to conclude that either: (i) the negative gradient of $|U_\theta|$, suggested by ring-diagram analysis, if real, should be smaller than the one considered here, (ii) the pumping velocities from DNS are unrealistically low, which cannot be excluded because the setup in Käpylä *et al.* (2006a) differs from the solar conditions, or (iii) the inclusion of the α effect in the flux transport model, enabling a dynamo process in the Sun's surface shear layer, is crucial. For the latter option we have to think of a short-wavelength dynamo wave traveling equatorwards. Magnetic quenching would cause a corresponding latitudinal modulation of U_θ, α, γ and $\eta_{\rm T}$. Here, quenching of $\eta_{\rm T}$ is perhaps the most promising effect as the value of $\eta_{\rm T}$ "decides" to what degree the field is frozen in the fluid.

Acknowledgments

We thank A. Muñoz-Jaramillo for pointing out the negative radial gradient of $|U_\theta|$ in the ring-diagram analysis and A. Kosovichev for his valuable comments.

References

Gonzáles Hernández, I., Komm, R., Hill, F., Howe, R., Corbard, T., & Haber, D. 2006, *ApJ*, 638, 576

Guerrero, G., Rheinhardt, M., Brandenburg, A., & Dikpati M. 2011, *MNRAS Letters*, in press, (arXiv:1107.4801)

† Note that γ_θ slightly differs from GDP08 insofar it is here $\neq 0$ at the surface. In the underlying DNSs a γ_θ vanishing at the surface could be an artifact due to the boundary conditions.

Guerrero, G. & de Gouveia Dal Pino, E. M. 2008, *A&A*, 485, 267
Hathaway, D. H. & Rightmire, L. 2010, *Science*, 327, 1350
Käpylä, P. J., Korpi, M. J., Ossendrijver, M., & Stix, M. 2006a, *A&A*, 455, 401
Komm, R. W., Howard, R. F., & Harvey, J. W. 1993, *Solar Physics*, 147, 207
Ulrich, R. K. 2010, *ApJ*, 725, 658
Zhao, J., Couvidat, S., Bogart, R. S., Parchevsky, K. V., Birch, A. C., Duvall, T. L., Beck, J. G., Kosovichev, A. G., & Scherrer, P. H. 2011, *Solar Physics*, pp 163–+

Session 3

Comparative Solar Minima from Sun to Earth

Comparative Magnetic Minima:
Characterizing quiet times in the Sun and stars
Proceedings IAU Symposium No. 286, 2011
C. H. Mandrini & D. F. Webb, eds.

doi:10.1017/S174392131200467X

Helioseismology - a clear view of the interior

Yvonne Elsworth, Anne-Marie Broomhall, and William Chaplin

School of Physics & Astronomy, University of Birmingham,
Edgbaston Park Road, West Midlands, Birmingham, UK, B15 2TT
email: y.p.elsworth@bham.ac.uk

Abstract. Helioseismology is a very powerful tool that allows us to explore the interior of the Sun. Here we give particular emphasis to the justification for the likely location of the zone that is most sensitive to cycle-related changes. For the low degree modes we find that more than one timescale for changes in the oscillations is discovered. We also note the successive cycles have differing sensitivities to the activity. We end with a warning of the risk of being misled with short datasets such as are seen with stellar data.

Keywords. Sun: helioseismology, Sun: interior, Sun: magnetic fields, Sun: oscillations

---

1. Introduction

The Sun's natural resonant oscillations, known as solar p modes are trapped in cavities below the surface of the Sun and their frequencies are sensitive to properties, such as temperature and mean molecular weight, of the solar material in the cavities. The basic principle governing the fundamental period of oscillation has been known for a long time and is that the period is related to the sound travel time. This period is similar to the dynamical timescale and is of order one hour for the Sun. The period scales as the inverse square root of the mean density. On the Sun we see not the fundamental but a higher overtone with a typical period of five minutes. It is no accident that this is a similar timescale to the turn-over time for a granule. Different modes of oscillations see different cavities with the lower boundary being very sensitive to the degree (or ℓ) of the mode and the upper boundary being mainly a function of frequency. A consequence of this is that the high degree modes are sensitive to just the outer regions of the Sun, while the low degree modes are sensitive to almost the complete volume. There are modes that are not global but only local. We do not consider them further here. In order to observe the high degree modes, in general one has to image the surface. However, no imaging is required for the low degree modes. Although one might think that, without imaging, one would observe just the lowest degree mode with $\ell = 0$, this is not true as there is some degree of inherent Doppler imaging.

By using a combination of measurements from high and low degree modes it is possible to build up maps of the sound speed and density in the interior of the Sun. It is also possible to determine the rotation profile with depth and latitude. Such images (see discussion in later section) point to the existence of several zones of high radial shear. One such zone lies at the base of the convection zone and is known as the Tachocline. Another radial shear zone is located at the surface.

It has been known since the mid 1980s that p-mode frequencies vary throughout the solar cycle with the frequencies being at their largest when the solar activity is at its maximum(e.g. Woodard & Noyes 1985; Pallé *et al.* 1989; Elsworth *et al.* 1990; Libbrecht &

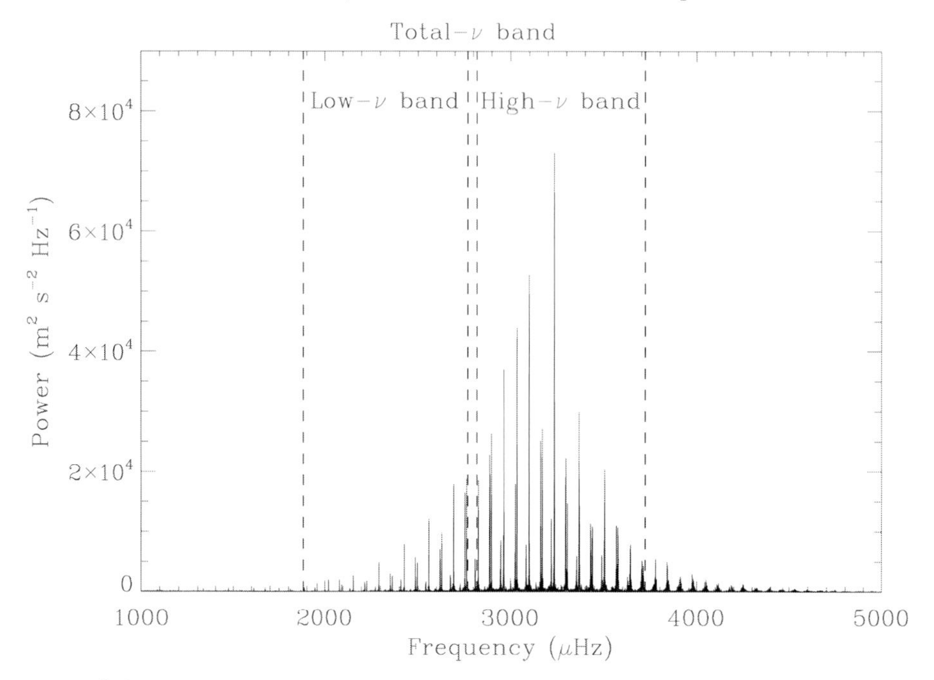

Figure 1. A frequency-power spectrum of BiSON data. The time series used to create this spectrum was 182.5 d in length. The vertical lines mark the low-, high- and total-frequency bands used when calculating the solar cycle frequency shifts.

Woodard 1990; Jiménez-Reyes *et al.* 2003; Chaplin *et al.* 2007; Jiménez-Reyes *et al.* 2007). [Please note here that the references listed are not exhaustive and while the references do not always include the earliest example we have tried to include useful ones whose reference lists themselves are informative.] For a low-degree mode at about 3000 μHz, the change in frequency between solar maximum and minimum is about 0.4 μHz. By examining the changes in the observed p-mode frequencies throughout the solar cycle we can learn about solar-cycle-related processes that occur beneath the Sun's surface.

The Birmingham Solar-Oscillations Network (BiSON) (Chaplin *et al.* 1996) makes Sun-as-a-star (unresolved) Doppler velocity observations, which are sensitive to the p modes with the largest horizontal scales (or the lowest angular degrees, ℓ). Consequently, the frequencies measured by BiSON are of the truly global modes of the Sun. These modes travel to the Sun's core, however, their dwell time at the surface is longer than at the solar core because the sound speed inside the Sun increases with depth. Therefore, the low-ℓ modes are most sensitive to variations in regions of the interior that are close to the surface and so are able to give a picture of the influence of near-surface activity.

BiSON is a network of autonomous ground-based observatories that are strategically positioned at various longitudes and latitudes in order to provide nearly continuous coverage of the Sun. The network began with just one station and gradually expanded over the course of 17 years, finally becoming a six-site network in 1992. There are four sites in the southern hemisphere and two in the northern hemisphere. The quality of the early data is poor compared to more recent data because of limited time coverage.

2. Determining variations in p-mode frequencies throughout the solar cycle

BiSON is in a unique position to study the changes in oscillation frequencies that accompany the solar cycle as it has now been collecting data for over 30 years. Here, we have analyzed the mode frequencies observed by BiSON during the last two solar cycles in their entirety i.e. from 1986 April 14 to 2010 October 8. To obtain solar cycle frequency shifts the BiSON data must be split into shorter subsets. However, the longer the time series the more precisely the mode frequencies can be obtained. Therefore, the length in time of the subsets must represent a balance between being long enough to accurately and precisely extract mode frequencies and short enough to resolve solar cycle variations. Here we show the results obtained when the BiSON data were divided into 182.5-day-long independent subsets. Figure 1 shows a frequency-power spectrum of 182.5 d of BiSON data. We now describe how solar cycle frequency shifts are obtained from BiSON data.

Estimates of the mode frequencies were extracted from each subset by fitting a modified series of Lorentzian models to the data using a standard likelihood maximization method, which was applied in the manner described in Fletcher *et al.* (2009). We first defined a frequency range of interest. For this range, a reference frequency set was determined by averaging the frequencies in subsets covering the minimum activity epoch at the boundary between cycle 22 and cycle 23. It should be noted that the main results described here are insensitive to the exact choice of subsets used to make the reference frequency set. Frequency shifts were then defined as the differences between frequencies given in the reference set and the frequencies of the corresponding modes observed at different epochs (Broomhall *et al.* 2009). This gives one value of the frequency shift for each epoch and frequency range. Note that it is, in general, not possible to get a spot reading of the frequency shift. It may require several months of observations to obtain a frequency shift value.

For each subset in time, three weighted-average frequency shifts were generated over different frequency intervals, where the weights were determined by the formal errors on the fitted frequencies: first, a "total" average shift was determined by averaging the individual shifts of the $\ell = 0$, 1, and 2 modes over fourteen overtones (covering a frequency range of $1.88 - 3.71$ mHz, see Figure 1); second, a "low-frequency" average shift was computed by averaging over seven overtones whose frequencies ranged from 1.88 to 2.77 mHz; and third, a "high-frequency" average shift was calculated using seven overtones whose frequencies ranged from 2.82 to 3.71 mHz. The lower limit of this frequency range (i.e., 1.88 mHz) was determined by how low in frequency it was possible to accurately fit the data before the modes were no longer prominent above the background noise. The upper limit on the frequency range (i.e., 3.71 mHz) was determined by how high in frequency the data could be fitted before errors on the obtained frequencies became too large due to increasing line widths causing modes to overlap in frequency.

Figure 2 shows mean frequency shifts of the p modes observed by BiSON (also see Broomhall *et al.* 2009; Salabert *et al.* 2009; Fletcher *et al.* 2010). A scaled version of the 10.7 cm flux is also plotted in Figure 2. The flux has been scaled by fitting a linear relationship between the 10.7 cm flux and the frequency shifts observed in the total-frequency band. The eleven-year cycle is seen clearly.

What causes the observed frequency shifts? Broadly speaking, the magnetic fields can affect the modes in two ways. They can do so directly, by the action of the Lorentz force on the gas. This provides an additional restoring force, the result being an increase of frequency, and the appearance of new modes. Magnetic fields can also influence matters

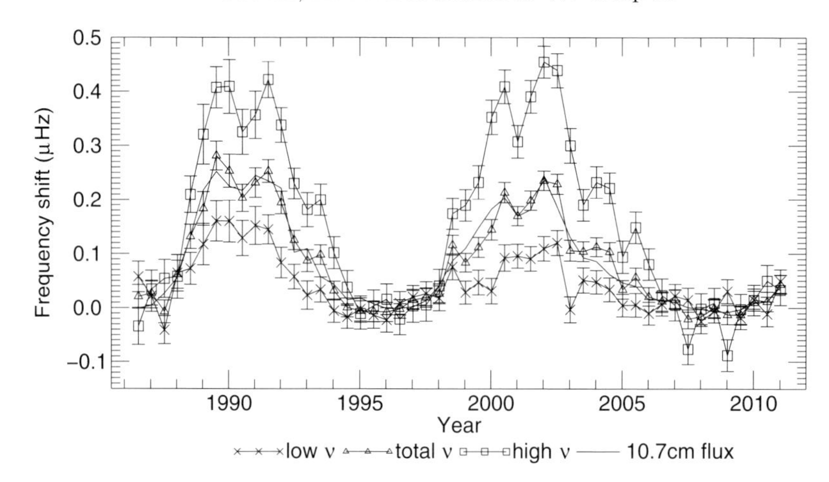

Figure 2. Average frequency shifts of "Sun-as-a-star" modes with frequencies between 1.88 and 3.71 mHz, 1.88 and 2.77 mHz, and 2.82 and 3.71 mHz (see legend). The results were obtained from 182.5 d time series. Also plotted is a scaled version of the 10.7 cm flux.

indirectly, by affecting the physical properties in the mode cavities and, as a result, the propagation of the acoustic waves within them. This indirect effect can act both ways, to either increase or decrease the frequencies.

The observed frequency shifts show dependencies on both frequency and ℓ. We now look in more detail at these dependencies and discuss what they tell us about the origin of the perturbation.

3. Dependence of solar cycle frequency shifts on frequency

Solar cycle frequency shifts, $\delta\nu_{n,\ell}$, have well-known dependencies on both angular degree, ℓ, and frequency, $\nu_{n,\ell}$, (see e.g. Libbrecht & Woodard 1990; Elsworth *et al.* 1994; Chaplin *et al.* 1998; Howe *et al.* 1999; Chaplin *et al.* 2001; Jiménez-Reyes *et al.* 2001). Figure 2 shows that the high-frequency band is much more sensitive to the solar cycle than the low-frequency band. The frequency dependence of the frequency shifts is a telltale indicator that the observed eleven-year signal must be the result of changes in acoustic properties in the few hundred kilometres just beneath the visible surface of the Sun, a region that the higher-frequency modes are much more sensitive to than their lower-frequency counterparts because of differences in the upper boundaries of the cavities in which the modes are trapped (Libbrecht & Woodard 1990; Christensen-Dalsgaard & Berthomieu 1991). We now go into more detail.

Stellar p modes are trapped in cavities in the solar interior. While the lower turning point of these cavities is mostly dependent on ℓ, the position of the upper turning point is mostly dependent on frequency. As p modes travel towards the centre of the Sun, refraction causes the oscillations to follow a curved path which takes them back to the surface. The lower turning point, therefore, depends on the direction of travel and consequently ℓ. On the other hand, the upper turning point is relatively independent of ℓ as by the time the oscillations reach the photosphere all waves travel in an approximately radial direction (for low and intermediate ℓ). Instead the position of the upper turning point is dependent on frequency. The modes are reflected because of the sharp decrease

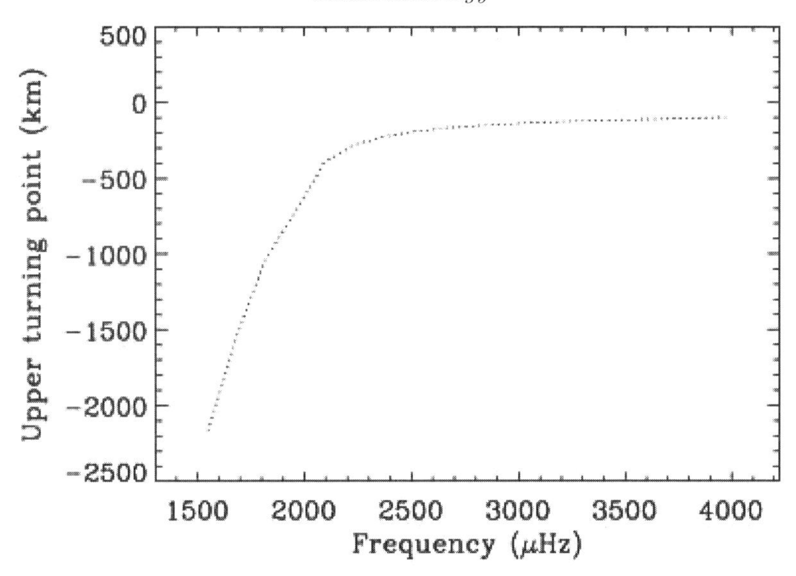

Figure 3. Taken from Chaplin *et al.* (2001). Upper turning point as a function of frequency for radial modes, as determined from model S of Christensen-Dalsgaard. Here, we define the zero-point location as that corresponding to where optical depth reaches unity at a wavelength of 500 nm.

in density of the plasma at the photosphere. If the density scale height is smaller than the lengthscale of the mode then the pressure changes required to make the mode cannot be maintained over the length of time that matches the wave period. As the density scale height decreases rapidly with altitude in the outer regions of the Sun the depth at which modes with different periods are reflected varies.

For a given ℓ the upper turning point of low-frequency modes is deeper than the upper turning point of high-frequency modes. At a fixed frequency lower-ℓ modes penetrate more deeply into the solar interior than higher-ℓ modes. Therefore, higher-frequency modes, are more sensitive to surface perturbations.

Libbrecht & Woodard (1990) discuss the origin of the perturbation. If the perturbations were to extend over a significant fraction of the radius, asymptotic theory implies that the fractional mode frequency shift would depend mainly on $\nu_{n,\ell}/\ell$. However, the observed $\delta\nu_{n,\ell}$ is not well described by a function of $\nu_{n,\ell}/\ell$. This implies that the relevant structural changes occur mainly in a thin layer. Thompson (1988) found that the effect of perturbing a thin layer in the propagating regions of modes, such as the HeII ionization zone, is an oscillatory frequency dependence in $\delta\nu_{n,\ell}$, which is not observed. This implies that the dominant frequency dependence is not the direct result of, for example, changes in the magnetic field at the base of the convection zone. If the perturbation was confined to the centre of the Sun the size of the frequency shift would increase with decreasing ℓ as lower ℓ modes penetrate deeper into the solar interior than high-ℓ modes. In fact, $\delta\nu_{n,\ell}$ increases with increasing ℓ.

Libbrecht & Woodard (1990), therefore, conclude that the oscillations are responding to changes in the strength of the solar magnetic activity near the Sun's surface. As modes below approximately 1800 μHz experience almost no solar cycle frequency shifts it is reasonable to conclude that the origin of the perturbation is concentrated in a region above the upper turning points of these modes. Figure 3 shows that the upper turning

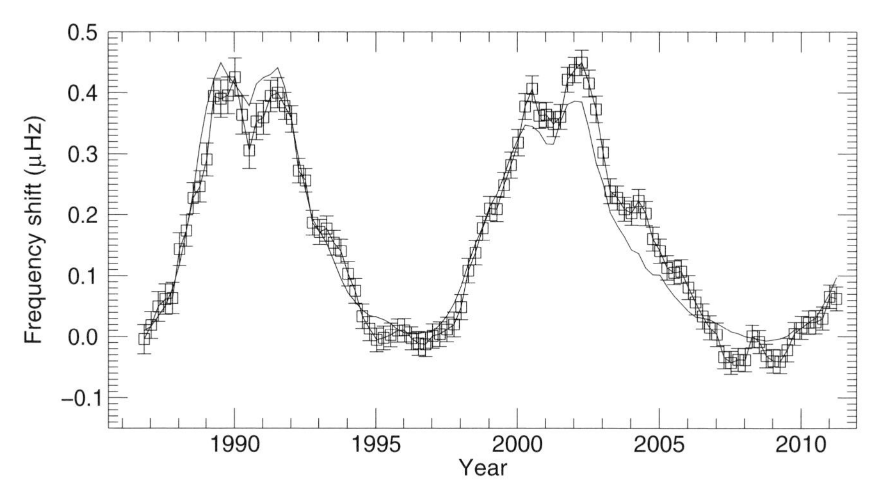

Figure 4. Oscillation frequency shifts in the high frequency band with the 10.7 cm flux overplotted as a thin line. The 10.7 and the frequency shifts are scaled to match in cycle 22

point of a mode with a frequency of $\sim$1800 μHz is about 1 Mm below the surface (which is defined as where optical depth reaches unity at a wavelength of 500 nm). Note that the upper turning point predicted by a model is strongly dependent on the properties of the model at the top of the convection zone.

4. Evidence for shorter-term variations in solar p-mode frequencies?

Over the past twenty years it has become apparent that significant (quasi-periodic) variability in activity is also seen on shorter timescales, between 1 and 2 years (e.g. Benevolenskaya 1995; Mursula *et al.* 2003; Valdés-Galicia & Velasco 2008). Is the periodicity the result of modulation of the main solar dynamo responsible for the eleven-year cycle or is it caused by a separate mechanism? Helioseismology can help to answer this question by looking for evidence of shorter-term variations in the p-mode frequencies, which, in turn, can provide information on solar-cycle-related processes that occur beneath the Sun's surface.

Shorter-term variations are indeed visible on top of the general eleven-year trend in p-mode frequencies (Broomhall *et al.* 2009; Fletcher *et al.* 2010; Broomhall *et al.* 2010). This is seen very clearly in Figure 4 which shows just the BiSON high-frequency data and the 10.7 cm radio signal. In this case the 10.7 is scaled to match the signal in cycle 22. The reason for this scaling will become apparent shortly. The period of this structure is approximately 2 years. Fletcher *et al.* (2010) showed that the signal was visible in both BiSON and GOLF data and Broomhall *et al.* (2010) demonstrated that the quasi-biennial signal was also evident in Variability of solar IRradiance and Gravity Oscillations (VIRGO; Fröhlich *et al.* 1995) data. VIRGO consists of three sun photometers (SPMs), that observe at different wavelengths, namely the blue channel (402 nm), the green channel (500 nm), and the red channel (862 nm). The results are similar for each individual channel.

In order to extract mid-term periodicities, we subtracted a smooth trend from the average total shifts by applying a boxcar filter of width 2.5 years. This removed the

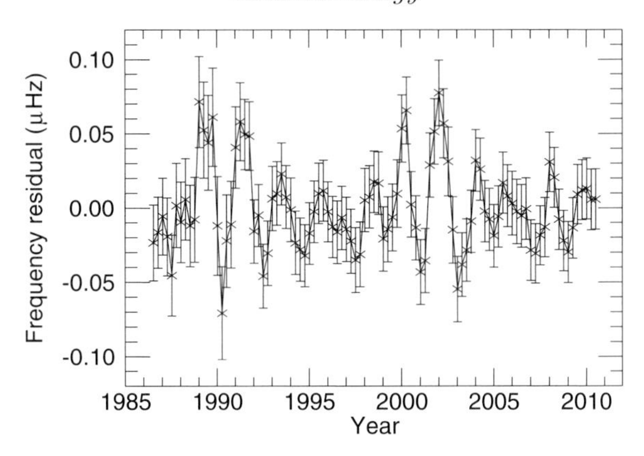

Figure 5. Rapidly varying component of the oscillation frequency shifts in the high frequency band

dominant eleven-year signal of the solar cycle. Note that, although the width of this boxcar is only slightly larger than the periodicity we are examining here, wider filters produce similar results. The resulting residuals, which can be seen in Figure 5, show a periodicity on a timescale of about 2 years.

The envelope of the two-year signal is modulated by the eleven-year signal. However there must be some addititive component because it is still present when the eleven-year signal is at minimum. Furthermore, the two-year signal must have its origin in significantly deeper layers than the eleven-year signal. Since the 2 year signal shows far less dependence on mode frequency, the origin of the signal must be positioned below the upper turning point of the lowest frequency modes examined (as the depth of a mode's upper turning point increases with decreasing frequency). The upper turning point of modes with frequencies of 1.88 mHz occurs at a depth of approximately 1000 km, whereas the influence of the eleven-year cycle is concentrated in the upper few 100 km of the solar interior. Put together, this all points to a phenomenon that is separate from, but nevertheless susceptible to, the influence of the eleven-year cycle.

One possibility is magnetic activity seated near the bottom of the layer extending 5% below the solar surface ($\sim 35,000$ km). This region shows strong rotational shear (see Figure 6), like the shear observed across the deeper-seated tachocline where the omega effect of the main dynamo is believed to operate (Corbard & Thompson 2002; Antia *et al.* 2008). The presence of two different types of dynamo operating at different depths has already been proposed to explain the quasi-biennial behaviour observed in other proxies of solar activity (Benevolenskaya 1998a,b). When the eleven-year cycle is in a strong phase, buoyant magnetic flux sent upward from the base of the envelope by the main dynamo could help to nudge flux processed by this second dynamo into layers that are shallow enough to imprint a detectible acoustic signature on the modes. The 2 year signal would then be visible. When the main cycle is in a weak phase, the flux from the second dynamo would not receive an extra nudge, and would not be buoyant enough to be detected in the proxies but could potentially be seen in the modes. While p-mode frequencies respond to conditions beneath the surface activity proxies respond to changes at or above the surface. The 2 year signals observed in other solar activity proxies are only detectible during phases of moderate to high activity (see, e.g. Vecchio & Carbone 2008; Hathaway 2010, and references therein). That the signal

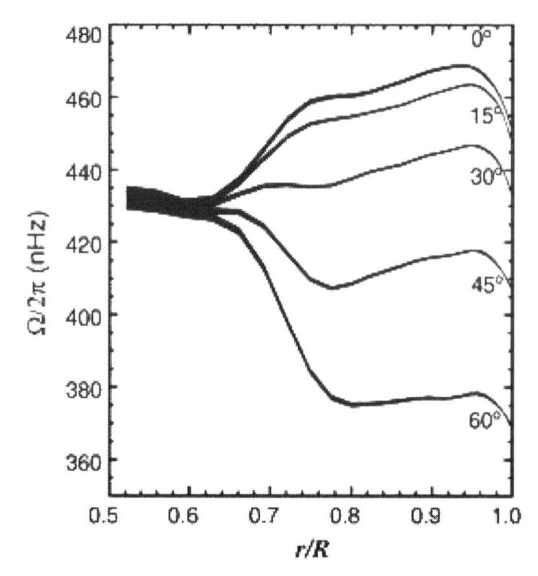

Figure 6. Taken from Howe *et al.* (2007). Mean rotation profile for RLS inversions of GONG data shown as slices ad constant latitude. Thickness of curves demonstrates the 3σ errors.

was also visible in the high-frequency modes during the recent extended minimum may also, therefore, point to behavioural changes in the main dynamo and its influence on the 2-year signal. We are currently exploring the latitudinal dependence of the two-year signal.

5. Comparison between cycles

It is recognised that the minimum out of which we are just emerging was unusually deep and long. It is interesting to ask if the unusual behaviour occured just at that minimum or whether there are other indications of changes in the structure of the Sun. To explore this we look at the oscillations data for the last two complete cycles. We now see the reason for scaling the 10.7 data for cycle 22 alone. Figure 7 shows the high and low frequency bands for cycles 22 and 23 and the initial part of the rise into cycle 24. The high frequency band, as indicated previously, shows a much stronger response to the activity cycle than does the low frequency band. However, the low frequency band is very different in amplitude relative to the high frequency band. In the cycle which has just finished, the sensitivity of the low frequency band to the solar activity is much lower. It is possible that we are seeing evidence that the conditions in the solar interior have started to change. It may be that the upper turning point of the modes is moving deeper in the Sun away from the zones of activity. Alternatively, the radial location of the activity may be different.

6. Stellar Context

For the Sun we can measure both high and low degree modes and thereby obtain a very detailed picture of the conditions within the Sun. However, for most stars this is impossible. We must do what we can with just the low-degree modes. It is therefore

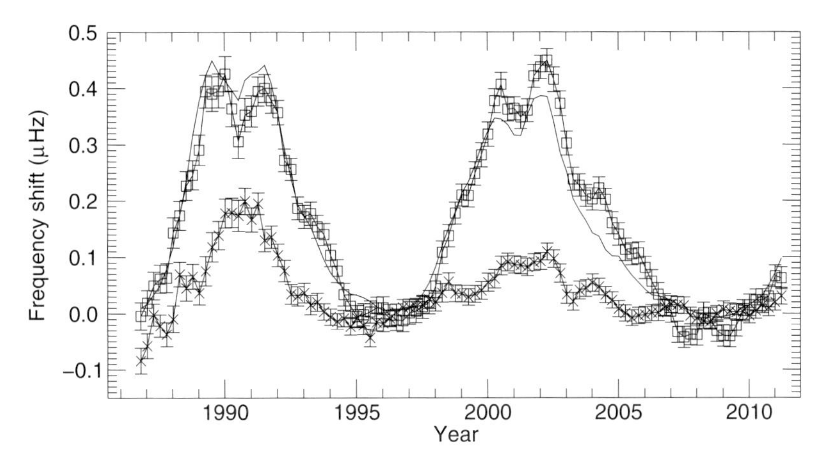

Figure 7. High and low frequency bands plus 10.7

useful to consider the inferences that we would draw from low-degree modes alone. We expect that magnetic activity will influence both the frequencies and the heights of the modes (Chaplin 2011). Effects that can be ascribed to activity were reported by Garcia for CoRoT observations of HD49933 (García *et al.* 2010). So we have clear expectations but what are the pitfalls. It is important to remember that the modes of oscillations are stochastically excited and we never see the underlying spectrum. This realisation noise is a big problem and a limiting factor in the inferences that can be drawn. It is very easy to be misled (Broomhall *et al.* 2011) into thinking that real changes are present when they are not.

7. Conclusions

We have seen that solar oscillations provide a window into the interior of the Sun. The global, low-degree modes have been measured for over two solar cycles and have the potential to allow us to compare one cycle with another. In the oscillations data, not only was the last minimum deep but there is clear evidence that the behaviour of the Sun at the peak of the last cycle was different from its behaviour in the previous cycle. We know that other indices are also seeing changes with, perhaps, the 10.7 cm radio flux and the sunspot number becoming somewhat decoupled and with the changes in the irradiance. This is a great opportunity to link the surface to the interior of the Sun. Furthermore the two-year signal seen in the oscillations and at times of high activity in some of the activity indices may be able to help us throw some light on the magnetic field generation in the Sun.

References

Antia, H.M., Basu, S., & Chitre, S.M., 2008. *ApJ*, 681, 680
Benevolenskaya, E. E., 1995. *Sol. Phys.*, 161, 1
Benevolenskaya, E. E., 1998a. *ApJ*, 509, L49
Benevolenskaya, E. E., 1998b. *Sol. Phys.*, 181, 479
Broomhall, A., Chaplin, W.J., Elsworth, Y., Fletcher, S. T., & New, R., 2009. *ApJ*, 700, L162

Broomhall, A., Fletcher, S. T., Salabert, D., Basu, S., Chaplin, W. J., Elsworth, Y., Garcia, R. A., Jimenez, A., & New, R., 2010. *ArXiv e-prints*

Broomhall, A. M., Chaplin, W. J., Elsworth, Y., & New, R., 2011. *MNRAS*, 413, 2978

Chaplin, W. J., Appourchaux, T., Elsworth, Y., Isaak, G.R., & New, R., 2001. *MNRAS*, 324, 910

Chaplin, W. J., Elsworth, Y., Howe, R., Isaak, G. R., McLeod, C. P., Miller, B. A., van der Raay, H. B., Wheeler, S. J., & New, R., 1996. *Sol. Phys.*, 168, 1

Chaplin, W. J., Elsworth, Y., Isaak, G. R., Lines, R., McLeod, C. P., Miller, B. A., & New, R., 1998. *MNRAS*, 300, 1077

Chaplin, W. J., Elsworth, Y., Miller, B. A., Verner, G. A., & New, R., 2007. *ApJ*, 659, 1749

Christensen-Dalsgaard, J. & Berthomieu, G., 1991. *Theory of solar oscillations.* Solar Interior and Atmosphere, 401–478

Corbard, T. & Thompson, M. J., 2002. *Sol. Phys.*, 205, 211

Elsworth, Y., Howe, R., Isaak, G.R., McLeod, C.P., Miller, B.A., New, R., Speake, C.C., & Wheeler, S. J., 1994. *ApJ*, 434, 801

Elsworth, Y., Howe, R., Isaak, G. R., McLeod, C. P., & New, R., 1990. *Nature*, 345, 322

Fletcher, S. T., Broomhall, A., Salabert, D., Basu, S., Chaplin, W. J., Elsworth, Y., García, R. A., & New, R., 2010. *ApJ*, 718, L19

Fletcher, S. T., Chaplin, W. J., Elsworth, Y., & New, R., 2009. *ApJ*, 694, 144

Fröhlich, C., Romero, J., Roth, H., Wehrli, C., Andersen, B. N., Appourchaux, T., Domingo, V., Telljohann, U., Berthomieu, G., Delache, P., *et al.*, 1995. *Sol. Phys.*, 162, 101

García, R. A., Mathur, S., Salabert, D., Ballot, J., Régulo, C., Metcalfe, T. S., & Baglin, A., 2010. *Science*, 329, 1032

Hathaway, D.H., 2010. *Living Reviews in Solar Physics*, 7, 1

Howe, R., Christensen-Dalsgaard, J., Hill, F., Komm, R., Schou, J., Thompson, M. J., & Toomre, J., 2007. *Advances in Space Research*, 40, 915

Howe, R., Komm, R., & Hill, F., 1999. *ApJ*, 524, 1084

Jiménez-Reyes, S. J., Chaplin, W. J., Elsworth, Y., García, R. A., Howe, R., Socas-Navarro, H., & Toutain, T., 2007. *ApJ*, 654, 1135

Jiménez-Reyes, S. J., Corbard, T., Pallé, P. L., Roca Cortés, T., & Tomczyk, S., 2001. *A&A*, 379, 622

Jiménez-Reyes, S. J., García, R. A., Jiménez, A., & Chaplin, W. J., 2003. *ApJ*, 595, 446

Libbrecht, K.G. & Woodard, M. F., 1990. *Nature*, 345, 779

Mursula, K., Zieger, B., & Vilppola, J. H., 2003. *Sol. Phys.*, 212, 201

Pallé, P. L., Régulo, C., & Roca Cortés, T., 1989. *A&A*, 224, 253

Salabert, D., García, R. A., Pallé, P.L., & Jiménez-Reyes, S.J., 2009. *A&A*, 504, L1

Thompson, M.J., 1988. In E. J. Rolfe, editor, *Seismology of the Sun and Sun-Like Stars*, volume 286 of *ESA Special Publication.* 321–324

Valdés-Galicia, J.F. & Velasco, V.M., 2008. *Advances in Space Research*, 41, 297

Vecchio, A. & Carbone, V., 2008. *ApJ*, 683, 536

Woodard, M. F. & Noyes, R. W., 1985. *Nat.*, 318, 449

Discussion

Janet Luhmann: (Comment) An alternative approach used by Lean, Sheeley and Wang was to use coronal EUV emission to "paint" potential source surface model lines. This may give you insight into magnetic geometry.

Yvonne Elsworth: Thanks a lot for this point. The research in this field will be followed in the future.

Jeff Linsky: (Comment) Fontenla's models in 2011 are different from those in 2009. Lyman alpha wings are not properly modelled, we have encouraged him to consider Lyman alpha broadening.

Yvonne Elsworth: I appreciate this point. Indeed the wings of Lyman alpha are important for the continuum levels in the UV. We will use the updated models in the future.

Comparative Magnetic Minima:
Characterizing quiet times in the Sun and stars
Proceedings IAU Symposium No. 286, 2011
C. H. Mandrini & D. F. Webb, eds.

doi:10.1017/S1743921312004681

Reconstruction of magnetic field surges to the poles from sunspot impulses

Nadezhda V. Zolotova[1] **and Dmitri I. Ponyavin**[1]

[1]Institute of Physics, St. Petersburg State University,
Ulyanovskaya ul.1, Petrodvorets, St. Petersburg, Russia, 198504
email: `ned@geo.phys.spbu.ru`

Abstract. The time-latitude diagram of the photospheric magnetic field of the Sun during 1975-2011 (Kitt Peak NSO, SOLIS NSO, SOHO MDI data) is analyzed using Gnevysvev's idea of impulsed structure of sunspot cycle and a flux transport concept. It is demonstrated that poleward migrations of magnetic trailing polarities are closely associated with the impulses of sunspot activity. We use a fitting procedure to reconstruct the sunspot impulses and poleward magnetic field surges. We compare our results for Cycle 22 model with the time-latitude diagram of the photospheric magnetic field of the Sun.

Keywords. Sun: activity, Sun: magnetic fields, (Sun:) sunspots.

---

1. Introduction

1.1. *Sunspot impulses*

Gnevyshev(1938) proposed a hypothesis of spatio-temporal organization of sunspot activity over the solar surface based on impulses. The scale of these impulses is some tens of degrees in latitude and from 0.5 to 2 years in duration (Antalová & Gnevyshev 1983). The times at which impulses appear in both hemispheres may not be the same. The 11-year cycle consists of two or more superposed impulses, which peak at different times at different latitudes and do not always obey Spörer's law (Zolotova & Ponyavin 2011a). Additionally, the Gnevyshev gap can be observed separately in each hemisphere (Temmer *et al.* 2006; Norton & Gallagher 2010).

By means of a Gaussian random field approximation we modeled the solar cycle shape in the northern and southern hemispheres (Zolotova & Ponyavin 2011b; Zolotova & Ponyavin 2012). We specified the distribution parameters and overlapping proportions. Then, we designed different lengths, magnitudes of cycles, the presence of the Gnevyshev gap or a single-peaked activity cycle separately for each hemisphere. We have shown that even weak sunspot impulses can change the shape of the cycles — the declining phase becoming longer than the ascending one (the Waldmeier Effect). It was demonstrated that using only the convolution of activity over latitude (e. g., area or sunspot number data) makes it difficult to recognize impulses. Even a monotonic decay of activity may be consistent with them. The overlapping of impulses hides the internal structure of the cycle. Finally, why the solar cycle consists of impulses is still a puzzle.

1.2. *Magnetic field surges*

Leighton (1964) proposed that the polar magnetic field reversals are the result of trailing polarity transport by supergranular diffusion. On the contrary, Howard & LaBonte (1981) suggested that the transport of magnetic field poleward does not occur by diffusion, but by directed flow.

While the role of diffusion is annihilating the leading flux at low latitudes, that of the meridional flow is to transport the net surplus of trailing flux to the poles, with only a minor help from diffusion (Wang *et al.* 1989). These authors suggest that if supergranular diffusion did not exist, meridional flow would convect equal amounts of leading and trailing flux to the poles, producing no net change in the polar fields. A high diffusion itself (without meridional flow) is able to reverse the polar fields (Leighton 1964; Baumann *et al.* 2004).

Howard & LaBonte (1981) noticed that the polar field formation is not continuous but episodic by the movement of the magnetic field from the sunspot latitudes to the poles. Further, Wang *et al.* (1989) determined the magnetic field surges as poleward-moving streams (waves) of either polarity. In flux-transport models the surges are a result of meridional flow on the diffusion background (Wang *et al.* 1989; Baumann *et al.* 2004).

In this paper, we reconstruct magnetic field surges from sunspot impulses without diffusion assignment. Impulses are derived from observational data. Using the meridional flow and latitudinal separation between the leading and trailing spots, we model the poleward magnetic field surges.

2. Impulse–surge relation

Figure 1 shows the time-latitude diagram of the photospheric magnetic field of the Sun from 1975 to 2011 (Kitt Peak NSO, SOLIS NSO, SOHO MDI data — Scherrer *et al.* 1995; Keller 1998) with imposed sunspot impulses for Cycles 21 to 23. Black contours delineate impulses of sunspot activity and color gradation of impulses indicates sunspot density. To reconstruct sunspot impulses we used the RGO/USAF/NOAA daily data set of sunspot positions (http://solarscience.msfc.nasa.gov/greenwch.shtml). Sunspot impulses were derived independently for each cycle, from Cycle 21 to 23.

It is seen that each sunspot impulse produces the latitude migration of unbalanced flux (poleward magnetic field surge) of new polarity in each hemisphere, whereas waves of old polarity occur in gaps between impulses.

We reproduced magnetic field surges as a result of meridional transport of a net surplus to the poles. The surplus is the difference between distributions of leading and trailing

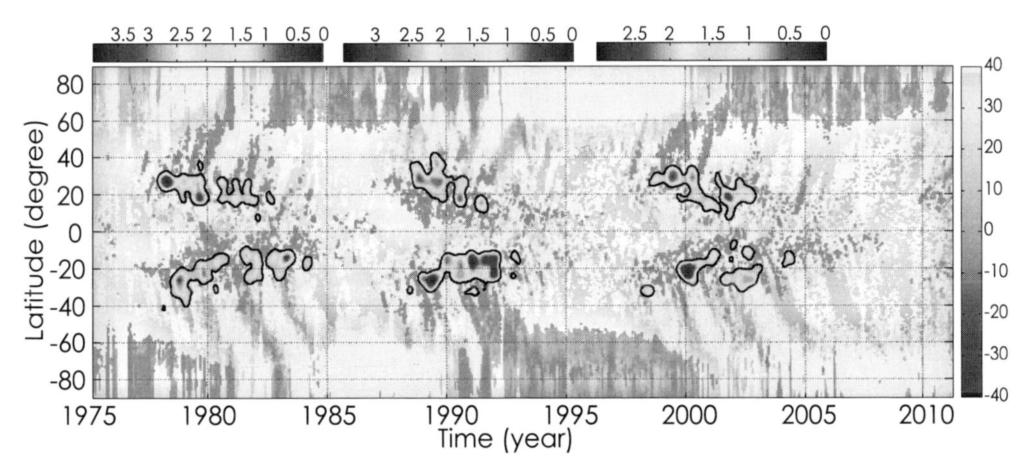

Figure 1. A magnetic butterfly diagram with imposed sunspot impulses from 1975 to 2011. Impulses are inside of the black contours.

polarities of bipolar sunspots. Thus, for surge reconstruction we required: sunspot impulses, meridional flow, and the value of the latitudinal separation between leading and trailing sunspots.

In models, it is usual to consider the effective turbulent diffusivity as a constant value (typically 600 km^2/sec) within a wide range (50–1500 km^2/sec) (see e.g., Wang *et al.* 1989; Baumann *et al.* 2004). However, there are not experimental measurements of diffusion on the Sun. Fortunately, the use of distributions permits modelling without diffusion assignment.

We defined the meridional flow pattern using an inverse proportionality to the cycle shape. To specify the north-south asymmetry of the meridional flow (Hathaway & Rightmire 2010) we modulated the meridional flow profile back in the past separately by hemisphere:

$$v = \begin{cases} (v_m - A_n/1000)\sin(\pi l/l_0), & for \quad 0^\circ < l < 90^\circ; \\ (v_m - A_s/1000)\sin(\pi l/l_0), & for \quad -90^\circ < l < 0^\circ, \end{cases} \tag{2.1}$$

where v_m is the maximal value of the meridional flow (we use 2.74^0 in latitude per solar rotation, corresponding to ~14 m/sec), A_n and A_s are the average sunspot areas in each hemisphere, l is the latitude, and l_0 is the latitude at which the meridional flow vanishes.

We determined the latitudinal separation Δl between leading and trailing sunspots using:

$$\Delta l = \Delta d \tan(0.5l). \tag{2.2}$$

where $\Delta d = 10^\circ$ is the longitudinal separation and $\alpha = 0.5l$ is the tilt angle.

Figure 2a shows sunspot impulses from 1985 to 1997. By means of latitudinal separation Δl we calculated the distributions of leading and trailing polarities of bipolar sunspots and the surplus. Applying the meridional flow pattern (Fig. 2b) to the surplus, we reconstructed poleward magnetic field surges of new and old polarities (Fig. 2c). Notice the absence of magnetic polarity cancellation across the equator. Figure 2d shows the observed photospheric magnetic field of the Sun during Cycle 22. It follows from

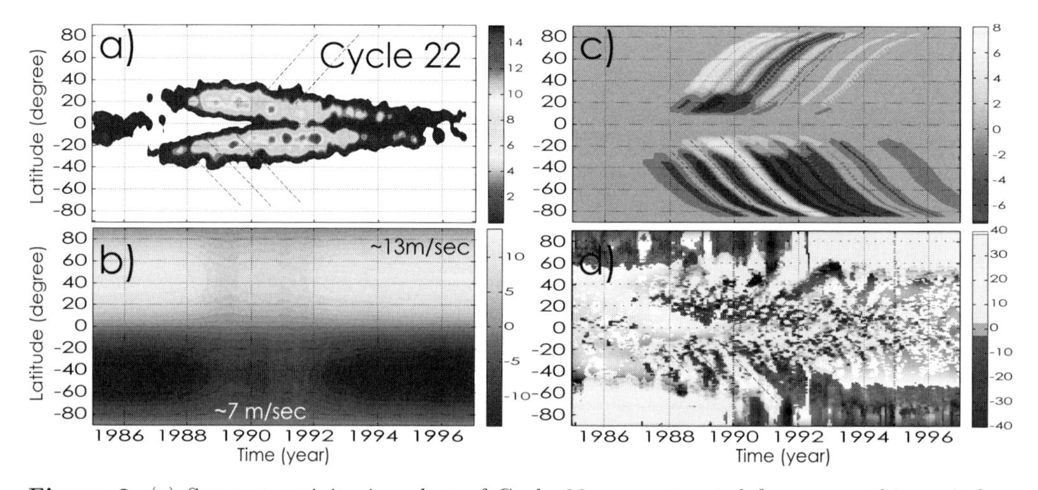

Figure 2. (a) Sunspot activity impulses of Cycle 22, reconstructed for a smoothing window $dx \times dy = 0.3 \times 0.2$ and iteration steps $i = 10$. (b) Pattern of meridional flow. (c) Reconstructed magnetic field surges. (d) Magnetic butterfly diagram for Cycle 22. Red (see the online version of the paper) dashed lines mark surges of old polarity.

Comparative Magnetic Minima:
Characterizing quiet times in the Sun and stars
Proceedings IAU Symposium No. 286, 2011
C. H. Mandrini & D. F. Webb, eds.

doi:10.1017/S1743921312004693

The Ni I lines in the solar spectrum

Mariela C. Vieytes[1,2], Pablo J. D. Mauas[1,3] and Juan M. Fontenla[4]

[1]Instituto de Astronomía y Física del Espacio
[2]Universidad Nacional de Tres de Febrero, email: mariela@iafe.uba.ar
[3]Universidad de Buenos Aires, email: pablo@iafe.uba.ar
Buenos Aires, Argentina
[4]Laboratory for Atmospheric and Space Physics, University of Colorado, Boulder, USA
email: juan.fontenla@lasp.colorado.edu

Abstract. The stratosphere is the region where the ozone chemistry is important for the balance of energy, and radiation in the near UV plays a fundamental role in the creation and destruction of ozone. However, the radiation in this range of wavelength has not been very well modeled. One of the most important elements, according to its abundance in the solar atmosphere, that contribute to the emission and absorption of radiation in the spectral range between 1900 and 3900 Å, is neutral nickel (Ni I). In this work we improve the atomic model of this element, taking into account 490 lines over the spectrum. We solve these lines in NLTE using the Solar Radiation Physical Modeling (SRPM) program and compare the results with observation of the quiet sun spectrum.

Keywords. radiative transfer, line: profiles, Sun: spectral irradiance

---

1. Introduction

It is well known that 90 % of the ozone is located and produced in the stratosphere by photodissociation of molecular oxygen by near UV Solar radiation in the Schuman-Runge bands and by ozone photodissociation in the Hartley-Huggins bands (1900 to 3900 Å). A change in the amount of solar radiation reaching this region can affect the production of ozone. Climate changes can be induced by ozone mixing ratio, changes that can affect the biosphere. Changes in ozone affect the amount of solar radiation transmitted through the stratosphere and produce tropospheric effects. Also, these changes affect the amount of biologically harmful radiation reaching the Earth's surface (Brasseur, 1997). For these reasons, there is a need for an accurate calculation of the Solar radiation that reach the top of the Earth atmosphere in the near UV range.

This spectral range has not been well modeled or observed. In order to build a semi-empirical model in the range of interest observations are essential. An important limitation is that observations from ground are difficult because of atmospheric absorption, and observations from space have limited spectral resolution. Also, another problem to obtain a realistic model is the availability of accurate atomic data for NLTE calculation.

Considering its abundance in the solar atmosphere, Ni I is an important species with quite strong lines in this range. A study that computes NLTE populations of this species is the one by Bruls (1993). In this work the author studied the behavior of the Ni I 6769.64 Å line used in MDI on board of SOHO and GONG for observing Doppler velocity and the vector magnetic field. An atomic model with only nineteen levels was built, and the level populations were computed using the model C by Vernazza *et al.* (1981).

To take into account the strong Ni I lines in the near UV range we increased the number of energy levels in the atomic model, and update the atomic data to calculate the populations in full NLTE using the latest available semi-empirical models for the Solar atmosphere (Fontenla *et al.* 2011).

In the present work we calculated the Quiet Sun spectrum by these new solar models and compared with observations of several line profiles at various wavelengths.

2. The new Ni I atomic model and spectral synthesis calculations

To carry out the calculations we used the Solar Radiation Physical Modeling code developed by Fontenla *et al.* (2011, and references therein). The code solves the radiative transfer, statistical equilibrium and momentum balance equation, assuming a one dimensional planeparallel atmosphere.

We used the latest semi-empirical models built by Fontenla *et al.* (2011) for features observed on the solar disk. These features were defined by their emitted intensity in Ca II K images taken with Precision Photometric Solar Telescope (PSPT) at Osservatorio Astronomico di Roma (Fontenla *et al.* 2009). The Quiet Sun spectrum was built by weighted averaging the three models listed in Table 1 and intend to reproduce the low solar activity state during 2008-2009.

We built a new Ni I atomic model with 61 energy levels, with the corresponding sublevels. With an energy level we mean the lower energy with the same configuration and term. The atomic data were obtained from NIST atomic spectra database (Ralchenko *et al.* 2011). We selected this number of levels to reproduce most of the near UV lines and strong visible lines present in the Solar spectrum. The populations for the different neutral atoms and ions were calculated in full NLTE. The atomic species taken into account were that from Table 2 in Fontenla *et al.* (2011), and the only change included was the new Ni I atomic model that increased the number of energy levels from 10 to 61.

The observations we used to compare our results are irradiance observations in the spectral range from 1969 to 3100 Å by Hall & Anderson (1991) and Anderson & Hall (1989). These observations are stratospheric balloon measurements, with all the complications to calibrate them correctly because of the atmospheric absorption. For this reason, these observations are here scaled so that the continuum matches the values in the calculated spectra. For comparison with these data the synthetic line profiles were convolved with a Gaussian function to simulate the resolution of the observations.

For comparing lines in the visible spectra, from 3290 to 7000 Å, we use the FTS Solar Atlas by Brault & Neckel (1999). These observations show the solar intensity at disk center obtained at the Kitt Peak Observatory and were calibrated in absolute units by a correction procedure whose accuracy may be questioned. More reliable absolute calibration exists for the Thuilliet *et al.* (2003), and the Harder *et al.* (2010) data. However, these data has much lower spectral resolution.

Figure 1 shows a comparison between several line profiles along the spectra with increasing wavelength. With solid lines we indicate our calculation, with dashed lines the spectra calculated by Fontenla *et al.* (2011) present only in the visible lines, and with doted lines the observations. The differences between the two synthetic profiles are in the NLTE calculations. In our case the populations were calculated in full NLTE with 61 levels. The spectra calculated by Fontenla *et al.* (2011) considered NLTE calculation only for 10 levels, and then they approximated the departure coefficients of the rest of the levels with the value obtained for the last level calculated in NLTE.

The figure shows a better agreement between our synthetic line profiles and the observed ones. It is important to note that several of these lines are quite strong, especially those within the range of interest shown in the first four panels. The last panel shows the Ni I 6769.64 Å line used in MDI on board of SOHO and GONG observations.

Table 1. Quiet Sun feature components and their respective models. We follow the feature and model index designations as in Fontenla *et al.* (2011)

Feature	Description	Photosphere-Chromosphere Model Index	Relative area on the Solar Disk
B	Quiet Sun inter-network	1001	80 %
D	Quiet Sun network lane	1002	19 %
F	Enhanced network	1003	1 %

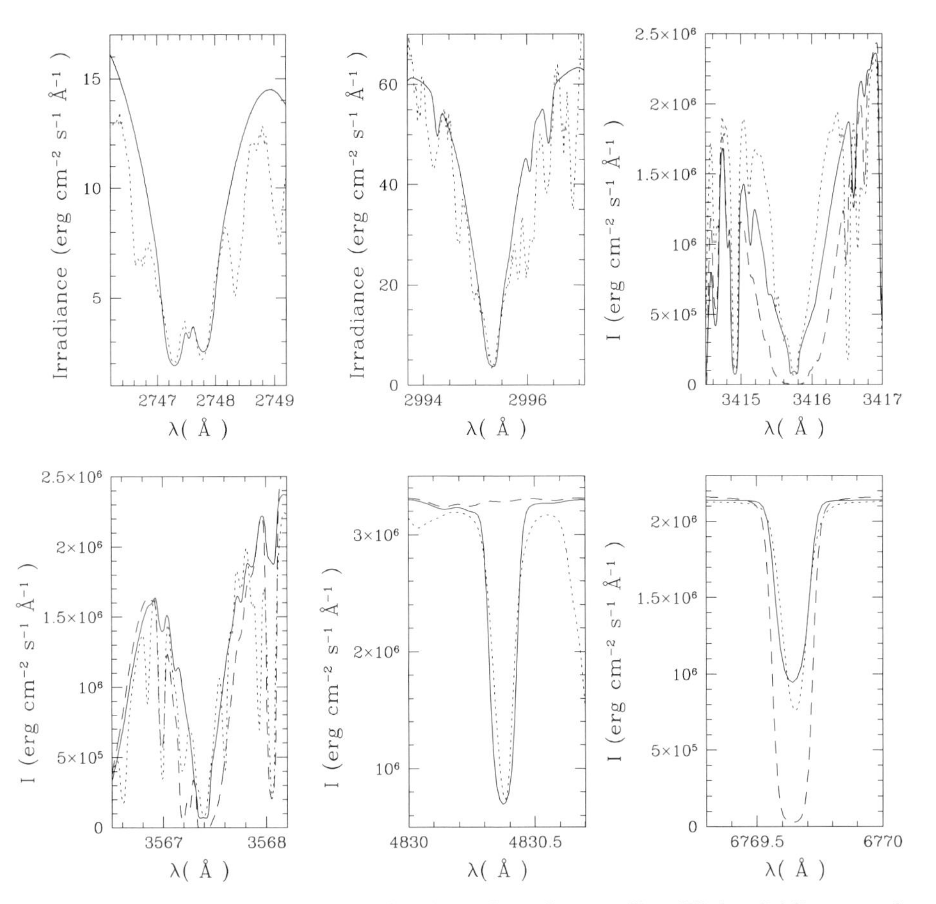

Figure 1. Comparison between observed and synthetic line profiles. With solid lines are shown our calculation, with dashed lines the spectra calculated by Fontenla *et al.* (2011) present only in the visible lines, and with doted lines the observations.

3. Summary

In this work we improved the Ni I atomic model to include the majority of the lines in the near UV range that are listed in the NIST atomic database. Several of these lines are quite strong and have not been taken into account in solar irradiance calculations.

There are large differences in the synthetic line profiles using different approximations. The results shown in this work stress the importance of full NLTE calculation and the use of a sufficient number of levels to obtain a realistic synthetic line profiles.

We reproduce very well the Ni I 6769.64 Å used in MDI and GONG observations, allowing a deeper study of the line formation, specially in different phases of the solar cycle, to investigate the impact of the solar activity in its behavior.

We do not show in this work details on the line source functions, departure coefficients or depths of formation, but these subjects will be included in a forthcoming paper.

References

Anderson, G. & Hall, L. 1989, *JGR*, 94, D56435

Brasseur, G. (ed.) 1997, *NATO ASI Series: The Stratosphere and its role in the Climate System*, 54, 1

Brault, J. & Neckel, H. 1999, *Sol. Phys.*, 184, 421

Bruls, J. 1993, *A&A*, 269, 509

Fontenla, J., Curdt, W., Haberreiter, M., Harder, J., & Tian, H. 2009, *ApJ*, 707, 482

Fontenla, J., Harder, J., Livingston, W., Snow, M., & Woods, T. 2011, *JGR*, 116, D20108

Hall, L. & Anderson, G. 1991, *JGR*, 96, D712927

Harder, J., Thuillier, G., Richard, E., Brown, K., Lykke, M., Snow, W., McClintock, J., Fontenla, J., Woods, T., & Pilewskie, P. 2010, *Sol. Phys.*, 263, 3

Ralchenko, Yu., Kramida, A. E., & Reader, J., and NIST ASD Team 2011, *NIST Atomic Spectra Database (ver. 4.1.0), [Online]. Available: http://physics.nist.gov/asd [2011, November 4]. National Institute of Standards and Technology, Gaithersburg, MD*

Thuillier, G., Herse, M., Labs, D., Foujols, T., Peetermans, W., Guillotay, D., Simon, P., & Mandel, H. 2003, *Sol. Phys.*, 214, 1

Vernazza, J., Avrett, E., & Loeser, R. 1981, *ApJS*, 45, 635

Discussion

Jeff Linsky: Oxygen abundance in the Sun has been very controversial because of its overlap with NiI line. It would be good to reevaluate the oxygen abundance

Mariela Vieytes: Thanks for your suggestion, I will study this problem with the new calculations we have done to improve the NiI atomic model.

Comparative Magnetic Minima:
Characterizing quiet times in the Sun and stars
Proceedings IAU Symposium No. 286, 2011
C. H. Mandrini & D. F. Webb, eds.

doi:10.1017/S174392131200470X

Towards the reconstruction of the EUV irradiance for solar Cycle 23

Margit Haberreiter

Physikalisch-Meteorologisches Observatorium Davos/World Radiation Center
7260 Davos Dorf, Switzerland
email: margit.haberreiter@pmodwrc.ch

Abstract. We present preliminary reconstructions of the EUV from 26 to 34 nm from February 1997 to May 2005, covering most of solar cycle 23. The reconstruction is based on synthetic EUV spectra calculated with the spectral synthesis code Solar Modeling in 3D (SolMod3D). These spectra are weighted by the relative area coverage of the coronal features as identified from EIT images. The calculations are based on one-dimensional atmospheric structures that represent a temporal and spatial mean of the chromosphere, transition region, and corona. The employed segmentation analysis considers coronal holes, the quiet corona, and active regions identified on the solar disk. The reconstructed EUV irradiance shows a good agreement with observations taken with the CELIAS/SEM instrument onboard SOHO. Further improvement of the reconstruction including more solar features as well as the off-limb detection of activity features will be addressed in the near future.

Keywords. Sun: corona, Sun: UV radiation, line: formation

1. Introduction

The solar EUV irradiance is the main energy input for the upper Earth's atmosphere with important effects on the ionosphere and thermosphere. The solar energy output changes on short time-scales of minutes to hours as well as longer times-scales such as the 27-day solar rotation cycle or the 11-year solar cycle. There is also indication that the EUV irradiance might shows a secular trend (Didkovsky *et al.* 2010).

The EUV radiation incident on the upper Earth's atmosphere leads to a change in its temperature and density (see e.g. Solomon *et al.* 2010). In order to understand the effects of the changing EUV radiation on the Earth's atmosphere a continuous data set covering the short-term and long-term variations is essential. However, as space instruments are limited with regard to their temporal and spectral coverage, reliable models are needed to fill the gaps of the observational data sets.

Several reconstruction approaches involve the use of proxies to describe the EUV variability. Lean *et al.* (2011) employ the two and three component NRLSSI model based on the Mg II and $F_{10.7}$ index to characterize and forecast the EUV variations. A further example for an empirical model is SOLAR2000 (Tobiska *et al.* 2000), a model based on an extensive number of irradiance proxies.

There is also ongoing work to determine which proxies, or spectral lines, are the best representatives for the variations of the entire EUV spectrum (see e.g. Kretzschmar *et al.* 2009, Dudok de Wit *et al.* 2009).

Proxy models have been quite successful in describing the EUV variations, however, in order to understand the complete physical processes driving the irradiance variations, it is important to model the variations of the full solar spectra. Warren (2006) utilizes differential emission measure distributions derived from spatially and spectrally resolved solar observations and full-disk solar images. The reconstruction presented here follows the same principle as used by Haberreiter *et al.* (2005) and Shapiro *et al.* (2011). It includes the calculation of synthetic spectra with a radiative transfer code for various activity features on the solar disk. Weighting the spectra by filling factors derived from the relative area coverage of these activity features or from proxy data then yields the time dependent irradiance spectrum.

In the following section the spectral synthesis code Solar Modeling in 3D (SolMod3D) is introduced. Then, in Section 3 the reconstruction approach is described briefly. Finally, in Section 4 our results are compared with the EUV irradiance observations carried out with the CELIAS/SEM instrument (Hovestadt *et al.* 1995) onboard the SOHO mission.

2. Spectral Synthesis

The spectral synthesis of the EUV is carried out with the SolMod3D code. It is a state-of-the-art radiative transfer code in full non-local thermodynamic equilibrium (NLTE). SolMod3D allows for the spherical line-of-sight integration which is very important for the correct calculation of the coronal emission. The code has already been successfully employed for the calculation of the quiet Sun EUV spectrum (Haberreiter 2011) and for solar limb studies (Thuillier *et al.* 2011).

The final goal is to reconstruct the EUV from 10 to 100 nm. Here, however, we focus on the wavelength range from 26 to 34 nm as observed with the SOHO/SEM instrument onboard SOHO. The calculations of the spectra are based on semi-empirical structures that represent a temporal and spatial mean of the photosphere, chromosphere, transition region and corona. For the photosphere, chromosphere, and transition region the full NLTE radiative transfer is solved based on the latest semi-empirical atmospheric structures by Fontenla *et al.* (2009). The coronal lines are calculated as optically thin lines. Currently we employ four coronal structures; three represent different features of the quiet Sun, i.e. the quiet corona, coronal network and active coronal network as described in Haberreiter (2011). For the bright coronal regions a fourth structure is used. Details of this structure will be presented in a future publication. The coronal holes are currently represented with a contrast of 0.5 with respect to the quiet corona.

It is important to note that these atmosphere structures constitute a temporal and spatial mean. As such, these structures fail to reproduce the very rich short-term variations of the solar atmosphere, in particular the corona. However, as we are mainly interested in the daily EUV variations, 1D structures are considered to be a suitable representation of the solar atmosphere. The advantage of the 1D atmosphere structures is that it allows us to take into account an extensive atomic data set (14,000 atomic levels and 170,000 spectral lines), essential for a realistic calculation of the EUV spectrum.

3. Reconstruction

First, intensity spectra are calculated for different features on the solar disk as described above. These spectra are then weighted by filling factors derived from the analysis of solar images. For the study presented in this paper we use the 3-component analysis of images taken with the Extreme UV Imaging Telescope (Delaboudinière *et al.* 1995, EIT) onboard SOHO carried out by Barra *et al.* (2009). The segmentation analysis is based on the so-called *fuzzy clustering*, a robust and fast technique that allows to monitor and track active regions on the solar disk. The analysis provides the disk-integrated relative area coverage of coronal holes, the quiet corona, and active coronal regions detected on the solar disk. However, extended active regions located beyond the solar limb are not yet

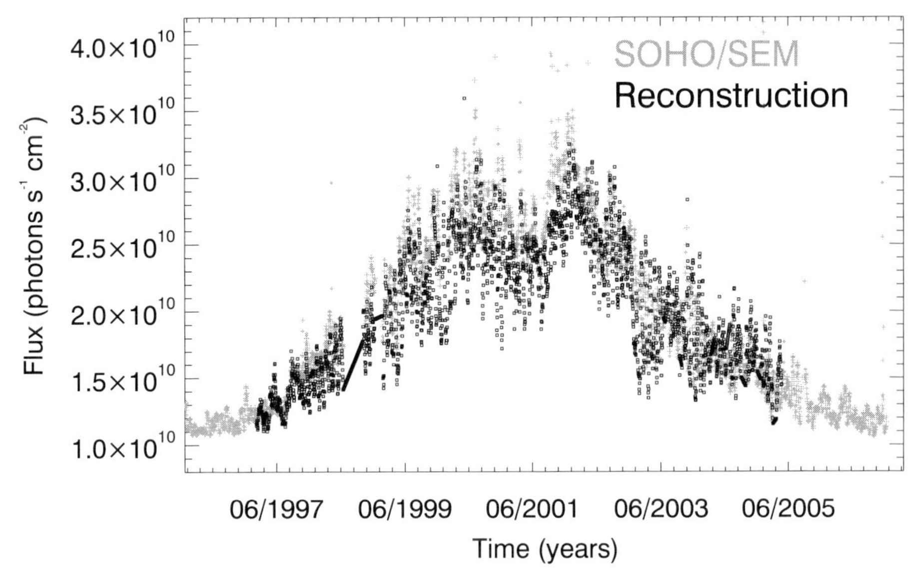

Figure 1. Preliminary results of the reconstruction of the EUV photon flux at 1 AU for the wavelength range from 26 to 34 nm (black squares) compared with the SOHO/SEM observations for the same wavelength range (grey crosses).

taken into account. To compensate the missing radiation, we assume here an additional 20% increase of the irradiance due to the flux coming from the extended corona. The detailed feature analysis of the off-limb features will be included in future work.

4. Results

Fig. 1 shows the reconstruction of the EUV from 26 to 34 nm from Feb 14, 1997 to May 1, 2005 compated with the SOHO/SEM observations covering the same wavelength range. The average quiet corona is represented as 75% of the quiet corona, 22% of the coronal network, and 3% of the active coronal network. This combination has already been used by Haberreiter (2011) for modeling the EUV for solar minimum conditions. For the coronal holes we use a contrast of 0.5 with respect to the quiet Sun intensity. We tested the effect of the contrast of coronal holes and conclude that the contrast of the coronal holes shows a minor effect on the overall EUV variability. The variations in the EUV are mainly driven by the varying area coverage of the bright active regions.

As the segmentation analysis of Barra *et al.* (2009) includes only on-disk features, the missing radiation from the extended corona has to be compensated by assuming that 20% of the on-disk radiation account for the radiation of the extended corona. The off-limb contribution will be studied in detail in the near future. A refinement of the features in the segmentation analysis will also be addressed in upcoming studies. Finally, for longer wavelength ranges we also need to account for the detailed center-to-limb variations of the contribution from the chromospheric features. Here, the goal is to implement PSPT images (see e.g. Ermolli *et al.* 2007). Given the limitations of these preliminary results the agreement with the SOHO/SEM observations is very promising.

5. Conclusion

We have presented a preliminary version of the reconstruction of the EUV variations from Feb 14, 1997 to May 1, 2005 for the wavelength range between 26 and 34 nm and compared it with observations taken with the CELIAS/SEM instrument. This work is based on the calculation of spectra in the EUV with the SolMod3D code and the filling factors for three coronal components. The good agreement of the reconstruction with the observed irradiance variations shows that our approach is suitable for the study of the EUV variability. Nevertheless, further analysis is required, in particular the study of the off-limb contribution to the EUV irradiance and the consideration of additional coronal activity features.

Acknowledgements

We thank Veronique Delouille and Cis Verbeeck for kindly providing the segmentation analysis of the EIT data. SOHO is a project of international cooperation between ESA and NASA. MH acknowledges support by the Holcim Foundation for the Advancement of Scientific Research.

References

Barra, V., Delouille, V., Kretzschmar, M., & Hochedez, J. 2009, *A&A*, 505, 361
Delaboudinière, J.-P., Artzner, G. E., Brunaud, J., *et al.* 1995, *Solar Phys.*, 162, 291
Didkovsky, L. V., Judge, D. L., Wieman, S. R., & McMullin, D. 2010, in: S. R. Cranmer, J. T. Hoeksema, & J. L. Kohl (eds.), *SOHO-23: Understanding a Peculiar Solar Minimum*, Astronomical Society of the Pacific Conference Series, Vol. 428, p. 73
Dudok de Wit, T., Kretzschmar, M., Lilensten, J., & Woods, T. 2009, *GRL*, 36, 10107
Ermolli, I., Criscuoli, S., Centrone, M., Giorgi, F., & Penza, V. 2007, *A&A*, 465, 305
Fontenla, J. M., Curdt, W., Haberreiter, M., Harder, J., & Tian, H. 2009, *ApJ*, 707, 482
Haberreiter, M. 2011, *Solar Phys.*, 274, 473
Haberreiter, M., Krivova, N. A., Schmutz, W., & Wenzler, T. 2005, *Adv. Sp. Res*, 35, 365
Hovestadt, D., Hilchenbach, M., Bürgi, A., *et al.* 1995, *Solar Phys.*, 162, 441
Kretzschmar, M., Dudok de Wit, T., Lilensten, J., *et al.* 2009, *Acta Geophysica*, 57, 42
Lean, J. L., Woods, T. N., Eparvier, F. G., *et al.* 2011, *JGR*, 116, 1102
Shapiro, A. I., Schmutz, W., Rozanov, E., Schoell, M., Haberreiter, M., Shapiro, A. V., & Nyeki, S. 2011, *A&A*, 529, A67
Solomon, S. C., Woods, T. N., Didkovsky, L. V., Emmert, J. T., & Qian, L. 2010, *GRL*, 37, 16103
Thuillier, G., Claudel, J., Djafer, D., *et al.* 2011, *Solar Phys.*, 268, 125
Tobiska, W. K., Woods, T., Eparvier, F., *et al.* 2000, *Journal of Atmospheric and Solar-Terr. Phys.*, 62, 1233
Warren, H. P. 2006, *Ad. Sp. Res*, 37, 359

Discussion

Jeffrey Linsky: I would like to draw your attention to a recently published paper by Juan Fontenla in which he updated the atmospheric models of the solar atmosphere (Fontenla *et al.* 2011). In particular the formation of the Lyman continuum has been improved. These latest set of solar atmosphere models has been successfully employed to reproduce the Lyman continuum of stars with various effective temperatures (Linsky *et al.* 2012).

Margit Haberreiter: We thank Jeff Linsky for this comment and will follow up on the recommended publications.

Comparative Magnetic Minima:
Characterizing Quiet Times in the Sun and Stars
Proceedings IAU Symposium No. 286, 2011
C. H. Mandrini & D. F. Webb eds.

doi:10.1017/S1743921312004711

Polar magnetic fields and coronal holes during the recent solar minima

Giuliana de Toma

High Altitude Observatory
National Center for Atmospheric Research
P.O. Box 3000, Boulder, CO 80307-3000
email: detoma@ucar.edu

Abstract. The slow decline of solar Cycle 23 combined with the slow rise of Cycle 24 resulted in a very long period of low magnetic activity during the years 2007–2009 with sunspot number reaching the lowest level since 1913. This long solar minimum was characterized by weak polar magnetic fields, smaller polar coronal holes, and a relatively complex coronal morphology with multiple streamers extending to mid latitudes. At the same time, low latitude coronal holes remained present on the Sun until the end of 2008 modulating the solar wind at the Earth in co-rotating, fast solar wind streams. This magnetic configuration was remarkably different from the one observed during the previous two solar minima when coronal streamers were confined near the equator and the fast solar wind was mainly originating from the large coronal holes around the Sun's poles. This paper presents the evolution of the polar magnetic fields and coronal holes during the past minimum, compare it with the previous minima, and discuss the implications for the solar wind near the Earth. It also considers the minimum of Cycle 23 in an historical perspective and, in particular, compares it to the long minima at the turn of the 19th century.

Keywords. Sun: sunspots, corona, solar wind

1. Introduction

The solar minimum between Cycles 23 and 24 was the longest and deepest minimum in about 100 years. The northern hemisphere reached solar minimum conditions already in 2006. In 2007–2008, sunspots continued to emerge, albeit at a very low level, preferentially in the southern hemisphere. In 2009 magnetic activity remained very low, with most of the active regions emerging on the Sun belonging to the new Cycle 24. The years 2008 and 2009 were extremely quiet years, with spots present on the Sun less than 30% of the time (Figure 1). Because magnetic flux emergence was so low for an extended period of time, it is not easy to identify a single point in time as the minimum between Cycle 23 and 24 but the periods in August 2008 and August 2009 stand out as particularly quiet times when spots were absent from the Sun for over 30 consecutive days.

This extremely quiet minimum differed in many ways from other recent minima for which we have a good observational record and changed our idea about solar minima. Not only were sunspot activity, CMEs, and flares extremely low during the past minimum, the solar wind had anomalously low densities and magnetic fields, both at the poles and in the ecliptic (e.g., McComas *et al.* 2008, Issautier *et al.* 2008, Smith & Balogh 2008, Jian *et al.* 2010). The interplanetary magnetic field measured at 1 AU continued to decline in 2006–2009 and by 2009 had decreased by 30% compared to 1996, reaching its lowest value ever recorded in 50 years of observations (e.g., Wang *et al.* 2009, Tsurutani *et al.* 2011). However, in spite of the low magnetic activity, the corona and heliosphere remained relatively complex throughout most of the minimum phase.

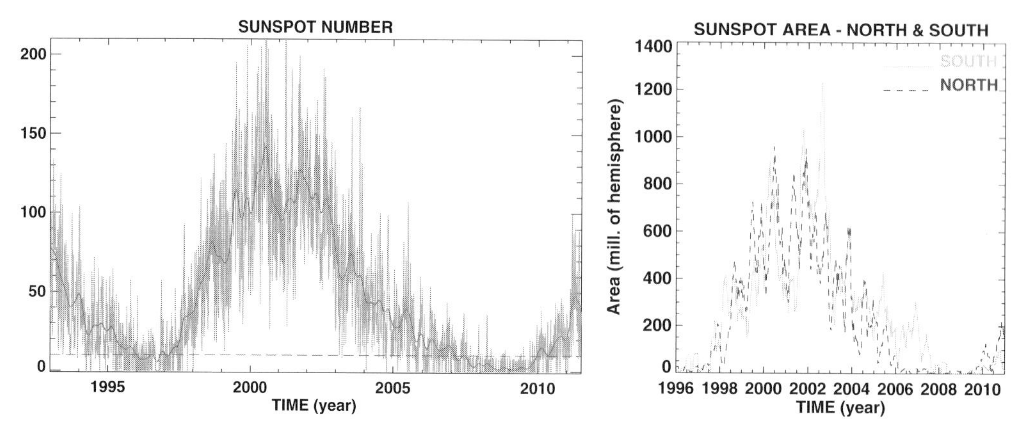

Figure 1. (Left) Daily sunspot number and 3-rotation averages. The minimum between Cycles 23 and 24 corresponds to a long period of extremely low magnetic flux emergence with sunspot activity lower than in 1996. (Right) Sunspot area averaged over 3 rotations for the northern and southern hemisphere. The north reached solar minimum conditions before the south, and Cycle 24 activity started first in the north and later in the south.

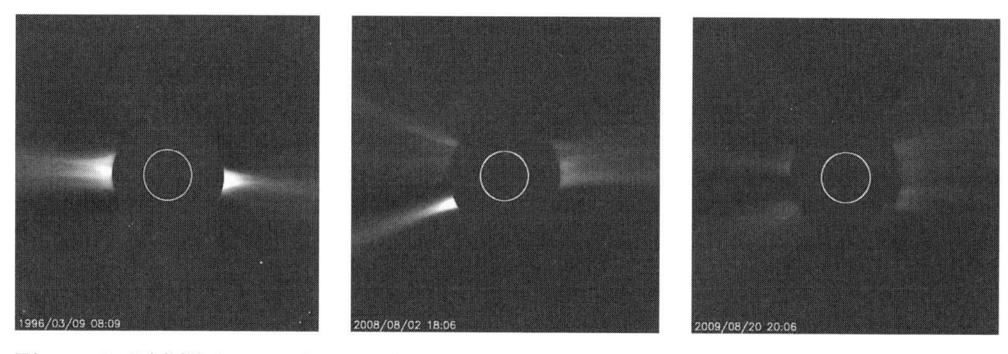

Figure 2. LASCO images showing the solar corona in visible light in March 1996 (left) August 2008 (center) and August 2009 (right). Even during the extremely quiet periods in August 2008 and August 2009 the shape of the solar corona was more complex than in 1996.

Sections 2 and 3 discuss the properties and evolution of the solar corona and solar wind during the minimum of Cycle 23 and the role of weak polar magnetic fields at the photosphere in determining the complexity of the corona and heliosphere during this very quiet time. Section 4 examines other low solar minima in the historical sunspot record, while conclusions are given in Section 5.

2. The Solar Corona during the Minimum of Cycle 23

The shape and structure of the solar corona and heliosphere were noticeably different during the minimum of Cycle 23 compared to other recent minima, and, interestingly, were not as simple as in 1996 or 1986 (e.g., Gibson *et al.* 2009, Luhmann *et al.* 2009, de Toma *et al.* 2010, Petrie *et al.* 2010). The solar corona was still quite complex in 2007 and 2008, even if sunspot activity was below the levels seen in 1986 and 1996, and slowly evolved toward a simpler structure in 2009, but it never reached a "dipolar" configuration with coronal streamers confined to a narrow band around the solar equator. Multiple streamers and pseudo-streamers (Wang *et al.* 2009, Riley *et al.* 2010, Gibson *et al.* 2011) remained present on the Sun even during times of extremely low magnetic activity (Figure 2). The heliosphere was also more complex for a significant fraction of the Cycle 23 minimum (e.g., Tokumaru *et al.* 2009, 2010) and only in 2009 did the heliospheric current sheet assume a relatively "flat" configuration, more typical of solar minimum (de Toma *et al.* 2010b).

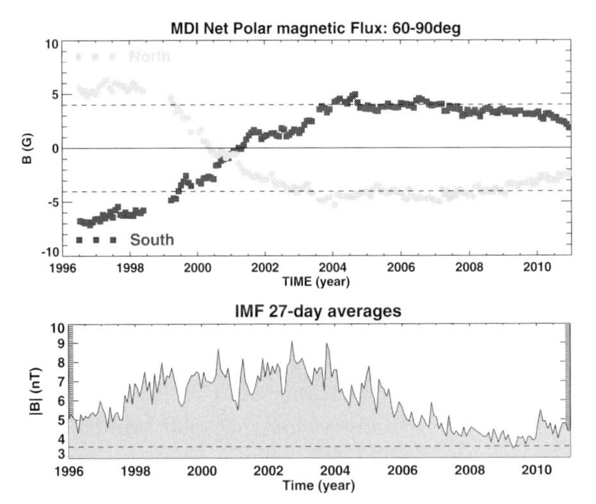

Figure 3. (Top) Polar magnetic flux as measured by SOHO/MDI. The strong decrease in the net polar magnetic flux during the past minimum was also observed at ground-based solar observatories, such as at Kitt Peak and Wilcox Solar Observatory. The polar magnetic fields remained almost constant between 2004 and 2008, and in mid 2009, with the increase in Cycle 24 activity, started to slowly decline. (Bottom) Interplanetary magnetic field (IMF) averaged over 27 days. In contrast to the polar fields, the IMF continued to slowly decline in 2006–2009.

The more complex morphology of the solar corona during the past minimum was largely because of the weak polar magnetic fields at the photosphere (e.g., Sheeley 2008). One of the striking observational features of the late declining phase and extended minimum of Cycle 23 was the low value of the polar magnetic flux. In 2004–2009, the net polar magnetic flux above 60° latitude was remarkably stable and about 40% lower than observed during the previous two minima as illustrated in Figure 3 (see also Svalgaard & Cliver 2007, Sheeley 2008, Wang *et al.* 2009). Thus, the continuous decline observed in the interplanetary magnetic field was not related to changes in the polar regions, but rather was caused by changes in the "open" magnetic flux from lower latitude sources, as we will discuss below.

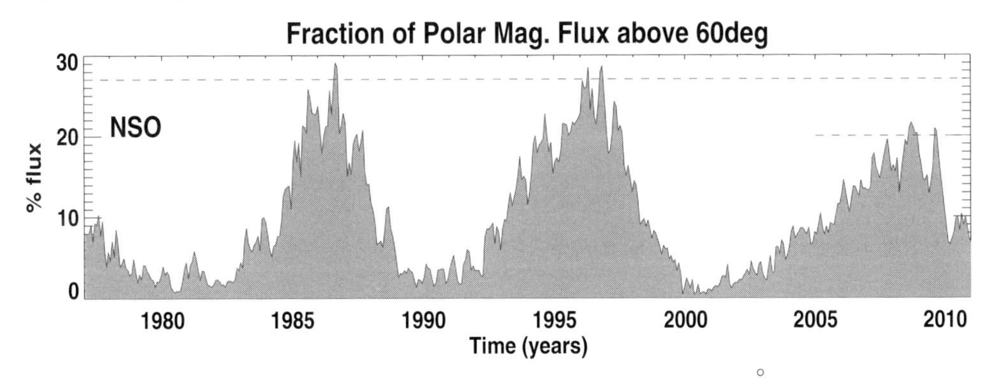

Figure 4. Fraction of the total magnetic flux in the polar region above 60° latitude as measured at NSO/Kitt Peak using KPVT and SOLIS magnetograms. The ratio of the net polar magnetic flux to the total magnetic flux during the minimum of Cycle 23 was significantly lower than during the previous two minima.

The lower value of the polar magnetic field resulted in global coronal field with a weaker polar dipole moment (l=1, m=0) while higher order moments did not change significantly. Figure 4 gives the ratio of the net polar magnetic flux –which is directly related to the strength of the polar dipole component– to the total unsigned magnetic flux at the photosphere, as measured at NSO/Kitt Peak. This ratio always reaches its maximum value near solar minimum, when polar magnetic fields are strongest and activity at mid and low latitudes is low. During the minimum of Cycle 23 the ratio of the polar to the total magnetic flux decreased significantly compared to the previous two minima. As a result, even if the polar dipole remained the dominant coronal field component (Petrie *et al.* 2010), the ratio of the polar dipole to the equatorial dipole and to the quadrupole moment decreased during the minimum in 2008–2009 relative to 1996 and 1986 minima, when the polar dipole was larger (e.g., Wang *et al.* 2009, Petrie *et al.* 2010, DeRosa *et al.* 2011). This, in turn, resulted into a more complex coronal configuration and in a tilted and warped heliospheric current sheet and allowed mid- to low-latitude streamers and pseudo-streamers to remain present even during times of extremely low activity (Gibson *et al.* 2011 and references therein).

2.1. *Coronal Holes*

The special solar minimum configuration at the end of Cycle 23 with very weak polar fields, and a weak polar dipole, had significant implications for the organization and evolution of coronal holes on the Sun and, consequently, for the solar wind. Figure 5 shows EUV synoptic maps of the Sun in the Fe XII 195Å lines where coronal holes appear as dark regions in the corona. The top two panels show the EUV corona in 1996 during the minimum of Cycle 22 when polar fields were strong. The bottom two panels show the corresponding coronal hole maps derived using a combination of magnetograms and four different EUV wavelengths (de Toma & Arge 2005, de Toma 2010). In 1996, coronal holes were organized in two large polar coronal holes that extended to or below 50° latitude. The area of each polar hole was about 7–8% of the Sun's surface while low-latitude coronal holes were small or absent (Harvey & Recely 2002, de Toma 2010).

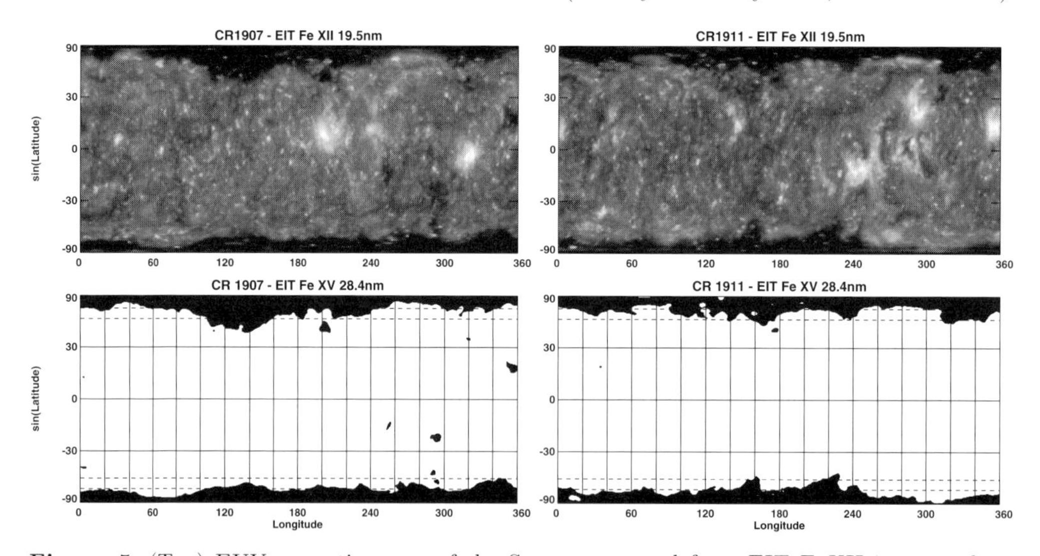

Figure 5. (Top) EUV synoptic maps of the Sun constructed from EIT Fe XII images where coronal holes appear as dark regions in the corona. The maps are in sin(latitude) and thus have equal area pixels. The two Carrington Rotations shown are CR 1907 in March 1996 and CR 1911 in July 1996 during the minimum of Cycle 22. (Bottom) Corresponding coronal hole maps. The dashed lines mark 50° and 60° latitude.

Figure 6 shows EUV synoptic maps of the Sun and the corresponding coronal hole maps for the minimum of Cycle 23. Polar coronal holes were significantly smaller, by 20–40%, than observed during the previous minimum (de Toma 2010) and mostly confined above

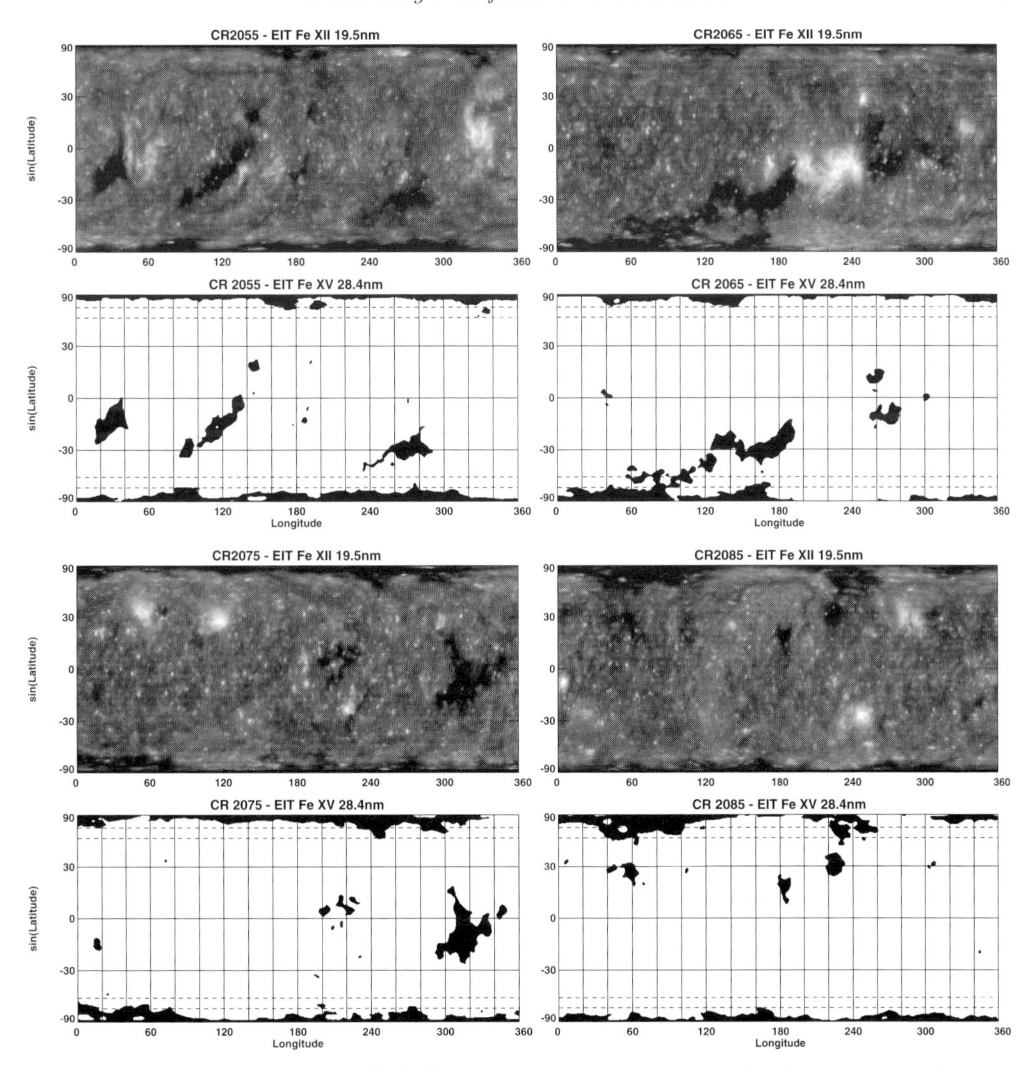

Figure 6. EUV synoptic maps of the Sun and corresponding coronal hole maps as in Figure 5 but for CR 2055 in April 2007, CR 2065 in January 2008, CR 2075 in October 2008, and CR 2085 in July 2009 during the extended minimum at the end of Cycle 23.

60° latitude. At the same time, isolated coronal holes of significant size were present on the Sun between $\pm 30°$ latitude. These low-latitude coronal holes, commonly seen during the declining phase of the solar cycle, persisted into the minimum phase of Cycle 23 and continued to be important sources of fast solar wind at the Earth in 2006–2008 (Gibson *et al.* 2009, Luhmann *et al.* 2009, Lee *et al.* 2009, Abramenko *et al.* 2010, de Toma 2010). They were long-lived and produced weak to moderate recurrent geomagnetic storms (Tsurutani *et al.* 2011). As a result, the solar wind speed maintained some of the 9-, 13.5-, and 27-day periodicities typically observed during the declining phase (Emery *et al.* 2009, 2010). The same periodicities were also seen in other observations, including auroral and geomagnetic indices, radiation belt electron flux (Gibson *et al.* 2009), and even propagated down into the upper atmosphere (Thayer *et al.* 2008, Lei *et al.* 2010). At the end of 2008, these large, low-latitude coronal holes started to close down and finally disappeared in early 2009 (de Toma 2010). In 2009, small and transient, mid-latitude coronal holes formed in the remnants of the first Cycle 24 active regions. These small

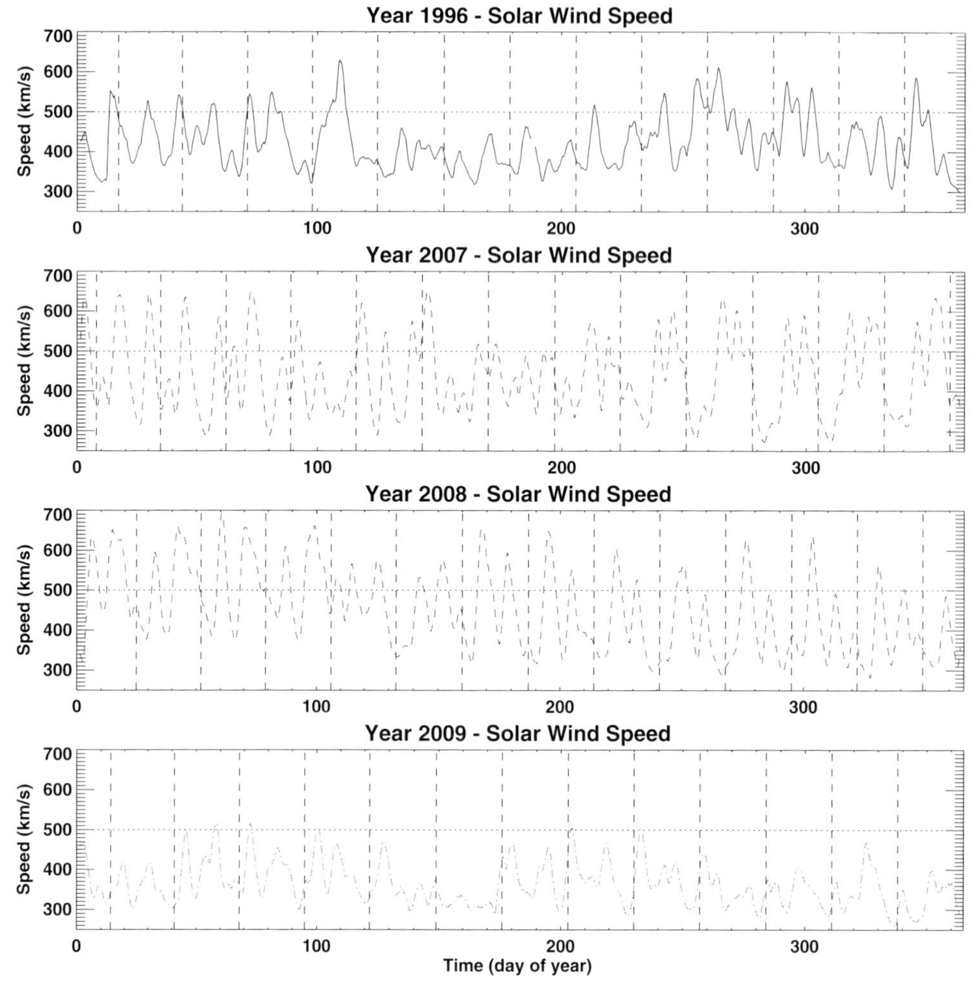

Figure 7. Solar wind velocity time series for the years 1996, 2007, 2008, and 2009. Hourly solar wind speeds have been averaged over 3 days.

and short-lived coronal holes were not important sources of the solar wind at the Earth, which originated mostly from the edges of the polar coronal holes during this time.

The persistence of large low-latitude coronal holes in 2007 and 2008 was possible because of the particular magnetic flux distribution at the Sun. Not only polar fields were weaker than in 1996 but the predominantly unipolar regions around the poles were confined to a smaller area at high latitudes. This gave rise to smaller polar coronal holes and made it easier for quasi-unipolar magnetic regions at low-latitude to "open" into the heliosphere. Numerical experiments using potential field source surface extrapolations where the photospheric polar magnetic fields are artificially modified (e.g., Wang *et al.* 2009, Luhmann *et al.* 2009, de Toma & Arge 2010) show that increasing the flux density near the poles in 2008 results in smaller low-latitude coronal holes, while decreasing it in 1996, makes it possible to form low-latitude coronal holes. de Toma & Arge (in preparation) find that not only the strength of the polar fields is important, but that the magnetic flux distribution near the poles and the latitudinal extent of the quasi-unipolar polar regions also play an important role in determining the distribution and size of coronal holes on the Sun at both high and low latitudes.

3. Solar Wind during the Minimum of Cycle 23

Figure 7 shows the time series of solar wind speed for the years 1996, 2007, 2008, and 2009. In 2007 and 2008, the yearly mean solar wind speed was 440 km/s and 448 km/s, respectively, similar to the 423 km/s observed in 1996. Nonetheless, in 2007 and 2008 the solar wind was more structured than in 1996 with regular high-speed streams, and fast solar wind was more commonly seen at the Earth. Speeds above 500 km/s were observed 29% and 33% of the time in 2007 and 2008 in hourly averaged solar wind data compared to 17% of the time in 1996. This fast wind originated from the large and long-lived, isolated coronal holes discussed above and gave origin to the prominent high-speed tail in the velocity distribution indicated by the arrow in Figure 8 (de Toma 2010). In 2009, with the disappearance of the low-latitude coronal holes, there was an abrupt change in the solar wind speed at the Earth. The mean solar wind velocity dropped to an yearly value of 365 km/s, a 20% decrease from 2008. Solar wind speeds above 500 km/s were seen at Earth only 4% of the time and the high-speed velocity tail in the solar wind distribution disappeared. At the same time, the interplanetary magnetic field continued to slowly decline, as these low-latitude source regions of fast wind closed down.

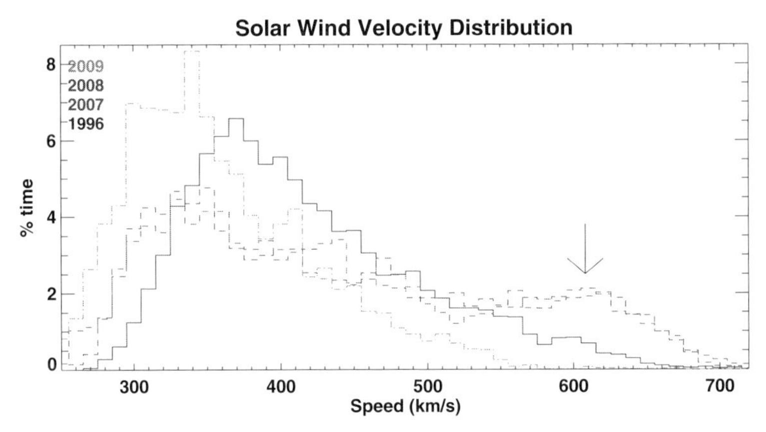

Figure 8. Solar wind velocity distributions for hourly solar wind speeds for the year 1996, 2007, 2008, and 2009. In 2007 and 2008, the velocity distribution is almost bimodal with a secondary peak at about 600 km/s. This high-velocity tail disappears completely in 2009 and the peak of the distribution shifts towards lower speeds.

The changes in the solar wind speed during the minimum of Cycle 23 had a very important effect on the Earth's radiation belts which remained elevated in 2007 and 2008, when solar wind speeds above 500 km/s were relatively common (Gibson *et al.* 2009), and almost completely vanished in 2009, when velocities dropped below 500 km/s (D. Baker, private communication). Geomagnetic activity also reached its lowest value in 2009 (e.g., Emery *et al.* 2010, Tsurutani *et al.* 2011) when the solar wind magnetic field and speed reached their minimum level.

4. Solar Minima in the Historical Record

While this minimum clearly differed from the other solar minima during the Space Age, the historical sunspot record indicates that this was not an unusual minimum and that very long and deep minima have occurred several times in the past. In particular, the minima at the end of the 19th century and beginning of the 20th century were as long or longer than the quiet period in 2007–2009. Unfortunately, we do not have direct solar wind observations for that period and only very limited solar observations besides sunspots.

Figure 9. Photographs of the total solar eclipses in December 1889 in French Guiana (left) and in June 1901 in Sumatra (center). The right panel shows a composite image for the 1901 eclipse obtained by combining multiple frames with different exposures (courtesy of M. Druckmüller).

Routine monitoring of the solar corona in broad-band visible light started in the '60s at Mauna Loa Solar Observatory (Fisher *et al.* 1981) while coronal observations in the Fe XIV emission line go back to the '40s (Billings 1966). Before this time, we only have solar eclipses available to infer the morphology of the corona (e.g., Judge *et al.* 2010). However, total eclipses are rare and short in duration and give us only a snapshot of the solar corona. They occur approximately once in 18 months, and not always at easy-to-reach locations, therefore, they cannot give us information about coronal evolution on time scales of less than a year, yet, they are useful to gain insight to the coronal morphology at different phases of the solar cycle. At the High Altitude Observatory in Boulder, Colorado, we have a collection of old eclipse photographs, *http : //mlso.hao.ucar.edu/mlso_eclipses.html*, that goes back to 1869. Figure 9 shows two of such photographs for the years 1889 and 1901, during the solar minima at the end of Cycle 12 and 13. These minima with 212 and 287 days without spots had similar sunspot activity to the years 2008 and 2009. It is, thus, interesting to examine the coronal shape at the times of these old eclipses. Multiple photographs were taken with different exposure times which can be combined to obtain composite images of the solar corona to better display the streamer structure (see the right panel of Figure 9 and Judge *et al.*, 2010). The eclipse images in 1889 and 1901 in Figure 9 reveal a solar corona far from a simple dipolar structure and show that multiple coronal streamers existed for at least part of these long solar minima, suggesting weak polar magnetic fields. Interestingly, the *aa* index for these minima indicates a low level of geomagnetic activity, and a Fourier analysis of the index shows periodicities at 7-, 9-, and 13.5–days, which might indicate the presence of recurrent, low-latitude coronal holes. Thus, these minima at the turn of the 19th century may have been similar to the recent solar minimum.

5. Conclusions

The long and deep minimum at the end of Cycle 23 was the lowest and longest minimum in about a century. It differed in many ways from the other solar minima observed during the Space Age for which we have a good observational coverage both at the Sun and in the heliosphere. This gave us an opportunity to advance our understanding of solar minima and, in particular, of deep minima. It also changed our ideas about the corona and heliosphere at solar minimum.

In 2007–2009, the regions around the solar poles dominated by one magnetic polarity were covering a smaller area than in 1996, had generally weaker magnetic flux density, and were less unipolar (de Toma & Arge, in preparation). All these factors contributed to give weaker polar fields and, in turn, a weaker polar dipole moment. This special magnetic configuration during the Cycle 23 minimum led to a coronal morphology far from the typical dipolar structure observed during the minima in 1986 and 1996. Polar coronal holes were smaller and quasi-unipolar magnetic regions at lower latitudes were able to open to the heliosphere and survive longer on the Sun, forming large and persistent, low-latitude coronal holes during the years 2007–2008. At this time, the ratio of the polar dipole to higher order moments was significantly lower than during the previous minimum in 1996. As low activity continued, little new magnetic flux emerged on the Sun while existing low-latitude magnetic fields continued to disperse and decay under the effect of differential rotation, meridional circulation, and random convective motions. This resulted in a decrease of the higher order components of the Sun's global magnetic field. In 2009, low-latitude coronal holes finally closed down and the heliosphere relaxed to a simpler two-sector structure, even if not as simple as during the preceding minima, when polar fields were stronger (Gibson *et al.* 2011). This had important implications for the solar wind impinging on the Earth. The low-latitude coronal holes present on the Sun in 2007 and 2008 were sources of recurrent, high-speed streams that triggered moderate to low geomagnetic activity. In 2009, these low-latitude sources of fast solar wind went away and the solar wind at the Earth was coming mostly from the edges of the relatively small polar coronal holes. This is when the magnetic field carried by the solar wind, the solar wind speed, and geomagnetic activity all reached their lowest level, showing a time delay between the minimum observed at the photosphere and in the heliosphere (Tsurutani *et al.* 2011).

The differences between Cycle 23 minimum and other recent minima discussed above show that solar minima are not all the same and that the Sun can have different magnetic configurations even during very quiet times, with significant consequences for the heliosphere. What ultimately determines the structure of the corona and heliosphere, even at solar minimum, is the magnetic flux distribution at the Sun. Therefore, sunspot number, while a very good indicator of the global activity level, cannot describe the 3D corona and heliosphere.

While Cycle 23 minimum was markedly different from previous minima observed during the Space Age, the historical records of sunspots, solar eclipses, and geomagnetic activity indicate that the recent minimum was not peculiar and similar long and deep minima have occurred in the past. In particular, the low minima at the turn of the 19th century appear to resemble the recent minimum.

Finally, the extremely low magnetic activity reached during the end of 2008 and 2009 gave us some insight on what physical conditions could have existed during grand minima (Schrijver *et al.* 2011). The length of Cycle 23 minimum, combined with the very low magnetic flux emergence, allowed magnetic fields present on the Sun the time to diffuse and decay to smaller and smaller scales in a manner that we had not observed before with modern instrumentation (McIntosh *et al.* 2011).

Acknowledgements

I deeply thank my collaborators C.N. Arge, S. Gibson, B. Emery, J. Kozyra, P. Judge, and J. Burkepile for the long and useful discussions on the prolonged solar minimum at the end of Cycle 23. I also thank the IAU organizing committee for partially supporting my trip to the IAU Symposium 286 in Mendoza.

The National Center for Atmospheric Research is sponsored by the National Science Foundation. This work was partially supported by the LWS NASA Grant # NNH05AA49.

References

Abramenko, V., Yurchyshyn, V., Linker, J., Mikić, Z., Luhmann, J. G., & Lee, C. O., 2010, *ApJ*, 712, 813

Billings D. E., 1966, *A Guide to the Solar Corona*, New York, Academic Press

DeRosa, M. L., Brun, A. S., & Hoeksema, J. T., 2011, *Astrophysical Dynamics: From Stars to Galaxies*, Proceedings of the International Astronomical Union, IAU Symposium 271, 94

de Toma, G. & Arge, C. N., 2005, *ASP Proceedings of the NSO Workshop 22: Large-Scale Structures and Their Role in Solar Activity*, 251

de Toma, G. & Arge, C. N., 2010, *Twelfth International Solar Wind Conference*, AIP Conference Proceedings vol. 1216, 679

de Toma, G., Gibson, S. E., Emery, B. A., & Kozyra, J. U., 2010, *Twelfth International Solar Wind Conference*, AIP Conference Proceedings vol. 1216, 667

de Toma, G., Gibson, S. E., Emery, B. A., & Arge, C. N., 2010b, *SOHO 23: Understanding a Peculiar Solar Minimum*, ASP Conference Series vol. 428, 217

de Toma, G. 2010, *Solar Phys.*, doi 10.1007/s11207-010-9677-2

Harvey K. L. & Recely F., 2002, *Solar Phys.*, 211, 31

Emery, B. A., Richardon, I. G., Evans, D. S., & Rich, F. J., 2009, *J.A.S.T.P.*, 71, 1157

Emery, B. A., Richardson, I. G., Evans, D. S., Rich, F. J., & Wilson, G. R., 2010, *Solar Phys.*, doi 10.1007/s11207-011-9758-x,

Fisher, R. R., Lee, R. H., MacQueen, R. M., & Poland, A. I, .1981, *Applied Optics*, 20, 1094

Gibson, S. E., Kozyra, J. U., de Toma, G., Emery, B. A., & Onsager, T., Thompson B. J., 2009, *J. Geophys. Res.*, 114, A09105

Gibson, S. E., de Toma, G., Emery, B. A., Riley, P., Zhao, L., Elsworth, Y., Leamon, R. J., Lei, J., McIntosh, S., Mewaldt, R. A., Thompson, B. J., & Webb, D., 2011, *Solar Phys.*, in press

Issautier, K., Le Chat, G., Meyer-Vernet, N., Moncuquet, M., Hoang, S., MacDowall, R. J., & McComas, D. J., 2008, *GRL* 35, L19101

Jian, L. K., Russell, C. T., & Luhmann, J. G., 2010, *Solar Phys.*, doi 10.1007/s11207-011-9737-2

Judge, P. G., Burkepile, J., de Toma, G., & Druckmüller, M., 2010, *SOHO 23: Understanding a Peculiar Solar Minimum*, ASP Conference Series vol. 428, 171

Lee, C. O., Luhmann, J. G., Zhao, X. P., Liu, Y., Riley, P., Arge, C. N., Russell, C. T., & de Pater, I., 2009, *Solar Phys.*, 256, 345

Lei, J., Thayer, J. P., Wang, W., & McPherron, R. L., 2010, *Solar Phys.*, doi 10.1007/s11207-010-9563

Luhmann, J. G., Lee, C. O., Li, Y., Arge, C. N., Galvin, A. B., Simunac, K., Russell, C. T., Howard, R. A., & Petrie, G., 2009, *Solar Phys.*, 256, 285

McComas, D. J., Ebert, R. W., Elliott, H. A., Goldstein, B. E., Gosling, J. T., Schwadron, N. A., & Skoug, R. M., 2008, *GRL*, 35, L18193

McIntosh, S. W., Leamon, R. J., Hock, R. A., Rast, M. P., & Ulrich, R. K., 2011, *ApJ Lett.*, 730, L3

Petrie G. J. D., Canou, A., & Amari, T., 2010, *Solar Phys.*, doi 10.1007/s11207-010-9687-0

Riley, P., Lionello, R., Linker, J. A., Mikić, Z., Luhmann, J. G., & Wijaya, J., 2010, *Solar Phys.*, doi 10.1007/s11207-010-9698-x

Schrijver, C. J., Livingston, W. C., Woods, T. N., & Mewaldt, R. A., 2011, *J. Geophys. Res.*, 38, CiteID L06701

Sheeley, N. R. 2008, *ApJ*, 680, 1553

Smith, E. J. & Balogh, A., 2008, *GRL*, 35, L22103

Svalgaard, L. & Cliver, E. W., 2007, *ApJ Lett.*, 661, L203

Thayer J. P., Lei J., Forbes J. M., Sutton E. K., & Nerem S. M., 2008 *J. Geophys. Res.*, 113, CiteID A06307

Tokumaru, M., Kojima, M., Fujiki, K., & Hayashi, K., 2009, *GRL*, 36, CiteID L09101

Tokumaru, M., Kojima, M., & Fujiki, K., 2010, *J. Geophys. Res.*, 115, CiteID A04102

Tsurutani B. T., Echer, E., Guarnieri, F. L., & Gonzalez, W. D., 2011, *J.A.S.T.P.*, 73, 164
Wang Y. M., Robbrecht E., Sheeley N. R., 2009, *ApJ*, 707, 1372

Discussion

Jon Linker: Is it possible from X-ray and EUV emission to look at the area of coronal holes and to add up the coronal hole magnetic flux to determine the difference between the past minima?

Giuliana de Toma: Yes, we have not done it yet but this can be done combining coronal hole maps with synoptic magnetic maps. One important caveat is that to best observe the poles we need a favorable b-angle, so we cannot have a good view of both poles at the same time. Within the limits of how accurate coronal hole maps are, the flux balance for the open field can be determined and it is something worth doing.

Leif Svalgaard: Comment: The current minimum is very similar to the one of 100 years ago. This would suggest to me that TSI was also similar.

Giuliana de Toma: I agree.

Eric Priest: You talked about the spectrum of solar wind speeds. Can the differences in velocity in the solar wind be explained by how coronal holes spread with height?

Giuliana de Toma: Yes, it can. The very low speeds seen at the Earth in 2009 were because the solar wind during this time was coming mostly from the edges of the polar coronal holes that were smaller than in 1996, so there was a larger expansion factor.

Eugene Rozanov: Why do you compare 2008 and 1996? Perhaps 1996 is better comparable with 2009.

Giuliana de Toma: It depends on which observables you compare. At the photosphere, we do not see significant magnetic flux emergence in 2008 and 2009. Already in 2007, sunspot activity level was below the level observed in 1996. However, in the corona and heliosphere, there is a slow evolution throughout 2007–2009, with the solar wind reaching the lowest values in speed and magnetic field in 2009. This delay is not just in the Cycle 23 minimum but was also noticed during previous minima. The heliosphere tends to reach minimum conditions later than the photosphere.

Arnab Choudhuri: You are a co-author in one highly cited paper on solar cycle predictions. Theoretical issues were left out in your talk. What sort of lesson can the dynamo theorists learn from these observations?

Giuliana de Toma: This was the first time that physical models of the Sun were used to make solar cycle predictions. Because the predictions were so different, Cycle 24 will give us the opportunity to discriminate between low-diffusivity and high-diffusivity models and to test which class of models works best. This will help to constrain some important ingredients of the solar dynamo for which we do not have direct observations. It will be interesting to see what happens with Cycle 24. Of course, there is the possibility that Cycle 24 will end up being an average cycle, in which case it will be difficult to say anything conclusive about the two classes of models. Right now, we see a strong asymmetry between the two solar hemispheres. If this continues, it is possible that we will have a strong cycle in the northern hemisphere and a weak cycle in the southern hemisphere. This asymmetry will also be interesting to explain for dynamo modelers.

David Webb: This asymmetry is very interesting. How does this compare to previous cycles?

Giuliana de Toma: There seems to be a time lag between the two hemispheres. This has been going on since 2006, longer than we have have observed in previous cycles.

Sacha Brun: Comment: There is evidence that during the Maunder minimum the southern hemisphere was more active. We looked at this in detail. We find that when the dipole and quadrupole have the same size, one hemisphere is more active. So clearly the dipole vs. quadrupole amplitude plays a key role in this asymmetry.

Sacha Brun: Since sunspot number was so much smaller in 2008 than in 1996, how do explain the complex magnetic topology? What features modify the global field?

Giuliana de Toma: What is important for the coronal structure is ultimately the balance of the magnetic flux at high and low latitudes. During the recent minimum, the low-latitude magnetic flux was weak, but so were the polar magnetic fields. In 2007 and 2008, the relative strength of the polar dipole relative to the equatorial dipole, the quadrupole and the higher order moments was lower than in 1996 or 1986 and this is what allowed the corona to retain some complexity. In 2009, as magnetic flux at low latitudes decayed, the polar dipole moment became more prominent but never as important as in 1996.

Sacha Brun: You said that sunspot number is not a good indicator for the solar corona. Is there another index that we can use?

Giuliana de Toma: This is a hard question. I do not know if there is a good proxy for the corona. To infer the coronal structure, you really need the 3D magnetic field distribution since the organization of the field spatially is as important as the strength of the field for the corona and heliosphere.

Jürgen Schmitt: You pointed out that the last sunspot minimum was not unusual compared to previous minima around 1900 and should not be confused with a Maunder minimum. So how would the Sun look like when it is really in a Maunder Minimum and how does this differ from what we saw in 2008–2009?

Giuliana de Toma: Nobody can know for sure what the Sun looked like during the Maunder minimum. However, there is increasing evidence from the comparison with other stars in grand minima that solar irradiance was not very different during the Maunder minimum from what we observe during a normal minimum. The work of Schrijver and co-authors also points into this direction. In 2008 and 2009, the magnetic flux on the Sun had enough time to diffuse to a smaller scale to give us some idea of how the Sun would look like after a long period of inactivity. As argued by Schrijver *et al.*, the fact that TSI was about the same in 2008–2009 as it was during the previous minima, suggests that the Sun was not much dimmer during the Maunder minimum.

Comparative Magnetic Minima:
Characterizing quiet times in the Sun and stars
Proceedings IAU Symposium No. 286, 2011
C. H. Mandrini & D. F. Webb, eds.

doi:10.1017/S1743921312004723

Global magnetic fields: variation of solar minima

Andrey G. Tlatov[1] **and Vladimir N. Obridko**[2]

[1]Kislovodsk mountian astronomical station of the Pulkovo observatory,
Gagarina str. 100, Kislovodsk, Russia
email: tlatov@mail.ru

[2]Pushkov Institute of Terrestrial Magnetism, Ionosphere & Radio Wave Propagation, Russian Academy of Sciences,
Troitsk, Moscow Region Russia,
email: obridko@mail.ru

Abstract. The topology of the large-scale magnetic field of the Sun and its role in the development of magnetic activity were investigated using H_α charts of the Sun in the period 1887-2011. We have considered the indices characterizing the minimum activity epoch, according to the data of large-scale magnetic fields. Such indices include: dipole-octopole index, area and average latitude of the field with dominant polarity in each hemisphere and others. We studied the correlation between these indices and the amplitude of the following sunspot cycle, and the relation between the duration of the cycle of large-scale magnetic fields and the duration of the sunspot cycle.

The comparative analysis of the solar corona during the minimum epochs in activity cycles 12 to 24 shows that the large-scale magnetic field has been slow and steadily changing during the past 130 years. The reasons for the variations in the solar coronal structure and its relation with long-term variations in the geomagnetic indices, solar wind and Gleissberg cycle are discussed.

We also discuss the origin of the large-scale magnetic field. Perhaps the large-scale field leads to the generation of small-scale bipolar ephemeral regions, which in turn support the large-scale field. The existence of two dynamos: a dynamo of sunspots and a surface dynamo can explain phenomena such as long periods of sunspot minima, permanent dynamo in stars and the geomagnetic field.

Keywords. Solar, large-scale magnetic field, solar activity

1. Introduction

The structure of the large-scale magnetic field on the Sun is determined by the distribution of unipolar regions. The pattern of these unipolar regions is evident in the magnetograms and it reflects the results of interaction of all phenomena in the atmospheric layers. Alternatively the unipolar regions can be identified on the H_α synoptic charts McIntosh (1979). The prominences and H_α filaments, filament channels, $CaIIK$ plages and strong magnetic fields in spots can be used as markers for tracing the evolution and migration of these large-scale magnetic regions. At present, the summary series of H_α charts cover the period from 1887 up to now, comparable to the length of sunspot group series (McIntosh 1979, Makarov & Sivaraman 1989; Vasil'eva 1998), but one can consider the data since 1915 as the most reliable ones. Using filament distribution charts of the Meudon Observatory and daily spectrograms in the H_α and $CaIIK$ lines of the Kodaikanal Observatory (India), the Atlas of H_α charts of large-scale magnetic field of the Sun for the period 1915-1964 was constructed (Makarov & Sivaraman 1989). McIntosh (1979) has constructed H_α synoptic charts for the period 1964-1974. Vasil'eva (1998) reconstructed large-scale field topology during 1887-1900 and 1904-1914. There are similar data for the period 1975-1991 (SGD, 1975-1991). Beginning from 1979 up to the present time, the bulletin Solnechnye Dannye publishes monthly Synoptic Charts on the basis of the observations at the Kislovodsk Solar Station of the Pulkovo Observatory (S.D.: 1979-2011 *http://www.solarstation.ru*). Thus, there is a continuous data stock on the evolution of the large-scale magnetic field for more than 12 cycles of solar activity (Fig. 1). Other data on large-scale magnetic field data come from solar magnetographs. One of the most stable and long-term data series is from Wilcox Solar Observatory. Figure 2 shows a comparison of axisymmetric modes of low harmonics, obtained according to the H_α and magnetograph data. You can note the good agreement between the two types of data.

2. Configuration of Large-Scale Magnetic Fields during the Minimum Activity Epoch

Long-term variations of large-scale magnetic fields in different latitudes can be studied by means of a series of H_α synoptic charts. These charts show boundaries of magnetic field polarities. Routine data obtained at several observatories were used to extract tracers. These are observations of filaments, filament channels, and solar prominences. Thus, in contrast to observations with magnetographs, the spatial resolution of the data is constantly increasing owing to the telescope upgrade, for example. The H_α charts contain boundaries on the spherical surface separating positive and negative magnetic polarities. There are various indices for the characterization of large-scale field.

2.1. *Dipole-octopole index of a large-scale magnetic field* $\mathbf{A(t)}$

The photospheric large-scale magnetic field of the Sun can be represented as a function of latitude θ and longitude φ, using decomposition in spherical harmonics:

$$B_r = \sum_l \sum_m P_l^m (g_l^m \cdot cos(m\phi) + h_l^m \cdot sin(m\phi)) \cdot [(l+1)(R_\odot/r)^{l+2} + l(r/R_s)^{l-1}(R_\odot/R_s)^{l+2}]$$

where P_l^m- are the Legendre polynomials, R_s - radius of the "source surface" (usually $2.5 \cdot R_\odot$). The coefficients of decomposition g_l^m and h_l^m are surface integrals. Here $B_r(\theta, \phi)$ is the value of the surface magnetic field. In the case of H_α charts we took only the sign of the magnetic field, keeping the absolute value constant: +1G or −1G, according to whether the field was positive or negative. Nine harmonics were taken into account. It is possible to restore H_α charts of a magnetic field using the factors of decomposition g_l^m and

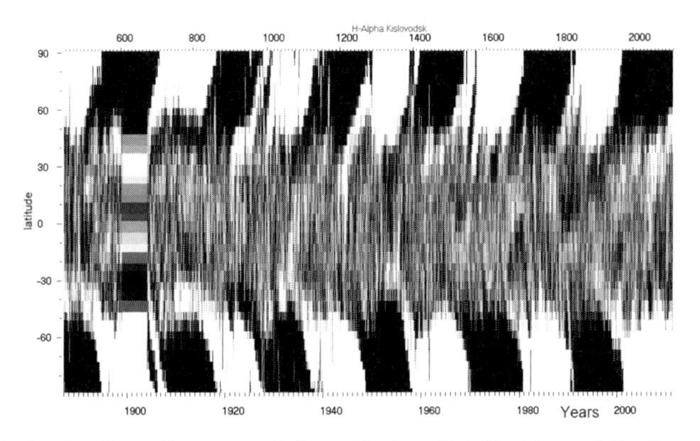

Figure 1. The latitude-time diagram of the polarity distribution of the large-scale magnetic field in the period 1887-2011.

h_l^m. It was supposed that the magnetic field is a potential field from the photosphere to the "source surface", which was assumed to be situated at $2.5 \cdot R_{\odot}$, where $R_{\odot}$ is the radius of the Sun. Let us consider the time behavior of low-l harmonics. We introduce a parameter $\mathbf{A(t)}$ characterizing the intensity of the dipole and octupole magnetic moments:

$$\mathbf{A(t)} = \sum_{\mathbf{m,l=1}} (\mathbf{g_l^m g_l^m} + \mathbf{h_l^m h_l^m}) + \sum_{\mathbf{m,l=3}} (\mathbf{g_l^m g_l^m} + \mathbf{h_l^m h_l^m})/\mathbf{3}$$

In Figure 3a (top) the distribution of $\mathbf{A(t)}$ is given for 1915-2011. The 11-yr cycle of a large-scale magnetic field of the Sun leaps to the eye. The procedure of smooth sliding by a window of 2 years was applied to eliminate noise. In Figure 3c (bottom) the solar activity in Wolf numbers, $\mathbf{W(t)}$ is given for comparison.

The $\mathbf{A(t)}$ index demonstrates well-marked 11-year activity cycles, and with respect to the 11-year Wolf number data $\mathbf{W(t)}$ one can see a phase shift in $\mathbf{A(t)}$ of 5 - 6 years. Namely, the $\mathbf{A(t)}$ index cycles precede the sunspot cycles $\mathbf{W(t)}$ (Figure 3a). It should be pointed out that the $\mathbf{A(t)}$ parameter includes only the dipole and octopole components of the background magnetic field, that is, modes $l = 1$ and 3. Even-order modes $l = 2, 4$ and higher-order modes have significantly smaller contributions. Lower and odd-order modes in synoptic charts characterize the configuration of the global solar magnetic fields in the minimum activity epoch.

The maximum value of index $\mathbf{A(t)}$ found near the activity minimum (A_{max}) can be used to predict the value of the Wolf number $\mathbf{W(t)}$ at the maximum of the following cycle (W_{max}). The correlation coefficient between these parameters is $R = 0.96$. The relationship between A_{max} and W_{max} can be obtained from the linear regression analysis, and we found $W_{max} = 1320 \cdot A_{max} - 54$. The prediction for the $24th$ activity cycle is, therefore, $W_{24} = 100 \pm 9$ (Figure 3a). The timing when the dipole - octopole index takes the maximum value can slightly vary with regard to the minimum activity epoch. For the prediction of the maximum Wolf number in the $24th$ cycle, we adopted the moment when index A takes the maximum value within the current activity minimum. Between the $18th$ and the $19th$ cycles, the maximum of index A took place at the beginning of 1952, which was two years before the Wolf number reached the minimum (Figure 3a).

2.2. *Area of Dominant Polarity at Middle and High Latitudes* Apz

During the minimum activity periods the polar regions are covered with unipolar magnetic fields, and the polarity of the north pole is opposite to that of the south pole. At the same time, the middle and high-latitude zones of each hemisphere are occupied with a field of predominant polarity, which has opposite signs in the two hemispheres. The area covered with the field of dominant polarity can be calculated by using the H_{α} synoptic charts, and can be used as an index (**Apz**) characterizing the topological organization of large-scale magnetic fields in the polar zones. Figure 3b represents the index **Apz**

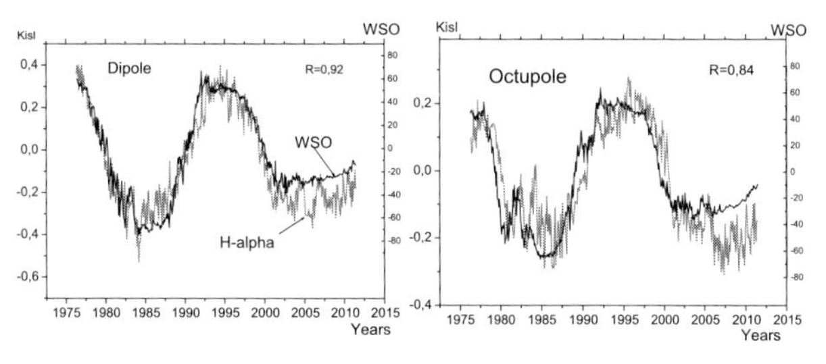

Figure 2. Comparison of axisymmetric modes of low harmonics, obtained according to the H_{α} and magnetograph data.

calculated for latitudes above 30° and summed over both hemispheres, expressed in units of $10^{16} m^2$. Local maxima in the **Apz** index are generally found at the minimum activity epochs. An exception is the maximum in **Apz** between cycles 18 and 19, which took place earlier than the sunspot number minimum. The relationship between the cycle amplitudes in **Apz** and Wolf numbers can be presented as $W_{max} = 3.92 \cdot Apz_{max} - 364.3$, $(R = 0.91)$. According to the value of **Apz** in the current minimum, the amplitude of the $24th$ cycle can be predicted as $W_{24} = 123 \pm 15$ (Figure 3b). This value derived from the linear regression using the data over eight activity cycles is close to the activity amplitude of the $23rd$ cycle. At the same time, the value of the **Apz** index itself in the $24th$ cycle is lower than in the minimum of the $23rd$ cycle.

2.3. *Average Latitude of Large-Scale Magnetic Fields*

The average latitude of large-scale magnetic fields of either positive or negative polarity can be considered as another characteristic, revealing the role of the topology of large-scale magnetic fields. The parameter may be represented by $\overline{\theta^p} = \int_{-\pi}^{\pi} \theta^p ds^p$, where θ^p is the latitude of a small area ds^p,Sp is the total area, and index p specifies either positive (+) or negative (−) polarity. Generally the average latitudes of large-scale magnetic fields migrate equatorward. The separation of the opposite polarities $\theta_{\pm} = abs(\theta_p - \theta_n)$ in the minimum activity epochs is in the range of $25^o - 40^o$. During the years 1915 - 2008, this value was largest between cycles 18 and 19 and smallest between cycles 19 and 20 (Figure 3c). The relationship between the value of $\theta_{\pm}$ in the activity minimum and the amplitude of the following sunspot cycle (W_{max}) is found as $W_{max} = 7.7 \cdot \theta_{\pm} - 111$, $(R = 0.89)$. The amplitude of the $24th$ cycle predicted from this relationship is $W_{24} = 120 \pm 17$ (Figure 3c).

3. Duration of Cycles Seen in Large-Scale Magnetic Fields and the Prediction of Activity Cycles

According to the hypothesis of the extended activity cycle (Wilson *et al.* 1988), the duration of an activity cycle is longer than of the sunspot cycle. Namely, the cycle may

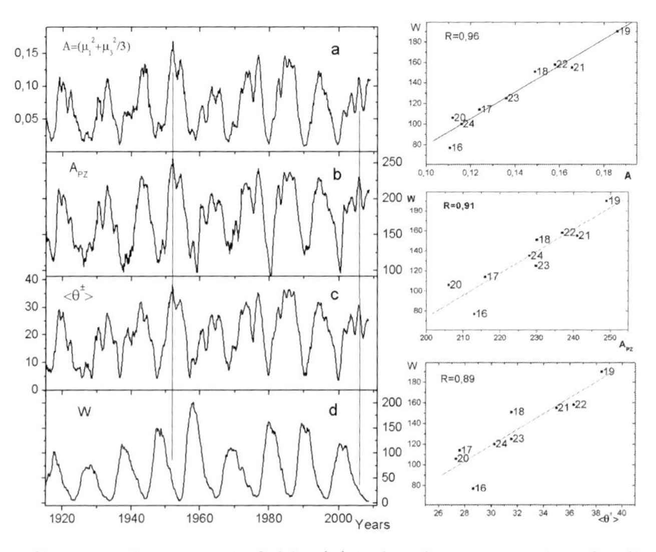

Figure 3. Indices of large-scale magnetic fields: (a) index **A**, representing the dipole and octopole components, (b) the area of polar zones **Apz** in latitudes above 30°, in units of $10^{16} m^2$, the separation between the average latitudes of positive and negative large-scale fields, $\theta_{\pm}$, expressed in degrees. The Wolf sunspot number W in panel (d) for comparison.

begin 1 - 3 years before the approach of sunspot minimum (Harvey 1992). The question is still unresolved, however, when a new activity cycle begins. Let us consider the time interval between the polarity reversal of global magnetic fields derived from the H_α synoptic charts and the following sunspot minimum. The moments of polarity reversal T_r^n is taken from the papers of several authors (Makarov *et al.* 2003; Tlatov 2009). In case the polarity reversal was not simultaneous on the two hemispheres, we chose the reversal of the hemisphere that took place later. There is a relation between the amplitude of the following sunspot cycle W_{n+1} as a function of the time interval between the polarity reversal of the large-scale magnetic fields and the sunspot cycle minimum, $\Delta T_R = T_{min}^{n+1} - T_r^n$. The shorter the time interval between the polarity reversal and the beginning of the next sunspot cycle is, the higher is the amplitude of the following sunspot cycle (Tlatov 2009):

$$W_{max}^{n+1} = 320(\pm 51) - 38.2(\pm 9.6) \cdot \Delta T_R,\ R = 0.78, \tag{3.1}$$

There is also a relation between the amplitude of the new sunspot cycle W_{max}^{n+1} and the time interval $\Delta T_2 = T_{max}^{n+1} - T_r^n$. Here T_{max}^{n+1} is the time of the new sunspot maximum and T_r^n is the time of the polarity reversal in the previous cycle. The relationship can be represented as:

$$W_{max}^{n+1} = 352(\pm 40) - 24.9(\pm 4.3) \cdot \Delta T_2,\ R = 0.87, \tag{3.2}$$

The relations between the amplitude of the following sunspot cycle and the time intervals given in Equations (3.1) and (3.2) contain information on the moments of activity minimum and maximum of the following cycle. On the other hand, another relationship given below (Equation (3.3)) involves the parameters available in the current cycle on the right-hand side:

$$W_{max}^{n+1} = 83(\pm 11) - 0.09(\pm 0.02) \cdot W_{max}^{n+1} \cdot \Delta T_3 \cdot abs(\Delta T_3),\ \Delta T_3 = T_{max}^n - T_r^n,\ R = 0.86, \tag{3.3}$$

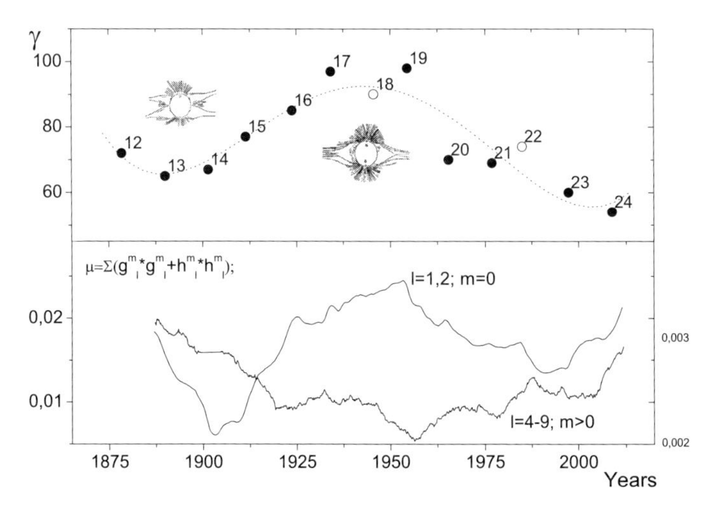

Figure 4. (Upper panel) Distribution of parameter γ for the structure of the corona of the minimal type. Solar activity cycles numbers are given. The shapes of the solar corona in cycles 13, 19 are shown. (Lower panel) The amplitudes of the harmonics were smoothed with a "running window" 11 years long. At the bottom of this panel we give the amplitudes of the axisymmetric modes dipole and quadruple ($l = 1, 2; m = 0$) and for harmonics, characterizing the sector structure of the higher harmonics ($l > 3; m > 0$).

We can demonstrate a number of other relationships between the time intervals which include the moments of polarity reversal of the large-scale magnetic fields:

$$W^{n+1}_{max} = 450(\pm 101) - 20.7(\pm 6.4) \cdot W^{n+1}_{max} \cdot \Delta T_4, \ \Delta T_4 = T^{n+1}_{max} - T^{n}_{r}, \ R = 0.72, \quad (3.4)$$

$$\Delta T_2 = 1.12(\pm 1.3) + 1.54(\pm 0,24) \cdot \Delta T_R, \ R = 0.88, \quad (3.5)$$

Earlier in the article by Makarov *et al.* (2003), the following relation has been found:

$$T^{n+1}_{r} - T^{n}_{r} = 5.8(\pm 0.4) + 304(\pm 33)/W^{n+1}_{max}, \ R = 0.95, \quad (3.6)$$

Equations (1)-(6) allow us to derive the recurrent formulae for the times $T^{n+1}_{max}, T^{n}_{rev}, T^{n+1}_{min}$ and the amplitude W^{n+1}_{max} in terms of the parameters $T^{n+1}_{rev}, T^{n}_{min}, T^{n}_{max}, W^{n}_{max}$. However, due to low correlation coefficients found, their use for the prediction for the forthcoming cycle is not promising. If we assume that the minimum of the $24th$ activity cycle took place at $T^{24}_{min} = 2009$ and $T^{23}_{r} = 2001.7$, then the amplitude of this cycle can be evaluated as $W^{24} = 42 \pm 27$. Equation (3.2) then gives the timing of the maximum of the new activity cycle. If we assume the amplitude of the $24th$ cycle as $W^{24} = 80$, the maximum of the $24th$ cycle is expected at $T^{24}_{max} = 2012.6 \pm 0.7$. Equation (3.3) then gives the prediction $W^{24} = 92 \pm 21$.

4. Long-term Changes of Coronal Shape and Geomagnetic Disturbance

The presence of long-term trends in the solar corona structure can be caused by changes in the configuration of the global magnetic field of the Sun. The role of active region formation during a solar activity minimum is significant. It has long been known that large coronal streamers typically lie above the polarity-inversion lines of the large-scale magnetic field marked by filaments and prominences (Vsekhsvyatskiy *et al.* 1965). For

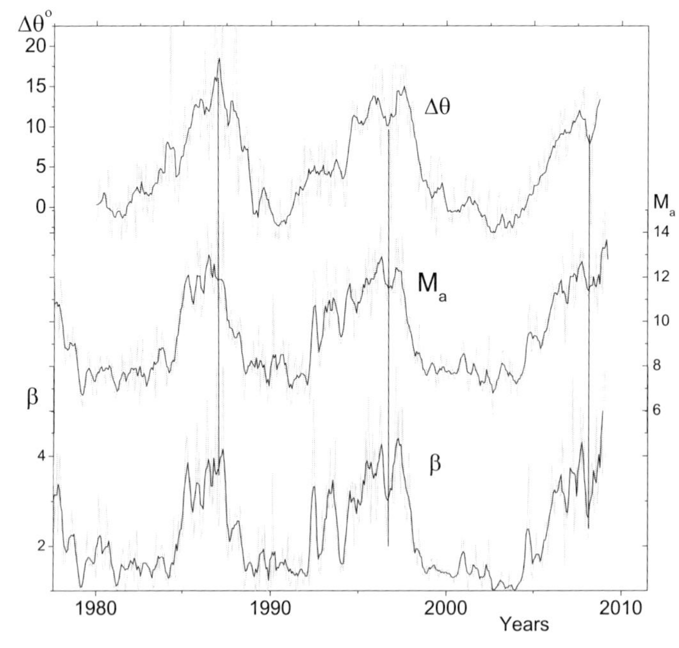

Figure 5. The comparison between a) the deviations of coronal rays $\Delta\theta$ with the solar wind parameters according to OMNI2 database. b) Alfven mach number M_a. c) Plasma beta β. Data of the solar wind parameters are smoothed in 20 Bartels rotations.

this reason, investigations of the large-scale corona shape give valuable information on the structure of the large-scale fields during a long time interval. During the activity minimum, the properties of the global magnetic field of the Sun manifest themselves in the most pronounced way. The magnetic field of the Sun is determined by large-scale structures. The northern and southern hemispheres of the Sun have magnetic fields of opposite polarity. The strength of the polar magnetic field is significantly higher than the fields in middle and low latitudes in the activity minimum period.

To analyze the corona shape at the eclipses during a solar activity minimum epoch in activity cycles 12 to 24, a corresponding index should be chosen. We need the index that characterizes the shape of the corona of the minimal type and is applicable to images and drawings of different qualities. The corona of the solar minimum is characterized by pronounced polar ray structures and large coronal streamers (Pasachoff *et al.* 2008). Let us introduce the parameter γ that characterizes the angle between high-latitude boundaries of the large coronal streamers at a distance of $2 \cdot R_{\odot}$. The γ parameter is the sum of the angles at the eastern and western limbs: $\gamma = 180 - (\gamma_W + \gamma_E)$. A parameter γ can be introduced for these eclipses. This parameter reduces parameter γ to the minimum phase. The application of this procedure is effective for recognizing the shape of the corona close to the minimum activity with the phase $\Phi < 0.4$. Figure 4 presents variations in parameter γ during the last 13 activity cycles.

The link between the coronal shape, geomagnetic disturbances and solar wind parameters can be clearly seen in 11-year activity cycle. Global magnetic field can lead to non-radial spreading of the coronal streamers and probably, to the spreading of the solar wind. In minimal activity when the value of the global field reaches its maximum, coronal rays are deflected aside from the helio-equator.

The solar corona structure corresponds to the configuration of solar magnetic fields. Since the magnetic field of the Sun is subjected to cyclic variations, the corona shape also changes cyclically. The coronal rays are distinctive structures in the solar corona, which propagate at a small angle to the radial direction from the Sun and display the electron density in K-corona enhanced by the factor of 3 to 10. The angle $\Delta\theta$ that describes the deviation of the rays from the radial position varies with the phase of the solar cycle and the latitude (Eselevich & Eselevich 2002; Tlatov 2010).

The regular observations with the SOHO/LASCO and Mark-III/IV coronagraphs at the Mauna Loa solar observatory make it possible to analyze the structure of the solar

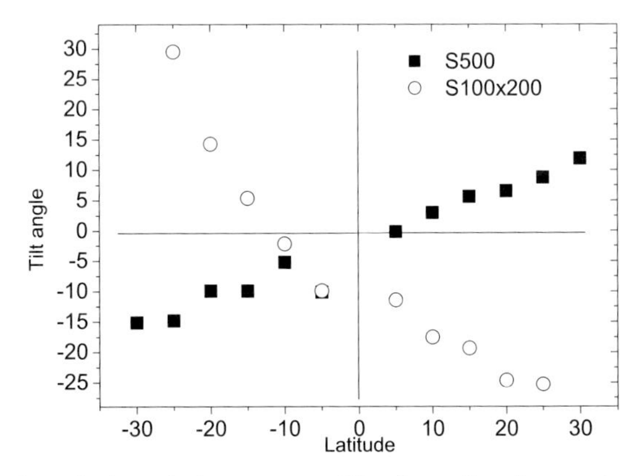

Figure 6. Latitudinal variation of the average tilt of small ephemeral regions (filled squares) and active regions (open circles). In both hemispheres, the following polarity in ERs is situated at lower latitude as compared with the leading polarity. This orientation is opposite to the tilt of ARS Tlatov (2010).

corona for the time comparable with the duration of the solar cycle. These data substantially complement extended series of observations of the corona in spectral lines carried out with extra-eclipse coronagraphs, since they make it possible to analyze coronal structures at sufficiently large distances from the solar limb, and also occasional observations of the "white" light corona during total eclipses. The coronagraph – polarimeter Mark-III detected the structure of the solar corona at the heights $\sim 1.15 : 2.45 R_{\odot}$ in 1980-1999. In 1998, at Mauna Loa observatory the new low-noise coronagraph Mark-IV, with a liquid-crystal modulator of polarization and a CCD, was mounted. To decrease the radial gradient and consequently to increase the contrast, we applied to the Mark data an artificial vignetting function. The LASCO-2 coronagraph telescope on board of SOHO satellite has been working since 1996 and covers the distance $1.5 R_{\odot} : 6 R_{\odot}$ above the solar limb. Thereby, here we have analyzed the structure of the corona for 1980-2008 on the basis of the data obtained at ground-based observations with Mark-III/IV coronagraphs and for 1996-2009 with the SOHO/LASCO-2 data.

In order to determine the deviation of coronal rays, we developed a technique of the identification of coronal streamers in two-dimensional images of the corona obtained with SOHO/LASCO-2 and Mark-III/IV in automatic mode. The analysis is based on discrimination of central parts of bright coronal structures propagating, as a rule, at some angle to the radial direction, discrimination of the points of the local maxima, and determination of the parameters of the approximating line section. Figure 5a presents the behavior of the shape of the coronal rays. But is it possible to establish the link directly between the coronal shape and solar wind parameters, measured from the Earth? Data bases OMNI1 and OMNI2 (http://nssdc.gfc.nasa.gov/omniweb) contain the information concerning hourly average value of key parameters of the solar activity, interplanetary atmosphere and geomagnetic disturbance since 1964. We collated changes of $\Delta\theta$ parameter with solar wind parameter and magnetic Mach number (Fig. 5). In minimal activity epoch there is a correspondence between the non-radial parameter $\Delta\theta$ and solar wind parameters. Comparing parameters $\beta = 8\pi nkT/B^2$ and $M_a = v/v_a$ we can make a conclusion that non-radial corona influences the relation of the solar wind, which grows in minimum solar activity and therefore magnetic field in minimal activity squeezes the solar wind flux towards helio-equator. These results complement the results of comparison of the three activity minima (Jian *et al.* 2011). This solar minimum has the slowest, least dense, and coolest solar wind, and the weakest magnetic field.

5. The fine structure of large-scale solar magnetic field

In the work by Tlatov *et al.* (2010) it was established that the orientation of the magnetic axis of ephemeral areas (ER) depends on the character of large-scale field (Fig. 6). However, large-scale magnetic field has zonation of polarity, as it is shown on Fig 8. We applied distribution of directions of magnetic axis ER onto zonality. There is a correspondence of distributions. It means that small-scale dipoles are oriented in

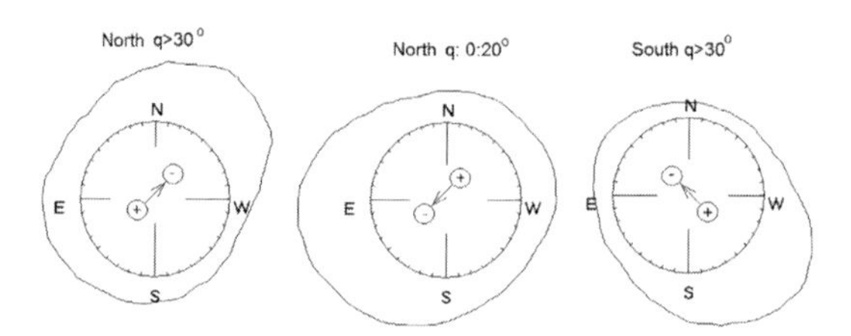

Figure 7. The distribution of the magnetic bipolar axis in ephemeral regions in different zones of the large-scale magnetic field.

accordance with the large-scale magnetic field distribution. Number of ER is equal to several hundreds, and consequently, ER can influence the formation of the large-scale magnetic field (Fig 7). The diagram shows that dipoles ER can strengthen large-scale magnetic field in near-surface dynamo process (Fig. 8). In the absence of sunspots the presence of ER would result in monotonous growth of the global magnetic field. During regular 11-year cycles the emergence of the sunspots lead to the reversal of the large scale field. But during the grand minima such as Maunder minimum other type of dynamo is possible. In contrast to traditional tachocline the superficial dynamo ER is present. As the result of ER dynamo action the global field is regenerated and the tachocline dynamo of the sunspots is turned over. It is possible that alternation of the activity cycles with the periods of deep minimum of the sunspots, such as Maunder minimum, is a consequence of the predominance of one or another type of dynamo on the Sun.

Constant magnetic activity of some types of stars Hall & Lockwood (2004) can be accounted for the same mechanism. It is obvious that in this case surface dynamo is prevailing, in some separate cases it can lead to the appearance of the gigantic spots on the poles.

In the Earth's dynamo surface type is also prevailing, and changes of the character is analogous to the sunspots cycles, and takes place periodically.

So, we can conclude that the large-scale field plays the key role in solar magnetism. This follows from the connection between the field configuration and solar activity cycles. Large-scale field, apparently, can exist even without spots due to generation of ephemeral areas. During the periods of spots' absence it can strengthen because of surface dynamo, which leads to the growth of the global magnetic field. Two-component dynamo can account for different phenomena, observed within the sun and stars, and effects of geomagnetism as well.

6. Overview

The indices presented above characterize the state of the solar atmosphere in the descending phase and at the minimum activity epoch. Their amplitude is related to the amplitude of the following cycle of activity. Some of these indices are related to the large-scale magnetic fields, and others characterize activity at high latitudes. Before a high cycle of

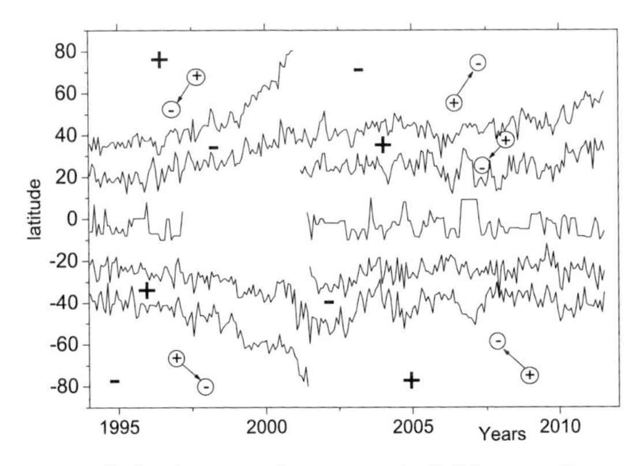

Figure 8. Zonal structure of the large-scale magnetic field according to H_α synoptic charts. Solid lines represent migration trajectories of magnetic neutral lines (or the zone boundaries with the opposite polarities on either of their sides in the order $+/-$ or $-/+$) of the large-scale magnetic field derived from H_α synoptic charts in the northern and southern hemispheres for the period 1994-2011. In the zones of different polarity of the magnetic field are shown the predominant direction of the magnetic axis of ER.

activity, in each hemisphere of the Sun a part of the large-scale field of prevailing polarity increases (index **Apz**), and the separation between the average latitudes of the magnetic fields of opposite polarities (in the opposite hemispheres) takes a maximum value (index $\theta_{\pm}$). These physical conditions provide high values of the dipole-octopole index **A**. On a unipolar background magnetic field arise polar faculae and calcium bright points.

An important parameter with promising prediction capability is the time interval between the polarity reversal of the large-scale magnetic fields and the sunspot cycle minimum, $\Delta T_R = T_{min}^{n+1} - T_r^n$. The smaller this time interval is, the more powerful the sunspots cycle can be expected. Probably, ΔT_R may be interpreted as a latent period of activity cycle development. As we have seen so far, the solar activity cycle can be interpreted as a process beginning with the polarity reversal of the large-scale magnetic fields. The total duration of the solar cycle was found to be 14.3 years for the 18th cycle and 18.6 years for the 14th cycle.

The analysis of the corona shape has revealed a long-term modulation of the global magnetic field of the Sun. Possibly, a secular modulation exists of the global solar magnetic field that is most pronounced during the solar activity minimum epoch. During the secular cycle of the global magnetic field of the Sun, the relation between the axisymmetric low order components and sectorial components of the magnetic field changes. The largest amplitude of the dipole component occurred during the interval 1944–1955. At the turn of the 19th to 20st and 20th to 21st centuries the solar corona shape and, possibly, the global magnetic field correspond to the configuration close to the octopole one (see Fig. 4).

Perhaps the large-scale field leads to the generation of small-scale bipolar ephemeral regions (Figs. 6, 7, 8), which in turn support the large-scale field. The existence of two dynamos: the dynamo of sunspots and the surface dynamo can explain phenomena such as long periods of sunspot minima, permanent dynamo in stars and the geomagnetic field.

Acknowledgements

This paper was supported by the Russian Fund of Basic Researches, and Program of the Russian Academy of Science.

References

Eselevich, V. G. & Eselevich, M. V. 2002, *Solar.Phys.*, 208, 5

Hall, J. C. & Lockwood, G. W. 2004, *ApJ*, 614, 942

Harvey, K. L. 1992, In: Harvey, K. L. (ed.), *The Solar Cycle, ASP Conf. Ser.* 27, 335

McIntosh, P. S. 1979, *Annotated Atlas of H_α Synoptic Charts, NOAA, Boulder*

Makarov, V. I. & Sivaraman, K. R. 1989, *Solar. Phys.*, 119, 35

Makarov, V. I., Tlatov, A. G., & Sivaraman, K. R. 2003, *Solar. Phys.*, 214, 41

Jian, L. K., Russell, C. T., & Luhmann, J. G. 2011, *Solar. Phys.*, Doi 10.1007/s11207-011-9737-2

Pasachoff, J. M., Rušin, V., Druckmüller, M., Druckmülerová, H., Bělik, M., Saniga, M., Minarovjech, M., Marková, E., Babcock, B. A., Souza, S. P. & Levitt, J. S. 2008 , *ApJ*, 682, 638

Tlatov, A. G. 2009, *Solar. Phys.*, 260, 465

Tlatov, A. G., Vasil'eva, V. V., & Pevtsov, A. A. 2010, *ApJ*, 717, 357

Tlatov, A. G. 2010, *ApJ*, 714, 805

Vasil'eva, V. V. 1998, in: V. I. Makarov & A. V. Stepanov (eds.), *New Cycle of Activity of the Sun : Observational and Theoretical Aspects* (St.-Petersburg, Pulkovo), p. 213

Vsekhsvyatskiy, S. K., Nikolskiy, G. M., Ivanchuk, V. I., & Nesmyanovich, A. T. *et al.* 1965, *Solar Corona and Corpuscular Radiation in the Interplanetary Space*, (Kiev, Naukova Dumka), 293

Wilson, P. R., Altrock, R. C., Harvey, K. L., Martin, S. F., & Snodgrass, H. B. 1988, *Nature* 333,748

Comparative Magnetic Minima:
Characterizing Quiet Times in the Sun and Stars
Proceedings IAU Symposium No. 286, 2011
C. H. Mandrini & D. F. Webb, eds.

doi:10.1017/S1743921312004735

The 3D solar minimum with differential emission measure tomography

Alberto M. Vásquez[1,2], Richard A. Frazin[3], Zhenguang Huang[3], Ward B. Manchester IV[3] and Paul Shearer[3]

[1]Instituto de Astronomía y Física del Espacio (CONICET-UBA),
CC 67 - Suc 28, (C1428ZAA) Ciudad de Buenos Aires, Argentina
email: albert@iafe.uba.ar

[2]Facultad de Cs. Exactas y Naturales, Universidad de Buenos Aires, Argentina

[3]Deptartment of Atmospheric, Oceanic and Space Sciences,
University of Michigan, Ann Arbor, MI 48109, USA

Abstract. Differential emission measure tomography (DEMT) makes use of extreme ultraviolet (EUV) image series to deliver two products: a) the three-dimensional (3D) reconstruction of the coronal emissivity in the instrumental bands, and b) the 3D distribution of the *local differential emission measure* (LDEM). The LDEM allows, in turn, construction of 3D maps of the electron density and temperature distribution. DEMT is being currently applied to the space-based EUV imagers, allowing reconstruction of the inner corona in the height range 1.00 to 1.25 $R_\odot$. In this work we applied DEMT to different Carrington Rotations corresponding to the last two solar Cycle minima. To reconstruct the 2008 minimum we used data taken by the Extreme UltraViolet Imager (EUVI), on board the Solar TErrestrial RElations Observatory (STEREO) spacecraft, and to reconstruct the 1996 minimum we used data taken by the Extreme ultraviolet Imaging Telescope (EIT), on board the Solar and Heliospheric Observatory (SOHO). We show here comparative results, discussing the observed 3D density and temperature distributions in the context of global potential magnetic field extrapolations. We also compare the DEMT results with other observational and modeling efforts of the same periods.

Keywords. Sun: corona, Sun: fundamental parameters, Sun: UV radiation, methods: data analysis

1. Introduction

The increasing capability of today's computers allows for increasingly complex global MHD models of the solar corona. This calls for the development of global 3D constraints for the fundamental coronal parameters, such as the electron density and temperature. Such developments are of interest in themselves, in order to observationally study the 3D structure of coronal features.

The DEMT technique, described in detail in Frazin *et al.* (2009) (FVK09 hereafter), allows reconstruction of the 3D distribution of the coronal DEM. In a first stage, time series of EUV images are used to reconstruct de 3D distribution of the *filter band emissivity* (FBE) via *solar rotational tomography* (SRT), for each band of the telescope independently. In a second stage, the values of the FBEs reconstructed in each tomographic grid cell are used to invert for a parametrized local-DEM (LDEM) pertaining only to the plasma contained in that cell.

In previous works we used DEMT to study two rotations of the solar Cycle 23 minimum phase, namely the Carrington Rotation (CR-)2068 period (Vásquez *et al.* (2011), V11 hereafter), which was the object of coordinated study efforts (known as the *Whole Heliosphere Interval*, WHI), and CR-2077 (Vásquez *et al.* (2010), V10 hereafter), which corresponds to the absolute minimum of solar Cycle 23. Both works used STEREO/EUVI data. In the present work we revisit those two periods, updating the ionization equilibrium calculations used to compute the temperature responses of each instrumental band, and including comprehensive 3D comparative results. In addition, we show results for the solar minimum of the previous solar Cycle 22, using SOHO/EIT data of CR-1914.

Fig. 1 shows the normalized temperature responses of the coronal bands of both the STEREO/EUVI-B and the SOHO/EIT instruments, computed using CHIANTI v6.0.1, assuming the Feldman *et al.* (1992) abundance set, and the Bryans *et al.* (2009) ionization equilibrium calculations.

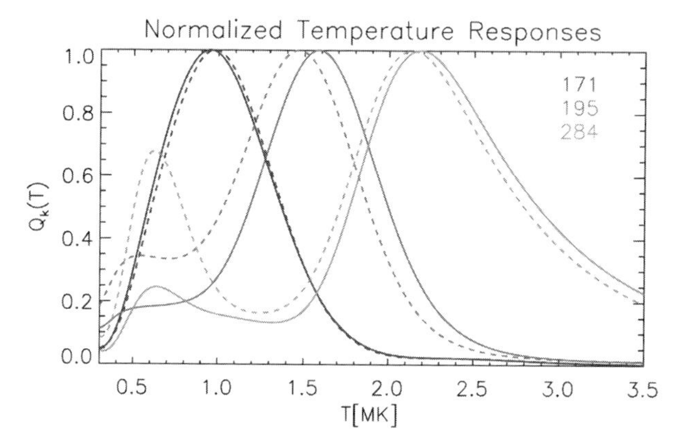

Figure 1. Temperature responses of STEREO/EUVI (solid) and SOHO/EIT (dashed).

2. Reconstructing the 3D FBE and Determining the LDEM

In this work, the inner corona (1.0-1.25 $R_\odot$) is discretized on a 25×90×180 (radial×latitudinal×longitudinal) spherical tomographic grid. As an example of the results of the first stage of DEMT, Fig. 2 shows the FBE maps for CR-2077 and CR-1914 at a selected height of the tomographic grid, and for the three bands of EUVI and EIT. The saturated regions correspond to *zero density artifacts* (ZDAs) in the reconstructions, which are locations of reconstructed zero emissivity, and are the result of coronal dynamics acting during the data acquisition while SRT assumes a static coronal structure.

The overploted thick-solid curves indicate the open/closed boundary of PFSS models, based on MDI synoptic magnetograms for each period, while the solid-thin white (black) curves are contour levels of positive (negative) magnetic strength. The PFSS models are computed using the finite-difference iterative solver FDIPS by Tóth *et al.* (2011), on a 150 × 180 × 360 spherical grid, covering 1.0 to 2.5 $R_\odot$. There is an overall good agreement between the location of the brighter-FBE structures and the PFSSM-closed regions, especially clear in the 195 and 284 Å bands. This is consistent with the closed plasma having enhanced densities and temperatures, implying larger emissivities.

To find the temperature distribution of the plasma within each tomographic cell, we assume a parametrization for the LDEM and use the FBE values to derive the parameters by minimizing the quadratic deviations between the tomographic and the predicted FBE in all bands. The instruments used in the present work (EUVI and EIT) have three bands each one, and we parametrize the LDEM as a single-normal distribution in each tomographic cell. With the new Atmospheric Imaging Assembly (AIA) instrument, on board the Solar Dynamics Observatory (SDO) spacecraft, the number of bands is increased, and other parametrizations can be explored (see Nuevo *et al.*, 2012, in this volume).

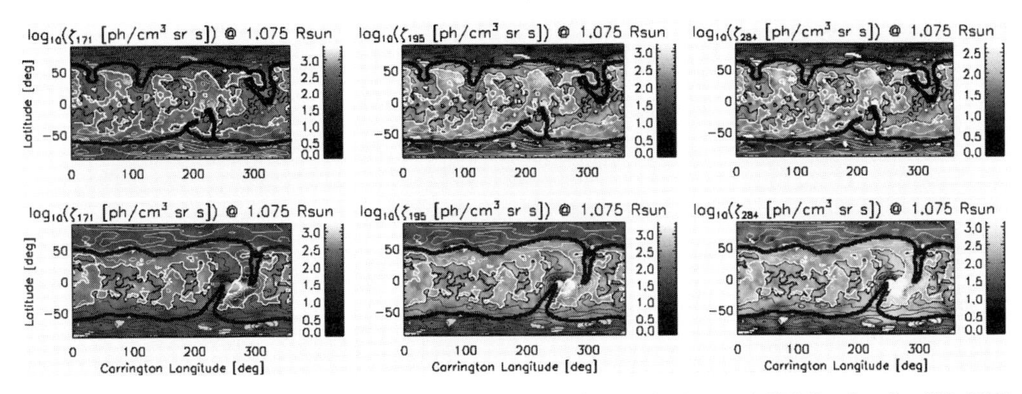

Figure 2. Examples of tomographically reconstructed emissivities at 1.075 $R_\odot$ for the CR-2077 (top panels) and CR-1914 (bottom panels) periods, in the 171 (right), 195 (middle), and 284 Å (left) coronal bands. The CR-2077 reconstructions are based on STEREO/EUVI-B data, while the CR-1914 reconstructions are based on SOHO/EIT data. The saturated regions correspond to ZDAs in the reconstructions (see text). The thick-solid curves indicate the open/closed boundary of the PFSS models for each period, while the solid-thin white (black) curves are contour levels of positive (negative) magnetic strength. All figures in this paper were designed to be seen in color in the online version, where their visibility is highly improved.

Using only three bands, a single-normal function is able to accurately predict the three tomographic FBEs in most regions. This is quantitatively shown in detail in Fig. 3 where, for three heights of the tomographic grid, we display the mean relative quadratic deviation (χ^2, see Eq. (1.4) in Nuevo *et al.*, 2012, in this volume for details) between the tomographic and predicted FBEs, as synthesized from the LDEM in each tomographic cell for the three periods we analyzed. We thresholded the lower values at $\chi^2 = 10^{-3}$, or an average agreement between the synthetic and tomographic FBEs better than $\sim 3\%$, which are the dark-green regions in the maps. It is readily seen that in most of the coronal volume the agreement is as good as, or superior, to this value.

The black regions correspond to ZDAs, where no LDEM is computed. The white regions indicate tomographic grid cells where $\chi^2 > 5 \times 10^{-2}$, or an average disagreement of about 20% between the tomographic and synthetic FBEs. We consider those cells to have *non-satisfactory-fits* (NSF) of their LDEM, i.e. regions where the assumed parametrization is not able to reasonably reproduce the tomographic FBEs. The NSFs usually occur around ZDAs, as well as in near-polar fainter regions. These regions are characterized by unusual relative values of the FBE in the three bands, making the single-normal function unable to reproduce them. A clear example of this is readily seen in the case of the CR-1914 period, where NSFs extend over a large range of longitudes in the northern hemisphere near-polar region. This is due to the data being contaminated by a filter leak in the 284 Å band near the north pole. The abnormally enhanced 284 Å signal makes a single-normal function unable to reproduce the three bands FBE with accuracy.

3. Global 3D Maps of Solar Minimum DEMT Results

Once the LDEM is found at each tomographic grid cell, its zeroth through second moments (Eqs. (1.1)-(1.3) in Nuevo *et al.*, 2012, in this volume) give the squared electron density N_e^2, mean electron temperature T_m, and squared electron temperature spread W_T^2, respectively. In Fig. 4 we show the results corresponding to CR-2077 at three different heights of the tomographic grid, for all latitudes and longitudes. In Figs. 5 and 6 we show the corresponding results for CR-2068 and CR-1914, respectively.

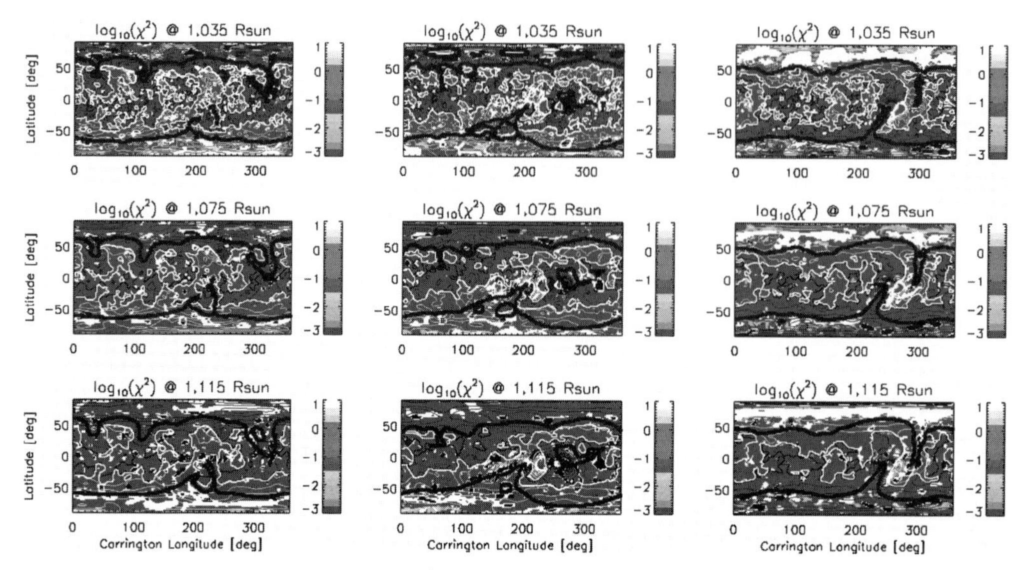

Figure 3. Mean cuadratic deviation (χ^2) between the tomographic and predicted FBE at three heights of the tomographic grid, for CR-2077 (right panels), CR-2068 (middle panels), and CR-1914 (left panels).

Several characteristics are common to all three tomographic reconstructions and PFSS models. Firstly, there is an overall good agreement in the location of the PFSS model open/closed (O/C) boundary and the location of the DEMT electron density and temperature transition between enhanced values within the closed regions and lower values in the open regions. Exceptions to this are usually related to low latitude incursions of open regions, especially when they are narrow and more S-N oriented. We believe this is the result of smearing of the results due to the fact that our current implementation of SRT does not consider coronal differential rotation. There is also the issue of coronal dynamics, but these effects should be minimized at the minimum of the cycle. Secondly, the largest temperature regions are systematically located within closed regions, along and around polarity inversion lines. Across the O/C boundary the change in temperature is quite sharp, which is shown in detail in Figs. 11 and 12.

To globally quantify the differences between open and closed regions, Figs. 7 through 9 show histograms of N_e, T_m, and W_T at 1.075 $R_\odot$, separately for the open and closed regions. We find a similar 2:1 ratio for the closed-to-open electron density contrast, with the electron density median value being about 10% larger for the CR-1914 period, compared to the CR-2077 and CR-2068 periods. We find the median T_m to be 30% larger in the 1996 closed region, compared to the 2008 periods. Differences in temperature could be partly due to the different instruments used. In inverting simulated DEM synthetic data (not included in this work), we found no systematic differences in the EUVI and EIT results. On the other hand, stray light issues are different for both instruments and have not yet been taken into account. Still, we have applied DEMT to the CR-2077 dataset using a preliminary blind deconvolution of the EUVI images (Shearer & Frazin, 2011) and found that the median temperature results are not greatly affected in the streamer region. All of these elements support the idea that the difference found in the temperature of the closed regions is a physical one, with the 2008 period being about 30% cooler than the 1996 one. In all the periods we found a median $T_m < 1$ MK in open regions, which is consistent with other studies revealing sub-MK electron temperatures at high latitudes.

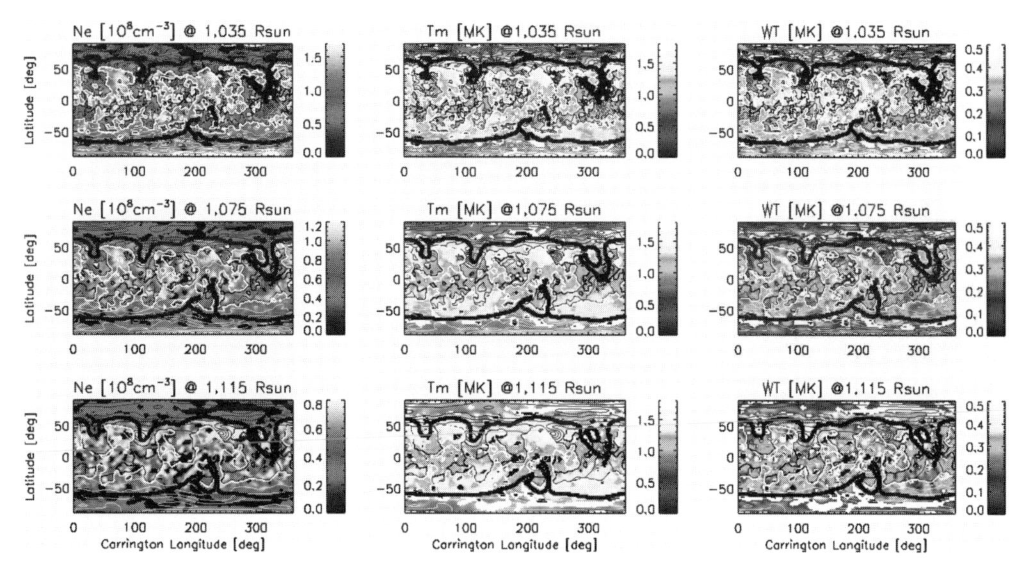

Figure 4. Reconstructed maps of N_e (left panels), T_m (middle panels), and W_T (right panels) for CR-2077, at heights 1.035 (top), 1.075 (middle), and 1.115 $R_\odot$(bottom), based on STEREO/EUVI data. In the temperature maps the black regions correspond to ZDAs, while the white regions to NSFs (see text), while both regions are shown as black in the density maps. The thick-solid curves indicate the open/closed boundary of the PFSS models for the period, while the solid-thin white (black) curves are contour levels of positive (negative) magnetic strength.

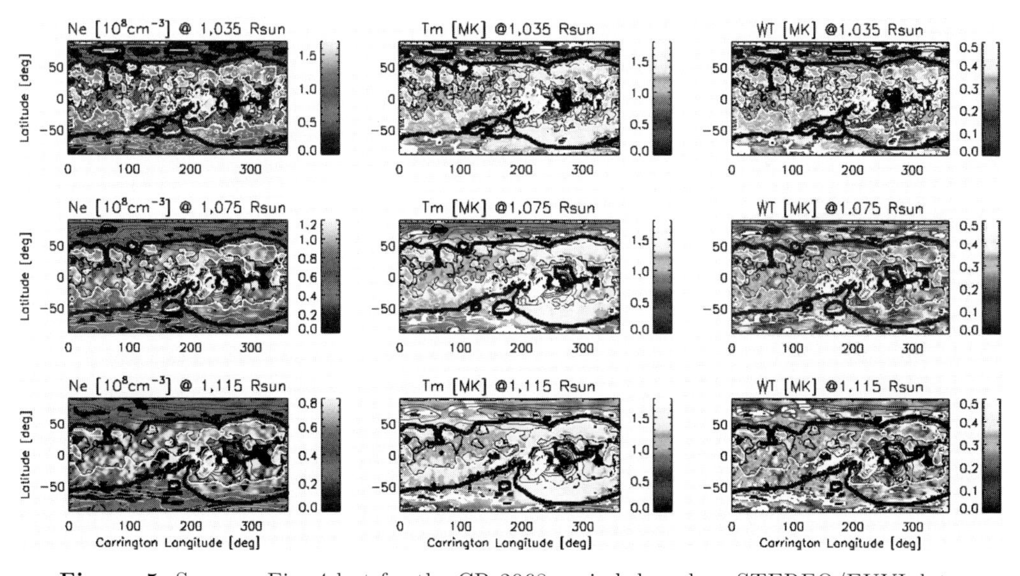

Figure 5. Same as Fig. 4 but for the CR-2068 period, based on STEREO/EUVI data.

Using the LDEM N_e and T_m values to estimate the gas total thermal pressure as $p_{\rm th} \sim 2k_B N_e T_m$ and the magnetic strength of the PFSS model to compute the magnetic pressure $p_B \equiv B^2/8\pi$, Fig. 10 shows plasma $\beta \equiv p_{\rm th}/p_B$ maps at 1.075 $R_\odot$. The CR-2077 and CR-1914 periods, both corresponding to a deep solar minimum, show more N/S symmetry than CR-2068. This is due to this last period belonging to the declining phase of Cycle 23 but still showing multipolar components from the AR complex, which clearly affected the overall structure as the polar fields were particularly weak during

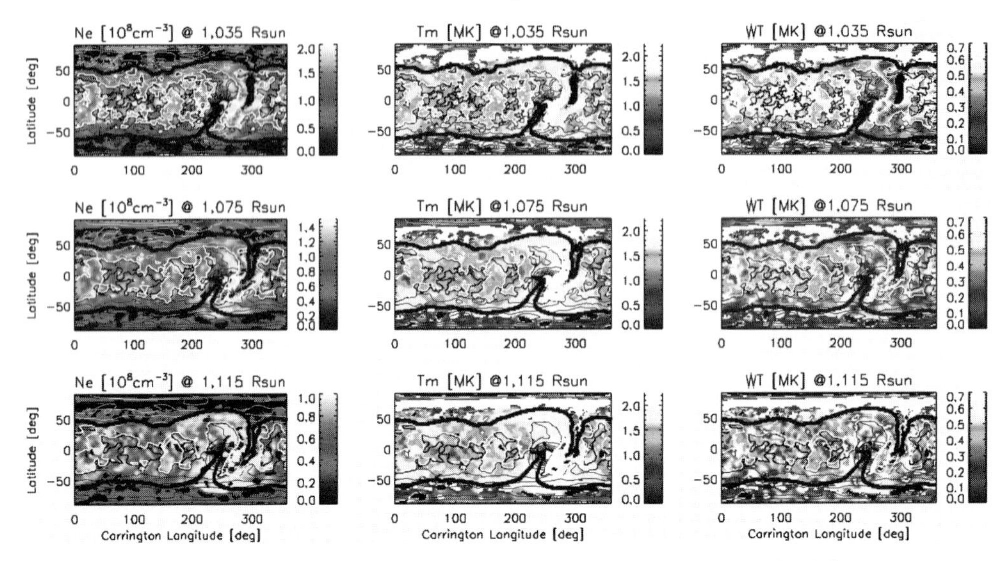

Figure 6. Same as Fig. 4 but for the CR-1914 period, based on SOHO/EIT data.

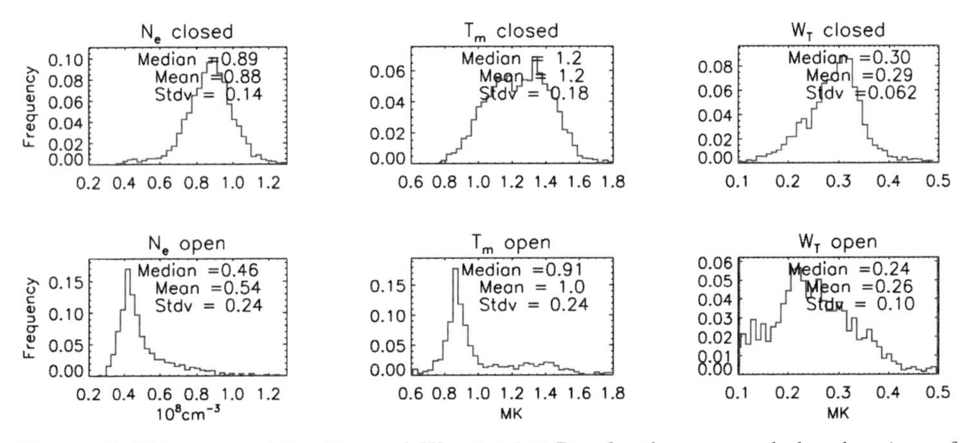

Figure 7. Histograms of N_e, T_m, and W_T at 1.075 $R_\odot$, for the open and closed regions of CR-2077.

this period. We find $\beta > 1$ in streamer cores, similar to the Li *et al.* (1998) study of the July 1996 streamer based on SXT and UVCS data, and a PFSS extrapolation, which estimated $\beta \sim 5$ at 1.15 $R_\odot$.

4. DEMT Results as a Function of Latitude and Height

To analyze the large scale corona latitudinal dependence of the DEMT results during both minima we select, for the periods CR-2077 and CR-1914, the ranges of more quiet longitudes and average the results in longitude. For CR-1914 we select the range of longitudes [340°,150°]. For CR-2077 we select the ranges [0°,20°], [50°,100°], [130°,170°], [250°,280°], and [340°,360°], which correspond to the more quiet and simply organized corona. In both periods we are then averaging over a longitude range spanning about 180°. The results for CR-2077 and CR-1914 are displayed in the top three panels of Figs. 11 and 12, respectively.

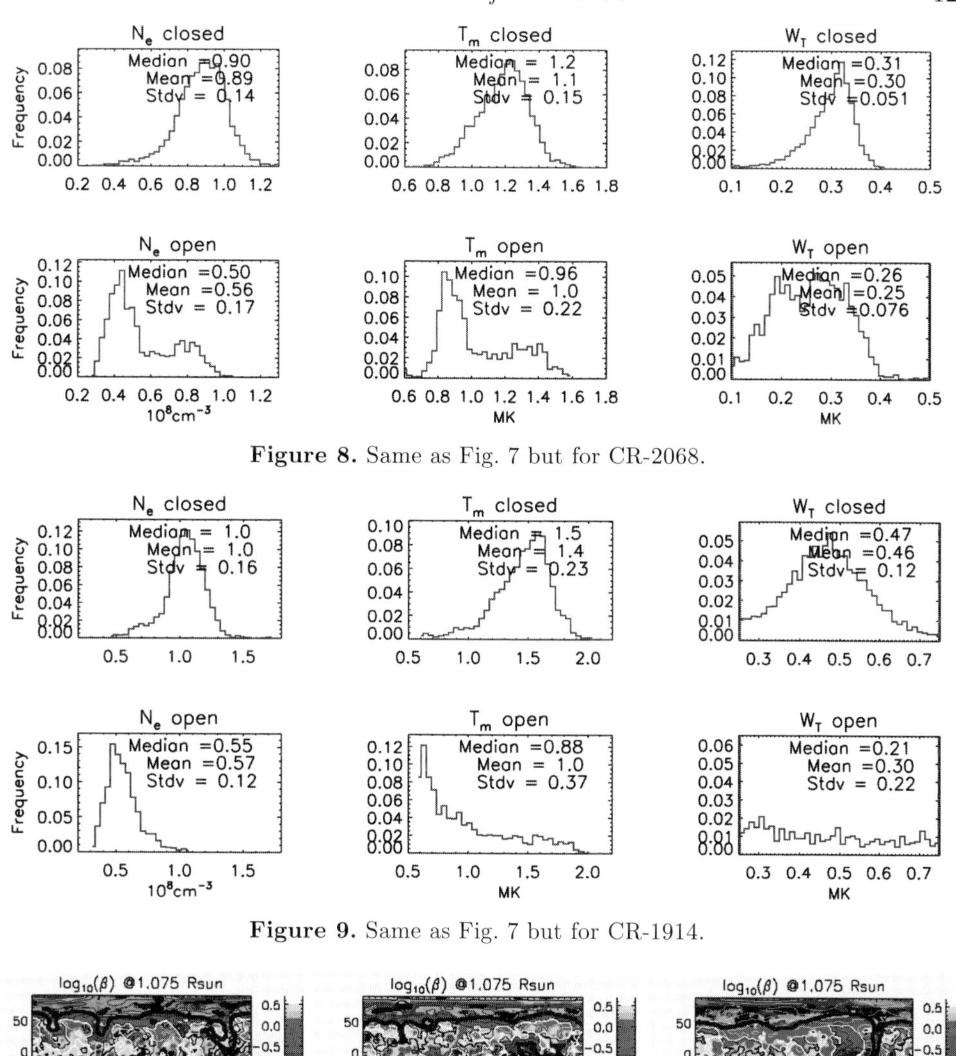

Figure 8. Same as Fig. 7 but for CR-2068.

Figure 9. Same as Fig. 7 but for CR-1914.

Figure 10. Plasma-β maps for CRs 2077 (right), 2068 (middle), and 1914 (left), at 1.075 $R_\odot$.

In both periods the electron density N_e is always more enhanced in the center of the closed regions, corresponding to the *streamer belt* core region, with a contrast of order 2:1 compared to the subpolar regions. About 70% of the transition of density between the streamer and subpolar regions occurs across a ~20° layer around the O/C boundary.

In both periods the mean electron temperature T_m is larger at mid-latitudes, within the closed regions, showing always a quite sharp transition to lower values across the O/C boundary. In both minima, within the closed region of large scale streamers covering both hemispheres, as for example from longitudes 0°to 150° for CR-2077 and the same range of longitudes for CR-1914, the temperature is lower at its central latitudes than near the O/C boundary. This latitudinal variation of temperature is similar to that found by Wilhelm *et al.* (2002). They performed a 1998 streamer study based on SUMER data and found larger T_e at mid latitudes and lower values in the streamer central latitudes (1.4

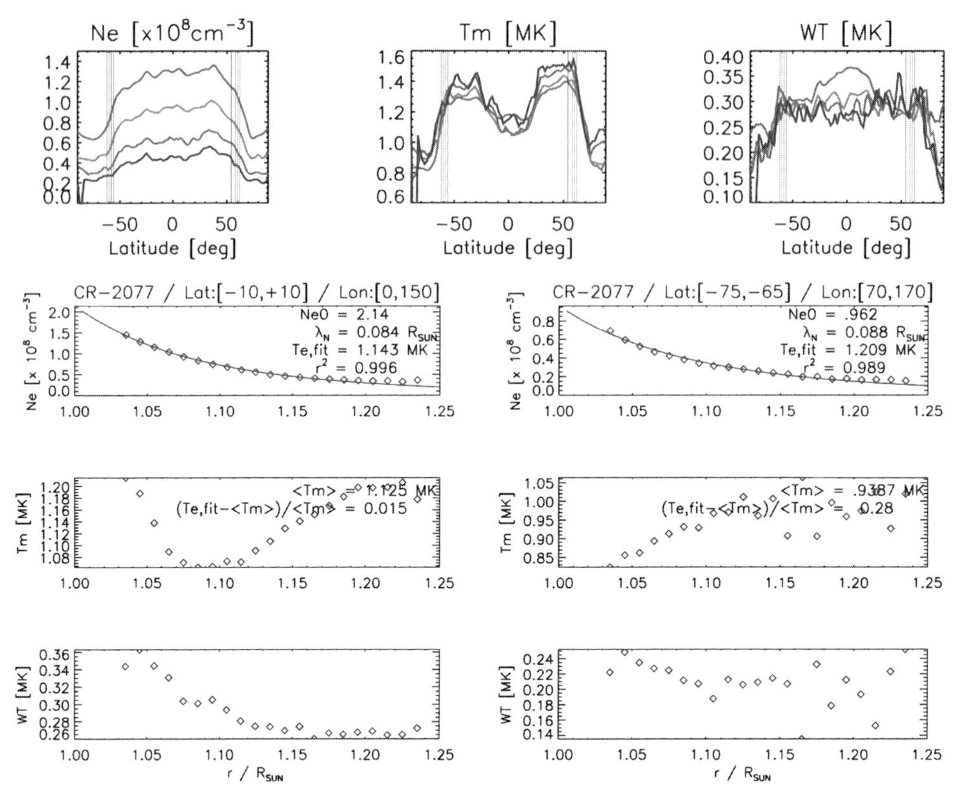

Figure 11. Averaged DEMT results for CR-2077. Top three panels: longitudinal average of the latitudinal variation for the quiet longitudes (see text) at four heights: 1.035 (red), 1.075 (green), 1.115 (magenta), and 1.155 R$_\odot$ (blue). For each height, using the same color code, the vertical lines indicate the average latitude of the O/C boundary in both hemispheres. Bottom Panels: average dependence with height for the streamer (left) and subpolar (right) regions. The colours correspond to the on line version of the paper.

and 1.0 MK, respectively). In CR-2068, the temperatures are also found to be enhanced in both streamer and pseudo-streamer regions, seen between longitudes 250°and 350°, in the southern and northern mid latitudes, respectively, surrounding the equatorial coronal hole seen in Fig. 5 (see also Figs. 4 and 8 in Gibson *et al.*, 2011).

The temperature spread W_T also presents a quite sharp transition to lower values right outside the O/C boundary, but its behavior within the streamer is a bit different than for T_m. At lower heights (1.035 and 1.075 R$_\odot$) there is an indication of an anti-correlation between both quantities, with W_T achieving larger values where T_m shows a minimum in the central part of the streamer core (see top right panel in Figs. 11 and 12).

Figs. 11 and 12 also show the average dependence with height of LDEM results in both the streamer central latitudes and in the subpolar regions, averaged over different ranges of longitudes and latitudes, as indicated in each plot. In each analyzed region we applied an isothermal hydrostatic (HS) fit to the LDEM $N_e(r)$ data and found the electron temperature $T_{e,\text{fit}}$ from the fit scale height, assuming $T_e = T_\text{H}$. Taking the LDEM $\langle T_m \rangle$ as a measure of the true T_e, one can use the expression of the HS fit height scale temperature to solve for the proton temperature (Vásquez *et al.* 2011), to find $\frac{T_\text{H}}{\langle T_m \rangle} \approx 1+2\left(\frac{T_{e,\text{fit}}-\langle T_m \rangle}{\langle T_m \rangle}\right)$.

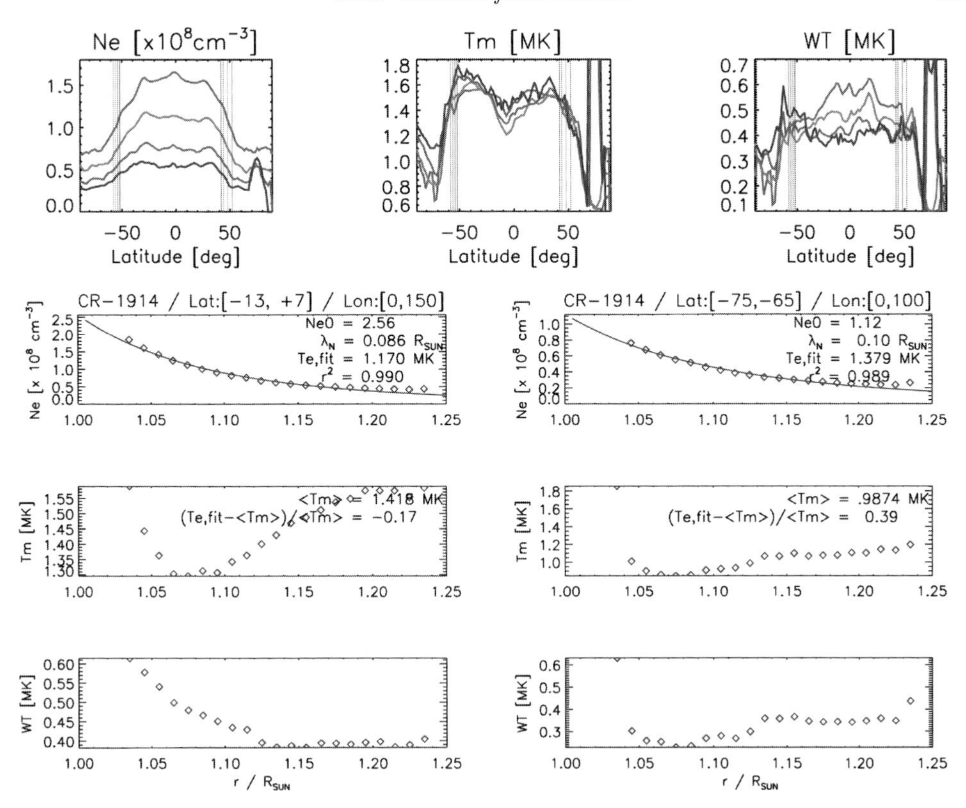

Figure 12. CR-1914. Top three panels: latitudinal variation of the results at four heights. Bottom Panels: average dependence with height for the streamer (left) and subpolar (right) regions.

The variation with height of the LDEM N_e in the CR-2077 streamer is consistent with the isothermal HS fit, as the correlation coefficient of the fit r^2 is high and also $T_{e,\mathrm{fit}} \sim \langle T_m \rangle$. We find the same for the CR-2068 streamer (not shown here due to space limitations). The CR-1914 streamer results present a different behavior, with also a high correlation coefficient but with $T_{e,\mathrm{fit}}$ being about 17% lower compared to $\langle T_m \rangle$. This may indicate a HS situation with excess electron temperature compared to protons. In the open subpolar regions all three periods show similar characteristics: a high value of the correlation coefficient (a bit lower than in the streamers though), and $T_{e,\mathrm{fit}} > \langle T_m \rangle$, with relative differences between 30 and 40 %. This is consistent with either proton excess temperatures or the presence of pressure mechanisms other than thermal that are linear with the electron density (see Vásquez *et al.* 2011 for a detailed CR-2068 analysis).

5. Concluding Remarks

We have performed DEMT analysis of three solar minimum rotations, two corresponding to the minimum phase at the end of solar Cycle 23, the third one corresponding to the minimum at the end of solar Cycle 22. In the case of the solar Cycle 23 periods, we are revisiting previous analysis (see Vásquez *et al.* 2010, 2011), updating the ionization equilibrium calculations assumed in the emission model to that of Bryans *et al.* (2009), and also providing here more comprehensive comparative figures of the 3D distribution of the results, along with detailed information on the degree of success of the parametrized LDEM to accurately predict the tomographically inverted emissivities. We analyzed all three periods in the context of PFSS magnetic extrapolations corresponding to each period, which were also improved compared to previous analysis. Some conclusions that arise from this analysis are as follows,

(*a*) The streamer:subpolar electron density ratio is of order 2:1 in all three periods. About ~70% of transition occurs over a ~20° latitude width layer around the O/C boundary. Note that the DEMT electron density is proportional to $1/\sqrt{[\mathrm{Fe}]}$, where [Fe] is the iron abundance, which has been assumed uniform in this work.

(*b*) In streamers, the LDEM $N_e(r)$ and $T_m(r)$ are quite consistent with isothermal hydrostatic equilibrium over the range of heights 1.0 to 1.2 $\mathrm{R}_\odot$.

(*c*) The largest temperatures are always found within streamers and pseudo-streamers, along and around polarity inversion lines. These regions are located at mid latitudes, near the O/C boundary, while the central latitudes of the streamer show relatively lower temperatures, still above 1 MK.

(*d*) Open subpolar regions are characterized by sub-MK electron temperatures. There is in general a quite sharp latitudinal gradient of temperature at the O/C boundary.

(*e*) In open regions surrounding streamers, we also find HS fits to present low residuals, but characterized by $T_{e,\mathsf{fit}} > T_m$. This is consistent with either pressure mechanisms other than thermal (but linear in density) being present and/or $T_\mathsf{H} > T_e$ (up to ~50% larger, see also Vásquez *et al.*, 2011)

(*f*) The CR-1914 minimum streamer is 30% hotter than for CR-2077. Based on the PFSS models, the CR-2077 streamer had an 8% larger basal area than both the CR-2068 and the CR-1914 streamers.

Future improvements of DEMT include the use of SDO/AIA data for more extensive temperature constraints on the LDEM inversion (a currently undergoing effort, see Nuevo *et al.*, 2012, in this volume) and validation of the DEMT results with Markov chain Monte Carlo (MCMC) simulations. We are also carrying on a study of DEMT inversion of simulated DEM with different available EUV imagers (EIT, EUVI, AIA), in order to learn how to interpret differences of DEMT results obtained with data from different instruments.

Regarding future improvements of the tomographic inversion of emissivities, the consideration of coronal differential rotation has the potential to allow for better resolution of N-S oriented elongated structures. The effects of dynamics in static tomographic reconstructions can be mitigated by using time-dependent approaches, as the Kalman-filtering method (Frazin *et al.*, 2005; Butala *et al.*, 2010). Local DEM analysis can then be applied to produce a time-dependent version of the DEMT technique. Stray light contamination due to the point-spread-function (PSF) affects all EUV imagers. Efforts to perform PSF blind-deconvolution of both STEREO/EUVI and SDO/AIA imagers are being developed by Shearer & Frazin (2011). Future inclusion of this procedures in the tomographic data pre-processing pipeline will greatly improve DEMT analysis.

DEMT major sources of uncertainty include the tomographic inversion regularization level, uncertainties in the assumed atomic data to compute the temperature response of the EUV imagers, and the coronal dynamics effects. All these factors combined lead to uncertainties of up to about 30% in the LDEM moments. A comprehensive study of DEMT uncertainties will be the subject of a future publication.

Acknowledgements

A.M.V. acknowledges the IAU for their support to attend the IAU Symposium 286 at the City of Mendoza, Argentina.

References

Bryans, P., Landi, E., & Savin, W. 2009, *ApJ*, 691, 1540-1559
Butala, M. D., Hewett, R. J., Frazin, R. A., & Kamalabadi, F. 2010, *Solar Phys.*, 262, 495-509
Feldman, U., Mandelbaum, P., Seely, J. L., Doschek, G. A., & Gursky, H. 1992, *ApJ*, 81, 387-408
Frazin, R. A., Kamalabadi, F., & Weber, M. A. 2005, *ApJ*, 628, 1070-1080
Frazin, R. A., Vásquez, A. M., & Kamalabadi, F. 2009, *ApJ*, 701, 547-560
Gibson, S. E., de Toma, G., Emery, B., Riley, P., Zhao, L.,Elsworth, Y., Leamon, R. J., Lei, J., McIntosh, S., Mewaldt, R. A., Thompson, B. J., & Webb, D. 2011, *Sol. Phys.*, 274, 5-27
Li, J., Raymond, J. C., Acton, L. W., Kohl, J. L., Romoli, M., Noci, G., & Naletto, G. 1998, *ApJ*, 506, 431-438
Nuevo, F. A., Vásquez, A. M., Frazin, R. A., Huang, Z., & Manchester IV, W. B. 2012, this volume
Shearer, P. & Frazin, R. A. 2011, *Phys. Rev. Lett.*, submitted
Tóth, G., van der Holst, B., & Huang, Z. 2011, *ApJ*, 732, 102-108
Vásquez, A. M., Frazin, R. A., & Manchester IV, W. B. 2010, *ApJ*, 715, 1352-1365
Vásquez, A. M., Huang, Z., Manchester IV, W. B., & Frazin, R. A. 2011, *Sol. Phys.*, 274, 259-284
Wilhelm, K., Inhester, B., & Newmark, J. S. 2002, *Astron. Astrophys.*, 382, 328-341

Discussion

LINKER: Have you tried to take your emission measure and T values to compare with XRT?

VÁSQUEZ: No, we have not yet done so, but that is something worth exploring. Thank you.

GIBSON: Were your bimodal population seen in the density and temperature histograms of the open regions related to the bimodal winds described by Giuliana (de Toma) in her talk?

VÁSQUEZ: Actually, slide 12 displays the frequency histograms for the N_e, T_m, and W_T values at 1.075 $R_\odot$, for the magnetically open and closed regions separately, covering the whole range of longitudes. To establish a correlation with the solar wind speed, it would be necessary to study the DEMT results along field lines in the MHD model and compare with the velocities of the model along the lines. We will also look into this.

Comparative Magnetic Minima:
characterizing quiet times in the Sun and stars
Proceedings IAU Symposium No. 286, 2011
C. H. Mandrini & D. F. Webb

doi:10.1017/S1743921312004747

The role of streamers in the deflection of coronal mass ejections

F. P. Zuccarello[1,2,3], A. Bemporad[4], C. Jacobs[1], M. Mierla[5,6], S. Poedts[1], and F. Zuccarello[3]

[1]Centre for Plasma Astrophysics, K.U. Leuven, Belgium.

[2]INAF - Osservatorio Astrofisico di Catania, via S. Sofia 78, 95123 Catania, Italy.

[3]Dipartimento di Fisica e Astronomia, Sezione Astrofisica, Via S. Sofia 78, 95123 Catania, Italy.

[4]Istituto Nazionale di Astrofisica (INAF), Osservatorio Astronomico di Torino, Strada Osservatorio 20, 10025 Pino Torinese, Torino, Italy.

[5]Institute of Geodynamics of the Romanian Academy, Bucharest, Romania.

[6]Royal Observatory of Belgium, Brussels, Belgium.

Abstract. On 2009 September 21, a filament eruption and the associated Coronal Mass Ejection (CME) was observed by the STEREO spacecraft. The CME originated from the southern hemisphere and showed a deflection of about 15° towards the heliospheric current sheet (HCS) during its propagation in the COR1 field-of-view (FOV). The aim of this paper is to provide a physical explanation for the strong deflection of the CME. We first use the STEREO observations in order to reconstruct the three dimensional (3D) trajectory of the CME. Starting from a magnetic configuration that closely resembles the potential field extrapolation for that date, we performed numerical magneto-hydrodynamics (MHD) simulations. By applying localized shearing motions, a CME is initiated in the simulation, showing a similar non-radial evolution, structure, and velocity as the observed event. The CME gets deflected towards the current sheet of the larger northern helmet streamer, due to an imbalance in the magnetic pressure and tension forces and finally it gets into the streamer and propagates along the heliospheric current sheet.

Keywords. Sun: coronal mass ejections (CMEs), methods: numerical, Sun: corona, Sun: magnetic fields

1. Introduction

Since the Skylab and *Solar Maximum Mission* (SMM) era (e.g. MacQueen, Hundhausen & Conover 1986), the occurrence of latitudinal deflections of coronal mass ejections (CMEs) towards the equator is a well known phenomenon, as well as similar deflections of flare associated shock waves (e.g. Fengsi & Dryer 1991). Later on, in the *SOlar and Heliospheric Observatory* (SOHO) era, many detailed investigations of deflections have been performed: statistical results show that during solar minima, CME deflections occur preferentially towards the equator, while during periods of intense solar activity both deflection towards the equator and towards the poles are observed, depending on the location and total area of coronal holes (Cremades, Bothmer & Tripathi 2006). Recently, Lopez *et al.* (2011, IAU Symposium 286, poster contribution) investigated the deflection of CMEs during the two previous solar minima. The authors found that between 60-75 % of the studied events exibit a deflection towards the nearest streamer independently of which solar minimum is considered. This indicates that the same physical mechanism could be responsable for the observed deflection of CMEs.

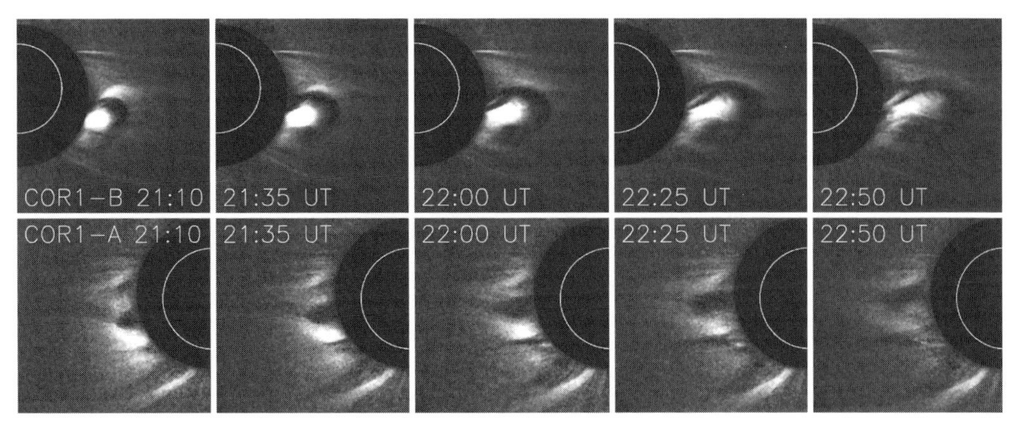

Figure 1. Sequence of COR1 images acquired during the CME propagation by STEREO-B (top) and -A (bottom). The COR1-B images show a three-part CME composed of a bright leading edge, the dark cavity and the bright core. The three-part components are not visible in the corresponding COR1-A images (bottom panels), where a more diffuse structure is observed. This figure is published in Zuccarello *et al.* (2012).

The aim of this paper is to provide a physical explanation for the deflection of the CME observed on 2009 September 21. In order to do this, we use *Solar TErrestrial RElations Observatory* (STEREO; see Kaiser *et al.* 2008) observations to reconstruct the three dimensional dynamics of the CME and combine these with MHD simulations of this event.

2. Observations

On 2009 September 21, a small prominence eruption leading to a CME occurred. This eruption has been observed by both the Extreme UV Imagers (EUVI) and the COR1 coronagraphs aboard the twin STEREO spacecraft.

The CME was better observed in COR1-B images as a classical three-part structure event, with a bright leading-edge, a dark cavity and a bright core (Fig. 1, top panels). The CME entered in the instrument field-of-view (FOV) around 19:45 UT being observed (as a three-part structure) until 00:10 UT on September 22, while the erupting core was visible until $\sim$ 01:35 UT. The CME core first appeared above the COR1 occulter at a projected latitude of $\simeq 25°$ South. The core expanded northward until $\sim$ 22:30 UT, when the top of the core was at a projected latitude of 15° South, i.e. closer to the equatorial plane. Interestingly, between 22:30 UT and 23:00 UT the CME core underwent a further and faster migration toward the equator, eventually approaching it. The CME was finally observed on 2009 September 22 by the COR2-B instrument as a faint three-part structured bubble expanding along the equatorial plane.

The CME was much more diffuse in COR1-A images and the three-part components were not clearly observed as compared with COR1-B images (Fig. 1, bottom panels). This is likely due to the large separation angle between the STEREO-A and -B spacecraft, making the CME, which expanded closer to the STEREO-B plane of the sky, very faint in the STEREO-A data.

3. Simulations

Using the two vantage points of the STEREO spacecraft, we reconstructed the 3D trajectory of the CME (see Zuccarello *et al.* (2012) for more details). We found that

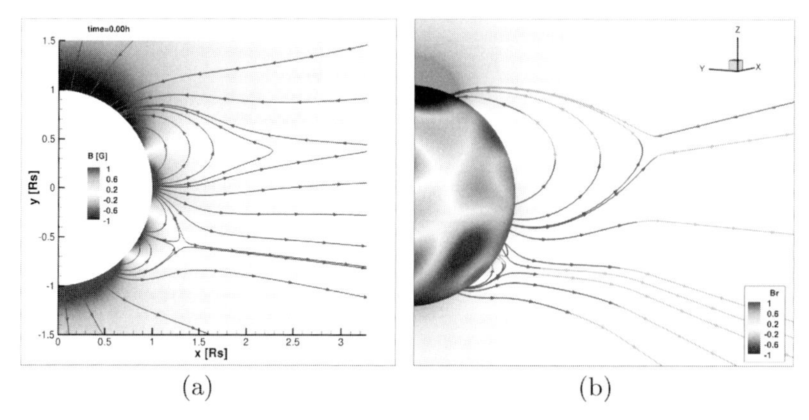

Figure 2. Magnetic field distribution (colour scale) and selected field lines for (a) the steady state of the simulation and (b) the PFSS extrapolation obtained from the MDI data on 2009 September 19 at 12:04 UT. This figure is published in Zuccarello *et al.* (2012). A color version of the figure is available in the online version.

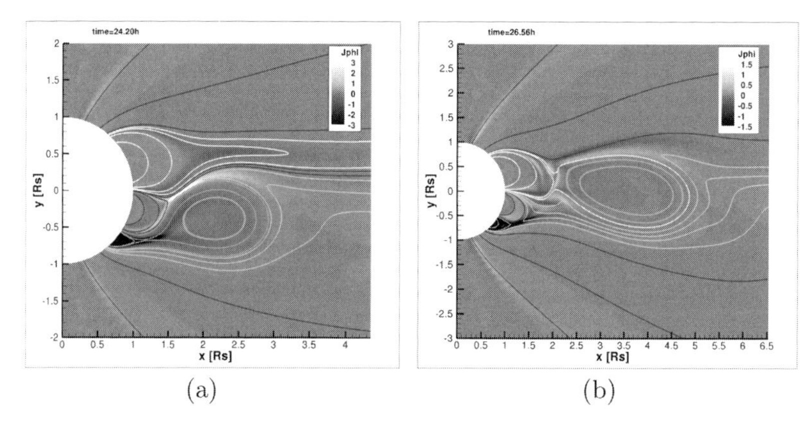

Figure 3. Snapshots of the time evolution of the azimuthal component of the current density (gray color scale) overplotted with some selected field lines. Different flux systems are identified by different colors (see text for more details). This figure is published in Zuccarello *et al.* (2012). A color version of the figure is available in the online version.

during its propagation the CME undergoes a longitudinal deflection not larger than $10°$, mainly travelling along a meridional plane. Therefore, the ideal MHD equations are solved numerically on a spherical, axisymmetric (2.5 D) domain covering the region between the solar north and south pole, i.e. $(r, \vartheta) \in [1\,R_\odot, 30\,R_\odot] \times [0, \pi]$.

Figure 2 shows the stationary solution for the MHD simulation (a) and the potential field source surface (PFSS) extrapolation for the 2009 September 19 (b). The initial magnetic configuration of the simulation presents a morphology similar to the reconstructed potential magnetic field. We would like to note that the key properties of the reconstructed field, i.e. the asymmetry between the two outer arcades, the northward shift of the cusp of the helmet streamer and the southern pseudostreamer, are all reproduced.

In order to form the prominence and drive the eruption, we apply localized shearing motions along the polarity inversion line of the southern loop system (Zuccarello *et al.* 2012). Figure 3(a) shows the magnetic configuration of the system after 21.84 hr. The grey scale denotes the azimuthal component of the current density, while the different colours of the field lines indicate different flux systems. Regions of high current density indicate the reconnection location. As a consequence of the applied shearing motions, the

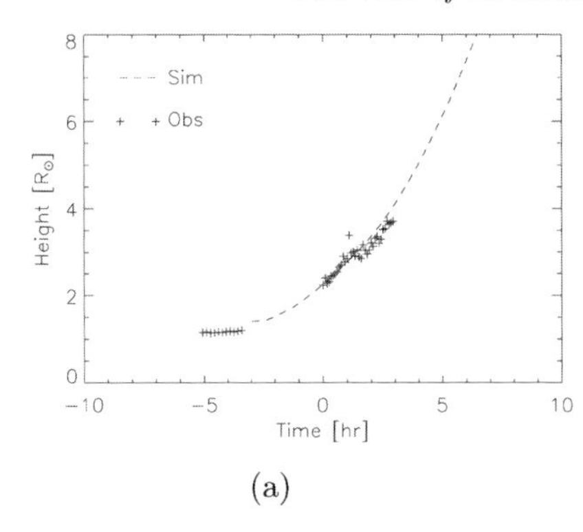

(a)

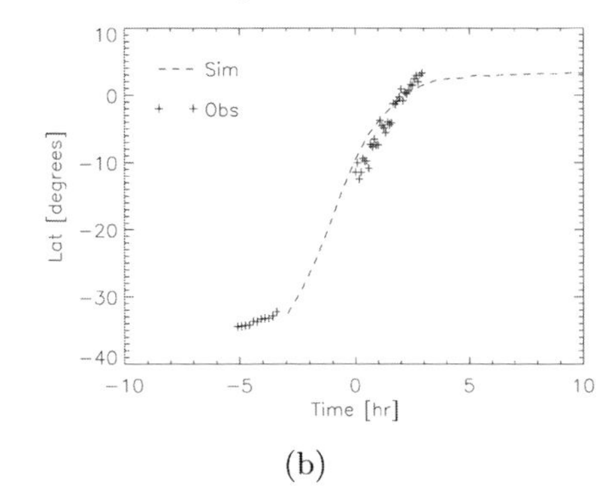

(b)

Figure 4. Comparison between the simulation (dashed line) and the observation (plus signs). (a) Altitude versus time and (b) latitude versus time. Time zero is 20:00 UT on 2009 September 21, corresponding to the time at which the CME was at 2.25 $R_\odot$. This figure is published in Zuccarello *et al.* (2012).

magnetic pressure increases and the southern side arcade starts to expand. During this expansion, the null point in the pseudostreamer is pushed northwards and one elongated current sheet is formed between the expanding southern arcade and the open field region at the north side of it, eventually initiating the magnetic reconnection. As a consequence of this reconnection, the magnetic flux of the expanding southern arcade (orange field lines) is transferred partially to the central arcade (red field lines), that becomes bigger, and partially to the open flux of the northern helmet streamer (blue field lines). The result of this process of interchange reconnection is visualized in the figure by the cyan field lines, i.e. originally closed field lines belonging to the southern arcade and that now belong to the southern coronal hole.

The pinching at the flanks of the southern arcade resulted in the formation of the flux rope (pink field lines) and during this reconnection process more and more magnetic flux is transferred from the southern arcade to the flux rope. The reconnection at the upper part of the expanding arcade results in a magnetic pressure imbalance between the north and the south part of the side arcade that, as a consequence, is deflected towards the equator. At a certain moment, due to the ongoing reconnection inside the southern arcade and the continuous growth of the central arcade, the newly formed open flux of the southern coronal hole (cyan field lines) will reconnect with the flux of the central arcade definitely separating the flux rope from its formation location and further contributing to the deflection of the CME toward the heliospheric current sheet (see Fig. 3(b)).

4. Discussion

In order to compare the early stages of the dynamics of the event, Fig. 4(a) shows the height-time plot for both the simulation (dashed line) and the reconstructed trajectory of the CME (plus signs). For the purpose of comparison with the observations, we set the origin of the time axis at the moment at which the core of the CME has an altitude of 2.25 $R_\odot$ in both the simulation and the observation. The simulated flux rope has a height-time evolution that is comparable with the altitude reconstruction of the CME. For both the simulated and the observed CMEs it takes about 6 hr to reach an altitude of 4 $R_\odot$ and both CMEs are slow.

In order to further compare the dynamics of the simulated and observed CME, in Fig. 4(b) we show the latitude-time plot for both the simulation (dashed line) and the reconstructed CME (plus signs). The prominence has a latitude of about 35° south and at time -5 hr (15:00 UT on 2009 September 21) it is evolving in the EUVI-B FOV. In about one and half hour it reaches a latitude of about 31° south and disappears from the EUVI FOV. The simulated flux rope starts from a location of about 33° south and experiences a deflection of about 20° in 3 hr, approaching an altitude of 2.25 $R_\odot$. At this altitude the core of the observed CME is visible in the COR1 FOV and its latitudinal deflection can be followed for another three hours. At 2.25 $R_\odot$ the core of the observed CME has a latitude of about 12° south and quickly approaches a latitude of about 4° north. This latitudinal behavior is well reproduced by the simulation.

Concluding, this study shows that during solar minima, as a consequence of the global magnetic field structure, even CMEs originating from high latitude can be easily deflected towards the heliospheric current sheet, eventually resulting in geoeffective events.

Acknowledgements

This research was funded by projects GOA-2009-009 (K.U.Leuven), G.0304.07 (FWO Vlaanderen), 3E090665 (FWO Vlaanderen), and C 90347 (ESA Prodex 9). The work of MM was supported from the contract TE - 73/11.08.2010. For the simulations we used the infrastructure of the VSC - Flemish Supercomputer Center, funded by the Hercules foundation and the Flemish Government - department EWI. STEREO is a NASA project. SOHO is a joint ESA and NASA project.

References

Cremades, H., Bothmer, V., & Tripathi, D. 2006, *Adv. Space Res.*, 38, 461.

Fengsi, Wei & Dryer, Murray 1991, *Solar Physics*, Volume 132, Issue 2, pp. 373-394

Kaiser, M. L. *et al.* 2008, *Space Sci. Rev.*, 136, 5

Lopez, F. M., Cremades, H., & Balmaceda, L. 2003, in: C. H. Mandrini & D. Webb (eds.), *Comparative Magnetic Minima: characterizing quiet times in the Sun and stars*, Proc. IAU Symposium No. 286 (Mendoza: ASP), *in press.*

MacQueen, R. M., Hundhausen, A. J., & Conover, C. W. 1986, *J. Geophys. Res.*, 91, 31

Zuccarello, F. P., Bemporad, A., Jacobs, C., Mierla, M., Poedts, S., & Zuccarello, F. 2012, *ApJ*, 744, 66.

Discussion

Janet Luhmann: There were two STEREO viewpoints. Why did you just pick one viewpoint, instead of both viewpoints?

Francesco Zuccarello: Due to the axial symmetry, our simulations are appropriate to describe events that are seen almost on the plane-of-sky (POS) for the coronagraph. This event was seen on the limb from STEREO B; this is the reason why we selected the B viewpoint. However, the simulation can reproduce the reconstructed 3D height and latitudinal time evolution of the CME.

Comparative Magnetic Minima:
Characterizing quiet times in the Sun and stars
Proceedings IAU Symposium No. 286, 2011
C. H. Mandrini & D. F. Webb, eds.

doi:10.1017/S1743921312004759

Magnetic clouds along the solar cycle: expansion and magnetic helicity

Sergio Dasso[1,2], Pascal Démoulin[3], and Adriana M. Gulisano[1,4]

[1]Instituto de Astronomía y Física del Espacio, CONICET-UBA, CC. 67, Suc. 28, 1428 Buenos Aires, Argentina
email: dasso@df.uba.ar

[2]Departamento de Física, Facultad de Ciencias Exactas y Naturales, Universidad de Buenos Aires, 1428 Buenos Aires, Argentina

[3]Observatoire de Paris, LESIA, UMR 8109 (CNRS), F-92195 Meudon Principal Cedex, France

[4]Instituto Antártico Argentino (DNA), Cerrito 1248, Buenos Aires, Argentina

Abstract. Magnetic clouds (MCs) are objects of extreme importance in the heliosphere. They have a major role on releasing magnetic helicity from the Sun (with crucial consequences on the solar dynamo), they are the hugest transient object in the interplanetary medium, and the main actors for the Sun-Earth coupling. The comparison between models and observations is beginning to clarify several open questions on MCs, such as their internal magnetic configuration and their interaction with the ambient solar wind. Due to the decay of the solar wind pressure with the distance to the Sun, MCs are typically in expansion. However, their detailed and local expansion properties depend on their environment plasma properties. On the other hand, while it is well known that the solar cycle determines several properties of the heliosphere, the effects of the cycle on MC properties are not so well understood. In this work we review two major properties of MCs: (i) their expansion, and (ii) the magnetic flux and helicity that they transport through the interplanetary medium. We find that the amount of magnetic flux and helicity released via MCs during the last solar minimum (years 2007-2009) was significantly lower than in the previous one (years 1995-1997). Moreover, both MC size and mean velocity are in phase with the solar cycle while the expansion rate is weakly variable and has no relationship with the cycle.

Keywords. magnetic fields, magnetic clouds, coronal mass ejections (CMEs), solar wind, interplanetary medium, plasmas

---

1. Introduction

The study of the heliosphere has advanced greatly in the last few years. The comparison between models and observations is clarifying several issues of this system. The synergy of combining modeling with different observational techniques has led to a very important progress in our understanding of the heliosphere, with important consequences on Sun-Earth connection and space weather.

1.1. *Magnetic clouds*

Interplanetary Coronal Mass Ejections (ICMEs) are the biggest transient structures in the solar wind (SW). An important set of ICMEs is known as Magnetic Clouds (MCs), a term introduced by Burlaga *et al.* (1981). A MC is characterized by an enhanced magnetic field strength with respect to ambient values, a smooth and large rotation of the magnetic field vector, and low proton temperature (e.g., Burlaga *et al.* 1981; Klein & Burlaga 1982; Burlaga 1995).

When a MC is traveling in the Earth direction, it can be geoeffective. In particular, depending on the orientation and strength of its magnetic field, a MC can initiate intense

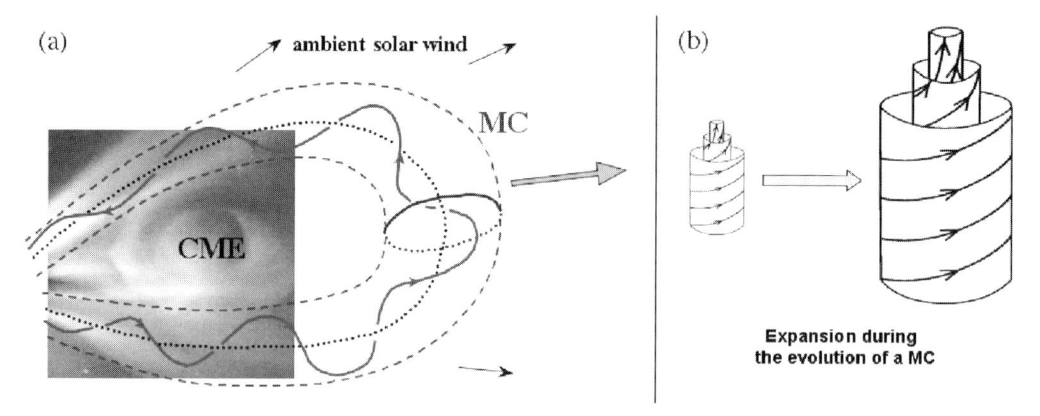

Figure 1. **(a)** Cartoon of a CME-MC interacting with the surrounding solar wind plasma while evolving in the interplanetary medium (the CME image is from SOHO/LASCO). **(b)** Due to the decrease of the solar wind pressure with the distance to the Sun, MCs are globally expanding objects.

geomagnetic perturbations as consequence of reconnection processes in the terrestrial magnetopause (e.g., Gonzalez *et al.* 1999, and references therein). It is worth noting that the field strength depends both on the solar source region and on the MC expansion during its transit from the Sun to Earth (because of the conservation of magnetic flux). Moreover, fast MCs generally drive a shock wave, which can increase the geoeffectiveness of the event.

MCs can be modeled locally using an helical cylindrical geometry as a first approximation (e.g., Farrugia *et al.* 1995). Several models have been used for describing the magnetic structure of the local cylinder of MCs. The most frequently used, is the linear force-free field (e.g., Burlaga *et al.* 1981; Burlaga 1988; Lepping *et al.* 2003), which corresponds to the first eigen-function of the curl operator for cylindrical symmetry.

A cylindrical non-linear force-free and uniformly twisted field (e.g., Farrugia *et al.* 1999) has been used as a possible alternative to describe MCs. Also several non force-free models have been applied, using different shapes for their cross sections (e.g., Hu & Sonnerup 2001; Cid *et al.* 2002; Hidalgo *et al.* 2002; Vandas & Romashets 2003; Démoulin & Dasso 2009b). All these models are physically different. Comparisons of these models with observations and synthetic data have been done (e.g., Riley *et al.* 2003; Dasso *et al.* 2005b) but it is not yet clear which of them gives the best representation of MCs.

MCs are evolving while traveling in the solar wind. Indeed, it is very frequent to observe an 'in situ' velocity profile of MCs with a clear signature of expansion (i.e. with a faster velocity in the front than at the rear, e.g. Gulisano *et al.* 2010). Figures 1a and 1b show a cartoon of a solar eruption, launching a flux rope, and the consequent expansion of the associated MC. Several models have considered the expansion effects on the decay of the observed magnetic profile (e.g., Shimazu & Vandas 2002; Berdichevsky *et al.* 2003; Démoulin & Dasso 2009a; Nakwacki *et al.* 2008, 2011).

The large scale magnetic structure of MCs is helical, then these objects transport large amounts of magnetic flux (F) and helicity (H) from the Sun to the outer heliosphere. The content of F and H in MCs have been quantified using several models (e.g., Dasso *et al.* 2005b; Lynch *et al.* 2005). Nakwacki *et al.* (2008) have estimated the bias produced by the MC expansion on F and H. By comparing static and expanding models, they have concluded that these quantities do not differ more than 25% when using both kind of approaches.

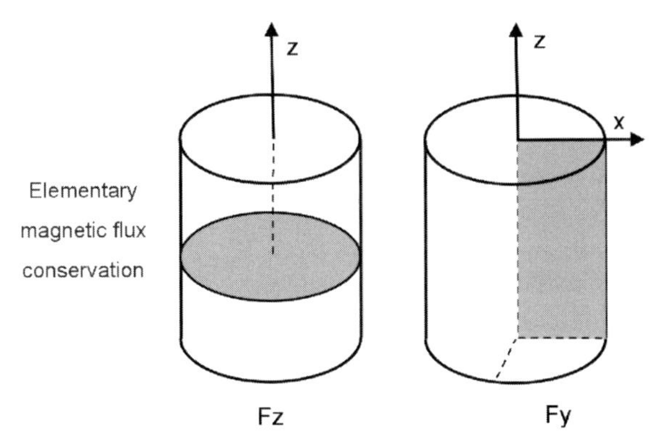

Figure 2. Scheme of surfaces associated with F_z (axial flux) and F_y (azimuthal flux).

The twisted magnetic structures associated with a large amount of H transported by MCs have important consequences also on the propagation of energetic particles in the solar wind. In particuliar, Masson *et al.* (2012) have analyzed the Earth-arrival time for high energy particle released from the Sun during relativistic solar particle events. They have shown that this Earth-arrival time is different when the interplanetary medium presents structures as the Parker's model for the SW and when it presents transient structures as ICMEs/MCs.

1.2. *Eruptive events along the solar cycle*

During the last solar maximum (solar cycle 23) an important number of large active regions with long lifetimes was observed. These ARs produced several huge magnetic clouds, which transported very large amounts of magnetic flux and helicity into the heliosphere. Some of the magnetic clouds launched during this period were associated with a complex topology of the coronal magnetic field (e.g. Schmieder *et al.* 2011).

On the other hand, the last solar minimum (minimum of solar cycle 24) presented the slowest and least dense solar wind, and the weakest interplanetary magnetic field, when compared with the last three minima periods (Jian *et al.* 2011). Coronal holes near the ecliptic produced recurrent fast streams of fast solar wind during this last minima (de Toma 2011). There was also an north-south asymmetry of the AR locations, with more ARs located in the South (de Toma 2012, this volume). This implies an expected predominance of positive helicity for ICMEs which erupted in this period. A comparison of a variety of outward-traveling transients observed in the solar corona during these two last solar minima, as achieved during the observing campaigns of the Whole Sun Month (WSM, from 10 August to 8 September 1996) and of the Whole Heliosphere Interval (WHI, from 20 March to 16 April 2008) was made by Cremades *et al.* (2011). They found more ejecta produced by active regions during WHI than during WSM.

2. Global magnetic quantities

2.1. *Magnetic field and cross section*

MCs have low plasma β (e.g., Lepping *et al.* 2003), they present a configuration in a quasi-equilibrium (a state near to a force-free field), and they are formed by a flux rope magnetic structure. Then, the magnetic forces dominate the dynamics inside the structure of MCs for a given time and their magnetic configuration can be modeled using the cylindrical linear force-free field $\vec{B^L} = B_0[J_1(\alpha_0 r)\vec{\phi} + J_0(\alpha_0 r)\vec{z}]$ (Lundquist 1950).

There are evidences both from MHD simulations (Riley *et al.* 2003; Manchester *et al.* 2004) and from observations (Owens *et al.* 2006; Savani *et al.* 2011; Nakwacki *et al.* 2011) that MCs propagating in the interplanetary medium have a trend to develop oblate shapes for their cross sections. This effect is mainly due to the interaction between the MC and the solar wind (Vandas *et al.* 1995; Riley *et al.* 2003). However, the aspect ratio between the major axis (which is expected to be perpendicular to the Sun-ward direction) with respect to the minor one seems to be moderate (typically $\leqslant 2$). In particular, from a reconstruction of the magnetic cross section using a Grad-shafranov formalism, Liu *et al.* (2008) found that the MC core ($\sim$ the inner half part of the flux rope) is almost circular, while the flux rope periphery is more oblate [see Figure 5 of Liu *et al.* 2008]. This result can be interpreted as a consequence of magnetic tension, which will favor a circular shape when the magnetic field is strong enough. The deformation of the external part can be interpreted as a consequence of the interaction with the ambient solar wind.

2.2. *Magnetic flux and helicity*

Magnetic flux (F) and helicity (H) are ideal magnetohydrodynamical (MHD) invariants. Magnetic helicity quantifies the twist and linkage of magnetic field lines and it is approximately conserved in the solar atmosphere and the heliosphere because of the high magnetic Reynolds number (Berger 1984). Since MCs are large twisted flux tubes, they transport important amounts of H through the solar wind (a review on magnetic helicity in MCs can be found in Dasso 2009).

A quantification of F and H can be done using a specific model for describing the MC magnetic configuration. A local frame of coordinates linked to the cloud is very useful to make quantitative comparisons between models and observations, and to quantify F and H.

Figure 2 shows a cartoon showing the local system (for a precise definition of this system, see e.g., Dasso *et al.* 2005b). The magnetic fluxes are computed across two surfaces: across a surface perpendicular to $\hat{z}$ (F_z, the axial flux) and across a surface perpendicular to $\hat{y}$ (F_y, the azimuthal flux). Simple analytical expressions for F_z, F_y, and H, can be derived in function of the free parameters of the Lundquist model (α_0, B_0) and the radius of the cross section (R) (e.g., Dasso *et al.* 2005b) and thus, these quantities can be evaluated after fitting these parameters to observations of a given cloud.

Different methods and models, to quantify F and H from in situ observations, have also been used (e.g., Dasso *et al.* 2005b; Lynch *et al.* 2005; Nakwacki *et al.* 2008, 2011). Comparison of estimations for F and H obtained from different models, that fit relatively well to in situ observations of several samples of MCs, have been done (e.g., Gulisano *et al.* 2005; Dasso *et al.* 2005a; Nakwacki *et al.* 2008). These studies concluded that, for a given event, differences on estimations of F and H using different models are typically much lower than the dispersion of these quantities for different events. These results imply that helicity and fluxes are relatively well defined quantities for MCs.

Quantitative studies of F and H both in MCs and in the solar corona are used to relate MCs to their solar sources (using F and H approximative conserved properties). Moreover, these studies set constrains to coronal magnetic configurations and on flux rope formation/eruption models (Mandrini *et al.* 2005; Luoni *et al.* 2005; Attrill *et al.* 2006; Longcope *et al.* 2007; Qiu *et al.* 2007; Harra *et al.* 2007; Mandrini *et al.* 2007; Rodriguez *et al.* 2008; Möstl *et al.* 2008; Ravindra *et al.* 2011).

2.3. *Magnetic flux and helicity in MCs during the last two minima*

Lepping *et al.* (2011) studied some properties of MCs observed at one astronomical unit by the spacecraft Wind in the two last solar minima. In particular they study MCs during

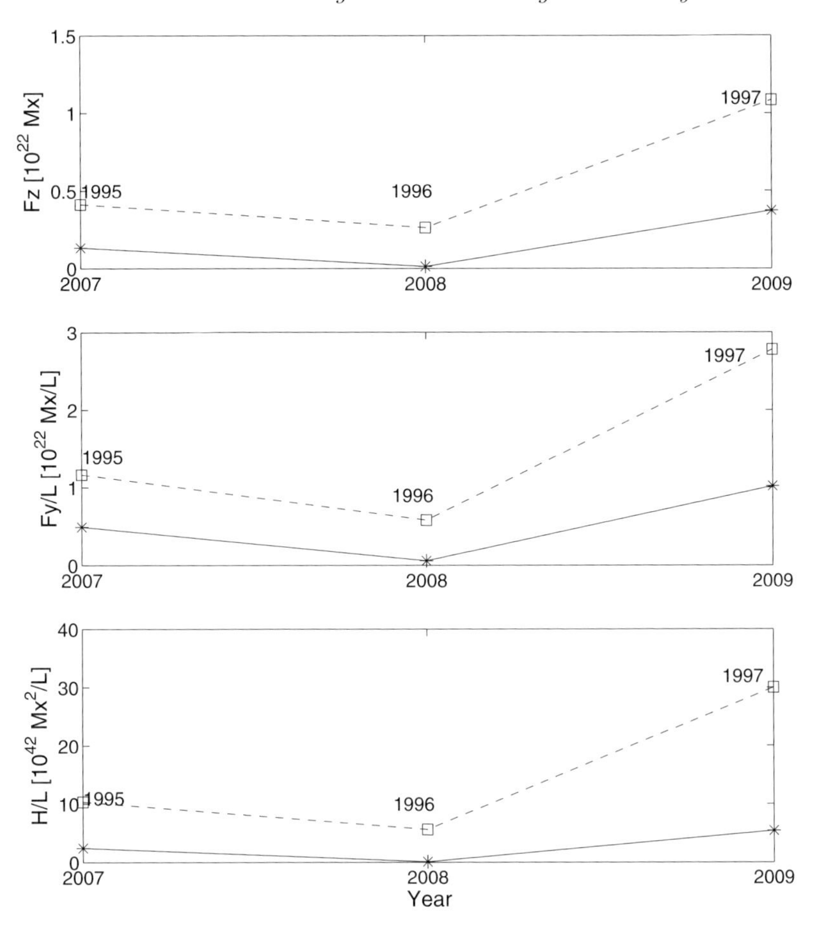

Figure 3. Comparison of magnetic quantities in MCs cumulated in each year of periods I and II (see Section 2.3). From upper to lower panels: axial flux (F_z), azimuthal flux per unit length (F_y/L), and magnetic helicity per unit length (H/L). All these magnetic quantities are lower in Period II than in period I.

the previous solar minimum (years 1995-1997, which we call period I) and during the last recent minimum (year 2007-2009, which we call period II). From a comparison of periods I and II, they find that period II presented: (1) a lower number of events (29 MCs for period I and 18 MCs for period II), (2) a significantly lower axial magnetic field (B_0 is in average lower by 33 %), (3) similar velocities, (4) lower duration of the 'in situ' observations, then (5) smaller MC size, and (6) $\sim$ 50% less axial flux (F_z), than during period I.

In our study we analyze the same set of MCs studied by Lepping *et al.* (2011), and from using the fitted free parameters of the Lundquist's model (B_0, α_0, and R, see Section 2.2), we compute the axial (F_z) and the azimuthal (F_y) fluxes, and the magnetic helicity (H). The results are shown in Figure 3. This figure follows the style used in Figure 1 of Lepping *et al.* (2011). In that paper the authors show the number of events in the two periods, we show here the cumulated magnetic quantities considering all the MCs observed in each year.

Figure 3 shows that, during each of the years of period II, MCs transported a significantly lower amount of magnetic flux and helicity, when compared with the corresponding years in period I. The fraction of total content of fluxes and helicity, considering all the

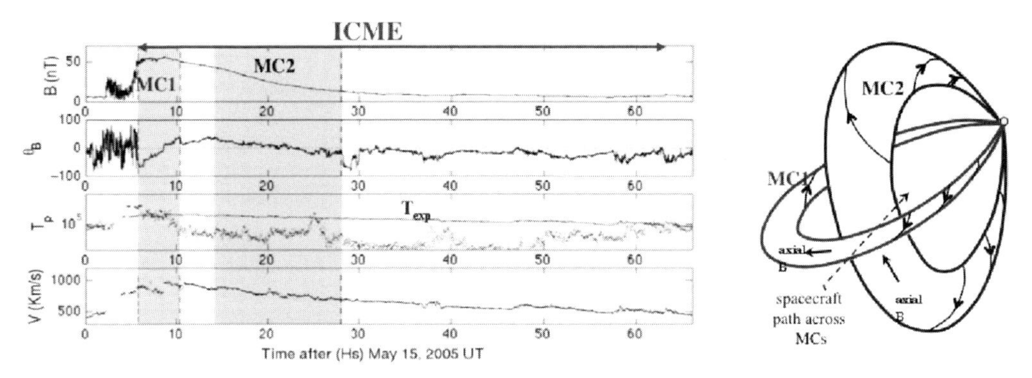

Figure 4. The interaction between two magnetic clouds affect their expansion rate. This figure shows 'in situ' observations (left) and a cartoon of two interacting MCs (right), observed in May 2005. MC1 is strongly compressed by the overtaking MC2, and the velocity profile of MC1 shows almost no expansion, while MC2 has a classical expansion profile (Dasso *et al.* 2009).

events for each period, is: $F_z^{II}/F_z^I \sim 0.29$, $F_y^{II}/F_y^I \sim 0.35$, $H^{II}/H^I \sim 0.17$. Moreover, a typical MC in period II transported less flux and helicity than a typical MC in period I. The mean (median) values of F and H ratios for MCs in period II with respect to period I are: $< F_z^{II} > / < F_z^I > \sim 0.47$ (0.35), $< F_y^{II} > / < F_y^I > \sim 0.56$ (0.51), $< H^{II} > / < H^I > \sim 0.28$ (0.15). Then, a lower total magnetic flux and helicity released from the Sun in period II as MCs comes from a combination of (i) less number of events and (ii) weaker MCs during period II with respect to period I.

3. Expansion

3.1. *Self-Similar expansion*

The expansion of MCs can be quantified from the 'in situ' observations of the proton velocity profile. The expansion can be described using a model developed by Démoulin *et al.* (2008), which is derived from a few basic hypothesis:

(1) The motion of parcels of fluid inside a MC can be split in two: (a) a global motion describing the position $\vec{r}_{CM}(t) = D(t)\hat{\mathbf{v}}_{CM}$ of the center of mass (CM) with respect to a fixed heliospheric frame and (b) an internal expansion where the fluid elements are expressed relative to the CM.

(2) During the MC crossing the spacecraft, the MC center travels with an almost uniform speed ($\vec{v}_{CM}$, with $|\vec{v}_{CM}| = V_0$), then

$$D(t) = D_0 + V_0(t - t_0)\,. \tag{3.1}$$

(3) The local MC frame ($\hat{\mathbf{x}}$, $\hat{\mathbf{y}}$, $\hat{\mathbf{z}}$) (see Figure 2), defines the three principal directions of expansion.

(4) The flux rope expansion is self similar with different expansion rates in each of the three principal directions.

These assumptions imply that the position at time t of an element of fluid is described as:

$$\vec{r}(t) = x(t)\,\hat{\mathbf{x}} + y(t)\,\hat{\mathbf{y}} + z(t)\,\hat{\mathbf{z}} \tag{3.2}$$
$$= x_0\,e(t)\,\hat{\mathbf{x}} + y_0\,f(t)\,\hat{\mathbf{y}} + z_0\,g(t)\,\hat{\mathbf{z}}\,, \tag{3.3}$$

with $x(t), y(t), z(t)$ being the fluid coordinates in the local frame, and x_0, y_0, z_0 the

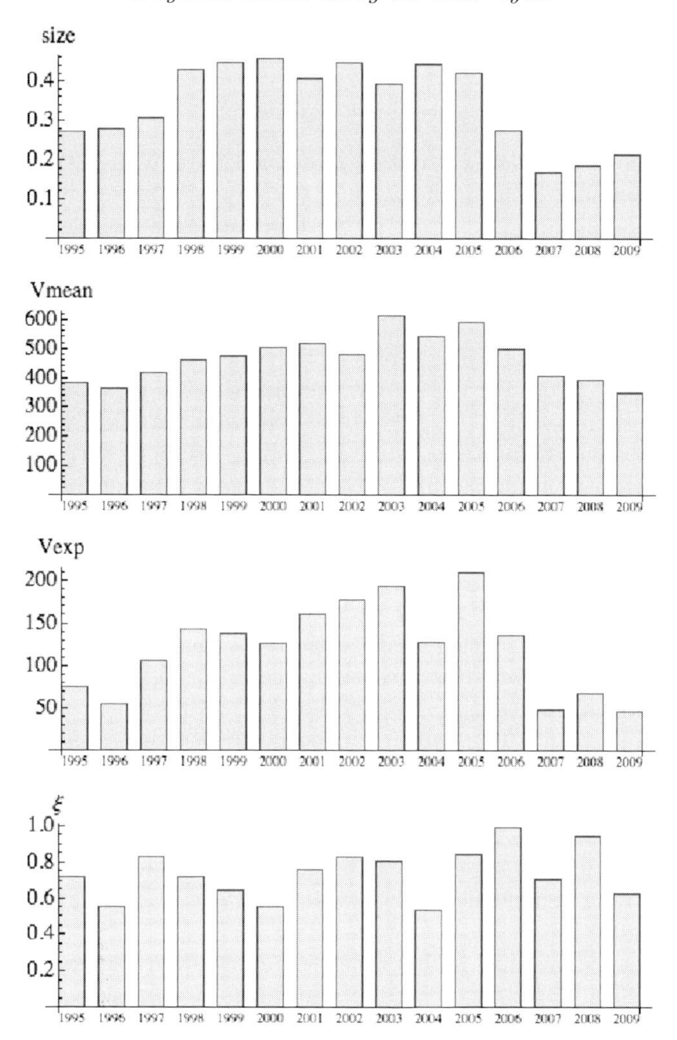

Figure 5. Distribution of the MC properties along the solar cycle. From the top to the bottom panels: size, mean velocity, expansion velocity, and expansion rate (ζ).

reference positions at time t_0. The functions $e(t)$, $f(t)$, and $g(t)$, provide the specific time dependences for the self-similar evolution.

Based on observations of MCs at different distance from the Sun (e.g., Liu *et al.* 2005; Wang *et al.* 2005; Leitner *et al.* 2007; Gulisano *et al.* 2010), it is possible to approximate $e(t)$ as:

$$e(t) = (D(t)/D_0)^l \, . \tag{3.4}$$

Similar expressions can be used for $f(t)$ and $g(t)$ (replacing the exponent l by m and n, respectively).

Neglecting the evolution of the spacecraft position while it observes the MC, the measured velocity profile (V_x) along the direction $\hat{\mathbf{v}}_{CM}$ is expected to be (Démoulin *et al.* 2008):

$$V_x = -V_0 + V_0 \frac{t - t_0}{D_0/V_0 + t - t_0} \zeta \tag{3.5}$$

$$\approx -V_0 + \frac{V_0^2}{D_0}\zeta(t-t_0)\,, \tag{3.6}$$

$$\text{with}\quad \zeta = l\sin^2\gamma + n\cos^2\gamma\,, \tag{3.7}$$

and with the angle between $\hat{\mathbf{z}}_{\rm cloud}$ and $\hat{\mathbf{v}}_{CM}$ defined as γ. Thus, with an isotropic expansion in two directions ($l = n$), the slope of the observed linear velocity profile allows us to determine the expansion rate, ζ, of the flux rope (for anisotropic expansion and acceleration effects on the velocity profile see Démoulin *et al.* 2008; Nakwacki *et al.* 2011).

3.2. *Expansion along the cycle*

We analyze a set of 75 magnetic clouds observed by the spacecraft ACE at one astronomical unit from 1995 to 2009. For each event, we quantify the mean velocity (V_{mean}), the size (S), the expansion velocity (V_{exp}), and the expansion rate (ζ). S is derived from the mean velocity and the MC duration, while V_{exp} and ζ are both derived from a linear fit of the observed velocity profile. Figure 5 shows the mean value of these quantities inside bins of one year. We found that S, V_{mean} and V_{exp} are in phase with the solar cycle (in agreement with Jian *et al.* 2011), with larger and faster MCs near the solar maximum, and smaller and slower MCs observed in periods of solar minimum.

The expansion rate, ζ, has only weak variations and it does not follow the solar cycle. The dispersion of ζ from its theoretically expected value of $\sim 0.7 - 0.9$ (Démoulin & Dasso 2009a) can be interpreted as a consequence of the perturbation of the observed cloud by fast solar wind streams, or by another MC (Gulisano *et al.* 2010). Thus, this result is in agreement with MCs which can be perturbed during all the phases of the solar cycle.

Moreover, for MCs at a distance D from the Sun there is a relationship between ζ and V_{exp} given by $V_{exp} = \zeta V_{mean} S/D$. Then, from the non-dependence of ζ with the solar cycle (lower panel of Figure 5) and from the increasing behaviour of V_{mean} and S with solar activity, a strong dependence of V_{exp} with the cycle is expected, as indeed present (third panel of Figure 5).

The above results show that MCs properties, apart from the expansion rate ζ, are cycle dependant. Such dependence needs to be taken into account when a typical MC is used, for example for building a magnetic helicity budget (e.g. Démoulin 2007, and references therein).

4. Summary and Conclusions

In this work we revised methods which quantify the content of magnetic flux and helicity in MCs from 'in situ' observations of the magnetic field, and also methods which quantify their expansion from 'in situ' observations of the proton velocity. We also studied the influence of the solar cycle on these MC properties.

We compared the amount of flux and helicity released from the Sun during the last two minima and found that, due to a less number of events and less flux and helicity contained in each typical event, the amount of H (F) released via MCs during the last solar minimum (years 2007-2009) was lower than 20% (35%) than during the previous one (years 1995-1997). MCs are one of the main agents in charge of releasing magnetic helicity from the Sun. Then, these quantitative results can be useful for better understanding the solar dynamo and then to improve our knowledge of the so weak last solar minimum itself.

We studied also the variability of the MCs expansion along the solar cycle, and found that the dimensionless expansion parameter (ζ) has no dependence with the solar cycle.

This results is coherent with previous ones because ζ mainly depends on the global decay of the solar wind pressure with the distance to the Sun (Démoulin & Dasso 2009a). However, because of the dependence of both the size and speed of MCs with the solar magnetic activity, the velocity of expansion (V_{exp}) has a strong dependence with the solar cycle (with a significantly lower expansion during solar minima). Moreover, as a consequence of MC interaction with streams of the fast solar wind during periods of solar minima (as occurred in the last minimum), MCs can be perturbed even during periods with low level of solar activity.

Acknowledgments

The authors acknowledge financial support from ECOS-Sud through their cooperative science program (N° A08U01). S.D. is a member of the Carrera del Investigador Científico, CONICET. S.D. and A.M.G. acknowledge partial support by Argentinean grants UBACyT 20020090100264 (UBA), PIP 11220090100825/10 (CONICET), PICT 2007-856 (ANCPyT). S.D. acknowledges support from the Abdus Salam International Centre for Theoretical Physics (ICTP), as provided in the frame of his regular associateship.

References

Attrill, G., Nakwacki, M. S., Harra, L. K., *et al.* 2006, *Solar Phys.*, 238, 117
Berdichevsky, D. B., Lepping, R. P., & Farrugia, C. J. 2003, *Phys. Rev. E*, 67, 036405
Berger, M. A. 1984, Geophys. Astrophys. Fluid Dyn., 30, 79
Burlaga, L., Sittler, E., Mariani, F., & Schwenn, R. 1981, *J. Geophys. Res.*, 86, 6673
Burlaga, L. F. 1988, *J. Geophys. Res.*, 93, 7217
Burlaga, L. F. 1995, *Interplanetary magnetohydrodynamics* (Oxford University Press, New York)
Cid, C., Hidalgo, M. A., Nieves-Chinchilla, T., Sequeiros, J., & Viñas, A. F. 2002, *Solar Phys.*, 207, 187
Cremades, H., Mandrini, C. H., & Dasso, S. 2011, *Solar Phys.*, 274, 233
Dasso, S. 2009, in IAU Symposium, Vol. 257, IAU Symposium, ed. N. Gopalswamy & D. F. Webb, pp. 379–389
Dasso, S., Gulisano, A. M., Mandrini, C. H., & Démoulin, P. 2005a, Advances in Space Research, 35, 2172
Dasso, S., Mandrini, C. H., Démoulin, P., Luoni, M. L., & Gulisano, A. M. 2005b, Adv. Spa. Res., 35, 711
Dasso, S., Nakwacki, M., Démoulin, P., & Mandrini, C. H. 2007, *Solar Phys.*, 244, 115
Dasso, S., Mandrini, C. H., Schmieder, B., *et al.* 2009, *J. Geophys. Res.*, 114, A02109
de Toma, G. 2011, *Solar Phys.*, 274, 195
de Toma, G. 2012, This volume
Démoulin, P. 2007, *Adv. Spa. Res.*, 39, 11, 1674
Démoulin, P. & Dasso, S. 2009a, *Astron. Astrophys.*, 498, 551
Démoulin, P. & Dasso, S. 2009b, *Astron. Astrophys.*, 507, 969
Démoulin, P., Nakwacki, M. S., Dasso, S., & Mandrini, C. H. 2008, *Solar Phys.*, 250, 347
Farrugia, C. J., Janoo, L. A., Torbert, R. B., *et al.* 1999, in Habbal, S. R., Esser, R., Hollweg, J. V., Isenberg, P. A. (eds.), *Solar Wind Nine, AIP Conf. Proc.*, Vol. 471, 745
Farrugia, C. J., Osherovich, V. A., & Burlaga, L. F. 1995, *J. Geophys. Res.*, 100, 12293
Gonzalez, W. D., Tsurutani, B. T., & Clúa de Gonzalez, A. L. 1999, Space Sci. Rev., 88, 529
Gulisano, A. M., Dasso, S., Mandrini, C. H., & Démoulin, P. 2005, JASTP, 67, 1761
Gulisano, A. M., Démoulin, P., Dasso, S., Ruiz, M. E., & Marsch, E. 2010, *Astron. Astrophys.*, 509, A39
Harra, L. K., Crooker, N. U., Mandrini, C. H., *et al.* 2007, *Solar Phys.*, 244, 95
Hidalgo, M. A., Cid, C., Vinas, A. F., & Sequeiros, J. 2002, *J. Geophys. Res.*, 107, 1002
Hu, Q. & Sonnerup, B. U. Ö. 2001, Geophys. Res. Lett., 28, 467

Jian, L. K., Russell, C. T., & Luhmann, J. G. 2011, *Solar Phys.*, 274, 321
Klein, L. W. & Burlaga, L. F. 1982, *J. Geophys. Res.*, 87, 613
Leitner, M., Farrugia, C. J., Möstl, C., *et al.* 2007, *J. Geophys. Res.*, 112, A06113
Lepping, R. P., Berdichevsky, D. B., Szabo, A., Arqueros, C., & Lazarus, A. J. 2003, *Solar Phys.*, 212, 425
Lepping, R. P., Wu, C.-C., Berdichevsky, D. B., & Szabo, A. 2011, *Solar Phys.*, 274, 345
Liu, Y., Luhmann, J. G., Huttunen, K. E. J., *et al.* 2008, Astrophys. J. Lett., 677, L133
Liu, Y., Richardson, J. D., & Belcher, J. W. 2005, *Planetary Spa. Sci.*, 53, 3
Longcope, D., Beveridge, C., Qiu, J., *et al.* 2007, *Solar Phys.*, 244, 45
Lundquist, S. 1950, *Ark. Fys.*, 2, 361
Luoni, M. L., Mandrini, C. H., Dasso, S., van Driel-Gesztelyi, L., & Démoulin, P. 2005, JASTP 67, 1734
Lynch, B. J., Gruesbeck, J. R., Zurbuchen, T. H., & Antiochos, S. K. 2005, *J. Geophys. Res.*, 110, A08107
Manchester, W. B. I., Gombosi, T. I., Roussev, I., *et al.* 2004, *J. Geophys. Res.*, 109, A02107
Mandrini, C. H., Nakwacki, M., Attrill, G., *et al.* 2007, *Solar Phys.*, 244, 25
Mandrini, C. H., Pohjolainen, S., Dasso, S., *et al.* 2005, *Astron. Astrophys.*, 434, 725
Masson, S., Démoulin, P., Dasso, S., & Klein, K.-L. 2012, *Astron. Astrophys.*, 538, A32
Möstl, C., Miklenic, C., Farrugia, C. J., *et al.* 2008, Annales Geophysicae, 26, 3139
Nakwacki, M., Dasso, S., Démoulin, P., Mandrini, C. H., & Gulisano, A. M. 2011, *Astron. Astrophys.*, 535, A52
Nakwacki, M., Dasso, S., Mandrini, C. H., & Démoulin, P. 2008, JASTP, 70, 1318
Owens, M. J., Merkin, V. G., & Riley, P. 2006, *J. Geophys. Res.*, 111, 3104
Qiu, J., Hu, Q., Howard, T. A., & Yurchyshyn, V. B. 2007, *Astrophys. J.*, 659, 758
Ravindra, B., Yoshimura, K., & Dasso, S. 2011, *Astrophys. J.*, 743, 33
Riley, P., Linker, J. A., Mikić, Z., *et al.* 2003, *J. Geophys. Res.*, 108, 1272
Rodriguez, L., Zhukov, A. N., Dasso, S., *et al.* 2008, *Annales Geophysicae*, 26, 213
Savani, N. P., Owens, M. J., Rouillard, A. P., *et al.* 2011, *Astrophys. J.*, 732, 117
Schmieder, B., Démoulin, P., Pariat, E., *et al.* 2011, *Adv. Spa. Res.*, 47, 2081
Shimazu, H. & Vandas, M. 2002, *Earth, Planets, and Space*, 54, 783
Vandas, M., Fischer, S., Dryer, M., Smith, Z., & Detman, T. 1995, *J. Geophys. Res.*, 100, 12285
Vandas, M. & Romashets, E. P. 2003, *Astron. Astrophys.*, 398, 801
Wang, C., Du, D., & Richardson, J. D. 2005, *J. Geophys. Res.*, 110, A10107

Discussion

Axel Brandenburg: Are there different signs of the helicity? In which hemisphere did you see positive helicity? Is the sign of the helicity of the ICME the same as in the active region?

Sergio Dasso: Yes, there are magnetic clouds with positive global helicity (right-handed) and there are others with negative one (left-handed). Active regions with positive helicity are typically observed in the southern solar hemisphere. The sign of the helicity of the clouds is the same as in the source active regions (e.g. Mandrini *et al.* 2005; Dasso *et al.* 2007; Longcope *et al.* 2007; Dasso *et al.* 2009; Nakwacki *et al.* 2011), with just a few exceptions in complex regions (e.g. Schmieder *et al.* 2011). In fact, the last minimum was more active in the southern hemisphere and, consequently, there were more clouds with positive helicity.

Comparative Magnetic Minima:
Characterizing quiet times in the Sun and Stars
Proceedings IAU Symposium No. 286, 2011
C. H. Mandrini & D. F. Webb, eds.

doi:10.1017/S1743921312004760

Coronal transients during two solar minima: their source regions and interplanetary counterparts

Hebe Cremades[1] **Cristina H. Mandrini**[2] **and Sergio Dasso**[3]

[1]UTN - Facultad Regional Mendoza/CONICET
Rodríguez 243, Ciudad, Mendoza M5502AJE, Argentina
email: hebe.cremades@frm.utn.edu.ar

[2]Instituto de Astronomía y Física del Espacio (CONICET-UBA) and FCEN (UBA)
CC. 67 Suc. 28, 1428, Buenos Aires, Argentina

[3]Instituto de Astronomía y Física del Espacio (CONICET-UBA) and FCEN-UBA
CC. 67 Suc. 28, 1428, Buenos Aires, Argentina

Abstract. We have investigated two full solar rotations belonging to two distinct solar minima, in the frame of two coordinated observational and research campaigns. The nearly uninterrupted gathering of solar coronal data since the beginning of the SOHO era offers the exceptional possibility of comparing two solar minima for the first time, with regard to coronal transients. This study characterizes the variety of outward-travelling transients observed in the solar corona during both time intervals, from very narrow jet-like events to coronal mass ejections (CMEs). Their solar source regions and ensuing interplanetary structures were identified and characterized. Multi-wavelength images from the space missions SOHO, Yohkoh and STEREO, and ground-based observatories were studied for coronal ejecta and their solar sources, while in situ data registered by the ACE spacecraft were inspected for interplanetary CMEs and magnetic clouds. Instrumental aspects such as dissimilar resolution, cadence, and fields of view are considered in order to discern instrumentally-driven disparities from inherent differences between solar minima.

Keywords. Sun: corona, Sun: coronal mass ejections (CMEs), Sun: activity, solar wind

---

1. Introduction

The Whole Sun Month (WSM) and the Whole Heliosphere Interval (WHI) consisted of two series of coordinated efforts carried out almost 12 years apart during two consecutive solar minima, covering the periods August 10 - September 8, 1996 and March 20 - April 16, 2008 respectively. The characterization and modeling of the large-scale solar minimum corona in connection to in-situ observations of the solar wind and interactions with Earth are among their main goals. The campaigns led to a wealth of in-depth studies of the solar corona's configuration during those solar minima (e.g., Gibson *et al.* 1999, Riley *et al.* 1999, Gibson *et al.* 2009, Gopalswamy *et al.* 2009, Landi & Young 2009). Only the two latter papers directly deal with coronal mass ejections (CMEs), key drivers of space weather; while only Gibson *et al.* (2009) have explored differences between two solar minima, i.e. between WSM and WHI, though focusing on solar wind high-speed streams.

This survey attempts to gain insight into intrinsic dissimilarities of the solar ejective aspect, during two solar rotations of two distinct but consecutive solar minima, whilst distinguishing from instrumentally-produced effects (see Cremades *et al.* 2011). It considers all distinguishable types of ejecta, from the most impressive wide and bright CMEs to the narrowest and faintest events, as well as their possible in situ counterparts.

2. Identification of events

The proposed approach to compare the ejective aspects of these two particular solar rotations requires the inspection of the coronagraph data available during those time intervals. These catalogs were consulted and taken as a basis: The SOHO LASCO CME Catalog at the CDAW Data Center, the list by O. C. St. Cyr at the SOHO LASCO NRL website, the COR1 CME Catalog at NASA GSFC, and the COR2-based CACTUS list of detections at the Royal Observatory of Belgium. Moreover, this study tends to complement independent surveys by Sterling (2010), Webb *et al.* (2010), and Webb *et al.* (2011). In particular, we have considered all kinds of "clustered" outward-traveling material in the white-light corona as coronal ejecta. This broad criterion includes not only bright, significant CMEs but also extremely faint ones, as well as thin and narrow jets. Data from the SOHO/LASCO C2 coronagraph were inspected for ejective activity both during WHI and WSM. The images have a pixel size of 11.2 arc sec and a spatial coverage of 2.2-6.0 Rs. During WSM, the practical cadence rarely overcame two images per hour, while the field of view (FOV) was frequently cropped from the full 1024×1024 pixels to 1024×576 centered on the Sun. During WHI, inspected data also included STEREO SECCHI COR1 and COR2 coronagraphs. The inner COR1 coronagraph covers 1.4-4.0 Rs with a pixel size of 7.5 arc sec, while the outer COR2 extends from 2.0 to 15 Rs with a pixel size of 14.7 arc sec. During WHI, COR1 and COR2 recorded images with a cadence of 20 and 30 minutes, respectively.

A total of 45 (in LASCO C2) and 143 (in LASCO C2, COR1A&B, and COR2A&B) were found respectively in each of these solar rotations. The number of ejective events identified by us in data of each instrument exceeded the numbers reported by the respective catalogs, likely because of the strict selection criteria we systematically followed. Out of the 143 events identified during WHI, 84 were observed by LASCO C2, 131 by COR1A and/or B, and 85 by COR2A and/or B.

3. Coronal events and candidate sources

The identified ejecta for both investigated time periods were classified according to their white-light appearance, into: i) CMEs, bright, significant, "conventional" events as defined by Hundhausen *et al.* (1984); ii) Faint CMEs, similar but weak compared to the background corona, hence frequently not reported by catalogs; iii) Jets, very narrow and fast ejecta (see e.g., St. Cyr *et al.* 1997); and iv) Streamer-swelling events, persistent outward flows of material, commonly at equatorial streamers.

Table 1 contains the number of ejective events identified during WSM and WHI, sorted according to the above categories. The number of all ejecta types during WSM is significantly lower than those during WHI, except for the "Streamer Swelling" kind. The latter seems to have been characteristic of Cycle 22's solar minimum, which portrayed well-formed polar coronal holes (CHs), an almost lack of low-latitude ones, and a streamer belt confined to equatorial latitudes. The situation during Cycle 23's solar minimum was radically different, implying a more complex global coronal structure, likely hindering persistent outflows of the streamer-swelling type. Jets are usually very fast events, prone to occur at polar position angles, and best detected at low heights and if travelling in the plane of the sky. LASCO C2's lower cadence and frequently cropped FOV at the poles thus made their detection difficult during WSM; while COR1's lowest threshold

Table 1. Types of ejecta identified during both time intervals.

	CME	Faint CME	Jet	Streamer Swelling
WSM	14	7	16	8
WHI	29	30	84	0

altitude and stereoscopic view allowed the detection of a large amount of jets during WHI. LASCO C2 Faint CMEs during WHI were double those during WSM, also likely due to the poorer cadence during WSM. Many of the faint CMEs are evident only when viewed in consecutive images, because the human eye is very sensitive to the motion of an organized structure rather than by the structure itself. Thus, having fewer images of a faint event implies that there is less of a chance of detecting it.

From the above analysis it could be generalized that the "conventional" CME rate during WHI was double that during WSM also because of instrumental differences between both periods. However, WSM cadence was good enough to detect fast CMEs, while the cropped FOV at the poles should not be a limitation since CMEs at WSM times typically travelled close to equatorial latitudes. The comparison between observations registered by the same instrument (LASCO C2) thus reveals an inherent difference in this category of ejecta, independent of instrumental issues.

Candidate source regions could be recognized for 27 out of the 45 ejective events identified during WSM, while candidate sources for the WHI period could be deduced for 89 of the 143 events. Low-coronal data were inspected for eruptive signatures: images from SOHO/EIT and Yohkoh/SXT were used during WSM, and from SOHO/EIT and STEREO SECCHI/EUVI during WHI. The latter extended the Sun's longitudinal coverage to almost 230°. $H\alpha$ data from the Global High-Resolution $H\alpha$ network were inspected for flares, filament disappearances, and erupting prominences.

After careful inspection of eruptive signatures, four types of source regions could be discerned: active regions (ARs), quiescent filaments, bipoles in quiet Sun locations, and bipoles within or at the boundary of coronal holes, both polar and low latitude. Table 2 summarizes the productivity of each of these during both time periods. Active regions appear to have played a major role as sources of white-light ejecta during WSM. Out of the 18 identified transients from ARs, nine seemed to originate from AR 7981, one in AR 7982, and eight in unnumbered ARs. However, during WHI all AR sources were numbered (10987-10990). On the quiet Sun, quiescent filament disappearances represent a small fraction of the identified candidate sources in our survey. The number of bipoles within/next to coronal holes as source candidates drastically increased during WHI with respect to WSM. This can be attributed to the low operational cadence of SOHO/EIT and Yohkoh/SXT, far from enough to detect such a fast episode as the launch of a jet, except for a few fortunate cases. Bipoles not associated with CHs were scarce, accounting for $\sim 18\%$ of jet sources. Jets are mainly produced by bipoles, and associated with ARs only in exceptional cases. Unidentified source regions represent doubtlessly the largest fraction. A smaller amount of unidentified sources during WHI is likely not only related to instrumental differences between both periods, but also to the fact that STEREO could survey a larger portion of the solar sphere, thus being able to observe part of the Sun's far side.

Table 2. Source region types observed during both time intervals.

	Active Region	Quiescent filament	CH-related bipole	Quiet Sun bipole	Unknown
WSM	18	1	5	1	20
WHI	17	12	49	11	54

4. Interplanetary Structures

In situ data from OMNI and from the Advanced Composition Explorer (ACE), respectively, were inspected during WSM and WHI to find solar wind structures potentially associated with the identified transients. The criteria to select interplanetary structures required the existence of magnetic field higher than the surroundings, low proton temperature, low plasma β, and large and coherent rotation of the magnetic field vector. We could ascertain seven candidates to transient interplanetary structures in the OMNI (Wind) data for the WSM period, while only five could be discerned in the ACE data for WHI. The identified structures consist of small flux ropes and even one magnetic cloud (MC) candidate during WHI. Unfortunately, it was not possible to identify their potential source regions at the Sun. The small magnitude of the events and the lack of obvious eruptions in the appropriate time windows, hindered possible associations. It is worth noting that the identified small flux ropes could have been locally generated within the solar wind due to magnetic reconnection across the heliospheric current sheet (Moldwin *et al.* 2000), thus not strictly solar in origin. Still, Feng *et al.* (2007) found continuous size and energy distributions between large and small flux ropes, suggesting a broad range of CMEs, from large and bright to weak ones, hard to detect with coronagraphs.

5. Conclusion

This analysis provides insight into the ejective aspect of two solar rotations during two consecutive solar minima. The high detection rate of ejective events is notably higher for the WHI period, explained by the improved cadence, resolution, and longitudinal coverage achieved by the SOHO and STEREO missions during WHI. However, the elevated number in the case of "conventional" CMEs cannot be accounted for by instrumentally-driven disparities, given the high contrast and extension exhibited by this type of events. Source region identification of the analyzed ejective events was less ambiguous during WHI, while in WSM poor spatial coverage and lower cadence introduced large uncertainties. Large active regions were present in both periods, though with no apparent impact on geomagnetic activity: geoactivity parameters were exceptionally quiet during WSM, while the connection with Earth during WHI was due to recurrent high speed streams (Gibson *et al.* 2009). Investigation of in situ data yielded no significant MC or interplanetary CME at 1 AU, though some candidate flux rope structures could be recognized in both rotations.

References

Cremades, H., Mandrini, C. H., & Dasso, S. 2011, *Solar Phys.*, 274, 233

Feng, H. Q., Wu, D. J., & Chao, J. K. 2007, *J. Geophys. Res.*, 112, A02102

Gibson, S. E., Biesecker, D., Guhathakurta, M., Hoeksema, J. T., Lazarus, A. J., Linker, J., Mikic, Z., Pisanko, Y., Riley, P., & Steinberg, J., *et al.* 1999, *ApJ*, 520, 871

Gibson, S. E., Kozyra, J. U., de Toma, G., Emery, B. A., Onsager, T., & Thompson, B. J. 2009, *J. Geophys. Res.*, 114, A09105

Gopalswamy, N., Thompson, W. T., Davila, J. M., Kaiser, M. L., Yashiro, S., Mäkelä, P., Michalek, G., Bougeret, J.-L., & Howard, R. A. 2009, *Solar Phys.*, 259, 227

Hundhausen, A. J., Sawyer, C. B., House, L., Illing, R. M. E., & Wagner, W. J. 1984, *J. Geophys. Res.*, 89, 2639

Landi, E. & Young, P. R. 2009, *ApJ*, 707, 1191

Moldwin, M. B., Ford, S., Lepping, R., Slavin, J., & Szabo, A. 2000, *Geophys. Res. Lett.*, 27, 57

Riley, P., Gosling, J. T., McComas, D. J., Pizzo, V. J., Luhmann, J. G., Biesecker, D., Forsyth, R. J., Hoeksema, J. T., Lecinski, A., & Thompson, B. J. 1999, *J. Geophys. Res.*, 104, 9871

St. Cyr, O. C., Howard, R. A., Simnett, G. M., Gurman, J. B., Plunkett, S. P., Sheeley, N. R., Schwenn, R., Koomen, M. J., Brueckner, G. E., & Michels, D. J. 1997, in: Wilson, A. (ed.), *31st ESLAB Symp.* (SP-415, ESA, Noordwijk), p. 103

Sterling, A. 2010, in: Corbett, I. F. (ed.), *Whole Heliosphere Interval: Overview of JD16, Highlights of Astronomy* (15, IAU, Cambridge Univ. Press), p. 498

Webb, D. F., Gibson, S. E., & Thompson, B. J. 2010, in: Corbett, I. F. (ed.), *Whole Heliosphere Interval: Overview of JD16, Highlights of Astronomy* (15, IAU, Cambridge Univ. Press), p. 471

Webb, D. F., Cremades, H., Sterling, A. C., Mandrini, C. H., Dasso, S., Gibson, S. E., Haber, D. A., Komm, R. W., Petrie, G. J. D.., McIntosh, P. S., Welsch, B. T., & Plunkett, S. P. 2011, *Solar Phys.*, 274, 57

Comparative Magnetic Minima:
Characterizing quiet times in the Sun and Stars
Proceedings IAU Symposium No. 286, 2011
C. H. Mandrini & D. F. Webb, eds.

doi:10.1017/S1743921312004772

Coronal ejections from convective spherical shell dynamos

J. Warnecke[1,2] P. J. Käpylä[1,3] M. J. Mantere[3] and A. Brandenburg[1,2]

[1]Nordita, Roslagstullsbacken 23, SE-10691 Stockholm, Sweden, email: joern@nordita.org
[2]Department of Astronomy, Stockholm University, SE-10691 Stockholm, Sweden
[3]Department of Physics, PO BOX 64, FI-00014 Helsinki University, Finland

Abstract. We present a three-dimensional model of rotating convection combined with a simplified model of a corona in spherical coordinates. The motions in the convection zone generate a large-scale magnetic field which is sporadically ejected into the outer layers above. Our model corona is approximately isothermal, but it includes density stratification due to gravity.

Keywords. MHD, Sun: magnetic fields, Sun: coronal mass ejections (CMEs), turbulence

1. Introduction

The Sun sheds plasma into the heliosphere via coronal mass ejections (CMEs). There has been significant progress in the study of CMEs in recent years. In addition to improved observations from spacecrafts like SDO or STEREO, there have also been major advances in the field of numerical modeling of CME events. One of the main motivations for understanding the generation and dynamics of CMEs is to have more reliable predictions for space weather. CMEs can have strong impacts on Earth and can affect microelectronics aboard spacecrafts. However, an important side effect of CMEs is that through them the Sun sheds magnetic helicity from the convection zone which may prevent the solar dynamo from being quenched at high magnetic Reynolds numbers (Blackman & Brandenburg, 2003).

In many CME models, footpoint motions of the magnetic field in the photosphere are taken from two-dimensional observations at the surface. This is an approximation to the full three-dimensional field generated by a turbulent dynamo. An alternative way of modeling CMEs would be to perform a 3-D convection simulation to generate the magnetic field and the photospheric motions self-consistently. However, convection zone and corona have very different timescales. In solar convection the dominant timescale varies from minutes to days. While this is short compared with the dynamo cycle, timescales in the solar corona can be even shorter because the Alfvén speed is large.

In earlier work (Warnecke & Brandenburg 2010, Warnecke *et al.* 2011) we have established a two-layer model with a unified treatment of the convection zone and the solar corona in a single three-dimensional domain. In those models, magnetic fields were produced by turbulence from random helical forcing mimicking the effects of convection and rotation. This model was able to produce recurrent plasmoid ejections which are similar to observed eruptive features on the Sun. In Warnecke *et al.* (2012) we have developed this approach further and considered self-consistent convection instead, where differential rotation arises from the interaction of rotation and convection. Here we present some preliminary results of this study. We find the formation of a large-scale magnetic field, which is eventually ejected into the corona. This mechanism could play an important role for the formation of CMEs and flares.

2. The model

As in Warnecke & Brandenburg (2010) and Warnecke *et al.* (2011) a two-layer model is used. Our convection zone is similar to that in Käpylä *et al.* (2010, 2011). The domain is a segment of the Sun and is described in spherical polar coordinates (r, θ, ϕ). We mimic the convection zone starting at radius $r = 0.7\,R$ and the solar corona until $r = 1.5\,R$, where R denotes the solar radius, used from here on as our unit length. In the latitudinal direction, our domain extends from colatitude $\theta = 15^\circ$ to 165° and in the azimuthal direction from $\phi = 0^\circ$ to 90°. We solve the equations of compressible magnetohydrodynamics,

$$\frac{\partial \boldsymbol{A}}{\partial t} = \boldsymbol{U} \times \boldsymbol{B} + \eta \nabla^2 \boldsymbol{A}, \tag{2.1}$$

$$\frac{\mathrm{D} \ln \rho}{\mathrm{D}t} = -\boldsymbol{\nabla} \cdot \boldsymbol{U}, \tag{2.2}$$

$$\frac{\mathrm{D}\boldsymbol{U}}{\mathrm{D}t} = \boldsymbol{g} - 2\boldsymbol{\Omega} \times \boldsymbol{U} + \frac{1}{\rho}\left(\boldsymbol{J} \times \boldsymbol{B} - \boldsymbol{\nabla} p + \boldsymbol{\nabla} \cdot 2\nu\rho\mathbf{S}\right), \tag{2.3}$$

$$T\frac{\mathrm{D}s}{\mathrm{D}t} = \frac{1}{\rho}\boldsymbol{\nabla} \cdot K\boldsymbol{\nabla} T + 2\nu\mathbf{S}^2 + \frac{\mu_0\eta}{\rho}\boldsymbol{J}^2 - \Gamma_{\mathrm{cool}}, \tag{2.4}$$

where the magnetic field is given by $\boldsymbol{B} = \boldsymbol{\nabla} \times \boldsymbol{A}$ and thus obeys $\boldsymbol{\nabla} \cdot \boldsymbol{B} = 0$ at all times. The vacuum permeability is given by μ_0, whereas magnetic diffusivity and kinematic viscosity are given by η and ν, respectively. $\mathsf{S}_{ij} = \frac{1}{2}(U_{i;j} + U_{j;i}) - \frac{1}{3}\delta_{ij}\boldsymbol{\nabla} \cdot \boldsymbol{U}$ is the traceless rate-of-strain tensor, and semicolons denote covariant differentiation, $\boldsymbol{\Omega} = \Omega_0(\cos\theta, -\sin\theta, 0)$ is the rotation vector, K is the radiative heat conductivity and $\boldsymbol{g} = -GM\boldsymbol{r}/r^3$ is the gravitational acceleration. The fluid obeys the ideal gas law with $\gamma = 5/3$ being the ratio of specific heats. We consider a setup in which the stratification is convectively unstable below $r = R$, whereas the region above is stably stratified and isothermal due to a cooling term Γ_{cool} in the entropy equation. The Γ_{cool} term is r dependent and causes a smooth transition to the isothermal layer representing the corona.

The simulation domain is periodic in the azimuthal direction. For the velocity we use stress-free conditions at all other boundaries. For the magnetic field we adopt radial field conditions on the $r = 1.5\,R$ boundary and perfect conductor conditions on the $r = 0.7\,R$ and both latitudinal boundaries. Time is expressed in units of $\tau = (u_{\mathrm{rms}}k_{\mathrm{f}})^{-1}$, which is the eddy turnover time in the convection zone. We employ the Pencil Code†, which uses sixth-order centered finite differences in space and a third-order Runge-Kutta scheme in time; see Mitra *et al.* (2009) for the extension to spherical coordinates.

3. Results

In this work we focus on a run which has fluid Reynolds number $\mathrm{Re} = 3$, magnetic Reynolds number $\mathrm{Re}_M = 32$ and Coriolis number $\mathrm{Co} = 7$. We define the fluid and magnetic Reynolds number as $\mathrm{Re} = u_{\mathrm{rms}}/\nu k_{\mathrm{f}}$ and $\mathrm{Re}_M = u_{\mathrm{rms}}/\eta k_{\mathrm{f}}$, respectively, and the Coriolis number as $\mathrm{Co} = 2\Omega_0/u_{\mathrm{rms}}k_{\mathrm{f}}$. After around 100 turnover times, the onset of large-scale dynamo action due to the convective motions is observed. The magnetic field reacts back on the fluid motions and causes saturation after around 200 turnover times. The saturation is combined with an oscillation of the magnetic field strength in the convection zone. The field reaches its maximum strength of about 60% of the equipartition field strength, $B_{\mathrm{eq}} = (\mu_0\overline{\rho\boldsymbol{u}^2})^{1/2}$, which is comparable with the values obtained in the forced turbulence counterparts both in Cartesian and spherical coordinates (Warnecke & Brandenburg 2010, Warnecke *et al.* 2011). The magnetic field in rotating convection

† `http://pencil-code.googlecode.com`

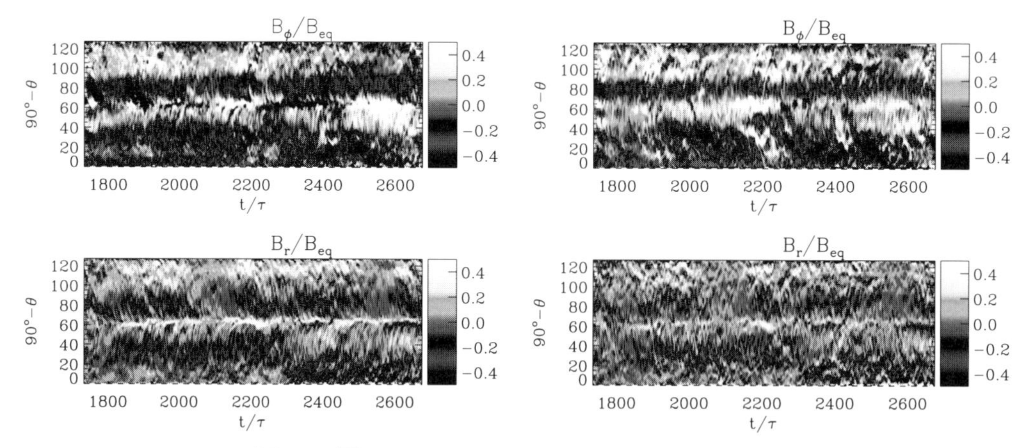

Figure 1. Variation of $\overline{B}_\phi$ and $\overline{B}_r$ in the convection zone at $r = 0.89R$ (left panel) and $r = 0.79R$ (right). Dark blue shades represent negative and light yellow positive values. The dotted horizontal lines show the location of the equator at $\theta = \pi/2$. The magnetic field is normalized by the equipartition value.

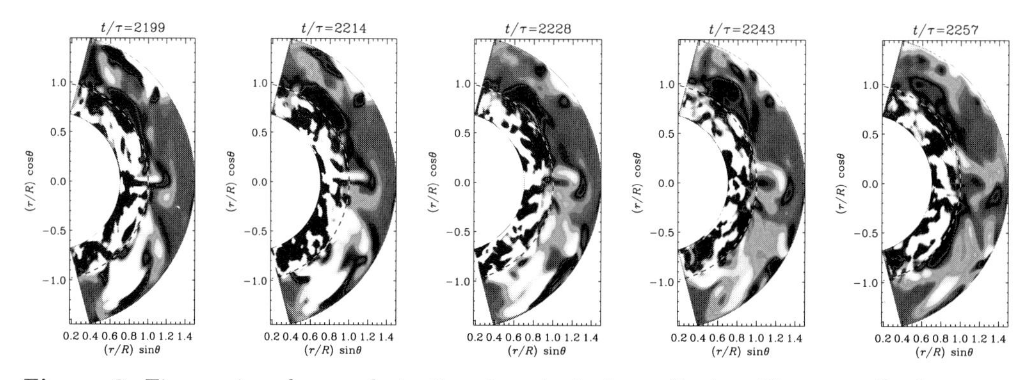

Figure 2. Time series of coronal ejections in spherical coordinates. The normalized current helicity, $\mu_0 R\overline{\boldsymbol{J}\cdot\boldsymbol{B}}/\langle\overline{\boldsymbol{B}^2}\rangle_t$, is shown in a color-scale representation at different times; dark blue represents negative and light yellow positive values. The dashed horizontal lines show the location of the surface at $r = R$. Adapted from Warnecke *et al.* (2012).

seems to show certain migration properties. In Figure 1, we show the azimuthal ($\overline{B}_\phi$) and radial ($\overline{B}_r$) magnetic fields versus time (t/τ) and latitude ($90° - \theta$) for two different heights. The magnetic field emerges through the surface and is ejected as isolated structures. The dynamical evolution is clearly seen in the sequence of images of Figure 2, where the normalized current helicity ($\mu_0 R\overline{\boldsymbol{J}\cdot\boldsymbol{B}}/\langle\overline{\boldsymbol{B}^2}\rangle_t$) is shown. If one focuses on the region near the equator ($\theta = \pi/2$), a small yellow (i.e. positive) feature with a blue (negative) arch emerges through the surface to the outer atmosphere, where it leaves the domain through the outer boundary. This ejection does not occur as a single event—it rather shows recurrent behavior. We do not, however, find a clear periodicity in the ejection recurrence, like in earlier work. Even though the ejected structures are much smaller than in Warnecke *et al.* (2011), their shape is similar. However, the detection of an ejection with the aid of the current helicity is much more difficult in convection-driven simulations than in forced turbulence. In Figure 2, one can see large structures diffusing through the surface into the upper atmosphere at higher latitudes. These structures disturb the emergence of ejections and hamper their detection. These larger diffusive structures are also visible in Figure 3, where the normalized current density is averaged over two narrow

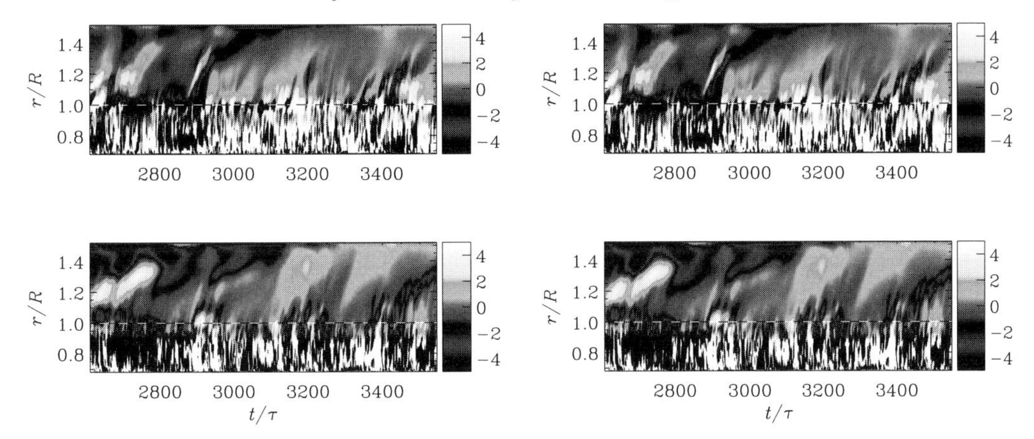

Figure 3. Dependence of the dimensionless ratio $\mu_0 R\overline{\boldsymbol{J}\cdot\boldsymbol{B}}/\langle\overline{\boldsymbol{B}^2}\rangle_t$ on time t/τ and radius r in terms of the solar radius. The top panels show a narrow band in θ in the northern hemisphere and the bottom ones in the southern hemisphere. We have also averaged in latitude from 4° to 20° (left panel) and 33° to 46° (right). Dark blue shades represent negative and light yellow positive values. The dotted horizontal lines show the location of the surface at $r = R$.

latitude bands on each hemisphere. The formation of these diffuse structures in the corona seem to get suppressed when the stratification of the system is increased. Nevertheless, ejections are still visible, for example around $t/\tau = 1900$, $t/\tau = 2200$ (see Figure 2) and $t/\tau = 2400$.

In Warnecke *et al.* (2011) the sign of the current helicity in the northern (southern) hemisphere was negative (positive) in the interior and positive (negative) in the coronal part. However, in Figure 3 the current helicity shows different behaviors in the northern and southern hemispheres, and one cannot tell clearly the leading sign. This has to do with the much lower values of relative kinetic helicity $h_{\rm rel}(r,t) = \langle\boldsymbol{\omega}\cdot\boldsymbol{u}\rangle/\omega_{\rm rms}u_{\rm rms}$ in the convection simulations. Values of up to $h_{\rm rel} = \mp 0.4$ are reached at some radii in the two hemispheres. In the forced turbulence simulations of our earlier work, we studied purely helical systems with nearly $h_{\rm rel} = \mp 1$.

In summary, we have been able to advance our two-layer model approach by including self-consistent convection, which generates the magnetic field and eventually drives ejections. Unlike our earlier work, the ejections occur now non-periodically and through smaller structures, which is now closer to the behavior displayed by the Sun. Furthermore, detailed investigations covering a wider range of magnetic and kinetic Reynolds number as well as rotation rates show promising results; see Warnecke *et al.* (2012) for details.

References

Blackman, E. G. & Brandenburg, A. 2003, ApJ, 584, L99
Käpylä, P. J., Korpi, M. J., Brandenburg, A., Mitra, D., & Tavakol, R. 2010, *AN*, 331, 73
Käpylä, P. J., Korpi, M. J., Guerrero, G., Brandenburg, A., & Chatterjee, P. 2011, A&A, 531, A162
Mitra, D., Tavakol, R., Brandenburg, A., & Moss, D. 2009, ApJ, 697, 923
Warnecke, J. & Brandenburg, A. 2010, A&A, 523, A19
Warnecke, J., Brandenburg, A., & Mitra, D. 2011, A&A, 534, A11
Warnecke, J., Käpylä, P. J., Mantere, M. J., & Brandenburg, A. 2012, *Sol. Phys.* (in press, arxiv:1112.0505)

Discussion

Daniel Gómez: Why do you use the current helicity and not the magnetic helicity?

Jörn Warnecke: The spectra of current helicity and magnetic helicity are related to each other by a k^2 factor, where k is the wavenumber. The current helicity integrated over all wavenumbers is equal to the integrated magnetic helicity spectrum weighed toward large k. Current helicity can therefore be regarded as a proxy of magnetic helicity at small scales. Furthermore, unlike magnetic helicity, current helicity is gauge-invariant and therefore a physically meaningful quantity.

Janet Luhmann: There are different flavors of CMEs. Which ones do you model here: streamers, blow outs? Comment: There is very clear cycling of the helicity of the activity clouds. It would be interesting to follow these results to complement your model.

Jörn Warnecke: We do not model specific CMEs. Rather, the objective of this project is to link dynamo-generated magnetic fields to driving ejecta from its surface. Our model is not sufficiently realistic to make meaningful comparisons with detailed observations. Nevertheless, the visual similarity with actual CMEs inspires us to look more carefully at certain features that might relate our ejecta with CMEs.

Francesco Zuccarello: Why are some CMEs deflected?

Jörn Warnecke: Even though the boundary condition for the magnetic field is an open one (vertical field condition), the condition for the velocity is a closed one (stress free). So the mass associated with CMEs cannot get out and must fall back. We are currently working on open boundary conditions allowing mass to escape. This would also allow a solar wind to develop.

Jon Linker: For CMEs on the Sun associated with active regions: you do sometimes see a CME that occurs right after an active region emerges, but very often the active region hangs around and decays before it produces a CME. It sort of appears in your model that you see the eruption of the CME through the photosphere and the whole thing is blown away completely. I am wondering if there needs to be more restraining (coronal) field there to begin with, otherwise many of these so called CMEs might just be emergence of an active region.

Jörn Warnecke: It is right, that we do not include constraining coronal fields in our models. In addition our density stratification and the radiation effects of the photosphere and chromosphere are very simplified. Nevertheless, it is interesting to interpret our results as the emergence of an active region, and we shall look more closely into this.

Comparative Magnetic Minima:
Characterizing quiet times in the Sun and stars
Proceedings IAU Symposium No. 286, 2011
C. H. Mandrini & D. F. Webb, eds.

doi:10.1017/S1743921312004784

Dynamic evolution of interplanetary shock waves driven by CMEs

P. Corona-Romero[1,2] and J. A. Gonzalez-Esparza[2]

[1]Posgrado en Ciencias de la Tierra, Universidad Nacional Autonoma de Mexico,
Av. Universidad 2000, Mexico City, Mexico.
email: piter.cr@gmail.com

[2]Insituto de Geofisica Michoacan, Universidad Nacional Autonoma de Mexico
Tzintzuntzan 310, Morelia, Mexico.
email: americo@geofisica.unam.mx

Abstract. We present a study about the propagation of interplanetary shock waves driven by super magnetosonic coronal mass ejections (CMEs). The discussion focuses on a model which describes the dynamic relationship between the CME and its driven shock and the way to approximate the trajectory of shocks based on those relationships, from near the Sun to 1 AU. We apply the model to the analysis of a case study in which our calculations show quantitative and qualitative agreements with different kinds of data. We discuss the importance of solar wind and CME initial conditions on the shock wave evolution.

Keywords. Sun: coronal mass ejections (CMEs), Shock waves, Sun: activity, Methods: analytical.

1. Introduction

A shock wave is an energetic compressive perturbation which propagates faster than the characteristic speed of the medium. An interplanetary shock wave transfers energy to the solar wind increasing its entropy and kinetic energy. From a macroscopic perspective, a shock wave can be regarded as a discontinuity which separates two distinct fluids whose properties are related by the so called jump conditions (Landau & Lifshitz, 2005). Interplanetary shocks associated with CMEs are an important issue due to their impact on the Earth magnetic conditions and increase our contingency plans.

In Heliophysics there are two main conditions in which a shock wave evolves: driven and decaying. During the first one, the shock energy transferred to the medium is restored by a driver. On the other hand, during the decaying, the shock constantly loses energy by transferring it to the medium. Corona-Romero & Gonzalez-Esparza (2011) studied the dynamical evolution of shocks associated with CMEs using numerical simulations and analytical models. They found that the CME-shock evolution presents three different dynamic phases: driving, decoupling, and decaying. These dynamic phases are defined by the evolution of the linear momentum between the CME and the shock wave, i.e. the linear momentum flux through the plasma sheath (ambient wind modified by the shock propagation).

Figure 1 shows a sketch of the CME-shock system at the driving (a) and decaying (b) phases. During the driving phase the CME transfers momentum to the sheath, which results in a strong compression of the plasma sheath (darker gray region in panel a). On the other hand, during the decaying phase the CME is no longer transferring momentum to the sheath. Thus, during the decaying phase, the compression on the sheath decreases (panel b) and the distance between the CME and the shock increases. The decoupling process bounds these two opposite behaviors, during which the dynamic disconnection between driver (CME) and shock wave occurs by sheath relaxation.

2. CME-shock propagation

We can approximate the trajectory of shocks and CMEs on the basis of the three phases introduced above. In Figure 2 we present a comparison between the data and the calculations of the speeds (left panel) and heliocentric positions (right panel) for the May 13 2005 CME-shock event. The speed-position temporal evolution was studied previously by Bisi *et al.* (2010, and references there in) by combining different measurements such as white light coronagraph images (diamonds), the frequency drift of kilometric type II spectrum (triangles), interplanetary scintillation (squares and asterisks), and in situ spacecraft observations (crosses). Though the data do not refer to the same structure, they are related to the same phenomena (CME−shock event) and give us an approximation of the CME and shock propagations.

At the earlier moments of the CME and shock propagation, during the driving phase, the CME-shock configuration can be approximated by a quasi-spherical figure which drives a bow shock (Ontiveros & Vourlidas, 2009). In this phase we approximate the CME-shock configuration by quasi-stationary bow shock. In this case we observe two characteristics: (1) the standoff distance (shortest distance between the CME and the shock, see Figure 1a) is a function of the magneto-sonic Mach number and the CME geometry, and (2) the standoff distance is almost constant, i.e. the CME and the shock share the same speed. These characteristics can be appreciated in Figure 2a, where CME and shock speeds appear similar (between the solid and dashed vertical gray lines) and, thus, the distance between them is constant. The duration of this period highly depends on the CME initial kinetic energy and on the solar wind conditions between them (Corona-Romero & Gonzalez-Esparza, 2011).

It is well accepted that fast CMEs arrive at 1 AU with slower speeds than their initial coronagraph counterparts (Gopalswamy, 2000), i.e. in general fast CMEs decelerate in the interplanetary medium. Once a fast CME decelerates, it is not able to transfer momentum

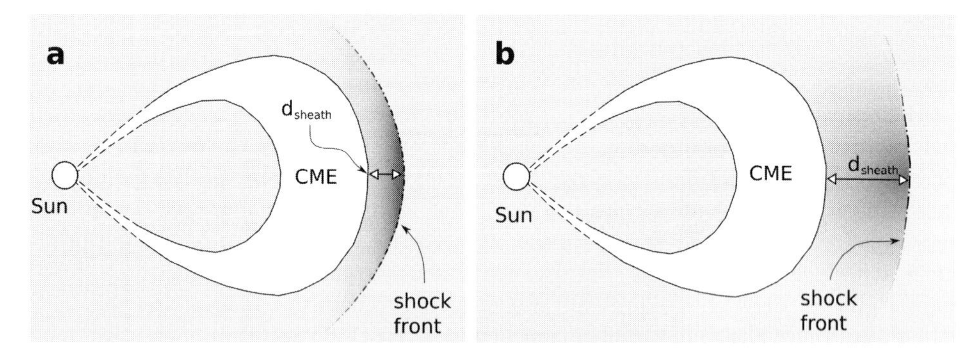

Figure 1. Sketch of a fast CME-shock system propagating through the interplanetary medium showing the driving (a) and decaying (b) configurations. From left to right, the Sun (white circle), the solar wind (light gray), the CME (white croissant-like shape), the plasma sheath or shocked solar wind (darker gray) and the shock front (dashed-dotted black line), respectively. The gray color shades represent the density of the ambient solar wind, being the darker regions (plasma sheath) the denser ones. (a) The CME drives the shock front (the dashed-dotted black line is clearly defined), the sheath is highly compressed and the standoff distance (d_{sheath}) is constant. (b) The shock is getting dissipated (the dashed-dotted black line is vanishing), the compression on the sheath material decreases and the standoff distance grows.

to the sheath, circumstance which triggers the decoupling process. In this process, with the absence of the CME momentum flux, the plasma sheath begins to relax; such a relaxation propagates throughout the sheath reaching the shock front. In Figure 2 (left panel) we see that along the decoupling process (limited by the dashed and dotted-dashed vertical gray lines) the shock preserves its driving speed and together with the CME deceleration produces an increasing in the plasma sheath's volume (with the standoff distance growing). The duration of this phase is proportional to the quotient of the driving standoff distance and the magneto-sonic speed inside the plasma sheath.

When the relaxation has propagated all over the plasma sheath width (the standoff distance) the shock speed begins to decrease, denoting the start of the decaying phase. During the latter phase, the CME is still decelerating, tending to reach the ambient solar wind speed and the shock decays into a magneto-sonic perturbation. In Figure 2, (a) and (b) panels, we can identify two characteristics in the decaying phase: (1) the standoff distance keeps increasing with time (the shock is faster than the CME) and (2) both velocities are lower than those of their driving counterparts.

3. Comparison with observations

According with our dynamic model, during the driving phase and the decoupling process the shock front propagates with the CME leading edge speed. On the other hand, during the decaying phase the shock heliocentric position (r_{shock}) and speed (v_{shock}) decay as $r_{shock} \propto t^{2/3}$ and $v_{shock} \propto r_{shock}^{-1/2}$, respectively, with t the time (Cavaliere & Messina, 1976). The previous expressions have been proposed by Smart & Shea (1985)

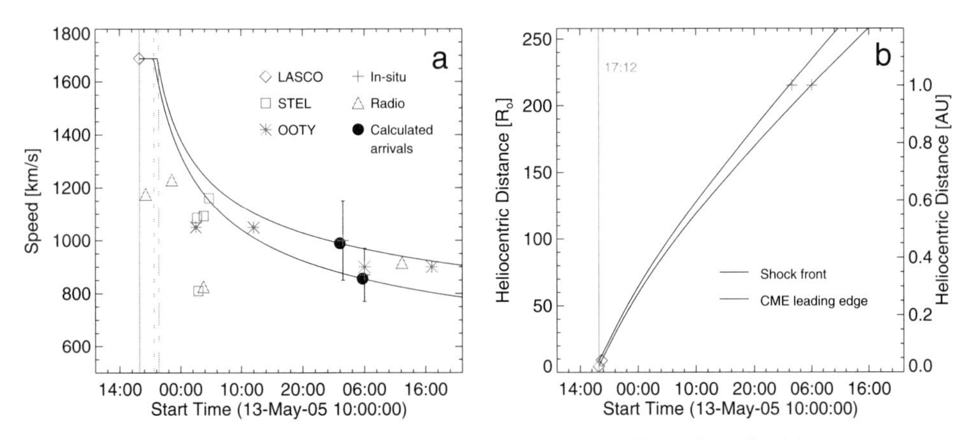

Figure 2. Comparison between the results of the analysis and the data for May 13 2005 event. (a) Calculated CME and shock speeds (black lines) vs. time. (b) Calculated CME and shock heliocentric distances vs. time (black lines). The different symbols in panels (a) and (b) indicate data from coronagraph (LASCO) images, speeds inferred from radio emission, in situ and interplanetary scintillation (STEL and OOTY) measurements. The driving phase lasts 2.38 hours (dashed vertical gray line, panel a) and the decoupling process 0.78 hours (dotted-dashed vertical gray line, panel a). To calculate the CME's nose trajectory we follow the description by Corona-Romero & Gonzalez-Esparza (2011) with $\Delta t = 0.50\ h$, $a = 5.01$, $c = 8$, $r_0 = 4.57\ R_o$ and $v_{0cme} = 1689\ km/s$. The ambient wind conditions at 1 AU were $n_p = 4\ cc$, $T_p = 130 \times 10^3\ K$, $|B| = 6\ nT$, and $v_{sw} = 420\ km/s$, respectively. The shock front propagation was calculated according the respective solution by Cavaliere & Messina (1976). The driving standoff distance and the CME radius were calculated using the models of Farris & Russell (1994) and Bothmer & Schwenn (1998), respectively.

and corroborated by Pintér & Dryer (1990) in order to calculate transit times and to describe the propagation of interplanetary shocks associated with solar flares.

In this case study we find reasonable agreement between the data and the calculated positions and speeds. In Figure 2 we see that in situ (gray crosses) arrival speeds (panel a) and travel times (panel b) are quantitatively similar to their analytical counterparts (black circles). On the other hand, the interplanetary scintillation (OOTY and STEL) and the radio burst data show qualitative agreements with the predicted trajectory of the sheath (limited by the solid and dashed black lines).

The shock wave is not driven by the CME when it arrives to the Earth's orbit. According to the data, the CME speed at the solar corona ($\approx 1690\ km/s$) is considerably slower than its 1 AU counterpart ($\approx 800\ km/s$). On the other hand, the shock front speed at 1 AU ($\approx 1100\ km/s$) is faster than the CME, but it is still slower than the initial speed of the CME. Both conditions are consistent with the signatures for the decaying phase.

Transit times (panel b) and arrival speeds (panel a) matched with in situ data and speed profiles were consistent with the different sets of data (panel b). According to our results, the driving phase and decoupling process last 3.2 hours approximately, while the CME-shock transit time to 1 AU is larger than 30 hours. Thus, the CME-shock evolution, in this case study, spends most of its transit time in the decaying stage.

4. Summary

We described the dynamical phases (driving, decoupling and decaying) that might be present in the fast CME-interplanetary shock propagation. We also analyzed a case study by applying the dynamical phase descriptions. Our results showed quantitative and qualitative agreements with different kinds of data. The three-phases dynamical evolution described above may help to approximate the trajectory, transit times, and arrival speeds of interplanetary shocks associated with fast CMEs. In order to do so we require the initial conditions of the CME and solar wind, which are determinant on the shock evolution.

Acknowledgments

Pedro Corona Romero acknowledges the grant from CONACyT and the IAU support. J. A. Gonzalez Esparza thanks the support of the CONACyT 48494 and PAPIIT IN105310-3 projects.

References

Bisi, M. M., Breen, A. R., Jackson, B. V., Fallows, R. A., Walsh, A. P., Mikić, Z., Riley, P., Owen, C. J., Gonzalez-Esparza, A., Aguilar-Rodriguez, E., Morgan, H., Jensen, E. A., Wood, A. G., Owens, M. J., Tokumaru, M., Manoharan, P. K., Chashei, I. V., Giunta, A. S., Linker, J. A., Shishov, V. I., Tyul'Bashev, S. A., Agalya, G., Glubokova, S. K., Hamilton, M. S., Fujiki, K., Hick, P. P., Clover, J. M., & Pintér, B. 2010, *Solar Phys.*, 265, 49

Bothmer, V. & Schwenn, R. 1998, *Ann. Geophysicae*, 16, 1

Cavaliere, A. & Messina, A. 1976, *ApJ*, 209, 424

Corona-Romero, P. & Gonzalez-Esparza, J. A. 2011, *JGR*, 116, A05104

Farris, M. H. & Russell, C. T. 1994, *JGR*, 99, 17681

Gopalswamy, N., Lara, A., Lepping, R. P., Kaiser, M. L., & Berdichevsky, D., St. Cyr, O. C. 2000, *GRL*, 27, 145

Landau, D. L. & Lifshitz, M. E. 2005, *Fluid Mechanics, Course of Theoretical Physics Vol. 6* (Edit. Elsevier).

Ontiveros, V. & Vourlidas, A. 2009, *ApJ*, 693, 267
Pintér, S. & Dryer, M. 1990, *Bull. Astron. Inst. Czechosl.*, 41, 137
Smart, D. F. & Shea, M. A. 1985, *JGR*, 90, 183

Discussion

DANIEL GÓMEZ: In MHD there are three types of shocks. What happens if the CME decelerates?

PEDRO CORONA-ROMERO: In general fast CMEs are associated with fast MHD shocks. When a fast CME begins to decelerate, its role as shock driver ends. Thus, the shock begins to decay and its speed decreases. The important point is to understand how a shock dissipates: Does it change from a fast to a slow shock and finally to an MHD wave? Does a fast shock directly dissipate into an MHD wave? Does turbulence play an important role? But those are more likely statistical physics topics than MHD ones.

Comparative Magnetic Minima:
Characterizing quiet times in the Sun and Stars
Proceedings IAU Symposium No. 286, 2011
C. H. Mandrini & D. F. Webb, eds.

doi:10.1017/S1743921312004796

Dynamical evolution of anisotropies of the solar wind magnetic turbulent outer scale

M. E. Ruiz[1], S. Dasso[1,2], W. H. Matthaeus[3], E. Marsch[4] and J. M. Weygand[5]

[1]Instituto de Astronomía y Física del Espacio (CONICET-Universidad de Buenos Aires), C.C. 67, Sucursal 28, 1428, Buenos Aires, Argentina. email: meruiz@iafe.uba.ar

[2]Departamento de Física, Facultad de Ciencias Exactas y Naturales, Universidad de Buenos Aires, Pabellón 1 (1428), Buenos Aires, Argentina. email: dasso@df.uba.ar

[3]Bartol Research Institute, Department of Physics and Astronomy, University of Delaware, Newark, DE, USA. email: whm@udel.edu

[4]Max-Planck-Institut für Sonnensystemforschung, Max-Planck-Straße 2, Katlenburg-Lindau, Germany. email: marsch@mps.mpg.de

[5]Institute of Geophysics and Planetary Physics, University of California, Los Angeles, CA, USA. email: jweygand@igpp.ucla.edu

Abstract. The evolution of the turbulent properties in the solar wind, during the travel of the parcels of fluid from the Sun to the outer heliosphere still has several unanswered questions. In this work, we will present results of an study on the dynamical evolution of turbulent magnetic fluctuations in the inner heliosphere. We focused on the anisotropy of the turbulence integral scale, measured parallel and perpendicular to the direction of the local mean magnetic field, and study its evolution according to the aging of the plasma parcels observed at different heliodistances. As diagnostic tool we employed single-spacecraft correlation functions computed with observations collected by Helios 1 & 2 probes over nearly one solar cycle. Our results are consistent with driving modes with wave-vectors parallel to the direction of the local mean magnetic field near the Sun, and a progressive spectral transfer of energy to modes with perpendicular wave-vectors. Advances made in this direction, as those presented here, will contribute to our understanding of the magnetohydrodynamical turbulence and Alfvénic-wave activity for this system, and will provide a quantitative input for models of charged solar and galactic energetic particles propagation and diffusion throughout the inner heliosphere.

Keywords. Solar wind, turbulence, magnetohydrodynamics.

1. Introduction

The solar wind (SW) is a turbulent plasma and the fluctuations of their bulk properties are ussually studied within the framework of magnetohydrodynamic (MHD) turbulence.

As originally proposed by Belcher & Davis (1971), it is often considered that the main source of the interplanetary fluctuations is near the Sun, below the alfvénic critical point. Near 0.3 astronomical unit (AU) from the Sun, fluctuations are observed to be highly alfvénic with the sign of the cross helicity indicating mainly outward propagation from the Sun. Nevertheless, Coleman's (1968) point of view of an evolving turbulent SW is essential for the evolution of the turbulence throughout the heliosphere. Velocity gradients at large scales can drive local nonlinear instabilities that inject kinetic energy only (i.e., injection of zero cross helicity turbulence, leading to a decrease of cross helicity with heliodistance and to a state of well developed turbulence (Roberts *et al.* 1992).

Solar wind turbulence is anisotropic with respect the mean magnetic field $\mathbf{B}_0$ as has been suggested in many studies (Robinson & Rusbridge 1971; Shebalin *et al.* 1983;

Oughton *et al.* 1974; Goldreich & Sridhar 1995). The simplest models commonly used for the description of anisotropic SW fluctuations are the slab model, where fluctuations have wavevectors parallel to $\mathbf{B}_0$, and the 2D model where fluctuations have wavevectors perpendicular to $\mathbf{B}_0$. These models, although very simplified, provide a useful parametrization of anisotropy in SW turbulence, in the sense that all wavevector contributions can be grouped into these two categories.

The solar wind three-dimensional structure is highly dependent upon the solar cycle (see, e.g. McComas *et al.* (2003)). During low solar activity, the SW presents a bi-modal structure, with a regular fast wind at high latitudes and a much more variable slow wind at low latitudes. On the contrary, near solar maximum, the wind structure is more complex, being highly variable at all heliolatitudes and arising from very diverse sources. Thus, it is reasonably to expect a stronger velocity shear during solar maximum than during minimum.

Then, one would like to know if the solar cycle, through this shear mechanism, assuming identical turbulent initial conditions near the Sun, can affect the SW turbulence properties?

In order to explore this possible scenario, we look at anisotropies in the correlation scale of the turbulence, and examine its evolution with heliodistance for different periods of time, thereby taking into account the dynamical age of the turbulence as well as the magnetic field direction and SW speed. To address this question, we employ Helios 1 (H1) & Helios 2 (H2) observations made over almost a complete 11-year solar cycle. These are unique spacecrafts since they have systematically explored the inner heliosphere providing us with the youngest samples of wind observed *in situ*.

In the following, we briefly present the theoretical background for this work, describe the data processing, analyze the anisotropy in the correlation scale, and present and discuss the results.

2. Correlation lengths: analysis, results and discussion

The magnetic autocorrelation function R is commonly employed in studies of turbulent magnetofluids like the SW (e.g., Tu & Marsch (1995)).

Assuming homogeneity in space and time, R can be defined as:

$$R(\mathbf{r}, \tau) = \langle \mathbf{b}(\mathbf{x}, t).\mathbf{b}(\mathbf{x} + \mathbf{r}, t + \tau) \rangle \tag{2.1}$$

where $\mathbf{r}$ and τ are the spatial and temporal lags respectively.

Observations made by only one probe allows us to calculate R in the SW as $R(-\mathbf{V}_{\mathbf{sw}}\tau, \tau)$, with $\mathbf{V}_{\mathbf{sw}}$ the SW velocity.

However, due to the superalfvénic character of the SW, it is possible to construct spatial correlation functions by means of the *frozen-in flow* Taylor hypothesis: $R_{bb}(-\mathbf{V}_{\mathbf{sw}}\tau, \tau = 0) = R_{bb}(\mathbf{r} = \mathbf{0}, \tau)$ (Taylor 1938). Then the intrinsic lag time dependence in equation 2.1 can be ignored and R becomes a function of $\mathbf{r}$ alone, giving $R(\mathbf{r}) = \langle \mathbf{b}(\mathbf{0}).\mathbf{b}(\mathbf{r}) \rangle$.

As mentioned before, anisotropies will develop with respect to $\mathbf{B}_0$ and therefore correlation functions will not behave in the same way in different directions. A measure of this correlation anisotropy can be given by means of the integral scale λ computed from R. If we consider a spatial lag $\mathbf{r}(\theta)$ making an angle θ with respect to $\mathbf{B}_0$, the correlation lengths in the $\hat{\mathbf{r}}(\theta)$ direction can be defined as:

$$\lambda(\theta) = \frac{\int_0^\infty \langle \mathbf{b}(\mathbf{0}).\mathbf{b}(r\hat{\mathbf{r}}(\theta)) \rangle}{\langle b^2 \rangle} \tag{2.2}$$

Conventionally, λ can be viewed as an anisotropic measure of the size of the energy-containing eddies in turbulence.

We applied the concepts summerized above to H1 and H2 data (time cadence of 40 seconds), which correspond to observations almost in the ecliptic plane, between 0.3 AU and 1.0 AU. We analyze observations in a range from December 1974 to June 1981. We group the observations into intervals (I) of 24-hour length, and use the same procedure as in Ruiz *et al.* (2011) to construct an appropiate data set. For each interval, this data set contains a correlation function (R^I) and its respective correlation length (λ^I), the distance from the Sun to the spacecraft (D^I), the SW speed and the angle (θ^I) between the direction of the mean magnetic field $\langle \mathbf{B}_0^I \rangle$ and the SW velocity. The number of intervals analyzed for H1 (H2) for solar maximum and solar minimum are 165 (181) and 333 (261), respectively.

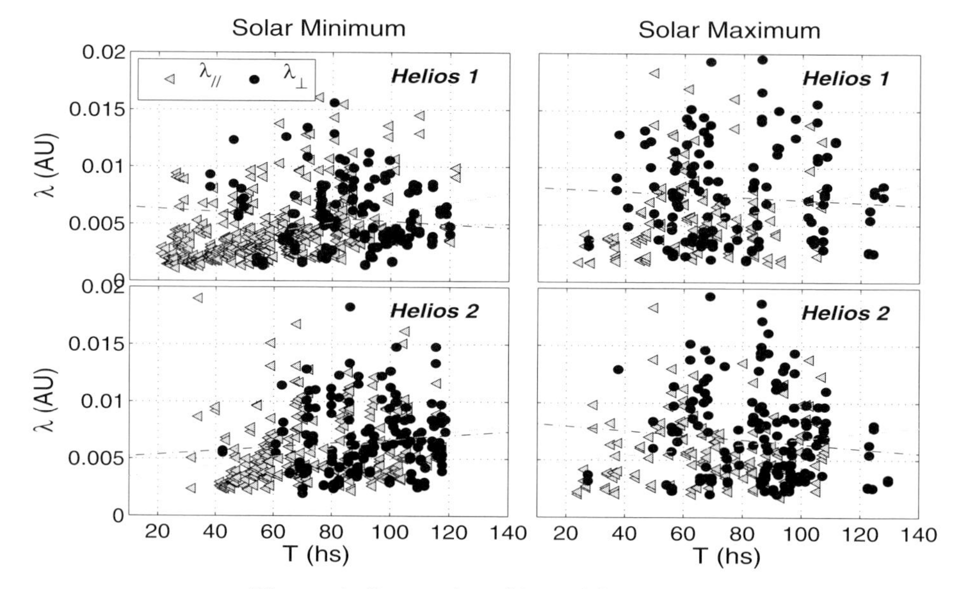

Figure 1. Scatterplot of $\lambda_{\parallel}$ and $\lambda_{\perp}$ vs. age.

Table 1. Fitted values of the slope (m, reported in 10^{-5}AU hs^{-1}), assuming $\lambda = mT + b$.

	Solar Minimum		Solar Maximum	
Dates	December 1974-December 1977 (H1) January 1976-December 1977 (H2)		November 1978-June 1981 (H1) December 1978-June 1981 (H2)	
	$m_{\parallel}$	$m_{\perp}$	$m_{\parallel}$	$m_{\perp}$
Helios 1	4.3±1.3	-1.4±1.0	4.4±3.1	-1.2±2.4
Helios 2	3.5±1.1	1.7±1.1	2.5±1.9	-2.0±2.4

Since the degree and type of anisotropy can vary with heliodistance (Ruiz *et al.* 2011) and with the wind speed (Dasso *et al.* 2005; Weygand *et al.* 2011), we perform an evolution analysis of the anisotropy in correlation lengths in terms of the turbulent age of the plasma, which is simply the nominal time it takes a SW parcel moving at speed V_{sw}^I to travel from the Sun to the spacecraft located at D^I. Then, in each interval we compute this time as $T^I = D^I/V_{sw}^I$, and establish two angular channels, the parallel ($0° < \theta < 40°$) and the perpendicular channel ($50° < \theta < 90°$) into which we grouped observations, thus obtaining in particular the parallel ($\lambda_{\parallel}$) and perpendicular ($\lambda_{\perp}$) correlation lengths.

The relative order between $\lambda_{\parallel}$ and $\lambda_{\perp}$ can be interpreted qualitatively in terms of the relative abundance of the two basic components of the MHD scale fluctuations, slab and

2D. Accordingly, preponderance of the slablike component is identified by $\lambda_{\parallel}/\lambda_{\perp} < 1$, while preponderance of the 2D component is identified by $\lambda_{\parallel}/\lambda_{\perp} > 1$ (Matthaeus *et al.* (1990)).

Figure 1 shows the evolution of $\lambda_{\parallel}$ and $\lambda_{\perp}$ with the turbulent age T, for both spacecrafts H1 (upper panels) and H2 (bottom panels), differentiating observations that correspond to solar minimum (left panels) respect to those observed in the solar maximum (right panels). Single lines, clear for $\lambda_{\parallel}$ and dark for $\lambda_{\perp}$, are linear fits to the data. They reveal a tendency of $\lambda_{\parallel}$ to grow with T. Instead, for $\lambda_{\perp}$ we find that the slope is not well defined and their values are marginally inside the error bar; it can be positive (H1 at minimum), negative (H1 at maximum) or even zero (H2 at minimum and maximum), as indicated in Table 1. Furthermore, the mean evolution of $\lambda_{\parallel}$ and $\lambda_{\perp}$ indicates that for T smaller than $\sim$ 70-100 hours, the slab component of the fluctuations is dominant ($\lambda_{\parallel} < \lambda_{\perp}$), while for T larger than $\sim$ 70-100 hours, the 2D component is stronger ($\lambda_{\parallel} > \lambda_{\perp}$). The inversion of the relative order of both populations occurs between $T \simeq 70$ hs and $T \simeq 100$ hs, and this fact is independent of the stage of the solar cycle.

The evolution of the anisotropy of correlation lengths with the aging of the solar wind ($T = D/V_{sw}$) is consistent with the injection of alfvénic fluctuations near the Sun and a subsequent spectral transfer from modes with wavectors parallel to $\mathbf{B}_0$ ($k_{\parallel}$) to modes with $k_{\perp}$. There is no observable dependence of this evolution with the solar cycle. During both periods of the cycle, the fluctuations evolve towards a larger abundance of the 2D-type population, with a similar age of transition between the two populations.

Acknowledgements

SD is member of the Carrera del Investigador Cientifico, CONICET. MER is a fellow of CONICET. MER and SD acknowledge support from the Argentinean grants: UBACyT 20020090100264 (UBA), PIP 11220090100825/10 (CONICET), PICT 2007-00856 (ANCPyT). SD acknowledges support from Abdus Salam International Centre for Theoretical Physics (ICTP) in the frame of his regular associateship. JW and WHM acknowledge partial support by NASA Heliophysics Guest Investigator Program grant NNX09AG31G, and NSF grants ATM-0752135 (SHINE) and ATM-0752135.

References

Belcher, J. W. & Davis L. Jr. 1971, *J. Geophys. Res.*, 76, 3534

Coleman, P. J. Jr. 1968, *ApJ*, 153, 371

Dasso, S., Milano, L. J., Matthaeus, W. H., & Smith, C. W. 2005, *ApJ* 635, L181

Goldreich, P. & S. Sridhar 1995, *ApJ*, 438, 763

Matthaeus, W. H., Goldstein, M. L., & Roberts, D. A. 1990, *J. Geophys. Res.*, 95, 20673

McComas, D. J., Elliott, H. A., Schwadron, N. A., Gosling, J. T., Skoug, R. M., & Goldstein, B. E. 2003, *GRL*, 30(10), 100000-1

Oughton, S., Priest, E. R., & Matthaeus, W. H. (1994), *J. of Fluid Mech.*, 280, 95

Roberts, D. Aaron, Ghosh, S., Goldstein, M. L., & Mattheaus, W. H. (1991), *Phys. Rev. Lett.*, 67, 3741

Robinson, D. C. & Rusbridge, M. G. 1971, *Phys. of Fluids*, 14, 2499

Ruiz, M. E., Dasso, S., Matthaeus, W. H., Marsch, E., & Weygand, J. M. 2011, *J. Geophys. Res.* 116 (A15), 10102

Shebalin, J. V., Matthaeus, W. H., & Montgomery, D. 1983, *J. of Plasma Phys.*, 29, 525

Taylor, G. 1938, *The Spectrum of the turbulence*, in *Proc. R. Soc. London Ser. A, 164*, p. 476

Tu, C. Y. & Marsch, E. 1995, *MHD structures, waves and turbulence in the solar wind: observations and theories*, Dordrecht: Kluwer

Weygand, J. M., Matthaeus, W. H., Dasso, S., & Kivelson, M. G. 2011, *J. Geophys. Res.*, 116, A08102

Comparative Magnetic Minima:
Characterizing quiet times in the Sun and stars
Proceedings IAU Symposium No. 286, 2011
C. H. Mandrini & D. F. Webb, eds.

doi:10.1017/S1743921312004802

Interplanetary conditions: lessons from this minimum

J. Luhmann[1], C. O. Lee[1], P. Riley[2], L. K. Jian[3], C. T. Russell[3] and G. Petrie[4]

[1]Space Sciences Laboratory, 7 Gauss Way, University of California, Berkeley, CA 94720, USA
[2]Predictive Science Inc., 9990 Mesa Rim Rd., San Diego, CA 92121 USA
[3]Inst. of Geophysics and Planetary Physics, Slichter Hall, UCLA, Los Angeles, CA 90095, USA
[4]National Solar Observatory, 950 N. Cherry Ave., Tucson, AZ 85719, USA

Abstract. Interplanetary conditions during the Cycle 23-24 minimum have attracted attention because they are noticeably different than those during other minima of the space age, exhibiting more solar wind stream interaction structures in addition to reduced mass fluxes and low magnetic field strengths. In this study we consider the differences in the solar wind source regions by applying Potential Field Source Surface models of the coronal magnetic field. In particular, we consider the large scale coronal field geometry that organizes the open field region locations and sizes, and the appearance of the helmet streamer structure that is another determiner of solar wind properties. The recent cycle minimum had an extraordinarily long entry phase (the decline of Cycle 23) that made it difficult to identify when the actual miminum arrived. In particular, the late 23rd cycle was characterized by diminishing photospheric fields and complex coronal structures that took several extra years to simplify to its traditional dipolar solar minimum state. The nearly dipolar phase, when it arrived, had a duration somewhat shorter than those of the previous cycles. The fact that the corona maintained an appearance more like a solar maximum corona through most of the quiet transitional phase between Cycles 23 and 24 gave the impression of a much more complicated solar minimum solar wind structure in spite of the weaknesses of the mass flux and interplanetary field. The extent to which the Cycle 23-24 transition will affect Cycle 24, and/or represents what happens during weak cycles in general, remains to be seen.

Keywords. solar wind sources, coronal field structure

---

1. Introduction

The Cycle 23-24 solar minimum has generated considerable interest because of its distinctive behavior compared to the three previous cycles of the space age. As has been pointed out by many recent authors and speakers, the solar field has been both weaker overall and exhibits a lower axial dipole moment, resulting in several important consequences for solar irradiance and the interplanetary medium (e.g. see SOHO 23: Understanding a Peculiar Solar Minimum, ASP Conf. Ser. V.428, 2010). Emissions at EUV wavelengths were reduced to the point where the Earth's thermosphere practically disappeared (Solomon *et al.* 2011). Observations both in the ecliptic at 1 AU and at higher latitudes on the Ulysses spacecraft indicated that the interplanetary magnetic flux and the solar wind mass flux were each lower by ~30% than their values in the previous solar minimum in 1995-6 (McComas *et al.* 2008, Smith & Balogh 2008). These parameters affected other heliospheric conditions including the galactic cosmic ray flux (e.g. Schwadron *et al.* 2010). The altered solar wind dynamic pressure also had significant impacts on planetary plasma interactions and geomagnetic activity. Here we focus on the coronal field topology that in part determined the interplanetary conditions during the Cycle 23-24 transition. In particular we use Potential Field Source Surface (PFSS) models to illustrate how the large scale coronal field features, including coronal streamer structure and ecliptic solar wind sources, compare to those during earlier solar minima.

2. Interplanetary Conditions at Cycle Minima

A review of interplanetary parameters over the previous three cycles through the Cycle 23-24 minimum can be found in the recent paper by Jian *et al.* (2011). These authors discuss the now well-documented weak interplanetary field strengths (reported by both Smith & Balogh 2008 and Lee *et al.* 2009, among others) and the low solar wind densities associated with the low solar wind mass flux mentioned above. The recorded high galactic cosmic ray fluxes are an expected result of the low interplanetary field, given past cycle correlations showing field strengths are anti-correlated with the fluxes (Cane *et al.* 1999). Of special interest here are the results of their analysis of solar wind structures through Cycle 23 including the bracketing minima. In particular, the stream interaction regions (SIRs) were examined for both their occurrence rates and average properties and found to be more numerous but weaker than in the previous minimum.

Figure 1 provides an overview of a selection of the interplanetary parameters analyzed by Jian *et al.*, from the OMNI 1 AU data base. The measurements shown are 27 day averages, which minimize variations introduced by corotating features. What is most notable from a cursory examination of these time series is the lack of clear solar cycle modulations in anything but the interplanetary magnetic field. The origin of the interplanetary field modulation during the minima on either side of Cycle 23, and in particular its relationship to the solar magnetic field, was explored in some detail by Lee *et al.* (2011) using complementary magnetograph observations. These authors demonstrated that the interplanetary field strength could be obtained from photospheric magnetic field maps by using the Potential Field Source Surface Model of the coronal field to estimate open flux, and then assuming that open flux was uniformly distributed over the heliosphere. Their analysis could explain the interplanetary flux magnitude difference between the previous (Cycle 22-23) and recent (Cycle 23-24) minima, but what about the greater number of weak SIRs? The answer may lie in the solar wind source distribution.

From the experience of the three previous solar cycle minimum periods in 1975-76, 1985-6 and 1995-6, and especially from the last one for which we have the comprehensive solar information from SOHO imagers available, certain expectations were established. The picture that emerged from the combination of observations and modeling was one in which the solar corona took on an approximately dipolar global appearance, with most of the ecliptic solar wind coming from the open fields near the low latitude boundaries of the polar coronal holes (e.g. Linker *et al.* 1999). Moreover, within the minimum there were periods when the coronal dipole equator was practically in the ecliptic plane. At other times it could be tilted with respect to the solar rotation axis, and/or have warps due to the presence of one or more active regions at lower latitudes, typically the last of the preceding cycle at the equatorward end of the butterfly diagram wings. The latter promoted the existence of sometimes large polar coronal hole extensions, an example of which was the elephant trunk feature of the first Whole Sun Month study campaign (Gibson *et al.* 1999). Both dipolar coronal tilting and polar coronal hole extensions bring higher speed polar hole wind into the ecliptic plane every solar rotation. The canonical solar minimum picture of interplanetary conditions dominated by corotating high speed streams from polar coronal holes and their stream interaction regions established by Hundhausen (1979; also Zhao and Hundhausen 1981) became the expected state for

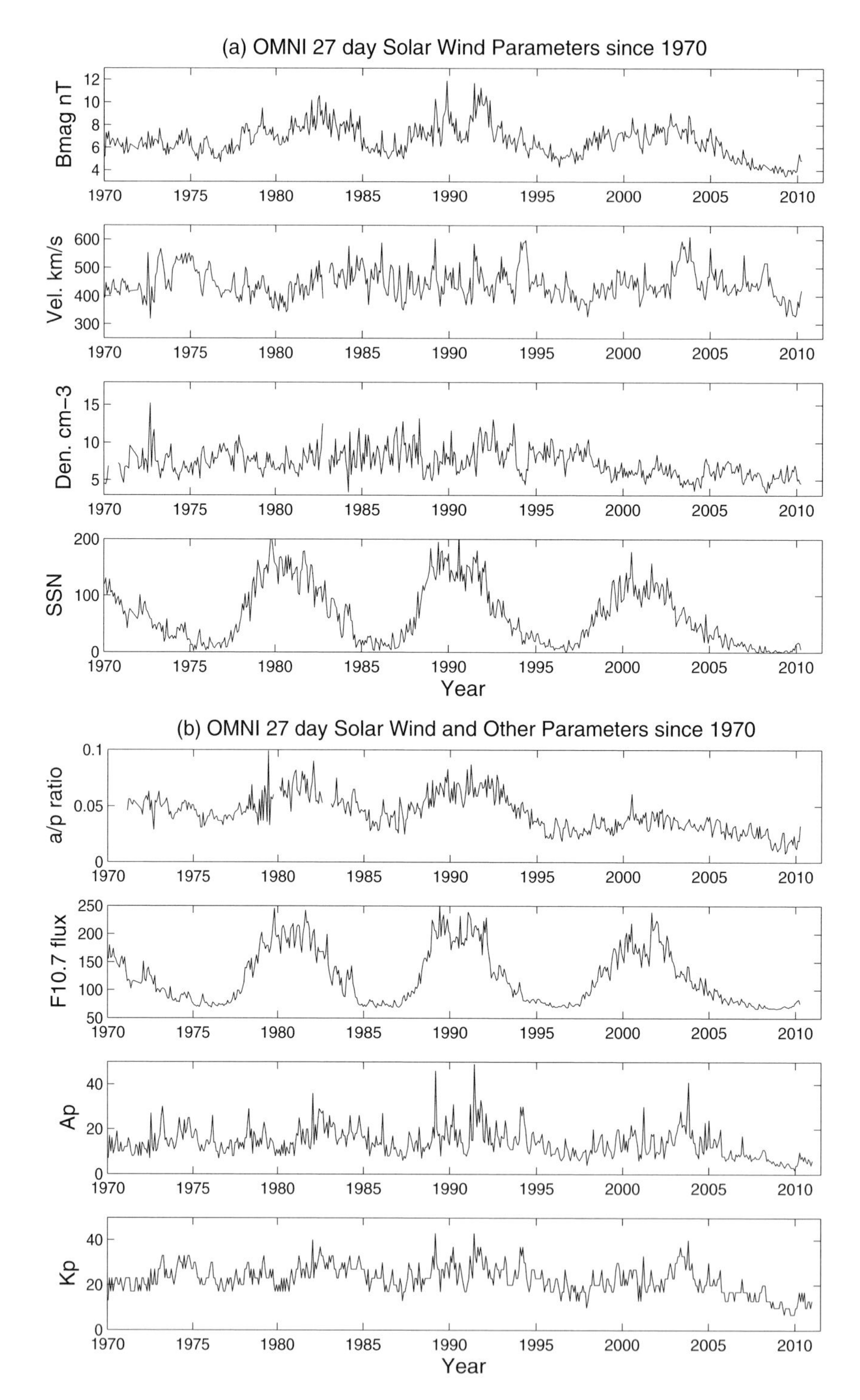

Figure 1. OMNI 27-day averaged solar wind plasma and field data are shown for the 3+ cycles of the space age, together with sunspot number, F10.7cm radio flux (a solar EUV proxy), and the Ap geomagnetic index. Of all of the interplanetary parameters, the magnetic field and solar wind alpha-to-proton (a/p) ratio are the only ones that show a clear solar cycle modulation.

solar quiet periods. However the recent minimum period seems to require a somewhat modified picture.

3. Sources of the Solar Minimum Interplanetary Conditions

A number of previous authors have noted the unusually wide coronal streamer belt (e.g. McComas *et al.* 2008) and the ubiquity of low latitude coronal holes during the Cycle 23-24 minimum (Abramenko *et al.* 2010, deToma, in this proceedings). The reasons for these can be found in the weakening solar polar magnetic fields relative to the mid-to-low latitude magnetic fields of decayed active regions. Luhmann *et al.* (2009) discussed how this combination produces a large scale coronal field that has significant higher order multipole contributions compared to the axial dipole influence. Instead of a single, wide helmet streamer belt overlying essentially all the smaller scale closed coronal fields, the corona includes large topologically distinct pseudostreamers (e.g. Wang *et al.* 2009). The cartoon in Figure 2 (top) illustrates the major features the large scale field structure. The formation of pseudostreamers can be understood as the result of two effects. One is the location of the source surface, the provisional radial distance at which the coronal fields are opened by the solar wind. A smaller source surface will always produce a more structured coronal hole pattern than a large source surface (e.g. see Lee *et al.* 2011 for an illustration). The other effect is the existence on the solar surface of sufficiently large and strong dipolar magnetic regions with surface dipole moments having a direction opposite to the polar dipole moment and its overlying large scale field. As Luhmann *et al.* (2003) illustrated, the opposing surface dipole moment allows a part of the helmet streamer to essentially break away from the main streamer belt. The weak polar fields and inferred small source surface of Cycle 23 (see Lee *et al.* 2011) conspired to regularly produce pseudostreamers. Thus the multi-rayed coronal field that prevailed throughout the long Cycle 23 decline more closely resembled a solar maximum configuration than a dipole, even though the photospheric field was weak overall. Figure 2 (bottom) displays a sampling of large scale coronal field pictures contructed using GONG magnetogram-based Potential Field Source Surface (PFSS) models for Carrington Rotations in and around the Cycle 23-24 minimum.

PFSS models have been shown to reproduce coronal hole geometries and streamer features during quiet to moderately active solar conditions (e.g. Levine *et al.* 1977). The average polarity of the interplanetary magnetic field can also be inferred to remarkable accuracy by assuming the source surface neutral line separates outward and inward going fields (Hoeksema 1984, Wang *et al.* 2009). Lee *et al.* (2011) and also deToma (this proceedings) modeled coronal holes for Cycle 23 Carrington Rotations that generally capture their locations, shapes and sizes seen in EUV images. In the above mentioned paper by Luhmann *et al.* (2009) a numerical experiment was done with the PFSS model to demonstrate that weakened polar fields relative to the same mid-to-low latitude fields produced a situation where the ecliptic solar wind source mapping (via open field lines) that usually connects to polar coronal hole boundaries instead connects to low to mid latitude coronal holes. Riley & Luhmann (2012) further show that part of this source mapping change involves frequent ecliptic connections to the ubiquitous pseudostreamer boundaries in the years ~2006-9. Although these latter authors use an MHD model, their results can also be seen in the PFSS versions of the solar wind source mapping. This combination of successful long-term and recent applications of PFSS models suggests that we can further apply them toward comparing solar minima.

Figures 3 and 4 illustrate the overall solar cycle evolution of coronal holes and their near-ecliptic connections since the early 1970s according to PFSS models. This is an

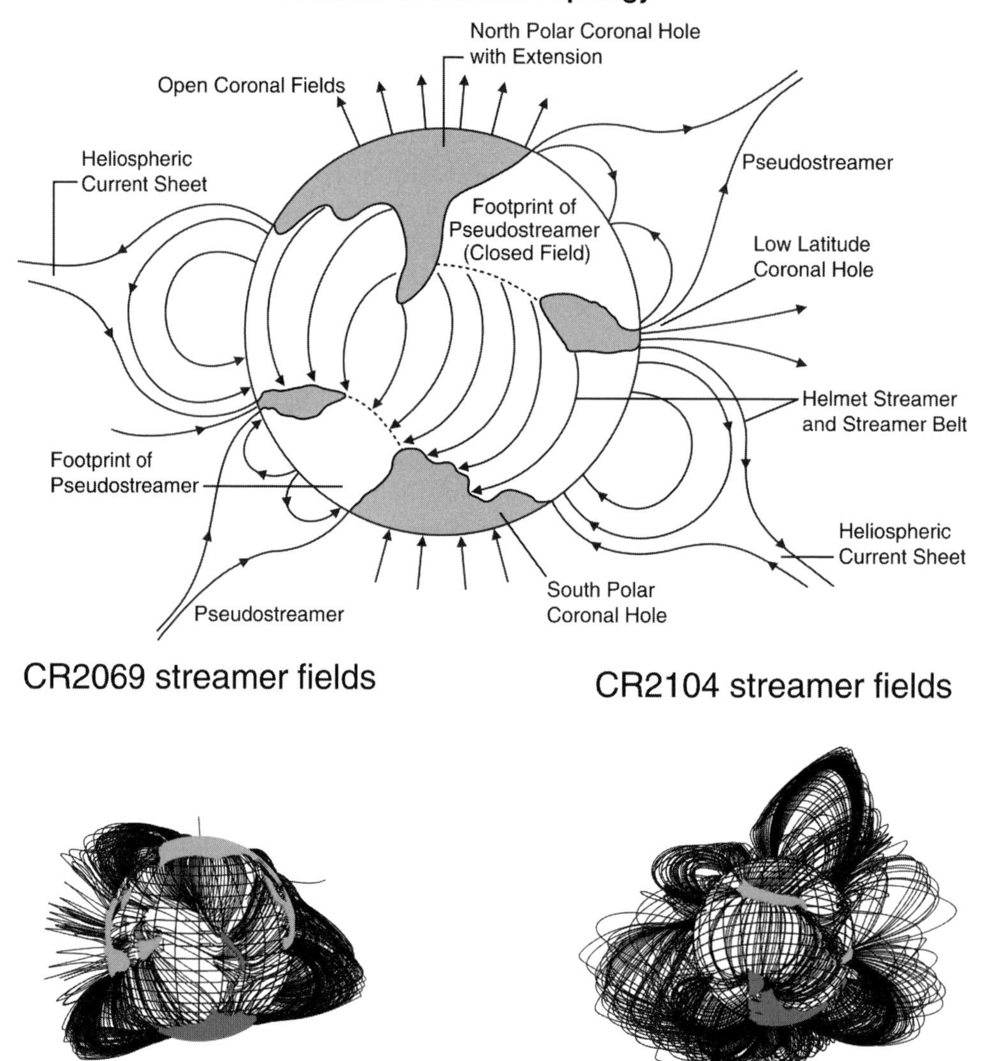

Figure 2. (top) Illustration of the main features of the large scale coronal field. Frequently occurring departures from dipolar topology including numerous coronal holes at mid-to-low latitudes and pseudostreamers separate from the helmet streamer belt. (bottom) A sampling of magnetogram-based PFSS models of large scale coronal fields for the Cycle 23 declining phase and the Cycle 23-24 minimum.

update of similar plots presented in Luhmann *et al.* (2002), where the ability of the ecliptic mapping results to mimic the solar cycle trends in average solar wind parameters observed upstream of Earth was demonstrated. These updated plots display the results of Mt. Wilson Observatory (MWO) magnetogram-based models up to 2007, transitioning to GONG magnetogram-based models thereafter (e.g. after Carrington Rotation (CR) 2050). Carrington Rotations are used as the time coordinate for consistency with the synoptic map origins of the results. For overall perspective and context, Figure 3 (top) shows the sunspot number for the periods of the modeled open field footprints in Figure 3 (bottom). Here one sees how the coronal hole distributions evolve over the solar cycle (with greatest model accuracy around solar minima). The polar coronal holes and their extensions persist except around solar maximum, when they disappear for several rotations as polar field reversal takes place. One also sees the band of mid-to-low latitude coronal holes and polar hole extensions associated with the active region belt. These features follow the butterfly diagram of the Hale Cycle, which reaches its midpoint in equatorward drift at solar maximum when the polar coronal holes disappear. It is important to note that the polar holes do not tilt toward low latitudes as many cartoons suggest, but instead shrink and then disappear (sometimes not exactly at the same time in both hemispheres), leaving the mid-to-low latitude coronal holes associated with active regions to dominate the solar maximum period.

Figure 4 (top) shows only those open field footpoints that map to within ~10 degrees of the equator, color coded by magnetic polarity (radially outward or inward field). These indicate the sources of the near-ecliptic solar wind according to the PFSS model. The corresponding magnetic neutral line on the source surface, approximating the latitude of the heliospheric current sheet at its origin at a few solar radii, is displayed in Figure 4 (bottom) as a complement to Figure 4 (top). These plots collectively illustrate the differences in the Cycle 23-24 solar minimum solar wind sources compared to the preceding few cycle minima. What is most striking is the long duration of the decline into the Cycle 23-24 minimum. Whereas in the previous few cycles a significant fraction of the transition between solar cycles was occupied by polar hole dominated conditions, in the Cycle 23-24 minimum the mid-to-low latitude active region related coronal holes do not disappear except for a few Carrington Rotations in late 2009. The consequences this has had for interplanetary conditions over the last ~5 years are considered below.

4. Implications

It is now generally-agreed that the boundaries of coronal streamers, which are also the boundaries of coronal holes, produce mainly slow solar wind including a transient component. In addition, analyses of the ion composition of the slow solar wind suggests it may have at least two main sources (Wang 2012, Kasper *et al.* 2007), perhaps the polar hole boundaries and the low-to-mid latitude coronal holes discussed above. Do the coronal conditions in the Cycle 23-24 minimum favor either polar coronal hole boundary or mid-to-low latitude coronal hole mappings more than in the previous cycle minima? To address this question we use the PFSS model results in Figure 4a to obtain statistics of near-ecliptic mapped footpoint field latitudes for the ~3+ solar cycles shown. We distinguish between polar hole boundary and low-to-mid latitude coronal hole boundary sources by separately counting the points at latitudes above and below 50 degrees. As each Carrington Rotation in our model plots involves the same number of source points mapped from a source surface equatorial band, this produces a consistent picture of the comparative ecliptic mappings from cycle to cycle. The results in Figure 5 (top) more quantitatively illustrate what is inferred from looking at Figure 4a. By comparing the fraction of the total near-ecliptic mappings that go to the polar hole boundaries (latitudes >50 degrees) with the fraction going to mid-to-low latitudes, one can see how much of each cycle is invested in one or the other source. Interestingly, the duration of the period when polar hole boundary sources dominates is similar for all the cycle minima shown, including the Cycle 23-24 minimum. However, Cycle 23 was distinguished by its markedly longer (by several years) declining phase, during which the low-to-mid latitude mapping dominance persisted. These results suggest that our perception of the Cycle 23-24 minimum solar wind structure is affected more by the long decline of Cycle 23 than

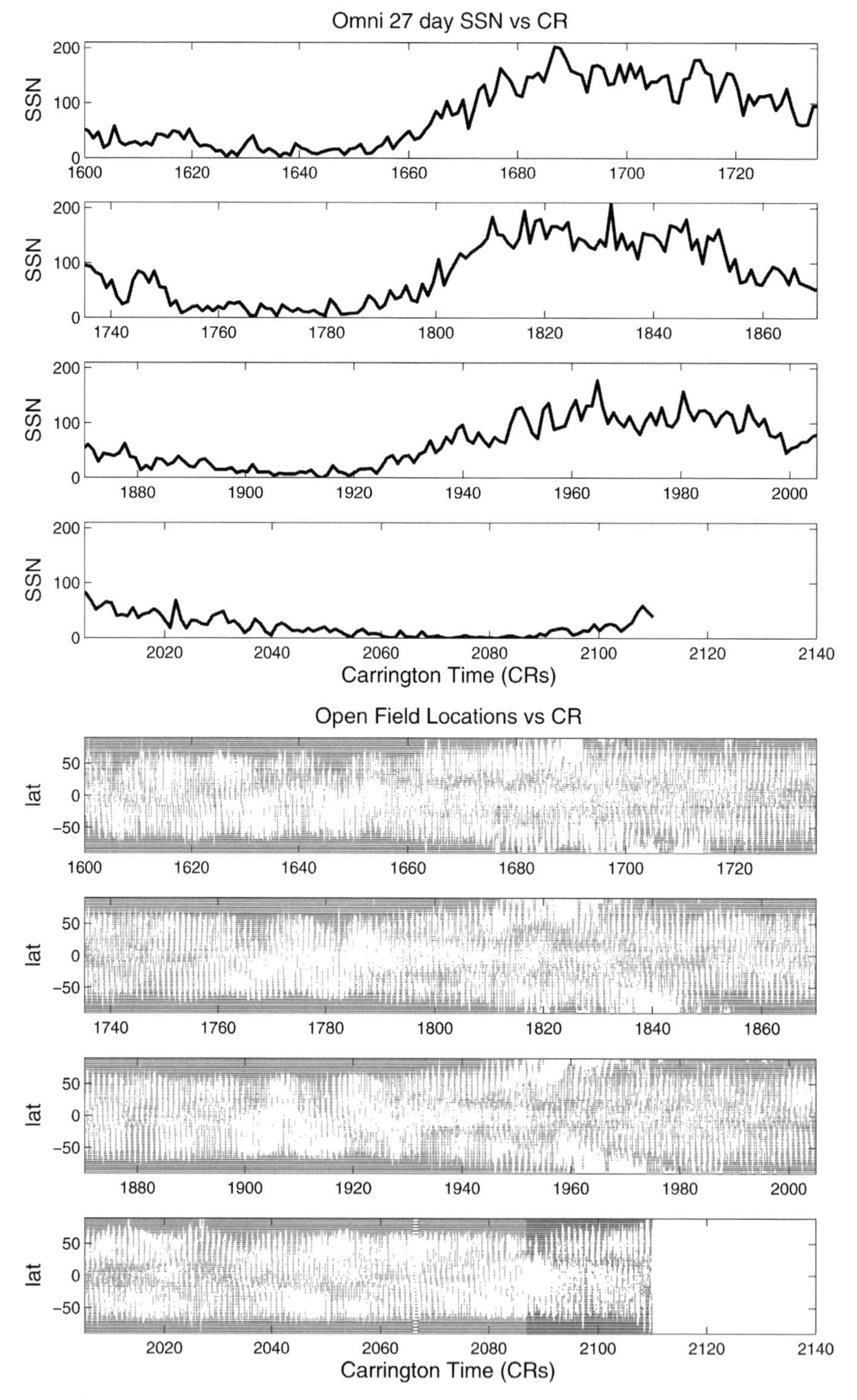

Figure 3. (top) Sunspot number and (bottom) footpoint locations of open coronal magnetic fields on the photosphere according to PFSS models. The models are MWO magetogram based prior to Carrington Rotation 2050, and GONG magnetogram based thereafter. See the text for a discussion of the features in the open field patterns and the online version for color figures.

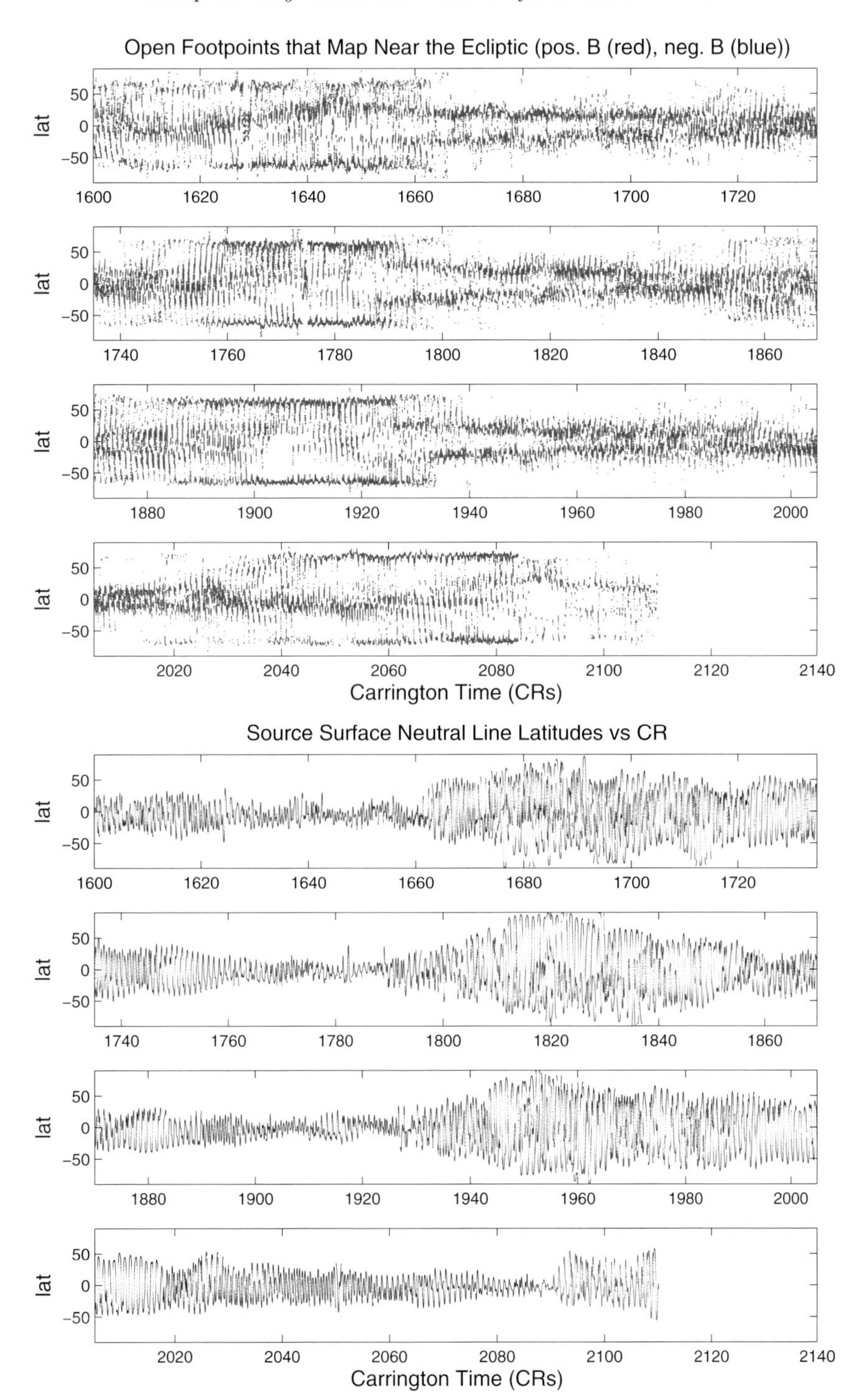

Figure 4. (top) A subset of the open fields shown in Figure 3 (bottom) that corresponds to locations that map to the low heliolatitudes near the ecliptic. These are in principle the source regions for the interplanetary conditions prevailing around each cycle minimum. The points are color coded (see the online version) according to the solar field polarity at the open field footpoints (red=outward, blue=inward). (bottom) Magnetic neutral line latitude on the source surface for the PFSS models used in Figure 4 (top)

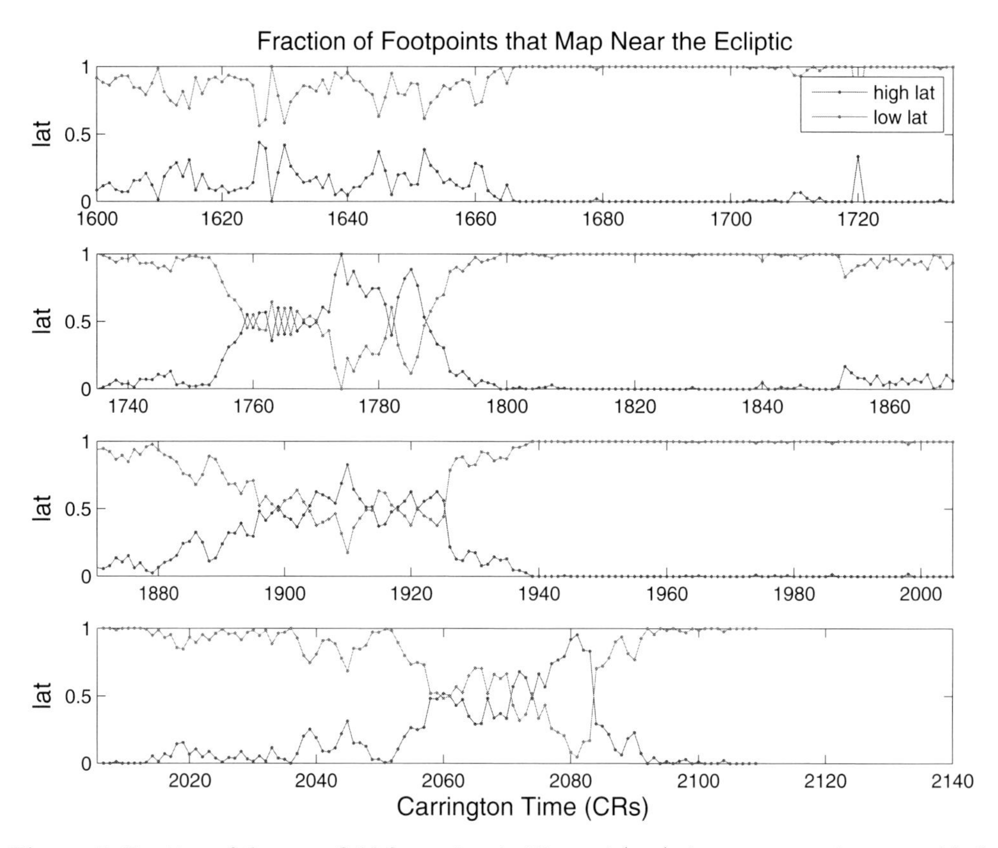

Figure 5. Fraction of the open field footpoints in Figure 4 (top) that map to polar coronal hole boundaries (defined here as latitudes >50 deg.) compared to the fraction mapping to mid-to-low latitude sources. The periods where polar hole boundary mapping dominates can be viewed as the classical dipolar coronal phases of the solar minima. See color figure in the online version of the paper.

anything special about the dipolar phase. Of course the weak mass flux and interplanetary field are present throughout this period regardless of the coronal source complexity.

As some authors have suggested, the long, slow decline may have contributed to the relative weakness of the photospheric field, including the polar fields, at the end of Cycle 23. This long decline moreover maintained a complexity of the coronal field, and thus the solar wind sources, by limiting the polar field strength buildup. However observations also indicate that less flux emerged from the subphotospheric convection zone during Cycle 23 in the first place (Penn & Livingston 2010), leading to a chicken-egg situation. It may be that when flux emergence is weak, for reasons of either weak internal generation or weak delivery to the surface, the subsequent flux redistribution process –by association or as a consequence– further weakens the surface field, including the polar region fields. Moreover –if by association or as a consequence– it takes longer for the new cycle field emergence to get started, even more time is available for the previous cycle surface fields to diminish in strength. The result for the interplanetary conditions includes a long-lived period of weak but complex coronal field and related solar wind sources prior to the onset of the next cycle. What has happened on the Sun and in the interplanetary medium in the past decade is part of a highly interconnected evolutionary process. Will it affect the subsequent cycles? If some internal mechanism delivers a new cycle of emerged active region fields that are robust (e.g. not heavily influenced by what went on during the prior cycle), the next cycle could in principle evolve mainly according to its own dynamics, However, if the next cycle, perhaps through memory and recirculation of flux (e.g. Dikpati 2011), again produces weak emerged fields, the result could be a continuation of the present conditions of weak solar wind and interplanetary fields from weak but complex coronal sources. Whether the solar surface and interplanetary field and mass flux would be reduced yet further remains to be seen.

Acknowledgments

The UCB and PSI contributions to this work were made possible by the NSF Science and Technology Center award to CISM through Boston University, under grant ATM0120950. The UCLA effort was enabled by NASA support of the STEREO IMPACT investigation through a subaward from UCB. OMNI data were obtained from the Space Physics Data Facility at Goddard Space Flight Center.

References

Abramenko, V., Yurchyshyn, V., Linker, J., Mikic, Z., Luhmann, J. G., & Lee, C. O. 2010, *ApJ*, 712, 813

Cane, H. V., Wibberenz, G., Richardson, I. G., & von Rosenvinge, T. T. 1999, *Geophys. Res. Lett.*, 26, 565

Dikpati, M. 2011, *Space Sci. Rev.*, 143, DOI 10.1007/s11214-011-9790-z

Gibson, S. E. *et al.* 1999, *ApJ*, 520, 871

Hoeksema, J. T. 1984, *Structure and evolution of the large scale solar and heliospheric magnetic fields, Ph.D. thesis, Stanford Univ., Stanford, Calif.*

Hundhausen, A. J. 1979, *Rev. Geophys.*, 17, 2034

Jian, L. K., Russell, C. T., & Luhmann, J. G. 2011, *Sol. Phys.*, 274, 321

Kasper, J., Stevens, M. L., Lazarus, A. J., Steinberg, J. T., & Ogilvie, K. W. 2007, *ApJ*, 660, 901

Lee, C. O., Luhmann, J. G., Hoeksema, J. T., Sun, X., Arge, C. N., & dePater, I. 2011, *Sol. Phys.*, 269, 367

Lee, C. O., Luhmann, J. G., Zhao, X. P., Li, Y., Riley, P., Arge, C. N., Russell, C. T., & dePater, I. 2009, *Sol Phys.*, 256, 345

Levine, R. H., Altschuler, M. D., Harvey, J. W., & Jackson, B. V. 1977, *ApJ*, 215, 636

Linker, J. A., Mikic, Z., Biesecker, D. A., Forsyth, R. J., Gibson, S. E., Lazarus, A. J., Lecinski, A., Riley, P., Szabo, A., & Thompson, B. J. 1999, *J. Geophys. Res.*, 104, 9809

Luhmann, J. G., Li, Y., Arge, C. N., Galvin, A. B., Simunac, K., Russell, C. T., Howard, R. A., & Petrie, G. 2009, *Sol. Phys.*, 256, 285

Luhmann, J. G., Li, Y., & Zhao, X. P., S. Yashiro 2003, *Sol. Phys.*, 213, 367

Luhmann, J. G., Li, Y., Arge, C. N., Gazis, P. R., & Ulrich, R. 2002, *J. Geophys. Res.*, 107(A8), 1154

McComas, D. J., Ebert, R. W., Elliott, H. A., Goldstein, B. E., Gosling, J. T., Schwadron, N. A., & Skoug, R. M. 2008, *Geophys. Res. Lett.*, 35, L18103

Penn, M. & W. Livingston 2010, *eprint arXiv:1009.0784*

Riley, P. & J. G. Luhmann 2012, *Sol. Phys.*, 277, 355

Schwadron, N. A., Boyd, A. J., Kozarev, K., Golightly, M., Spence, H., Townsend, L. W., & Owens, M. 2010, *Space Weather*, 8, S00E04

Smith, E. J. & A. Balogh 2008, *Geophys. Res. Lett.*, 35, L22103

Solomon, S. C., Qian, L., Didkovsky, L. V., Viereck, R. A., & Woods, T. N. 2011, *J. Geophys. Res.*, 116, A00H07

Wang, Y. M. 2012, *Space Sci. Rev.*, in press

Wang, Y. M., Robbrecht, E., & Sheeley, N. R. 2009, *ApJ*, 707, 1372

Zhao, X. P. & Hundhausen, A. J. 1981, *J. Geophys. Res.*, 86, 5423

Discussion

Arnab Choudhuri: The last minimum was not dipolar. If you try to model the streamer belt located farther South would it be correct?

Janet Luhmann: The problem is that a dipole with a planar neutral line just cannot fit most of the time. One needs to add the higher order harmonic contributions to get the observed warps and structures, it has to be more than a dipole translation to describe the corona during this recent minimum.

Jeffrey Linsky: Does the total mass flux integrated over all latitudes of the solar wind change between solar maximum and minimum and, in particular, was it lower during this minimum?

Janet Luhmann: Ulysses measurements suggested this minimum's mass flux was ∼30% lower than the previous minimum, as other speakers also noticed. This reduction seems to be mainly in the density, rather than in the velocity. Dave Webb and others have published estimates of solar cycle variations based on previous observations (as per comment by Dave Webb). I do not have those numbers handy.

Axel Brandenburg: Could you clarify what was plotted when you showed the fine structure of the solar wind? You mentioned HI intensity, but does black then mean a strong decrease against some finite mean value?

Janet Luhmann: The movie shown was made from difference images, so the contrasts indicate large increases (white) and decreases (black) in the densities detected in the white light of the HI images.

Margit Haberreiter: How does the unusual shape of the magnetic field affect the shielding from cosmic ray flux?

Janet Luhmann: The solar wind transients that result over a wide heliolatitude band probably merge with the stream interaction regions before they get far out into the heliosphere, so hard to say what effect these might have on galactic cosmic rays. It's usually merged interaction regions that seem to affect modulation form what I understand.

Eric Priest: What is the definition of a pseudo streamer?

Janet Luhmann: It is a closed coronal field structure outside of the main (circumsolar) streamer "belt". It can have a circular or elongated footprint, depending on the solar distribution.

Mark Giampapa: What is the origin of the dominance of higher order multipole moments in the field and does it have anything to do with the "blobby" turbulent-looking outflow?

Janet Luhmann: The Sun somehow did not produce polar fields (by the usual transport processes of convection plus diffusion) as strong as in the past. As a result, the decayed active region fields that dominate lower latitudes had a much greater influence over the global coronal field than in the previous minimum. The latitude extent of the wind "blobbiness" this minimum, as seen by HI, is one result.

Comparative Magnetic Minima:
Characterizing quiet times in the Sun and Stars
Proceedings IAU Symposium No. 286, 2011
C. H. Mandrini & D. F. Webb, eds.

doi:10.1017/S1743921312004814

The floor in the solar wind: status report

E. W. Cliver[1]

[1]Space Vehicles Directorate, Air Force Research Laboratory Sunspot, NM, USA
email: ecliver@nso.edu

Abstract. Cliver & Ling (2010) recently suggested that the solar wind had a floor or ground-state magnetic field strength at Earth of ~2.8 nT and that the source of the field was the slow solar wind. This picture has recently been given impetus by the evidence presented by Schrijver *et al.* (2011) that the Sun has a minimal magnetic state that was approached globally in 2009, a year in which Earth was imbedded in slow solar wind ~70% of the time. A precursor relation between the solar dipole field strength at solar minimum and the peak sunspot number (SSN_{MAX}) of the subsequent 11-yr cycle suggests that during Maunder-type minima (when SSN_{MAX} was ~0), the solar polar field strength approaches zero - indicating weak or absent polar coronal holes and an increase to nearly ~100% in the time that Earth spends in slow solar wind.

Keywords. Sun, magnetic fields, solar wind

1. Introduction

Does the solar wind, at the current stage of the Sun's life, have a ground state or floor? Was such a ground state approached during the Maunder Minimum? Interest in the notion of such a floor has increased because of the recent prolonged and deep solar minimum and the prospect that we might be entering a Gleissberg-type minimum such as that of ~1900 or perhaps even a Maunder-type minimum (Eddy, 1976).

The concept of a floor in the solar wind magnetic field strength (B) was introduced separately by Svalgaard and Cliver (2007) and Owens *et al.* (2008). Svalgaard and Cliver posited a floor in B of ~4.6 nT based on their long-term geomagnetic-based reconstruction of B and the correlation of B with the sunspot number (SSN) while Owens *et al.* (2008) obtained a B value of ~4 nT at solar minimum when the rotation-averaged coronal mass ejection rate was extrapolated to zero. Both papers viewed the floor as the state of the near-Earth solar wind in the absence of 11-yr cyclic activity.

Shortly after these floor values were proposed, the solar wind B dropped to its lowest values observed during the space age, ~4.2 nT in 2008 and ~3.9 nT in 2009, prompting downward revisions of the proposed levels by both groups of authors: to ~3.7 nT by Crooker and Owens (2010) and to ~2.8 nT by Cliver and Ling (2010). Here I will focus on the work by Cliver and Ling (2010) which also suggested a novel source for such a ground state - the omnipresent slow solar wind. The chain of reasoning that led to this suggestion began with the paper on sunspot cycle prediction by Schatten *et al.* (1978).

2. Precursor relations and the floor

Schatten *et al.* (1978) argued that because (in the dynamo model for solar activity) poloidal fields at solar minimum provide the "seed" for the toroidal fields of the subsequent maximum, the peak sunspot number of a cycle should be related to the strength of the solar polar fields at the preceding minimum. It was difficult to test this idea directly at the time however, because reliable measurements of the solar polar fields had been made at only one solar minimum up to that point. A direct use of this method (i.e., not based on proxy input parameters such as geomagnetic indices, e.g., Wang and Sheeley, 2009) was made by Svalgaard *et al.* (2005) who predicted that the present cycle would be the smallest in ~100 years with a peak sunspot number (SSN_{MAX}) of 75 ± 8. This prediction (Fig. 1(a)), which was based on only two cycles and a force fit through the origin, was close to the value of 90 eventually adopted by the NOAA/NASA/ISES prediction panel (http://www.swpc.noaa.gov/SolarCycle/SC24/index.html).

Because the floor of ~4.6 nT in the solar wind proposed by Svalgaard and Cliver (2007) was thought to apply to the Maunder Minimum, the precursor relation in Fig. 1(a) implies that this value, which was approached at every 11-yr minimum during the space age, is relatively independent of the solar polar field strength. This misconception was based in part on Ulysses observations through November 2006 (Balogh & Smith, 2006) which showed little change in the radial field strength between fast latitude scans near the minima following solar cycles 22 and 23. As Svalgaard and Cliver (2007) wrote, "While the radial IMF strengths [B_R] are essentially identical for these minima, the solar polar magnetic fields are ~40% weaker at present (60 μT vs. 100 μT for ~1995; Svalgaard *et al.* (2005) and current data from Wilcox Solar Observatory [WSO]). It seems hard to escape the conclusion that the polar fields do not determine the magnitude of the IMF at solar minimum." Subsequently, Smith and Balogh (2008) reported that B_R measured by Ulysses dropped by ~35% between the minima following cycles 22 and 23 and, as noted above, B dropped below 4 nT in the ecliptic plane in 2009. Clearly solar wind B was responding to the change in the solar polar field strength. The relationship between solar minimum values of the solar dipole moment [DM] and near-Earth B for the last four minima can be seen in Fig. 2 (taken from Cliver and Ling, 2010). Linear extrapolation of the regression line indicates that when the Sun's dipole moment goes to zero [implying from Fig. 1(a) that the subsequent solar cycle will have a SSN_{MAX} value ~ 0 (i.e., Maunder-type minimum conditions; Eddy, 1976, Hoyt and Schatten, 1998)], the solar wind will have a floor of ~2.6 nT.

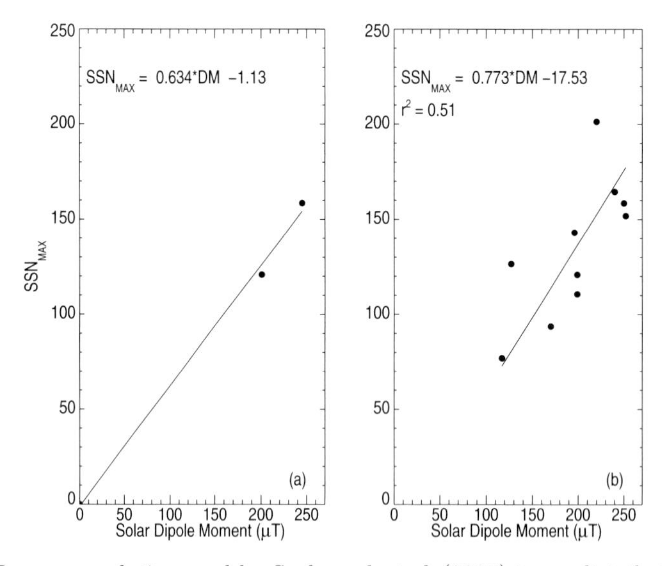

Figure 1. (a) Precursor relation used by Svalgaard *et al.* (2005) to predict that cycle 24 would be the smallest in ~100 years with a peak sunspot number of ~75. (b) Same as (a) but with more data points based on the correlation between B_{MIN} and DM in Fig. 2 and a long-term reconstruction of solar wind B (Svalgaard & Cliver, 2010)

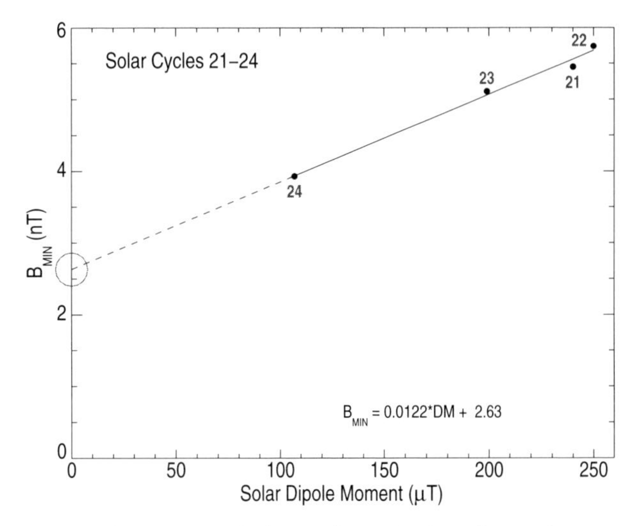

Figure 2. Solar magnetic field strength (B_{MIN}) at the last four solar minima (cycles 21-24) plotted vs. the solar dipole-field strength. The linear extrapolation suggests the existence of a floor in B_{MIN} at a value of ~2.6 nT (red circle, see the online version of this paper) (from Cliver and Ling, 2010).

The relationship between B_{MIN} and DM, in combination with the long-term geomagnetic-based reconstruction of solar wind B (Svalgaard and Cliver, 2010), can be used to infer DM going back in time. Thus it is possible to substantiate (Fig. 1(b)) the prediction of Svalgaard *et al.* that was based on only two cycles. For an input of 119.3 μT (Svalgaard *et al.*, 2005), the equations in (a) and (b) both give a prediction of ~75. Note that for zero SSN_{MAX}, the dipole moment is zero within the uncertainty (23 ± 32 μT) justifying the fit through the origin in Fig. 1(a). Alternatively, one could use B_{MIN} as a proxy for DM in a precursor relationship, as was done by Cliver and Ling (2010), who found that a peak sunspot number of ~0 corresponded to a floor in B of ~2.9 nT, near the value of ~2.6 nT inferred from Fig. 2. Cliver and Ling therefore suggested that the floor had a value of ~2.8 nT, significantly below the Svalgaard *et al.* (2007) value of ~4.6 nT and also under Crooker and Owens' revised value of ~3.7 nT. The reduced floor value of 2.8 nT has several advantages. It permits the predicted SSN_{MAX} values to go to ~0 in a precursor relation based on B_{MIN}. It makes the ~50% drop in solar wind B_{MIN} (measured above the floor) between the minima preceding cycles 23 and 24 comparable to the corresponding ~45% drop in DM (Cliver & Ling, 2010). Finally, a floor at ~2.8 nT accommodates most of the 10,000 year ^{10}Be-based reconstruction of B from Steinhilber *et al.* (2010) except for several sharp dropouts where B goes to ~0 nT (Fig. 3).

3. Is the slow solar wind the source of the floor?

Cliver and Ling (2010) noticed that the revised floor value of ~2.8 nT corresponded roughly to the relatively constant contribution of the slow solar wind to annual averages of B (Fig. 4, taken from Cliver and Ling, 2010). They attributed the fact that the contribution of the slow solar wind was generally less than 2.8 nT to the accounting scheme of Richardson *et al.* (2000, 2002) which was designed to apportion geomagnetic activity to the three basic wind types [coronal mass ejections (CMEs), high-speed streams (HSSs), and slow solar wind (SSW)]. In this scheme, co-rotating interaction regions and shock-sheath regions are attributed to HSSs and CMEs, respectively. At solar minimum, when

CMEs essentially disappear and high-stream speeds are less frequent, a more accurate accounting of the SSW contribution can be obtained. During the magnetic minimum year of 2009, Earth was imbedded in SSW ~70% of the time and extrapolations of B to a value of 2.8 nT indicate that Earth would be immersed in SSW, with a speed of ~300 km s^{-1}, over 90% of the time at the floor (Cliver & Ling, 2010). In other words, the slow solar wind is the floor.

4. Is the "minimal solar activity" identified by Schrijver *et al.* the source of the slow solar wind and the floor?

Cliver and Ling (2010) proposed that "the floor corresponds to a baseline (non-cyclic or ground state) open solar flux of $\sim 8 \times 10^{13}$ Wb which originates in persistent small-scale (supergranular or granular) field." Recently, Schrijver *et al.* (2011), from an analysis of a series of Ca II K line data taken at Kitt Peak from 1974-present for the quietest regions on the solar disk, reported "a baseline activity level that is independent of the global sunspot cycle ... regularly observed locally in the quiet-Sun network", i.e., a floor in solar magnetic activity. Quoting from their abstract, "We argue that there is a minimum state of solar magnetic activity associated with a population of relatively small magnetic bipoles which persists even when sunspots are absent ... The minimal solar activity ... was approached globally after an unusually long lull in sunspot activity in 2008-2009. Therefore, the best estimate of magnetic activity ... for the least-active Maunder Minimum phases appears to be provided by direct measurement in 2008-2009." The existence of such a floor in solar magnetic activity is a requirement for a floor in the solar wind. Moreover, the suggestion that 2009 provided our best glimpse thus far of the magnetic state of the Sun during the Maunder Minimum is consistent with the notion that the solar wind during the Maunder Minimum was characterized by slow solar wind. If, in fact, the precursor relation in Fig. 1 extends down to $DM = 0$ ($SSN_{MAX} \sim 0$), then we would expect the polar coronal holes and high-speed stream component of the solar wind (in which Earth was imbedded for ~25% of the time during 2009) to disappear during Maunder-type minima, leaving only slow solar wind.

Schrijver *et al.* (2011) obtained a minimum total unsigned solar flux of 15×10^{22} Mx from SOHO MDI (Scherrer *et al.*, 1995) observations during 2009. Approximately 5% of

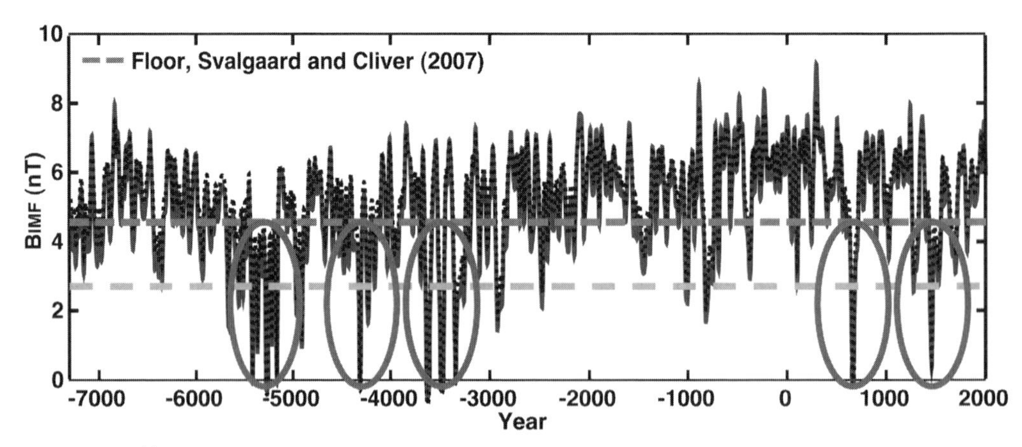

Figure 3. ^{10}Be-based reconstruction of 9,300 years of solar wind B adapted from Steinhilber *et al.* (2010). The positions of the old (dashed green line) and new (dashed blue line) floors are shown. The red ovals encompass the sharp dropouts to ~0 nT. The B data are 40-yr running means. For color figures see the online version of this paper.

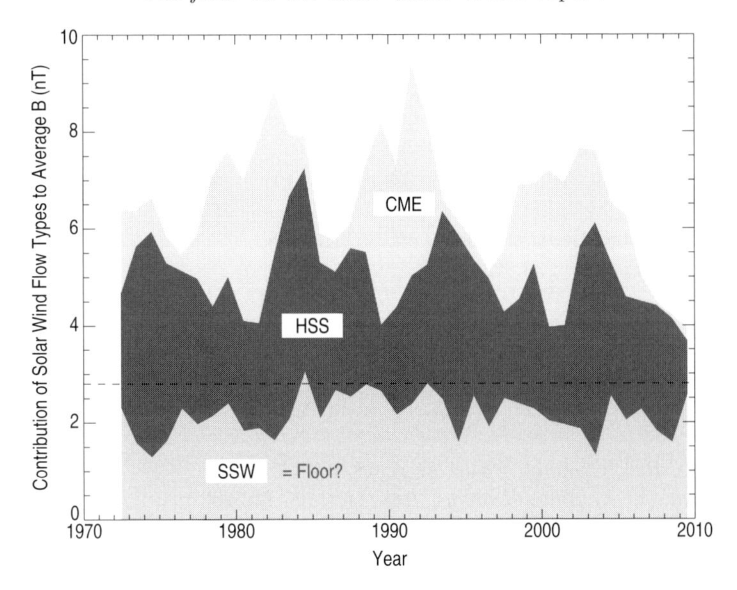

Figure 4. Cumulative distribution of the contributions of SSW, HSSs, and CMEs to average B from 1972-2009. The dashed line is drawn at the floor value of ~2.8 nT (from Cliver and Ling, 2010). See color figure in the online version of this paper.

this flux would need to escape the Sun to account for the floor in B deduced by Cliver and Ling (2010).

5. Questions

Open questions abound. Does the "Livingston-Penn" effect (a secular decrease in the maximum magnetic field strength of sunspots; Livingston & Penn, 2009, and references therein) change the nature of spots at extended minima, thereby affecting the linear extrapolation in the precursor relation between SSN_{MAX} and DM or B_{MIN}? Did the solar cycle continue during the Maunder Minimum at a time when Fig. 1 suggests that the solar dipole field essentially disappeared? The concentration of ^{10}Be in ice cores suggests that it did (Beer *et al.*, 1998; Berggren *et al.*, 2009), while the auroral record (Siscoe, 1980) indicates a general diminution in solar wind activity during this time. Can the "minimal solar activity" of Schrijver *et al.* (2011) maintain a solar wind in the absence of the solar polar fields and, if so, would that solar wind be primarily slow solar wind? When the Maunder Minimum was rediscovered, both Parker (1975) and Eddy (1976) hypothesized that the Sun would resemble a giant coronal hole with high speed wind in all directions. The most recent minimum, which presumably has given us our best glimpse of Maunder Minimum conditions thus far, suggests quite the opposite.

Acknowledgements

We thank Sarah Gibson, Cristina Mandrini, Hebe Cremades, and David Webb for organizing a very timely and stimulating conference.

References

Balogh, A. & Smith, E. J. 2006, AGU Fall Meeting, Abstract No. SH44A-05
Beer, J., Tobias, S., & Weiss, N. 1998, *Solar Phys.* 181, 237

Berggren, A.-M., *et al.* 2009, *Geophys. Res. Lett.* 36, L11801
Cliver, E. W. & Ling, A. G. 2010, *Solar Phys.* DOI: 10.1007/s11207-010-9657
Crooker, N. U. & Owens, M. J. 2010, in: Cranmer, S., Hoeksema, T., & Kohl, J. (eds.) *Proc. of SOHO 23: Understanding a Peculiar Minimum CS-428* (San Francisco: ASP), p. 279
Eddy, J. A. 1976, *Science* 192, 1189
Hoyt, D. V. & Schatten, K. H. 1998, *Solar Phys.* 181, 491
Livingston, W. & Penn, M. 2009, *Eos, Trans. Am. Geophys. Union* 90, 257
Owens, M. J., *et al.* 2008, *Geophys. Res. Lett.* 35, L20108
Parker, E. N. 1975, *Sci. Am.* 233, 42
Richardson, I. G., Cliver, E. W., & Cane, H. V. 2000, *J. Geophys. Res.* 105, 18203
Richardson, I. G., Cane, H. V., & Cliver, E. W. 2002, *J. Geophys. Res.* 107, 1187
Schatten, K. H., Scherrer, P. H., Svalgaard, L., & Wilcox, J. M. 1978, *Geophys. Res. Lett.* 5, 411
Scherrer, P. H., *et al.* 1995 *Solar Phys.* 162, 129
Schrijver, C. J., Livingston, W. C., Woods, T. N., & Mewaldt, R. A. 2011, *Geophys. Res. Lett.* 38, L06701
Siscoe, G. L. 1980, *Rev. Geophys. Space Phys.* 18, 647
Smith, E. J. & Balogh, A. 2008, *Geophys. Res. Lett.* 35, L22103
Steinhilber, F., Abreu, J. A., Beer, J., & McCracken, K. G. 2010, *J. Geophys. Res.* 115, A01104
Svalgaard, L., Cliver, E. W., & Kamide, Y. 2005, *Geophys. Res. Lett.* 32, L01014
Svalgaard, L. & Cliver, E. W. 2007, *ApJ (Lett.)* 661, L203
Svalgaard, L. & Cliver, E. W. 2010, *J. Geophys. Res.* 115, A09111
Wang, Y.-M. & Sheeley, N. R., Jr. 2009, *ApJ (Lett.)* 649, L11

Discussion

Arnab Choudhuri: In our paper of 2007 (Jiang *et al.*, MNRAS, 381, 1527) we made some effort to use the polar faculae as a proxy for the next activity cycle. We found two outliers. We would be happy to discuss this with you.

Ed Cliver: Yes, of course. Comparison of Jiang *et al.* (2007) with Figure 1(b) revealed that the two outliers based on polar faculae (for the minima preceding Cycles 16 and 20) were not the same as for ours (Cycles 15 and 19). While the outliers in the Jiang *et al.* study would have resulted in over-predictions, the outliers in ours corresponded to under-predictions of the cycle peak SSN.

Sacha Brun: Why do you extrapolate between the floor IMF and the sunspot number, as we expect the Sun to drive a dynamo field independent of sunspot emergence, one may want to look instead at the polar field during quiet phases?

Ed Cliver: That comparison/extrapolation is shown in Figure 2 and indicates a solar wind Bmin value of $\sim$ 2.6 nT when the solar polar field strength approaches zero at solar minimum.

Comparative Magnetic Minima:
Characterizing quiet times in the Sun and Stars
Proceedings IAU Symposium No. 286, 2011
C. H. Mandrini & D. F. Webb, eds.

doi:10.1017/S1743921312004826

Probing the heliosphere with the directional anisotropy of galactic cosmic-ray intensity

Kazuoki Munakata[1]

[1]Physics Department, Shinshu University,
3-1-1 Asahi, Matsumoto, Nagano 390-8621, Japan
email: kmuna00@shinshu-u.ac.jp

Abstract. Because of the large detector volume that can be deployed, ground-based detectors remain state-of-the-art instrumentation for measuring high-energy galactic cosmic-rays (GCRs). This paper demonstrates how useful information can be derived from observations of the directional anisotropy of the high-energy GCR intensity, introducing the most recent results obtained from the ground-based observations. The anisotropy observed with the global muon detector network (GMDN) provides us with a unique information of the spatial gradient of the GCR density which reflects the large-scale magnetic structure in the heliosphere. The solar cycle variation of the gradient gives an important information on the GCR transport in the heliosphere, while the short-term variation of the gradient enables us to deduce the large-scale geometry of the magnetic flux rope and the interplanetary coronal mass ejection (ICME). Real-time monitoring of the precursory anisotropy which has often been observed at the Earth preceding the arrival of the ICME accompanied by a strong shock may provide us with useful tools for forecasting the space weather with a long lead time. The solar cycle variation of the Sun's shadow observed in the TeV GCR intensity is also useful for probing the large-scale magnetic structure of the solar corona.

Keywords. Galactic cosmic-rays, Cosmic-ray anisotropy, Cosmic-ray density gradient, Cosmic-ray precursors of ICME, Sun's shadow in 10 TeV cosmic-ray intensity

1. Introduction

Galactic cosmic-rays (GCRs) are extremely high-energy nuclei that travel close to the speed of light in space. They are ubiquitous in the Milky Way and make up a substantial fraction of the total energy of the Galaxy, equivalent to the energy in large-scale magnetic fields and thermal gases. Being charged particles, they are deflected when crossing the magnetic field in the space, and the amount of the total deflection in an average magnetic field magnitude is dependent on both their momentum and path lengths. The cosmic-ray flux at energies high enough to undergo minimal deflection is so small that cosmic-ray sources in the Galaxy far away from us have been proved difficult to be observed directly by measuring the directional anisotropy of GCR intensity. The significant deflection and the pitch angle scattering by the irregular magnetic field, on the other hand, produce the diffusive streaming of GCRs which has been observed as a GCR anisotropy at the Earth with an amplitude of $\sim$0.1-1.0 %. The present paper demonstrates how useful information can be derived from the ground-based observations of the anisotropy.

Ground-based detectors of GCRs use the atmosphere as an active component and measure secondary particles produced from the interaction between the atmospheric nuclei and the primary cosmic-rays which are mostly protons. Because of the large detector volume that can be deployed at ground-based stations, neutron monitors (Simpson, Fonger & Treiman 1953) and muon detectors (Fujimoto *et al.* 1984) remain state-of-the-art instrumentation for measuring $>$ 1 GeV cosmic-rays. These instruments excel at recording time variations of the cosmic-ray flux with a great statistical accuracy which is high enough for measuring the typically small directional anisotropy. Their energy range is highly complementary with the upper range of energies measured by cosmic-ray detectors flown in space.

There are several reasons that particles at these energies are interesting from a space weather perspective. First, they travel nearly at the speed of light. Cosmic-ray particles that interact with a shock or the interplanetary coronal mass ejection (ICME) and escape into the upstream region will race ahead of the much slower shock, bringing advance warning of a disturbance approaching the Earth. Second, the particles have large mean free paths of the pitch angle scattering in the irregular magnetic field. This is important because precursory signatures of an approaching disturbance will be wiped out by scattering on scales larger than a mean free path. Third, the particles have Larmor radii that are large compared to Earth's magnetosphere, but are smaller than or comparable to the scale size of a typical disturbance. For instance, a typical energy of primary GCRs for neutron monitors would be 10 GeV which corresponds to a Larmor radius of 0.02 AU in a 10 nT field. A typical energy for muon detectors is 50 GeV which corresponds to a Larmor radius of 0.1 AU. This is significant because it implies that kinetic anisotropies, such as the diamagnetic drift anisotropy expressed by a product of the Larmor radius and the spatial gradient of the cosmic-ray density, are responding to the large-scale structure of the solar wind disturbance. Instrumentation and methods for using cosmic-rays in space weather applications have advanced dramatically in recent years. The existing muon detector network has been improved to consist of four multi-directional muon detectors in Japan (Nagoya), Australia (Hobart), Brazil (São Martinho) and Kuwait (Kuwait), now forming the Global Muon Detector Network (GMDN). For the detail information of the GMDN and data analyses, readers can refer to Okazaki *et al.* (2008). The most recent results derived from observations with the GMDN are introduced in the following sections 2 and 3, while the solar cycle variation of the Sun's shadow observed in an extremely high-energy GCR intensity is briefly reported in section 4.

2. Observations of the spatial gradient of cosmic-ray density in the heliosphere

The GCR intensity recorded at the Earth changes in the solar activity- and magnetic-cycles reflecting the solar cycle variations of the modulation parameters such as the sunspot number, the magnitude of the interplanetary magnetic field (IMF) and the tilt-angle of the heliospheric current sheet (HCS). The best-known example of such variation is the 11-year variation of the count rate of neutron monitors which represents the temporal variation of the omnidirectional intensity of GCRs. While the omnidirectional intensity measured by a single detector represents the temporal variation of the GCR density at the single location of the detector, the directional anisotropy of the GCR intensity tells us about the spatial distribution of the GCR density around the detector. Since the magnitude of the GCR anisotropy due to the diffusive streaming is proportional to the spatial gradient of the GCR density, one can derive the gradient vector in three dimensions from the anisotropy precisely observed with the global network of detectors. The spatial density gradient is important, because it tells us the average feature of the large-scale magnetic field which is governing the spatial distribution of GCRs in the heliosphere and is still difficult to derive directly from any other in-situ and/or ground-based measurements.

An example of such observations by the GMDN is shown in Fig. 1 which displays the solar cycle variation of the GCR gradient since 2001 (Kozai *et al.* 2011). The negative

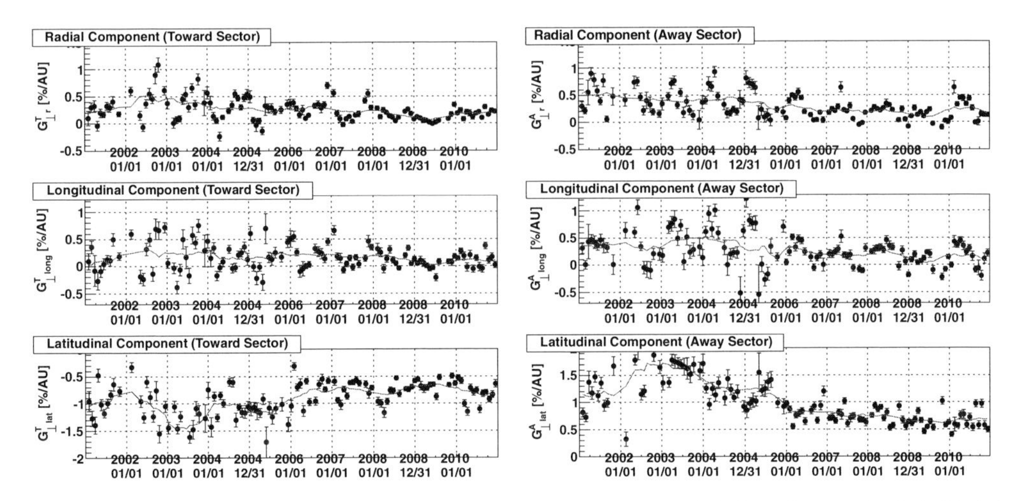

Figure 1. Derived density gradient perpendicular to the IMF. Solid circles in the left (right) panels display the average three components (r, θ, ϕ) of $\boldsymbol{G}_{\perp}(t)$ in the heliocentric polar coordinate system in the Toward (Away) IMF sector in every rotation. The thick curve in each panel represents the central moving average of the solid circles over 14 rotations. From now on, colour figures refer to the online version of this paper.

(positive) latitudinal gradient $G_{\perp lat}(t)$ in Toward (Away) IMF sector in the bottom panel of Fig. 1 indicates the local maximum of the GCR density on the HCS separating IMF sectors. This is qualitatively in accord with the "drift model" prediction for the density distribution during the "neagtive" polarity period of the solar polar magnetic field (also referred as the $A < 0$ epoch) (Kóta & Jokipii 1983). In Fig. 1, it is also clear that $\mid G_{\perp lat}(t) \mid$ decreases significantly from 2003 to 2008-2009 according to the solar activity decrease. The long-term variation of the GCR gradient has attracted less attention than the 11-year variation of the neutron monitor's count rate, but it also provides us with an important information of the global distribution of GCRs reflecting the large-scale magnetic structure of the heliosphere.

Shown in Fig. 2 is another example of the GCR gradient observed in association with the arrival of the ICME accompanied by a strong interplanetary shock (Kuwabara *et al.* 2009). The best-fit density in the top left panel shows a clear signature of a large Forbush decrease (FD) indicating that the Earth entered in the GCR depleted region formed downstream of the shock and inside the ICME. Based on numerical simulations of the cross-field diffusion of GCRs into the magnetic flux rope which is modeled with an expanding straight "cylinder" as an idealized representation of a local section of a flux rope, the best-fitting analyses between the observed and model density gradients are performed and the cylinder geometry, represented by the best-fit parameters including the radius, the expansion speed and the orientation of the cylinder axis, is derived (Munakata *et al.* 2006, Kuwabara *et al.* 2004, 2009). The ICME structure in this event is also determined independently from a magnetic flux rope analyses of the *ACE* (Advanced Composition Explorer) magnetic field and solar wind data and is compared with the one from the cosmic-ray analyses. It is seen in the bottom panels that two geometries are very consistent with each other. From March 2001 to May 2005, 11 ICME events that produced FDs > 2 % were observed, and clear variations of the density gradient due to ICME passage were observed in 8 of 11 events. In five of the eight events, signatures of magnetic flux rope structure (large, smooth rotation of magnetic field) were also seen, and the ICME geometry and orientation deduced from the two methods were very

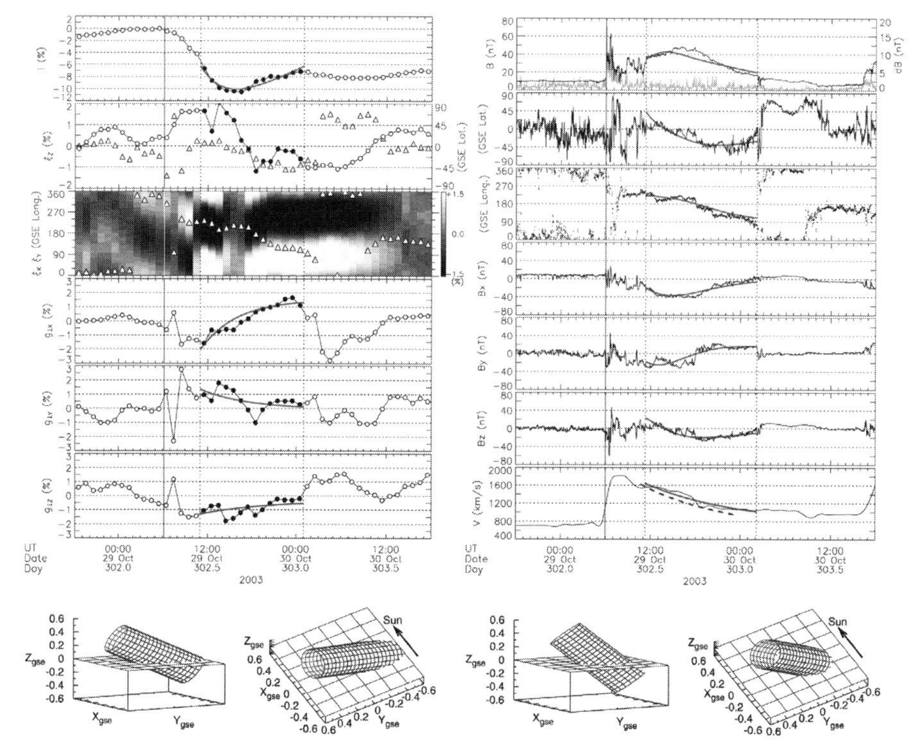

Figure 2. Observation and modeling of ICME geometry on 29 October 2003 (left) from cosmic-ray density gradients observed by the muon detector network and (right) from magnetic flux rope model based upon *ACE* IMF measurements. Each left panel from the top shows hourly value of the observed GCR density, north-south anisotropy, the component anisotropy in the ecliptic plane in a gray scale format, and the three components of the density gradient in GSE coordinates. Latitude and longitude of the hourly mean IMF are plotted over the anisotropy as triangle marks in the second and third panels, respectively. The vertical solid line shows the arrival time of the shock, while the dotted lines show ICME/flux rope boundaries. In bottom panels, the cylinder geometry in GSE coordinates deduced from both methods is compared.

similar in three events. This suggests that the cosmic-ray based method can be used as a complementary method for deducing ICME geometry especially for events where a large FD is observed.

3. Cosmic-ray precursors of the interplanetary disturbance

While the relationship between ICMEs and FDs of cosmic-ray intensity is now well established (Cane 1993, Cane *et al.* 1994, 1996), it is less generally recognized that cosmic-ray decreases are often accompanied by strong enhancements of the cosmic-ray anisotropy (Lookwood 1971, Duggal & Pomerantz 1976; Nagashima *et al.* 1992), some of which extend into the region upstream of the approaching shock. Such precursory anisotropies provide a key mechanism by which information about the presence of a disturbance can be transmitted to remote locations. Because cosmic-rays are fast and have large scattering mean free paths in the solar wind, this information is carried rapidly to the Earth and can be useful for space weather forecasting. Precursory anisotropies have generally been interpreted as kinetic effects related to interaction of ambient cosmic-rays with the approaching shock (Barnden 1971; Nagashima *et al.* 1994; Belov *et al.* 2001; Leerungnavarat *et al.* 2003). Precursory decreases may result from a "loss-cone"

(LC) effect, in which the station is magnetically connected to the cosmic-ray depleted region downstream of the shock. Precursory increases may result from particles that have received a small energy boost by reflecting from the approaching shock (Dorman *et al.* 1995). The LCs are typically visible at the Earth 4-8 hours ahead of the shock arrival associated with major geomagnetic storms (Munakata *et al.* 2000; Rockenbach *et al.* 2011).

An example of a LC precursor preceding a large geomagnetic disturbance in December 2006 is shown in Fig. 3 (Fushishita *et al.* 2010). This figure displays the intensity maps observed with the GMDN during 36 hours between the flare onset (02:14 UT, December 13) and the SSC (14:14 UT, December 14). Each square panel is the observation in one hour indicated by the day-of-year (DOY) above the panel. Each muon detector in the GMDN consists of two identical horizontal layers of unit detectors, vertically separated by a certain distance. By counting pulses of the twofold coincidences between a pair of detectors on the upper and lower layers, one can record the rate of muons from the corresponding incident direction. If one has a square $n \times n$ array of unit detectors aligned to the north-south (or east-west) direction in the i-th muon detector, therefore, one can record muon rates $I^{obs}_{i(k,l)}(t)(k, l = -n+1, ..., 0, ..., n-1)$ in total $(2n-1) \times (2n-1)$ directional channels at the time t, where positive (negative) k and l represent eastern (western) and northern (southern) inclined incident, respectively, with $k = l = 0$ corresponding to the vertical incident. $I^{obs}_{i(k,l)}(t)$ is divided by the statistical count error in each pixel and plotted as a function of k and l respectively on the horizontal and vertical axes in a color-coded format in each square panel of Fig. 3, which is called the "2D significance map" of the muon intensity. Panels in Fig. 3a, c and e display the 2D maps observed by São Martinho (Brazil), Hobart (Australia) and Kuwait (Kuwait), respectively. Red (blue) color in each pixel denotes the excess (deficit) intensity relative to the ominidirectional intensity in an entire field of view (FOV) of each detector. Also shown by white curves in each panel are contour lines of the pitch-angle measured from the sunward IMF direction and calculated for cosmic-rays incident to each pixel. It is seen that the zero pitch-angle region is first captured in the FOV of São Martinho in DOY 347.354-347.563 (08:00-13:00 UT, December 13), and then by Hobart in DOY 347.771-347.979 (18:00-23:00 UT, December 13) and again by São Martinho in DOY 348.354-348.563 (08:00-13:00 UT, December 14) according to the Earth's spin. A striking feature of this event is that a LC signature (i.e. intensity deficit around the zero pitch-angle) is seen first by São Martinho more than one day before the SSC. This suggests that the LC precursor already existed only 7 hours after the CME eruption at 02:54 UT on December 13, when the interplanetary shock driven by an ICME was located at 0.4 AU from the Sun (Liu *et al.* 2008).

Based on a pitch angle distribution deduced from numerical simulations of high-energy particle transport across the shock (Leerungnavarat *et al.* 2003), the best-fit analyses to the data in Fig. 3 are carried out and the best-fit parameters denoting the model anisotropy are derived. The 2D maps reproduced from the best-fit parameters are shown in Figs. 3b, 3d and 3f. The best-fit analyses indicate that the lead time of this LC precursor is as long as 16 hours and the maximum intensity deficit at the LC center exceeds -6 % which is almost twice the size of the FD observed with the GMDN. This implies that the maximum intensity depression behind the shock is much larger than the FD size recorded at the Earth in this event. The precursor observed with the long lead time, like the event in Fig. 3, is of particular importance for the possible space weather forecast using cosmic-ray measurements. For an accurate observation of such event, however, one needs further improvement of the GMDN. First, an incomplete

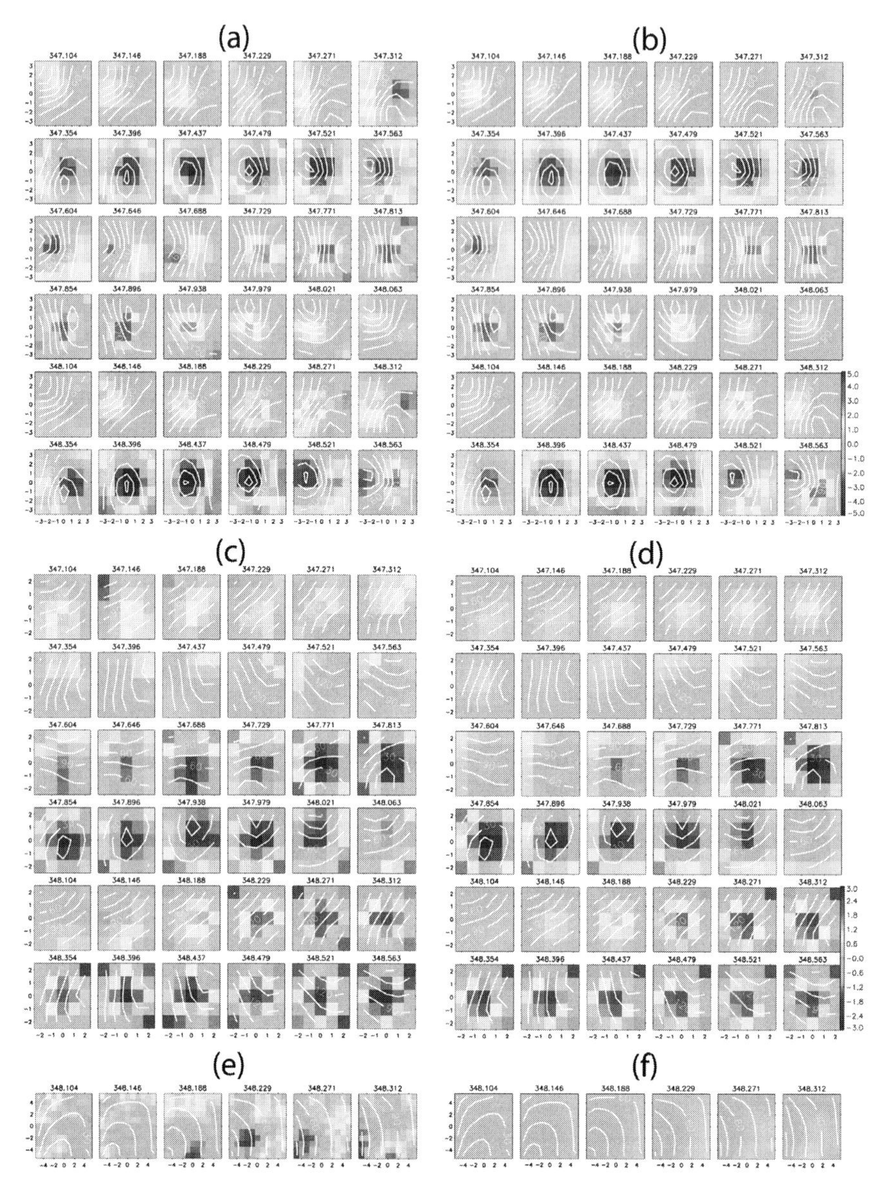

Figure 3. 2D significance maps observed by the GMDN prior to the SSC onset on December 14, 2006. The color scale is set ± 5 (five times the statistical error) for São Martinho and ± 3 (three times the statistical error) for Hobart and Kuwait as indicated by color bars at the right bottom corners of panels b and d. All maps during 36 hours between the flare onset and the SSC are shown in a-d for São Martinho and Hobart, while maps by Kuwait are shown only for 6 hours when the weak LC signature is visible with this small detector. The integers attached along the x- and y-axes in each small panel indicate k and l, representing respectively the east-west and north-south inclinations of the viewing direction (see text).

sky-coverage of the GMDN allowed us to analyze the LC anisotropy only for half a period, when the sunward IMF direction was in the FOV of the GMDN. Second, the insufficient detection areas in the network increased the statistical error and introduced a non-uniformity into the response of the GMDN to the LC anisotropy. Such a non-uniform response also introduced large fluctuations in the obtained best-fit parameters.

These problems can be solved by expanding the detection area of the smallest detector in Kuwait and also by installing new detector(s) to expand the FOV of the GMDN, as a preparation for the next solar maximum expected at around the year of 2013.

4. Solar cycle variation of the Sun's shadow observed in 10 TeV GCR intensity

The anisotropy of GCRs with extremely high energies also provides us with unique information of the large-scale magnetic field of the Sun. The Larmor radius of a 10 TeV (10 tera-electron-volt or 10^{13} eV) proton in 5 nT IMF is as large as 40 AU and GCRs with this energy travel nearly straight in the interplanetary space between the Sun and Earth. The Sun and Moon shield GCRs arriving from the directions behind them and cast tiny "shadows" in the GCR intensity observed at the Earth. The Sun's shadow is of particular interest because it reflects the large-scale solar magnetic field near the Sun, which is still difficult to observe with any direct and/or remote measurements. The Tibet Air Shower (AS) experiment at Yangbajing in Tibet, China has succeeded for the first time in observing a clear solar cycle variation of the Sun's shadow over an entire solar cycle. The Tibet AS experiment achieved the world highest count rate ($\sim$ 230 Hz) and the best angular resolution ($\sim 0.9°$) of the GCR incident direction. For the Tibet AS experiment, readers can refer to Amenomori *et al.* (2009).

For analyzing the Sun's shadow, an on-source event number ($N_{\rm on}$) is first defined as the number of events arriving from the direction within a circle of 0.9° radius from a certain point on the celestial sphere. The background or off-source event number ($N_{\rm off}$) is also calculated by averaging the number of events over the eight off-source windows which are located at the same zenith angle as the source direction but apart in the azimuthal direction. Both $N_{\rm on}$ and $N_{\rm off}$ are calculated on every 0.05° grid of the Geocentric Solar Ecliptic (GSE) longitude and latitude surrounding the optical Sun's center. The deficit intensity relative to the background event number is then estimated as $D_{\rm obs} = (N_{\rm on} - N_{\rm off})/N_{\rm off}$ at every grid by using the yearly mean $N_{\rm on}$ and $N_{\rm off}$. Shown in Fig. 4 are 2D maps of $D_{\rm obs}$ for every year from 1996 to 2009 (Amenomori *et al.* 2011). In each panel of this figure, $D_{\rm obs}$ is plotted as a function of the GSE longitude and latitude on the horizontal and vertical axes, respectively. It is remarkable in this figure that $D_{\rm obs}$ changes considerably in the solar activity cycle, i.e. the shadow becomes clear (with larger negative $D_{\rm obs}$) around 1996 and 2008 when the solar activity is close to the minimum, while it becomes faint (with smaller negative $D_{\rm obs}$) around 2000 when the activity was high. Since the Moon's shadow observed with the same apparatus during the same period is quite constant (Amenomori *et al.* 2009), the variation of the Sun's shadow in this figure is not likely of the instrumental origin.

For quantitative analyses of the temporal variation in Fig. 4, the average ($\bar{D}_{\rm obs}$) of $D_{\rm obs}$ within an on-source window around the origin of 2D map is calculated for every year. Fig. 5c shows the temporal variation of $\bar{D}_{\rm obs}$ together with the sunspot number and the tilt-angle of the HCS (Wilcox Solar Observatory 2010). It is seen in Fig. 5c that $\bar{D}_{\rm obs}$ shows a clear solar cycle variation with an amplitude as large as 50 % of $\bar{D}_{\rm opt}$ displayed by the horizontal line, which represents the deficit intensity expected when all cosmic-rays from the direction of the optical Sun disk are excluded from the observation. Fig. 5 shows a good correlation between the variations of $\bar{D}_{\rm obs}$ and the sunspot number or the HCS tilt-angle with the absolute correlation coefficient ($|r|$) exceeding 0.85. To examine the central position of the Sun's shadow, $D_{\rm obs}$ in Fig. 4 are projected onto the horizontal and vertical axes and then the central GSE longitude and latitude of the projection are obtained for every year. In the average GSE latitude, no significant deviation from

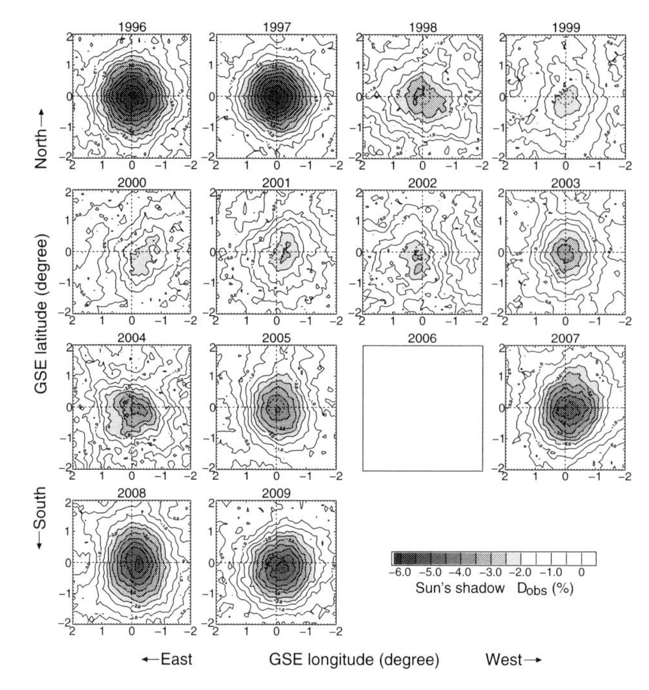

Figure 4. 2D maps of the observed Sun's shadow for every year between 1996 and 2009. Each panel displays a contour map of the yearly mean deficit intensity ($D_{\rm obs}$) as a function of the GSE longitude and latitude on the horizontal and vertical axes, respectively (see text). The data in 2006 are excluded from the plot since the AS array was scarcely operating due to upgrading the array and also due to instrumental troubles.

0.0° is found. The average GSE longitude in 1996-1997 is also only $+0.039° \pm 0.038°$ which is significantly smaller than the observed geomagnetic deflection of the Moon's shadow ($-0.120° \pm 0.024°$) (Amenomori *et al.* 2009). The average longitude in 2007-2009 during the next solar minimum period, on the other hand, is $-0.149° \pm 0.036°$ and significantly different from the longitude during the previous minimum in 1996-1997. This is due to the reversal of the solar dipole field during the solar maximum period around 2000. The magnetic deflection of the Sun's shadow in the solar corona is canceled by the geomagnetic deflection in the opposite sense in 1996-1997 solar minimum, while it is enhanced by the geomagnetic deflection in the same sense in 2007-2009 solar minimum. The detailed Monte Carlo (MC) simulations based on the coronal field model are now in progress aiming to clarify the physical implications of the observed solar cycle variation of the Sun's shadow. These simulations of the Sun's shadow enable us to examine coronal field model and probe the large-scale structure of the solar magnetic field as a function of time in the solar cycle.

Acknowledgements

The observations with the GMDN have been supported in part by Grants-in-Aid for Scientific Research from the Ministry of Education, Culture, Sports, Science and Technology in Japan, and by the joint research programs of the Solar-Terrestrial Environment Laboratory, Nagoya University. The observations with the Kuwait Muon Telescope are supported by the Kuwait University grant SP03/03. We thank N. F. Ness for providing the *ACE* magnetic field data via the ACE Science Center. The collaborative experiment

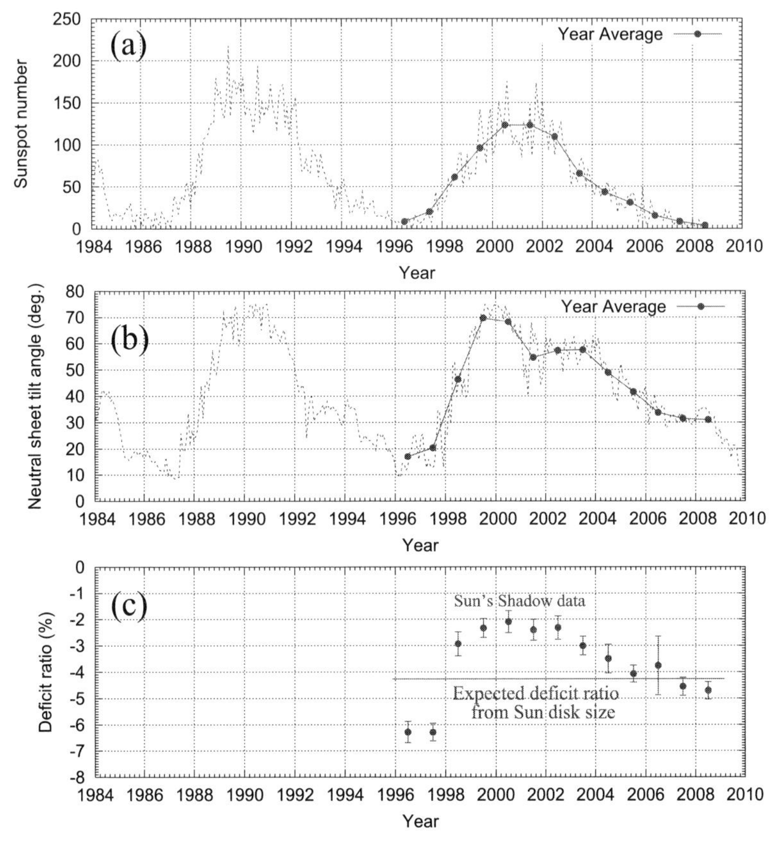

Figure 5. Solar cycle variations of the intensity deficit in the Sun's shadow. The solid circles in each panel from the top displays the yearly mean value of (a) the sunspot number, (b) the HCS tilt-angle in degree and, (c) yearly mean deficit intensity observed in the Sun's shadow ($\bar{D}_{\rm obs}$). The horizontal solid line in panel (c) indicates the deficit intensity expected when all cosmic-rays from the direction of the optical Sun disk are excluded from the observation.

of the Tibet Air Shower Arrays has been performed under the auspices of the Ministry of Science and Technology of China and the Ministry of Foreign Affairs of Japan. This work was supported in part by a Grant-in-Aid for Scientific Research on Priority Areas from the Ministry of Education, Culture, Sports, Science and Technology, by Grants-in-Aid for Science Research from the Japan Society for the Promotion of Science in Japan, and by the Grants from the National Natural Science Foundation of China and the Chinese Academy of Sciences.

References

Amenomori, M., Bi, X. J., Chen, D., Cui, S. W., & Danzengluobu, Ding, L. K. *et al.* 2009, *ApJ*, 692, 61

Amenomori, M., Bi, X. J., Chen, D., Chen, W. Y., & Cui, S. W., Danzengluobu *et al.* 2011, *Proc. of the 32nd Internat. Cosmic Ray Conf. (Beijing)*, 11, 242

Barnden, L. R. 1971, *Solar Phys.*, 18, 165

Belov, A. V., Bieber, J. B., Eroshenko, E. A., Evenson, P., Pyle, R., & Yanke, V. G. *et al.* 2001, *Proc. of the 27th Internat. Cosmic Ray Conf. (Hamburg)* (Schaltungsdienst Lange o.H.G., Berlin), 9, 3507

Cane, H. V. 1993, *J. Geophys. Res.*, 98, 3509

Cane, H. V., Richardson, I. G., von Rosenvinge, T. T., & Wibberenz, G. 1994, *J. Geophys. Res.*, 99, 21429
Cane, H. V., Richardson, I. G., & von Rosenvinge, T. T. 1996, *J. Geophys. Res.*, 101, 21561
Dorman, L. I., Iucci, N., & Villoresi, G. 1995 *Proc. of the 24th Internat. Cosmic Ray Conf. (Rome)*, 4, 892
Duggal, S. P. & Pomerantz, M. A. 1976, *J. Geophys. Res.*, 81, 5032
Fjimoto, K., Inoue, A., Murakami, K., & Nagashima, K. 1984, *Report of Cosmic-Ray Research Lab.*, No.9, Nagoya University
Fushishita, A., Kuwabara, T., Kato, C., Yasue, S., Bieber, J. W., & Evenson, P. *et al.* 2010, *ApJ*, 715, 1239
Kóta, J. & Jokipii, J. R. 1983, *ApJ*, 265, 573
Kozai, M., Munakata, K., Kato, C., Yasue, S., Kuwabara, T., & Bieber, J. W. *et al.* 2011, *Proc. of the 32nd Internat. Cosmic Ray Conf. (Beijing)*, 11, 301
Kuwabara, T., Munakata, K., Yasue, S., Kato, C., Akahane, S., & Koyama, M. *et al.* 2004, *Geophys. Res. Lett.*, 31, L19803-1
Kuwabara, T., Bieber, J. W., Evenson, P., Munakata, K., Yasue, S., & Kato, C. *et al.* 2009, *J. Geophys. Res.*, 114, A05109-1
Leerungnavarat, K., Ruffolo, D., & Bieber, J. W. 2003, *ApJ*, 593, 587
Liu, Y., Luhmann, J. G., Müller-Mellin, R., Schroeder, P. C., Wang, L., & Lin, R. P. *et al.* 2008, *ApJ*, 689, 563
Lookwood, J. A. 1971, *Space Sci. Revs*, 12, 658
Munakata, K., Bieber, J. W., Yasue, S., Kato, C., Koyama, M., & Akahane, S. *et al.* 2000, *J. Geophys. Res.*, 105, 27457
Munakata, K., Yasue, S., Kato, C., Kóta, J., Tokumaru, M., & Kojima, M. *et al.* 2006, *Adv. Geosci.*, 2, 115
Nagashima, K., Fujimoto, K., Sakakibara, S., Morishita, I., & Tatsuoka, R. 1992, *Planet. Space Sci.*, 40, 1109
Okazaki, Y., Fushishita, A., Narumi, T., Kato, C., Yasue, S., & Kuwabara, T. *et al.* 2008, *ApJ*, 681, 693
Rockenbach, M., Dal Lago, A., Gonzalez, W. D., Munakata, K., Kato, C., & Kuwabara, T. *et al.* 2011, *Geophys. Res. Lett.*, 38, L16108-1
Simpson, J. A., Fonger, W., & Treiman, S. B. 1953, *Phys. Rev.*, 90, 934
Wilcox Solar Observatory 2010, *WSO Computed "Tilt Angle" of the Heliospheric Current Sheet*, http://wso.stanford.edu/Tilts.html

Discussion

Ramón López: For this Sun shadow variation with the solar cycle, is the main driver the variation of the solar magnetic field?

Kazuoki Munakata: Yes, the strong magnetic field near the Sun has a major contribution in producing the Sun's shadow, we think.

David Webb: What is the energy of the particles in the shadow modeling?

Kazuoki Munakata: It is 10 TeV.

Comparative Magnetic Minima:
Characterizing quiet times in the Sun and stars
Proceedings IAU Symposium No. 286, 2011
C. H. Mandrini & D. F. Webb, eds.

doi:10.1017/S1743921312004838

Search for solar energetic particle signals in the Mexico City neutron monitor database

B. Vargas-Cárdenas[1]
J. F. Valdés-Galicia[2]

[1]Instituto de Geofísica, UNAM, México D.F., C.P. 04510.
email: `bernardo@geofisica.unam.mx`

[2]Instituto de Geofísica, UNAM, México D.F., C.P. 04510.
email: `valdes@geofisica.unam.mx`

Abstract. We performed a search for ground level solar cosmic ray enhancements on the full five minute database of the Mexico City neutron monitor using wavelet filters and two different statistical tests. We present a detailed analysis of the time series of November 2, 1992, where we found a previously unreported increment matching the onset time of the impulsive phase of the Ground Level Enhancement No. 54, thus providing evidence of an effective detection of high energy solar cosmic rays. This technique may help to find still undiscovered GLE signals in the Worldwide Neutron Monitor Database, to refine GLE spectra and, probably, to find a relationship between the latter and the solar cycle.

Keywords. Ground level enhancements, solar energetic particles

1. Introduction

There is an increasing interest in evaluating the sun's maximum capacity to accelerate particles. This has motivated different studies with data from high cutoff rigidity neutron monitors (Debrunner *et al.* 1997; Karapetyan 2008; Beisembaev *et al.* 2009; Chilingarian 2009). Therefore, we decided to search for GLE signals in the whole database of Mexico City neutron monitor (http://132.248.105.25/). Due to the site's high cutoff rigidity (8.27 GV) and high median response energy of the instrument to solar particles (10.7 GeV), any such detection would provide evidence for the production of high energy particles by solar eruptive phenomena. We found only one signal on November 2, 1992, whose significance could be validated by two different statistical tests, namely, Student's test (Karapetyan 2008; Beisembaev *et al.* 2009) and extremum statistics test (Chapman, Rowland & Watkins 2002; Chilingarian 2009). Some authors (Shea & Smart 2008; Belov, Eroshenko & Kryakunova 2009) have studied the distribution of Ground Level Enhancements (GLE's) along the solar cycle, but it's not clear what is the relationship between sunspot numbers and GLE frequency, intensity or spectra.

2. Data and Qualitative Analysis

GOES-7 satellite registered an X9.0 flare beginning at 2:31 with a peak at 3:08 and ending at 3:28 UT of November 2 (Fig. 1). There were two coronal mass ejections associated with this flare; one starting at 2:53 with a type II radio burst and the other at 3:02 UT with a type IV burst. Although there was no sudden commencement, the Kp index increased to a value of 5.5, and a sudden ionospheric disturbance of importance 3 and a widespread index equal to 5 started at 2:32, peaked at 3:04 and ended at 9:40 UT.

The direction of the IMF at 3:40 UT was 41°, a little eastward from the nominal Parker spiral, favoring the asymptotic directions of North American stations. To validate our claimed detection of a GLE signal on event 54, we also analyzed pressure corrected five minute count rates of Calgary, Deep River, Goose Bay, Hermanus, Inuvik, LARC, Lomnicky Stit, Oulu, Potchefstroom, SANAE and Tsumeb neutron monitors, as suggested by the Australian Antarctic Data Center (http://data.aad.gov.au/aadc/gle/); all these data were downloaded from the FTP site of IZMIRAN (ftp://cr0.izmiran.rssi.ru/). Dates and onset times of the events were taken from Miroshnichenko & Pérez-Peraza (2008), and Shea *et al.* (1995). We used the baseline intervals suggested by the Bartol Research Institute (http://neutronm.bartol.udel.edu/). Solar data were taken from Solar Geophysical Data 580 and 581. All these neutron monitor data show the signature of the gradual phase of the event, beginning approximately at 4:00 and ending about 9:00 UT, but Deep River and Goose Bay show signs of the impulsive phase.

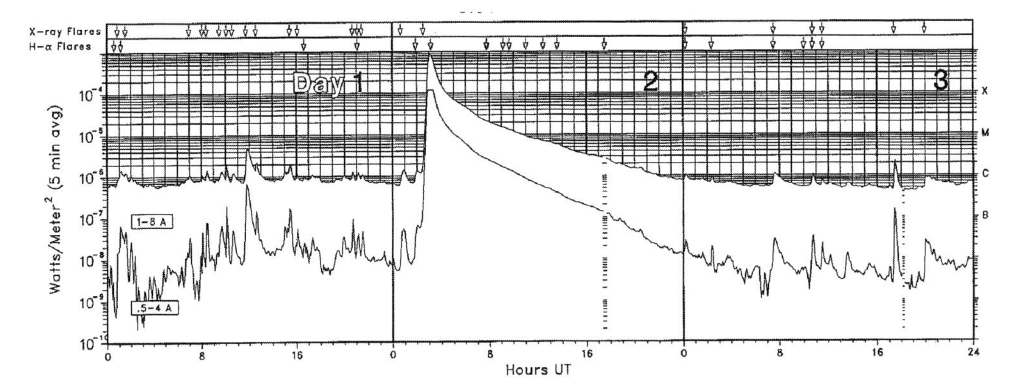

Figure 1. GOES-7 satellite hard and soft X ray data for November 1 to 3, 1992 taken from Solar Geophysical Data.

3. Results

Fig. 2 is a plot of Deep River and Goose Bay data flattened by wavelet based Daubechies filters (Daubechies 1988, 1993); both the impulsive and the gradual phases of the event are clearly distinguished. Deep River shows an increase in its count rates of 1.2% at 3:35 UT while Goose Bay has a 1.4% increase 3:50 UT; this corresponds to the impulsive phase of the event.

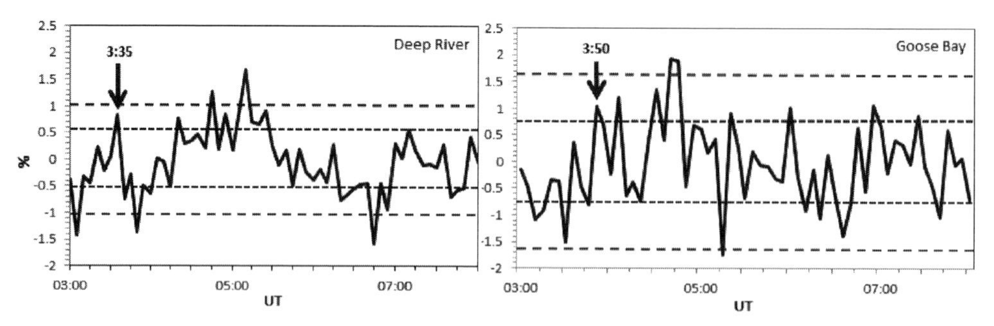

Figure 2. Flattened Deep River (left) and Goose Bay (right) neutron monitors data for November 2, 1992. Dashed lines indicate σ and 2σ thresholds. The two components of the event are evident on both plots.

A second increase is observed between 4:30 and 6:30 UT, approximately; this is the signature of the gradual phase. Mexico City neutron monitor did not see the lower energy gradual phase, but it shows a 2.1% increase at 3:45 UT falling above the 2σ threshold (Fig. 3), which should correspond to the impulsive phase. The difference in the timing of this peak between the three mentioned stations could be due to a high anisotropy in the high energy particle flux, but their coincidence in recording the event may be explained by the location of their corresponding asymptotic cones of acceptance, which are close to each other (Smart, Shea & Flckiger 2000).

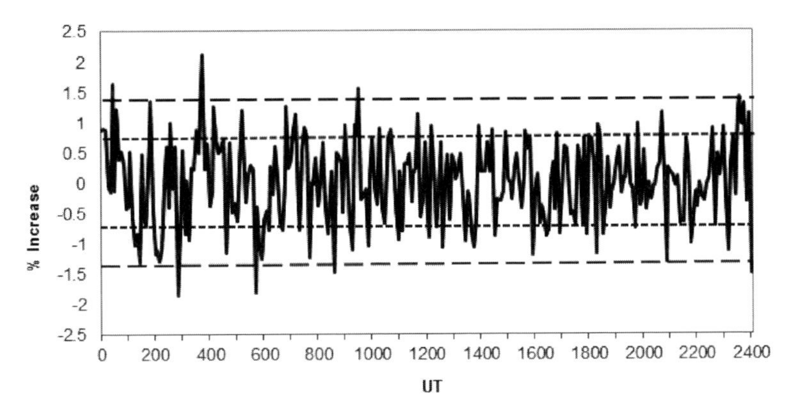

Figure 3. Mexico City data for November 2, 1992. Dashed lines represent σ and 2σ thresholds. The 2.1% increment observed at 3:45 UT corresponds to the onset time of GLE 54.

Mexico City's cone of acceptance had a better connection with the IMF direction, hence the higher value of its increase in the count rates, irrespective of the higher cut off rigidity of the station. Fig. 3 represents Mexico City's time series corresponding to November 2, 1992, also flattened by Daubechies filters.

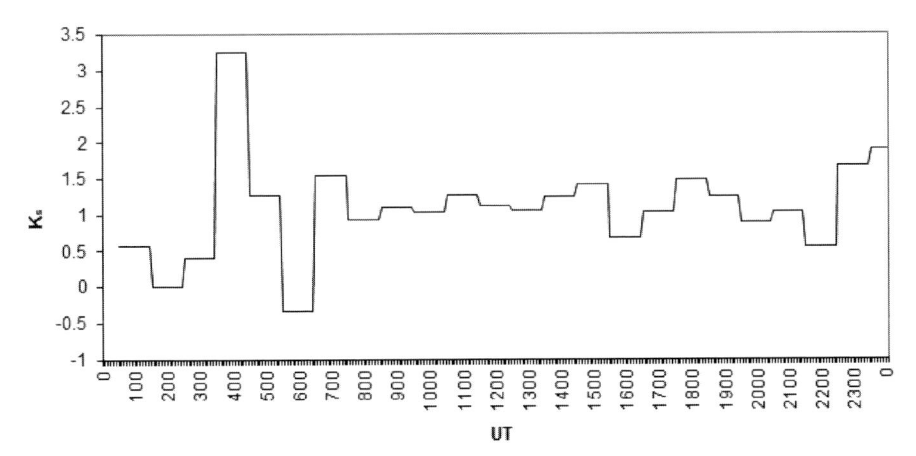

Figure 4. One hour Student's test for the Mexico City November 2, 1992 time series showing the significance of the 3:45 increment.

The 2.1% increment seen at 3:45 UT matches the reported onset time of GLE 54 and exceeds in 0.75 units the 2σ threshold. This is the only positive peak that exceeds the 2σ limit in the relevant time period. This result, together with the Goose Bay and Deep River flattened time series (Fig. 2) encouraged us to do the statistical analysis explained in Section 3. These could only be applied to Mexico City data as it is the only station without any trace of the gradual phase of the event. Fig. 4 shows Student's test for this time series. For the period including the peak at 3:45 UT we get a statistical significance of 99.8%. Fig. 5 shows the plot of normalized residuals of the Mexico City neutron monitor data.

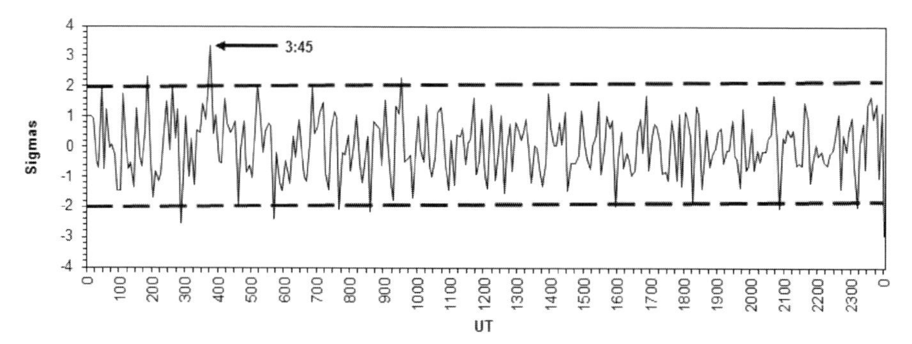

Figure 5. Residuals for the November 2, 1992 time series showing the significance of the 3:45 increment according to the extremum statistics test. Dashed lines represent the 2σ threshold.

The residual corresponding to 3:45 UT is clearly standing out well above the 2σ level and no other residual falls above it. The extremum statistics test gives a statistical significance of 96% for this peak. This is thus further evidence that the register in question might not correspond to a fluctuation of the isotropic galactic cosmic ray flux. As a consequence of the statistical tests performed to the Mexico City neutron monitor data during November 2, 1992, we can accept the hypothesis that we have found an additional cosmic ray signal superimposed to the galactic cosmic ray intensity with a statistical confidence that is beyond 96%. As there is no other compatible source, this signal should correspond to particles produced during the impulsive phase of GLE 54. The instrument's mean counting rate of 76,470 counts/hour = 6,373 counts/5 min with a corresponding standard deviation of 0.88% are an additional validation of the statistical results presented in this work. We calculate that under such disturbed geomagnetic conditions, the cutoff rigidity of the Mexico City site got lowered to a value of 7.5 GV; likewise, we have computed the median response energy of the instrument for the combined galactic and solar spectrum, using the upper limit spectum for protons (Miroshnichenko & Pérez-Peraza 2008) and got a value of 10.7 GV. Thus, we can infer that the Sun emitted particles with energies of at least 7.5 and most probably 10.7 GeV during the X9.0 flare that occurred on November 2, 1992.

4. Conclusions

We found evidence of the detection of solar particles during the event on 2 November 1992 (GLE 54) by the Mexico City neutron monitor, validated by two independent statistical tests, supporting the hypothesis that particles with energies around 10 GeV were produced by the Sun. Mexico City data, as well as those from other middle and low latitude stations, can contribute substantially to refine solar energetic particle spectra and therefore to improve the corresponding acceleration models. Furthermore, they may help to evaluate the solar capability to produce high energy protons. In particular, these results may help to refine the upper limit spectrum for protons (Miroshnichenko 1996; Miroshnichenko and Pérez-Peraza 2008). Also, a more accurate knowledge of GLE spectra could lead to find a relationship with the solar cycle. Finally, we conjecture that there may be more solar particle event signals still undiscovered in the Worldwide Neutron Monitor Network's database that require refined statistical analysis to be revealed.

References

Beisembaev, R., Dorbzhev, V., Dryn, E., & Kryakunova, O. 2009, *Adv. in Sp. Res.*, 43, 509

Belov, A. V., Eroshenko, E. A., Kryakunova, O. N., Kurt, V. G., & Yanke, V. G. 2009, *Proceedings of the 31st Int. Cosmic Ray Conf.*, Lodz

Chapman, S. C., Rowlands, G., & Watkins, N. W. 2002, *Nuclear Processes in Geophysics*, 9, 409

Chilingarian, A. 2009, *Adv. in Sp. Res.*, 43, 702

Daubechies, I. 1988, *Communications on Pure and Applied Mathematics*, 41, 909

Daubechies, I. 1993, *SIAM Journal on Mathematical Analysis*, 24, 499

Debrunner, H., Lockwood, J. A., Barat, C., Btikofer, R., Dezalay, J. P., Flckiger, E., Kuznetsov, A., Ryan, J. M., Sunyaev, R., Terekhov, O. V., Trottet, G., & Vilmer, N. 1997, *Astrophys. J.*, 479, 997

Karapetyan, G. G. 2008, *Astroparticle Physics*, 30(5), 234

Miroshnichenko, L. I. 1996, *Radiation Measurements*, 26(3), 421

Miroshnichenko, L. I. & Pérez-Peraza, J. A. 2008, *International Journal of Modern Physics A*, 23(1), 1

Shea, M. A., Smart, D. F., Gentile, L. C., & Campbell, J. M. 1995, in: N. Iucci, E. Lamanna (Eds.), *Proceedings of the 24nd Int. Cosmic Ray Conf.* (International Union of Pure and Applied Physics, Rome), Vol. 4, 244

Shea, M. A. & Smart, D. F. 2008, in: R. Caballero, J. C. D'Olivo, G. Medina-Tanco, L. Nellen, F. A. Snchez, J. F. Valds-Galicia (eds.), *Proceedings of the 30th Int. Cosmic Ray Conf.* (Universidad Nacional Autnoma de Mxico, Mexico City), Vol. 1, 261

Smart, D. F., Shea, M. A., & and Flckiger, E. O. 2000, *Space Science Reviews* 93, 305

Solar Geophysical Data No. 580, Part I 1992, *National Geophysical Data Center, Boulder*, December 1992

Solar Geophysical Data No. 581, Part II 1993, *National Geophysical Data Center, Boulder*, January 1993

Discussion

JANET LUHMANN: Do you have any doubts to see a solar cycle dependence of the GLEs? Wouldn't you expect it to depend on the existence of very fast CME shocks in the corona which generally happen during active times?

BERNARDO VARGAS: We believe these are flare-accelerated particles and not CME-accelerated particles, since shocks cannot accelerate particles to these energies; I mean, nobody has explained how a shock could accelerate particles to these high energies in an efficient manner. Don't forget these particles are relativistic.

Comparative Magnetic Minima:
Characterizing quiet times in the Sun and Stars
Proceedings IAU Symposium No. 286, 2011
C. H. Mandrini & D. F. Webb, eds.

doi:10.1017/S174392131200484X

Extremely low geomagnetic activity during the recent deep solar cycle minimum

E. Echer[1], B. T. Tsurutani[2] and W. D. Gonzalez[1]

[1]National Institute for Space Research (INPE)
S. J. Campos, SP, Brazil
email: ezequiel.echer@gmail.com, gonzalez@dge.inpe.br

[2]Jet Propulsion Laboratory(JPL), California Institute of Technology (CALTECH)
Pasadena, CA, USA
email: bruce.tsurutani@jpl.nasa.gov

Abstract. The recent solar minimum (2008-2009) was extreme in several aspects: the sunspot number, R_z, interplanetary magnetic field (IMF) magnitude B_o and solar wind speed V_{sw} were the lowest during the space era. Furthermore, the variance of the IMF southward B_z component was low. As a consequence of these exceedingly low solar wind parameters, there was a minimum in the energy transfer from solar wind to the magnetosphere, and the geomagnetic activity *ap* index reached extremely low levels. The minimum in geomagnetic activity was delayed in relation to sunspot cycle minimum. We compare the solar wind and geomagnetic activity observed in this recent minimum with previous solar cycle values during the space era (1964-2010). Moreover, the geomagnetic activity conditions during the current minimum are compared with long term variability during the period of available geomagnetic observations. The extremely low geomagnetic activity observed in this solar minimum was previously recorded only at the end of XIX century and at the beginning of the XX century, and this might be related to the Gleissberg (80-100 years) solar cycle.

Keywords. Sun: activity, Sun: magnetic fields, solar-terrestrial relations, solar wind

1. Introduction

The recent solar cycle was the longest with the deepest minimum in sunspot number (R_z) values in the space era (Gibson *et al.* 2009; Hathaway 2010; Russell, Luhmann & Jian 2010; de Toma 2011; Echer *et al.* 2011; Tsurutani, Echer & Gonzalez 2011). The deep minimum is also reflected in several solar observations, such as solar irradiance (Fröhlich 2009) and radio flux at 10.7 cm (Hathaway 2010), solar polar magnetic field (Wang, Robbrecht & Sheeley 2009), and in solar wind parameters at 1 AU (Luhmann *et al.* 2009; Russell, Luhmann & Jian 2010; Tsurutani, Echer & Gonzalez 2011), and in the outer heliosphere (McComas *et al.* 2008; Smith & Balogh 2008). The polar coronal holes were smaller in this minimum relative to the previous minimum (Kirk *et al.* 2009; de Toma 2011). Furthermore, extreme flux enhancements were observed in cosmic rays (Heber *et al.* 2009; Tsurutani, Echer & Gonzalez 2011). As a consequence, there have been many geophysical effects, both in the magnetosphere and ionosphere/atmosphere (Gibson *et al.* 2009; Minamoto & Taguchi 2009; Kataoka & Miyoshi 2010; Ram, Lin & Su 2010; Russell, Luhmann & Jian 2010; Echer *et al.* 2011; Emery *et al.* 2011; Verkhoglyadova *et al.* 2011).

Transference of solar wind energy to the magnetosphere depends on the interplanetary magnetic field (IMF) magnitude B_o, IMF direction and solar wind speed V_{sw} (Gonzalez *et al.* 1994; Echer *et al.* 2005). We examine these parameters at 1 AU in addition to the solar wind-magnetosphere energy coupling ϵ parameter (Perreault & Akasofu 1978) and the *ap* geomagnetic activity index (Bartel 1950; Rostoker 1972). Annual averages of geomagnetic indices for periods previous to the space era are also analysed. The study of these parameters will indicate the cause of the low geomagnetic activity during solar minimum.

2. Methodology

In this work, we investigate the long term variation of solar wind and geomagnetic parameters. We used 27-day Bartel rotation averages of solar wind parameters from NASA/GSFC solar wind data base (omniweb.gsfc.nasa.gov/). Also, 27-day averages of the sunspot number (Eddy 1976; Echer *et al.* 2005) and the geomagnetic activity *ap* index (Rostoker 1972) have been used to assess solar and geomagnetic activity. In order to estimate the solar-wind magnetosphere energy coupling, the ϵ parameter (Perreault & Akasofu 1978) is computed. For this calculation, we used high resolution 1-hour solar wind data and then averaged over 27-day Bartel rotations. The ϵ parameter is calculated as:

$$\epsilon = 10^7 l_{\rm o}^2 V_{\rm sw} B_{\rm o}^2 sin^4(\theta/2) \; [1]$$

The interplanetary parameters used in Eq. (1) are the IMF magnitude B_o, the IMF clock angle θ in the plane perpendicular to the Sun-Earth line, and the solar wind speed V_{sw}. The factor l_o is an empirically determined scale factor with the physical dimension of length, 7 R_E, representing the magnetopause radius. The parameter ϵ is given in [W].

For the long term evaluation of geomagnetic activity, the annual averages of the *ap*, *AE*, and *Dst* geomagnetic indices (Bartel 1950; Rostoker 1972), the *aa* index and the sunspot number R_z have been used. The *ap* and *aa* index were obtained from the National Geophysical Data Center (www.ngdc.noaa.gov/) and the R_z from the Solar Influences Data Center (sidc.oma.be/). The *AE* and *Dst* indices were obtained from the World Data Center for Geomagnetism (wdc.kugi.kyoto-u.ac.jp/). The mean magnetic field of the Sun (Scherrer *et al.* 1977) data was obtained from the Stanford University Wilcox Solar Observatory (wso.stanford.edu).

3. Results

3.1. *Space era*

Fig. 1 shows, from top to bottom, the sunspot number (R_z), the solar wind speed V_{sw}, the 1 AU interplanetary magnetic field (IMF) magnitude B_o, the ϵ parameter, and the *ap* geomagnetic index.

The data extend from 01 January 1964 to 31 December 2010. This interval covers essentially the entire space era when interplanetary parameters are available. Solar minimum epochs as given by the 13-month smoothed sunspot number are shown in the R_z panel as vertical dashed lines. The average of the parameters for ± 6 months around solar minimum is indicated in Table 1.

Fig. 1 shows that the solar cycle that has just ended (in 2008) is the longest in the space era (12.6 years, extending from 1996 to 2008). The length of cycles 20 to 22 were 11.7, 10.3 and 9.7 years, respectively. The solar minimum R_z average for the last cycle is 1.7, which is considerably lower than the other minima: 12.2, 12.3 and 8.0 (Hathaway

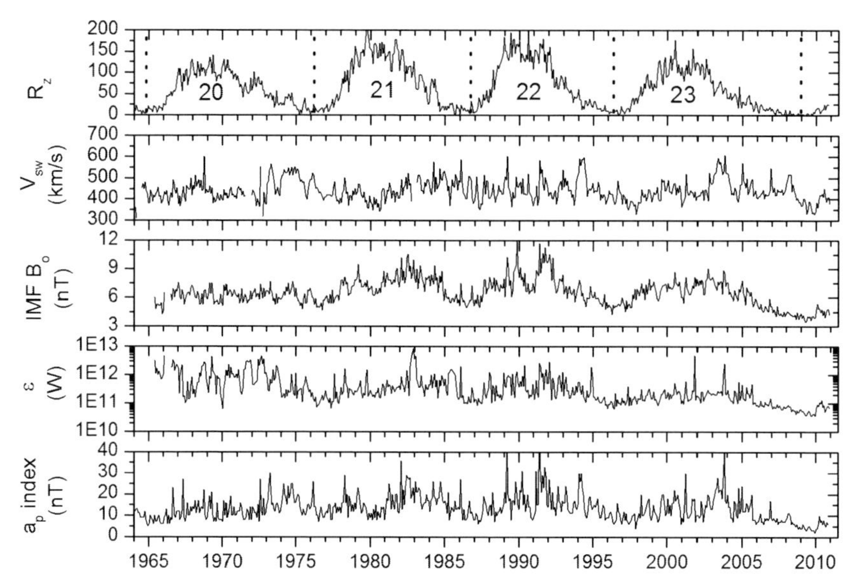

Figure 1. Solar rotation averages of sunspot number, R_z, solar wind speed V_{sw}, interplanetary magnetic field IMF magnitude B_o, ϵ parameter, ap index. Vertical dashed lines in the R_z panel indicates the time of solar minima.

Table 1. Average parameters for an interval of ± 0.5 year around solar cycle minima

Cycle	Date min (year/month)	R_z	B_o (nT)	V_{sw} (km/s)	ϵ (10^{11} W)	ap (nT)
19-20	1964/10	10.3	-	431	-	8.5
20-21	1976/03	12.5	5.8	459	1.61	13.5
21-22	1986/09	13.1	5.7	444	1.64	10.7
22-23	1996/05	8.7	5.1	419	1.09	9.5
23-24	2008/12	2.0	4.0	395	0.54	4.9

2010). It can be noted that sunspot number, IMF magnitude, solar wind speed and ap index have the lowest averages during the recent past minimum. The average R_z value for the period of 1 year around solar minimum is 2.0, lower than the previous four solar cycle minimum values of 8.7 to 13.1.

Comparison of the IMF magnitude at 1 AU to the R_z(Fig. 1) shows that the IMF is more-or-less in phase with the sunspot number, with maximum averages located close in time to the solar maxima. It can be noted that in the current cycle minimum, the IMF average is the lowest on record, about 4 nT. In the previous solar minima, the average IMF magnitude was always $\geqslant$ 5 nT.

The solar wind speed (Fig. 1 third panel) reached maximum values in the solar cycle declining phases in years 1973, 1984, 1994 and 2003. One possible explanation is that the high speed streams coming from coronal holes predominate in this phase of solar cycle (Tsurutani *et al.* 1995). High speed streams will cause a higher radial IMF component due to the high velocities and the stretching of the Parker spiral. The solar wind speed reached minimum values of $\sim$ 395 km/s during this last cycle minimum, slightly lower than previous cycles (always greater than 400 km/s). The solar rotation averaged ap index shows values of $\sim$ 5 nT (minimum of 2 nT) during the current solar minimum, contrasting to values of $\sim$ 8.5-13.5 nT in previous minima.

Since the energy transferred by the solar wind to the magnetosphere depends on V_{sw} and B_o, from the results of Table 1 for average V_{sw} and IMF B_o one expects that the average energy was also smaller during the recent solar cycle minimum as compared to the previous minima. Table 1 shows the average for each solar minimum period. The average power was $\sim$ 1.1 to 1.64 x 10^{11} W in the three previous solar minima, and 5.4 x 10^{10} W in the most recent minimum. Thus the energy input to the magnetosphere during this last solar minimum is only $\sim$ 30 % to 50 % of the value for previous solar cycle minima during space era.

The solar and interplanetary causes in the geomagnetic activity during the last two solar minima were studied (Tsurutani, Echer & Gonzalez 2011) . Minima in geomagnetic activity (MGA) at Earth occurring at the ends of solar cycles (SC) SC23 and SC22 have been identified. Figure 2 shows these MGA intervals. The two MGAs were present in 2009 and 1997, delayed from the sunspot number minima in 2008 and 1996 by 0.5 to 1 years. This is in agreement with geomagnetic activity occurring after sunspot minimum, as seen for example in the *aa* index (Gao 1986; Kane 2002; Hathaway 2010). For instance, it has been found that, for most of the solar cycles, the minimum in the *aa* index occurs more than three months later than the sunspot number minimum (Kane 2002).

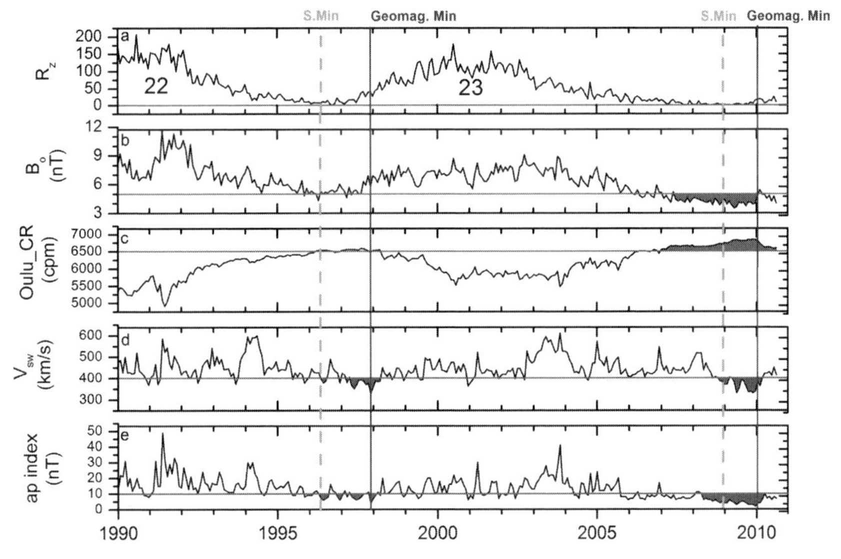

Figure 2. Solar, interplanetary, cosmic ray and geomagnetic activity data for SC22 and 23. A dashed vertical green line indicates the official sunspot minima. A vertical solid line shows the geomagnetic activity minima on Earth. Horizontal (arbitrary) red lines are shown for R_z =0, B_o =5 nT, CR = 6500 cpm, V_{sw} =400 km s1 and *ap* = 10 nT. The shaded regions in the *ap* panel are used to define the minimum geomagnetic activity (MGA) intervals, MGA22 and MGA23. (After Tsurutani, Echer & Gonzalez 2011).

A detailed comparison between the last two MGA was performed (Tsurutani, Echer & Gonzalez 2011). It was found that the solar wind speed was about the same for the more recent and the previous cycle (390 vs. 395 km/s), but the IMF B_o was much lower in the recent minimum (4.2 versus 5.5 nT). The geomagnetic activity was $ap = 5$ nT versus $ap =$ 8.7 nT. The cause of the lower IMF B_o was the Sun's magnetic field; using Wilcox Observatory/Stanford University observations of the Sun's mean magnetic field, it was found that the solar magnetic field was much weaker in the recent minimum (Tsurutani, Echer & Gonzalez 2011) . Furthermore, the polar magnetic fields of the Sun are weaker as well (Wang, Robbrecht & Sheeley 2009; Russell, Luhmann & Jian 2010; de Toma 2011).

Fig. 3 shows daily averages of sunspot number R_z, the solar mean magnetic field and the IMF B_o for 1975-2010. It can be seen that the solar magnetic field is much weaker in this current minimum than in the previous two solar minima. The IMF B_o at 1 AU is also a minimum as a consequence. The Sun's field is lower in the last cycle relative to the two previous ones.

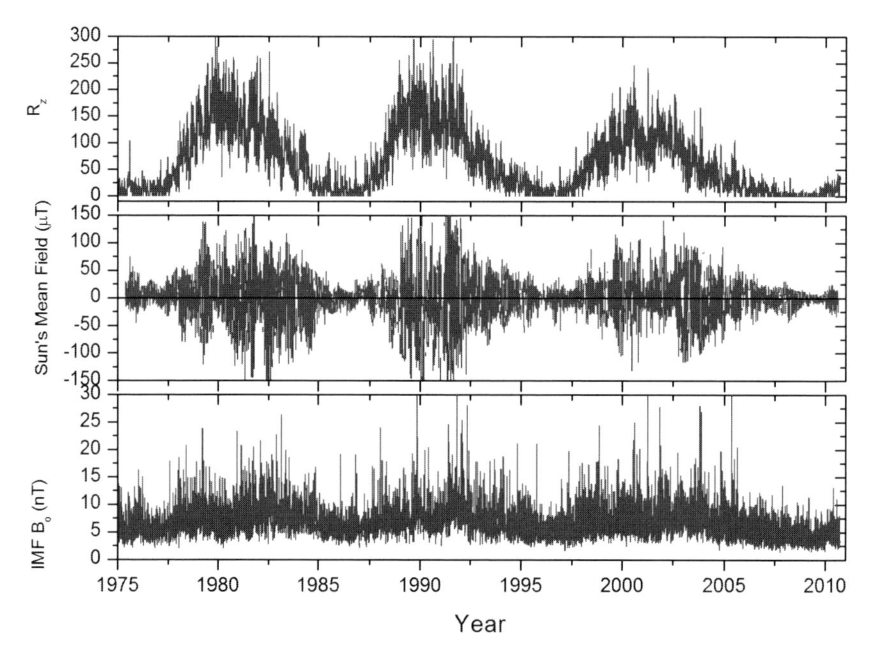

Figure 3. Daily averages of sunspot number, R_z, Stanford's Sun's mean field and the IMF B_o for 1975-2010.

The solar wind speed was very low in the MGA23, with no stream with peak speed higher than 600 km/s at Earth's orbit. As it is well known, high speed solar wind streams emanate from coronal holes (Zirker 1977; Tsurutani *et al.* 1995; Richardson 2006). When the coronal hole is located at low solar latitudes, streams from its center can reach the Earth. When the coronal hole is located at middle latitudes, only the solar wind emanating from the edges of the coronal hole can reach Earth. The velocities at the edges are lower due to superradial expansion of the solar wind. In this recent minimum, there is a sharp contrast between 2008 and 2009 (de Toma 2011). At the end of 2008, the large low latitude coronal holes were closing down and smaller, mid-latitude coronal holes were appearing. In 2009, the disappearance of the low-latitude coronal holes shifted the sources of the high speed solar wind to higher latitudes, mostly to the edges of the solar coronal holes (de Toma 2011). Furthermore, the variances of the IMF B_z were computed (Tsurutani, Echer & Gonzalez 2011). They found that the nested and normalized nested variances (e.g., Tsurutani *et al.* 1982) are lower in the recent minimum than in the previous one and are also lower than intervals of high speed streams in year 2003 (Tsurutani, Echer & Gonzalez 2011).

The location of the coronal holes relative to the ecliptic plane led to low solar wind speeds and low interplanetary magnetic field B_z variances (and normalized variances) at Earth. Since the coupling of energy transfer between the solar wind and the magnetosphere depends on the IMF B_o, the IMF B_z fluctuations and the velocity of the solar wind, there was reduced solar wind-magnetospheric energy coupling during 2009, leading to the lowest geomagnetic *ap* index in its recording history.

3.2. *Historical data*

It is important to assess how unprecedented is the very low geomagnetic activity observed in the recent solar minimum. We assess the geomagnetic activity using data from the historical period with instrumental record, since the XIX century.

Fig. 4 shows the whole interval of *ap* (since 1932), and *Dst* and *AE* indices (since 1957). Also shown for comparison is the R_z. It can be seen that the current minimum in geomagnetic *ap* index is the lowest in the history of *ap* (see also Minamoto & Taguchi 2009). This indicates that solar wind-magnetosphere energy transfer during the current minimum was very low.

The *AE* index also shows lowest values during this period. As it is known that long-term averages of *AE* have high correlation with the *ap* index (Echer *et al.* 2004), this would be expected. The *Dst* index shows very low (in magnitude) values, the lowest in the space era, and comparable to another period of low values during the 1960s (Minamoto & Taguchi 2009).

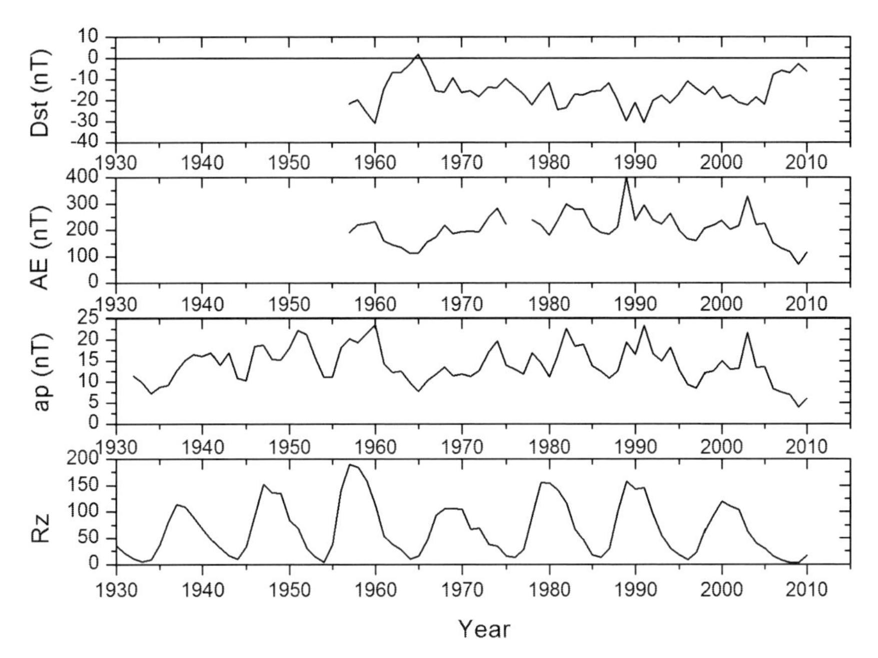

Figure 4. Yearly averages of geomagnetic indices: *Dst*, *AE*, and *ap*, and the sunspot number R_z.

Fig. 5 shows the annual averages of R_z and the *aa* index for 1868-2010. It can be seen that in the recent minimum the *aa* index was the lowest in the entire XX century and the first decade of XXI century, and it is comparable only to the low geomagnetic acitvity in the end of XIX century, when solar activity was lower. The recent minimum in R_z is also very low for the XX-XXI century, being compared only to low solar activity seen in the beginning of XX century (de Toma 2011).

The long term correlation between the geomagnetic *aa* index and R_z was found to be decreasing since the beginning of *aa* index recording in 1868 (Echer *et al.* 2004). Geomagnetic activity has a dual peak distribution, with one peak near solar maximum and other in the declining phase (Gonzalez, Gonzalez & Tsurutani 1990). The cause for the decrease of correlation between R_z and *aa* has been attributed to fact that the second *aa* peak is becoming stronger relative to the first one, which could be due to the fact

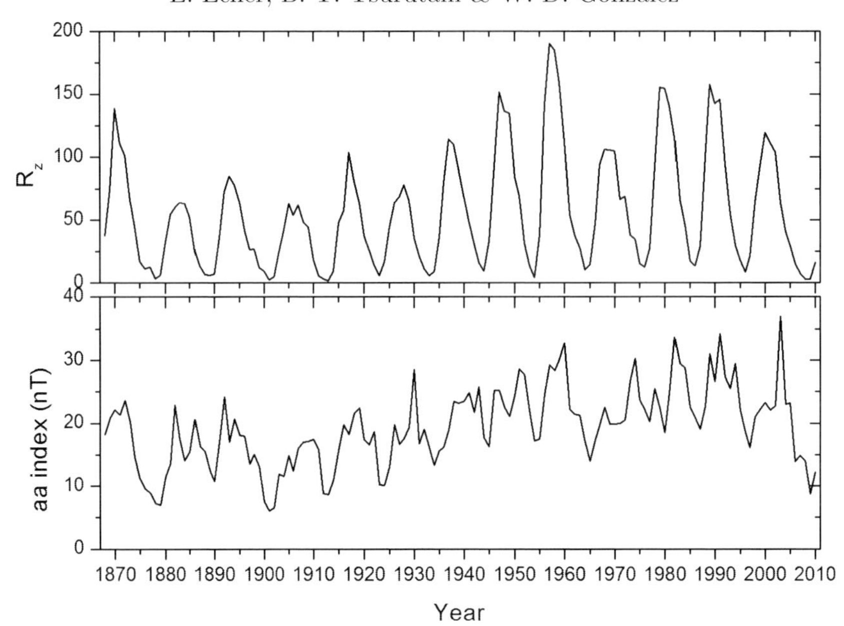

Figure 5. Yearly averages of sunspot number R_z (top) and *aa* index (bottom) for the interval 1868-2010.

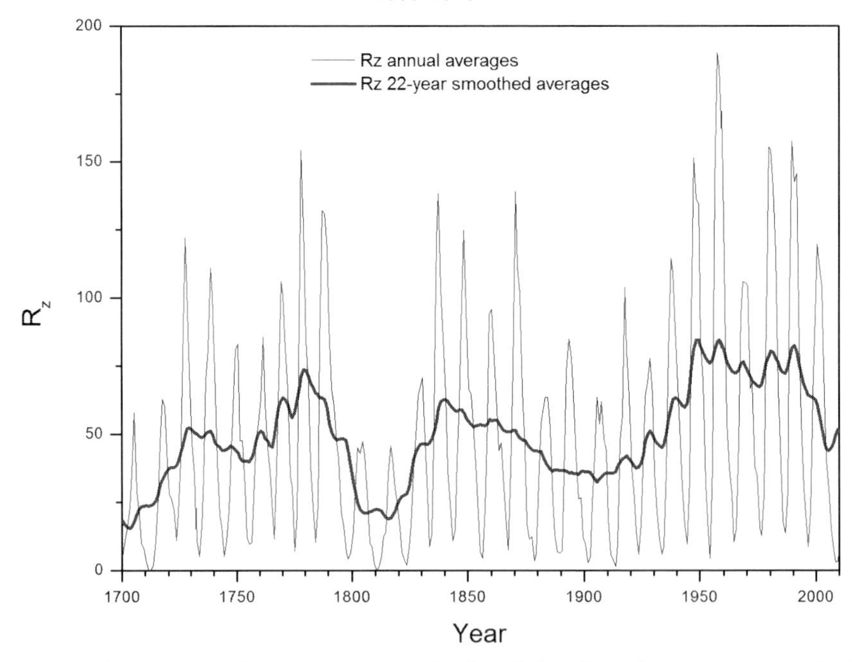

Figure 6. Yearly averages of sunspot number R_z (black lines), and a 22-year smoothed curve of R_z (blue lines) for the interval 1700-2010.

that solar open magnetic field structures (the *aa* 2nd peak) seemed to have increase their strength relative to closed magnetic activity structures (*aa* 1rst peak, R_z related) due to long term variability of the Sun's magnetic field (Stamper *et al.* 1999; Echer *et al.* 2004).

Fig. 6 shows the sunspot number and its 22-year smoothed curve for 1700-2010. This long term variation could be related to the Gleissberg cycle of solar activity (Hathaway 2010; Gonzalez *et al.* 2011). There are 8 to 10 solar cycles between consecutive minima of the modulation. The last two Gleissberg minima occurred approximately around 1900 and 2010 (maxima at 1840 and 1960), as can be seen in Figure 6. Thus the current minimum might be a result of the combination of a minimum in the Schwabe 11-year cycle and the Gleissberg minima, which could contributed to make this minimum deeper. The lowest values in *aa* index also occurred in the two periods of Gleissberg minima, around 1900 and 2010.

4. Conclusions

This past solar cycle minimum has been extreme in several aspects. The sunspot number (R_z) was the lowest in recent history. Associated with this characteristic, the duration of cycle 23 has been elongated in comparison to cycles 20 through 22. The solar wind speed V_{sw} was a minimum of 330 km/s (average 395 km/s) and the embedded IMF B_o was an exceptionally low ~4 nT. These low values of IMF B_o and V_{sw} caused a decrease in the energy transfer from solar wind to the magnetosphere, as estimated by the ϵ parameter, which was 30-50 % of previous solar cycle minima, causing the lowest value of space era in geomagnetic activity as quantified by *ap* index.

The solar wind magnetosphere energy transference depends on the solar wind speed V_{sw}, the IMF B_o and the IMF B_z component. The energy transfer, estimated by computing the ϵ parameter, was the lowest of the space era, and the cause is the combination of low solar wind speeds, low IMF magnitude and low IMFB_z variances (Alfvenic fluctuations). The low IMF B_o values at 1 AU are consequence of low solar fields, as observed for example in solar magnetograms. The V_{sw} is low because of the location of the CHs. The B_z component is low also because of the location of the CHs.

The recent solar minimum is also different from previous ones, being characterized by small midlatitude coronal holes and polar coronal holes, instead of big polar coronal holes extending equatorwards. We surmise that it is the superradial expansion from the small midlatitude coronal holes that gives the feeble high speed solar wind speeds. The minimum in the 11-year sunspot cycle was also coincident with a minimum in the Gleissberg cycle, which could be the cause of the deep minimum in solar activity. This result implies that in extended solar minima such as the Maunder minimum (1645-1715), exceptionally low solar magnetic fields and feeble solar winds and embedded magnetic fields are to be expected. The energy transference to Earth's magnetosphere and atmosphere is therefore expected to be a minimum during such epochs accounting for the lack of auroral sightings during that era.

Acknowledgements

EE would like to thank the CNPq (PQ-300211/2008-2) agency for financial support. WDG would like to thank FAPESP agency (2008/06650-9) for financial support. Portions of this work were performed at the Jet Propulsion Laboratory, California Institute of Technology under contract with NASA.

References

Bartels, J. 1950, *J. Geophys. Res.*, 55, 427

de Toma, G. 2011, *Solar Phys.*, 274, 195

Echer, E., Gonzalez, W. D., Gonzalez, A. L. C., Prestes, A., Vieira, L. E. A., Dal Lago, A., Guarnieri, F. L., & Schuch, N. J. 2004, *J. Atmos. Solar-Terr. Phys.*, 66, 1019

Echer, E., Gonzalez, W. D., Guarnieri, F. L., Dal Lago, A., & Vieira, L. E. A. 2005, *Adv. Space Sci.*, 35, 855

Echer, E., Tsurutani, B. T., Gonzalez, W. D., & Kozyra, J. U. 2011, *Sol. Phys.*, 274, 303
Eddy, J. A. 1976, *Science*, 192, 1189
Emery, B. A., Richardson, I. G., Evans, D. S., Rich, F. J., & Wilson, G. R. 2011, *Solar Phys.*, 274, 399
Fröhlich, C. 2009, *Astron. Astrophys.*, 501, L27
Gao, M. 1986, in: Y. Kamide & J. A. Slavin (eds.), *Solar-Wind Magnetosphere Coupling*, (Astrophysics and Space Science Library, 126), p. 149
Gibson, S. E., Kozyra, J. U., de Toma, G., Emery, B. A., Onsager, T., & Thompson, B. J. 2009, *J. Geophys. Res.*, 114
Gonzalez, W. D., Gonzalez, A. L. C., & Tsurutani, B. T. 1990, *Planet. Space Sci.*, 38, 181
Gonzalez, W. D, Joselyn, J. A., Kamide, Y., Kroehl, H. W., Rostoker, G., Tsurutani, B. T., & Vasyliunas, V. M. 1994, *J. Geophys. Res.*, 99, 5771
Gonzalez, W. D., Echer, E., Tsurutani, B. T., Clua de Gonzalez, A. L., & Dal Lago, A. 2011 *Space Sci. Rev.*, 158, 69
Hathaway, D. H. 2010, *Liv. Rev. Solar Phys.*, 7, 1
Heber, B., Kopp, A., Gieseler, J. , Muller-Mellin, R., Fichtner, H., Scherer, K., Potgieter, M. S., & Ferreira, S. E. S. 2009, *Astrophys. J.*, 699, 1956
Kane, R. P. 2002, *Ann. Geophys.*, 20, 1519
Kataoka, R. & Miyoshi, Y. 2010, *Space Weather*, 8, 1
Kirk, M. S., Pesnell, W. D., Young, C. A., & Hess Webber, S. A. 2009, *Sol. Phys.*, 257, 99
Luhmann, J. G., Lee, C. O., Li, Y., Arge, C. N., Galvin, A. B., Simunac, K., Russell, C. T., Howard, R. A., & Petrie, G. 2009, *Solar Phys.*, 256, 285
McComas, D. J., Ebert, R. W., Elliott, H. A., Goldstein, B. E., Gosling, J. T., Schwadron, N. A., & Skoug, R. M. 2008, *Geophys. Res. Lett.*, 35, L18103
Minamoto, Y. & Taguchi, Y. 2009, *Earth Planets Space*, 61, e25
Perreault, P. & Akasouf, S.-I. 1978, *Geophys. J. Royal Astron. Soc.*, 54, 547
Ram, S. T., Liu, C. H., & Su, S.-Y. 2010, *J. Geophys. Res.*, 115, A12340
Richardson, I. G. 2006, in: B. T. Tsurutani, R. McPherron, W. Gonzalez, G. Lu, J. H. A. Sobral & N. Gopalswamy (eds.), *Recurrent Magnetic Storms: Corotating Solar Wind Streams* (AGU Geophysical Monograph 167), p. 45
Rostoker, G. 1972, *Rev. Geophys. Space Phys.*, 10, 935
Russell, C. T., Luhmann, J. G., & Jian, K. J. 2010, *Rev. Geophys*, 48, RG2004
Scherrer, P. H., Wilcox, J. M., Svalgaard, L., Duvall, T. L., Dittmer, H., & Gustafson, E. K. 1977, *Solar Phys.*, 54, 353
Smith, E. J. & Balogh, A. 2008, *Geophys. Res. Lett.*, 35, L22103
Stamper, R., Lockwood, M., Wild, M. N., & Clark, T. D. G. 1999, *J. Geophys. Res.*, 104, 28325
Tsurutani, B. T., Smith, E. J., Pyle, K. R., & Simpson, J. A. 1982, *J. Geophys. Res.*, 87, 7389
Tsurutani, B. T., Gonzalez, W. D., Gonzalez, A. L. C., Tang, F., Arballo, J. K., & Okada, M. 1995, *J. Geophys. Res.*, 100, 21717
Tsurutani, B. T., Echer, E., & Gonzalez, W. D. 2011, *Ann. Geophys.*, 29, 83
Verkhoglyadova, O. P., Tsurutani, B. T., Mannucci, A. J., Mlynczak, M. G., Hunt, L. A., Komjathy, A., & Runge, T. 2011, *J. Geophys. Res.*, 116, A09325
Wang, Y.-M., Robbrecht, E., & Sheeley, N. R. 2009, *Astrophys. J.*, 707, 1372
Zirker, J. B. 1977, *Rev. Geophys. Space Phys.*, 15, 257

Discussion

Janet Luhmann: ICME fields at minimum are due to magnetic clouds. This cycle had north-leading ICME fields. How does this affect geoactivity, have you taken this into account? Also, there are lower solar wind densities during this minimum. Could this also affect the overall geoactivity?

Ezequiel Echer: ICME fields affect mainly the occurrence of intense storms, which are absent from 2007-2010 interval. However, for long term geomagnetic activity (as solar rotation averages), high speed streams have a stronger effect. The solar wind density can affect geomagnetic activity, but the main parameters (interplanetary magnetic field magnitude and southward direct Bz component, solar wind speed) are already low enough to cause this very low geomagnetic activity observed during the recent solar minimum.

Comparative Magnetic Minima:
Characterizing Quiet Times in the Sun and Stars
Proceedings IAU Symposium No. 286, 2011
C. H. Mandrini & D. F. Webb, eds.

doi:10.1017/S1743921312004851

A porcupine Sun? Implications for the solar wind and Earth

Sarah E. Gibson[1] **and Liang Zhao**[1]

[1]NCAR/HAO, 3080 Center Green Dr., Boulder, CO
email: sgibson@ucar.edu

Abstract. The recent minimum was unusually long, and it was not just the case of the "usual story" slowed down. The coronal magnetic field never became completely dipolar as in recent Space Age minima, but rather gradually evolved into an (essentially axisymmetric) global configuration possessing mixed open and closed magnetic structures at many latitudes. In the process, the impact of the solar wind at the Earth went from resembling that from a sequence of rotating "fire-hoses" to what might be expected from a weak, omnidirectional "lawn-sprinkler". The previous (1996) solar minimum was a more classic dipolar configuration, and was characterized by slow wind of hot origin localized to the heliospheric current sheet, and fast wind of cold origin emitted from polar holes, but filling most of the heliosphere. In contrast, the more recent minimum solar wind possessed a broad range of speeds and source temperatures (although cooler overall than the prior minimum). We discuss possible connections between these observations and the near-radial expansion and small spatial scales characteristic of the recent minimum's porcupine-like magnetic field.

Keywords. Sun: solar-terrestrial relations, Sun: solar wind; Sun: corona, Sun: magnetic fields

---

1. Introduction

The Whole Heliosphere Interval (WHI) (Thompson *et al.* 2011) was an internationally coordinated observing and modeling effort. Its science goals were to connect the origins and effects of solar structure and activity through the solar wind to the Earth and other planetary systems, and to characterize the three-dimensional solar minimum heliosphere. Results from WHI may be found in the 2011 Topical Issue of *Solar Physics*: "The Sun–Earth Connection near Solar Minimum."

In this paper, we will draw from analyses within that Topical Issue and elsewhere in order to summarize the evolution of the recent solar minimum, and compare it to the one immediately prior. In particular, we will discuss three representative properties: the global coronal magnetic field, the global solar wind structure in relation to the Earth's space environment, and the solar wind source temperature as encoded in its composition.

2. Comparing solar minima: Coronal magnetic field and solar wind/Earth interactions

The most recent minimum was different from the previous two minima, and the beginning of the recent minimum was unlike its end. In an overview article of the Solar Physics WHI Topical Issue, Gibson *et al.* (2011) described the prolonged recent minimum from Sun to Earth as it evolved from 2008 through 2009. In terms of magnetic flux emergence and associated closed-field quanitites (*e.g.*, sunspots, solar irradiance, coronal mass ejections) there was not much change during these years, with levels essentially staying as low or lower than the prior minimum. However, for quantities pertaining to or sensitive to the distribution of open flux at the Sun and in the heliosphere (*e.g.*, coronal

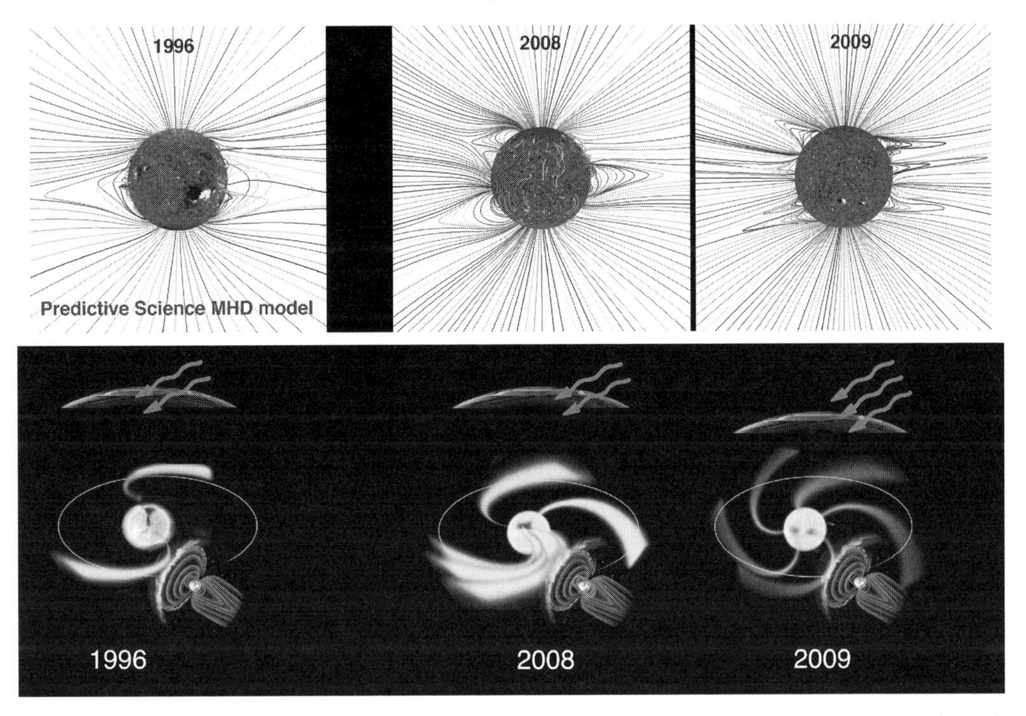

Figure 1. Top: Coronal magnetic field from the Magnetohydrodynamics on a Sphere (MAS) model for (left to right) Whole Sun Month Carrington Rotation (CR) 1913 in August-September 1996, August 1, 2008 eclipse prediction (using CR2071 and CR2072 data), and July 22, 2009 eclipse (using CR2084 and CR2085). Bottom: artist's conception for similar times of solar wind morphology (faster wind streams indicated in yellow emanating from coronal holes) and impact for the Earth's radiation belt (large relativistic electron population indicated by red) and cosmic rays (high levels indicated by number of squiggly red arrows). For color figures, see the online version of this paper.

holes, heliospheric current sheet (HCS) tilt, cosmic rays, solar wind velocity at the Earth, the Earth's radiation belts) there was significant evolution, from levels that had not yet reached those of the prior solar minimum in early 2008, to space-age-record-setting levels by mid 2009.

Figure 1 (top row) illustrates the magnetic morphology of the corona. Last minimum reached a predominantly dipole state, with closed field straddling a fairly flat HCS. 2008 saw a warped HCS, and both open and closed field at a broader range of latitudes. In 2009 the HCS flattened, but because of a weak polar magnetic field the dipole was not as dominant as in 1996. Both closed and open field were still present at low-to-mid latitudes, although the non-polar open-field coronal holes were no longer large and long-lived as they had been in 2008 (de Toma 2011). Moreover, much of the closed field in both 2008 and 2009 were so-called "pseudo" streamers, in that they were surrounded by open magnetic field of a single polarity (Riley & Luhmann 2012). The HCS-straddling closed field of the more traditional streamer was still present as it was in 1996, but, particularly in 2009, was narrower in latitudinal width, as were also the other pseudostreamer closed field regions.

Figure 1 (bottom row) shows an artist's conception of the solar wind for these time periods. During 1996, the Sun was not completely dipolar in that there were equatorward extensions of the Northern polar hole, which drove the Earth's space environment with fast solar wind streams for several months as they rotated past. However, as discussed in

Gibson *et al.* (2009), the presence of large, long-lived and low-latitude coronal holes in 2008 had a much stronger effect, leading to periodicities and elevated levels in the Earth's radiation belts and aurora (also the thermosphere, as discussed in Lei *et al.* (2010) and Gibson *et al.* (2011)). By 2009 those coronal holes had dissipated, and the Sun's magnetic configuration was essentially axiymmetric. It was still complex, with open and narrow-closed "porcupine"-like magnetic structures poking out in all directions, but the loss of the localized long-lived open field structures resulted in a loss of the periodic forcing of fast wind streams and associated enhanced radiation belts. Overall, the heliosphere reached the depths of its minimum, and cosmic rays were free to reach space-age record high levels, due at least in part to space-age record-low levels of interplanetary magnetic field strength (Mewaldt *et al.* 2010).

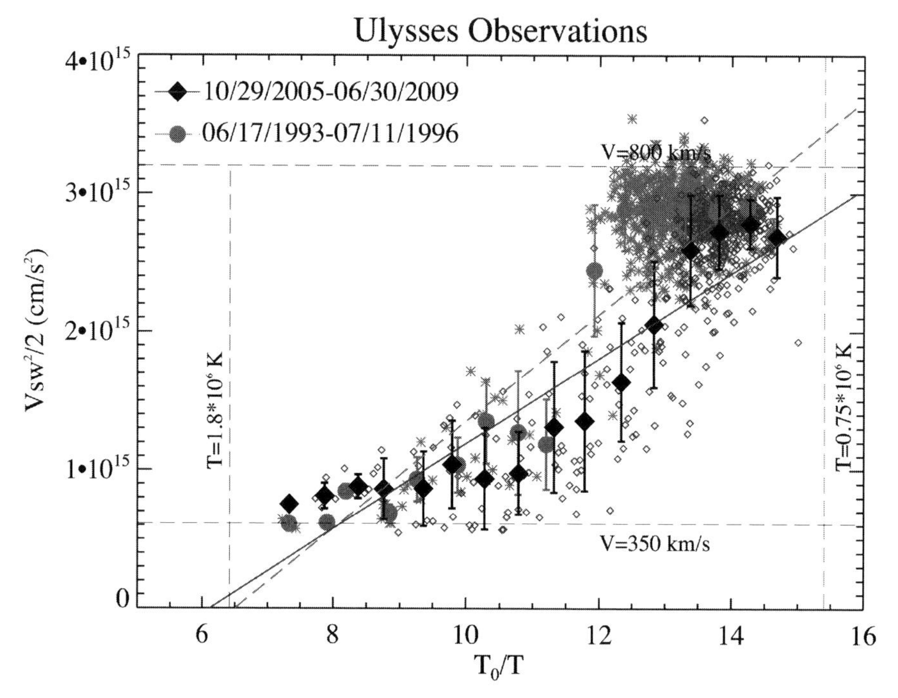

Figure 2. Ulysses measurements of solar wind source temperature (T) extracted from Oxygen charge-state ratio. T_o/T is plotted vs. solar wind velocity-squared, where $T_o \simeq 14MK$ (T_o is defined in Zhao & Fisk (2011) and has a weak dependence on temperature – see Fig. 13 of that paper for a comparable plot). The black diamonds are binned (bin size: $T_0/T = 0.5$) averages for the most recent minimum (shown with higher resolution by smaller blue diamonds), and the red large dots are the binned averages for the prior minimum (shown with higher resolution by the smaller red asterisks). The time intervals are of the same length and Ulysses orbital position for the two minima. ICMEs have been removed by eliminating points with high values of O^{7+}/O^{6+}. For color figures, see the online version of this paper.

3. Comparing solar minima: Solar wind composition

The solar magnetic field morphologies of the last two minima were also reflected in solar wind composition. For both minima the solar wind O^{7+}/O^{6+} ratio, related to the temperature of solar wind source regions, indicated slow wind from a hot source at HCS crossings, and fast wind from a cold source deep in polar holes. However, just as the closed field around the HCS was narrower in the more recent minimum, Zhao & Fisk

(2011) found that the width in the heliosphere of the slow, hot-sourced wind straddling the HCS was narrower.

Figure 2 shows the source temperature extracted from the Oxygen charge-state ratio plotted versus velocity-squared, illustrating an inverse correlation that ranges from hot-sourced, slow wind to cold-sourced, fast wind. To first order this relationship has not changed greatly from one minimum to the other. Note that the time intervals chosen are for the same Ulysses orbital positions and similar solar cycle levels.

However, a closer look reveal two interesting differences between the two minima. First, there is a general shift to the right in Figure 2 from the 1996 minimum (red/asterisks) to the more recent minimum (blue/diamonds). This indicates a a cooler source during the recent minimum, particularly for the slow, cold-sourced wind (upper right) coming from the polar coronal holes (von Steiger & Zurbuchen 2011).

Second, it is clear that the distribution in the most recent minimum is not as bimodal as the prior minimum. At times near the 1996 minimum, the heliosphere was largely filled with fast, cold-polar-hole-sourced wind, as is evidenced by the large clump of red points in the upper right of Figure 2. As the Ulysses orbit crossed the HCS, some slow, hot-sourced wind was seen (bottom left), but there are not many red points in between these two extremes. In contrast, the most recent minimum possessed a broad range of source temperatures and velocities, as seen in the more uniformly-spaced curve of blue dots.

4. Discussion and Conclusions

The magnetic field at the solar surface controls the distribution of open flux in the corona, the morphology and temperature of the coronal plasma, and the structure and impact of the solar wind at the Earth. The differences between the two solar minima were reflected in all of these properties.

The solar wind Oxygen composition indicates that, although the behavior in the depths of the coronal hole or right at the HCS were similar for the two minima, a population of warm/mid-speed wind was also present in the more recent minimum. It seems likely that this was related to the population of mixed closed/open field seen at mid-to-low latitudes in the recent minimum, and, in particular, pseudostreamers.

Riley & Luhmann (2012) found that a pseudostreamer in the solar wind during the recent minimum was associated with slow wind. They pointed out that this is not consistent with the empirical relationship often found between magnetic super-radial expansion and solar wind speed (Wang & Sheeley 1990). Figure 1 (top-middle panel) illustrates this point, *i.e.* that the field around a bipolar HCS-straddling streamer (streamers in the West and Southeast) expands more than the field around a unipolar pseudostreamer (streamer in the Northeast). Riley & Luhmann (2012) argued that this supported models where interchange reconnections at a boundary layer between open and closed fields are the source of the slow solar wind, as these predict similar properties for HCS-straddling streamers and pseudostreamers, unlike expansion-based slow wind models.

We cannot tell from our Figure 2 which wind is associated with pseudostreamers. However, we do note that a porcupine-like field by definition has low expansion everywhere, since the "quills" of narrow streamers and surrounding open field lines stick out near-radially. Expansion-factor models would predict that this should lead to faster solar wind overall, and yet 2009 saw the lowest wind velocities at the Earth and Ulysses saw slow and moderate speed wind at mid-latitudes during 2008 and the beginning of 2009.

The temperature of the slow wind is another interesting issue. Arguably for both minima, but particularly for the recent one, the slope in Figure 2 is flatter in the slow-wind portion (bottom-left). Indeed, it appears nearly horizontal in the black-diamonds of the recent minimum. The smaller, narrower loops of the recent minimum may be associated, either directly, or indirectly (e.g., as common consequences of a weakened magnetic field), with the cooler temperatures seen overall. Is it possible that the range in slow-wind temperatures from hot to warm also arises from a range from larger loops to smaller loops? Is it indicative of a time evolution of the size of such loops? Or is it it due to the changing Ulysses orbital position, from HCS-streamer intersecting to higher-latitude pseudostreamers? What is the temperature (and speed) of wind emitting from the small coronal holes at mid-latitude during 2009? These open questions are well worth pursuing, and the recent minimum an excellent opportunity for pinning down the origins and characteristics of the solar wind.

Acknowledgements

We thank Giuliana de Toma, Larisza Krista, and Alysha Reinard for helpful discussions. We thank Predictive Science Inc. and Janet Kozyra for the material shown in Figure 1, and acknowledge the Ulysses Solar Wind Composition Instrument (SWICS) for solar wind data. LZ is supported by the NASA Living With a Star Heliophysics Postdoctoral Fellowship Program, administered by the University Corporation for Atmospheric Research (NCAR), which is supported by the National Science Foundation.

References

de Toma, G. 2011, *Solar Phys.*, 274, 195 doi:10.1007/s11207-010-9677-2

Gibson, S. E., Kozyra, J. U., de Toma, G., Emery, B. A., Onsager, T., & Thompson, B. J. 2009, *Journ. Geophys. Res.*, 114, A09105, doi:10.1029/2009JA014342

Gibson, S. E. *et al.* 2011, *Solar Phys.*, 274, 5, doi:10.1007/s11207-011-9921-4

Lei, J., Thayer, J. P., Wang, W., & McPherron, R. L. 2010, *Solar Phys.*, 274, 427, doi:10.1007/s11207-010-9563-y

Mewaldt, R. A. *et al.* 2010, *Astrophys. J.* (Letters), 723, L1, 723, doi:10.1088/2041-8205/723/1/L1

Riley, P. & Luhmann, J. 2012, *Solar Phys.*, 421, in press, doi:10.1007/s11207-011-9909-0

Thompson, B. *et al.* 2011, *Solar Phys.*, 274, 29, doi:10.1007/s11207-011-9891-6

von Steiger, R. & Zurbuchen, T. H. 2011, *Journ. Geophys. Res.*, 116, A01105, doi:10.1029/2010JA015835

Wang, Y.-M. & Sheeley, N. 1990, *Astrophys. J.*, 355, 726, doi:10.1086/168805

Zhao, L. & Fisk, L. 2011, *Solar Phys.*, 274, 379, 10.1007/s11207-011-9840-4

Discussion

JON LINKER: Have you looked at Magnesium in the solar wind?

SARAH GIBSON: The Mg^{10+}/O^{6+} ratio behaves basically like that of O^{7+}/O^{6+}, in that last minimum there was a bimodal distribution of basically low-speed/high Mg^{10+}/O^{6+}; high-speed/low Mg^{10+}/O^{6+}, but this minimum the middle is filled in with middle speed, middle Mg^{10+}/O^{6+}. Also it is lower amplitude over all speeds, as was true for the O^{7+}/O^{6+}.

LEIF SVALGAARD: You shouldn't call it "Heliospheric Current Sheet" so close to the Sun.

JON LINKER: Like it or not, that is what the community calls it.

Comparative Magnetic Minima:
Characterizing quiet times in the Sun and stars
Proceedings IAU Symposium No. 286, 2011
C. H. Mandrini & D. F. Webb. eds.

doi:10.1017/S1743921312004863

Modeling of the atmospheric response to a strong decrease of the solar activity

Eugene V. Rozanov[1,2], Tatiana A. Egorova[1], Alexander I. Shapiro[1] and Werner K. Schmutz[1]

[1]Physikalisch-Meteorologisches Observatorium, World Radiation Center
Dorfstrasse 33, CH-7260, Davos, Switzerland
email: t.egorova@pmodwrc.ch

[2]Institute for Atmospheric and Climate Science, ETH Zurich,
CH-8092, Zurich, Switzerland
email: e.rozanov@pmodwrc.ch

Abstract. We estimate the consequences of a potential strong decrease of the solar activity using the model simulations of the future driven by pure anthropogenic forcing as well as its combination with different solar activity related factors: total solar irradiance, spectral solar irradiance, energetic electron precipitation, solar protons and galactic cosmic rays. The comparison of the model simulations shows that introduced strong decrease of solar activity can lead to some delay of the ozone recovery and partially compensate greenhouse warming acting in the direction opposite to anthropogenic effects. The model results also show that all considered solar forcings are important in different atmospheric layers and geographical regions. However, in the global scale the solar irradiance variability can be considered as the most important solar forcing. The obtained results constitute probably the upper limit of the possible solar influence. Development of the better constrained set of future solar forcings is necessary to address the problem of future climate and ozone layer with more confidence.

Keywords. Climate, ozone, solar irradiance, energetic particles

1. Introduction

The warming of the Earth's climate due to anthropogenic greenhouse gases is evident and can become dangerous for mankind in the nearest future (IPCC 2007). The ozone layer was endangered by the anthropogenic emission of chlorine and bromine containing species but is expected to fully recover in the middle of the 21st century due to the limitations introduced by Montreal protocol and its amendments (WMO 2011). On the other hand the current unusually long solar minimum hints to gradual decrease of the solar activity in the future similar to the Dalton minimum of the solar activity (Abreu *et al.* 2008; Lockwood *et al.* 2009). What are the implications of this would-be decrease of solar activity for future climate and ozone layer changes? Should we expect any compensation or enhancement of the anthropogenic effects on the atmosphere in the future? This question can be addressed only if we take into account all relevant direct effects of anthropogenic and solar forcing as well as complex feedbacks between different atmospheric processes (Gray *et al.* 2010). An expected decline of the solar activity will be accompanied by a decrease of the both total solar irradiance (TSI) and spectral solar irradiance (SSI) leading directly to the global cooling, ozone depletion and cooling in the tropical stratosphere followed by a deceleration of the polar night jets and cooler winters over northern land masses (Egorova *et al.* 2004, Gray *et al.* 2010). The significance of these effects depends on the magnitude of the applied solar irradiance forcing. The published estimates of the TSI change from Maunder minimum to present are highly uncertain and cover the range from 0.5 to 5 W/m^2 (Gray *et al.* 2010; Feulner 2011) depending on the treatment of secular changes in quiet Sun contribution (e.g., Shapiro *et al.* 2011; Schrijver *et al.* 2011). The simulations of the past climate change driven by different TSI reconstructions (e.g., Stott *et al.* 2003; Feulner 2011) have shown that the application of larger solar forcing is more promising to understand the unexplained global warming in the first half of 20th century (IPCC 2007). The decrease of the solar activity can also change the pattern of energetic particle precipitation. The expected decline of the solar magnetic activity will lead to less intensive deflection of galactic cosmic rays (Barnard *et al.* 2011) followed by higher ionization rates in the lower atmosphere. A weak solar magnetic activity in the future will not be favorable for coronal mass ejections leading to substantial decrease of powerful solar proton events (Barnard *et al.* 2011). Substantial decrease of solar wind pressure followed by weaker geomagnetic activity can be also expected if we apply observed correlation between solar magnetic activity and different geomagnetic indexes. Therefore, the effects of solar irradiance can be partially compensated by an expected decrease of the geomagnetic activity leading to less intensive production of nitrogen and hydrogen oxides followed by less intensive ozone destruction and relative warming inside polar vortices (Rozanov *et al.* 2005; Semeniuk *et al.* 2011). On the other hand an increase of galactic cosmic rays flux will facilitate ozone destruction and cooling in the polar lower winter stratosphere leading to opposite effects, i.e. to an acceleration of the polar night jets and warmer winters over Europe (Calisto *et al.* 2011; Semeniuk *et al.* 2011). The multitude of factors affecting climate requires the application of proper models, which are able to treat different anthropogenic and solar related forcing mechanism and atmospheric feedbacks. Therefore, in order to understand the resulting changes in the atmosphere we apply the chemistry-climate model SOCOL in time-slice mode driven by anthropogenic and solar forcing. In the following sections we describe the model, the set-up of performed experiments and discuss the results.

2. Model description and experimental set-up

The CCM SOCOL consists of the global circulation model MA-ECHAM4 and the chemistry-transport model MEZON. MA-ECHAM4 (Manzini *et al.* 1997) is a spectral model with T30 horizontal truncation resulting in a grid spacing of about 3.75; in the vertical direction the model has 39 levels in a hybrid sigma-pressure coordinate system spanning the model atmosphere from the surface to 0.01 hPa. The chemical-transport part MEZON (Egorova *et al.* 2003) exploits the same vertical and horizontal resolution and treats 41 chemical species of the oxygen, hydrogen, nitrogen, carbon, chlorine and bromine groups, which are coupled by 140 gas-phase reactions, 46 photolysis reactions and 16 heterogeneous reactions in/on aqueous sulfuric acid aerosols, water ice and nitric acid trihydrate (NAT). The solar irradiance variability represented in the model is accounted in the radiation and photochemical modules to calculate the response of the energy budget, heating rates and photolysis frequencies (Egorova *et al.* 2004). The original version of the CCM SOCOL was described by Egorova *et al.* (2005). Evaluation of the CCM SOCOL (Egorova *et al.* 2005; Eyring *et al.* 2007) revealed model deficiencies in the chemical-transport part and led to the development of the CCM SOCOL v2.0. A comprehensive description of the CCM SOCOL v2.0 is presented by Schraner *et al.* (2008). CCM SOCOL v2.0 participated in the SPARC CCMVal-2 intercomparison campaign (SPARC CCMVal 2010) and showed substantial improvement of transport and chemical diagnostics; however some shortcomings in the simulation of gas transport still remains. This version also includes additional source of nitrogen and hydrogen oxides due to ionization of the neutrals in the atmosphere by different precipitating energetic particles. The ionization rates due to the Galactic Cosmic Rays (GCR) have been parameterized using the recently developed CRAC:CRII (Cosmic Ray induced Cascade: Application for Cosmic Ray Induced Ionization) model extended toward the upper atmosphere (Usoskin *et al.* 2010). The model is based on a Monte-Carlo simulation of the atmospheric cascade and reproduces the observed data within 10% accuracy in the troposphere and lower stratosphere (Usoskin *et al.* 2010). The results of the CRAC:CRII model have been parameterized (Usoskin & Kovaltsov 2006) to give ion pair production rate as a function of the air pressure, geomagnetic cutoff rigidity and solar modulation potential (SMP).

Table 1. Description of the boundary conditions for the performed model experiments.

Experiment	Anthropogenic forcing[1]	TSI[2]	SSI[3]	SMP[4]	Ap[5]	SPE[6]
REF	1995-2005	1367.77	0.0098	741.0	15.05	1995-2005
ANT	2045-2055	1367.77	0.0098	741.0	15.05	1995-2005
APS	2045-2055	1363.87	0.0082	216.0	6.1	N/A

Notes:
[1] Anthropogenic forcing includes greenhouse gases, ozone destroying substances, CO and NO_x emissions.
[2] Total solar irradiance (TSI) is in W/m^2. [3] SSI stands for spectral solar irradiance at 205 nm in (W/m^2/nm).
[4] Solar modulation potential (SMP) is in MV. [5] Ap is geomagnetic index. [6] SPE is averaged intensity of solar proton events.

The ionization rates due to solar proton events (SPE) have to be prescribed. For the satellite era (1963 to 2008) we applied daily averaged ionization rates as functions of pressure between 888 hPa and $8*10^{-5}$ hPa (Jackman *et al.* 2008). The ionization rates were introduced to the model over the polar cap from 60° to 90° geomagnetic latitudes. The area of ionization by high energy (more than 10 Mev) protons is located well inside our model domain and their effects should be properly accounted for. The representation of energetic electron effects is more complicated task, because their energy deposition occurs mainly outside our model domain (i.e., above 80 km). Therefore, we have parameterized the influx of NO_x produced by energetic electrons above the model top proposed by Baumgaertner *et al.* (2009) on the basis of the empirical relation with geomagnetic Ap index using their "average excess" NO_x mode. The effect of the energetic electrons on the mesosphere and stratosphere is confined to the polar vortex. Therefore minimum absolute latitude of 55° has been used, i.e. the products of the ionization by energetic electrons can enter our model domain only over the high latitudes. Further downward propagation of ionization products depends on the presence of polar vortices and appropriate vertical transport, which guarantees that the ionization products properly affect lower mesosphere and stratosphere. The applied parameterization does not include high (more than 50 KeV) energy electrons which deposit their energy below 80 km. The lack of this process in our model can be justified by smaller contribution of high energy electron precipitation events to the total NO_x production by particles (Sinnhuber *et al.* 2011) and the absence of proper parameterization. The GCR and SPE ionization rates cannot be directly used in CCM SOCOL which has no explicit treatment of ion chemistry, therefore it is necessary to convert the ionization into the NO_x and HOx production rates. Following Porter *et al.* (1976), we assumed that 1.25 NO_x molecules are produced per ion pair, and 45 % of this NO_x production is assumed to yield ground state atomic nitrogen, while 55 % is assumed to go into $N(^2D)$ with instantaneous conversion to NO. The production of HOx has been studied by Solomon *et al.* (1981) with a 1-D time-dependent model of neutral and ion chemistry. They parameterized the number of odd hydrogen particles produced per ion pair as a function of altitude and ionization for daytime, polar summer conditions of temperature, air density and solar zenith angle. We implement these parameterizations in the CCM SOCOL to take into account the production of NO_x and HOx induced by GCR and SPE from the ground up to 0.01 hPa level. The errors associated with this approach are within 10-20% (Egorova *et al.* 2011) which is comparable with the accuracy of ionization rate calculations. For this study, we have carried out three 20-year long runs of CCM SOCOL v2.0 in time slice mode (see Table 1). The reference run (REF) has been driven by boundary conditions for the source gases, aerosol loading, solar irradiance, sea surface temperatures and sea ice concentration identical to the CCMVal-2 experiments (Morgenstern *et al.* 2010) representing the climatology around year 2000. The same procedure was applied for the solar modulation potential (proxy for GCR), ionization rates by SPE and Ap index (proxy for low energy electrons). For the second run (ANT) we applied the boundary condition identical to CCMVal-2 REF-B2 experiments (Morgenstern *et al.* 2010) representing the climatology around year 2050 keeping the solar activity related forcing applied for reference simulation. This run represents the future atmospheric state due to anthropogenic forcing only. For the third run (APS) we applied anthropogenic forcing identical to ANT run, but the solar forcing was prescribed for the case of expected strong decrease of the solar activity similar to the Dalton minimum. All solar related forcing was taken as an average over the Dalton minimum period. For the solar total and spectral solar irradiance (TSI and SSI) we have used the latest reconstruction presented by Shapiro *et al.* (2011). The solar modulation potential for the Dalton minimum was taken from Steinhilber *et al.* (2010). The frequency of the SPE was set to zero according to strong solar minimum case of Barnard *et al.* (2011). The geomagnetic Ap index values for the Dalton minimum was set according to Steinhilber (2011, private communications). The first 10 years of all runs are considered as a spin-up time which is necessary for the adaptation to the new boundary conditions and reaching quasi-equilibrium state. In the next sections we analyze the results obtained from the last 10-years of all model experiments. The comparison of the 10-mean climatology allows estimating the potential contribution of the solar forcing to the future climate change and statistical significance of the obtained results.

3. Results

Annual and zonal mean difference of total odd nitrogen (NO_y) in the future relative to present for ANT and APS runs is illustrated in Fig. 1. The anthropogenic changes of NO_y are not very large. The project decrease of NO_x emissions (IPCC 2007) leads to the NO_y decrease in the northern lower troposphere. Slight NO_y increase in the stratosphere is the result of gradually increasing anthropogenic production of N_2O (the source gas for NO_y). Negative tendencies in the tropical lower stratosphere and polar mesosphere are caused by relative deceleration of the meridional circulation in the future (SPARC CCMVal 2010, chapter 4). The application of the solar forcing substantially changes the results. Introduced decrease of Ap index leads to lower production of NO_x by precipitating electrons and significant decrease of NO_y in the mesosphere/upper stratosphere. As expected (Semeniuk *et al.* 2011) this effect is more pronounced in the polar areas leading to NO_y decrease by up to 80% in the upper mesosphere and up to 20% in the upper stratosphere. The applied decrease of the solar UV irradiance results in slower NO photolysis and suppressed destruction of NO_y via cannibalistic reaction ($N+NO=N_2+O$) leading to enhancement of NO_y in the extra-polar stratosphere. Enhanced ionization by GCR leads to additional production of NO_y below 20 km more pronounced in relatively clean southern troposphere. These results are consistent with the estimates of NO_y response to energetic electron precipitation and GCR published by Semeniuk *et al.* (2011)

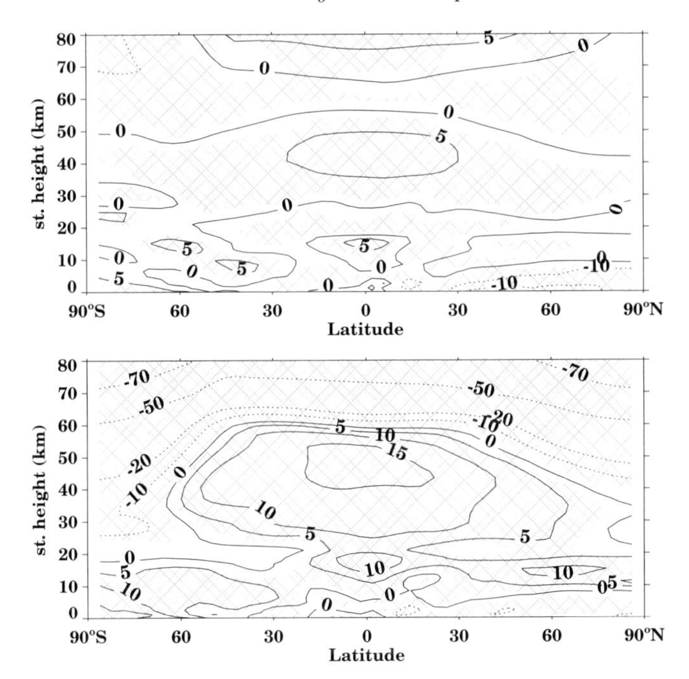

Figure 1. Annual and zonal mean difference (%) of NO_y in the future relative to present for ANT (anthropogenic forcing, upper panel) and APS (anthropogenic and solar forcing, lower panel) runs. Hatching represent the areas where the statistical significance exceed 90% level.

and Calisto *et al.* (2011) and shows that the solar forcing can plays leading role in the future evolution of NO_y. Annual and zonal mean difference of ozone in the future relative to present for ANT and APS runs is shown in Fig. 2. The anthropogenic effects on ozone consist of significant ozone increase in the troposphere caused by the enhancement of ozone precursors (i.e., carbon monoxide) and in the lower and upper stratosphere caused by substantial decline of the halogen loading in the future regulated by the Montreal protocol and its amendments (WMO 2011). The later effect is particularly visible in the southern lower stratosphere, where the ozone hole is not so deep in the considered year 2050. The ozone decrease in the tropical lower stratosphere reflects an increase of meridional circulation in the warmer climate (SPARC CCMVal 2010). The decrease of the ozone in the mesosphere can be attributed to the enhancement of the HO_x production caused by the increase of the stratospheric water vapor (SPARC CCMVal 2010). The application of the solar forcing changes the situation dramatically only in the polar mesosphere, where the decrease of NO_y (see Fig. 1) suppresses ozone depletion by catalytical oxidation. In the rest of the atmosphere the solar forcing tends to compete with anthropogenic. Global cooling due to decrease of TSI leads to deceleration of meridional circulation and partial compensation of the ozone depletion is the tropical lower stratosphere. In the middle and upper stratosphere the decrease of solar UV irradiance leads to less intensive ozone production and some compensation of the halogen loading effects. More intensive NO_y production in the troposphere due to GCR causes some compensation of the anthropogenic ozone increase in the upper troposphere. Thus, we can see that the applied solar forcing works in the direction opposite to anthropogenic effects.

Fig. 3 shows the contribution of the solar activity changes to the future total column ozone calculated as a difference between the results of APS and ANT runs. The anthropogenic influence results in ubiquitous increase of the total column ozone in the year

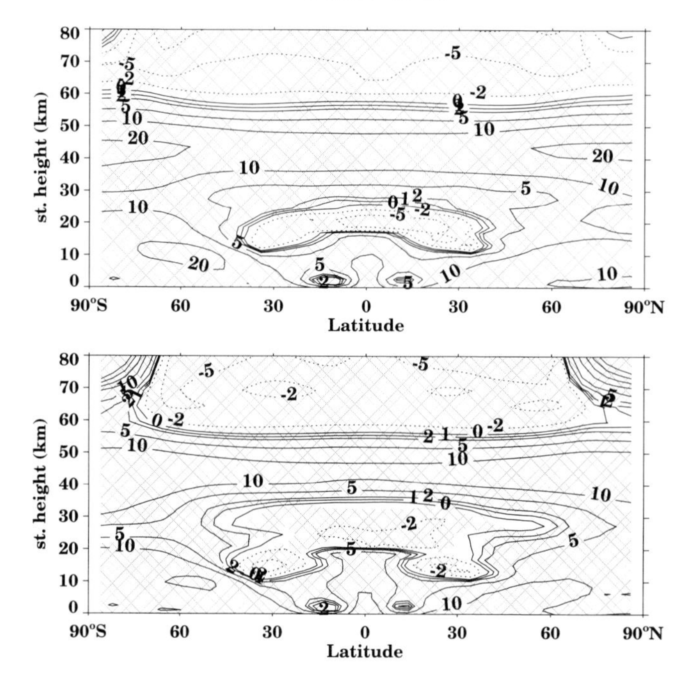

Figure 2. Annual and zonal mean difference (%) of ozone in the future relative to present for ANT (anthropogenic forcing, upper panel) and APS (anthropogenic and solar forcing, lower panel) runs. Hatching represent the areas where the statistical significance exceed 90% level.

2050 compare to present day (not shown). This increase is the most pronounced in the polar regions reaching 70 DU during late spring time, but is also statistically significant over the middle and tropical latitudes. These results are in a good agreement with multi-model assessment (SPARC CCMVal 2010). The influence of solar related forcing works in the direction opposite to anthropogenic effects leading to the total ozone depletion in global scale. The solar influence is the most important outside of the polar area leading to substantial compensation of the anthropogenic influence. Over the high latitudes solar influence is marginally significant, but even in this case it can compensate about 30-40% of anthropogenically induced total ozone recovery. Annual and zonal mean

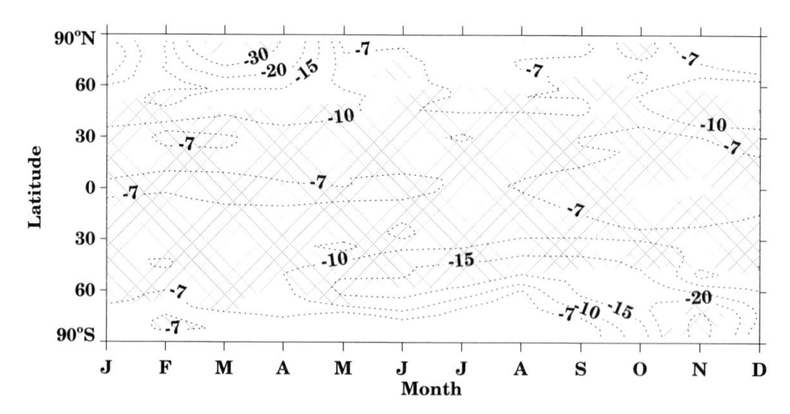

Figure 3. Zonal mean difference (DU) of the future total column ozone between APS (anthropogenic and solar forcing) and ANT (anthropogenic forcing) runs. Hatching represent the areas where the statistical significance exceed 90% level.

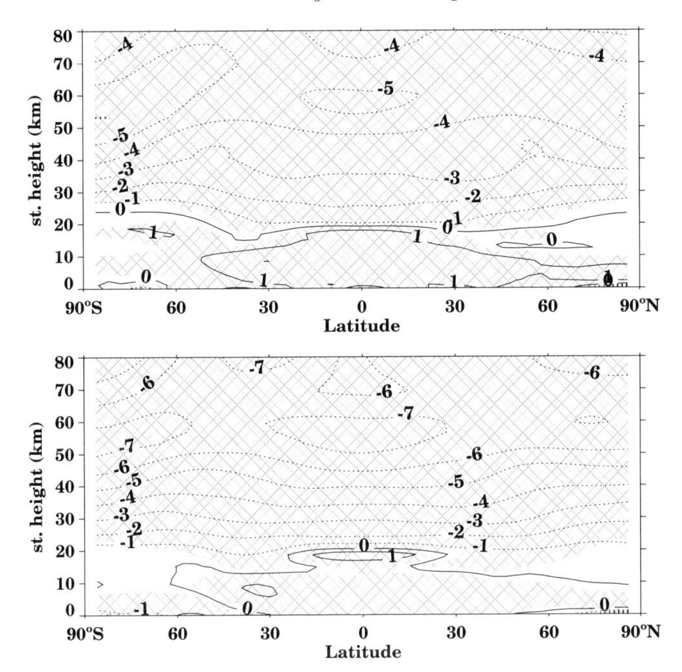

Figure 4. Annual and zonal mean difference (K) of the temprature in the future relative to present for ANT (anthropogenic forcing, upper panel) and APS (anthropogenic and solar forcing, lower panel) runs. Hatching represent the areas where the statistical significance exceed 90% level.

temperature change in the future relative to present for ANT and APS runs is illustrated in Fig. 4. The projected increase of greenhouse gas emissions leads to the pronounced cooling in the entire stratosphere reaching 5 K near the stratopause. The ozone recovery (see Fig. 2) caused by smaller halogen loading in the future atmosphere partially compensate the cooling due to greenhouse gases. The compensation is almost complete in the polar lower stratosphere, where the ozone recovery is the most remarkable. The increase of downward infrared radiation caused by greenhouse gases produces tropospheric warming exceeding 1K. The solar influence slightly alters the pattern of the temperature response. Substantial drop of solar UV irradiance leads to decrease of radiative heating, ozone concentration and additional (up to 2K) cooling in the stratosphere. The ozone increase in the polar mesosphere (see Fig. 2) caused by lower geomagnetic activity slightly compensates the cooling there, but cannot compete with the effects of solar UV irradiance. The introduced reduction of TSI affects surface energy budget and leads to global mean cooling at the surface by about 0.5 K followed by tropospheric cooling with about the same magnitude. The geographical distribution of the annual mean solar contribution to the 2 meter temperature over the land masses is illustrated in Fig. 5, which shows that the solar contribution is statistically significant almost everywhere and not homogeneously distributed. The most pronounced cooling (up to 1.2 K) appears over India, Central Asia, Siberia and Antarctica. The cooling over the other geographical regions is in the range from 0.5 to 0.7 K. This space inhomogeneity can be explained by the influence of the other introduced solar forcing mechanisms (solar UV irradiance and energetic particles) which cannot affect global mean temperature but are able to redistribute the pattern of surface temperature response (Egorova *et al.* 2004; Calisto *et al.* 2011). The comparison of the temperature changes in the future for ANT and APS runs shows that the solar forcing can compensate about 50% of the climate warming due to

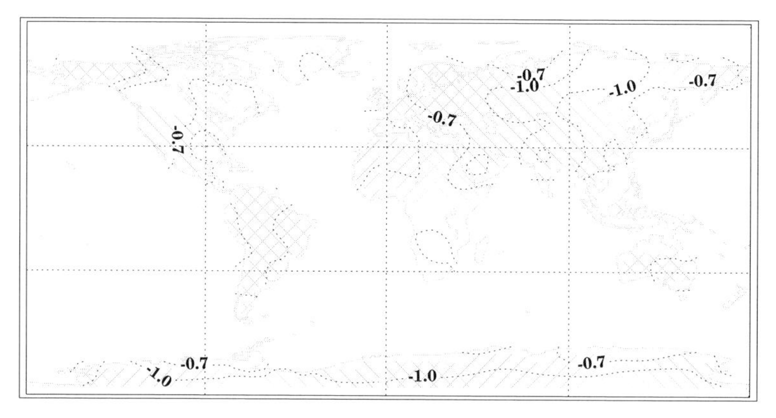

Figure 5. Annual mean difference (K) of the future 2 meters temperature between APS (anthropogenic and solar forcing) and ANT (anthropogenic forcing) runs. Hatching represent the areas where the statistical significance exceed 90% level.

pure anthropogenic factors, while in some geographical locations solar forcing can even dominate.

4. Summary

We have simulated the present and future climate and atmospheric state using CCM SOCOL v2.0 in time slice mode. To estimate the consequences of a potential strong decrease of the solar activity we simulated the future using pure anthropogenic forcing as well as in combination with different solar activity related factors. We have taken into account possible changes of the total solar irradiance, spectral solar irradiance, energetic electron precipitation, solar protons and galactic cosmic rays suggesting that the solar activity will become similar to the Dalton minimum around year 2050. The comparison of the model simulations shows that introduced strong decrease of solar activity can lead to some delay of the ozone recovery and partially compensate greenhouse warming acting in the direction opposite to anthropogenic effects. On the other hand, the anthropogenically induced cooling in the stratosphere is enhanced by solar forcing. The model results also show that all considered solar forcings are important in different atmospheric layers and geographical regions; however, the solar irradiance variability can be considered as the most important for global problems. The obtained results constitute probably the upper limit of the possible solar influence. Deeper understanding and construction of better constrained set of future solar forcings is necessary to address the problem of future climate and ozone layer state with more confidence. The development of more reliable solar forcing data sets requires in turn maintaining and extending of all relevant satellite and ground based observations as well as further theoretical investigations.

Acknowledgements

The research leading to this paper was partially supported by the Swiss National Science Foundation under grant CRSI122-130642 (FUPSOL).

References

Abreu, J., Beer, J., Steinhilber, F., Tobias S., & Weiss, N. 2008, *Geophys. Res. Lett.*, 35, L20109

Barnard, L., Lockwood M., Hapgood, M., Owens, M., Davis, C., & Steinhilber, F. 2011, *Geophys. Res. Lett.*, 38, L16103

Baumgaertner, A., Jöckel, P., & Brühl, C. 2009, *Atmos. Chem. Phys.*, 9, 2729
Calisto, M., Usoskin, I., Rozanov, E., & Peter, T. 2011, *Atmos. Chem. Phys.*, 11, 4547
Egorova, T., Rozanov, E., Zubov, V., & Karol, I. 2003, *Izvestiya, Atmospheric and Oceanic Physics*, 39, 277
Egorova, T., Rozanov, E., Manzini, E. *et al.* 2004, *Geophys. Res Lett.*, 31, L06119
Egorova, T., Rozanov, E., Zubov, V., Manzini, E., Schmutz, W., & Peter, T 2005, *Atmos. Chem. Phys.*, 5, 1557
Egorova, T., Rozanov, E., Ozolin, Y., Shapiro, A. V., Peter, T., & Schmutz, W. 2011, *Journal of Atmospheric and Solar-Terrestrial Physics*, 73, 356
Eyring, V., Waugh, D. W., Bodeker, G. *et al.* 2007, *J. Geophys. Res.*, 112, D16303
Feulner, G. 2011, *Geophys. Res. Lett.*, 38, L16706
Gray, L., Beer, J., Geller, M. *et al.* 2010, *Rev. Geophys.*, 48, RG4001
IPCC: Intergovernmental Panel on Climate Change 2007, *Cambridge University Press*, 489
Jackman, C., Marsh, D., Vitt, F. *et al.* 2008, *Atmos. Chem. Phys.*, 8, 765
Lockwood, M., Rouillard, A., & Finch, I. 2009, *Astrophys. J.*, 70, 937
Manzini, E., McFarlane, N. A., & McLandress, C. 1997, *J. Geophys. Res.*, 102, 25751
Morgenstern, O., Giorgetta, M. A., Shibata, K., *et al.* 2010, *J. Geophys. Res.*, 115, D00M02
Porter, H., Jackman, C., & Green, A. 1976, *J. Chem. Phys.*, 65, 154
Rozanov, E., Callis, L., Schlesinger, M., Yang, F., Andronova, N., & Zubov, V. 2005, *Geophys. Res. Lett.*, 32, L14811
Shapiro, A. I., Schmutz, W., Rozanov, E., Schoell, M., Haberreiter, M, Shapiro, A. V., & Nyeki, S. 2011, *Astron. Astrophys.*, 539, A67
Schraner, M., Rozanov, E., Schnadt Poberaj, C. *et al.* 2008, *Atmos. Chem. Phys.*, 8, 5957
Semeniuk, K., Fomichev, V., McConnell, J., Fu, C., Melo, S., & Usoskin, I. 2011, *Atmos. Chem. Phys.*, 11, 5045
Sinnhuber, M., Kazeminejad, S., & Wissing J. M. 2011, *J. Geophys. Res.*, 116, A02312
Schrijver, C.,Livingston, W., Woods, T., & Mewaldt, R. 2011, *Geophys. Res. Lett.*, 38, L06701
Solomon, S., Rusch, D., Gerard, J., Reid, G., & Crutzen, P. 1981, *Planetary Space Science*, 29, 885
SPARC CCMVal: Eyring, V., Shepherd, T., & Waugh, D. (eds) 2010, *WCRP-132/WMO/TD-1526/SPARC*, 5
Steinhilber, F., Abreu, J., Beer, J., & McCracken, K. 2010, *J. Geophys. Res.*, 115, A01104
Stott, P., Jones, G., & Mitchell, J. 2003, *J. Clim.*, 16, 4079
Usoskin, I., Kovaltsov, G., & Mironova, I. 2010, *J. Geophys. Res.*, 115, D10302
Usoskin, I. & Kovaltsov, G. 2006, *J. Geophys. Res.*, 111, D21206
WMO (*World Meteorological Organization*): Scientific Assessment of Ozone Depletion: 2010. 2011, *Global Ozone Reaserch and Monitoring Project Report No* 52, 516

Discussion

Leif Svaalgard: Cosmic rays and clouds. Some believe in this effect some not. Is that included in your model?

Eugene Rozanov: This effect is not included in the model, because we do not have physically based parameterization to do it.

Axel Brandenberg: It is not true that there is no data on cosmic rays influence on clouds. There are lab experiments and mechanisms that have been developed for the effects of cosmic rays forcing.

Eugene Rozanov: There are some data and mechanisms, but it is not enough to properly include them in the model. Physically based parameterizations are still not available.

Eric Priest: How important is UV variability?

EUGENE ROZANOV: It is very important. It explains most of the ozone changes and non-homogeneity of the surface air temperature response. The change of TSI gives rather homogeneous pattern of the temperature response. For example, the band of cooling over the Russia is formed by UV changes.

JANET LUHMANN: Will the polar ice change affect the results of your model?

EUGENE ROZANOV: Yes. The model I used here has simplified representation of the see ice. Better representation of the see ice is necessary. Similar experiments are on-going with more complicated version of the model.

MARK GIAMPAPA: What are the errors in your estimates from your model?

EUGENE ROZANOV: The results inside hatched areas are statistically significance at more than 90% level. The uncertainties depend on the accuracy of the forcing projections, which can hardly be estimated.

LEIF SVAALGARD: Before the Dalton minimum and other similar periods there were several volcanic eruptions, which could produce colder climate. Why not use volcanic forcing for the future runs?

EUGENE ROZANOV: We tried to estimate possible compensation of greenhouse effect and ozone recovery by solar influence. Volcanic forcing is important for the prediction of future climate, but this question is out of scope of this particular work.

NOT IDENTIFIED WOMAN: I would urge you to consider volcanic eruptions in your simulation of the future climate.

EUGENE ROZANOV: There is no reliable scenario for future volcanic eruption frequency and strength. Anyway, if there is enhanced volcanic activity in the future, the compensation of the greenhouse warming and ozone recovery will be also enhanced.

Comparative Magnetic Minima:
Characterizing quiet times in the Sun and stars
Proceedings IAU Symposium No. 286, 2011
C. H. Mandrini & D. F. Webb, eds.

doi:10.1017/S1743921312004875

Coronal Mass Ejection deflection in the corona during the two last solar minima

Fernando Marcelo López[1], Hebe Cremades[2] and Laura Balmaceda[3]

[1]Universidad Nacional de San Juan - FCEFyN
Av. Ignacio de la Roza 590 (O), J5402DCS, San Juan, Argentina
email: fermlop@gmail.com

[2]UTN - Facultad Regional Mendoza/CONICET
Rodriguez 243, Ciudad, Mendoza M5502AJE, Argentina
email: hebe.cremades@frm.utn.edu.ar

[3]Instituto de Ciencias Astronómicas, de la Tierra y el Espacio/CONICET
Av. España Sur 1512, J5402DSP, San Juan, Argentina
email: labalmaceda@gmail.com

Abstract. In the framework of the IAU Working Group on Comparative Solar Minima, we investigate the latitudinal deflection of Coronal Mass Ejections (CMEs) with respect to the location of their uniquely identified solar source regions. Data compiled during the Whole Sun Month (WSM) and Whole Heliosphere Interval (WHI) campaigns allowed for comparisons between the two last solar minima.

The analysis of the coronal streamers' distribution during these intervals led to study of the dependence of CME deflection on the angular separation between their source regions and the nearest streamer. All performed analyses consider exclusively projected structures on the plane of the sky, disregarding longitudinal deflections as well.

The results of the present study indicate that for both minima most of the events (62.5% for WSM, 84.2% for WHI) are deflected towards the nearest streamer, following the boundary conditions imposed by the heliospheric current sheet.

Most of the deflections found in the WHI period could be explained by the more complex structure in the global distribution of magnetic field present during that minimum. On the other hand, the low number of events detected during the WSM period hinders the statistical comparison between both campaigns.

Keywords. Sun: corona, Sun: coronal mass ejections (CMEs), Sun: activity

1. Introduction

Coronal Mass Ejections (CMEs) are known to be an important source of geomagnetic storms, which makes the study of their direction of propagation in the corona and interplanetary medium crucial, in order to determine the potential arrival of events to the terrestrial magnetosphere. Therefore, the deflection suffered by CMEs in the corona is one of the fundamental parameters to be considered at the time of forecasting a potentially geoefective event.

The deflection of CMEs in the solar corona and its relationship with coronal features have been previously studied. Cremades *et al.* (2006) found a connection between CME deflection and coronal hole (CH) location and area. Similarly, Gopalswamy *et al.* (2009) found that CMEs are significantly affected when the eruptions occur close to a CH, with the CH acting as a magnetic wall constraining the CME trajectory. Not in conflict with these results, Gui *et al.* (2011) suggest that CMEs tend to deflect to regions with lower magnetic energy density. Recent detailed case studies aimed to understand the fundamentals of CME deflection include those of Shen *et al.* (2011) and Zuccarello *et al.* (2012). In view of hitherto existing evidence of the influence of the corona's environment on early CME trajectory, and the essentially different coronal magnetic configuration present during the past two solar minima, this report deals with CME deflections during a full rotation of these minima and their connection to streamers, white-light coronal features that denote global magnetic field configuration.

The next Section describes the event selection process, followed by the main results about the types of CME-streamer interaction and the relationship between CME deflection and distance to streamer. Main findings are discussed and summarized in Section 5.

2. Events Selection

The WSM (10 August to 8 September 1996) and the WHI (20 March to 16 April 2008) campaigns represent a unique opportunity to compare different aspects of the corona/heliosphere during the last two solar magnetic minima. Using data from the Solar and Heliospheric Observatory (SOHO), Yohkoh, and Solar-Terrestrial Relations Observatory (STEREO), Cremades *et al.* (2011) identified all distinguishable coronal transients during these time intervals, from wide CMEs up to very narrow jets. Here, we analyze a subset of those events fulfilling the following criteria: 1) their angular width is higher than 10°, and 2) their solar source region can be uniquely identified on images of the Extreme-UV Imaging Telescope (SOHO/EIT) or Extreme UltraViolet Imager (STEREO/EUVI). Following these criteria, the original set of events reduced to a total of 8 events for WSM and 19 for WHI. Furthermore, this analysis considers only streamers and disregards pseudostreamers. The first separate open field lines of opposite polarities and are thus associated with the heliospheric current sheet (HCS), while the latter overlie loop arcades separating field lines of the same polarity.

3. Types of CME - Streamer Interaction

When inspecting the interaction between CMEs and streamers, we found three main classes, according to their behaviour on images of the corona and Carrington maps from the Large Angle and Spectrometric Coronagraph (LASCO C2). The kinds of interaction we found are as follows: 1) "Regular Case", when there is a well-defined, narrow streamer and the CME is not "channeled" through it; 2) "Channeled CME", when the CME travels

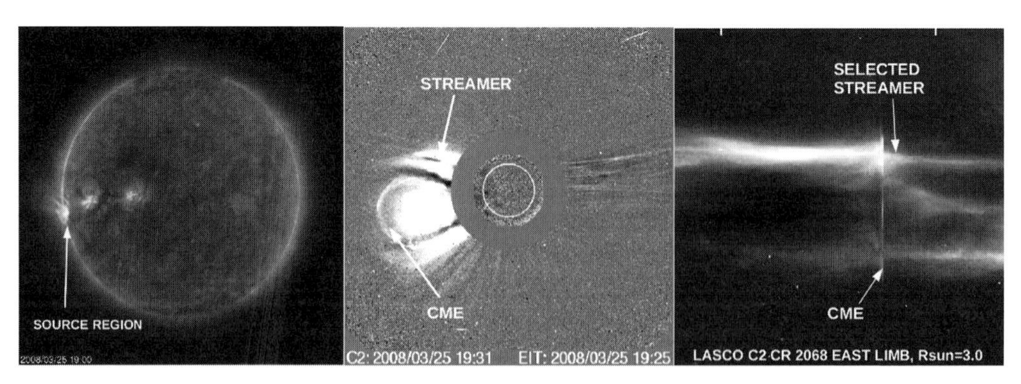

Figure 1. Example of the kind of interaction "Regular Case": EUV image with the associated source region (left), LASCO C2 running difference image (centre), and the corresponding LASCO C2 Carrington map obtained at 3 solar radii from the Sun's centre (right).

within the streamer, i.e. in the direction determined approximately by the central PA of the streamer; and 3) "Spread-Out Streamer" when the streamer is diffuse and wide.

Fig. 1 presents an example of the first kind of interaction mentioned above. To the left, an EUV image shows the associated source region, while the ensuing CME can be appreciated in a LASCO C2 running difference image (centre). The corresponding LASCO C2 Carrington map taken at 3 solar radii from the Sun's centre allows the visualization of the streamers' behaviour in time (right).

4. Relationship Between CME Deflection and Distance to Streamers

After classifying the events according to the three types defined above, we study the dependence of CME deflection on the angular distance between the associated source region

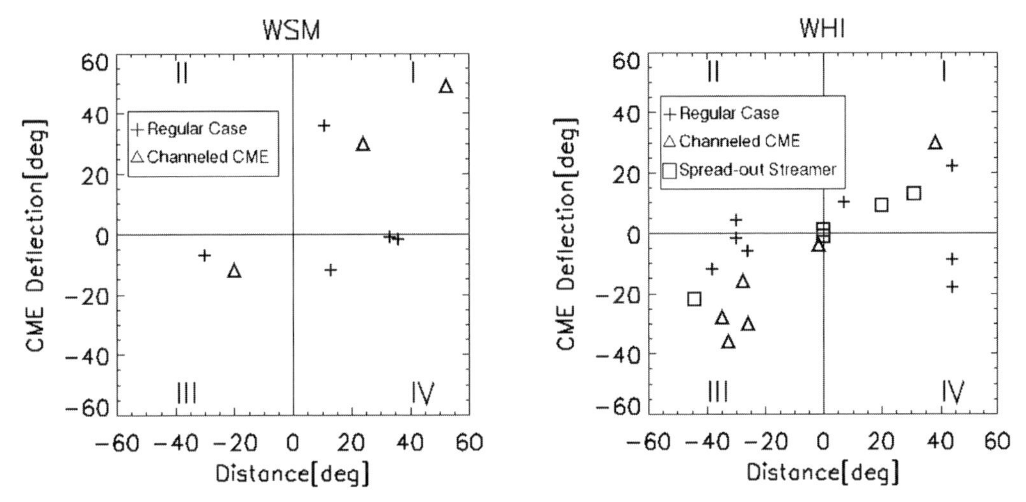

Figure 2. Relationship between the deflection of CMEs and the distance from their source region during the WSM (left) and WHI (right).

and the nearest streamer. This angular distance is given by the difference of the respective position angles, i.e. distance=ST_PA - SR_PA. The CME deflection angle is given by the difference between those position angles of the CME and its associated source region, i.e. CME deflection=CME_PA - SR_PA). The CME central position angles were obtained from the SOHO/LASCO's catalog of CMEs (http://cdaw.gsfc.nasa.gov/CME_list/index-.html), whereas the streamer position angles were determined using LASCO C2 Carrington maps. The source region position angles were taken from Cremades *et al.* (2011). Most of the low-coronal sources of the events here analyzed were located at heliographic longitudes higher than 50°. With the source regions close to the solar limb, projection effects are kept to a minimum.

Fig. 2 compares WSM and WHI with regard to CME deflection as a function of the angular distance between source regions and streamers. The three different types of interaction are represented by different symbols. In order to analyze the direction of CME deflection, we divide the plots in Fig. 2 into quadrants (I, II, III and IV) centred at distance=0 and CME deflection=0. When the deflection and distance angles are of the same sign, the CME deflects towards the streamer (quadrants I and III in Fig. 2). On the other hand, if the signs are opposite, the CME deflects in the direction opposite to that of the streamer (quadrants II and IV in Fig. 2). It is evident from the figure that the general behaviour follows these rules. In order to quantify the goodness of the linear

Table 1. Correlation coefficients between the CME deflection and distance angles during WSM and WHI for the three types of interaction.

	Regular Case	Channeled CME	Spread-Out Streamer
WSM	r=0.08	r=0.99	-
WHI	r=0.12	r=0.96	r=0.97

relationship between CME deflection and distance angles, we calculated the correlation coefficient (r) for each of the three types of interaction (see Table 1).

5. Discussion

From the definitions of the CME deflection and distance angles, it arises that CME deflections away from their source regions towards the nearest streamer and the distance angle must have the same sign. On the contrary, the signs are opposite when deflections are such that the CME moves away from the streamer.

Most of the events here analyzed (16 out of 19, i.e. 84.2% for WHI; and 5 out of 8, i.e. 62.5% for WSM) are indeed deflected towards the nearest streamer, following the boundary conditions imposed by the HCS. This occurs in spite of the essentially different configuration of the HCS during both of the inspected minima.

From the correlation analysis it is found for both time intervals, that those events classified as "Regular Case" do not present a linear relationship between the deflection and distance angles, even showing deflections in the direction opposite to the streamer. On the other hand, events classified as "Channeled CME" or "Spread-Out Streamer" present very good correlation coefficients, with all deflections happening in the streamer direction.

The high amount of deflections towards the nearest streamer is especially evident for the WHI period. For the WSM, the relationship between both angles follows the same trend, though with higher scattering. It is very likely that the lack of events during the WSM period is hindering the analysis.

The different coronal structure exhibited by the Sun during the two analyzed periods of minimum yielded also different absolute CME deflections. However, this is logical given the essentially distinct configuration of the HCS and the solar corona. Still, during both periods most CMEs deflected towards their respective nearest streamer, showing a similar behavior and thus indicating the same physical principle in action.

Acknowledgements

The authors acknowledge useful discussions during the Symposium with F. Zuccarello. L.B. and H.C. are members of the Carrera del Investigador Científico of CONICET.

References

Cremades, H., Bothmer, V., & Tripathi, D. 2006, *Adv. Space Res*, 38, 461.

Cremades, H., Mandrini, C. H., & Dasso, S. 2011, *Solar Phys.*, 274, 233.

Gopalswamy, N., Mäkelä, P., Xie, H., Akiyama, S., & Yashiro, S., 2009, *J. Geophys. Res.*, 114, A00A22.

Gui, B., Shen, C., Wang, Y., Ye, P., Liu, J., Wang, S., & Zhao, X., 2011, *Solar Phys.*, 271, 111.

Shen, C., Wang, Y., Gui, B., Ye, P., & Wang, S., 2011, *Solar Phys.*, 269, 389.

Zuccarello, F., P., Bemporad, A., Jacobs, C., Mierla, M., Poedts, S., & Zuccarello, F., 2012, *ApJ*, 744, 66.

Comparative Magnetic Minima:
Characterizing quiet times in the Sun and Stars
Proceedings IAU Symposium No. 286, 2011
C. H. Mandrini & D. F. Webb, eds.

doi:10.1017/S1743921312004887

High-speed streams in the solar wind during the last solar minimum

Georgeta Maris[1], **Ovidiu Maris**[2], **Constantin Oprea**[1], **Marilena Mierla**[1,3,4]

[1]Institute of Geodynamics of the Romanian Academy,
RO-020032, Bucharest, Romania
email: gmaris@geodin.ro

[2]Institute for Space Sciences, RO-077125 Bucharest, Romania

[3]Royal Observatory of Belgium, Brussels, Belgium

[4]Research Center for Atomic Physics and Astrophysics,
Faculty of Physics, University of Bucharest, Romania

Abstract. The paper presents a statistical analysis of the fast solar wind streams during the last prolonged minimum. Defining a minimum phase as the period with the monthly relative sunspot numbers (smoothed values) having a value of less than 20, we considered for this analysis the interval February 2006 September 2010. The High-Speed Streams (HSSs) in the solar wind were determined by their main parameters: duration, maximum velocity, velocity gradient. A comparative analysis of the HSS dynamics during the last solar minimum with the previous solar minimum (1996-1997) concludes the paper.

Keywords. Sun: activity, (Sun:) solar wind

---

1. Introduction

Solar wind plays an important role in the heliospheric structure and dynamics and it is "the medium" through which all the solar perturbations are propagating towards the Earth. Lots of space missions were and still are dedicated to recording its composition, velocity, density and temperature, inside and, especially, outside of the terrestrial magnetosphere (WIND, ACE, Ulysses and SOHO). Solar wind plasma consists of electrons, protons, helium and heavier nuclei, which are carrying alongside them the solar magnetic field resulting in an interplanetary magnetic field. The average velocity of the solar wind plasma is 350 km/s with a minimum of about 200 km/s and a maximum over 1000 km/s. The high speed streams (HSSs) in the solar wind were measured by many recording earth-orbiting or solar-orbiting spacecraft. The fast solar wind (>650 km/s) is characterized by a high temperature, a low density and a low mass flux, while the slow solar wind (<400 km/s) is cooler, denser and has a larger mass flux. There are also other differences in composition, in protons and electrons temperature anisotropy, etc. between slow and fast solar wind (Schwenn 2006).

In this paper we present a statistical analysis of the HSSs during the last prolonged solar minimum in comparison with their statistics during the previous minimum.

2. Data and Methods of Analysis

2.1. *HSSs and their Parameters*

A lot of HSSs definitions were given by different authors and HSSs catalogues covering solar cycles nos. 20 - 23 were set up (Lindblad & Lundstedt 1981; 1983; Lindblad *et al.*

1989; Mavromichalaki *et al.* 1988; Mavromichalaki & Vassilaki 1998; Maris & Maris 2009; 2012). We used the same definition and selection procedure of the streams as Lindblad and Lundsted because it allows for a more precise determination of the HSS beginning and end: a 'high-speed stream' is a solar wind flow having $\Delta V1 > 100 km/s$ that lasted for two days, where $\Delta V1$ is the difference between the smallest 3-hr mean plasma velocity for a given day ($V0$) and the largest 3-hr mean plasma velocity for the following day ($V1$).

From among the HSS parameters we used their maximum velocity V_{max} (in km/s), the duration d, in days, and the maximum gradient of the velocity:

$$\Delta V_{max} = V_{max} - V_0,$$

where V_0 - the minimum (pre-stream) solar wind velocity (in km/s).

2.2. *Minimum Phase of the 11-yr Solar Cycle*

The 11-yr solar cycle (SC), generally defined through the sunspot relative number, is the most evident periodicity that can be observed in all the solar atmospheric phenomena. The period of one solar cycle is defined as the interval between two successive minima of the Wolf curve. The length of approximately 11 years was statistically determined. It certainly represents the clearest periodicity observed also in all eruptive solar phenomena (flares, CMEs, solar active prominences etc.) that follow the 11-yr cycle evolution with their power and frequency of occurrence. It is best seen in the frequency indices' variability of some active solar phenomena.

We emphasize that the phases of the 11-yr solar cycle (minimum, ascendant, maximum and descendant phases) are not "rigid" components; they could have different durations and intensities in different cycles. Furthermore, maximum and minimum phases of the 11-yr cycle are referring to some intervals of time during which solar activity has a maximum or minimum level, respectively, unlike the momentary maximum or minimum of the 11-yr SC, that are referring to the maximum or minimum monthly (or annual) values, respectively, reached by some solar index during the cycle. We emphasize the main characteristics of the 11-yr solar cycle phases:

- The ascendant phase - the only interval of any cycle during which the influences of the solar cycles adjacent to it (the precedent and the following ones) are not present;
- The maximum phase - a phase with significant activity of the majority of solar phenomena having superimposed on it, in its second part, the beginning of the polar solar magnetic field reversal;
- The descendant phase - the most "complex" phase, when in the solar atmosphere and in the tachocline (underneath the convective zone) there are two magnetically opposite dipoles;
- The minimum phase - the apparent "quiet" phase when the old cycle (the old dipole) is using up its reserves (it's dying out) but in the sub-surface levels the new cycle is beginning as new active regions with the opposite polarity distribution.

We analyzed the HSS activity level during different phases of the 11-yr solar cycle in some previous papers (Maris & Maris 2005; 2010; 2011). We have chosen the phase limits considering the maximum and minimum structure of the curve for the smoothed sunspot relative number, W, in each SC, the sunspot relative number being the standard parameter of the 11-year SC dynamics. We have considered the minimum phases as the intervals with the sunspot relative number $W < 20$. According to this definition of the 11-year cycle minimum, we obtained for SCs 20-23 the phases specified in Table 1.

The lengths of the minimum phases are more or less equal for SCs 20 - 23, but the length of SC 24 minimum is more than double of the previous cycle minimum (that

Table 1. Minimum phase durations of the solar cycles nos. 20 - 24 (calendar months) and the HSS numbers.

Sc no.	Minimum Phase(mm.yyyy)	Duration (months)	HSS number	HSS no./month
20	01.1964 – 10.1965	22	34	1.5
21	04.1975 – 02.1977	23	66	2.9
22	02.1985 – 02.1987	25	69	2.8
23	05.1995 – 05.1997	25	79	3.2
24	02.2006 – 09.2010	57	283	5.0

being SC 23). The number of the HSSs recorded during the minimum phases (column 4 in Table 1), as well as the number of HSSs recorded per month (column 5 in Table 1) are not significantly different for SCs nos. 21 - 23. The small values recorded in the minimum of cycle no. 20 (34, and 1.5, respectively) can be, at least partially, explained by the limited ways of recording the solar wind plasma outside the magnetosphere during that era. In complete contrast, after 1996, the ACE and SOHO space missions, along with WIND and Ulysses, have ensured for a continuous monitoring of the solar wind plasma in regions outside the terrestrial magnetosphere.

3. HSS Statistics during SC 23 and SC 24 Minima

Figs. 1 - 3 present the statistical distributions of the HSSs in intervals of their duration, maximum velocity and maximum velocity gradient, respectively for SCs 23 (left panels) and 24 (right panels) minima.

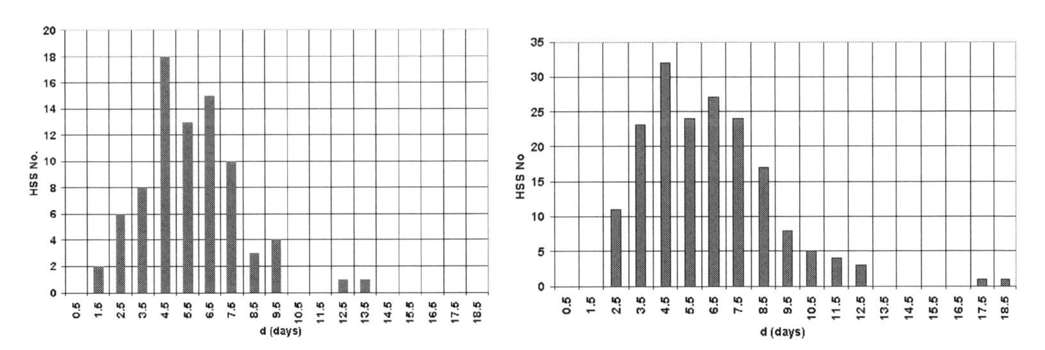

Figure 1. HSSs distributions by their durations during the minima of SC 23 (left panel) and SC 24 (right panel).

In Fig. 1 it is clearly seen that during SC 23 minimum, the majority of HSSs had durations of 4 - 8 days with peaks at 4.5 and 6.5 days, and during SC 24 minimum, the majority of HSSs had larger durations, of 3 - 9 days, with the same peaks at 4.5 and 6.5 days.

In Fig. 2 it is clearly seen that during SC 23 minimum, the maximum number of HSSs had $500 \leqslant V_{max} \leqslant 600 km/s$ whereas during SC 24 minimum, the maximum number of HSS moved to the higher values of maximum velocity, $600 \leqslant V_{max} \leqslant 700 km/s$.

We can remark in Fig. 3 that the maximum number of HSSs had $100 \leqslant \Delta V_{max} \leqslant 300 km/s$ during the SC 23 minimum, whereas the maximum number of HSSs is rather equally distributed in the three division of 100km/s of the 100 - 400 km/s interval during the SC 24 minimum.

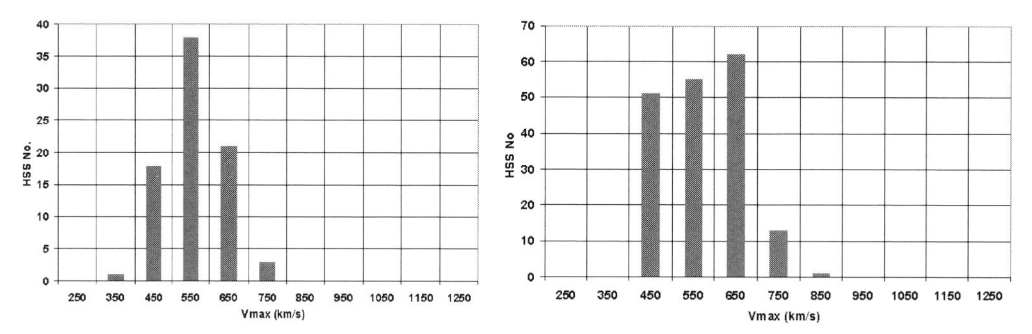

Figure 2. HSSs distributions by their maximum velocity during the minima of SC 23 (left panel) and SC 24 (right panel).

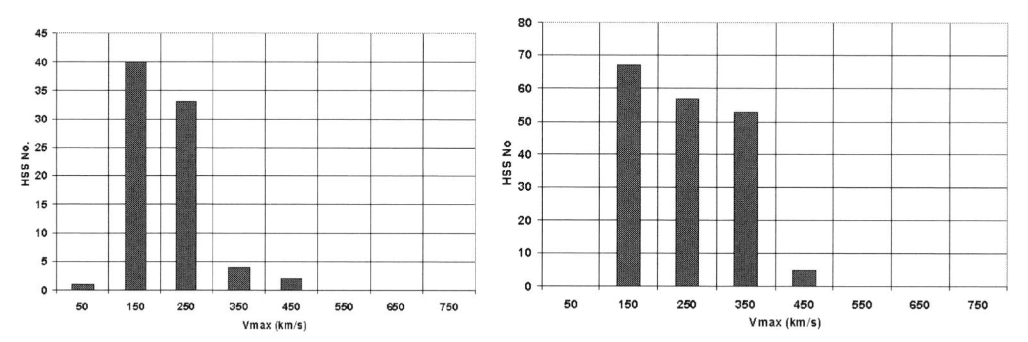

Figure 3. HSSs distributions by their maximum gradient velocity during the minima of SC 23 (left panel) and SC 24 (right panel).

A similar analysis for the minimum phases of all SCs 20 - 24 did not show any significant differences between odd and even SCs.

4. Summary

- The durations of the minimum phases for SCs 20-23 are quite equal (they lasted for 22 - 25 months), but the last minimum phase exceeded two times the previous minimum phases (57 months);
- The HSS number per minimum is slowly growing from one cycle to the next one, but it remains about 3/month for SCs 21 - 23. Frequencies of 1.5 HSSs/months and about 5 HSSs/months are opposite exceptions for the minimum phases of SC 20 and SC 24, respectively;
- Minimum phases situated after SCs with lower W_{max} (nos. 20 and 23) are more active in HSS occurrence.

Consequently, the minimum phases of the 11-yr solar cycles are not quiet intervals but they are periods with significant activity in the solar corona where especially more equatorial coronal holes could appear and could be sources of more energetic HSSs.

Acknowledgments

The work of CO and MM was supported from the project TE no. 73/11.08.2010.

References

Lindblad, B. A. & Lundstedt, H. 1981, *Sol. Phys.*, 74, 197-206

Lindblad, B. A. & Lundstedt, H. 1983, *Sol. Phys.*, 88, 377-382

Lindblad, B. A., Lundstedt, H., & Larsson B. 1989, *Sol. Phys.*, 120, 145-152

Maris, O. & Maris, G. 2005, *Adv. Space Res.*, 35, 2129-2140

Maris, O. & Maris, G. 2009, at: `http://www.spaceweather.eu/`, in *Chap. Data Acces*, or at: `http://spacescience.ro/new1/HSS_Catalogue.html`

Maris, G. & Maris, O. 2010, Proc. IAU Symposium 264 *Solar and Stellar Variability Impact on Earth and Planets*, A.G. Kosovichev, A.H. Andrei and J.-P. Rozelot (Eds.), Cambridge Univ. Press., 359-362, doi: 10.1017/S1743921309992924

Maris, G. & Maris, O. 2011, *3rd School and Workshop on Space Plasma Physics*, I. Zhelyazkov, T. Mishonov (Eds.), AIP Conf. Proc. 1356, 177-191, doi: 10.1063/1.3598104

Maris, G. & Maris, O. 2012, Chapter 7 in: *Advances in Solar Terrestrial Physics*, G. Maris & C. Demetrescu (Eds.), Research Signpost Publ., Trivandrum, Kerala, India, ISBN: 978-81-308-0483-5, 97-134, in press

Mavromichalaki, H., Vassilaki, A., & Marmatsouri, E. 1988, *Sol. Phys.* 115, 345-365

Mavromichalaki, H. & Vassilaki, A. 1998, *Sol. Phys.* 183, 181-200

Schwenn, R. 2006, *Living Rev. Solar Phys. 3*, `http://www.livingreviews.org/lrsp-2006-2`.

Comparative Magnetic Minima:
characterizing quiet times in the Sun and stars
Proceedings IAU Symposium No. 286, 2011
C. H. Mandrini & D. F. Webb, eds.

doi:10.1017/S1743921312004899

Geomagnetic effects on cosmic ray propagation for different conditions

Jimmy J. Masías-Meza[1], Xavier Bertou[1] and Sergio Dasso[2]

[1]Centro Atómico Bariloche (CNEA-CONICET), U.N. de Cuyo, Bariloche, Rio Negro, Argentina
email: masiasmj@ib.cnea.gov.ar
email: bertou@cab.cnea.gov.ar

[2]Instituto de Astronomía y Física del Espacio (UBA-CONICET) y Departamento de Física (FCEN-UBA), Buenos Aires, Argentina
email: dasso@df.uba.ar

Abstract. The geomagnetic field (B_{geo}) sets a lower cutoff rigidity (R_c) to the entry of cosmic particles to Earth which depends on the geomagnetic activity. From numerical simulations of the trajectory of a proton (performed with the MAGCOS code) in the B_{geo}, we use backtracking to analyze particles arriving at the Auger Observatory location. We determine the asymptotic trajectories and the values of R_c in different incidence directions. Simulations were done using several models of B_{geo} that emulate different geomagnetic conditions.

Keywords. cosmic rays, Sun: magnetic fields

1. Introduction

It is well known that the transport of galactic cosmic rays (GCRs) with energies $E < 20$ GeV is modulated by interplanetary conditions (Ferreira *et al.* 2004). In addition, the trajectory of these particles in the Earth's environment can be affected by the geomagnetic field, as will be addressed in the next sections.

Since 2005, a new detection mode has been implemented in the Auger Observatory (Malargüe City). It allows the detection of transient cosmic ray effects, e.g. Forbush decreases (Abreu *et al.* 2011), as well as long period modulations, e.g. daily.

Using the MAGCOS code (http://cosray.unibe.ch∼laurent/magnetocosmics), we do numerical simulations of the trajectory of a proton and analyze the main properties of arrival at Malargüe, such as the asymptotic trajectories and the value of the rigidity cutoff (R_c) in different incidence directions.

The International Geomagnetic Reference Field (IGRF, Finlay *et al.* 2010) is a semi empirical description of the Earth's magnetic field, updated every 5 years since 1955, and supported by data provided by satellites, observatories and surveys around the world. This model is mainly of dipolar topology and includes the secular variation of the main dipole moment. The Tsyganenko models (TSY, Tsyganenko 2002) include observations from major external magnetospheric sources: ring current, magnetotail current system, magnetopause currents, and large-scale system of field-aligned currents (in calm and storm conditions). In these models, the magnetosphere is deformed by the solar wind, with stretched field lines in the day-side and with a long tail in the night-side. Both models (IGRF and TSY) will be used in this study.

2. Cutoff rigidity calculations and asymptotic directions

As mentioned earlier, we simulate trajectories by the backtracking method; that is, the initial location is always with Malargüe coordinates. So, we determine the "transmittance function" as follows: for each value of rigidity we put a 0 when the trajectory leads to an asymptotic direction (i.e. the incidence direction before entering to the magnetosphere), and a 1 if the particle is bent back to the Earth's surface; where case "0" is considered to be an allowed trajectory for the particle, and case "1" is considered a forbidden one.

The "diffuse" region between the first and last values of allowed rigidity, is called "penumbra" (Dorman 2008).

For vertical incidence ($zenith = 0^o$), Figure 1 shows the transmittance function of protons for Malargüe city (approximate Auger Observatory location), using four different models of the geomagnetic field: Centered Dipole, Shifted Dipole, IGRF, and IGRF + TSY01 model (Tsyganenko 2002). The Shifted Dipole means that the magnetic dipole is translated from the Earth's center (Bartels 1936). In all cases, the real angle shift of the geomagnetic pole respect to the rotation axis is used. The "jump" of the penumbra from the centered dipole model to the shifted one might be produced by the loss in $|B_{geo}|$, as the dipole shift direction on the globe is almost in the opposite direction of Malargüe's.

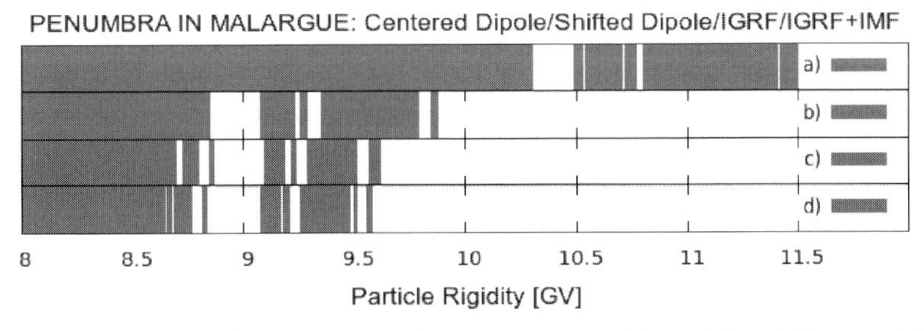

Figure 1. Transmittance function using four magnetic field models: a) Centered Dipole b) Shifted Dipole c) IGRF d) IGRF + TSY01.

The transmittance was simulated using steps of 0.001GV in rigidity. To obtain an effective cutoff rigidity R_c, we employ $R_c = R_L + 0.001GV * N$, where R_L is the first low rigidity for which there is an allowed trajectory, and N is the number of allowed rigidities in the penumbra region. For vertical incidence, the vertical effective cutoff rigidity at Malargüe is obtained at $R_c^{Mlg} = 9.13$ GeV (IGRF 2010 + TSY01).

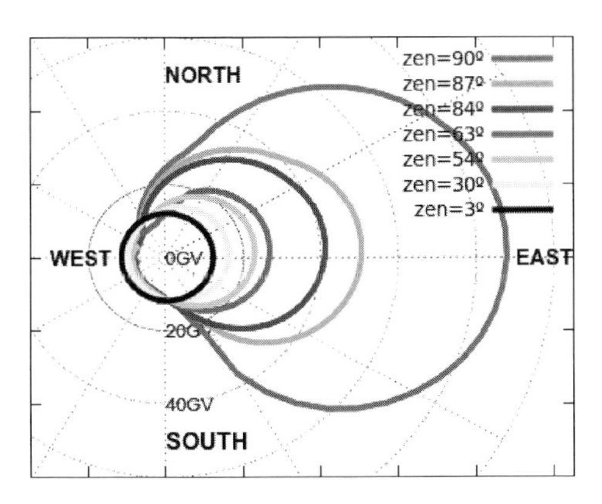

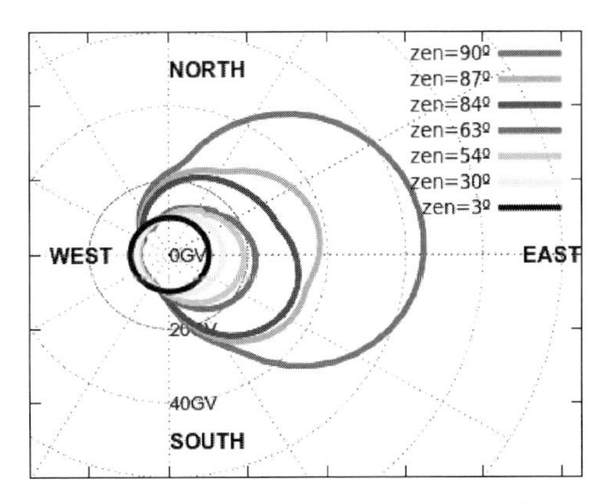

Figure 2. Effective cutoff rigidity (radius) in function of zenith (color) and azimuth (polar angle) incidence, using the Centered-Dipole model (left) and IGRF + TSY01 model (right).

In this manner, we can obtain an effective cutoff rigidity R_c for different zenith and azimuth incidences. This is performed by the four models, obtaining significant changes from the centered dipole to the shifted dipole. This can be seen in Figure 2, where only calculations with Centered Dipole and IGRF+TSY01 are shown.

For particle rigidities above R_c^{Mlg}, asymptotic directions can be determined (Figure 3). We can see that as the particle rigidity decreases, the asymptotic trajectories get closer to the equator region. Computations for Rome Neutron Monitor ($R_c = 6.27$GV) location shows the same tendency.

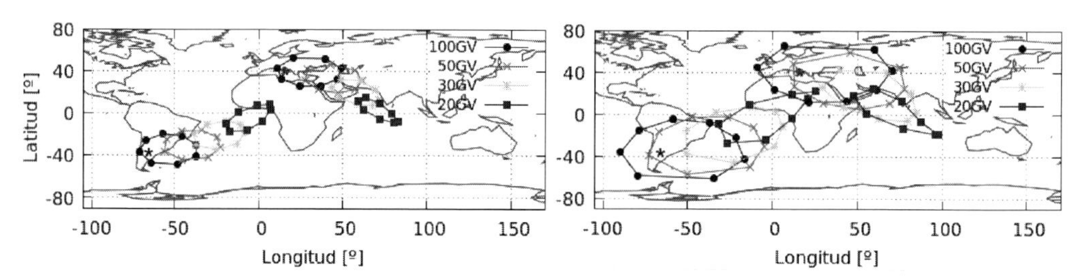

Figure 3. Asymptotic directions (projected on the Earth's surface) for two zenith incidence values: 15° (left) and 30° (right) and eight incidence azimuth values (45°,90°,..,360°), for two locations: Auger Observatory and Rome NM, using the IGRF+TSY01 model.

3. Simulations in geomagnetic storm conditions

Some properties of geomagnetic disturbances can be characterized by the Dst index (see, e.g., Dasso *et al.* 2002). Typical values of Dst, for different levels of perturbations, are -30nT (small), -50nT (Moderate) and -100nT (Intense), while the typical duration for these geomagnetic storms is $\sim$ 10 h (Gonzalez *et al.* 1994).

The values obtained for R_c^{Mlg} in different Dst conditions during a day are plotted in Figure 4. Increasing modulation for increasing Dst index is seen from 0nT to -600nT. For illustration, we evaluate the TSY01 Model, to visualize the topology of the geomagnetic field, for $Dst = 0$nT and $Dst = -250$nT (Figure 5).

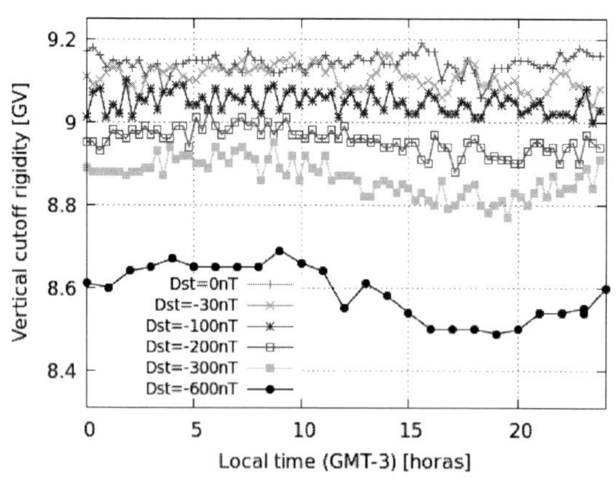

Figure 4. Effective cutoff rigidity as a function of time along the day, using the IGRF+TSY01 model, for different geomagnetic storm levels.

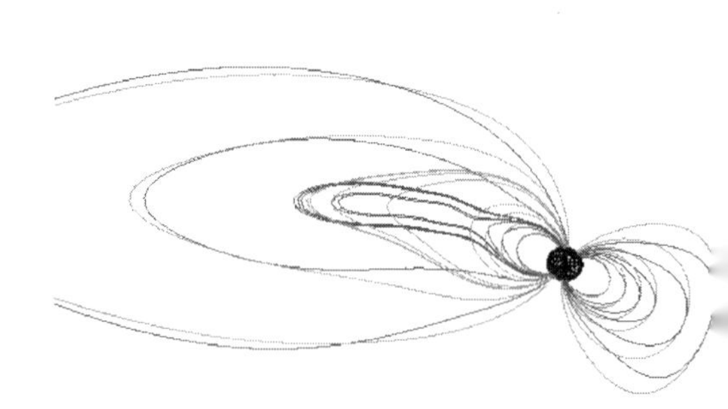

Figure 5. B_{geo} topology with $Dst = 0$nT (green) and $Dst = -250$nT (red), with IGRF+TSY01 model. Earth size is at scale.

With the method described above, we determine the transmittance for vertical incidence into Malargüe City (35.3°S, 69.3°W) for the last 20 years to see its evolution. In

Figure 6, we show the computations for Dst = 0nT (calm) and Dst = −500nT (very intense storm). The linear trend of the $R_c^{M\,lg}$ with time alone is −0.03GV/year, while the linear trend (not shown) with Dst is −0.001GV/nT.

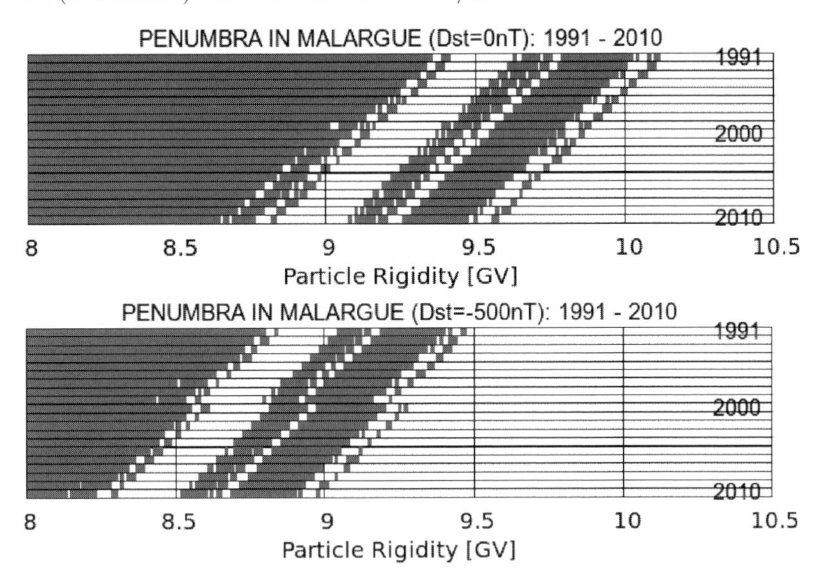

Figure 6. Transmittance functions for the last twenty years, for quiet (Dst = 0nT) and very intense storm (Dst = −500nT) conditions.

4. Conclusions

We have determined the cutoff rigidity for different azimuth and zenith incidence directions, using trajectory computations at the site of the Pierre Auger Observatory. We identified the asymptotic direction for particles arriving at Malargüe and found that they do not change during the day. We determined the variation of the cutoff rigidity along one day for different geomagnetic storm conditions. We computed the variation of the transmittance function in the last 20 years, finding a −0.03GV/year decrease in $R_c^{M\,lg}$, compatible with results in Smart & Shea (2008). Simulations, considering different values of the Dst index, show a −0.001GV/nT decrease of R_c with Dst. All these results can be used to interpret cosmic ray modulation using Auger data for calm and storm periods.

References

Abreu, P. and the Pierre Auger collaboration 2011, *JINST* 6, No. P01003.

Bartels, J. 1936, *J. Geophys. Res.*, 41, 225

Dasso, S., Gómez, D., & Mandrini, C. H. 2002, *J. Geophys. Res.*, 107(A5), 1059, doi:10.1029/2000JA000430

Dorman, L., Cosmic rays in magnetospheres of the Earth and other planets, *Springer*, 2008

Ferreira, S. E. S. & Potgieter, M. S. 2004, *Advances in Space Research*, 34, 115

Finlay, C. C. *et al.* 2010, *Geophys. J. Int.*, 183, 1216

Gonzalez, W. D. & Vasyliunas, V. M. *et al.* 1994, *J. Geophys. Res.*, 99, 5771

Smart, D. F. & Shea, M. A. 2008, *Proceed. of the 30th Internat. Cosmic Ray Conference*, 1, 737

Tsyganenko, N. A. 2002, *J. Geophys. Res.*, 107 (A8), doi:10.1029/2001JA000219

Comparative Magnetic Minima: Characterizing Quiet Times in the Sun and Stars
Proceedings IAU Symposium No. 286, 2011
C. H. Mandrini & D. F. Webb, eds.

doi:10.1017/S1743921312004905

The 3D solar corona Cycle 24 rising phase from SDO/AIA tomography

Federico A. Nuevo[1,2], Alberto M. Vásquez[1,2], Richard A. Frazin[3], Zhenguang Huang[3], and Ward B. Manchester IV[3]

[1]Instituto de Astronomía y Física del Espacio (CONICET-UBA), CC 67 - Suc 28, (C1428ZAA) Ciudad de Buenos Aires, Argentina
email: federico@iafe.uba.ar

[2]Facultad de Cs. Exactas y Naturales, Universidad de Buenos Aires, Argentina

[3]Department of Atmospheric, Oceanic and Space Sciences, University of Michigan, Ann Arbor, MI 48109, USA

Abstract. We recently extended the differential emission measure tomography (DEMT) technique to be applied to the six iron bands of the Atmospheric Imaging Assembly (AIA) instrument aboard the Solar Dynamics Observatory (SDO). DEMT products are the 3D reconstruction of the coronal emissivity in the instrument's bands, and the 3D distribution of the local differential emission measure, in the height range 1.0 to 1.25 $R_\odot$. We show here derived maps of the electron density and temperature of the inner solar corona during the rising phase of solar Cycle 24. We discuss the distribution of our results in the context of open/closed magnetic regions, as derived from a global potential field source surface (PFSS) model of the same period. We also compare the results derived with SDO/AIA to those derived with the Extreme UltraViolet Imager (EUVI) instrument aboard the Solar TErrestrial RElations Observatory (STEREO).

Keywords. Sun: corona, Sun: fundamental parameters, methods: data analysis

---

1. Introduction

The DEMT technique is described in detail in Frazin *et al.* (2009). We briefly summarize here its key elements, focusing on the determination of the *local differential emission measure* (LDEM), pertaining only to the plasma contained within each tomographic grid cell (see below). The inner corona (1.0-1.25 $R_\odot$) is discretized on a $25\times90\times180$ (radial$\times$latitudinal$\times$longitudinal) spherical computational grid. A time series of EUV images covering a full solar rotation is tomographically inverted, for each band k separately, to obtain the *filter band emissivity* (FBE), $\zeta_i^{(k)}$, at every tomographic cell i. The tomographic inversion assumes a static corona, so that solar dynamics occurring in the Sun produce artifacts in the reconstructions. Such artifacts include smearing and negative values of the reconstructed FBEs, or zero when the solution is constrained to positive values. These are called *zero density artifacts* (ZDAs).

The FBE values at each cell are then used to perform the LDEM analysis. In this work we parametrize the LDEM as a combination of one or two normal distributions, $\xi_i(T) = \mathcal{N}(T; \lambda_i = [T_0, \sigma_T, a]_i)$ (in a similar way to Aschwanden & Boerner 2011), to treat three EUVI and four AIA bands (see Section 2), respectively. In the case of AIA, one of the normal distributions has its centroid fixed at the maximum sensitivity temperature of the 335 Å band. To find λ_i we minimize the sum of the quadratic deviations between the synthesized and tomographic FBEs: $\Phi(\lambda_i) = \sum_{k=1}^{K} [\zeta_i^{(k)} - \int dT\ \xi(T; \lambda_i)\ Q_k(T)]^2$, where $Q_k(T)$ is the temperature response of the k-th band. Fig. 1 shows the temperature responses of the coronal bands we used, for both the STEREO/EUVI and the

SDO/AIA instruments, computed using CHIANTI v6.0.1, assuming the Feldman *et al.* (1992) abundance set, and the Bryans *et al.* (2009) ionization equilibrium calculations. Using the LDEM we derive plasma electron parameters at each tomographic grid cell i,

$$N_{e,i}^2 = \int dT \; \xi_i(T), \tag{1.1}$$

$$T_{m,i} = (1/N_{e,i}^2) \int dT \; \xi_i(T) \; T, \tag{1.2}$$

$$W_{T,i}^2 = (1/N_{e,i}^2) \int dT \; \xi_i(T) \; (T - T_{m,i})^2, \tag{1.3}$$

and quantify the success of the LDEM in reproducing the K tomographic FBEs as,

$$\chi_i^2 = \frac{1}{K} \sum_{k=1}^{K} \left(1 - \zeta_i^{(k,\mathrm{synth})} / \zeta_i^{(k)}\right)^2, \tag{1.4}$$

where the $\zeta_i^{(k,\mathrm{synth})}$ is given by the k-th temperature integral within the functional Φ defined above. We consider a cell to have a non-satisfactory-fit (NSF) of the parametrized LDEM when $\chi^2 > 5 \times 10^{-2}$, or a mean quadratic deviation of order 20% between the tomographic and synthesized FBEs.

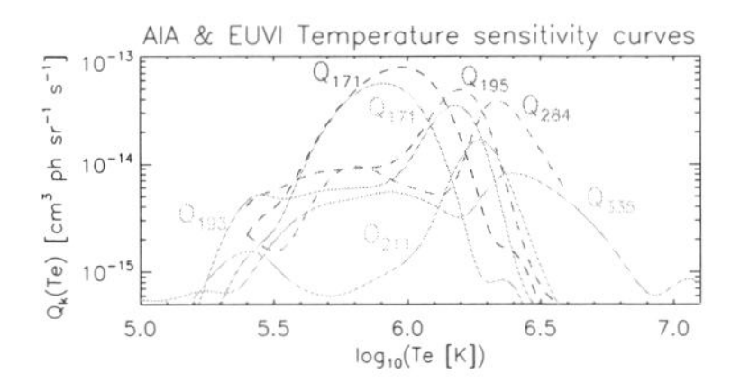

Figure 1. Temperature responses of the SDO/AIA (solid) and STEREO/EUVI-B (dashed). Figures in colour can be found in the online version.

2. Results

We performed two DEMT analyses of the period CR 2106 (2011, 20 January through 16 February), one using STEREO/EUVI data, and another one using SDO/AIA. For STEREO/EUVI we obtained the 3D FBE distribution for its three coronal bands (171, 195, and 284 Å), while for SDO/AIA we did the same for its four stronger bands (171, 193, 211, and 335 Å). Using each set of FBE tomographic reconstructions separately, we derived the LDEM moments from Eqs. (1.1)-(1.3).

Fig. 2 shows the DEMT results and χ^2 at 1.075 $R_\odot$, with similar maps obtained at all 25 tomographic grid height bins. Using a PFSS extrapolation of the same period based on GONG data, we overplot magnetic- strength B contour levels (thin black and white curves for negative and positive polarity), as well as the magnetically open/closed region boundaries (thick black curves). The PFSS model is computed using the finite-difference iterative solver FDIPS by Tóth *et al.* (2011), on a $150 \times 180 \times 360$ spherical grid, covering 1.0 to 2.5 $R_\odot$. While the morphology of the corona is clearly more complex compared to solar minimum, there is an overall good agreement between the magnetic-PFSSM and the density/temperature-DEMT structures.

The differences in the N_e, T_m and W_T maps obtained from both data sets are mainly due to the inclusion of AIA 335 Å band, and also to differences between the temperature

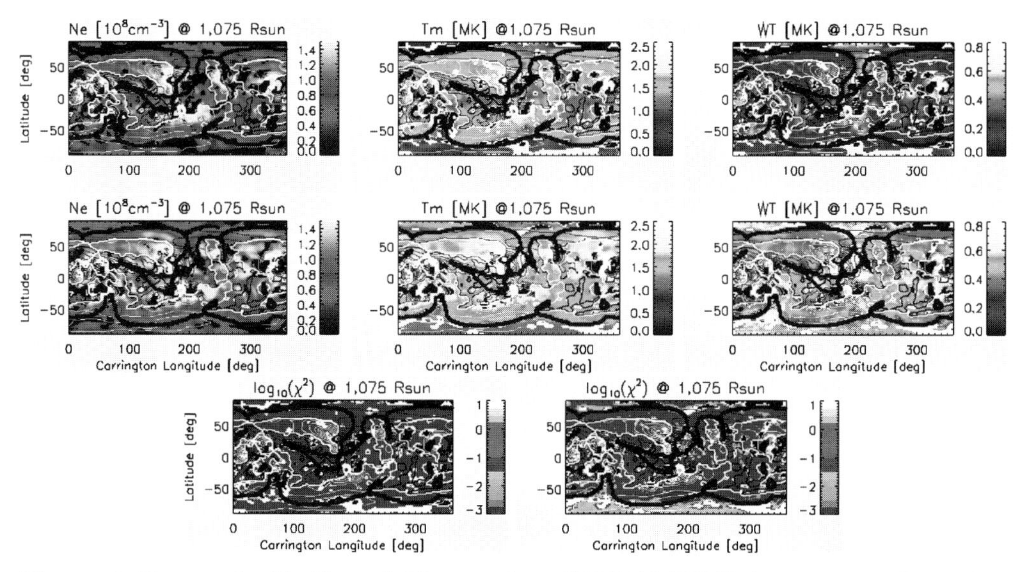

Figure 2. Top panels: N_e, T_m and W_T map from STEREO/EUVI data at 1.075 R$_\odot$ with a single-normal parametrization. Middle panels: N_e, T_m and W_T map from SDO/AIA data at 1.075R$_\odot$ with a double-normal parametrization. Bottom panels: χ^2 maps for STEREO/EUVI result (left) and SDO/AIA results (right). ZDAs and NSFs are indicated as black and white cells, respectively, in the T_m, W_T, and χ^2 maps, and as black cells in the N_e maps.

responses of the other three AIA bands with those of the EUVI instrument. Fig. 3 shows scatter plots and histograms comparing the DEMT results obtained with EUVI and AIA. The AIA results show systematically larger values, with N_e, T_m and W_T median values increasing by 9, 16, and 71%, respectively, when compared to the EUVI results.

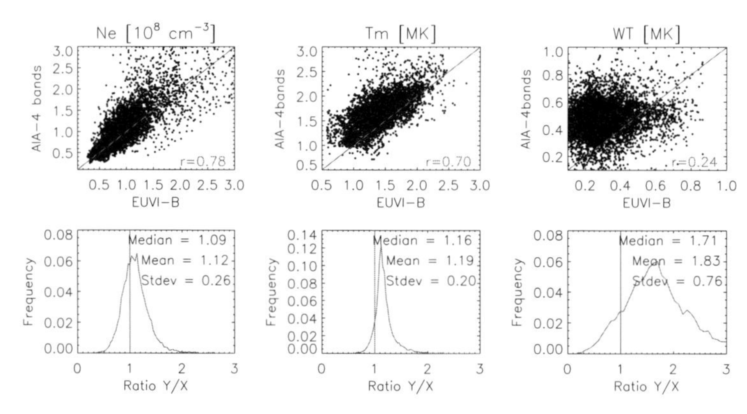

Figure 3. Comparison of LDEM results from EUVI and from AIA, at 1.075 R$_\odot$.

Fig. 4 shows statistics comparing the AIA DEMT results in open vs. closed regions (as determined from the PFSS model) at 1.075 R$_\odot$. The AIA DEMT results show that the streamer plasma is about 100% more dense and 50% hotter than in open regions. Also, when compared to EUVI DEMT CR-2077 and CR-2068 Solar Minimum results (see Vásquez *et al.* 2012, in this volume), the CR-2106 EUVI DEMT results indicate the streamer region is ~10% more dense and and ~20% hotter.

Fig. 5 shows the dependence with height of the AIA DEMT results for selected streamer regions, averaged over the range of latitudes and longitudes specified in each plot. We

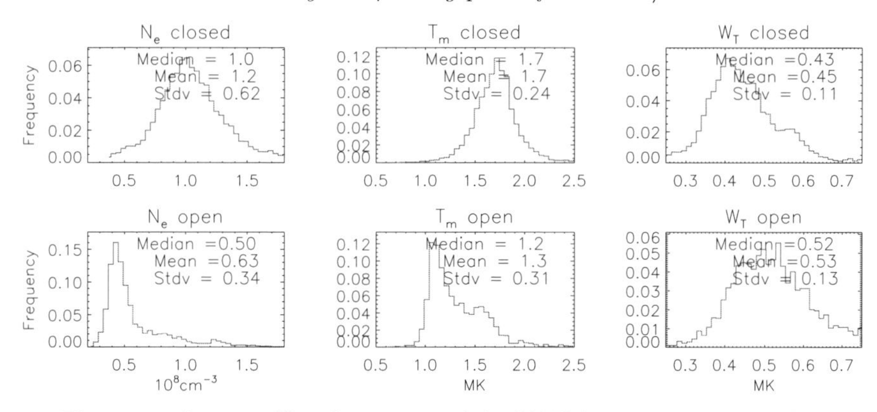

Figure 4. Open-vs-Closed statistics of the LDEM moments at 1.075 R$_\odot$.

apply hydrostatic fits to the LDEM $N_e(r)$ data, and find the electron temperature $T_{e,\mathsf{fit}}$ from the fit scale height, assuming $T_e = T_\mathsf{H}$. Taking then the LDEM $< T_m >$ as a measure of the true T_e, it follows $\frac{T_\mathsf{H}}{<T_m>} \approx 1 + 2(\frac{T_{e,\mathsf{fit}} - <T_m>}{<T_m>})$ (neglecting the He abundance). The CR-2106 streamer results show consistency with hydrostatic equilibrium, as the correlation coefficients are high, both in active and quiet zones. The fact that in those regions $T_{e,\mathsf{fit}} << T_m >$, may then be indicative of $T_e > T_\mathsf{H}$ (Vásquez *et al.* 2011).

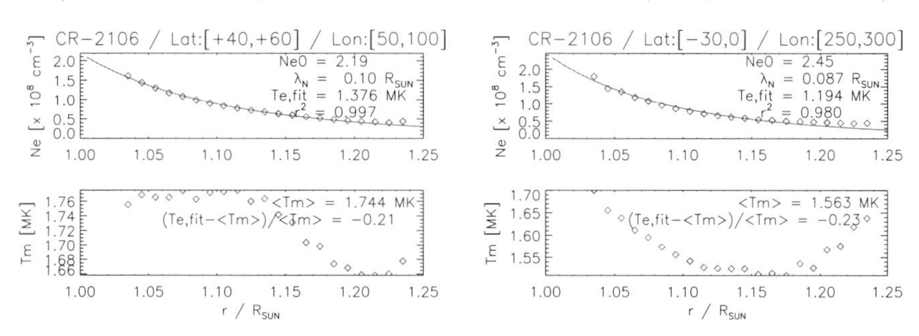

Figure 5. Average dependence with height of LDEM moments for selected streamer regions, and hydrostatic fits to the LDEM $N_e(r)$ data.

In a future publication we will develop a study of EUVI and AIA inversions of simulated DEM data to learn how to interpret the differences arising from the DEMT analysis based on both instruments.

Acknowledgements

F.A.N. and A.M.V. acknowledge CLAF and IAU for their support to attend IAU Symposium 286 in Mendoza, Argentina.

References

Aschwanden, M. & Boerner, P. 2011, *ApJ*, 732, 81-95
Bryans, P., Landi, E., & Savin, W. 2009, *ApJ*, 691, 1540-1559
Feldman, U., Mandelbaum, P., Seely, J. L., Doschek, G. A., & Gursky, H. 1992, *ApJ*, 81, 387-408
Frazin, R. A., Vásquez, A. M., & Kamalabadi, F. 2009, *ApJ*, 701, 547-560
Tóth, G., van der Holst, B., & Huang,Z. 2011, *ApJ*, 732, 102-108
Vásquez, A. M., Huang, Z., Manchester IV, W. B., & Frazin, R. A. 2011, *Sol. Phys.*, 274, 259-284
Vásquez, A. M., Frazin, R. A., Huang, Z., Manchester IV, W. B., & Shearer, P. 2012, this volume

Comparative Magnetic Minima: Characterizing quiet times in the Sun and Stars
Proceedings IAU Symposium No. 286, 2011
C. H. Mandrini & D. F. Webb, eds.

doi:10.1017/S1743921312004917

Earth–directed coronal mass ejections and their geoeffectiveness during the 2007–2010 interval

Constantin Oprea[1], **Marilena Mierla**[1,2,3] **and Georgeta Maris**[1]

[1]Institute of Geodynamics of the Romanian Academy,
RO–020032, Bucharest, Romania
email: const_oprea@yahoo.com

[2]Royal Observatory of Belgium, Brussels, Belgium

[3]Research Centre for Atomic Physics and Astrophysics,
Faculty of Physics, University of Bucharest, Romania

Abstract. In this study we analyse the coronal mass ejections (CMEs) directed towards the Earth during the interval 2007–2010, using the data acquired by STEREO mission and those provided by SOHO, ACE and geomagnetic stations. A study of CMEs kinematics is performed. This is correlated with CMEs interplanetary manifestations and their geomagnetic effects, along with the energy transfer flux into magnetosphere (the Akasofu coupling function). The chosen interval that is practically coincident with the last solar minimum, offered us a good opportunity to link and analyse the chain of phenomena from the Sun to the terrestrial magnetosphere in an attempt to better understand the solar and heliospheric processes that can cause major geomagnetic storms.

Keywords. Sun: coronal mass ejections (CMEs), (Sun:) solar-terrestrial relations

1. Introduction

Coronal mass ejections (CMEs) and their implication at the geomagnetic level has been a topic for many studies in the past decades (see e.g. Gopalswamy *et al.* 2006; Echer *et al.* 2008; Zhang *et al.* 2007). All these studies have tried to give a better insight over the connection mechanisms between the CMEs and the geomagnetic storms, in a continuous attempt to build as much as possible a profile in order to predict whether a particular CME event can cause or not a strong geomagnetic storm. Thus, we know that a big number of the frontside halo CMEs are geoeffective and that the geomagnetic storms which are associated with consecutive halos are among the most intense (Gopalswamy *et al.* 2006). The intensity of geomagnetic storms has a very strong dependence to the southward component of the interplanetary magnetic field, followed by the initial speed of the CME and the ram pressure (Srivastava & Venkatakrishnan 2004).

In this perspective we investigated the geoeffectivity of the CMEs directed towards the Earth in the time interval 2007–2010, period coincident with the Sun's minimum activity.

2. Data and Analysis Methods

In this study we analyse the CMEs which arrived to the Earth in the period 2007–2010 and produced geomagnetic storms. The intensities of the storms were from minimum to moderate values (with Dst varying between –30 nT and –80 nT). We have eleven such events.

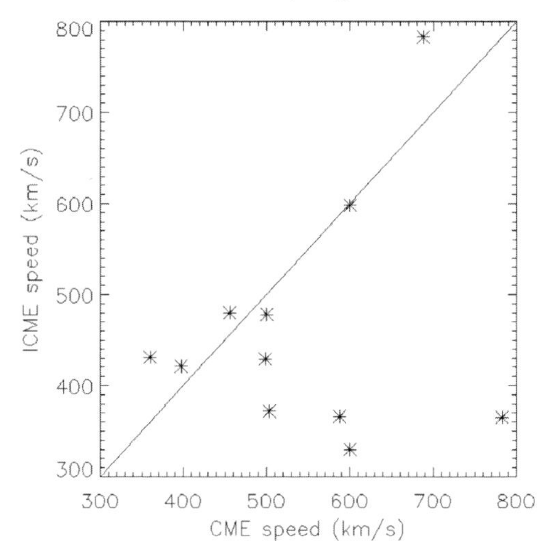

Figure 1. ICME speed versus reconstructed CME speed.

In order to select the CMEs which arrived to the Earth (ICMEs - or interplanetary CMEs) we used the data from ACE spacecraft and the data provided by Emilia Kilpua (Kilpua *et al.*, 2012). The corresponding solar sources (CMEs) were searched in a time interval up to maximum 6 days before the ICMEs, in images from LASCO (Brueckner *et al.* 1995) instrument onboard SOHO and COR instruments (Howard *et al.* 2008) onboard STEREO. The geomagnetic indices corresponding to each storm are taken from different geomagnetic stations around the globe.

Regarding the analysis methods, we derived the 3D reconstructed CMEs speeds by applying the forward-modelling (FM) technique (Thernisien *et al.* 2009) on the STEREO/COR2 and LASCO/C3 data and/or triangulation (Liu *et al.* 2010) on COR2 data. In order to investigate the behaviour of the ICMEs parameters and the impact of the ICMEs on the geomagnetic field we calculated the correlation coefficients between different ICME parameters and the geomagnetic index Dst. We also used the superposed epoch analysis (e.g. Mustajab 2011).

3. Speed analysis

Out of 21 events we have selected 11 events in the interval 2007–2010, when the STEREO separation angle was between 70 and 160 degrees and we applied the FM technique in order to derive their real (3D) speeds. Note that not all of these events produced geomagnetic storms. The calculated speeds were compared with in-situ speeds (the speeds of ICMEs recorded at ACE spacecraft). It was observed, that out of the 11 events, six were decelerated while travelling into the interplanetary space, four were accelerated and one CME kept a constant speed from the Sun to the Earth (see Figure 1). This means that the CMEs interacted with the surrounding solar wind and the drag forces accelerated or decelerated the events, depending on the speed of the solar wind.

4. ICME signatures versus Dst

From a total of 21 ICMEs observed in the interval 2007–2010, only 11 have produced geomagnetic storms (Dst<-30 nT). To understand the impact of the 11 ICMEs that arrived at the Earth on the geomagnetic field we computed the correlation coefficients of various interplanetary parameters (IP) versus the minimum Dst geomagnetic index.

The ICMEs parameters considered for this analysis were: the interplanetary magnetic field (IMF) magnitude (B), Z component of the IMF (Bz), Bs·V (where Bs=|Bz|when Bz<0 and Bs=0 when Bz⩾0), the plasma speed (V), the plasma temperature (T) and the proton density (ρ). Furthermore, we computed the correlation coefficient between the minimum Dst and the total energy injected into Earth's magnetosphere (eq. 2), which was found using the Akasofu coupling function (eq. 1) (Akasofu 1983). The Akasofu coupling function takes into consideration the processes of reconnection within the magnetosphere, as being the principal source of the injected energy (De Lucas *et al.* 2007).

$$\varepsilon = 10^7 V B^2 l_0^2 \sin^4(\frac{\theta}{2}), [J/s] \quad (1)$$

where: V and B are the physical units defined above, θ is the IMF clock-wise angle in the plane perpendicular to the Sun–Earth line, l_0 is the magnetopause radius (l_0 = 7R$_E$), $\theta = \tan^{-1}(\frac{B_y}{B_z})$.

$$W_\varepsilon = \int_{t0}^{tm} \varepsilon dt, [J] \quad (2)$$

where: W_ε is obtained by integrating ε over the main phase of each geomagnetic storm, from t_0 to t_m. All the measured units are given in International System.

We also computed the correlation coefficients between the IPs measured at the same time (t_0), one hour earlier (t_{-1}), two hours earlier (t_{-2}) and three hours earlier (t_{-3}) than the minimum Dst and the minimum Dst value. The computed coefficients show a poor correlation between the IPs and the minimum Dst index value. The best correlations were found for ICME speeds taken at two hours (r=-0.57) and three hours (r=-0.58) before the minimum Dst and for Bs·V taken at three hours (-0.55) before the minimum Dst. W$_\varepsilon$ had a rather low correlation coefficient of -0.51, although an important amount of energy was injected into the magnetosphere in the main phase of each geomagnetic storm.

We think that the poor correlation of the interplanetary structures parameters with the Dst index were due to the little number of geomagnetic storms occurred in the 2007–2010 interval. Also, the high speed streams (HSS) had an important implication in the disturbances at the geomagnetic level, taking in consideration that the geomagnetic storms produced by the ICMEs were, in some cases only contributions overimposed to already incipient geomagnetic disturbances produced by the HSS (Maris & Maris 2010).

5. Superposed epoch analysis

For a better understanding of the importance of the ICMEs parameters we used the superposed epoch analysis. In this analysis we considered as origin (t = 0) the time when the minimum Dst was observed. The period of the superposed epoch analysis was 24 hours before and 48 hours after this minimum. Then, the mean values of the interplanetary and geomagnetic parameters (the Bz component of the IMF, the Dst geomagnetic indexes and the Akasofu coupling function) at a given time, over the 11 events, were computed (Figure 2).

From Figure 2 we can see a better dependence between the Bz, the Akasofu coupling function and the Dst index than we observed in the case of the correlation coefficients computation. Bz is deacreasing on the main phase of the geomagnetic storms up to two hours before the minimum Dst value and then it starts to increase. The energy injected into the magnetosphere is increasing during the main phase of the storm. Even so, it can be observed that before the beginning of the main phase of a geomagnetic storm the Akasofu coupling function shows an increasing trend, meaning that for some reasons an

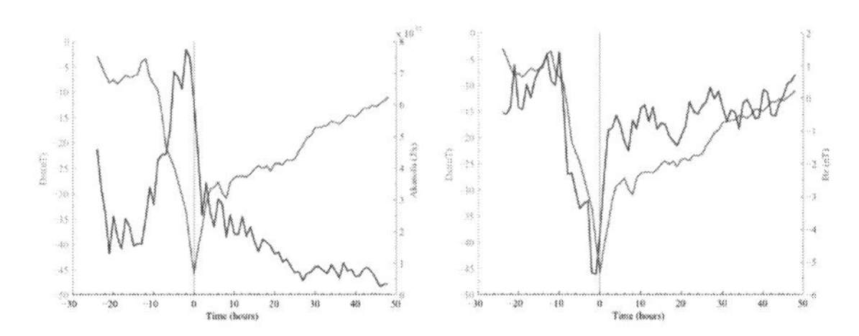

Figure 2. Plots of the mean values of: Dst and Akasofu coupling function (left), Dst and Bz (right).

important amount of energy was already injected into the magnetosphere at an earlier time. A possible explanation is the perturbations caused by the HSS in this 2007–2010 interval.

6. Summary

In this paper we studied the ICMEs manifestations from the 2007–2010 interval. In this period a number of 21 ICMEs were observed at ACE, from which only 11 caused weak and moderate geomagnetic storms (Dst<–30 nT).

Majority of the 11 events in the interval 2008-2009 for which the 3D speeds were calculated were decelerated while travelling into the interplanetary space, and four were accelerated. This suggests different kinds of their interaction with the ambient solar wind.

The correlation between different ICME parameters and Dst was very weak. A slightly better correlation was observed between the minimum Dst and the ICME speed measured three hours earlier than minimum Dst.

The highest energy was injected into the magnetosphere on the main phase of the geomagnetic storm. HSSs played also an important role to the production of the storms.

Acknowledgements

We acknowledge the use of SOHO, STEREO, ACE and geomagnetic data. We thank Emilia Kilpua for providing the ICMEs parameters. The work for this paper was supported from the project TE no 73/11.08.2010.

References

Akasofu, S.-I. 1983, *Space Sci. Revs*, 34, 173

Brueckner G. E., Howard R. A., Koomen M. J. *et al.* 1995, *Solar Phys.*, 162, 357

De Lucas, A., Gonzalez, W. D., Echer, E. *et al.* 2007, *Jour. Atmosph. and Solar-Terres. Phys.*, 69, 1851

Echer, E., Gonzalez, W. D., Tsurutani, B. T. & Gonzalez, A. L. C. 2008, *Jour. Geophys. Res.*, 113, A05221

Gopalswamy, N., Yashiro, S., & Akiyama, S. 2007, *Jour. Geophys. Res.*, 112, A06112

Howard, R. A., Moses, J. D., Vourlidas, A. *et al.* 2008, *Space Sci. Rev.* 136, 67

Kilpua, E. K. J., Mierla, M., Rodriguez, L., Zhukov, A. N., Srivastava, N., West, M. 2012, *Sol. Phys.*, in press

Liu, Y., Davies, J. A., Luhmann, J. G., Vourlidas, A., Bale, S. D., & Lin, R. P. 2010, *Astrophys. J.* (Letters), 710, L82

Maris, G. & Maris, O. 2010, *Highlights of Astronomy*, 15, 494

Mustajab, F. & Badruddin 2011, *Astrophys. Space Sci.*, 331, 91

Srivastava, N. & Venkatakrishnan, P. 2004, *Jour. Geophys. Res.*, 109, A10103

Thernisien, A., Vourlidas, A., & Howard, R. A. 2009, *Solar Phys.*, 256, 111

Zhang, J., Richardson, I. G., Webb, D. F. *et al.* 2007, *Jour. Geophys. Res.*, 112, A10102

Comparative Magnetic Minima:
Characterizing quiet times in the Sun and Stars
Proceedings IAU Symposium No. 286, 2011
C. H. Mandrini & D. F. Webb, eds.

doi:10.1017/S1743921312004929

Evolution of a very complex active region during the decay phase of Cycle 23

Mariano Poisson[1,2], Marcelo López Fuentes[1,2], Cristina H. Mandrini[1,2], Pascal Démoulin[3], Etienne Pariat[3]

[1]Instituto de Astronomía y Física del Espacio (CONICET-UBA), CC 67 Suc 28, 1428 Buenos Aires, Argentina

[2]Facultad de Ciencias Exactas y Naturales, UBA, Buenos Aires, Argentina

[3]Observatoire de Paris, LESIA, 92195 Meudon, France

Abstract. We study the emergence and evolution of AR NOAA 10314, observed on the solar disk during March 13-19, 2003. This extremely complex AR is of particular interest due to its unusual magnetic flux distribution and the clear rotation of the polarities of a δ-spot within the AR. Using SOHO/MDI magnetograms we follow the evolution of the photospheric magnetic flux to infer the morphology of the structure that originates the AR. We determine the tilt angle variation for the δ-spot and find a counter-clockwise rotation corresponding to a positive writhed flux tube. We compute the magnetic helicity injection and the total accumulated helicity in the AR and find a correlation with the observed rotation.

Keywords. Sun: photosphere, Sun: magnetic fields

1. Introduction

It has been shown in previous studies that magnetically complex solar active regions (ARs) present the higher rates of flare production and coronal mass ejections (CMEs) (Liu *et al.* 2005). Magnetic flux distributions that form δ-spots at photospheric level are among these cases (Linton *et al.* 1998). These structures are related to the emergence of magnetic flux tubes that have been highly distorted by the effect of convective turbulence beneath the photosphere (Fan 2009). For this reason, it is expected to find high magnetic helicity injection rates in these particular ARs. The accumulation of magnetic helicity in the solar atmosphere is related to the energy release mechanisms, since a large amount of free magnetic energy is present in highly stressed magnetic structures. Recent studies have pointed at these peculiar ARs as the sources of the stronger geomagnetic storms (Szajko *et al.*, this volume).

Here, we study how the complex photospheric evolution of AR 10314 is consistently related to a high magnetic helicity injection. In Section 2 we describe the observed photospheric motion of the 4 main polarities. In Section 3 we compute and analyze the evolution of the tilt angle of the δ-spot and the magnetic helicity injected in the AR. In Section 4 we discuss the flux tube morphologies that can qualitatively reproduce the observed evolution of the polarities and quantitatively explain the amount of magnetic helicity calculated in Section 3.

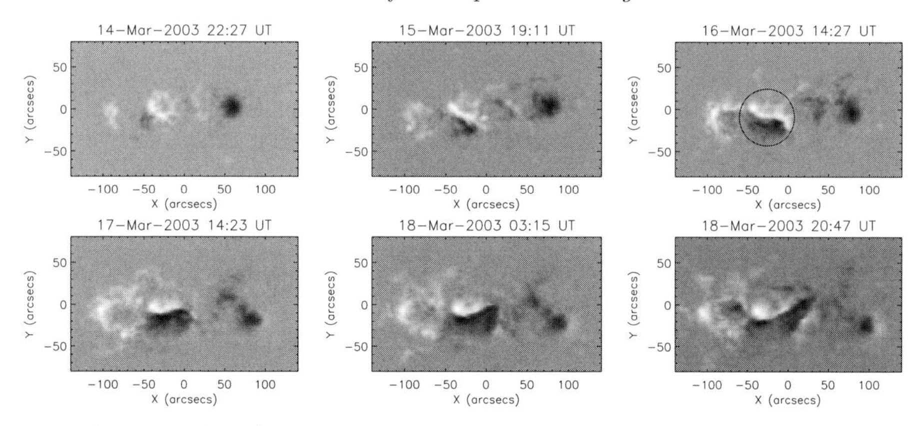

Figure 1. Selected SOHO/MDI magnetograms for the evolution AR 10314. The AR was observed from its first emergence on March 13 2003 to its disappearance on the west limb around March 20 2003. In the upper right panel we indicate with a circle the location of the δ-spot formed at the center of the AR. The main polarities of the δ-spot are observed to rotate one around the other during the studied evolution.

2. Photospheric evolution

We follow the photospheric evolution and the coronal activity of the studied AR using full disk longitudinal magnetograms taken with the Michelson Doppler Imager (MDI, see Scherrer *et al.* 1995) on board the Solar Heliospheric Observatory (SOHO).

The AR NOAA 10314 emerged near the solar disk center (S15W18) on March 13, 2003. It evolved from a dual-bipolar system to a more complex configuration with the appeareance of a δ-spot at its center (marked with a circle in the upper right panel of Figure 1). A feature that is of particular interest for this work is the constant rotation of the inversion line of the δ-spot (see the lower row panels of Figure 1).

3. Analysis

3.1. *Evolution of the δ-spot tilt*

Using MDI magnetograms we compute the flux weighted mean positions of the positive and negative polarities of the δ-spot (see López Fuentes *et al.* 2003). From these positions we compute the tilt angle as the angle that the vector **S** joining both polarities (from positive to negative) form with the East-West direction.

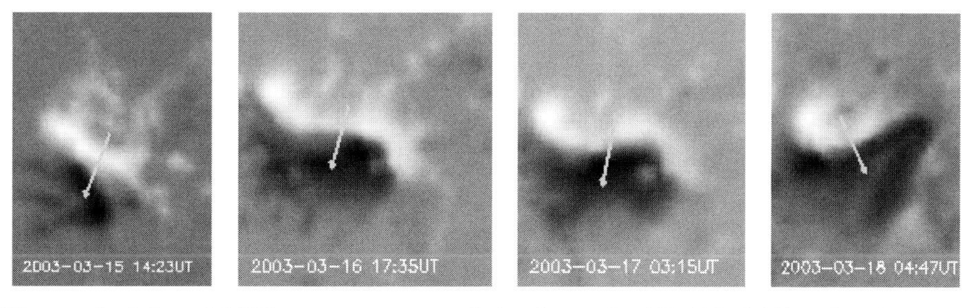

Figure 2. Selected MDI magnetograms of the δ-spot in AR 10314. The ploted vectors represent the relative position of the positive and negative polarities.

Figure 2 shows the tilt angle variation of the δ-spot by displaying the vector **S** for selected frames. Our results show a sustained counter-clockwise rotation of the polarities. From the observed rotation we are able to infer that the writhe of the associated magnetic flux tube, as well as the main helicity flux contribution are positive.

3.2. *Magnetic helicity injection*

We analyze the injection of magnetic helicity during the AR evolution following the procedure described in Pariat *et al.* (2005). We use the Differential Affine Velocity Estimator (DAVE) routine developed by Schuck (2005, 2006) in order to find the velocity field, from which we can in turn compute the magnetic helicity density. In figure 3 we show selected maps of magnetic helicity density. The density maps show a significant positive helicity injection in the area of the δ-spot region. The integration of the helicity density for the full AR confirms the dominance of positive helicity injection and is consistent with the observed counter-clockwise rotation of the tilt angle (Poisson *et al.* 2011).

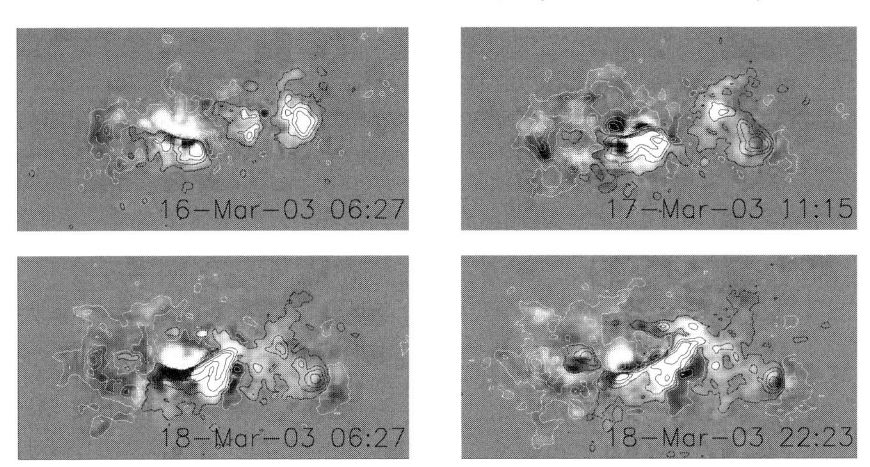

Figure 3. Magnetic helicity density maps for AR 10314. White and black intensities over the map correspond to positive and negative helicity flux contributions. Overlayed white and black contours correspond to positive and negative photospheric magnetic field strengths of 100, 600, 1200 and 1800 G.

4. Discussion

We devise two possible models for the magnetic structure that produced AR 10314 that are consistent with the photospheric field distribution and the evolution of the tilt angle. The first one (see Figure 4, left panel) corresponds to a single deformed flux tube that presents a kink instability, and the second one (Figure 4, right panel) is formed by two independent flux tubes.

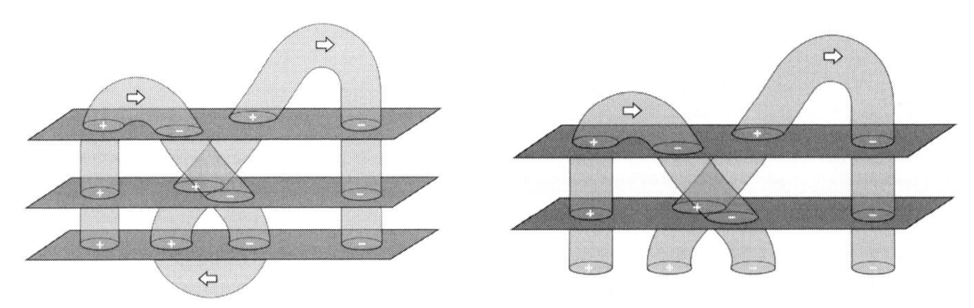

Figure 4. Possible structures of the magnetic flux tube that produced AR 10314. The planes indicate the relative position of the photosphere during the flux tube emergence, the arrows show the global direction of the magnetic field along the tube, and the circles with signs correspond to the location of the polarities observed at the photospheric plane for different stages of the evolution.

Although the method used to compute the magnetic helicity does not allow us to discard any of these two posibilities, the continuous rotation of the tilt angle and the constant distance between the main polarities of the δ-spot, strongly support the first scenario (single flux tube with a kink).

5. Conclusions

We study the evolution of AR 10314 from a dual-bipole to a complex configuration with the development of a δ-spot. We propose two different models of the flux tube morphology in order to explain the observed AR photospheric evolution. One of the models corresponds to a flux tube with a kink instability and the other to two independent flux tubes. We study the evolution of the tilt angle and observe a counter-clockwise rotation of the δ-spot with a constant distance between its main polarities. These results are consistent with the emergence of a positive writhed flux tube and support the single flux-tube model. Finally, we analyze the magnetic helicity injection during the AR evolution and find an overwhelming dominance of positive helicity, which is also consistent with the above results (for further analysis see Poisson *et al.* 2011).

Acknowledgements

MLF and CHM are members of the Carrera del Investigador Científico of the Consejo Nacional de Investigaciones Científicas y Técnicas (CONICET), Argentina. The authors would like to thank Dr. Peter Schuck for the permission to use the Differential Affine Velocity Estimator (DAVE) routine. SOHO is a project of international cooperation between ESA and NASA.

References

Fan, Y. 2009, Living Reviews in Solar Physics, 6, 4
Linton, M. G., Dahlburg, R. B., Fisher, G. H., & Longcope, D. W. 1998, *ApJ*, 507, 404
Liu, C., Deng, N., & Liu, Y. *et al.* 2005, *ApJ*, 622, 722
López Fuentes, M. C., Démoulin, P., Mandrini, C. H., Pevtsov, A. A., & van Driel-Gesztelyi, L. 2003, *A&A*, 397, 305
Pariat, E., Démoulin, P., & Berger, M. A. 2005, *A&A*, 439, 1191
Poisson, M., López Fuentes, M., Mandrini, C. H., Démoulin, P., & Pariat, E. 2011, submitted to Adv. Space Res.
Scherrer, P. H., Bogart, R. S., Bush, R. I. *et al.* 1995, *Sol. Phys.*, 162, 129
Schuck, P. W. 2005, *ApJ*, 632, L53
Schuck, P. W. 2006, *ApJ*, 646, 1358
Szajko, N., Cristiani, G., Mandrini, C. H., & Dal Lago, A. 2011, this volume

Comparative Magnetic Minima:
Characterizing quiet times in the Sun and Stars
Proceedings IAU Symposium No. 286, 2011
C. H. Mandrini & D. F. Webb, eds.

doi:10.1017/S1743921312004930

Very intense geomagnetic storms: solar sources, characteristics and cycle distribution

Natalia Szajko[1], Germán Cristiani[1], Cristina H. Mandrini[1] and Alisson Dal Lago[2]

[1]Instituto de Astronomía y Física del Espacio, Buenos Aires, Argentina
email: natisolsz@hotmail.com

[2]Instituto Nacional de Pesquisas Espaciais, São José dos Campos, São Paulo, Brazil

Abstract. We revisit previous studies in which the characteristics of the solar and interplanetary sources of intense geomagnetic storms have been discussed. We consider the very intense geomagnetic storms that occurred during Solar Cycle 23 by setting a value of $\mathrm{Dst}_{\mathrm{min}} \leqslant -200$ nT as threshold. We have identified and characterized the solar and interplanetary sources of each storm. After this, we investigate the overall characteristics of the interplanetary (IP) main-phase storm driver, including the time arrival of the shock/disturbance at 1 AU, the type of associated IP structure/ejecta, the origin of a prolonged and enhanced southward component (B_z) of the IP field, and other characteristics related to the energy injected into the magnetosphere during the storm.

Keywords. Sun: activity, Sun: coronal mass ejections (CMEs), interplanetary medium

---

1. Introduction

It is now well established that major geomagnetic storms are the consequence of a sequence of events that originate in the Sun and result in a geoeffective solar wind flow near Earth (see examples in Brueckner *et al.* 1998; Webb *et al.* 2000). Broadly speaking, the geoeffective solar wind disturbances can be separated in two types. One of them is associated to IP coronal mass ejections (ICMEs). ICMEs are the counterparts of CMEs in the IP medium. The other type is associated to the fast solar wind coming from solar coronal holes; this flow interacts with the preceeding slow solar wind in zones called corotating interaction regions (CIRs). Several recent studies have found that major geomagnetic storms may be driven by either ICMEs/MCs or CIRs (see Echer *et al.* 2008, and references therein).

Solar Cycle 23 is unique in the sense that it is the first of the space age during which the Sun has been imaged almost continuously. The Large Angle and Spectrometric Coronagraph (LASCO, Brueckner *et al.* 1995), on board the Solar and Heliospheric Observatory (SOHO), has provided a long-term set of CMEs for which several characteristic parameters have been catalogued in a comprehensive data base (http://cdaw.gsf.nasa.gov-/CME_list/, Gopalswamy *et al.* 2009). The combination of LASCO data with observations from other SOHO instruments, such as the Extreme-ultraviolet Imaging Telescope (EIT, Delaboudiniere *et al.* 1995) and the Michelson Doppler Imager (MDI, Scherrer *et al.* 1995), allows us to determine the solar CME source region and its magnetic characteristics. In addition to this, the plasma and magnetic field experiments on board the Advanced Composition Explorer (ACE) and Wind give the opportunity of full *in situ* data coverage in the same period of time.

For each very intense storm ($\mathrm{Dst}_{min} \leqslant$ -200 nT) during Cycle 23, we determine the time, angular width and plane-of-sky, expansion and radial velocities of the source CME, the type and heliographic location of the CME solar source region (including the characteristics of sunspot groups), and the time duration of the associated flare. After this, we investigate the overall characteristics of the IP main-phase storm driver including the arrival time of the shock/disturbance at 1 AU, the type of associated IP structure/ejecta, the origin of a prolonged and enhanced southward component (B_z) of the IP field, and other characteristics related to the energy injected into the magnetosphere during the storm (*i.e.* the solar wind maximum convected electric field, E_y). Our analysis, thus, complements and extends those of other works in the literature.

2. Selection criteria of the events

We use the Dst final values from the World Data Center for Geomagnetism (http://wdc.-kugi.kyoto-u.ac.jp/dstdir/index.html) to select the events in our set. The temporal extension of Solar Cycle 23 was taken from October 1996 to December 2008 (see *e.g.* http://www.ips.gov.au/solar). The hourly averaged Dst data are analyzed and plotted to select storms for which $\mathrm{Dst}_{\mathrm{min}} \leqslant -200$ nT. We found 19 cases that comply with our selection criterion and one case for which $\mathrm{Dst}_{\mathrm{min}} = -197$ nT. Taking into account that the data are averaged over one hour, we have decided to include this marginal case in our set. First and second columns of the Table 1 show the date and time when Dst reached its minimum and the corresponding value.

3. The solar sources: CMEs and their origin at the Sun

To identify the solar source event and the region from which it originates at the Sun's surface, we have proceeded by tracing back the possible solar candidate from Sun to Earth and, in several ambiguous cases, back from Earth to Sun. Our procedure to identify the solar source event is as follows. We first consider a time window between 24 hours (transit CME speed from Sun to Earth $\approx$ 1800 km s^{-1}) and 120 hours (transit CME speed $\approx$ 350 km s^{-1}) previous to the geomagnetic event to select a candidate CME. This time window roughly takes into account the range of plausible CME speeds measured in coronagraph data. In order to decrease the possible number of candidates, we first consider only frontside full halo CMEs and, in a second step, partial halo CMEs with a large "apparent" angular width (AW $\geqslant 150^\circ$). Once the source CME and the location on the Sun from which it originates are determined, we identify: the AR, the class (in soft X-rays and Hα) and duration of the associated flare using Solar Geophysical Data reports and GOES data, and the degree of magnetic complexity of the AR using MDI data. The flare duration is taken from the time of impulsive soft X-ray increase to the time when the flux returns either to its pre-flare level or another flare occurrs in a different or the same AR, being clearly distinguishable from the CME associated flare. From the third to the ninth columns of Table 1 we list the AR NOAA number and its heliographic location, the standard sunspot group classification at the time of CME occurrence, the X-ray and Hα classifications for the associated flare, the flare duration computed as discussed above, the time of first appearance in C2, the CME type, its velocity (second order fitting to C2 and C3 data), and the CME lateral expansion velocity (computed as discussed in Dal Lago *et al.*, 2003, 2004).

4. The associated interplanetary medium events

Since storms are driven by the solar wind magnetic field and plasma impinging on the Earth's magnetosphere, we use here *in situ* data from instruments aboard ACE to

Table 1. Very intense geomagnetic storms during Solar Cycle 23 and their solar and IP sources. (1): Low intensity bipolar region where a filament eruption occurs. (2): Location of AR9393 at the time of occurrence of two CMEs, candidate sources of the two-step storm. (3): The halo CME associated with this storm is related to the interaction between AR10695 and AR10696. The heliographic coordinates in this column are those of AR10696. (4): This duration corresponds to two consecutives flares in AR10501, one on 18 Nov 2003 - 07:52 UT (M3.2/2N) and the one included in the table. (5): No LASCO data from July to September 1998. ND in the fifth column means no Hα data.

Dst		Active Region		Flare		CME			IP medium
Time	Dst_{min} [nT]	AR location	Sunspot group	Importance X-ray/opt.	Durat. [hours]	Lasco C2	v [km/s]	v_{exp} [km/s]	B_S origin
04 May 1998-05:00 UT	-205	AR8210 (S15W15)	$\gamma\delta$	X1.1/3B	5.23	02 May 98-14:06 UT	697	1228	sheath
25 Sep 1998-09:00 UT	-207	AR8340 (N20E09)	β	M7.1/3B	16.16	No data(5)			sheath + MC
22 Oct 1999-06:00 UT	-237	No AR(1) (S20E05)	QS	C1.2/ND	3.70	18 Oct 99-00:06 UT	263	546	ICME
06 Apr 2000-23:00 UT	-287	AR8933 (N16W66)	β	C9.7/2F	5.22	04 Apr 00-16:32 UT	1232	1927	sheath
16 Jul 2000-00:00 UT	-301	AR9077(N22W07)	$\beta\gamma\delta$	X5.7/3B	4.60	14 Jul 02-10:54 UT	1534	2178	MC
2 Aug 2000-09:00 UT	-235	AR9114(N11W11)	$\beta\gamma$	C2.3/SF	5.18	09 Aug 00-16:30 UT	731	898	sheath + MC
17 Sep 2000-23:00 UT	-201	AR9165(N14W07)	$\beta\delta$	M5.9/2B	4.23	16 Sep 00-05:18 UT	1162		MC
31 Mar 2001-08:00 UT 31 Mar 2001-21:00 UT	-387 -284	AR9393(N16W10)(2) AR9393(N18E02)	$\beta\gamma\delta$	X1.7/1N M4.3/SF	3.78 3.83	29 Mar 01-10:26 UT 28 Mar 01-12:50 UT	965 582	1511	sheath + MC? MC?
11 Apr 2001-23:00 UT	-271	AR9415(S21W04)	$\beta\gamma\delta$	M7.9/2B	5.00	09 Apr 01-15:54 UT	1198	2679	sheath + MC
06 Nov 2001-06:00 UT	-292	AR9684(N06W18)	$\beta\gamma$	X1.0/3B	9.95	04 Nov 01-16:35 UT	1514	3670	sheath
24 Nov 2001-16:00 UT	-221	AR9704(S10W39)	$\beta\gamma\delta$	M9.9/3B	8.53	22 Nov 01-23:30 UT	1371	2800	sheath
30 Oct 2003-00:00 UT	-353	AR10486(S16E08)	$\beta\gamma\delta$	X17/4B	11.16	28 Oct 03-11:30 UT	2229	4800	MC
30 Oct 2003-22:00 UT	-383	AR10486(S15W02)	$\beta\gamma\delta$	X10/2B	5.28	29 Oct 03-20:54 UT	1670	4000	sheath
20 Nov 2003-20:00 UT	-422	AR10501(N03E08)	$\beta\gamma\delta$	M3.9/2N	8.42(4)	18 Nov 03-08:50 UT	1645	2900	MC
27 Jul 2004-13:00 UT	-197	AR10652(N08W33)	$\beta\gamma\delta$	M1.1/ND	6.68	25 Jul 04-15:14 UT	1366		sheath + MC
08 Nov 2004-06:00 UT	-373	AR10696(N10E08)	$\beta\gamma\delta$	M9.3/2N M5.9/ND M3.6/ND	6.90	06 Nov 04-02:06 UT	1111	1562	MC?
10 Nov 2004-10:00 UT	-289	AR10696/AR10695 (N10W10)(3)	$\beta\gamma\delta/\beta$	X2.0/ND	8.16	07 Nov 04-16:54 UT	1696	3077	sheath + MC
15 May 2005-08:00 UT	-263	AR10759(N12E12)	β	M8.0/2B	12.78	13 May 05-17:22 UT	1690		MC
24 Aug 2005-11:00 UT	-216	AR10798(S16W70)	$\beta\gamma\delta$	M2.9/SF	9.70	23 Aug 05-14:54 UT	2123	3100	sheath

identify the IP structures responsible for each goemagnetic storm. In particular, we have used plasma data from the Solar Wind Electron Proton and Alpha Monitor (SWEPAM, McComas *et al.* 1998) and magnetic field data from the Magnetic Fields Experiment (MAG, Smith *et al.* 1998).

Taking into account their magnetic and plasma signatures, we are able to identify various types of structures. These structures include ICMEs, ICMEs containing a flux tube with MC properties, the sheath between the CME driven shock and the ICME, and regions with clear signatures of interaction between ICMEs and high speed streams from coronal holes. Last column of Table 1 indicates the inferred origin of the prolonged and enhanced B_z.

5. Summary and discussion

We have analyzed the full set of very intense geomagnetic storms that occurred during Solar Cycle 23 in search of their solar and interplanetary origin.

When we investigate their distribution, which was double–peaked, we see that 15% of the events occurs during the cycle rising-phase. All these events are very intense storms. During the first cycle maximum, the number of very intense storms increases to 20%.

During the second cycle maximum, it reaches 25% with all of them, but one, being superstorms (following Gonzlez *et al.* 2002, we call superstorm one for which Dst_{min} $\leqslant$ -250 nT). However, 40% of the very intense storms with half of all superstorms (6), including one extreme event, occurs during the cycle descending phase. These results, then, show that the distribution of very intense storms, as that of intense storms along the cycle (see Echer *et al.* 2008, and references therein), presents two peaks: one during the cycle maximum and another one during the descending phase. These results have been discussed in Gonzalez *et al.* 2011, and references therein.

All CME sources of the IP disturbances causing very intense geomagnetic storms are either full halo CMEs or partial halos with a large angular width as, in principle, expected. The calculated average lateral expansion velocity of the CMEs is $\approx$ 2400 km s^{-1}. With the lateral expansion velocity, we can apply the phenomenological expression derived by Dal Lago *et al.* (2003) and compute the Sun-Earth line velocity, in all cases this velocity is much higher than the projected plane-of-the-sky speed. All the listed CMEs, but one, are really fast events with radial velocities reaching 4000 km s^{-1} with an average value of $\approx$ 2200 km s^{-1}. It is also evident from our analysis that all fast CMEs are strongly decelerated during their transit to 1 AU, while slow CMEs (see *e.g.* event number 3) are accelerated by the ambient solar wind.

The CME solar sources of all analyzed storms, but one, are ARs. The ARs where the CMEs originate show, in general, high magnetic complexity; δ spots are present in 74% of the cases, 10% are formed by several bipolar sunspot groups, and only 16% present a single bipolar sunspot group. This is not surprising as it is well-known that ARs contaning δ spots display a high level of activity (see Zirin & Liggett 1987).

All CMEs are associated to long duration events (LDEs), exceeding 3 hours in all cases, with around 75% lasting more than 5 hours. The associated flares are, in general, intense events, classified as M or X in soft X-rays; only three of them fall in the C class.

When we look for the location of the CME source regions producing very intense storms, we find that 75% are located at a distance smaller than half a solar radius from the solar disk center (one flare/CME is not observed by SOHO). If we separate this restricted set in Dst_{min} ranges, the ones in the ranges with lower absolute values have source regions at larger distances from Sun center. In four of these cases the AR is closer to the western limb, with two events at a longitude $\geqslant 60°$.

Finally, considering the IP structures responsible for a long and enhanced B_z, we find that 35% are MCs or ICME fields, 30% sheath fields, and 30% combined sheath and MC or ICME fields. Therefore, for this particular set, any of these structures is equally important. We have found no storm originated by CIR fields, only one storm is related to the compression of an ICME by a high speed stream coming from a coronal hole. Strictly speaking, this is not a CIR as it is not a region of interaction between slow and fast solar wind. We have also found that the linear relation between the maximum value of E_y and the storm intensity holds (with a correlation coefficient of 0.73).

References

Brueckner, G. E., Delaboudiniere, J.-P., Howard, R. A., Paswaters, S. E., St. Cyr, O. C., Schwenn, R., Lamy, P., Simnett, G. M., Thompson, B., & Wang, D., 1998, *Geophys. Res. Lett.*, 25, 3019

Brueckner, G. E., Howard, R. A., Koomen, M. J., Korendyke, C. M., Michels, D. J., Moses, J. D., Socker, D. G., Dere, K. P., Lamy, P. L., Llebaria, A., Bout, M. V., Schwenn, R., Simnett, G. M., Bedford, D. K., & Eyles, C. J. 1995, *Sol. Phys.*, 162, 357

Dal Lago, A., Vieira, L. E. A., Echer, E., Gonzalez, W. D., de Gonzalez, A. L. C., Guarnieri, F. L., Schuch, N. J., & Schwenn, R. 2004, *Sol. Phys.*, 222, 323

Dal Lago, A., Schwenn, R., & Gonzalez, W. D. 2003, *Adv. Spac. Res.*, 32, 2637

Delaboudiniere, J.-P., Artzner, G. E., Brunaud, J., Gabriel, A. H., Hochedez, J. F., Millier, F., Song, X. Y., Au, B., Dere, K. P., Howard, R. A., Kreplin, R., Michels, D. J., Moses, J. D., Defise, J. M., Jamar, C., Rochus, P., Chauvineau, J. P., Marioge, J. P., Catura, R. C., Lemen, J. R., Shing, L., Stern, R. A., Gurman, J. B., Neupert, W. M., Maucherat, A., Clette, F., Cugnon, P., & van Dessel, E. L. 1995, *Sol. Phys.*, 162, 291

Echer, E., Gonzalez, W. D., Tsurutani, B. T., & Gonzalez, A. L. C. 2008, *Jour. of Geophys. Res. (Space Physics)*, 113(A12), A05221

Gonzalez, W. D., Tsurutani, B. T., Lepping, R. P., & Schwenn, R. 2002, *Jour. of Atmos. and Solar-Terres. Phys.*, 64, 173–181

Gopalswamy, N., Yashiro, S., Michalek, G., Stenborg, G., Vourlidas, A., Freeland, S., & Howard, R., 2009, *Earth Moon and Planets*, 104, 295

McComas, D. J., Bame, S. J., Barker, P., Feldman, W. C., Phillips, J. L., Riley, P., & Griffee, J. W. 1998, *Space Science Reviews*, 86, 563

Scherrer, P. H., Bogart, R. S., Bush, R. I., Hoeksema, J. T., Kosovichev, A. G.,Schou, J., Rosenberg, W., Springer, L., Tarbell, T. D., Title, A., Wolfson, C. J., & Zayer, I. 1995, *Sol. Phys.*, 162, 129

Smith, C. W., L'Heureux, J., Ness, N. F., Acuña, M. H., Burlaga, L. F., & Scheifele, J. 1998, *Space Scien. Rev.*, 86, 613

Gonzalez, W. D., Echer, E., Tsurutani, B. T., Clúa de Gonzalez, A. L., & Dal Lago, A. 2011, *Space Science Reviews*, 158, 69

Webb, D. F., Lepping, R. P., Burlaga, L. F., DeForest, C. E., Larson, D. E., Martin, S. F., Plunkett, S. P., & Rust, D. M. 2000, *Jour. of Geophys. Res.*, 105, 27251

Zirin, H. & Liggett, M. A. 2004, *Sol. Phys.*, 113, 267

Session 4

Stellar Cycles

Comparative Magnetic Minima:
Characterizing quiet times in the Sun and Stars
Proceedings IAU Symposium No. 286, 2011
C. H. Mandrini & D. F. Webb, eds.

doi:10.1017/S1743921312004954

Stellar cycles: general properties and future directions

Mark S. Giampapa

National Solar Observatory
950 N. Cherry Ave., POB 26732, Tucson, AZ 85726-6732 USA
email: giampapa@nso.edu

Abstract. We discuss the general properties of stellar cycles with emphasis on their amplitudes as a function of stellar parameters, particularly those stellar characteristics relevant to dynamo-driven magnetic activity. We deduce an empirical scaling relation between cycle frequency and differential rotation based on previously established empirical relations. We also compare the recent Cycle 23 to cycles in solar-type stars. We find that the extended minimum of Cycle 23 resembled in its Ca II H & K emission at minimum the mean levels of activity seen in stars with no cycles.

Keywords. Stars:activity cycle, Sun:cycle, stars:chromosphere, stars:differential rotation

1. Introduction

Any review of stellar cycles must begin with a reference to the pioneering study of Olin Wilson, who asked the question, "Does the chromospheric activity of main-sequence stars vary with time, and if so, how?" (Wilson 1978). The question was clearly inspired by the threefold recognition of (1) the existence of the solar cycle as most prominently observed in the form of the ~11-year variation in sunspot numbers, (2) the spatial association of chromospheric Ca II resonance line emission with magnetic field regions, on the Sun, particularly near spots, and (3) the observation of Ca II H & K line emission in the spectra of late-type stars indicative of the presence of chromospheres, analogous to the solar chromosphere, in the outer atmospheres of stars. As we know from Wilson's dedicated efforts, and the long-term extension of this program by Sallie Baliunas and her colleagues at Mt. Wilson (Baliunas *et al.* 1995), late-type stars exhibit an array of chromospheric variability ranging from regular variations with multi-year periods similar to that of the Sun to irregular variability with no clear pattern to stars with essentially flat curves, i.e., no long-term cycle-like variability.

About 25% of the Mt. Wilson sample exhibits irregular variations with considerable scatter in H & K core emission although some of these "irregular stars" can display aperiodic variability. Typically, these objects are characterized by relatively higher mean levels of Ca II emission compared to the Sun and they generally belong to a somewhat younger stellar population, such as members of the ~600 Myr Hyades. Approximately 60% of the Mt. Wilson sample shows smooth cyclic variability similar to that of the Sun. These stars also have mean levels of Ca II emission more near the mean level of the Sun. Finally, about 15% of the sample objects are simply constant, or 'flat', in their time series of H & K emission suggesting either no cycle variability or a very long period cycle. While it might be expected that this group of stars would be characterized by an even lower mean level of Ca II chromospheric emission, continuing the implied trend, we will see that the 'flat case' is more complicated.

In the following, I will further review the Mt. Wilson survey results in the context of correlations, beginning with cycle amplitudes and stellar properties and continuing with empirical relationships between cycle periods, rotation, and differential rotation. A comparison of the Sun with other solar-type stars in the context of different stellar samples will be discussed as will the most recent observations of new cycle properties, such as multiple cycles and short-period cycles. The possible implications for dynamo models will be noted but without firm conclusions. The review concludes with a summary followed by a discussion of promising directions for future research.

2. Amplitudes of Stellar Cycles

We refer the reader to Fig. 2 in Saar & Brandenburg (2002), which gives the amplitude of stellar cycles in dwarf stars as the fractional change in the normalized chromospheric flux in the Ca II H & K resonance lines relative to the mean level of this index, as a function of $B - V$ color, or stellar effective temperature. Inspection of their Fig. 2 reveals that relative cycle amplitudes generally increase toward cooler stars with thicker convection zones (as a fraction of the stellar radius), attaining an apparent maximum in the mid-K spectral types. Thereafter, relative amplitudes decline though it is important to note that the regime of the M dwarf stars has not been extensively explored. The reason for this is primarily observational given the comparative faintness of these cool stars combined with the reduced efficiencies of detectors in the blue/near-UV region of the H & K lines near 400 nm.

Saar & Brandenburg (2002) find that cycle amplitudes do not exhibit a strong dependence on rotation or, in the case they considered, inverse Rossby number. However, we know from the rotation-activity connection that chromospheric emission does depend on rotation (e.g., Noyes *et al.* 1984). Therefore, the change in chromospheric emission as modulated by the cycle must behave in a similar way on rotation as does the mean level of emission. Since the fractional change in the chromospheric emission index, i.e., the amplitude of the cycle, is always less than one, Saar & Brandenburg (2002) suggest that only a portion of the mean H & K emission derives from a cycling dynamo, i.e., the "cycling portion" modulates the mean magnetic field-related emission arising from the non-cycling (small-scale) component of the dynamo, which must have a similar dependence on rotation as the larger-scale, cycling component.

We refer the reader, again from Saar & Brandenburg (2002), to their Fig. 4, where the observed relationship between two dimensionless quantities representing the logarithm of cycle amplitude versus the logarithm of the cycle frequency divided by the rotation frequency. Representing the data in this way reveals two relations, or "branches"—the "A" or "Active" branch and the "I" or "Inactive" branch. While a single relation using dimensional quantities can be constructed to fit the data (Baliunas *et al.* 1996), it has considerably more scatter than the representation in Fig. 4 of Saar & Brandenburg (2002). The other appeal is that the presence of the two branches suggests a discontinuity or jump between "active" and "quiet" stars that is reminiscent of the so-called Vaughan-Preston gap (Vaughan & Preston 1980). In this kind of "Chromospheric H-R Diagram" there is a distinctive gap separating active (and generally younger) stars from more quiescent and generally older stars. One prevailing thought as to the origin of the Vaughan-Preston gap is that it is the manifestation of two different dynamo modes. Another minority view is that the gap is the result of a discontinuity in the local star formation rate combined with the assumption of a smooth decline of chromospheric activity with age.

Saar & Brandenburg (1999) offer an interpretation in the context of α-ω kinematic dynamo models where the α effect (i.e., the poloidal flux generation) is sensitive to stellar rotation. Recalling briefly, this class of dynamo models requires a region of radial differential rotation where poloidal field lines interact with azimuthal flows to produce toroidal flux (the "ω effect"). The interaction of buoyant toroidal flux with cyclonic fluid motions in the convection zone produces poloidal fields that can then be acted upon again by the ω effect—this process is referred to as the "α effect." Saar & Brandenburg (1999) consider the hypothesis that the dynamo is partly unstable, giving rise to a transition between modes that, in turn, results in the Inactive (I) and Active (A) branches, respectively. The mode transition itself may be due to rotation-dependent changes in the degree of enhancement of the α effect, presumably due to interactions of toroidal flux with rotation and fluid motions in the convection zone.

3. Rotation, Differential Rotation, and Cycle Periods

Given that rotation and differential rotation are key physical parameters in models for the origin of magnetic flux generation in the Sun and cool stars, their inter-relationships along with their relationship to cycle parameters, such as cycle period, merit consideration. In Fig. 1 (from Saar 2011), we see differential rotation frequency, normalized to solar differential rotation, as a function of rotation frequency. These kinds of data are difficult to obtain since they depend on the detection of subtle period changes in long-term observations of the rotational modulation of active regions on the stellar surface. In Fig. 1 the slower rotators follow a power law with normalized $\Delta\Omega \sim \Omega^{0.68}$ for $\Omega < 3$ d^{-1}. The relation appears to reach a maximum and then declines toward rapid rotators that also are "saturated activity" stars, i.e., stars characterized by high levels of chromospheric and coronal emission levels that do not increase further with faster rotation rates. Their surfaces are thought to be completely covered, or "saturated," with magnetic active regions. However, if these objects are plotted in a graph of differential rotation versus $B-V$ color then a clear trend of declining differential rotation with redder $B-V$ color is revealed (Saar 2011, his Fig. 1). The result suggests a dependence of differential rotation on effective temperature or, in turn, relative convection zone depth, which implies a possible relationship with convective turnover times. This immediately evokes the Rossby number (or inverse Rossby number) as a relevant parameter since the Rossby number incorporates both rotation period and convective turnover time in a single

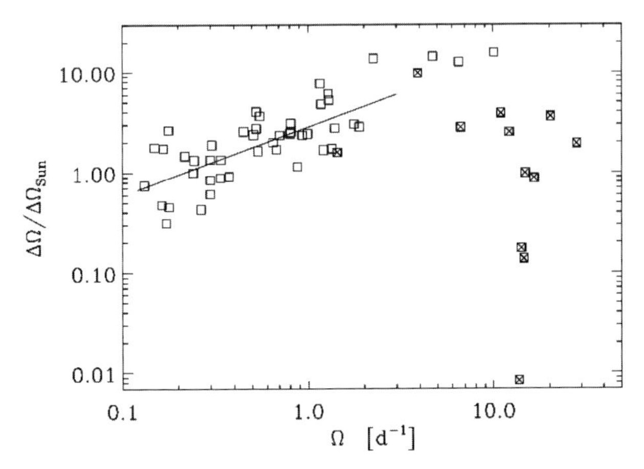

Figure 1. Normalized differential rotation frequency versus rotation frequency (from Saar 2011)

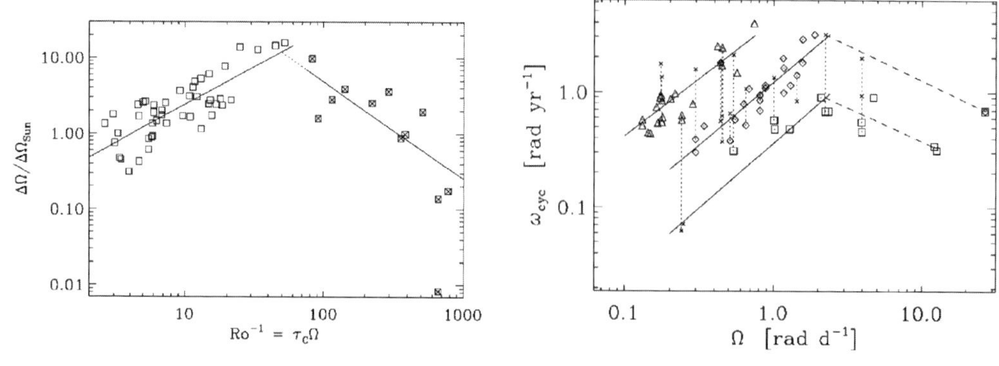

Figure 2. A comparison of the behavior of cycle frequency with inverse Rossby number and normalized differential rotation frequency with rotation (from Saar 2011)

dimensionless quantity. Saar (2011) points out that this suggests a relationship involving deep-seated dynamo action at rapid rotation in the especially high-activity stars.

When cycle frequency is examined as a function of rotation frequency, three parallel tracks emerge in the data, each with cycle frequency, $\omega_{cyc} \sim \Omega^{1.1}$ (Saar 2011). We compare this relation to that for normalized differential rotation but now as a function of inverse Rossby number rather than just rotation frequency (Fig. 2). The cycle frequency—rotation frequency diagram exhibits a rise to a peak followed by a decline though the decline is somewhat tentative given the scant data available. Nevertheless, stars with well-defined cycle periods show a correlation with rotation. This diagram appears very similar in shape and in location of the maximum to that in the differential rotation versus inverse Rossby number diagram. Saar (2011) claims that the similarity in shape of these relations suggests that differential rotation has left an "imprint" on cycle periods and their relationship to rotation. From an observational perspective, the way in which we measure differential rotation and cycle periods, i.e., varying concentrations of magnetic flux in time and latitude/longitude distribution on stars, could naturally lead to similar diagrams. In particular, detecting differential rotation in late-type stars requires the appearance of active complexes at different latitudes and asymmetrically distributed in longitude in order to see rotational modulation. Cycle modulation requires a globally averaged variation in chromospheric H & K emission over long periods of time, which may be correlated with the latitudinal variation in the appearance of active region complexes over cycle time scales. In other words, the similarity of these two diagrams could be evidence for a "butterfly diagram" in stars, analogous to what we see in the Sun, where emergent sites of active complexes move from high-latitudes (slower differential rotation) to lower latitudes (faster differential rotation) over the course of a cycle.

Recalling our discussion of cycle amplitudes, and the empirical relationships given above between rotation, differential rotation, and cycle frequency, we can examine further the dependence of cycle amplitudes on these parameters. In particular, Saar & Brandenburg (2002) find that cycle amplitude, A_{cyc}, is proportional to $(\omega_{cyc}/\Omega)^{\alpha}$, where α is 0.65 for the I-branch and 0.85 for the A-branch stars, respectively. We recall that $\omega_{cyc} \sim \Omega^{1.1}$. Hence, by substitution, $A_{cyc} \sim \Omega^{0.1\alpha}$. Thus, cycle amplitudes have a very weak dependence on rotation, as noted earlier by Saar & Brandenburg (2002; see their Fig. 3) from observations. We may explore the correlation of cycle amplitudes with differential rotation given the power-law relation between normalized differential rotation and rotation frequency cited previously. Again we find that cycle amplitudes depend only weakly on differential rotation with $A_{cyc} \sim (\Delta\Omega/\Delta\Omega_{\odot})^{0.1}$, for either the

I-branch or the A-branch. By contrast, we can use the aforementioned empirical relationships between differential rotation and rotation, and cycle frequency and rotation, to find $\omega_{cyc} \sim (\Delta\Omega/\Delta\Omega_{\odot})^{1.6}$. Hence, a star with twice the differential rotation frequency of the solar value would have a cycle frequency approximately three times higher and, therefore, a correspondingly shorter period.

4. The Sun in a Stellar Context

Recent observational work has expanded in different ways on the original Mt. Wilson program using stellar samples that either overlap or are completely different from the Mt. Wilson sample. In a recent preprint, Lovis *et al.* (2011) utilized data from the HARPS (High-Accuracy Radial velocity Planetary Searcher) survey—intended for Doppler searches for extrasolar planets—to examine the systematic chromospheric properties in their subsample, including possible magnetic cycle detection during the ~7 years the survey has been in operation. Their subsample includes stars that are broadly solar-type that also were selected for their low projected rotation velocities of v sin$i$ $<$ 3-4 km-s^{-1} and single status (i.e., no evidence the star is a member of a binary or multiple system).

An overview of the results from the subsample analyzed by Lovis *et al.* (2011; see their Fig. 14) are encapsulated in Fig. 3 showing the mean chromospheric emission index as a function of effective temperature, i.e., a chromospheric "H-R diagram" referred to earlier in the discussion of the Vaughan-Preston gap in Section 2. Please see the electronic version of this paper for the color-coded version of Fig. 3 discussed in the following, or Figs. 14 and 13, respectively, in Lovis *et al.* (2011). Stars with "large-amplitude" cycles are represented by red filled circles while stars with no large-amplitude cycle, to within their adopted criterion of a relative variation of 4%, are denoted by the blue dots. The black filled circles represent stars with no detected cycle but for which the existence of a cycle cannot be completely ruled out. The Sun is included at its mean value for the chromospheric index, among other stars with detected cycles. Interestingly, and perhaps suggestively, the mean Sun also appears in this diagram to reside very near a transition between stars with cycles and stars with no cycles.

The H-R diagram of the subsample studied by Lovis *et al.* (2011; their Fig. 13), given here as Fig. 4, yields further insight on a comparison of the two populations of stars with cycles and without 'large-amplitude' cycles. Lovis *et al.* (2011) note that the stars without cycles appear among the warmer stars and do not appear among the cooler K dwarfs. While this is indeed true, the salient point is that the stars without cycles are systematically slightly higher in luminosity (at a given effective temperature), suggesting that some evolution has occurred, i.e., they belong to a slightly older population. The fact that it is the warmer stars, which are also the relatively more massive stars, is consistent with membership in a slightly older population since they will evolve faster than the cooler dwarfs. Furthermore, rotational evolution in the form of spin-down will occur more rapidly among the more massive stars (Stauffer *et al.* 1991).

It is worthwhile to consider the Sun and its cycle in the above context of cycling and non-cycling solar-type stars. We display in Fig. 5 the time series of the 0.1 nm K-index parameter from the SOLIS Integrated Sunlight Spectrometer (ISS) of the National Solar Observatory at Kitt Peak.

The ISS obtains spectra of the Sun seen as a star. The K-index is a measure of the strength of the core of the Ca II K line, which includes the chromospheric emission. From inspection of the SOLIS ISS K-index parameter time series we find that the K-index minimum for Cycle 23 occurred during June 2008. We have the same time series

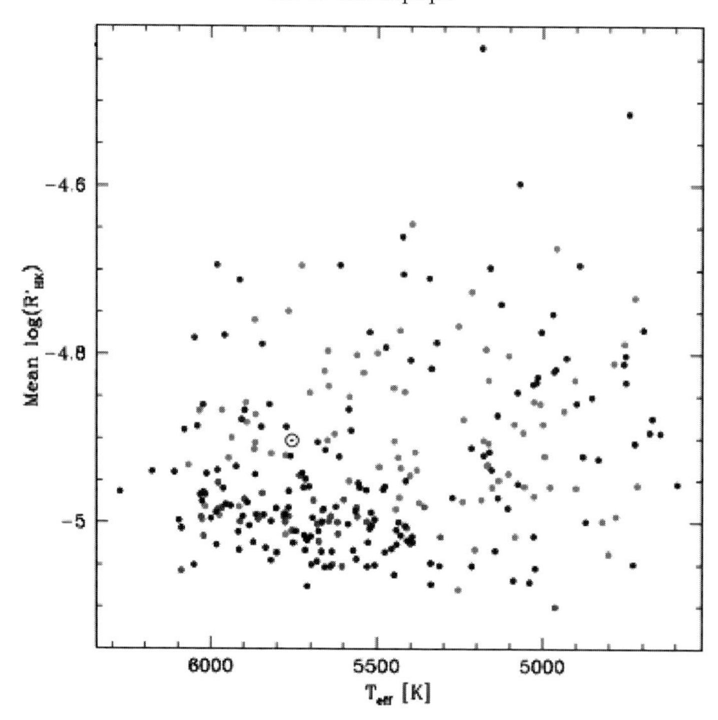

Figure 3. Normalized mean chromospheric emission for stars with various cycle properties as a function of effective temperature (from Lovis *et al.* 2011)

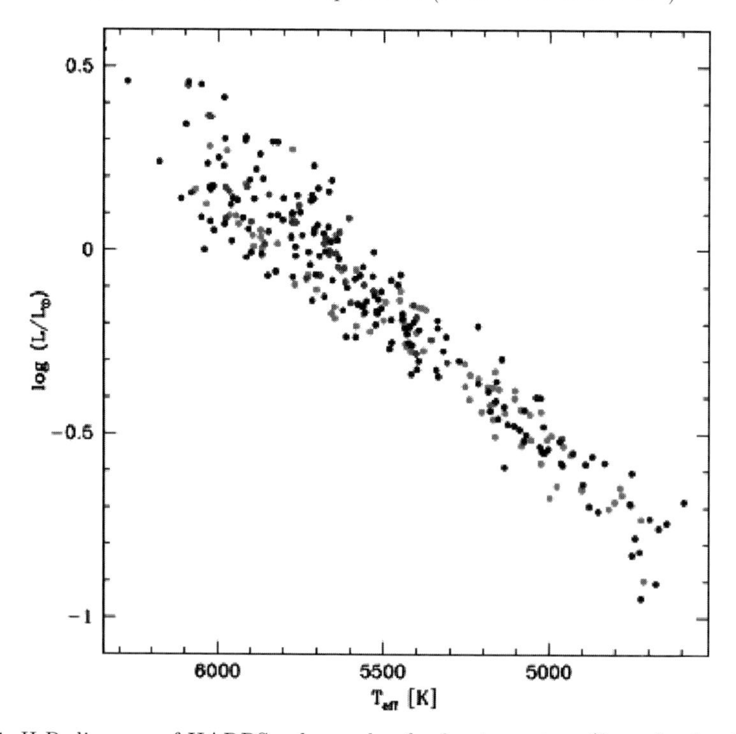

Figure 4. H-R diagram of HARPS subsample of solar-type stars (from Lovis *et al.* 2011)

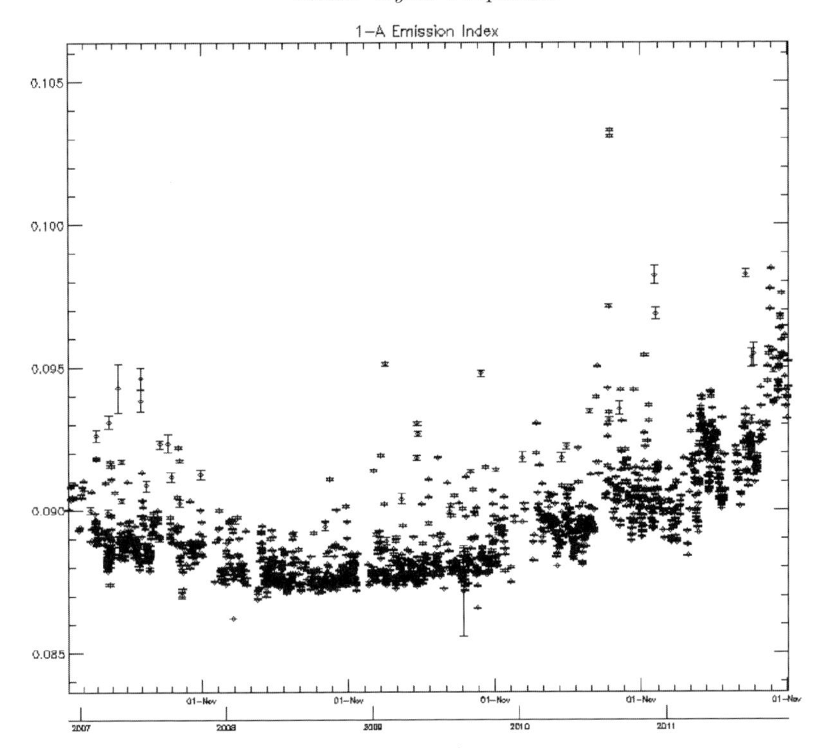

Figure 5. Time series of the Ca II K line 1.0 Å index from the SOLIS ISS instrument

for the Ca II H line. This enables us to combine the H and K series to derive the solar counterpart of the stellar HK index—a combined measure of the H & K line core strength—following a calibration procedure described by Giampapa *et al.* (2006). The resulting value of the corresponding stellar R'_{HK} index for the Cycle 23 minimum Sun is 1×10^{-5}. Referring back to Fig. 3, this places the Sun seen as a star at its Cycle 23 minimum in the very midst of other stars observed to not have a large-amplitude cycle, as defined by Lovis *et al.* (2011). In other words, a distant observer performing a similar survey would have concluded that the Cycle 23 minimum Sun was a candidate for a star in a Maunder minimum state of prolonged quiescence. It is worthwhile to examine the Cycle 23 'minimum Sun' in comparison with other stellar samples.

In their survey of the chromospheric activity of the solar-type stars in the solar-age and solar-metallicity cluster, M67, by Giampapa *et al.* (2006), the position of the Sun at the 2008 minimum in their Fig. 3 would be among the most quiet sun-like stars in M67, below the previous recorded solar minima. It is important to note that M67 includes a homogeneous sample of solar-type stars, in contrast to field star samples such as the HARPS sample, which tend to be inhomogeneous in age and chemical composition. We therefore may ask if the especially quiescent sun-like stars in M67 are in a Maunder-minimum state or in a prelude to a state of prolonged quiescence. Furthermore, is the Sun in a transition to a period of prolonged quiescence, perhaps even another Maunder Minimum?

A stellar candidate for such a transition to a Maunder-minimum-like state is HD 3851 from Baliunas *et al.* (1995), displayed in Fig. 6. The star is cooler than the Sun with a spectral type between K0 and K2, and with a cycle period of nearly 14 years. The amplitude of variation has declined significantly. In addition, there appears to be a long-term trend of a declining mean level of, in this case, the Mt. Wilson S-index, which is a

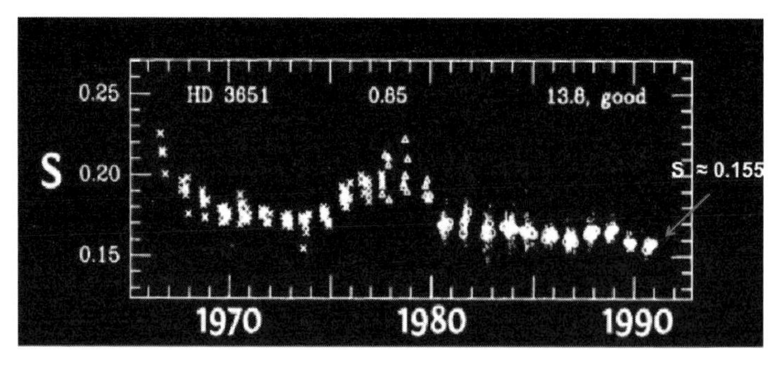

Figure 6. Example of a star that may be in transition to a Maunder-minimum state. The figure is provided courtesy of J. Hall and adapted from Baliunas *et al.* (1995)

measure of the strength of the H & K bands normalized to nearby continuum levels. As we saw previously (Fig. 3), low mean chromospheric emission is correlated with the lack of large-amplitude cycles. The low point shown in the last measurements of this object from the Mt. Wilson program is at S = 0.155. This corresponds to the same S-index value at the Cycle 23 minimum Sun in 2008. We envision a Maunder-minimum state as one in which there is either no or only a low-amplitude cycle at a low mean level of activity. But to confuse this picture further, in their monitoring of 57 sun-like stars, Hall & Lockwood (2004) classified ten stars as "flat" (over 6-10 years of observation of each target). Yet their mean level of chromospheric emission in the flat-activity stars spans the range of emission levels from below solar minimum values to well in excess of solar maximum levels. We note, parenthetically, that the respective distributions of S index values for the cycling and the non-cycling, flat stars in the Hall & Lockwood (2004) sample are statistically identical in the sense that there are no differences in the population from which they are drawn. Thus, flat time series in H & K are not always correlated with only low levels of activity. The distribution of HK values for the Hall & Lockwood (2004) sample, and for the Giampapa *et al.* (2006) M67 solar-type stars, appears to be unimodal. There does not appear to be a bimodal distribution as was reported in the earlier study of Baliunas & Jastrow (1990) that, as shown later by Wright (2004), was contaminated with evolved subgiant stars that were in a quiescent state as a consequence of their relative age.

5. Other Facets of Stellar Cycles

Baliunas *et al.* (1995) discovered examples of solar-type stars with multiple cycles. For example, 15 Sge (HD 190406; G1 V) appears to have a shorter, secondary cycle of 2.6 years superimposed on its much longer primary cycle of 16.9 years (also see Hall *et al.* 2007). In an investigation based on the results from the Mt. Wilson sample, Böhm-Vitense (2007) investigated the correlation between cycle periods and rotation periods, identifying both an "Active" (A) branch and an "Inactive" (I) branch (see her Fig. 1). Secondary cycles were found to be primarily on the extrapolation of the I-sequence to faster rotation periods. Böhm-Vitense (2007) therefore suggested that the operative dynamo for the I-sequence stars must also be operating in the A-branch stars, and that the secondary cycles arise from the same I-branch dynamo mechanism. The Sun in her diagram appears conspicuously between the two branches, thereby giving the impression that it is a "transition object." Perhaps so, but I would caution that in this sample the I-branch stars in her Fig. 1 are all K stars with slower rotation than the Sun. Therefore, this may not be a suitable sample for comparison with the Sun. This points to the critical importance of how a stellar sample is defined before conclusions can be drawn concerning the comparative behavior of the Sun with that of other stars.

In the case of the Sun, Fletcher *et al.* (2010) report the detection of frequency shifts in low-degree solar oscillation modes from BISON data, suggesting evidence of an ∼2 year period that may be interpreted as a secondary cycle. Jain *et al.* (2011) could not confirm this claim using GONG data but the issue may still be outstanding pending a more detailed comparison of the data analysis methods utilized in the two respective studies. I would remark, however, that in my own analysis of the multi-decadal Ca II K line time series obtained by W. C. Livingston utilizing the NSO McMath-Pierce solar telescope, I could not find evidence for an ∼1-2 year period arising from a possible cycle secondary to the primary 11-year solar cycle. In brief summary, the presence of secondary cycles has been interpreted as evidence of more than one dynamo operating in solar-type stars and, perhaps, in the Sun itself. However, more than one mode of a single dynamo mechanism may also be operative and distinguishing between these possibilities is a current challenge for dynamo theory.

In addition to secondary cycles, unusually short period cycles have been identified. In particular, Metcalfe *et al.* (2010) reported the detection of a 1.6 year cycle in a F8 V star with no evidence for a long-term trend indicative of a longer cycle. Therefore, this short cycle could be its primary cycle. Recalling the empirical scaling between cycle frequency and differential rotation given in §2, such a short period could be a manifestation of particularly strong differential rotation at a level of ∼ 3 times the solar differential rotation rate.

6. Summary and Future Directions

Some of the principal points I sought to convey to the reader include the following:

• Cycle amplitudes exhibit similar trends with cycle frequency (normalized by rotation frequency) for stars on the Active and Inactive branches, respectively. However, cycle amplitudes do not show any dependence on rotation frequency alone, or on differential rotation.

• Cycle frequencies have a relatively stronger correlation with differential rotation frequency than with rotation frequency. We deduced an empirical scaling relation between cycle frequency and differential rotation given by $\omega_{cyc} \sim (\Delta\Omega/\Delta\Omega_{\odot})^{1.6}$

• The Sun appears to reside at a boundary between stars with and without cycles. In fact, the recent Cycle 23 "minimum Sun" seen as a star looked like stars with no cycles, i.e., a possible 'Maunder-minimum star.'

• Relatively short period secondary cycles superimposed on longer period primary cycles have been detected in other solar-type stars and may exist in the Sun itself. Most stars with secondary periods tend to have higher mean levels of chromospheric emission and faster rotation as compared to the Sun.

• The existence of secondary cycles, and an Active and Inactive branch in cycle frequency (period)–rotation frequency (period) diagrams, have been interpreted as evidence of more than one dynamo mechanism operating in at least some stars and perhaps the Sun itself. Alternatively, multiple cycle periods may be manifestations of multiple modes of a single dynamo mechanism.

As for future directions of research, the future is *now* as the application of asteroseismic techniques to the detection of stellar cycles or possible secondary solar cycles becomes more extensive. In addition, starspot cycles inferred from *Kepler* and *CoRoT* data will be invaluable. Both these approaches will depend critically on the long term continuation of ground-based facilities and these revolutionary space missions. This will require both diligence and vigilance with funding agencies to maintain these crucial capabilities. In parallel efforts, it is extremely important to identify solar twins and homogeneous populations of stars in order to carry out comparative studies of stellar cycles and the solar cycle. Inhomogeneous stellar samples can lead to erroneous conclusions. The solar-age and solar-metallicity open cluster, M67, appears to be an excellent solar laboratory with stars that are spectroscopically identical to the Sun (e.g., Önehag *et al.* 2011). Finally, the cycle properties in the limit of thick convection zones as could be investigated through long-term studies of M dwarf stars remains an open and urgent question.

Acknowledgements

I am pleased to acknowledge discussions with S. Saar and J. Hall that materially contributed to this manuscript. I also would like to thank the organizers of IAU Symposium 286 in Mendoza, Argentina for their kind hospitality and for a productive and thoroughly enjoyable meeting. The National Solar Observatory is operated by AURA for the U. S. National Science Foundation under a cooperative agreement.

References

Baliunas, S. L. *et al.* 1995, *ApJ*, 438, 269
Baliunas, S. L. *et al.* 1996, *ApJ*, 460, 848
Baliunas, S. L. & Jastrow, R. J. 1990, *Nature*, 348, 520
Böhm-Vitense, E. 2007, *ApJ*, 657, 486
Giampapa, M. S. *et al.* 2006, *ApJ*, 651, 444
Hall, J. C. & Lockwood, G. W. 2004, *ApJ*, 614, 942
Hall, J. C., Lockwood, G. W., & Skiff, B. A. *AJ*, 133, 862
Fletcher, S. T. *et al.* 2010, *ApJ* (Letters), 718, L19
Jain, K. *et al.* 2011, *ApJ*, 739, 6
Lovis, C. *et al.* 2011, *A&A*, submitted
Metcalfe, T. S. *et al.* 2010, *ApJ* (Letters), 723, L213
Noyes, R. W. *et al.* 1984, *ApJ*, 279, 763
Önehag, A. *et al.* 2011, *A&A*, 528, 850
Saar, S. H. 2011, *The Physics of Sun and Star Spots, Proceedings of IAU Symposium No. 273*, eds. D. P. Choudhary & K. G. Strassmeier, p. 61
Saar, S. H. & Brandenburg, A. 1999, *ApJ*, 524, 295
Saar, S. H. & Brandenburg, A. 2002, *AN*, 323, 357
Stauffer, J. R. *et al.* 1991, *ApJ*, 374, 142
Vaughan, A. H. & Preston, G. W. 1980, *PASP*, 92, 385
Wilson, O. C. 1978, *ApJ*, 226, 379
Wright, J. T. 2004, *AJ*, 128, 1273

Discussion

Arnab Choudhuri: 1) You showed very interesting plots of differential rotators, are they all solar type with the equator moving faster or are there anti-solar cases in the sample? 2) You showed a star which was a candidate to enter a Maunder Minimum. The data went only until 1990, do you have data in the last 16 or 20 years?

Mark Giampapa: 1) The differential rotation is inferred by changes in rotation period coming from the rotational modulation of active regions asymmetrically distributed in longitude. So, there is no latitude information and, therefore, we don't know if there are anti-solar examples. 2) No, there are no subsequent data. The Mt. Wilson program ended due to lack of funding.

EMRE IŞIK: You said there was a flat activity star with a considerable HK flux but is not showing cycles. Are there more stars like this in the sample?

MARK GIAMPAPA: The sample of stars is from Jeff Hall's "S-cubed" monitoring program at Lowell Observatory in Flagstaff, Arizona. There are 57 solar-type stars in his sample, of these, 10 show 'flat activity', that is, no cycle-like or statistically significant variation.

LEIF SVALGAARD: During the Maunder Minimum we know from cosmic ray data that the dynamo was still going. If magnetic activity is still there, there should be Ca II emission going in a cyclic fashion with regard to your stellar cycling and non-cycling data. Do you have any comment?

MARK GIAMPAPA: The detection of cycle variation depends on the index or diagnostic being used., whether it is sunspot number, modulation of cosmic rays, or something else. In the flat activity stars, the time series of cyclic Ca II emission is not present within the small errors. Of course, we don't have direct information on the integrated Ca II emission during the Maunder Minimum. So, it is difficult to know what the limits of Ca II variability might have been.

JEFF LINSKY: If we observe the Sun from the poles, the solar cycle would have looked less intense. Could it be that stars with low stellar cycles are seen from the poles?

MARK GIAMPAPA: In Hall's sample of 57 stars, there are 10 stars that exhibit flat activity in their K-line time series. Statistically, it is an extremely low probability that in a sample of only 57 stars, 10 would be seen nearly pole-on. Also, one could still see a cycle modulation of the K-line emission for a star viewed pole-on though the amplitude would be smaller.

STEVE SAAR: (Comment) You have to be careful about what you use for your index, they have not been calibrated for effective metallicity or gravity because different metallicities or gravities can shift can shift the curve in the R'HK–effective temperature diagram, mainly due to changing the correction for the photospheric contribution to the H and K line cores.

MARK GIAMPAPA: Yes, I agree. Lovis *et al.* in their paper do make a correction for metallicity to the R'HK index.

Comparative Magnetic Minima:
Characterizing quiet times in the Sun and stars
Proceedings IAU Symposium No. 286, 2011
C. H. Mandrini & D. F. Webb, eds.

doi:10.1017/S1743921312004966

Investigating stellar surface rotation using observations of starspots

Heidi Korhonen[1,2,3]

[1]Niels Bohr Institute, University of Copenhagen, Juliane Maries Vej 30, DK-2100 Copenhagen, Denmark email: heidi.korhonen@nbi.ku.dk

[2]Finnish Centre for Astronomy with ESO (FINCA), University of Turku, Väisäläntie 20, FI-21500 Piikkiö, Finland

[3]Centre for Star and Planet Formation, Natural History Museum of Denmark, University of Copenhagen, Øster Voldgade 5-7, DK-1350, Copenhagen, Denmark

Abstract. Rapid rotation enhances the dynamo operating in stars, and thus also introduces significantly stronger magnetic activity than is seen in slower rotators. Many young cool stars still have the rapid, primordial rotation rates induced by the interstellar molecular cloud from which they were formed. Also older stars in close binary systems are often rapid rotators. These types of stars can show strong magnetic activity and large starspots. In the case of large starspots which cause observable changes in the brightness of the star, and even in the shapes of the spectral line profiles, one can get information on the rotation of the star. At times even information on the spot rotation at different stellar latitudes can be obtained, similarly to the solar surface differential rotation measurements using magnetic features as tracers. Here, I will review investigations of stellar rotation based on starspots. I will discuss what we can obtain from ground-based photometry and how that improves with the uninterrupted, high precision, observations from space. The emphasis will be on how starspots, and even stellar surface differential rotation, can be studied using high resolution spectra.

Keywords. stars: activity, stars: late-type, stars: rotation, stars: spots

---

1. Introduction

It is widely accepted that the global behaviour of the solar magnetic field can be explained by a dynamo action which is due to interaction between magnetic fields and fluid motions. The Sun is thought to have an $\alpha\Omega$-type dynamo, in which the poloidal field is created from the toroidal one by helical turbulence (α-effect), and the toroidal field is obtained by shearing the already existing poloidal field by differential rotation (Ω-effect). This kind of dynamo is also thought to work in other main-sequence stars with similar internal structure as the Sun has, i.e., stars with masses of $\sim$0.4–2.0 $M_\odot$.

Stellar rotation has a major impact on the over-all efficiency of the dynamo action, and thus on the level of observed magnetic activity. The relationship between rotation and activity was first studied in detail by Pallavicini *et al.* (1981) who investigated the correlation between the X-ray luminosity and projected rotation velocity, $v \sin i$. Since then several studies have shown that the magnetic activity increases with increasing rotation rate, until finding a rotation rate after which no increase, and even maybe a small decrease, occurs (e.g., Micela *et al.* 1985; Pizzolato *et al.* 2003; Wright *et al.* 2011). This so-called saturation limit of the magnetic activity is reached at certain spectral type dependent rotation period, which increases toward later spectral types (Pizzolato *et al.* 2003).

There is also evidence that the stellar cycle length correlates to some extend with the rotation rate. Faster rotation tends to create shorter activity cycles (e.g., Noyes *et al.* 1984; Saar & Brandenburg 1999; Oláh *et al.* 2000). In a diagram where the cycle frequency ($\omega_{\rm cycl}$) is plotted against the rotational frequency Ω the stars also seem to occupy three different regions: so-called inactive, active, and super-active regions (e.g., Saar & Baliunas 1992; Saar & Brandenburg 1999).

The studies of open clusters with known ages show that young clusters, with age of some tens of Myrs, have many rapidly rotating cool stars. With increasing age the amount of cool rapidly rotating stars decreases, and around 600 Myrs the stars with similar internal structure to the Sun's have slowed down to relation that is dependent on stellar mass (see e.g., Barnes 2003; Meibom *et al.* 2011). The observed spin-down of the stars over their lifetimes can be attributed to the magnetic breaking, which is driven by mass-loss through a magnetised stellar wind (e.g., Skumanich 1972). Therefore, many young cool stars (spectral classes G–M) are rapid rotators. The rapid rotation introduces significantly stronger magnetic activity than is seen in their older main sequence counterparts, like the Sun. On the other hand, the Sun was most likely in its youth a very active stars. All the energetic events caused by the magnetic activity can heat the possible planet forming disc around the star and have an impact on its composition, thus also affecting the possible planet and planetary system formation process. The phenomena caused by stellar activity can also have similar effects on stellar brightness and radial velocity as orbiting planets, making it at times difficult to distinguish between planets and activity signatures (see, e.g., Queloz *et al.* 2001).

It is clear that stellar rotation has a significant effect on magnetic activity of stars. In this review I will discuss how to measure stellar rotation using starspots, and especially concentrating on detailed studies of stellar surface differential rotation, which is a crucial parameter in the solar and stellar dynamos.

2. Methods for studying stellar rotation using starspots

Stars are point sources and studying their surface features is very demanding. For a long time it was not possible to obtain direct, spatially resolved, images of the stellar surface, except in some very rare cases of near-by giant and supergiant stars, like Betelgeuse (Gilliland & Dupree 1996). During the last years a breakthrough using long baseline infrared interferometers has occurred. Images with milli-arcsecond resolution can now be produced, and a variety of targets have been imaged with astonishing result, e.g., bulging stars rotating near their critical limit (Monnier *et al.* 2007), and images of the transiting disk in the ϵ Aurigae system (Kloppenborg *et al.* 2010). Infrared interferometric imaging has produced amazing results of stellar surfaces and the time is drawing near when even dark spots on cool stars can be imaged. In addition, helioseismology has given us an unprecedented view of the solar rotation, both at the surface and in the interior, but obtaining similar results for stars is still very much 'work in progress'.

The main methods for studying stellar rotation are similar to the ones used originally on the Sun, i.e., tracing the movement of magnetic features, mainly starspots. This can be done using photometry and high resolution spectroscopy through Doppler imaging techniques. The rest of the article is focused on these two methods. Still, it has to be remarked that stellar rotation has also been studied using spectral line profile shapes (e.g., Reiners & Schmitt 2003) and Ca II H&K emission variations (e.g., Donahue *et al.* 1996; Katsova *et al.* 2010).

2.1. *Photometry*

In many active stars the areas occupied by starspots are so large that they cause brightness variations which can be few tens of percent from the mean light level, thus easily

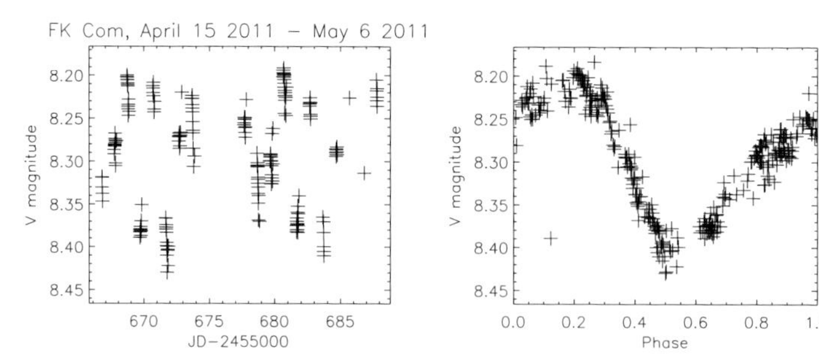

Figure 1. Example of ground-based observations of an active star FK Com. The plot gives the V magnitudes plotted against the Julian Date (left panel) and against the phase (right panel). The observations have been obtained in April–May 2011 using automatic photometric telescopes in Arizona. More information on the telescopes is given by Strassmeier *et al.* (1997).

observable even with small ground-based telescopes (see, e.g., Kron 1947). In comparison, biggest sunspots would only cause ∼0.01% decrease in the solar brightens and would require extremely precise instruments to detect.

Numerous studies of stellar rotation on active stars have been carried out based on photometry. Long-term monitoring campaigns have been carried out by several groups resulting in numerous papers on stellar rotation (e.g., Messina & Guinan 2003; Strassmeier *et al.* 1997). Most of the campaigns are primarily used for studying long-term stellar cycles, but also the rotation periods and their possible variations have been investigated.

In general, the observations are carried out by automatic telescopes and typically a couple of observations per night are obtained during as many nights as possible within the observing season. Depending on the exact stellar rotation period this results in a time series from which the period might be difficult to determine accurately. Fig. 1 gives an example of a ground-based light-curve of an active star, in this case the single yellow giant FK Com. These observations actually present a good phase coverage light-curve obtained during a dedicated campaign. The typical dataset has much less observations within similar time period.

If the period is not straight forward to determine from the often sparse ground-based observations, it is one of the easiest stellar properties to obtain from continuous space-based observations. With the recent launch of several photometric space missions, the high accuracy, high cadence, stellar light-curves are now also revolutionising the active star research. These missions include the Canadian micro-satellite MOST (for results on active stars see, e.g., Rucinski *et al.* 2004 and Siwak *et al.* 2011) and the French CoRoT satellite (e.g., Lanza *et al.* 2009; Silva-Valio *et al.* 2010; Huber *et al.* 2010). The real break-through happened, though, with the launch of the NASA's Kepler satellite. Kepler provides unprecedented accuracy light-curves of about 150 000 stars, among them several active stars (see Fig. 2 and, e.g., Frasca *et al.* 2011). Based on GALEX data a sample of about 200 active stars have been selected for Kepler observations in a Guest Observer programme (Brown *et al.* 2011). The space-based data has shown us that basically every stellar rotation has slightly different spot configuration. This casts serious doubts in the old practice of using ground-based data from several rotations together.

2.2. *Doppler imaging*

Doppler imaging is a method that can be used for detailed mapping of the stellar surface structure (e.g., Vogt *et al.* 1987; Piskunov *et al.* 1990). Hereby high resolution, high signal-to-noise spectra at different rotational phases are used to measure the rotationally modulated distortions in the line-profiles. These distortions are produced by the

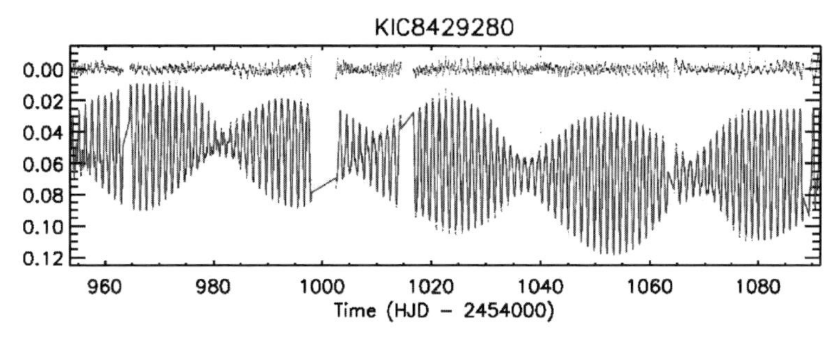

Figure 2. Example of a high precision continuous space-based observations of a young active star. The observations have been obtained using Kepler and have been published by Frasca *et al.* (2011).

inhomogeneous distribution of the observed characteristic, e.g., effective temperature, element abundance or magnetic field. Surface maps, or Doppler images, are constructed by combining all the observations from different phases and comparing them with synthetic model line profiles. Doppler imaging techniques were first used in the abundance mapping of Ap stars. Nowadays, Doppler imaging is more commonly used for temperature mapping of rapidly rotating late type stars (e.g., Korhonen *et al.* 2007; Skelly *et al.* 2010).

One has to keep in mind though, that for successful Doppler imaging a priori knowledge of the stellar rotation, usually based on earlier photometric investigations, is needed. If the rotation period of the Doppler imaging target is not know, it is very difficult to plan the observations so that a good phase coverage needed for Doppler imaging is obtained. Also, if the rotation period is not known accurately, comparison of the maps recovered at different epochs is not straight forward. The spectra and the variations seen in them can be used for estimating the rotation period of the target, but still the best results are obtained if also long-term photometric monitoring is carried out for accurate period determination.

Until autumn 2011 Doppler imaging has been applied to some 70 stars. Strassmeier (2009) gives a recent review on starspots and their properties. Here only a short summary on the main results we have learned about starspots using Doppler imaging is given.

Due to the enhanced magnetic activity in rapidly rotating cool stars, the spots are much larger than the spots observed in the Sun. The largest starspot recovered with Doppler imaging is on the active RS CVn binary HD 12545 which, in January 1998, had a spot that extended approximately 12×20 solar radii (Strassmeier 1999). The image of this starspot is shown in Fig. 3.

The lifetime of the large starspots can also be much longer than that of the sunspots, even years instead of weeks for sunspots (e.g., Rice & Strassmeier 1996; Hussain 2002). One has to keep in mind though, that the spatial resolution obtained with Doppler imaging is not good enough for distinguishing whether the large spots are one single spot or a group of spots. If they are groups of spots, the individual spots could exhibit much shorter lifetimes than the group itself. Based on numerical simulations Işık *et al.* (2007) have shown that in a rapidly rotating active star the expected spot lifetime is few months, depending on the spot latitude (mid-latitude spots live a shorter time than equatorial or high latitude spots) and spot size (larger spots live longer). For sub-giants even longer spot lifetimes are obtained.

The latitudes at which starspots often occur are very different from those for the sunspots. In rapidly rotating late type stars spots can appear at very high latitudes, unlike in the Sun. This can be explained by the increase in the Coriolis force induced

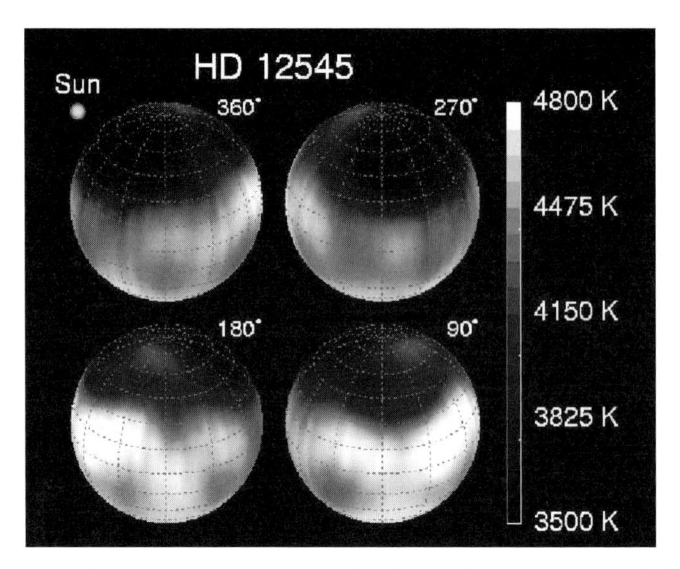

Figure 3. Doppler imaging results of HD 12545 by Strassmeier (1999).

by the rapid rotation (e.g., Schüssler & Solanki 1992; Granzer *et al.* 2000). Still, these calculations cannot explain the formation of the polar caps, i.e., spots that are located at the rotational poles of the star, except in very young stars. These polar caps are still often seen in the Doppler images of also older late type stars (e.g., Weber & Strassmeier 2001).

3. Surface differential rotation

Gaseous bodies, like the stars, can rotate differentially, i.e., different latitudes can have different rotation rates. In the Sun the rotation velocity of the photosphere depends strongly on the latitude; the rotation of the solar equator is approximately 30% shorter than the period at the poles. Helioseismological studies show that this latitude dependence persists throughout the convection zone (e.g., Thompson *et al.* 1996).

Differential rotation is one of the main elements in the dynamo models. Together with the helical turbulence and meridional flow it is responsible for the main features of the solar and stellar magnetic activity (see, e.g., Rüdiger *et al.* 1986; Brun & Toomre 2002). Therefore, it is important to also measure differential rotation on other stars than the Sun.

Usually it is assumed that the differential rotation law of the Sun can be generalised to stars, leading the surface rotation law to be expressed by

$$\Omega = \Omega_{\rm eq} + \beta \sin^2 \psi, \tag{3.1}$$

where ψ is the latitude, $\Omega_{\rm eq}$ is the equatorial angular velocity and β defines the magnitude of the differential rotation. The relative differential rotation coefficient is given by

$$\alpha = \frac{\Omega_{\rm eq} - \Omega_{\rm pol}}{\Omega_{\rm eq}} \quad \text{or } \alpha = \frac{-\beta}{\Omega_{\rm eq}} \tag{3.2}$$

where $\Omega_{\rm pol}$ is the polar angular velocity. Note though that the exact formulation, symbols and signs, vary from author to author. Therefore, one has to be careful when comparing different works.

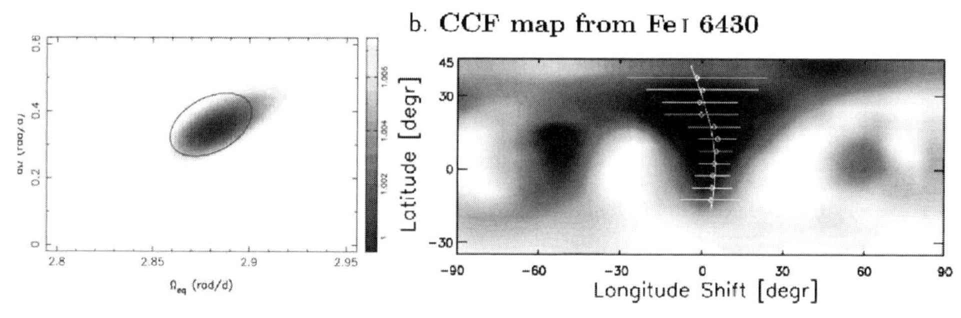

Figure 4. Examples of surface differential rotation determination using χ^2-landscape (left; Marsden *et al.* 2011) and cross-correlation (right; Kővári *et al.* 2007a) methods.

Measuring stellar differential rotation is not straight forward. Currently the best way to study stellar surface differential rotation in detail, both the strength and the sign, is by using surface maps obtained with Doppler imaging. One way, similar to using sunspots and other magnetic features to study the solar surface differential rotation, is to cross-correlate several surface maps obtained at different times with Doppler imaging. This allows to investigate the changes in the locations of the spots and how that depends on the latitude (see, e.g., Barnes *et al.* 2000; Weber *et al.* 2005; Kővári *et al.* 2007a). The other often applied method is so-called χ^2-landscape technique in which a solar-like surface differential law is implemented into the Doppler imaging code, and many maps are obtained to study which parameters give the best solution (e.g., Petit *et al.* 2002; Marsden *et al.* 2006; Dunstone *et al.* 2008). For examples of both methods, look at Fig. 4. Stellar surface differential rotation has also been studied by combining spot latitude information from Doppler images and spot rotation period from simultaneous photometry (Korhonen *et al.* 2007).

3.1. *Surface differential rotation with spectral type and rotation period*

Donahue *et al.* (1996) investigated the rotation periods and the change in the rotation period, i.e., the magnitude of the surface differential rotation, in 36 stars from Ca II H&K S-index measurements. They found that there was a power-law correspondence between these values, where $\Delta\Omega$ is proportional to the mean seasonal rotation, Ω, with the power of 0.7, independent of mass. Barnes *et al.* (2005) collected measurements from different sources and methods, and obtained results which also imply power-law correlation between the rotation and surface differential rotation, but with a power of only 0.15. On the other hand, recent investigation by Saar (2011) gives relation $\Delta\Omega \propto \Omega^{0.68}$, i.e. similar to the Donahue *et al.* (1996) results. It is clear that selecting homogeneous sample for these studies is difficult. Additionally, the surface differential rotation estimates used in some of these studies come from different methods and exhibit different systematic effects, therefore making comparison difficult.

The theoretical calculations by Kitchatinov & Rüdiger (1999) show that the absolute value of the surface differential rotation decreases initially as the rotation period decreases from the solar value, but changes to a slight increase for periods of few days. They also predict that the differential rotation for the giant stars is larger than that for the dwarfs. Surface differential rotation measurements of 10 young G2–M2 dwarfs were obtained using Doppler imaging by Barnes *et al.* (2005). These measurements show an increase in the magnitude of differential rotation towards earlier spectral types, which is consistent with the theoretical calculations. Similar results have also been obtained by Saar (2011).

Küker *et al.* (2011) carried out an theoretical study of rotation of G dwarfs. They computed model convection zones of different depth and investigated their large-scale gas motions, i.e. rotation and meridional flow, and compared the results to observations. Their calculations could easily produce the rotation laws of the slowly rotating Sun and several rapidly rotating G dwarfs. Their calculations failed to explain the extreme surface shear of HD 171488 (reported, e.g., by Marsden *et al.* 2005 and Jeffers & Donati 2009), except when using an artificially shallow convection zone. This high-lights the fact that even though theory and observations at times encouragingly agree, there are still many unexplained features in the solar and stellar magnetic activity.

3.2. *Anti-solar differential rotation*

In general, models for global circulation in outer stellar convection zones predict solar-type differential rotation, where the equator is rotating faster than the poles. However, Kitchatinov & Rüdiger (2004) have shown that anti-solar differential rotation could arise as a result of intensive meridional circulation.

Anti-solar differential rotation, where the polar regions rotate faster than the equator, has been suggested by observations of several active stars (e.g., Vogt *et al.* 1999; Weber 2007; Kővári *et al.* 2007b). Kővári *et al.* (2007b) investigated surface flow patterns on σ Gem from a series of observation spanning 3.6 consecutive rotation cycles, and found an anti-solar differential rotation with the surface shear of -0.022 ± 0.006 (approximately 1% of the shear in the Sun, but of opposite sign). Additionally, they found evidence of a poleward migration trend of spots with an average velocity of $\sim$300 m/s. The strong meridional flow hinted at σ Gem would support the hypothesis of Kitchatinov & Rüdiger (2004), which attributes the anti-solar differential rotation to strong meridional circulation. Similar trend is seen by Weber (2007) in the investigation of several active stars. One has to note though, that these meridional flow measurements can arise from artefacts in maps and have to be confirmed with data from several epochs.

3.3. *Temporal variations in surface differential rotation*

Intriguingly, temporal evolution of the surface differential rotation has been reported for two young single K stars, AB Dor and LQ Hya (see, e.g., Donati *et al.* 2003; Jeffers *et al.* 2007). And in this case the question is not about small-scale variations, like seen on the Sun, but changes that are as large as 50% of the mean $\Delta\Omega$. Donati *et al.* (2003) hypothesise that these temporal variations could be caused by the stellar magnetic cycle converting periodically kinetic energy within the convective zone into large-scale magnetic fields and vice versa, as originally proposed by Applegate (1992). They also remark that a definite demonstration of the temporal variation would require monitoring a few stars for a long time, and seeing both the differential rotation parameters and the activity proxies showing the same cyclical variations, or at least to exhibit strongly correlated fluctuations in the case of non-cyclic behaviour.

3.4. *Cautionary remark on surface differential rotation studies*

A recent study by Korhonen & Elstner (2011) used snapshots from dynamo models to investigate how well the cross-correlation method reproduces the latitudinal rotation rates used in the dynamo calculations. Their investigation showed that the cross-correlation method works well when the time difference between the maps is appropriate for recovering the surface differential rotation, and, importantly, if small-scale fields were included in the dynamo calculations. Using only the large-scale dynamo field the solution was dominated by the geometry of the dynamo field and the input rotation law was not recovered from the snapshots. Actually, the results in these cases showed much smaller surface differential rotation, than what was used in calculating the dynamo models. On the other hand, with additional injection of small-scale fields the input surface rotation law was well recovered.

This raises the question whether the large starspot seen in active stars can actually be created by small scale fields. If they are manifestations of the large-scale dynamo field, then according to the study by Korhonen & Elstner (2011) we would not even expect them to follow the surface differential rotation. Their results also show that the exact latitude dependence of the rotation changes during the stellar cycle. Which could explain the temporal variation of surface differential rotation discussed above.

4. Concluding remarks

On the whole photometry is the easiest and least time consuming way of carrying out stellar rotation studies. Long time series of observations can easily be obtained for even relatively faint targets. Still, the usual cadence for this kind of observations is few observations per night – at most. This usually results in data from several different stellar rotation been used together, and also in difficulties to pin-point the stellar rotation period accurately. Continuous high cadence space-based observations have shown us that the light-curves of active stars change from rotation to rotation. Thus the ground-based observations usually give us only an average spot configuration. The high precision light-curves from space missions, like CoRoT and Kepler, on the other hand provide us a unique opportunity to investigate stellar rotation and starspots with high temporal resolution. Actually, measuring the stellar rotation period from ground-based observations is often challenging, whereas from CoRoT and Kepler light-curves that is one of the easiest properties to determine.

The strength of the ground-based observations lies in investigating stellar cycles. Such long-term time series are virtually impossible to obtain from space due to the limited mission lifetimes. Also, instrumental artefacts, like trends, are often a problem in space-based instruments, again hampering accurate studies of stellar cycles. Whereas obtaining rotational periods with high precision can uniquely be done using space-based continuous photometry.

Surface differential rotation can be estimated using photometric observations. But the investigations suffer from the fact that the information on the spot latitudes is usually impossible to obtain, and thus no information on the sign of the differential rotation can be obtained. Also, the latitude range of the spots is unknown, therefore only a lower estimate of the magnitude of the differential rotation can be obtained. Although, from high precision space-based photometry it might be possible to obtain also the latitude information.

For obtaining the magnitude and sign of the surface differential rotation detailed surface maps are needed. Still, also the studies using Doppler images are not without problems. All the methods based on Doppler imaging suffer from the often restricted latitude range the starspots occur on, and from the possible artefacts in the maps. These artefacts can for example rise from incorrect modelling of the spectral line profiles. For example incorrect modelling of the line core would produce a signal which is always visible around zero velocity, i.e., would result in a polar spot. Additionally, in χ^2-landscape technique a predefined solar-type rotation law is assumed and thus no other latitude dependence can be recovered. On the other hand, in cross-correlation the time difference between the maps is crucial, with too small difference no change has time to occur, and with too long difference spot evolution due to flux emergence and disappearance can occur. Therefore, it is very demanding to obtain reliable measurements of the stellar surface differential rotation.

Even though, these concluding remarks offer more concerns than definite answers, it does not mean that the future of studying stellar rotation is bleak. Doppler imaging keeps on offering us an intriguing picture of stellar surface features and the space-based photometry is opening a real golden era for stellar rotation studies. In the future also asteroseismology and infrared/optical interferometry have a role to play in the new discoveries.

Acknowledgments

The author acknowledges the support from the European Commission under the Marie Curie IEF Programme in FP7.

References

Applegate, J. H. 1992, *ApJ* 385, 621
Barnes, J. R., Collier Cameron, A., James, D. J., & Donati, J.-F. 2000, *MNRAS* 314, 162
Barnes, J. R., Collier Cameron, A., Donati, J.-F., James, D. J., Marsden, S. C., & Petit, P. 2005, *MNRAS* 357, L1
Barnes, S. A. 2003, *ApJ* 586, 464
Brown, A., Korhonen, H., Berdyugina, S. V., *et al.* 2011, in: D. P. Choudhary, & K. G. Strassmeier (eds.), *Physics of Sun and Star Spots*, Proc. IAU Symposium No. 273 (Cambridge University Press), p. 78
Brun, A. S. & Toomre, J. 2002, *ApJ* 570, 865
Donahue, R. A., Saar, S. H., & Baliunas, S. L. 1996, *ApJ* 466, 384
Donati, J.-F., Collier Cameron, A., Semel, M., *et al.* 2003, *MNRAS*, 345, 1187
Dunstone, N. J., Hussain, G. A. J., Collier Cameron, A., *et al.* 2008, *MNRAS*, 387, 1525
Gilliland, R. L. & Dupree A. K. 1996, *ApJ* (Letters) 463, L29
Granzer, Th., Schüssler, M., Caligari, P., & Strassmeier, K. G. 2000, *A&A* 355, 1087
Frasca, A., Fröhlich, H.-E., Bonanno, A., Catanzaro, G., Biazzo, K., & Molenda-Zakowicz, J. 2011, *A&A* 532, A81
Huber, K. F., Czesla, S., Wolter, U., & Schmitt, J. H. M. M. 2010, *A&A*, 514, A39
Hussain, G. A. J. 2002, *AN*, 323, 349
Işık, E., Schüssler, M., & Solanki, S. K. 2007, *A&A* 464, 1049
Jeffers, S. V., Donati, J.-F., & Collier Cameron, A. 2007, *MNRAS* 375, 567
Jeffers, S. V. & Donati, J.-F. 2009, *MNRAS* 390, 635
Katsova, M. M., Livshits, M. A., Soon, W., Baliunas, S. L., & Sokoloff, D. D. 2010 *New Astron.* 15, 274
Kitchatinov, L. L. & Rüdiger, G. 1999, *A&A* 344, 911
Kitchatinov, L. L. & Rüdiger, G. 2004, *AN* 325, 496
Kloppenborg, B., Stencel, R., Monnier, J. D., *et al.* 2010, *Nature* 464, 870
Korhonen, H., Berdyugina, S. V., Hackman, T., Ilyin, I. V., Strassmeier, K. G., & Tuominen, I. 2007, *A&A*, 476, 881
Korhonen, H. & Elstner, D. 2011, *A&A* 532, A106
Kővári, Zs., Bartus, J., Strassmeier, K. G., *et al.* 2007a, *A&A* 463, 1071
Kővári, Zs., Bartus, J., Strassmeier, K. G., *et al.* 2007b, *A&A* 474, 165
Kron, G. E. 1947, *PASP* 59, 261
Küker, M., Rüdiger, G., & Kitchatinov, L. L. 2011, *A&A* 530, A48
Lanza, A. F., Pagano, I., Leto, G.,*et al.* 2009, *A&A* 493, 193
Marsden, S. C., Waite, I. A., Carter, B. D., & Donati, J.-F. 2005, *MNRAS* 359, 711
Marsden, S. C., Donati, J.-F., Semel, M., Petit, P., & Carter, B. D. 2006, *MNRAS* 370, 468
Marsden, S. C., Jardine, M. M., Ramirez Vélez, J. C., *et al.*, 2011, *MNRAS*, 413 1939

Meibom, S., Barnes, S. A., Latham, D. W., *et al.*, 2011, *ApJ* (Letters) 733, L9
Messina, S., & Guinan E. F. 2003, *A&A* 409, 1017
Micela, G., Sciortino, S., Serio, S., *et al.* 1985, *ApJ* 292, 172
Monnier, J. D., Zhao, M., Pedretti, E., *et al.* 2007, *Science* 317, 342
Noyes, R. W., Weiss, N. O., & Vaughan, A. H. 1984, *ApJ* 287, 769
Oláh, K., Kolláth, Z., & Strassmeier, K. G. 2000, *A&A* 356, 643
Pallavicini, R., Golub, L., Rosner, R, *et al.* 1981, *ApJ* 248, 279
Petit, P., Donati, J.-F., & Collier Cameron, A. 2002, *MNRAS* 334, 374
Piskunov, N. E., Tuominen, I., & Vilhu, O. 1990, *A&A* 230, 363
Pizzolato, N., Maggio, A., Micela, G., Sciortino, S., & Ventura, P. 2003, *A&A* 397, 147
Queloz, D., Henry, G. W., Sivan, J. P. *et al.* 2001 *A&A* 379, 279
Reiners, A., & Schmitt, J. H. M. M. 2003 *A&A* 398, 647
Rice, J. B. & Strassmeier, K. G. 1996, *A&A* 316, 164
Rüdiger, G., Krause, F., Tuominen, I., & Virtanen, H. 1986, *A&A* 166, 306
Rucinski, S. M., Walker, G. A. H., Matthews, J. M., *et al.* 2004, *PASP* 116, 1093
Saar, S. H. & Baliunas, S. L. 1992, in: K. L. Harvey (ed), *The Solar Cycle, Proceedings of the NSO/Sac Peak 12th Summer Workshop (Fourth Solar Cycle Workshop)* (ASP Conf. Series, Vol. 27), p. 150
Saar, S. H. & Brandenburg, A. 1999, *ApJ* 524, 295
Saar, S. 2011, in: D. P. Choudhary, & K. G. Strassmeier (eds.), *Physics of Sun and Star Spots*, Proc. IAU Symposium No. 273 (Cambridge University Press), p. 61
Silva-Valio, A., Lanza, A. F., Alonso, R., & Barge, P. 2010, *A&A*510, A25
Siwak, M., Rucinski, S. M., Matthews, J. M., *et al.* 2011, *MNRAS* 415, 1119
Skelly, M. B., Donati, J.-F., Bouvier, J., *et al.* 2010, *MNRAS* 403, 159
Skumanich, A. 1972, *ApJ* 171, 565
Schüssler M., Solanki, S. K. 1992, *A&A* 264, L135
Strassmeier, K. G., Bartus, J., Cutispoto, G., & Rodono, M. 1997, *A&AS* 125, 11
Strassmeier, K. G. 1999, *A&A* 347, 225
Strassmeier, K. G. 2009, *Astronomy & Astrophysics Review* 17, 251
Thompson, M. J., Toomre, J., Anderson, E. R., *et al.* 1996, Science 272, 1300
Vogt, S. S., Penrod, G., & Donald, Hatzes, A. P. 1987, *ApJ* 321, 496
Vogt, S. S., Hatzes, A. P., Misch, A., & Kürster, M. 1999, *ApJS*, 121, 546
Weber, M. & Strassmeier, K. G. 2001, *A&A* 373, 974
Weber, M., Strassmeier, K. G., & Washuettl, A. 2005, *AN*, 326, 287
Weber, M. 2007, *AN* 328, 1075
Wright, N. J., Drake, J. J., Mamajek, E. E., & Henry, G. W. 2011, *ApJ* in press, arXiv1109.4634

Discussion

Jeffrey Linsky: On the Sun we see spots and faculae or anti-spots, they rotate at the same period of the spots and they tend to limb brightening. Are you fitting the data but not representing well the real star by not considering anti-spots?

Heidi Korhonen: In Doppler imaging we also allow for hot regions in our maps. We also regularly see these regions and they are used in the differential rotation analysis together with the cool spots. It is true, though, that often in light-curve analysis hot spots are not included.

Katja Poppenhager: What do spot maps look like using the two methods Doppler and Keppler imaging?

Heidi Korhonen: Unfortunately the Keppler targets are quite faint; so, it is difficult to get good enough observations for Doppler imaging. We have some candidates for this and are working on them.

Dibyendu Nandy: Is it possible to determine spot migration like in a butterfly diagram using Doppler imaging?

Heidi Korhonen: In principle it is possible. The problem is that we do not have Doppler images for many stars with the required time base. Besides the stars that have many maps spanning enough years unfortunately have mainly high latitude spots, which do not change latitude much. We would very much like to find butterfly diagrams on other stars.

Mark Giampapa: Do you have a general impression of spot properties as a function of spectral type? For example, for different convection zone depths, how would the spot properties change?

Heidi Korhonen: Our sample of Doppler imaged targets is still quite small and the activity too affected by many other parameters, especially rotation. It is difficult to find good targets for comparison, but in general young stars show faster variability than the older ones and the spot contrast is smaller for cooler stars than for hotter stars.

Comparative Magnetic Minima:
Characterizing quiet times in the Sun and stars
Proceedings IAU Symposium No. 286, 2011
C. H. Mandrini & D. F. Webb, eds.

doi:10.1017/S1743921312004978

Modulated stellar and solar cycles: parallels and differences

K. Oláh[1], **L. van Driel-Gesztelyi**[1,2,3] **and K. G. Strassmeier**[4]

[1]Konkoly Observatory, Budapest, Hungary
email: olah@konkoly.hu
[2]Mullard Space Science Laboratory, University College London, UK
[3]Observatoire de Paris, LESIA, CNRS, UPMC Univ. Paris 06, Univ. Paris-Diderot, Meudon Cedex, France
[4]Leibniz Institute for Astrophysics Potsdam (AIP), Germany

Abstract. We present examples of activity cycle timescales on different types of stars from low-mass dwarfs to more massive giants, with wide-ranging rotation rates, and compare the observed cyclicities to the irradiance based solar cycle and its modulations. Using annual spectral solar irradiance in wavelength bands typical for stellar observations reconstructed by Shapiro *et al.* (2011), a direct comparison can be made between cycle timescales and amplitudes derived for the Sun and the stars. We show that cycles on multiple timescales, known to be present in solar activity, also show up on stars when the dataset is long enough to allow recognition. The cycle lengths are not fixed, but evolve - gradually during some periods but there are also changes on short timescales. In case the activity is dominated by spots, i.e., by cooler surface features, the star is redder when fainter, whereas other type of activity make the stars bluer when the activity is higher. We found the Sun to be a member of the former group, based on reconstructed spectral irradiance data by Shapiro *et al.* (2011).

Keywords. stars: activity, stars: spots, Sun: activity, sunspots

1. Introduction

The long-term variability of the Sun and other active stars' cycles is a well known feature originating from the quasi-periodic generation, evolution and decay of the magnetic field by a dynamo mechanism, which manifests itself in surface phenomena like spots and faculae. During a magnetic cycle the coverage by cooler and hotter areas of the stellar surface is changing. These changes are measured as brightness variability in case of stars, whereas for the Sun they can be traced through direct imaging. The time behavior of stellar surface phenomena is studied by searching for periodicities in the long-term brightness variation. On the Sun, the sunspot number and/or some proxy data (e.g. the 10.7 cm radio emission) characterize the solar cycle. For solar and stellar datasets, using time series analysis, cycles can be found. However, the amplitude of the variations depends on the observing wavelength of the datasets. In this paper we briefly describe the types of existing long-term datasets for active stars and the Sun and present a convenient and useful method to study the time-behavior of the cycles found. With the help of a recent annual spectral reconstruction of the solar irradiance variability back to centuries by Shapiro *et al.* (2011), we are able to make direct comparison between the long-term behavior of the solar and the stellar magnetic variability. Dynamo models indicate that the length of a cycle is determined by many factors (depth of the convection zone, rotational rate, strength of the magnetic field etc., see e.g., Brown *et al.* (2011)). In this paper we study the cycle features without addressing their origins.

2. Datasets and tools

As the 11-yr sunspot cycle is well documented for more than 2.5 centuries, the existence of stellar cycles were postulated not long after the first observations of the active stars begun, about half a century ago. Active stars exhibit brightness variation as the spotted, cooler regions of the stellar surface are moving in and out of view due to rotation, and long-term brightness changes superposed on the rotational modulation are also present in all cases. Systematic observations of stars proposed by Wilson (1968) started in 1966 with monitoring the relative fluxes of the cores of CaII H & K lines, which on the Sun vary with the area and intensity of chromospheric network and faculae, and thus can be used as indicators of magnetic activity in stellar chromospheres. This dataset allowed the determination of cycles on several active stars and even double cycles were found (see Baliunas *et al.* (1996)). Conventional photometric observations were carried out at several observatories but systematic observations of active stars only started with the advent of dedicated automated telescopes, some of those are working to date, such as e.g. the Vienna-Potsdam APT (Strassmeier *et al.* 1997).

Time series photometry spanning decades were gathered for a sample of active stars. We refer to Oláh *et al.* (2009) for details of the stellar data, but note that the data used here have been updated with data collected since then with the Vienna-Potsdam APT (Strassmeier *et al.* 1997). The data used are measured in V color, and differential magnitudes to the comparison stars are given in the plots.

A reconstruction of the annual mean spectral irradiance of the Sun from 1610 to the present, from 130 nm to 10μm for every year by Shapiro *et al.* (2011), allows us to study the Sun as if it was measured with broadband BVR filters generally used for stellar observations. The electronic tables of Shapiro *et al.* (2011) contain flux values of the Sun for the wavelength ranges of the above mentioned filters. Though a flux value like this is not identical to a flux which is gathered through a broadband filter with its special flux transmission, but is good enough approximation. The fluxes were transformed to magnitudes in the usual way ($-2.5\times\log(\text{flux})$), which in this case has a zero-point error, but that does not affect the variability we are searching for in the data.

To study the cycle lengths and their evolution the method Short Term Fourier Transform (hereafter STFT) is used here. In short, the datasets are multiplied by a Gaussian window around a certain epoch, then this part is Fourier-transformed, so the frequency spectrum around that epoch is generated. Then the epoch is shifted, so a series of Fourier spectra are obtained, and a two-dimensional power spectrum can be plotted. This process needs equidistant data in time, which is valid for the solar measurements and reconstructions. For the stars the situation is different due to gaps caused by weather conditions on daily basis, and longer interruptions because of the visibility of the stars from different geographical locations. The stellar data thus need pre-processing: the rotational signal is removed from the data and after that a moving averaging and spline smoothing is applied to obtain a continuous dataset. The details and tests of the method and the pre-processing of the stellar data can be found in the paper of Kolláth & Oláh (2009).

2.1. *The problem of the lengths of the datasets*

The longest datasets describing magnetic cycles of the closest active star, the Sun, include sunspot numbers, sunspot area, flux data determined from imaging observations, irradiance measurements, and various proxies describing the long-term changes. All these data are widely analyzed and a great number of attempts have been made to predict the activity of our star, due to its direct effect on life of Earth, see Kolláth *et al.* (2011). It has also been known since long, that the length of the Schwabe cycle is not constant, but

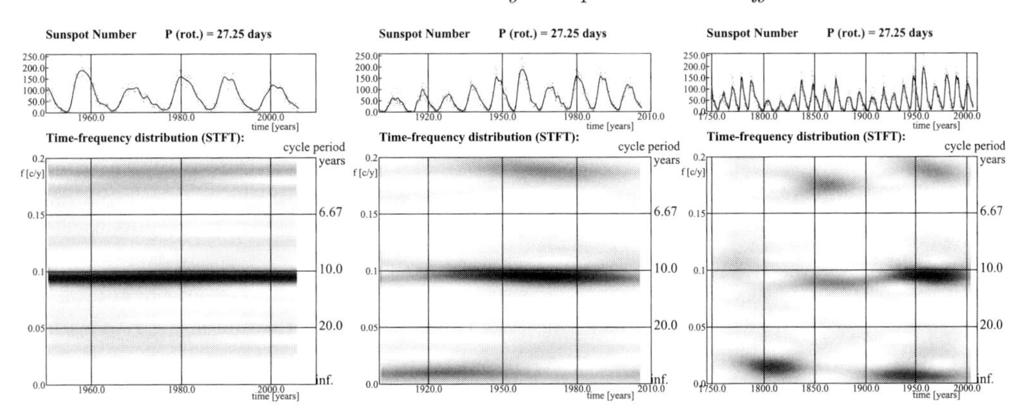

Figure 1. Time-frequency diagrams of solar activity using data for the last 50 years (left), 100 years (middle) and 250 years (right). The importance of the length of the dataset is demonstrated; from 50 years of data (which is longer than most of the stellar datasets) only the existence of the Schwabe cycle is certain. The full dataset shows an evolution of the length of the Schwabe cycle and reveals the century-long Gleissberg cycle and its time-variability.

varies between 8 and 15 years. The importance of the length of the datasets is demonstrated in Fig. 1. Time-frequency diagrams are plotted using the sunspot numbers for the last 50, 100 and 250 years, respectively. From the shortest dataset only the existence of an about 11-year long cycle is found. From the 100-year dataset an evolution in the Schwabe cycle appears, together with the sign of a changing long-term cycle. When more than 250 years of sunspot data are analyzed, a complicated variability of the Schwabe cycle and more details of the century-long Gleissberg cycle are recovered.

3. Stellar cycles

A comprehensive study of stellar cycles recovered from V-band photometry was published by Oláh *et al.* (2009) for 14 active stars, including a re-analysis of CaII data for further 6 stars. To illustrate the variety of stellar cycles, here we present time-frequency analysis of photometry for 3 stars of very different nature from the sample of Oláh *et al.* (2009), with the datasets now extended to 2011 (Fig. 2). The left panels are devoted to

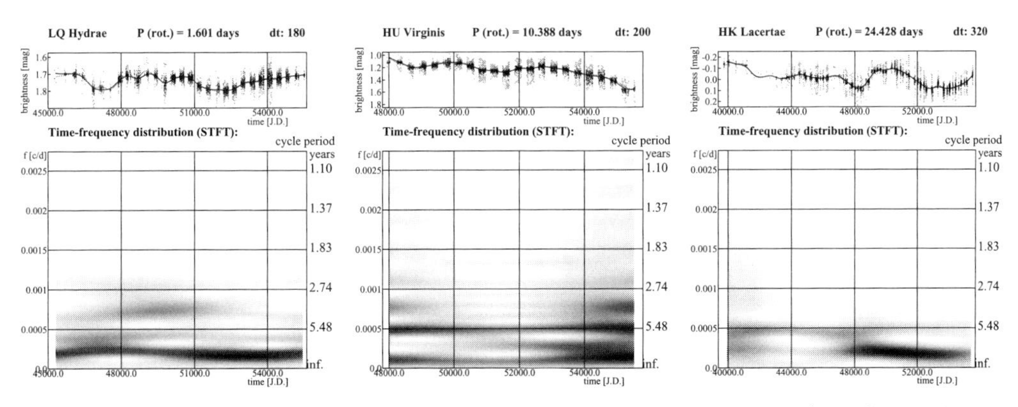

Figure 2. Long-term datasets (upper) and their time-frequency diagrams (lower) for LQ Hya (left), HU Vir (middle) and HK Lac (right). Multiple cycles showing changes both in time and amplitude are seen in all cases. In the upper panels the points represent original (red-grey) and processed (blue-black) observations, while the continuous line results from spline interpolations. For a colour figure, see the online version of the paper.

LQ Hya; this is a fast rotating, single star of K2V type, with a rotational period of 1.601 days and with an effective temperature of about 5100 K. The time-frequency diagram of the observations, spanning now for 28 years, clearly shows at least two, changing cycles, one on the timescale of 2-3 years and another, smoothly changing between 7 and 12 years. The middle panels of Fig. 2 display the results for HU Vir, the primary component of a synchronized binary star with an orbital and rotational period of 10.388 days. Its spectral type is K1IV-III, i.e., a subgiant with an effective temperature of 5000 K. The 22 years long dataset, now five years longer than in Oláh *et al.* (2009), confirms the previously known cycle of around 5.7 years and suggests a higher amplitude of the longer term variability. The rightmost panels of Fig. 2 show the results for the K0III giant HK Lac, which is the primary of a single-lined, synchronized binary with an orbital period of 24.428 days, and an effective temperature of 4750 K. We have the longest dataset available for this star, at present 53 years, which consists of both photometry and early photographic measurements. Cycles showing continuous changes are found around 5-6, and 10-13 years. A general decreasing trend is also evident from the data.

4. Solar cycles

The cycles of the Sun as conventionally treated are not fully comparable to stellar results, since the amplitudes of variability are wavelength-dependent, i.e. they are markedly different when measured from X-rays through visual-infrared to radio wavelengths. For the Sun, long series of data exist at various wavelengths, whereas long-term stellar measurements are made in a narrow vawelength range, in $BVRI$ bandpasses, i.e., between about 400-900 nm. The reconstructed spectral irradiance of Shapiro *et al.* (2011), from which typical fluxes in bandpasses similar to those used for stars can be extracted, allows a direct comparison between solar and stellar cycles both in time and in amplitude. Time-frequency diagrams for the Sun in near-B and -V Johnson passbands, are plotted in Fig. 3.

In the left panel of Fig. 3 the temporal evolution of the solar cycles are shown as if the measurements would have been made in B band. The amplitude of the Gleissberg cycle is naturally stronger during the Maunder (when the Schwabe cycle nearly disappeared) and Dalton (when the Schwabe cycle had very low amplitude) minima, whereas during the last 170 years the Schwabe cycle dominates. The right panel of Fig. 3 shows that in the V-band there is a domination of the Gleissberg cycle in the entire time-frequency diagram covering 400 years with a very weak signal of the Schwabe cycle in the latter two centuries. Note that the amplitudes are very small, the total change (min-max) is about 0.018 magnitude in B and less, than 0.005 magnitude in V. In case of ground-based

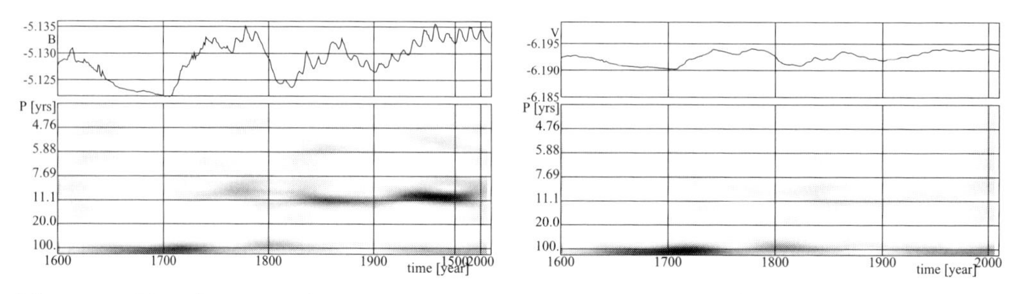

Figure 3. Time-frequency diagrams of the Sun, close to Johnson B (left) and V (right) bandpasses, similar to those traditionally used in stellar photometry. The relative amplitudes of the different cycles are well seen.

observations of stars such changes are equal to the measurement errors. For the more active stars in Fig. 2 one can see that the stellar cycle amplitudes can be of several tenths of a magnitude in the V band. From a stellar point of view, if the Sun was measured through the V filter for decades, its brightness would seem to be constant, suitable to be used as a comparison star.

The existence of approximate BVR data for the Sun allows another comparison to other active stars. As the area of the spots and faculae in view are changing due to stellar rotation and cycle phases, the stars also change their colors and, as inferred from this, their average surface temperatures. Spot-dominated stars are redder when fainter, while faculae-dominated stars are bluer when more activity (including spots) are present. Examples for both are found e.g., in Messina (2008), who identified e.g., V775 Her as spot-dominated and UX Ari as faculae-dominated. This latter type is a rare example of an active star being bluer when fainter.

Two color index vs. brightness diagrams of the Sun on Fig. 4 are based on the spectral reconstruction of Shapiro *et al.* (2011). The magnitudes have zero-point offsets. Therefore, we plotted the differences from the mean values of the actual data since we are looking for systematic changes only. The figure shows that the Sun belongs to the spot-dominated type of stars, although its variability is very small. Preminger *et al.* (2011), from full-disk images of the Sun through filters similar to the Strömgren b,y filters also concluded that the Sun is spot-dominated in the visible continuum. Note, that in Fig. 4 as well, as in all similar diagrams of stars, data are from different phases of rotations and (multiple) cycles. However, all these flux modulations have the same origin, i.e., they are manifestations of the magnetic activity of stars, on different timescales and with different amplitudes.

The dominance of spots or faculae originate from the variable contribution of these features. The color index variation reflects temperature changes as well; when e.g., the facular area decreases faster than the spotted one the star becomes fainter (decreased relative contribution from the brighter component) and redder (less hotter area), as measured in the visible. In the high energy bandpasses the Sun looks like other stars with comparable parameters - see Judge and Thompson (2012) and references therein. However, measuring the Sun in the conventional bandpasses used for the stars is not easy and this was not systematically done, except for those data presented by Preminger *et al.* (2011). The latter authors call attention to the problem of accurate determination of the quiet-Sun flux: at high level of the solar activity low contrast bright features modify the flux of the quiet regions but this remains unnoticed. We would like to emphasize that

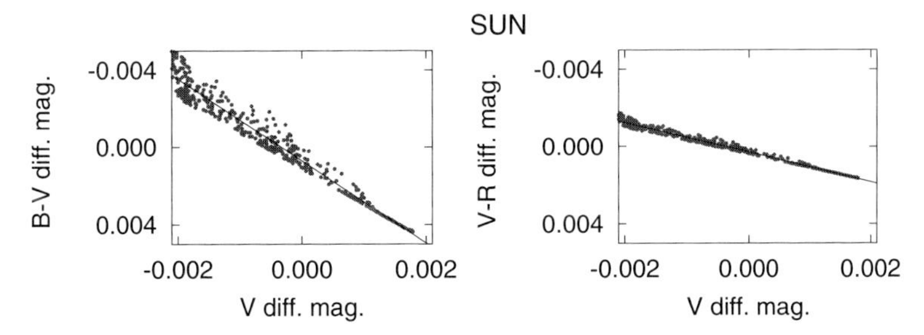

Figure 4. Solar brightness vs. temperature: relations between $B - V$ and $V - R$ color indices vs. V relative magnitudes of the Sun based on Shapiro and co-workers' (Shapiro *et al.* 2011) spectral reconstruction. The fluxes are transformed to magnitudes, and the deviation from the mean values are plotted. As the majority of active stars, the Sun seems to be redder when fainter, i.e. spot-dominated, although the amplitudes are small.

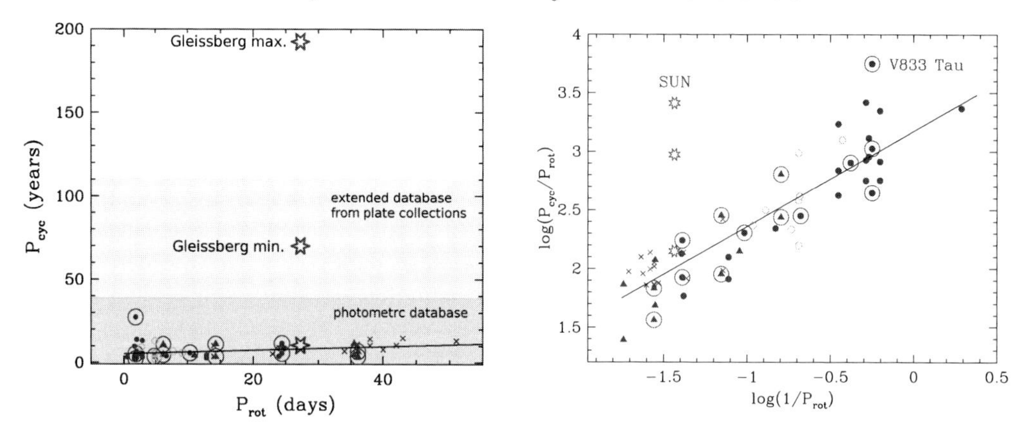

Figure 5. Stellar and solar cycles in the function of the rotational (or orbital) periods. Left: the dark grey area shows the length of a typical stellar dataset from photometry, and the light grey area at 110 years is the maximum length of stellar data extended using photographic data. Right: log-log representation of the cycle lengths normalised with the rotational periods as function of the inverse of the rotational periods. Solid symbols represent results from Oláh *et al.* (2009), circled ones among those are representative of different rotational rates. Open circles and crosses are from the literature (Frick *et al.* 2004; Messina & Guinan 2002).

when one compares the behaviors of the Sun and the stars, the same kinds of measurement should be used.

5. Discussion and Conclusions

Although it is interesting to study the multiple cycles and their changes of active stars and the Sun on their own, it is also important to see if there is any relation to physical parameters. The most evident measure at hand is the rotation of the objects. As discussed before, the lengths of the datasets are of crucial importance for identifying cycles. On the left panel of Fig. 5 we plot a rotation period vs. cycle length diagram for a sample of stars with the best cycle determinations currently exist. The two shaded areas show the lengths of the average photometric dataset (dark grey), and the maximum data record extended by archival photographic measurements (light grey). The moral of this figure is that the Gleissberg cycle of the Sun could at most be guessed from a dataset of the same length than the longest one available for stars. Additionally, the Sun, measured from a greater distance, with filters that are routinely used for stars, should have a light variation of at least 10-30 times higher than in fact has, for a detectable cycle!

An increase of the cycle lengths is seen towards longer rotational periods (Fig. 5, left panel). Baliunas *et al.* (1996) already found a loose relation between rotation and cycle length expressed in a way we plot the right panel of Fig. 5, i.e., logarithm of the cycle length normalized to the rotation period plotted as a function of the logarithm of the inverse rotation period. The slope of such a relation is related to the dynamo number (see Baliunas *et al.* (1996) for details). The relation is weak but it is the same for solar type, slowly rotating stars, dwarfs and giants, single stars and binary components. To quantify such a relation needs time - studying activity cycles of stars is only be possible with the help of routinely working automated telescopes, which have been successfully operating since decades and should continue to do so.

Acknowledgements

Thanks are due to Z. Kolláth and A. Shapiro for useful discussions, and for the referee, M. Güdel for useful remarks. Support from the organizers of the meeting is much appreaciated. This work was supported by the Hungarian Research grant OTKA K-081421.

References

Baliunas, S., Nesme-Ribes, E., Sokoloff, D., & Soon, W. 1996 *ApJ* 460, 848
Brown, B. P., Miesch, M. S., Browning, M. K. *et al.* 2011 *ApJ* 731, 69B
Frick, P., Soon, W., Popova, E., & Baliunas, S. 2004 *New Astron.* 9, 599
Judge, P. G. & Thompson, M. J. 2012 *this proceedings*
Kolláth, Z. & Oláh, K. 2009, *A&A* 501, 695
Kolláth, Z., Oláh, K., & van Driel-Gesztelyi, L. 2011, *this proceedings*
Messina, S. & Guinan, E. F. 2008, *A&A* 393, 255
Messina, S. 2008, *A&A* 480, 495
Oláh, K., Kolláth, Z., Granzer, T. *et al.* 2009, *A&A* 501, 703
Preminger, D., Chapman, G. A., & Cookson, A. M. 2011, *ApJ* 739, L45
Shapiro, A. I., Schmutz, W., Rozanov, E. *et al.* 2011, *A&A* 529A, 67S
Strassmeier, K. G., Boyd, L. J., Epand, D. H., & Granzer, Th. 1997, *PASP* 109, 697
Wilson, O. C. 1968, *ApJ* 153, 221

Comparative Magnetic Minima:
Characterizing quiet times in the Sun and Stars
Proceedings IAU Symposium No. 286, 2011
C. H. Mandrini & D. F. Webb, eds.

doi:10.1017/S174392131200498X

The solar wind in time

Jeffrey L. Linsky[1], **Brian E. Wood**[2] **and Seth Redfield**[3]

[1]JILA, University of Colorado and NIST,
Boulder, Colorado 80309-0440, USA
email: jlinsky@jilau1.colorado.edu

[2]Naval Research Laboratory, Space Science Division,
Washington, DC 20375, USA
email: brian.wood@nrl.navy.mil

[3]Astronomy Department, Wesleyan University, Middletown, CT 06459, USA
email: sredfield@wesleyan.edu

Abstract. We describe our method for measuring mass loss rates of F–M main sequence stars with high-resolution Lyman-α line profiles. Our diagnostic is the extra absorption on the blue side the interstellar hydrogen absorption produced by neutral hydrogen gas in the hydrogen walls of stars. For stars with low X-ray fluxes, the correlation of observed mass loss rate with X-ray surface flux and age predicts the solar wind mass flux between 700 Myr and the present.

Keywords. ISM:lines and bands, ISM:magnetic fields, solar wind, stars:winds, Sun: UV radiation, ultraviolet:ISM

---

1. Why mass loss rates are needed for the Sun and cool dwarf stars

Ultraviolet spectroscopy is an important tool for measuring mass loss rates ($\dot{M}$) of massive stars, late-type giants, and premain sequence stars when $\dot{M} > 10^{-10}$ $M_\odot yr^{-1}$. By comparison, the solar mass loss rate measured by space experiments is far smaller, about 2×10^{-14} $M_\odot yr^{-1}$. Until 10 years ago, there were no measured values of $\dot{M}$ for main sequence stars cooler than about spectral type A because there were no spectral or other diagnostics capable of measuring $\dot{M}$ far smaller than 10^{-10} $M_\odot yr^{-1}$.

While mass loss rates similar to the Sun do not significantly reduce a star's mass and thus nuclear evolution during its lifetime (see below), accurate measurements of $\dot{M}$ are needed for many reasons: (1) the torque exerted by even weak stellar winds reduces the stellar rotation rate and thus the magnetic dynamo and stellar activity, (2) knowledge of $\dot{M}(t)$ for the young Sun is important for understanding the evolution of the atmospheres of solar system planets like Mars that have lost their magnetic fields, and (3) the discovery of exopolanets close to their host stars highlights the need to predict the properties of stellar winds that impact the atmospheres of these exoplanets. Since we cannot measure $\dot{M}_\odot$ back in time, it is critical to find a reliable diagnostic to measure $\dot{M}$ for solar-mass stars with a range of ages and coronal properties in order to scale to the young Sun. Unfortunately, theoretical estimates of $\dot{M}$ are controversial, an empirical relationship between $\dot{M}$ and X-ray flux was not previously known, and no space probe will enter an astrosphere to directly measure $\dot{M}$ for a very long time.

2. The Lyman-α diagnostic for stellar mass loss rates

Baranov & Malama (1995) showed that charge exchange processes between solar wind protons and neutral hydrogen atoms in the local interstellar medium (ISM) produce a population of heated, decelerated hydrogen atoms concentrated in the interstellar upwind

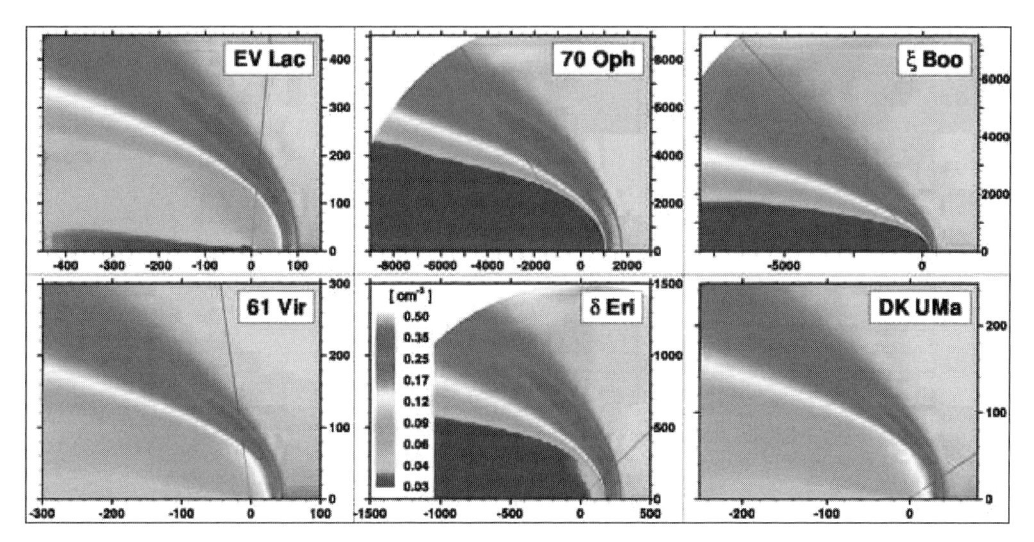

Figure 1. Theoretical models including charge exchange processes of the neutral hydrogen distribution around six stars. Units for the x and y axes are Astronomical Units, and the stars are moving to the right compared to the local ISM. High densities (red color, in the online version of this paper) indicate the location of hydrogen walls. Solid lines show the line of sight to the Earth (Wood *et al.* 2005).

direction, the so-called "hydrogen wall". These calculations and those in subsequent models of Zank *et al.* (1996), Izmodenov *et al.* (1999), and Müller *et al.* (2001) allowed Wood *et al.* (2001) to compute $\dot{M}$ for α Cen. They showed that along the line of sight to the star, the stellar Lyman-α line contains three absorption components due to interstellar hydrogen, the solar hydrogen wall, and the stellar hydrogen wall. The decelerated hydrogen atoms in the solar hydrogen wall produce absorption on the red side of interstellar absorption. Hydrogen atoms in stellar hydrogen walls are decelerated relative to the ISM as seen by the star and thus blue-shifted relative to the interstellar absorption as seen from the Earth.

Figure 1 shows maps of neutral hydrogen density computed for 6 stars studied by Wood *et al.* (2005). These models show the presence of hydrogen walls around the stars. Figure 2 shows the Lyman-α profiles for these stars observed with the *Space Telescope Imaging Spectrograph* on the *Hubble Space Telescope*. Also included are the interstellar hydrogen absorption profiles predicted from the observed deuterium Lyman-α interstellar absorption and additional astrosphere absorption based on models with a range of $\dot{M}$ values relative to the solar value. These models assume that the stellar wind speed is 400 km s^{-1}, but $\dot{M}$ scales roughly linearly with the wind speed.

The effect of increasing stellar mass loss rate is to increase the absorption on the blue side of the interstellar absorption. The theoretical Lyman-α profiles, which have been convolved with the instrumental profile, show good agreement with the observed profiles at a best fit value of $\dot{M}$ with about a factor of 2 uncertainty.

Figure 3 (left) shows the values of $\dot{M}$ obtained for 12 stars studied by Wood *et al.* (2005) and the Sun compared with the X-ray surface fluxes (F_X), obtained at different times than the $\dot{M}$ measurements. Since X-ray emission is variable, there should be some scatter in any correlation of such non-simulataneous data. Nevertheless, there appears to be a correlation of $\dot{M}$ per cm^2 of stellar surface with F_X over two orders of magnitude in $\dot{M}$ with the Sun consistent with the trend. The power-law fit to these data is

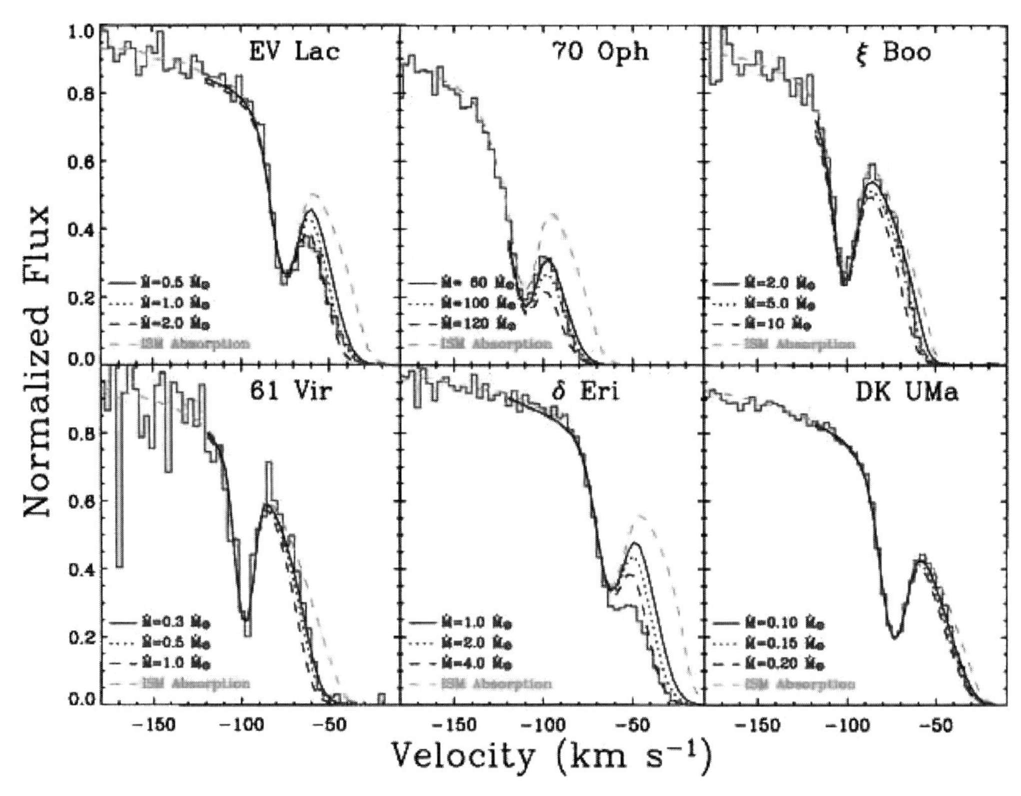

Figure 2. Observed Lyman-α profiles (red histograms) are compared to models with no mass loss (green (in the online version) dashed line marked ISM absorption) and models with different mass loss rates compared to the solar value (Wood *et al.* 2005). For a color figure, see the online version of the paper.

$$\dot{M} \propto F_X^{1.34 \pm 0.18}. \tag{2.1}$$

For $F_X > 8 \times 10^5$ erg cm^{-2} s^{-1}, the fit is no longer valid as all of the data points lie at lower values than predicted by an extrapolation of the fit. Why do the more active stars violate the fit shown by the less active stars?

Before we apply the $\dot{\mathrm{M}}$-F_X relation to estimating the properties of the solar wind in time, it is important to understand the limitations of this relation. We have already mentioned that the mass loss and X-ray data are not simulataneous and that we have assumed that all of these stars have outflow velocities of 400 km s^{-1} similar to the quiet Sun. Also, the relation is based on data for stars with a range of spectral types (G2 V to K5 V), and some are binaries. Two of the more active stars are M dwarfs, but one is a G8 V star (ξ Boo). Clearly $\dot{\mathrm{M}}$ values are needed for more stars to confirm the $\dot{\mathrm{M}}$-F_X relation for the less active stars and to determine whether the more active stars are inconsistent because they are more active or because they are M dwarfs. Nevertheless, eight G and K-type stars fit the relation within a factor of two despite the uncertainties, and we assume the relation to be valid as a working hypothesis. We have two approved *HST/STIS* programs to observe eight more stars and there may be other observed targets to fill in the diagram and test our conclusions.

Ayres (1997) found a relation between X-ray surface flux and age (t) for solar-type dwarf stars, $F_X \propto t^{-1.74 \pm 0.34}$. Combining this relation with the $\dot{\mathrm{M}}$-F_X relation leads to

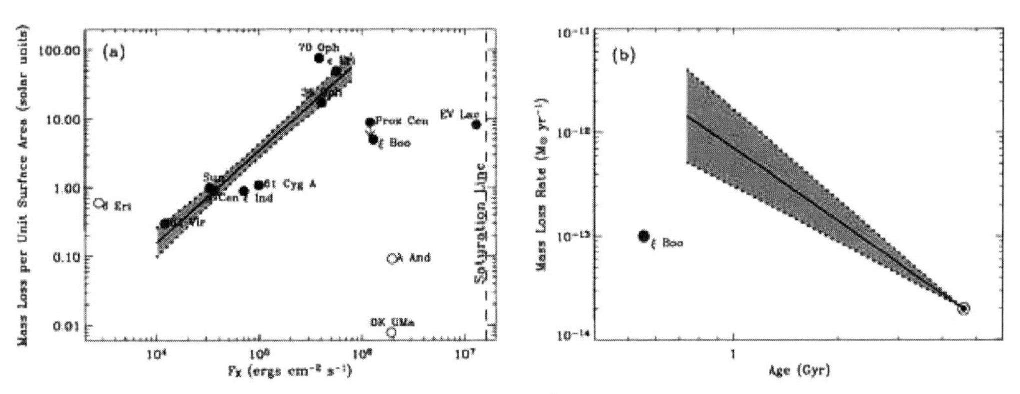

Figure 3. Left: Observed mass loss rates per cm^2 of stellar surface are compared to X-ray surface fluxes. Filed circles are for main sequence stars and unfilled circles are for subgiants and giants. The solid and dashed lines are for Eq. 2.1 and its uncertainty. Right: solar mass loss rate from 700 Myr to the present computed from Eq. 2.2 (Wood *et al.* 2005).

$$\dot{M} \propto t^{-2.33 \pm 0.55}. \tag{2.2}$$

This relation shown in the right hand side of Fig. 3 provides an estimate of $\dot{\mathrm{M}}_\odot$ back to an age of 700 Myr when F_X would have been larger than the upper limit of the $\dot{\mathrm{M}}$-F_X relation. This age is slightly larger than the age of the Hyades. The total mass loss for the Sun between 700 Myr and the present is 0.001 $M_\odot$ with an uncertainty of a factor of 10 (Wood *et al.* 2002). If $\dot{\mathrm{M}}$ remained at 1.5×10^{-12} ($\pm$ a factor of 3) $M_\odot$ yr^{-1} over the Sun's early life on the main sequence ($0 \leqslant t \leqslant 700$ Myr), then the additional mass loss would be 0.001 $M_\odot$ ($\pm$ a factor of 3). This amount of mass loss would not have modified the Sun's evolution, but the amount of mass loss before 700 Myr is uncertain.

Can theoretical models for stars with solar-type magnetic field topologies explain the $\dot{\mathrm{M}}$-F_X relation? Cranmer & Saar (2011) find that Alfvén wave-driven winds predict a factor of 30 increase in $\dot{\mathrm{M}}$ for a solar mass star as the rotation period decreases from 25 to six days, consistent with the increase in $\dot{\mathrm{M}}$ predicted by the $\dot{\mathrm{M}}$-F_X relation.

Why is the $\dot{\mathrm{M}}$-F_X relation not valid for the more active stars? A likely explanation is that the magnetic topology of rapidly rotating stars which have strong X-ray emission is different from that of slowly rotating stars. Zeeman-Dopler imaging studies show that rapidly rotating dwarf stars with mass greater than 0.5 $\mathrm{M}_\odot$ have strong torroidal fields, and low mass stars often have strong poloidal fields (Donati & Lanstreet 2009). The maximum F_X for which Eq. 2.1 is valid corresponds to a rotation period of six days for a solar-mass star. Zeeman-Doppler images of such stars show large polar spots surrounded by torroidal magnetic fields. Theoretical models for stars with different magnetic topologies are needed to understand why the $\dot{\mathrm{M}}$-F_X relation is not valid for these stars. We note that the solar wind mass flux is nearly the same in coronal holes and equatorial regions despite the difference in magnetic field topology (Phillips 1995).

References

Ayres, T. R. 1997, *J. Geophys. Res.*, 102, 1641

Baranov, V. B. & Malama, Y. G. 1995, *J. Geophys. Res.*, 100, 14755

Cranmer, S. R. & Saar, S. H. 2011, *ApJ*, 741, 54

Donati, J.-F. & Landstreet, J. D. 2009, *ARAA*, 47, 333

Izmodenov, V. V., Lallement, R., & Malama, Y. G. 1999, *A&A*, 342, L13

Müller, H.-R., Zank, G. P., & Wood, B. E. 2001, *ApJ*, 551, 495
Phillips, J. L. 1995, *Science*, 268, 1030
Wood, B. E., Linsky, J. L., Müller, H.-R., & Zank, G. P. 2001, *ApJ* Letters, 547, L49
Wood, B. E., Müller, H.-R., Zank, G. P., & Linsky, J. L. 2005, *ApJ*, 574, 412
Wood, B. E., Müller, H.-R., Zank, G. P., Linsky, J. L., & Redfield, S. 2005, *ApJ* Letters, 628, L143
Zank, G. P., Pauls, H. L., Williams, L. L., & Hall, D. T. 1996, *J. Geophys. Res.*, 101, 21,639

Discussion

DIBYENDU NANDI: We know that there is a strong power law relationship between magnetic flux and X-ray flux. Given that you find a connection between X-ray flux and stellar wind, perhaps this relationship can be utilized to infer the magnetic history of the Sun?

JEFFREY LINSKY: Yes, we will look for a correlation of the unsigned average magnetic flux with the mass loss rate and then the magnetic field history of the Sun.

JANET LUHMANN: Have you thought about the possible effect of a solar maximum type solar wind that is structured and bursty, rather than steady, laminar and uniform? Is your thinking affected by new IBEX results?

JEFFREY LINSKY: We find that the mass loss rate declines to a plateau when the X-ray surface flux becomes greater that 10^6 erg cm^{-2} s^{-1}, which corresponds to a stellar rotation rate of about 7 days for a solar mass star. At this rotation rate the stellar magnetic field is probably more dipole-like than for the Sun. We have not considered bursty and inhomogeneous flows.

We look forward to IBEX measurements of the speed of in-flowing helium atoms from the interstellar medium. This will tell us whether the heliosphere is located inside the Local Interstellar Cloud (LIC) or in a transition region between the LIC and another cloud. The answer to this question will determine the outer boundary condition of the solar wind.

Comparative magnetic minima:
Characterizing quiet times in the Sun and stars
Proceedings IAU Symposium No. 286, 2011
C. H. Mandrini & D. F. Webb, eds.

doi:10.1017/S1743921312004991

Stellar activity cycles in a model for magnetic flux generation and transport

Emre Işık

Department of Physics, Faculty of Science & Letters, Istanbul Kültür University
34156, Bakırköy, Istanbul, Turkey
email: e.isik@iku.edu.tr

Abstract. We present results from a model for magnetic flux generation and transport in cool stars and a qualitative comparison of models with observations. The method combines an $\alpha\Omega$-type dynamo at the base of the convection zone, buoyant rise of magnetic flux tubes, and a surface flux transport model. Based on a reference model for the Sun, numerical simulations were carried out for model convection zones of G- and K-type main sequence and subgiant stars. We investigate magnetic cycle properties for stars with different rotation periods, convection zone depths, and dynamo strengths. For a Sun-like star with $P_{\rm rot}$=9 d, we find that a cyclic dynamo can underly an apparently non-cyclic, 'flat' surface activity, as observed in some stars. For a subgiant K1 star with $P_{\rm rot}$=2.8 d the long-term activity variations resemble the multi-periodic cycles observed in V711 Tau, owing to high-latitude flux emergence, weak transport effects and stochastic processes of flux emergence.

Keywords. stars: activity, stars: interiors, stars: magnetic fields, (magnetohydrodynamics:) MHD, Sun: interior, Sun: magnetic fields

---

1. Introduction

Magnetically active stars are important to better understand the limits and the behaviour of the solar dynamo, in addition to how it used to operate during the early stages of solar evolution. The level of activity increases with the rotation rate and the fractional depth of the convection zone, up to a saturation level for ultra-fast rotators and fully convective stars. The distribution and coverage of stellar magnetic regions differ significantly from the solar patterns as the Rossby number (ratio of the rotation period to convective turn-over time) decreases (Strassmeier 2009). Temporal properties of magnetic activity are also of interest when comparing the solar cycle with stellar cycles. Multi-periodic stellar cycles have also been reported (Oláh *et al.* 2009). We present a model for the generation and transport of stellar magnetic fields (Işık, Schmitt & Schüssler 2007, 2011) and discuss possible magnetic flux transport mechanisms in active stars, in comparison with observations of stellar activity cycles.

2. The model

We have recently developed a three-part model for solar and stellar magnetic flux generation and transport (Işık, Schmitt & Schüssler 2007, 2011). In the first part, a one-dimensional mean-field $\alpha\Omega$ dynamo with a saturation mechanism related to buoyant flux removal is set up at the position of the overshoot region at the base of the convection zone (Schmitt & Schüssler 1989). For faster rotators, the strength of the α-effect is assumed to scale with the rotation rate. This is similar to the assumption that the differential rotation rate scales with the activity level and thus with the rotation rate, as suggested recently by Saar (2011), based on observations of a homogeneous sample of cool stars. In the

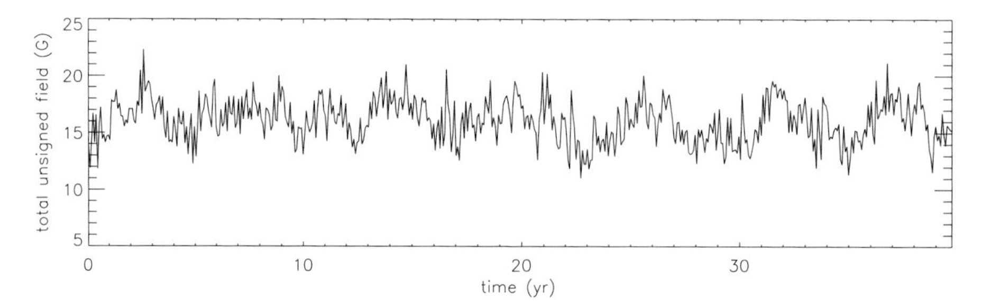

Figure 1. Variation of total unsigned magnetic flux density for the Sun-like model with $P_{\rm rot}$=9 d. The right axis shows the corresponding field strength.

second part, time-latitude distribution of the dynamo-generated toroidal field determines the probability distribution that a flux tube would become unstable at a given latitude and time. The tubes are assumed to have properties given by the criteria for magnetic buoyancy instability at a given latitude. The emergence latitudes and tilt angles of BMRs are determined by numerical simulations of flux tube rise through the convection zone. These results are used as input for the flux transport model at the surface (the third part), including effects of differential rotation, meridional flow, horizontal and radial diffusion. The number of emerging BMRs (per activity cycle) is scaled up with the rotation rate and their areas are determined randomly, following a probability distribution set by the Sun-like area distribution, $N(A) \propto A^{-2}$. The details of the model construction are described by Işık, Schmitt & Schüssler (2011).

3. Sun-like stars: disappearance of cycles

For the Sun-like models, the cycle period decreases with the rotation rate, owing to stronger α-effect. For $P_{\rm rot} \gtrsim 10$ d, the polar fields become stronger with increasing rotation rate, owing to increasing frequency of flux emergence and tilt angles, until another effect starts to set the stage: the increasing dynamo strength (the α-effect), thus the increasing cycle frequency. For $P_{\rm rot} \lesssim 10$ d polar regions undergo field reversals with increasingly high frequencies, so that the peak polar fields do not reach the field strengths expected from the emerging magnetic flux. For most of these cases the magnetic flux variations show clear cycles.

For $P_{\rm rot} = 9$ d, the level of total magnetic flux is considerably higher than the solar values, but the underlying periodic dynamo cycle is 'obscured' by the combined effects of rise and surface transport of magnetic flux (Fig. 1). The reason for the 'invisible' stellar cycle is that during 'magnetic minima' the magnetic flux at high latitudes become comparable to that of low latitudes during activity maxima. This intermediate rotator case represents a moderately active, but non-cyclic (or weakly cyclic) configuration (in fact, there is a weak cycle with $\sim$ 6-yr period). The model indicates a possibility that such stars may not be in a Maunder minimum state and are still observed as non-cyclic. Observational evidence on the existence of such stars already supports this possibility (Hall & Lockwood 2004).

4. K stars: from regular to fluctuating cycles

Işık, Schmitt & Schüssler (2011) have applied the Sun-like model to K-type stars with different radii. The first case is a K0-type main-sequence star, rotating 13 times faster than the Sun. Similar to the Sun-like star with the same rotation rate, large tilt

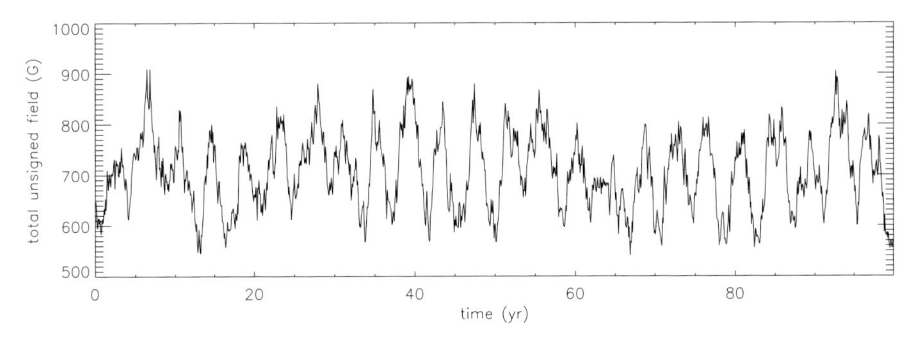

Figure 2. Variation of the surface-integrated unsigned magnetic field strength for the K1IV-type fast rotator with $P_{\rm rot}$=2.8 d.

angles lead to formation of activity belts, but in this case as shifted towards the poles, because of an additional geometric effect. Buoyant flux tubes experience a substantial Coriolis acceleration in the rotating frame and rise almost parallel to the rotation axis in fast rotators. The larger the fractional depth of the convection zone, the higher the emergence latitudes, because the latitude difference between the initial and final latitudes of the rising flux tube is larger.

The second model simulates a K1IV-type fast rotator, with the mass, radius, and the rotational period of the active subgiant component of the close binary system V711 Tau (=HR 1099). The deep-seated dynamo pattern and the surface emergence are highly separated in latitude, owing to rapid rotation and convection zone geometry, as explained above. The variation of the total unsigned magnetic field for 100 years is shown in Fig. 2. The random component in the model, ie., in latitudes and areas of BMRs at emergence, become significant for the polar regions, where the timescales for differential rotation, meridional flow, and the turbulent magnetic diffusion are too long to have an effect on the rapidly changing magnetic flux pattern. The ineffective surface transport and the highly confined, highly frequent BMR emergence thus lead to cycle-to-cycle (and long-term) fluctuations in total magnetic flux. Preliminary time-frequency analysis (Oláh 2011) indicates an interesting correspondence between the models and the long-term photometric observations of V711 Tau, as shown in Fig. 3. The dynamo cycle period in the model is about 4 years and the short-term cycle of V711 Tau is about 5 years. A longer-term 'cycle' shifts its period between 10-20 years in the model, whereas for V711 Tau a similar shifting cycle is observed between 9-18 years. A longer-term simulation for the K1IV model (Fig. 2) shows similar periods. However, it should be noted that (a) the average brightness on the one hand is compared with the total magnetic flux on the other, and (b) the relative amplitudes of the short- and long-term cycles are not similar in both cases.

5. Discussion

It is clear that the relationships between stellar dynamo mechanisms, emergence patterns, and surface transport processes are non-trivial. The effect of rapid rotation on rising magnetic flux tubes in stellar convection zones can lead to large departures of observable patterns from the deep-seated fields. Surface flux transport processes cause additional complications and indicate various possibilities when interpreting the observed activity patterns. Apart from several unknowns about active cool stars, we conjecture that three parameters are likely to have important effects in shaping stellar cycles: the cycle frequency, the range of latitudes of BMR emergence, and the tilt angles.

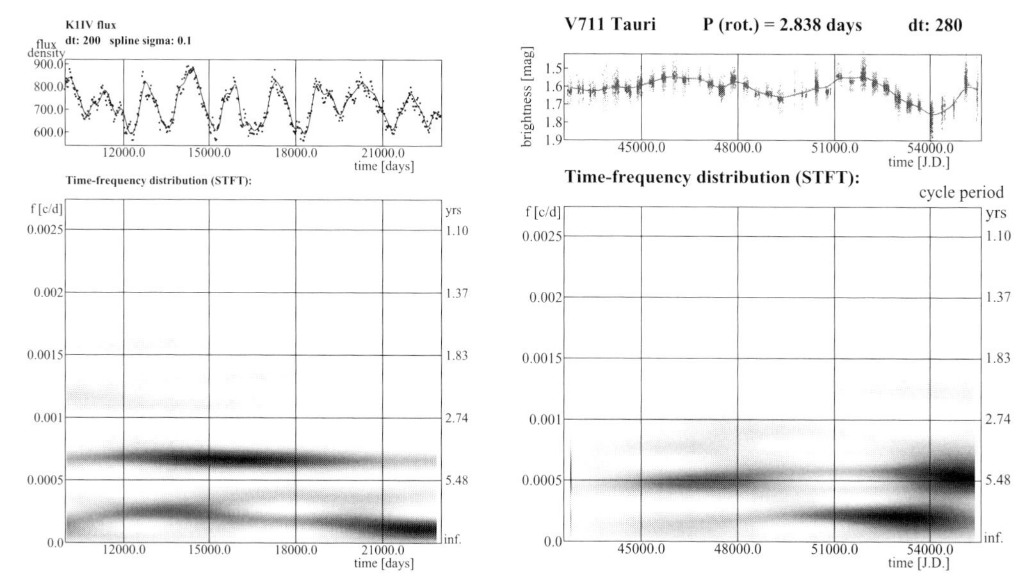

Figure 3. Time-frequency maps of the theoretical model of K1IV star with $P_{\rm rot} = 2.8$ d (*left panel*, based on total unsigned magnetic flux) and V711 Tau (*right panel*, based on stellar brightness). For V711 Tau, the data are from Oláh *et al.* (2009) extended to 2011 by new measurements from the Vienna APT (Strassmeier *et al.* 1997). For a colour figure, see the online version of the paper.

Multiple cycles observed in active stars (Oláh *et al.* 2009), such as V711 Tau in Fig. 3 (right panel), do not necessarily owe their existence to periodic events taking place in the stellar interior. In the theoretical model, the short-term cycle ($\sim$ 4 yrs) is simply determined by the mono-periodic dynamo in the interior. The long-term cycle signal in Fig. 3 (left panel) is caused by the stochastic nature of flux emergence. Although there are several degrees of freedom in modelling magnetic fields in stellar interiors, parallel investigation of nonlinear flux-transport dynamo models and observational results will certainly be useful when estimating the relevant physical mechanisms.

Acknowledgement

The author is grateful to Katalin Oláh for the data, time-frequency analysis and useful discussions.

References

Hall, J. C. & Lockwood, G. W. 2004, *ApJ*, 614, 942
Işık, E., Schmitt, D., & Schüssler, M. 2007, *AN*, 328, 1111
Işık, E., Schmitt, D., & Schüssler, M. 2011, *A&A*, 528, A135
Oláh, K., Kolláth, Z., Granzer, T., Strassmeier, K. G., & Lanza, A. F. *et al.* 2009, *A&A*, 501, 703
Oláh, K. 2011, *private communication*
Saar, S. H. 2010, *The Physics of Sun and Star Spots, IAU Symposium 273*, p. 61
Schmitt, D. & Schüssler, M. 1989, *A&A*, 223, 343
Strassmeier, K. G., Boyd, L. J., Epand, D. H., & Granzer, T. 1997, *PASP*, 109, 697
Strassmeier, K. G. 2009, *A&AR*, 17, 251

Discussion

ARNAB CHOUDHURI: You have shown some results which are non-axisymmetric. How do you get them?

EMRE ISIK: The mean-field dynamo model in the tachocline is axisymmetric. However, in order to implement the two-dimensional surface flux transport, I determine longitudes of emerging bipoles randomly.

Comparative Magnetic Minima:
Characterizing quiet times in the Sun and Stars
Proceedings IAU Symposium No. 286, 2011
C. H. Mandrini & D. F. Webb, eds.

doi:10.1017/S1743921312005005

Magnetic activity of cool stars in the Hertzsprung-Russell diagram

J. H. M. M. Schmitt

Hamburger Sternwarte, 21029 Hamburg, Gojenbergsweg 112, Germany
email: jschmitt@hs.uni-hamburg.de

Abstract.
I review the X-ray emission from cool stars with outer convection zones in comparison to the Sun with a focus on the properties of low-activity stars. I present the recent results of long-term X-ray monitoring which demonstrate the existence of X-ray cycles on stars with known calcium cycles. The evidence of a minimum stellar X-ray flux is presented and arguments are put forward for the view that the Sun in its extended minimum between 2008 - 2009 behaved very much like a Maunder-minimum Sun.

Keywords. Sun: activity, stars: activity, stars: coronae, X-rays: stars

1. Introduction

The periodic change in the frequency of sunspot appearance on the solar surface was the first observational evidence of the magnetic activity cycle of the Sun. As is well known, the frequency of sunspot occurrence on the solar surface varies with a period of approximately 11 years. After its discovery by the German amateur astronomer G. Schwabe in 1843, systematic sunspot observations have been carried out until now, and from historical records the sunspot appearance frequency can be reconstructed back to the first decade of the 17th century, when Galileo and Schreiner made their first telescopic sunspot observations. A sunspot frequency reconstruction going back to 1610 is shown in Fig. 1, rendering the solar sunspot record probably the longest available astronomical "light" curve. Obviously, the data quality of these sunspot observations is very much non-uniform, I do draw attention to the period between $\approx$ 1645 to 1715, the so-called *Maunder minimum*, named after E.W. Maunder, who, contemporaneously with G. Spörer, first identified this period as one with low sun spot numbers. Long considered an observational bias due to poor data coverage, Eddy (1976) convincingly demonstrated that "this 70-year period was indeed a time when solar activity all but stopped".

Modern multi-wavelength observations have shown the solar sunspot cycle to manifest itself in almost all spectral bands, in particular, these modern observations have shown that the observed cycle signatures are strongest and most readily apparent in coronal emission features. The coronal energy loss occurs predominantly at EUV and X-ray wavelengths, and in the X-ray range the Sun and its cycle look entirely different when compared to the optical band. A complete coverage of an activity cycle of the Sun has now been obtained by both the YOHKOH and SOHO satellites. As an example I show (in Fig. 2) the changing coronal appearance of the Sun in the emission line of Fe XV at 284 Å as measured by SOHO, which has been monitoring the solar EUV output since 1996. An inspection of the SOHO images shows a very profound change in the appearance of the EUV and X-ray Sun between solar minimum and solar maximum. The precise amplitudes of the solar X-ray and EUV variability is subject to some debate, and does in fact depend on the exact wavelength range considered. At harder X-rays, as measured for example by the GOES satellites, one finds variability amplitudes of two orders of magnitudes and more, while at softer wavelengths the variability amplitudes become smaller. Unfortunately, at soft X-ray wavelengths, where the XMM-Newton and *Chandra* observatories register the X-ray emission from other solar-like stars, no long term solar monitoring is available (yet).

A ground based proxy indicator of the solar soft X-ray emission is the so-called green coronal line at 5303 Å, which is produced by ions of Fe XIV, i.e., from plasma at temperatures very close to those ions responsible for the emission seen at EUV wavelengths in Fig.2. Historically, the green corona line has been one of the "coronium" lines, whose correct identification with highly ionized iron (rather than "coronium") led to the surprising realization that the corona of the Sun is actually much hotter than its photosphere (see discussion by Edlén 1945). Emission from this line has been monitored since the 1930es (see the solar green line "light" curve in Fig. 3), and the green line emission shows variability amplitudes of the factor 8 - 10 between solar maximum and solar minimum.

In this context it is important to consider what the Sun would look like as a spatially unresolved star. While sunspots can be – more or less – easily detected on the solar surface in spatially resolved observations (see, however, Svalgaard 2012), their effects on the total solar irradiance are quite small. While the sunspot cycle is clearly very obvious in Fig. 1-3, in total solar irradiance (TSI) the sunspot cycle is also clear but less obvious and of much smaller amplitude at a level of 0.1%. Successful TSI measurements only became available since the late 1970es, when spaceborne radiometers were first put into operation. Since then almost three complete solar cycles have been covered; an example of the total solar irradiance vs. time is shown in Fig. 4. Somewhat surprisingly, the total solar irradiance turned out to be larger at sunspot maximum when more sunspots appear on the solar surface (at least on average) than at solar minimum. It is, however, also clear that the TSI variability is much larger when the Sun is at cycle maximum. In fact, the passage of large sunspots across the solar surface does lead to noticeable dips in the range of a few tenth of percent at maximum, with the largest so far observed depression caused by sun spots was on the order of 0.3% observed in September 2003; at those times the TSI of the maximum Sun is actually below the TSI of the minimum Sun. In ground based observations of stars such variations in output could in principle be detected but only with some difficulties.

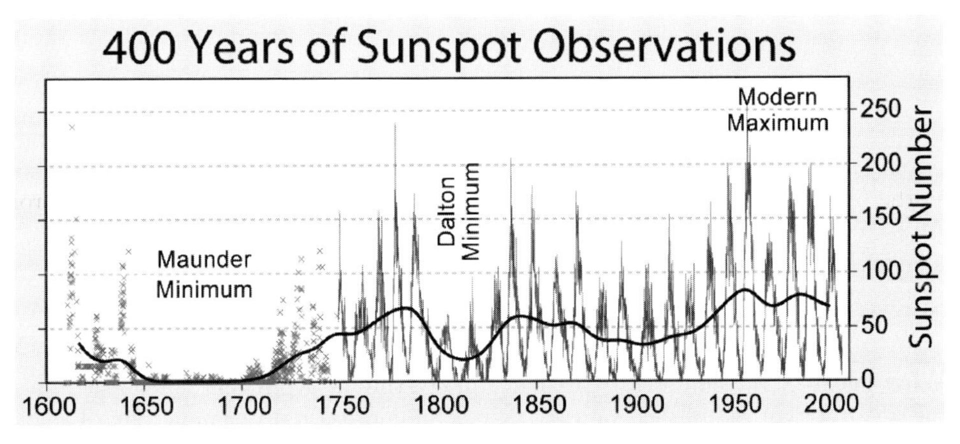

Figure 1. Sunspot number as function of time for the last four centuries (image created by Robert A. Rohde / Global Warming Art, taken from "http://www.globalwarmingart.com/wiki/File:SunspotNumbers.png". For a colour figure, see the online version of this paper.

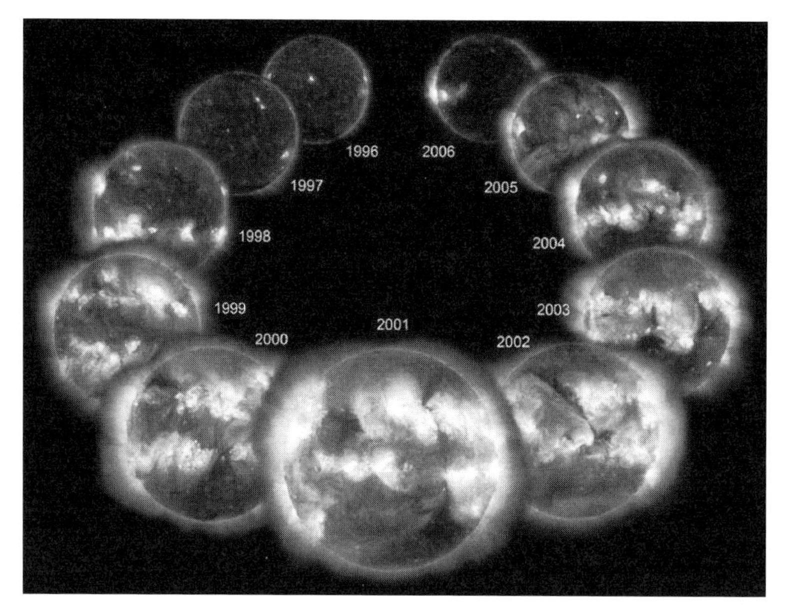

Figure 2. The changing face of the Sun as apparent in its EUV emission from Fe XV at 284 Å ; Image credit: SOHO - EIT Consortium, ESA, NASA. For a colour figure, see the online version of this paper.

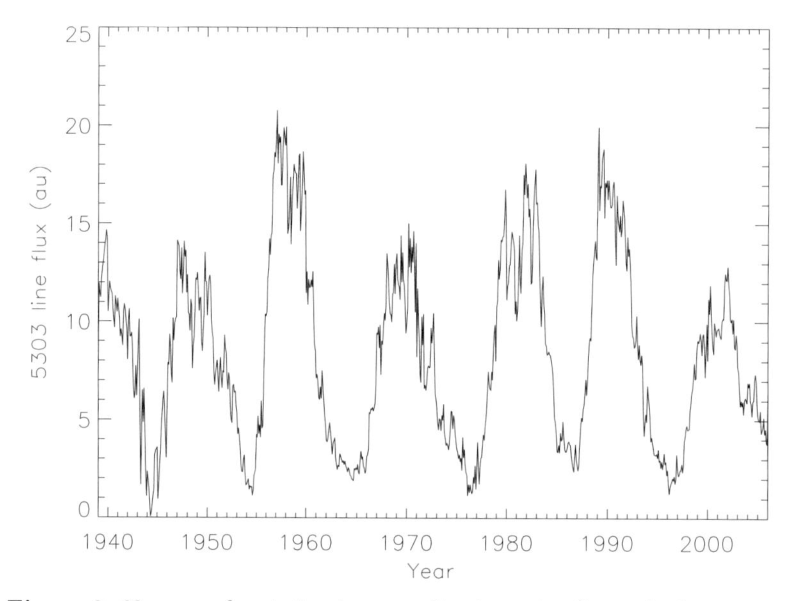

Figure 3. 60 years of variation in green line intensity (in a.u.); data source "ngdc.noaa.gov/stp/solar-data/solar-corona/index/Lomnicky/"

What makes X-ray emission particularly interesting as a physical diagnostic appears to be a correlation established between X-ray luminosity and magnetic flux. The close association of coronal X-ray emission with magnetic fields has been known, ever since spatially resolved X-ray images of the Sun had become available. By studying quiet and active Sun features as well as active stars, Pevtsov *et al.* (2003) were able to establish

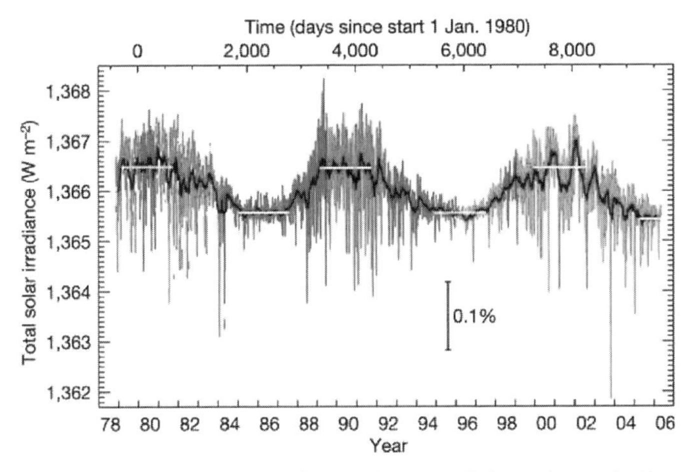

Figure 4. Total solar irradiance vs. time (taken from Foukal *et al.* 2006). For a colour figure, see the online version of this paper.

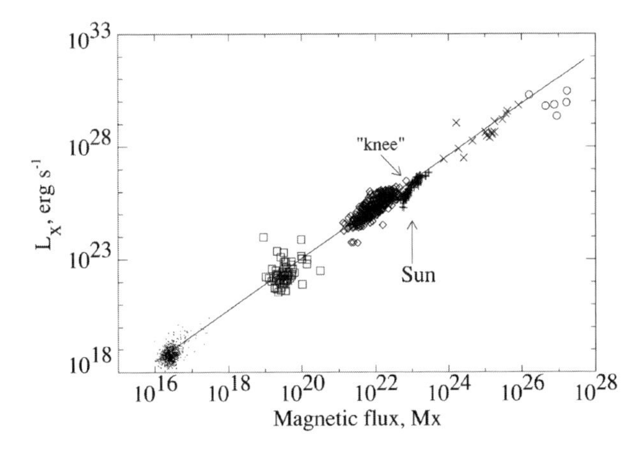

Figure 5. Correlation between X-ray luminosity and magnetic flux derived by Pevtsov *et al.* (2003)

an almost linear relationship between observed X-ray luminosity and magnetic flux that extends over 12 orders of magnitude, reproduced in Fig. 5. Thus, empirically X-ray emission appears to be a good proxy for magnetic fields or more precisely magnetic flux, a quantity that is very difficult to measure otherwise. Therefore the X-ray emission from other stars is used as a proxy for magnetic activity on those stars, although in many cases such magnetic fields have not been or cannot be measured directly.

2. X-ray emission from solar-like stars

Observations carried out with the *Einstein Observatory* between 1979 - 1983 were the first to demonstrate the wide-spread occurrence of X-ray emission from main sequence stars with outer convection zones (Vaiana *et al.* 1981). In particular, it appeared that that onset of outer convection as predicted by stellar structure theory is reflected by a rather vigorous onset of X-ray activity. This issue was studied by Schmitt *et al.* (1985), who found a large increase in detection rate when going from stars with $B - V \sim 0.2$ to stars with $B - V \sim 0.5$. Later the large stellar samples available from ROSAT observations allowed a precise delineation of this so-called “onset of convection”. Schröder & Schmitt

(2007) studied the detection statistics among bright stars of spectral type B and A (i.e., stars without outer convection zones) to stars of spectral type F5 (i.e., stars with convection zones; cf., Fig.6), using the complete data set from the ROSAT all-sky survey. While there is obviously a "bottom" of about 15% of X-ray detections among A-type stars, the precise nature of which is subject of some debate, there is also a sharp increase in X-ray detections going from spectral type A7 to F2, suggesting that stars with outer convection zones find it easier to produce X-ray emitting coronae. This finding is of course in line with the idea that magnetic dynamos operate in those convection zones and it is the magnetic fields generated by these dynamos which eventually cause the observed X-ray emission (see Fig.5).

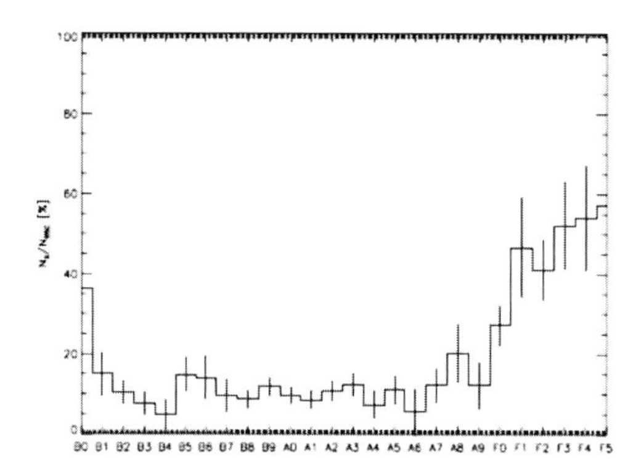

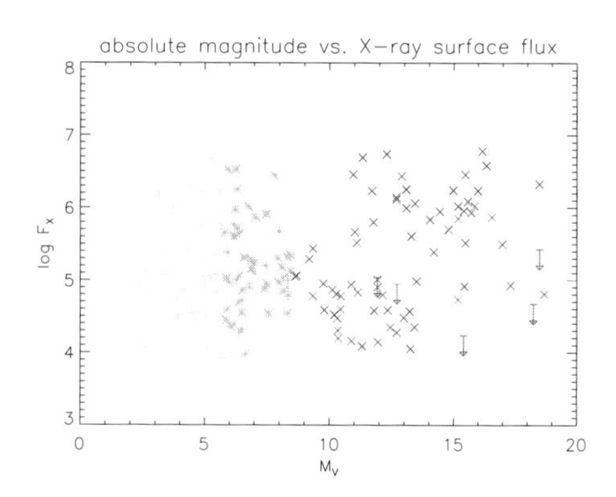

Figure 6. Left panel: detection rate of bright stars in the ROSAT all-sky survey vs. spectral type (taken from Schröder & Schmitt 2007). Right panel: X-ray surface flux of main sequence stars vs. absolute magnitude M_V (taken from Schmitt 1997). For a colour figure, see the online version of this paper.

What was left open by the *Einstein Observatory* observations was the question whether the X-ray emission for stars with outer convection was ubiquitous, i.e., whether X-ray emission is really found for all stars or whether there might possibly exist X-ray dark cool stars. To answer this question a far more systematic and, more importantly, complete study of X-ray emission from cool stars was required. To this end Schmitt (1997) investigated complete volume-limited samples of solar-like F and G dwarfs within 13 pc, K dwarfs within 10 pc and M dwarfs within 7 pc using data from both the ROSAT all-sky survey and ROSAT pointed observations. These samples are statistically meaningful and truly complete in the sense that all stars known within this volume were observed with sufficient sensitivity to detect solar-like X-ray emission levels (and even well below), while the coverage of the low-luminosity end of the X-ray luminosity distribution function had been rather limited with *Einstein Observatory* data. In Fig. 6 the results of Schmitt (1997) are plotted in the form of mean X-ray surface flux F_X vs. absolute magnitude M_V for a complete sample of nearby stars. As is clear from Fig. 6, none of the stars with colors $M_V < 5$ could be detected, but X-ray emission from **all** F-type stars was detected. Specifically, the detection rate for F-type stars is 100%, the detection rate for G-stars is more than 85 %, with all the upper limits resulting from stars for which only survey data (and no more sensitive pointing data) were available. The conclusion from these studies must then be that coronal formation is indeed a *universal phenomenon* on solar-like stars. A corona containing hot plasma is always formed at the interface between a turbulent outer convection zone and space. Truly X-ray dark solar-like (main sequence) stars do not exist at least within the immediate solar environment and must be very rare if they exist at all.

Another salient feature in Fig.6 is the existence of a rather well defined lower envelope $F_{X,lim}$ to the observed mean X-ray surface flux distribution. The apparent cutoff at surface fluxes of $F_{X,lim} \approx 10^4$ erg/cm^2/sec (in the ROSAT pass band of 0.1 - 2.4 keV) is not a question of lack of sensitivity since the non-detected A-type stars do indeed have upper limits well below $F_{X,lim}$. Because of the completeness of the samples both for the F and G stars as well as the K and M stars, one can state that among cool dwarfs stars with X-ray surface fluxes below $F_{X,lim}$ do not exist (at least in the considered volumes of space). It is interesting to compare this lower limit to the X-ray surface flux from solar emission features. Obviously, the solar minimum surface flux is at least similar if not identical to the minimum surface flux, and Fig. 6 suggests that the stars observed at their minimum flux levels should be interpreted as stars without any active regions present on their surfaces; a further discussion of this issue is given in Section 5.

3. X-ray cycles on other stars

Having established the universal character of stellar coronae around cool stars, the question arises to what extent are stellar coronae similar to the solar corona and to what extent are they different. In particular, the question arises whether other stars follow the same variability pattern in soft X-rays as the Sun. Earlier studies of X-ray emission from young stars in the Hyades cluster (Stern *et al.* 2003) showed a relatively good correlation between the X-ray fluxes measured at two different epochs (i.e., the *Einstein Observatory* data taken between 1979-1983 and the ROSAT survey observations taken in 1990-1991) separated by almost a decade, suggesting that the influence of cyclic variability must be relatively modest for this group of stars. Also, a comparison of the X-ray fluxes of stars, whose cyclic properties were known from the Mount Wilson HK program, showed that stars with calcium cycles tend to show less X-ray activity than stars with irregular calcium variability (Hempelmann *et al.* 1996).

With ROSAT the first systematic longterm monitoring program for time variable stellar X-ray emission was initiated in the 1990s (cf., Hempelmann *et al.* 2003). While the Sun is essentially monitored continuously by a small armada of spacecraft, such extensive X-ray monitoring is not feasible in a stellar context. First of all, for essentially all X-ray satellites there are viewing restrictions for any given direction, implying that a given source (star) can only be observed within some viewing interval typically lasting 4 to 8 weeks. And second, monitoring observations of stars are in competition with all other X-ray sources studied by any given X-ray observatory, and devoting megaseconds to such programs is simply not warranted. Typical stellar X-ray monitoring programs therefore consist of individual pointings at a given star, usually separated by 6 months. With ROSAT, the monitoring of the visual binary 61 Cyg A and B was begun (data available from 1992 to 1998), later, with XMM-Newton the monitoring of 61 Cyg A/B has been continued (since 2002) and three more stars, α Cen A and B and HD 80809, have been added to the monitoring list. Both 61 Cyg A and HD 80809 are stars with known calcium cycles, and both systems have been monitored in Ca H and K for a significant amount of time, while, unfortunately, no systematic longterm calcium monitoring is available for the α Cen system. The results of these monitoring observations have been published by Hempelmann *et al.* (2006), Robrade *et al.* (2007), Favata *et al.* (2004), Favata *et al.* (2008) and Robrade *et al.* (2012); as an example I show in Fig. 6 the XMM-Newton lightcurve of 61 Cyg A, covering the time frame from 2002 (when the 61 Cyg system became visible to XMM-Newton) until now. According to Baliunas *et al.* (1995) the calcium cycle of this star is 7 years, and from Fig.7 it is obvious and also in some ways expected that the X-ray emission from 61 Cyg A follows the Ca emission very well. A similar dependence between calcium and X-ray cycles is found for HD 80809, which is a little surprising since its X-ray luminosity is much larger than that of 61 Cyg A. Similarly, an X-ray cycle may exist for the case of α Cen A, however, since no calcium data is available for that star, it is too early to draw any firm conclusions on a possible X-ray cycle.

In this context it is surprising that variations as shown in Fig.7 can be measured at all. Naively one would assume that short term variability in the form of flares and rotational modulation would actually destroy the usually much smaller variability due to cyclic emission. Also it has to be kept in mind that the X-ray observations are by necessity snapshot observations, taken on a much shorter timescale than the rotational variation, while calcium data are often averaged over a whole rotational period. In actual fact, the individual XMM-Newton pointings of 61 Cyg A and B last typically 10 ksec. An examination of the whole data set available for 61 Cyg shows that, indeed, flares do occur, fortunately on a time scale shorter than 10 ksec and therefore "quiescent" emission periods with the corresponding emission levels can always be established, and it is these quiescent emission intervals that show cyclic variability (Robrade *et al.* 2012).

4. The extended solar minimum between 2008 - 2009

The last solar minimum in the years 2008 and 2009 received a lot of attention, both from the science community and the general public, because of the prolonged observed absence of sunspots. This minimum reminded of the Maunder-Minimum between 1650 - 1715, which is also characterized by an almost complete the absence of sunspots (cf., Fig. 1). Since the Maunder minimum period coincides with the time of the "little ice age" (cf., Eddy 1976), there is considerable interest in the question what the properties of the solar chromosphere and corona were at that time.

In the framework of the Mount Wilson Ca HK monitoring program the Sun has been regularly observed and its S-index was determined just like the S-indices of all other program stars (cf., Baliunas *et al.* (1995)). A very clear cycle is found in these data, in fact, the cycle of the Sun appears to be one of the best cycles (classified as "excellent" by Baliunas *et al.* (1995)) in the whole list of cycles presented in the paper by Baliunas *et al.* (1995). With our Hamburg Robotic Telescope (HRT, Hempelmann *et al.* 2005) we observed solar calcium spectra in the time frame between August 2008 to July 2009. Both sky spectra as well as Moon spectra were used and disk-averaged S-indices derived in the usual way. In Fig. 7 I display the resulting S-index "light" curve, taken from Mittag (2010). Interestingly but not surprisingly, the obtained S-values are consistently

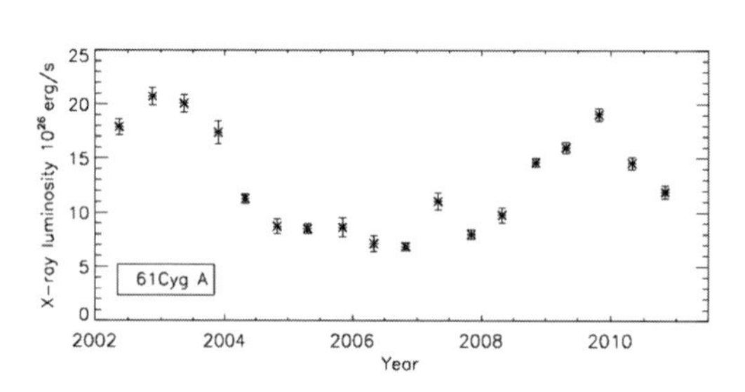

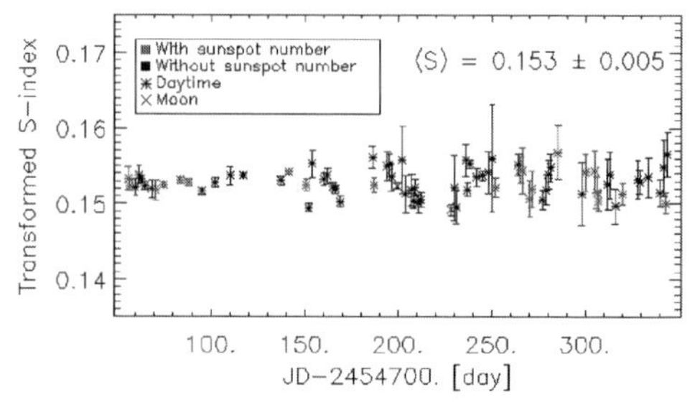

Figure 7. Left panel: XMM-Newton light curve of 61 Cyg A (taken from Robrade *et al.* (2012). Right panel: solar S-index variation in 2008-2009 (taken from Mittage 2010). For a colour figure, see the online version of this paper.

low, at times the S-index drops even to below S = 0.15 and the average S-index value (S = 0.153) in the observed period is below the minimum observed by Baliunas *et al.* (1995), i.e, S = 0.16; note the recorded S-indices were taken from both day-time spectra as well as lunar spectra and have been corrected for the effects of scattering. An inspection of contemporaneously obtained solar Ca K images also shows that plage regions were largely absent from the Sun during that period, and there are days without any clear signature of plage emission at all. We, therefore, conclude that not only was the sun spot number very low, but also the solar S-index was below the values measured between 1950 to 1990 and presented by Baliunas *et al.* (1995).

Is this observed low state of the Sun then already the state of the Sun during the Maunder minimum? There has been considerable debate in the literature on the definition of Maunder-Minimum stars and a recipe on how to find such objects. I do not want to delve into that discussion (see Saar 2012), I simply remark that, on the one hand, claims were published that the fraction of Maunder-Minimum stars, for example in the Mount Wilson sample, is quite substantial (Baliunas & Jastrow 1990), while, on the other hand, it was argued by other authors (cf., Wright 2004) that Maunder-Minimum stars in the Mount Wilson sample might simply not exist and have in fact been mistaken for subgiants. To clarify the issue and settle the position of the Sun within the stars, we used the catalog of Ca HK measurements published by Duncan *et al.* (1991), selected those stars with precise parallax information in order to classify the stars into main sequence stars, subgiants and giants, and study their distribution in an S-index vs. M_V diagram (cf., Fig.8); a far more extended discussion of this will be presented by Mittag *et al.* (2012). Fig.8 demonstrates that, indeed, subgiants and giants can have significantly smaller S-indices than their main sequence brothers, and therefore subgiants and giants, erroneously interpreted as main sequence objects, will result in make-believe Maunder minimum stars as pointed out by Wright (2004). However, as also apparent in Fig.8 and extensively discussed by Mittag *et al.* (2012), there is still a very well defined lower envelope in S-index, which empirically defines the lowest observed chromospheric activity state of a main sequence star. For solar-like stars with $B-V$ 0.64 this lower envelope to the observed S-index distribution is around S_{min} 0.145, and it is therefore clear that during its extended minimum between 2008 - 2009 the Sun came – in terms of S-index – very close to this empirically observed bottom of the S-index distribution for main sequence stars. In other words, the activity of the Sun as observed in Ca HK was essentially as low as solar-like stars at the effective temperature of the Sun can ever get, and therefore it is suggestive to interpret the minimum Sun between 2008 - 2009 as a close-up example of a Maunder minimum star.

5. A minimum X-ray flux for the Sun and the stars?

While long-term solar monitoring at soft X-ray wavelengths is not available yet, significant efforts have been undertaken in establishing solar reference spectra over a wide spectral range in order to facilitate a variety of studies of the Earth's climate and how it might be affected by the solar output in different energy bands. Woods *et al.* (2009) published Solar Irradiance Reference Spectra (SIRS) for the so-called Whole Heliospheric Interval (WHI) in Carrington Rotation 2068 covering the time span between March 20 to April 16 2008. In that time frame solar activity was already very low and I specifically use the data set from April 10 - April 16, where the Sun's activity was already very much depressed with rather low sun spot numbers and very low 10.7cm radio fluxes and therefore should - as pointed out by Woods *et al.* (2009) – represent the solar minimum conditions very well. For reference purposes I plot the solar surface flux in cgs units (per bin) for the WHI quiescent spectrum in Fig. 8.

I then proceed to compute integrated solar X-ray surface fluxes in various pass bands. Since the various X-ray missions tend to use their own specific band passes, I compute the X-ray fluxes in pass bands ranging from some low energy threshold (LET in keV) up to 2.4 keV; note the 'official' ROSAT band pass is 0.1 - 2.4 keV. The precise choice of the upper energy threshold is immaterial in this context since the solar input spectrum is very soft and contains very little flux at wavelengths below 10 Å (cf., Fig.9). As is clear from Fig.9, the choice of the lower energy threshold is crucial for the resulting surface flux value. The formal value of surface flux in the ROSAT 0.1 - 2.4 keV band for the WHI spectrum is 9062 erg/cm^2/s, which is very close to the estimated value of 10^4 erg/cm^2/s as the minimum X-ray surface flux for stars. Had the *Einstein Observatory* band pass of 0.2 - 4 keV band pass been chosen, a much smaller surface flux value of 3762 erg/cm^2/s would have resulted.

It is further useful to compare the resultant X-ray surface fluxes with previously published "old" values. Specifically I use the *Skylab* studies by Maxson & Vaiana (1977) as well as the YOHKOH SXT studies by Hara *et al.* (1994). Maxson & Vaiana (1977) produce (in their Fig. 10) a line of sight emission measure curve for coronal holes (CH) as well as an emission measure for large scale quiescent structure (LSS), while Hara *et al.* (1994) produce a two-temperature fit to their filter ratio data. Using CHIANTI, I calculate the equivalent soft X-ray fluxes in the 0.1-2.4 keV pass band, which are plotted in Fig.10; as is clear from Fig.10, both the *Skylab* and YOHKOH SXT data yield values lower than the SIRS, which is, however, not surprising since the YOHKOH SXT specifically refers to a coronal hole. On the other hand, all numbers agree reasonably well and do seem to point at the existence of a minimum X-ray flux also for the Sun. Converting the SIRS surface flux of 9062 erg/cm^2/s to a global X-ray luminosity, I find $L_X \approx 5.5 \times 10^{26}$ erg/sec, which I propose to use as a minimal solar X-ray luminosity.

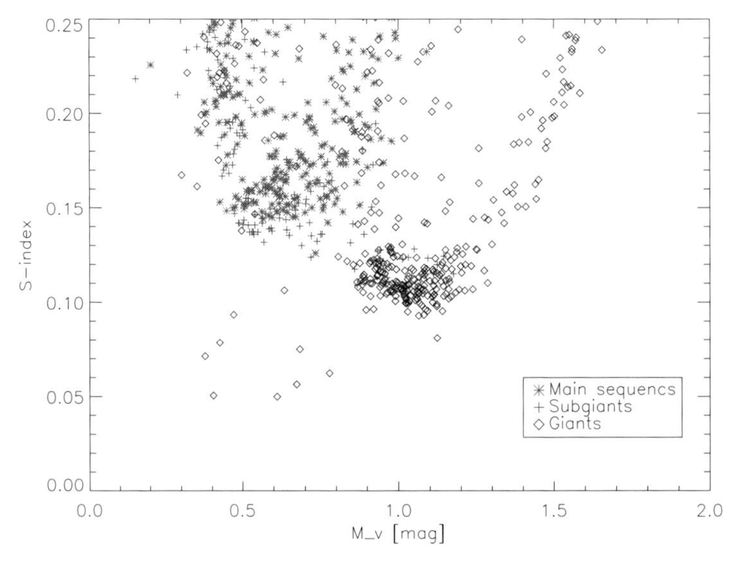

Figure 8. S-index (from Duncan *et al.* (1991)) for main sequence stars (asterisk), subgiants (plusses) and giants (diamonds), courtesy M. Mittag. For a colour figure, see the online version of this paper.

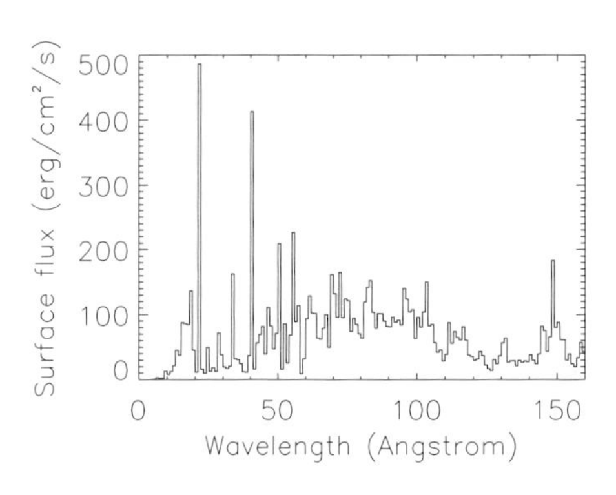

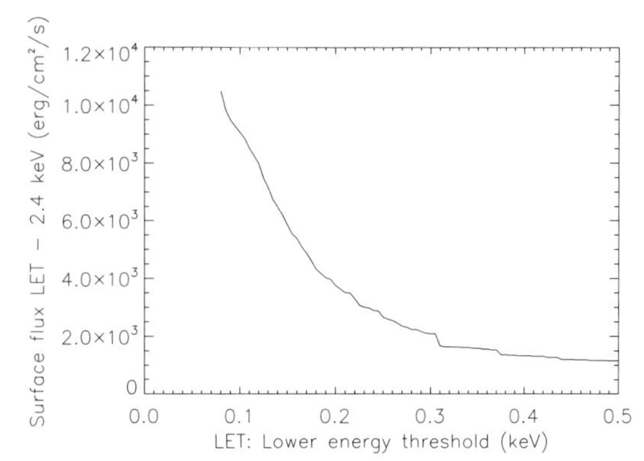

Figure 9. Left panel: solar surface flux (in cgs units per bin) vs. wavelength from SIRS. Right panel: X-ray surface flux as a function of passband; the ordinate is the lower energy threshold (see text for details).

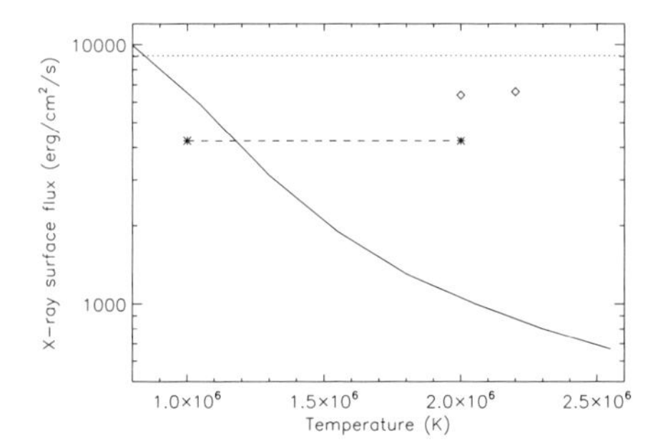

Figure 10. Comparison of calculated solar surface flux values: dotted line (SIRS), solid line represents the CH emission measure from Maxson & Vaiana (1977), diamonds correspond to LSS from Maxson & Vaiana (1977), and two-temperature CH from Hara *et al.* (1994).

The close agreement between the minimum solar flux and the lower envelope of the stellar flux distribution is certainly somewhat fortuitous since there is considerable uncertainty in converting the measured PSPC count rates to X-ray surface fluxes. This conversion depends, first, on the X-ray temperature (the precise value of which is not really known for these soft sources), and second, it depends on stellar size and distance. Therefore the scatter of data points near the lower flux threshold is not really surprising, rather, the real surprise is that the minimum flux corona as represented in the WHI spectrum fits so well to the minimum X-ray fluxes of stars. This agreement leads me then to argue that the X-ray corona of the Sun as observed in the extended last solar minimum is, first, compatible with the minimal X-ray coronae we observe on other stars and, second, that it is very likely to be compatible with the corona of the Sun during the time of the Maunder minimum. This view is consistent with the findings of Eddy (1976), who discusses reports on coronal sightings during total solar eclipses observed during the Maunder minimum period and finds that the solar wind was probably a fast wind and that the corona visible during the eclipse was by comparison small and unstructured, but still present.

References

Baliunas, S. & Jastrow, R. 1990, *Nature*, 348, 520
Baliunas, S. L., Donahue, R. A., Soon, W. H., *et al.* 1995, *Ap. J.*, 438, 269

Duncan, D. K., Vaughan, A. H., Wilson, O. C., *et al.* 1991, *Ap. J. Supp.*, 76, 383
Eddy, J. A. 1976, *Science*, 192, 118
Edlén, B. 1945, *MNRAS*, 105, 323
Favata, F., Micela, G., Baliunas, S. L. *et al.* 2004, *Astron. Ap.*, 418, L13
Favata, F., Micela, G., Orlando, S. *et al.* 2008, *Astron. Ap.*, 490, 1121
Foukal, P., Fröhlich, C., Spruit, H., & Wigley, T. M. L. 2006, *Nature*, 443, 161
Hara, H., Tsuneta, S., Acton, L. W. *et al.* 1994, *PASJ*, 46, 493
Hempelmann, A., Schmitt, J. H. M. M., & Stępień, K. 1996, *Astron. Ap.*, 305, 284
Hempelmann, A., Schmitt, J. H. M. M., Baliunas, S. L., & Donahue, R. A. 2003, *Astron. Ap.*, 406, L39
Hempelmann, A., Gonzalez Perez, J. N., Schmitt, J. H. M. M., & Hagen, H. J. 2005, 13th Cambridge Workshop on Cool Stars, Stellar Systems and the Sun, 560, 643
Hempelmann, A., Robrade, J., Schmitt, J. H. M. M., *et al.* 2006, *Astron. Ap.*, 460, 261
Maxson, C. W. & Vaiana, G. S. 1977, *Ap. J.*, 215, 919
Mittag M., 2010, PhD Thesis available in electronic form at "http://www.hs.uni-hamburg.de/DE/Ins/HRT/hrtpublication.html"
Mittag, M. 2012, in preparation
Pevtsov, A. A., Fisher, G. H., Acton, L. W., *et al.* 2003, *Ap. J.*, 598, 1387
Robrade, J., Schmitt, J. H. M. M., & Hempelmann, A. 2007, *Memorie della Societá Astronomica Italiana*, 78, 311
Robrade, J. 2012, in preparation
Saar, S. 2012, these proceedings
Schmitt, J. H. M. M., Golub, L., Harnden, F. R., Jr., *et al.* 1985, *Ap. J.*, 290, 307
Schmitt, J. H. M. M. 1997, *Astron. Ap.*, 318, 215
Schröder, C. & Schmitt, J. H. M. M. 2007, *Astron. Ap.*, 475, 677
Stern, R. A., Alexander, D., & Acton, L. W. 2003, The Future of Cool-Star Astrophysics: 12th Cambridge Workshop on Cool Stars, Stellar Systems, and the Sun, 12, 906
Svalgaard, L. 2012, these proceedings
Vaiana, G. S., Cassinelli, J. P., Fabbiano, G. *et al.* 1981, *Ap. J.*, 245, 163
Woods, T. N., Chamberlin, P. C., Harder, J. W. *et al.* 2009, *Geophys.Res.Letters J.*, 360, 1101
Wright, J. T. 2004, *A.J.*, 128, 1273

Discussion

Mark Giampapa: Images of SOHO for the Sun show bright maximum and nothing at the minimum. Can you explain why the peak to peak excursion for other stars are much smaller than for the Sun?

Jürgen Schmitt: The measured peak-to-peak variations over a cycle depend sensitively on the spectral range that is used. This also applies to solar data, for example the GOES hard X-ray fluxes. In the stellar case one is dominated by cooler plasma which is ensuing smaller variations.

Jeffrey Linsky: With the extended VLA we observe many radio stars with different emission mechanism, gyro-resonance or coherent emission, not thermal. Can you comment on that?

Jürgen Schmitt: I realize that the sensitivity of radio observations has been improved. It would of course be extremely valuable to have a stellar analog to the 10.7 cm radio flux monitoring that provides an excellent diagnosis for the solar cycle.

Comparative Magnetic Minima:
Characterizing quiet times in the Sun and stars
Proceedings IAU Symposium No. 286, 2011
C. H. Mandrini & D. F. Webb, eds.

doi:10.1017/S1743921312005017

Semi-empirical modelling of stellar magnetic activity

Adriana Valio†

CRAAM, Mackenzie University, Sao Paulo, Brazil
email: avalio@craam.mackenzie.br

Abstract. Since Galileo, for four hundred years, dark spots have been observed systematically on the surface of the Sun. The monitoring of the sunspot number has shown that their number varies periodically every 11 years. This is the well-known solar activity cycle that is caused by the periodic changes of the magnetic field of the Sun. Not only do spots vary in number on a timescale of a decade, but the total luminosity and other signatures of activity such as flares and coronal mass ejections also increase and decrease with the 11-year cycle. Still unexplained to the present date are periods of decades with almost an absence of activity, where the best known example is the Maunder Minimum. Other stars also exhibit signs of cyclic activity, however the level of activity is usually thousand times higher than the solar one. Obviously, this is due to the difficulty of observing activity at the solar level on most stars. Presently, a method has been developed to detect and study individual solar like spots on the surface of planet-harbouring stars. As the planet eclipses dark patches on the surface of the star, a detectable signature can be observed in the light curve of the star during the transit. The study of a different variety of stars allows for a better understanding of magnetic cycles and the evolution of stars.

Keywords. stars: activity, stars: spots, stars: rotation

---

1. Introduction

Sunspots are regions of strong magnetic fields on the surface of the Sun, of the order of hundreds to a few thousands Gauss. The magnetic field suppresses the overturning motion of the convective cells and thus hampers the flow of energy from the stellar interior outwards to the surface. Therefore, the region becomes cooler than the surrounding photosphere and thus appears darker.

Very likely, all cool stars with a convective envelope like the Sun, or even fully convective, will have spots on their surfaces. This is not a new idea, in 1667, the French astronomer Ismael Boulliau (1605-1694) introduced the concept of a starspot to explain the periodic light variability of *Omicron Ceti*, which turned out to be a Mira variable star, but nevertheless the concept was introduced. In the 1940s, starspots were observed by Kron in the eclipsing binary AR Lacertae as a significant light variability outside the eclipse (Kron 1947). Starspots are observable tracers of the internal dynamo activity, and their study provides a glimpse into the complex internal stellar magnetic field.

As dark spots cross the stellar disk due to rotation they modulate the total brightness of the star. The periodic variation generally follows the rotational period of the star. There exists an observational bias toward young stars that are fast rotators (need less observing time) and are also more active. Moreover, their light curve variability is stronger. This is the case of CoRoT-2, the second star discovered by the CoRoT satellite to harbour a planet (Alonso *et al.* 2008).

† aka Adriana V. R. Silva or Adriana Silva-Valio

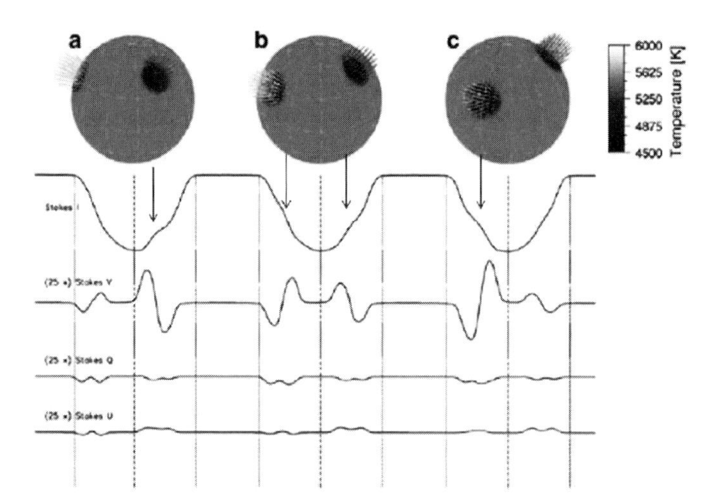

Figure 1. Zeeman-Doppler imaging: cool spots cause reduced line absorption at the wavelengths that are proportional to the Doppler velocity at their respective surface location.

The goal here is to spatially resolve the stellar disk and trace individual spots, just as Galileo, Scheiner, and others have done for the Sun 400 years ago. This type of study provides information on: i) stellar rotation and differential rotation; ii) spots location and physical characteristics such as size, temperature, and magnetic field intensity; and iii) magnetic cycles.

2. Starspot modelling

There are basically three methods for modelling starspots:

• Analysis of Doppler imaging from spectral lines.

• Light curve modulation: both considering a small number of spots (usually three) and the Maximum Entropy Method.

• Planetary transit mapping.

Doppler imaging is a tomographic technique based on high-resolution spectroscopy. It is applicable only to fast rotators, that is, $v \sin(i) \geqslant 20 - 25$ km/s. Cool spots on the stellar surface generate clearly detectable distortions in the profile of the absorption lines. These distortions move from blue to red wavelengths across the line profile as the spots rotate across the visible hemisphere due to the stellar rotation (see Figure 1). The span of migration within the line profile reveals the feature latitude. Thus, by detecting spots on different latitudes it is possible to determine the stellar differential rotation if any. Strassmeier has mapped the surface of 79 stars this way (Strassmeier 2009).

For slow rotators, such as solar analogues, the modelling of the irradiance variations is based on the information of the location and area of the active regions provided by the rotational modulation of the stellar flux. The assumption here is that the active regions do not change during their transit across the disk, thus the variability of the optical flux is due to the modulation of the visibility of the active regions. The quantities derived by the model are: i) the longitudinal distribution of the filling factor of active regions and ii) the variation of total area of active regions.

One of the caveats of this method is that the reconstruction of the stellar surface map is an ill-posed problem because of the non-uniqueness of the solution and its instability. Moreover, the rotational modulation of the flux is a function of the varia-

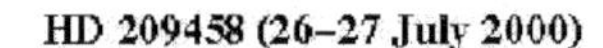

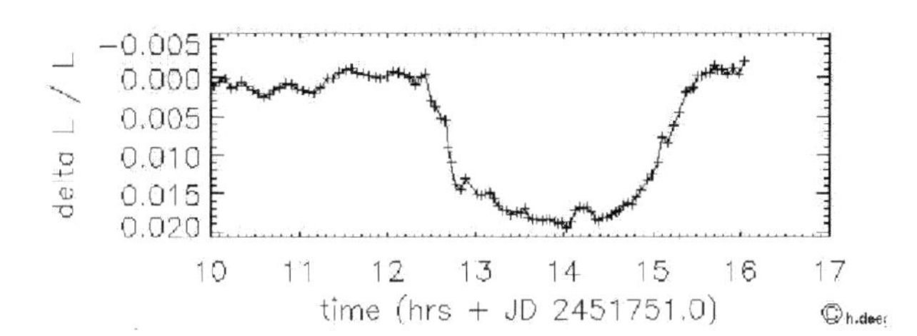

Figure 2. Example of the variation ("bump") caused by the occultation of a spot during a planetary transit.

tion of the surface brightness versus longitude, with no information on the latitude. Furthermore, the contrast of the surface brightness inhomogeneities must be assumed, that is the spot temperature cannot be derived from measurements in a single spectral band.

Earlier rotational modulation models were based on just two circular spots, where the coordinates and radii of the spots were adjusted to fit the light curve rotational modulation (Rodono *et al.* 1986). Later, a model with three active regions containing cool spots and bright faculae in a fixed proportion was applied by Lanza *et al.* (2003, 2004) to fit the total as well as the spectral solar irradiance variations. A total of 11 free parameters were needed to describe the model. However, very often the longitudinal distribution of the real active regions was too complex to be described with only three active regions, resulting in a poor agreement between the model and the observations.

The Maximum Entropy method is a regularization technique applied to a continuous spot distribution (Lanza *et al.* 2007). It yields better agreement with the Doppler imaging maps. In this case, the surface of the star is subdivided into 200 squared elements (each with 18° in latitude) with varying filling factors to reproduce the rotational modulation of the star. The downside is that there are a large number of free parameters (about 200) and the stellar rotation period is fixed, and not estimated by the method. The ratio between the area of faculae and of sunspots is also fixed. The regularized solutions are computed by minimizing a function that is a linear combination of the reduced χ^2 and a regularizing function that accounts for the a priori assumptions on the filling factor map of the spots. Lanza and collaborators applied this model to many stars, among them CoRoT-2, 4, 6, and 7.

Presently nearly 800 extrasolar planets have been discovered orbiting other stars. More than 200 (about 30%) of them transit their host star. During one of these transits, the planet may pass in front of a spot group and cause a detectable signal in the light curve of the stars, shown in Figure 2.

3. Spotted star with planetary transit

3.1. *Method*

The method discussed in this section simulates planetary transits, using the planet as a probe to study starspots (Silva 2003). This technique yields the individual spots physical characteristics such as:

- Size, or surface area coverage.
- Intensity, that can be converted to temperature assuming black body emission. This can be further translated to an estimate of the magnetic field provided that a solar model is used.
- Location, both the longitude and latitude of the spots.

Also, from observation of successive transits, presuming that the same spot is detected a few times on different transits, the stellar properties can be inferred such as:

- Rotation period.
- Differential rotation, if a mean period is known.
- Activity cycle, provided that a long enough observational time series is obtained.

The model creates a 2-D synthesized image of a star with a given limb darkening (linear or quadratic), whereas the planet is an opaque disk of radius r/R_s, where R_s is the stellar radius. The orbit is assumed to be circular and the orbital axis and stellar spin axis are parallel. Every 2 minutes (or a desired time interval), the planet is centred at its calculated position in a circular orbit (with semi-major axis a/R_s and inclination angle i). The light curve flux at every instant in time is the sum of all pixels in the image. The input parameters for the transit simulation are the orbital period, semi-major axis, inclination angle, and the ratio of the planet and stellar radii. The Southern hemisphere projection of the transit was arbitrarily chosen. The main program and auxiliary routines of this model can be found in www.craam.mackenzie.br/∼avalio/research.html.

The interesting feature of this model is to allow for the presence of sunspots on the stellar photosphere. The spots are considered circular and with a constant brightness. Also the foreshortenning effect of the spots close to the stellar limb is taken into account. Each spot is characterized by three parameters: radius (in units of planetary radius), intensity (measured with respect to the stellar central intensity), and location (longitude and latitude). To decrease the number of free parameters, so far I have only considered the latitude of the spots to be on the projected transit cord.

The spotted star with transiting planet model was applied to five stars observed by the CoRoT satellite: CoRoT-2 (Silva-Valio *et al.* 2010; Silva-Valio & Lanza 2011), CoRoT-4 and 6 (Valio & Lanza 2012), CoRoT-5, and CoRoT-8. The parameters of the star and its orbiting planet are given in Table 1 (where P_{rot} means the stellar rotation period). Images of the synthesized stars and the modelled spots during the first transit, with their respective planet at mid transit (depicted as a black circle) are shown in Figure 3.

3.2. *Results*

The number of transits observed varies from star to star, depending on the orbital period of the transiting planet. CoRoT-2 is the star with the larger number of transits, 77, because the planets orbits its host star every 1.743 d. Being a very young star, less than half a billion year, it is also very active, with a total of 392 spots detected, or an average detection of 5 spots per transit. CoRoT-4 was observed during a short run of less than 60 days, thus only 6 transits were detected and a total o 12 spots modelled. The other stars were detected during the long runs of 140 days, however not all transit observations are usable for the model. Therefore, 89, 71, and 33 spots were modelled for the stars CoRoT-5, CoRoT-6, and CoRoT-8, respectively.

Table 1. Stellar and planetary parameters

Star	CoRoT-2[1]	CoRoT-4[2]	CoRoT-5[3]	CoRoT-6[4]	CoRoT-8[5]
Spectral type	G7V	F8V	F9V	F9V	K1V
Mass (M_{sun})	0.97	1.10	1.0	1.055	0.88
Radius (R_{sun})	0.902	1.17	1.19	1.025	0.77
Teff (K)	5625	6190	6100	6090	5080
Prot (d)	4.54	8.87	26.6	6.35	21.7
Age (Gyr)	0.13-0.5	0.7-2.0	5.5-8.3	1.0-3.3	2.0-3.0
Mass (M_{jup})	3.31	0.72	0.467	2.96	0.22
Radius (R_{jup})	0.172	0.107	0.467	2.96	0.22
Orbital period (d)	1.743	9.203	4.038	8.886	6.212
Semi-major axis ($Rstar$)	6.7	17.47	9.877	17.95	17.61
Latitude (°)	-14.6	0	-47.2	-16.4	-29.4

Notes:
[1] Alonso *et al.* (2008)
[2] Aigrain *et al.* (2008)
[3] Rauer *et al.* (2009)
[4] Fridlund *et al.* (2010)
[5] Borde *et al.* (2010)

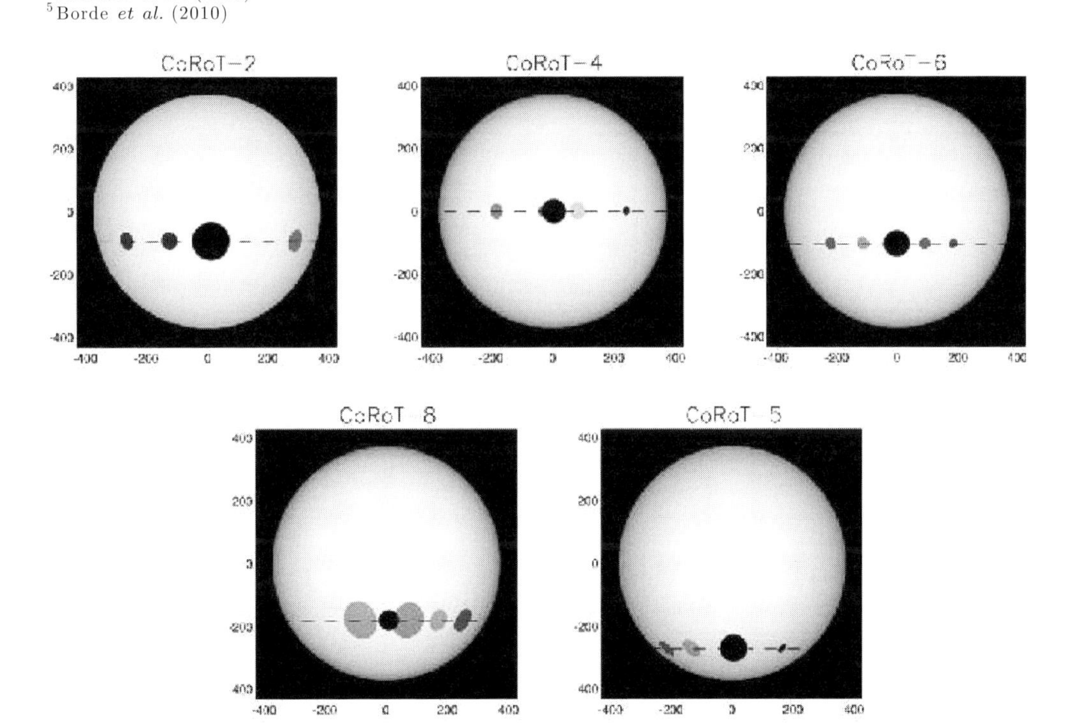

Figure 3. Five CoRoT stars with their transiting planet (black disk in the centre of the transit).

The average value of the radii of the spots and contrast with the rms are listed in Table 2, for all five stars. The last column displays the average values of the Sun for comparison. Assuming that the emission of both the surrounding photosphere and of the spot are black body emissions, the temperature of the spot may be estimated. These are the average temperatures values displayed in Table 2. Also, the 2-D histograms of the radius and contrast of the spots on each star are shown in Figure 4.

For each transit it is possible to estimate the area of the stellar surface covered by the spots, as has been done for CoRoT-2 (Silva-Valio *et al.* 2010), only within the latitude band obscured by the passage of the planet. The average total area of the spots in each

Table 2. Spots physical characteristics

Star	CoRoT-2	CoRoT-4	CoRoT-5	CoRoT-6	CoRoT-8	Sun
Radius ($\times 10^6$ m)	55 ± 19	51 ± 14	75 ± 17	48 ± 14	82 ± 21	12 ± 10
Area (%)	13 ± 5	6.0 ± 1.5	13 ± 3	8.2 ± 2.0	29 ± 10	< 1
Contrast (I_c)	0.45 ± 0.25	0.73 ± 0.14	0.48 ± 0.19	0.49 ± 0.18	0.52 ± 0.22	0.46 ± 0.19
T_{spot} (K)	4600 ± 700	5300 ± 500	5100 ± 600	5100 ± 500	4400 ± 600	4800 ± 400

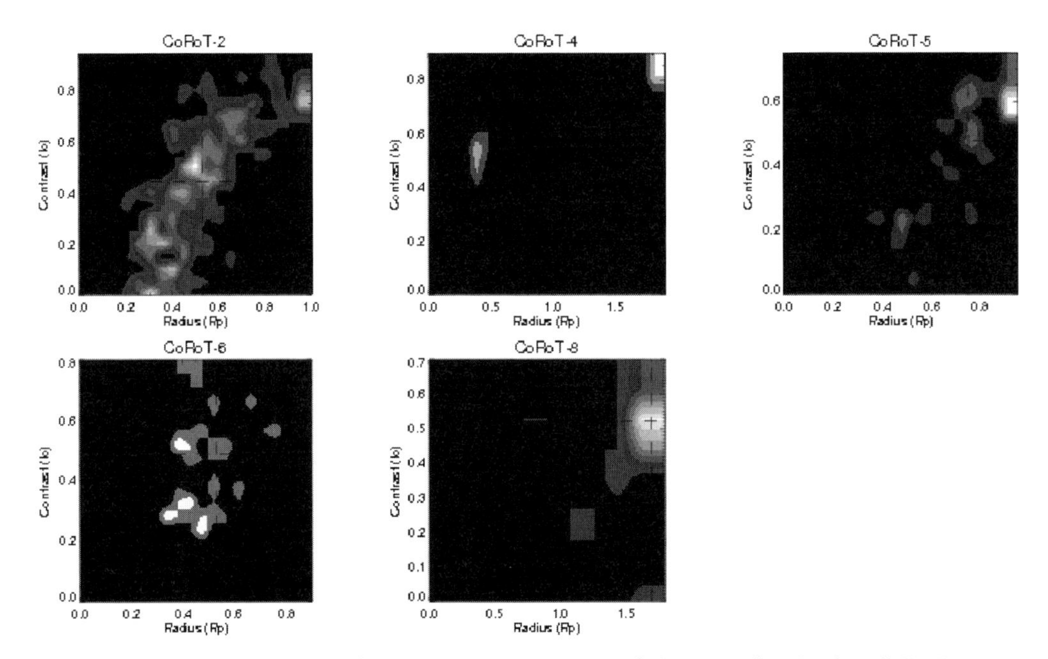

Figure 4. 2-D distributions of the radius and contrast of the spots for the five CoRoT stars.

transit is listed in Table 2 and shown, as a function of time in Figure 5. The star with the surface most covered by spots is CoRoT-8, with almost 30% coverage, note that this occurs at high latitudes ($-29°$).

Figure 6 presents the spotted stellar area as a function of the rotation period of the star (top right panel), orbital period (bottom left panel) and semi-major axis (bottom right panel) of the planet. The top left panel shows the rotation period of the star as a function of the orbital period, except for CoRoT-2 (with the closest hot Jupiter) there appears to be an anti-correlation between both periods.

A way to visualize how the spots are distributed on the stellar surface, restricted to the latitude bands covered by the planetary transit, of course, is to stack the longitude position of the spots on successive transits forming a "map" of the stellar surface. This was initially done considering the position of the spot as viewed from Earth. These are known as the topocentric longitude, and are limited to $\pm 90°$, where zero longitude corresponds to the line-of-sight at the the time of mid transit. Such maps for the 5 stars are shown in Figure 7, where the size of the circles are proportional to the modelled spot radius.

It is more interesting to know the position of the spots in a reference frame that rotates with the star that is called the rotational longitude. The conversion between topocentric

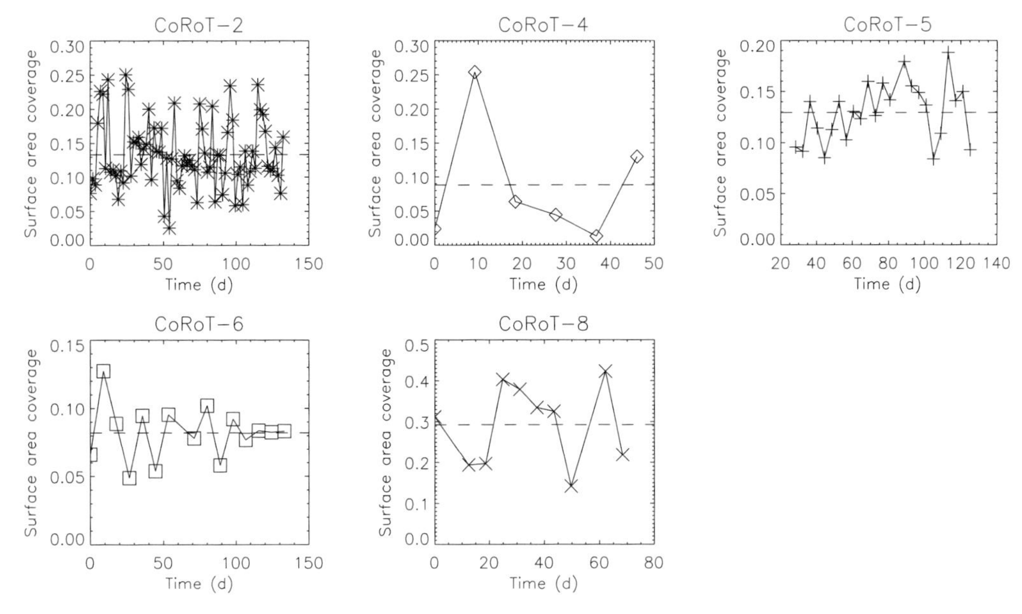

Figure 5. Spotted area coverage within the transit band latitudes as a function of time for the five CoroT stars.

and rotational longitudes is given by:

$$\beta_{rot} = \beta_{topo} - 360^\circ \frac{nP_{orb}}{P_{star}} \tag{3.1}$$

where P_{orb} is the orbital period (d), n is the transit number, and P_{star} is the stellar rotation period. The spotted surface maps of the five stars are displayed on Figure 8. The values on the top of each panel correspond to the rotation period of the star in days.

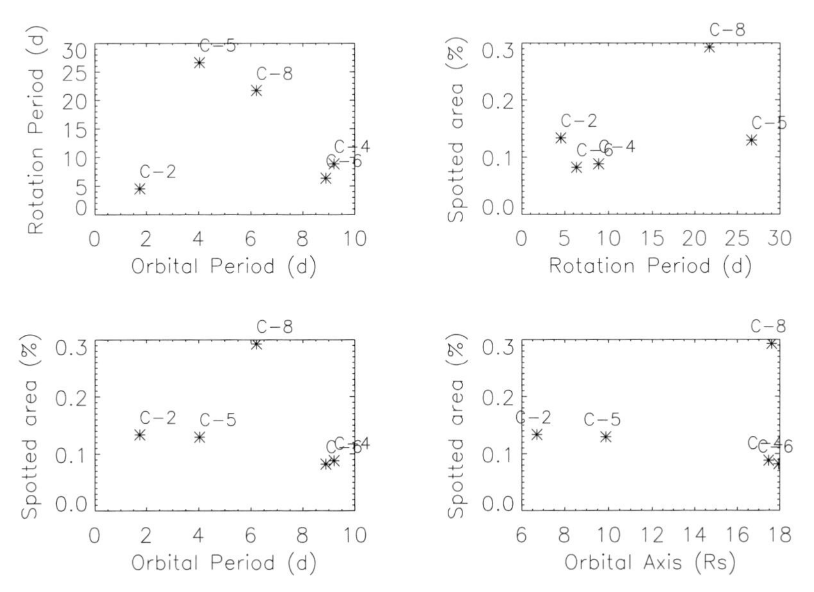

Figure 6. For the five CoroT stars: (top left) Stellar rotation period versus orbital period. (top right) Spotted surface area versus stellar rotation period. (bottom left) Spotted surface area versus planet orbital period and axis (bottom right).

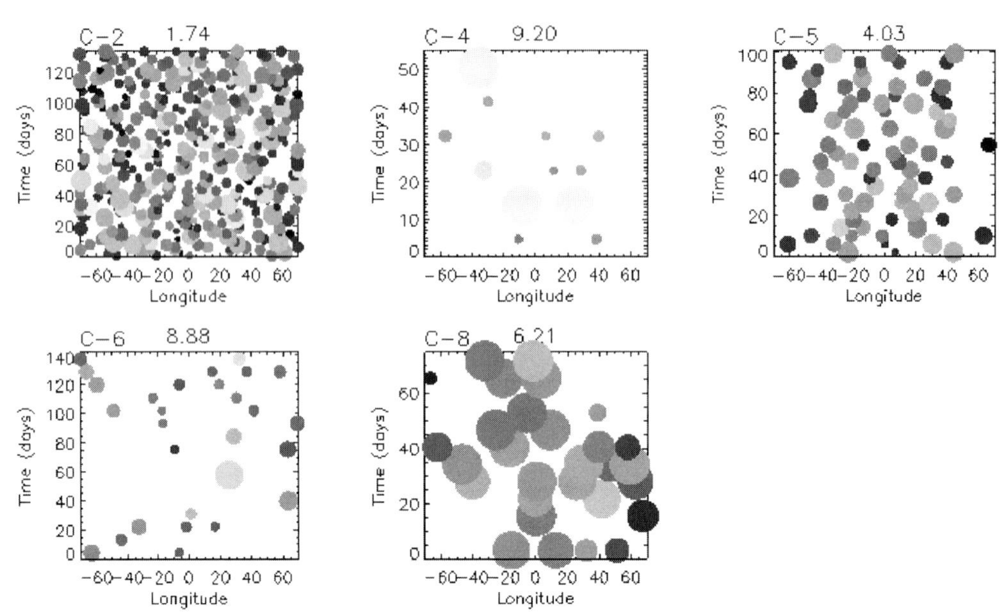

Figure 7. Surface maps of the spots for the five CoRoT stars as a function of their topocentric longitudes.

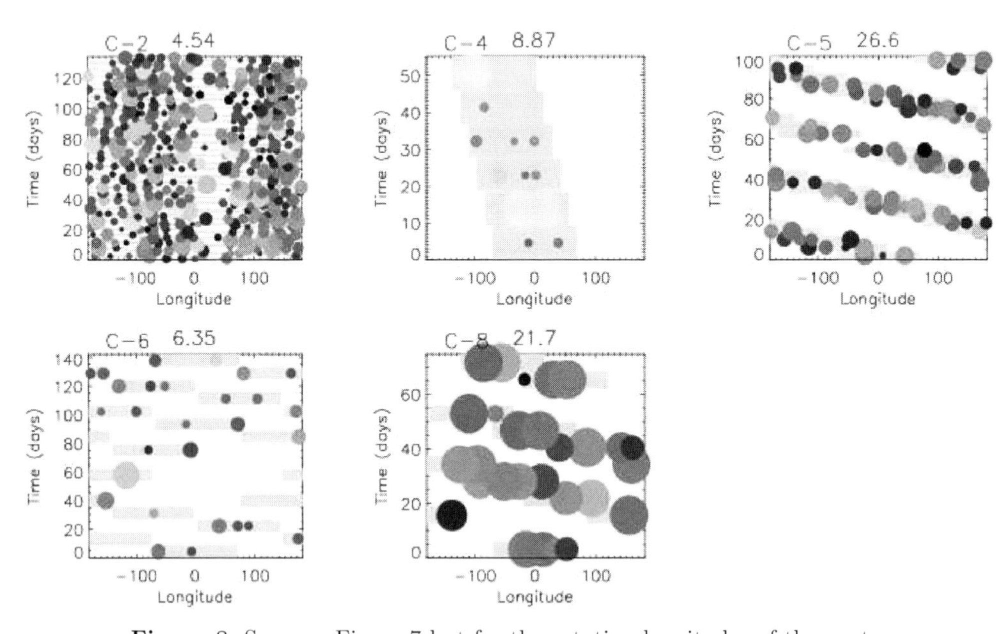

Figure 8. Same as Figure 7 but for the rotation longitudes of the spots.

As a last calculation, it is demonstrated how a magnetic activity cycle, similar to the 11 year solar cycle, may be inferred from the spotted surface area coverage for the star CoroT-2, reproduced again on the left panel of Figure 9. A running mean every 20 points is shown as the solid grey (red in the colour version) line. The power spectrum of the spotted area was calculated using the Scargle routine and is shown in the right panel. Several periods are identified with the orbital period, the stellar rotation, a 18 d periodicity observed in the spots coverage that is approximately equal to 10 times the

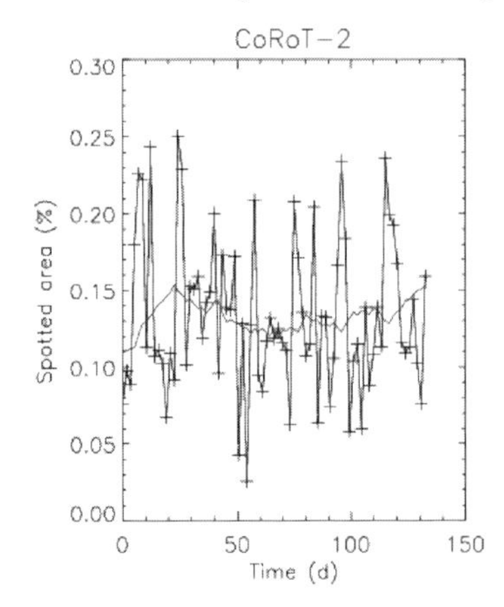

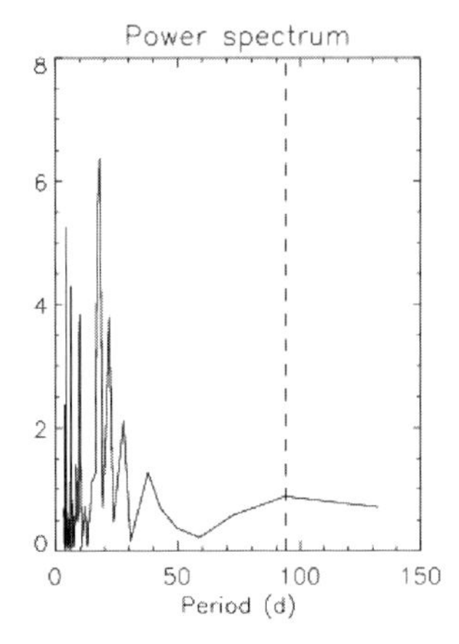

Figure 9. Left: Spotted surface area coverage of the spots on CoRoT-2 as a function of time for the 77 transits. Right: Power spectrum of the spotted area coverage.

orbital period, and the longest peak at around 94 days (dashed line). If this series was long enough such as to cover a significant fraction of the magnetic cycle of CoRoT-2, it would be detectable in the spotted area coverage of the star.

4. Discussion and conclusions

A technique that models the spots on the surface of a star by fitting the "bumps" detected on the transit light curve of stars with transiting planets was applied to five stars observed by the CoRoT satellite. This technique allows to determine the individual spots physical characteristics such as size and contrast.

The starspots modelled here are 4 to 7 times larger than regular sunspots, this happens probably due to the planet large size, and because spot groups are being detected (i.e., active regions) and not individual spots. The area covered by spots within the transit band of the star varies from 7 to 30%, on average. Moreover, there does not seem to be any correlation between the spotted area on the stellar photosphere and the rotation of the star, neither with the orbital parameters of the planet.

In conclusion, the modelling of small variations observed in the transit light curves yields:

- From single transits: Spots physical characteristics (size, temperature, location active longitudes, evolution/lifetime, surface area coverage, magnetic fields) (Silva 2003);
- From multiple transits: Stellar rotation (Silva-Valio 2008) and stellar differential rotation (Silva-Valio *et al.* 2010; Silva-Valio & Lanza 2011); and
- For longer observing period: Stellar activity cycles.

References

Aigrain, S., Collier Cameron, A., Ollivier, M. *et al.* 2008, *A&A*, 488, 43
Alonso, R., Auvergne, M., Baglin, A. *et al.* 2008, *A&A*, 482, 21
Borde, P., Bouchy, F., & Deleuil, M. 2010, *A&A*, 520, 66
Fridlund, M., Hbrard, G., & Alonso, R. 2010, *A&A*, 512, 14
Kron, G. E. 1947, *PASP*, 59, 261
Lanza, A. F., Rodono, M., Pagano, I., Barge, P., & Llebaria, A. 2003, *A&A*, 403, 1135
Lanza, A. F., Rodono, M., & Pagano, I. 2004, *A&A*, 425, 707
Lanza, A. F., Bonomo, A. S., & Rodono, M. 2007, *A&A*, 464, 741

Rauer, H., Queloz, D., Csizmadia, Sz. *et al.* 2009, *A&A*, 506, 281
Rodono, M., Cutispoto, G., Pazzani, V., *et al.* 1986, *A&A*, 165, 135
Silva, A. V. R. 2003, *ApJL*, 585, L147
Silva-Valio, A. 2008, *ApJL*, 683, L179
Silva-Valio, A., Lanza, A. F., Alonso, R., & Barge, P. 2010, *A&A*, 510, 25
Silva-Valio, A. & Lanza, A. F. 2011, *A&A*, 529, 36
Strassmeier, K. G. 2009, *A&ARv*, 17,251
Valio, A. & Lanza, A. F. 2012, *ApJ*, submitted

Discussion

Axel Brandenburg: If you say 4-7 times larger starspots than sunspots, one would like to know about selection effects?

Adriana Valio: The main selection effect is that these are planet hosting stars but solar type. The inferred starspot size is big because the probe, the planet - a hot Jupiter, is big. We are probably detecting the whole active region or sunspot group. Once I start analyzing transits of smaller planets, these values should decrease as we may identify individual spots.

Katja Poppenhager: You only constructed spots in the transit path (in line with the planet). What about spots at other latitudes? Are you planning to do some work including this in your model?

Adriana Valio: Presently, I don't model the spots on the stellar disk outside the transit latitude bands. But, as the spots don't change during transits (a few hours) and the light curve is normalized, that should not affect the results.

Francesco Zuccarello: How do you determine the stellar latitude where the planet transits in front of the star?

Adriana Valio: It is a geometric calculation from the semi-major axis and inclination angle of the planetary orbit.

Pablo Mauas: You could have spots just because of the interaction between the planet and the magnetosphere of the star. Perhaps the spots would not be there if it wasn't for the planet.

Adriana Valio: That may be true because there definitely is a planet-star interaction. This is clearly seen in the CoRoT-4 data, where an active longitude is observed at -30 deg. or 30 deg. from the sub-planetary point.

Comparative Magnetic Minima:
Characterizing quiet times in the Sun and Stars
Proceedings IAU Symposium No. 286, 2011
C. H. Mandrini & D. F. Webb, eds.

doi:10.1017/S1743921312005029

12 years of stellar activity observations in Argentina

P. J. D. Mauas[1], A. Buccino[1], R. Díaz[2], M. Vieytes[1], R. Petrucci[1], E. Jofre[3], X. Abrevaya[1], M. L. Luoni[1] and P. Valenzuela[1]

[1]Instituto de Astronomía y Física del Espacio,
C.C. 67 - Suc. 28, 1428, Buenos Aires, Argentina
email: pablo@iafe.uba.ar

[2]Institut d'Astrophysique de Paris, CNRS/UPMC, Paris, France. Observatoire de Haute-Provence, CNRS/OAMP, Saint-Michel l'Observatoire, France.

[3]Observatorio Astronómico de Córdoba, Argentina

Abstract. We present an observational program we started in 1999, to systematically obtain mid-resolution spectra of late-type stars, to study in particular chromospheric activity. In particular, we found cyclic activity in four dM stars, including Prox-Cen. We directly derived the conversion factor that translates the known S index to flux in the Ca II cores, and extend its calibration to a wider spectral range. We investigated the relation between the activity measurements in the calcium and hydrogen lines, and found that the usual correlation observed is the product of the dependence of each flux on stellar color, and it is not always preserved when simultaneous observations of a particular star are considered. We also used our observations to model the chromospheres of stars of different spectral types and activity levels, and found that the integrated chromospheric radiative losses, normalized to the surface luminosity, show a unique trend for G and K dwarfs when plotted against the S index.

Keywords. stars: activity, stars: late-type, stars: atmospheres

---

1. Introduction

In 1999 we started a program to systematically obtain spectra of late-type stars, to study in particular chromospheric activity. The stars were chosen to cover the spectral range from F to M, with different activity levels. Since we were particularly interested in the transition to completely convective stars, we included a larger number of M stars in our sample.

Our observations were made at the 2.15 m telescope of the Complejo Astronómico El Leoncito (CASLEO), which is located at 2552 m above sea level, in San Juan, Argentina. We obtained high-resolution echelle spectra with a REOSC spectrograph. The maximum wavelength range of our observations is from 3860 to 6690 Å, and the spectral resolution ranges from 0.141 to 0.249 Å per pixel ($R = \lambda/\delta\lambda \approx 26400$).

At present, we have about 5500 spectra of 150 stars, ranging from F to M with different activity levels. Altough most of the stars are single dwarfs, we have several binaries and a few subgiants. Currently, we have four observing runs per year. Details on the reduction and calibration procedures can be found in Cincunegui & Mauas (2004).

2. Flux-Flux calibrations

Since, unlike most surveys of this kind, the observations in the different spectral features are made simultaneously, our data provides an excellent opportunity to study the

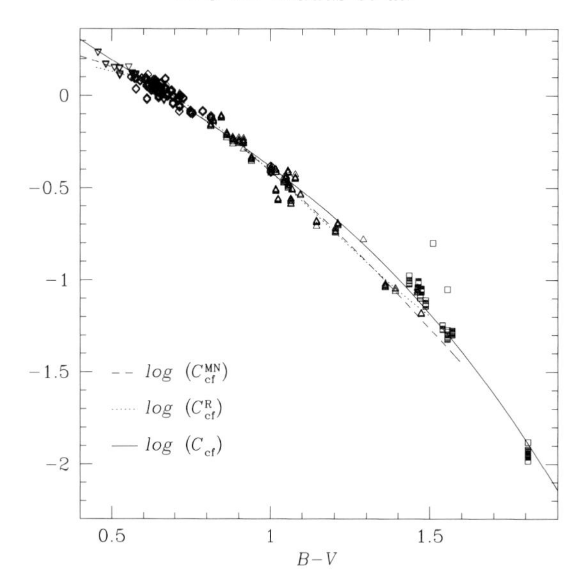

Figure 1. Conversion factor between Mount Wilson S index and F_{HK}. The dashed line corresponds to Noyes et al. (1984), the dotted line to Rutten (1984), and the full line to our derived factor (from Cincunegui, Díaz & Mauas 2007a).

correlation between different spectral features and activity indexes. To date, the most common indicator of chromospheric activity is the well-known S index, essentially the ratio of the flux in the core of the Ca II H and K lines to the continuum nearby (Vaughan et al. 1978). This index has been defined at the Mount Wilson Observatory, were an extensive database of stellar activity has been built over the last four decades. However, unlike our program, these observations are mainly concentrated on stars ranging from F to K, due to the long exposure times needed to observe the Ca II lines in the red in faint M stars. For this reason, the S index is poorly characterized for these stars. We first obtained the S index for our spectra, integrating with the corresponding profile (for details, see Cincunegui, Díaz & Mauas 2007a).

The Mount Wilson S index can be converted to the average surface flux in the Ca II lines through the relation:

$$F_{\rm HK} = F_{\rm bol} 1.34\, 10^{-4}\, S\, C_{\rm cf}\,, \tag{2.1}$$

where $C_{\rm cf}\,(B-V)$ is a conversion factor that depends on color. Two different expressions are widely used for this factor, the first one given by Middelkoop (1982) and corrected by Noyes et al. (1984) and the other one given by Rutten (1984). The deductions used in both works to derive $C_{\rm cf}$ involve complex calibration procedures.

Since we have simultaneous measurements of the S index and the core fluxes, we can calculate directly the correction factor as a function of the index and the flux, which is shown in Fig. 1 together with the other two expressions (for details and a whole discussion, see Cincunegui, Díaz & Mauas 2007a).

It is usually accepted that there is a tight relation between the chromospheric fluxes emitted in H-α and in the H and K Ca II lines, and that these two features can be used to study chromospheric activity. However, most works where this relation has been observed found it by using averaged fluxes for both the calcium and the hydrogen lines, which were not obtained simultaneously, and were even collected from different sources.

However, the situation is different when the individual stars are studied separately. In Fig. 2 we plot the individual simultaneous measurements of each flux for several stars of

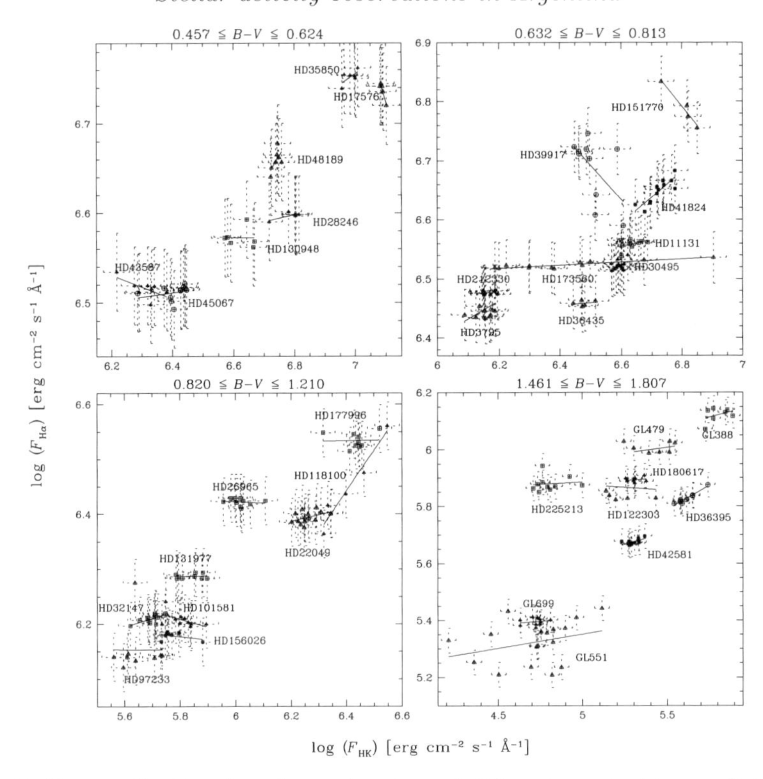

Figure 2. H-α vs. Ca II surface fluxes, for stars of different spectral types, divided into different color bins, as indicated (from Cincunegui, Díaz & Mauas 2007a).

different spectral types and different levels of activity, divided into color bins as stated. We also show the linear fits for each star. It can be seen that the behavior is different in each case: in some stars both fluxes are well correlated, although the slopes of the fits are not the same. In other stars the H-α flux seems to be almost independent of the level of activity measured in the Ca II lines, and there are even stars where the fluxes are anti-correlated.

In Díaz, Cincunegui & Mauas (2007) we studied the sodium D lines (D1: 5895.92 Å; D2: 5889.95 Å) in our stellar sample. We found a good correlation between the equivalent width of the D lines and the color index $(B-V)$ for all the range of observations. Since equivalent width is a characteristic of line profiles that do not require high resolution spectra to be measured, this fact could become a useful tool for subsequent studies. Finally, we constructed a spectral index (R'_D) as the ratio between the flux in the D lines and the bolometric flux. Once corrected for the photospheric contribution, this index can be used as a chromospheric activity indicator in stars with a high level of activity. Additionally, we found that combining some of our results, we obtained a method to calibrate in flux stars of unknown color.

3. Cycles in M-stars

One of the main goals of our program was to extend the studies of stellar cycles to M-stars, at and beyond the limit for full convectivity. The first star we studied was Proxima Centaury, a dMe 5.5 star with strong and frequent flaring activity (Cincunegui, Díaz & Mauas 2007b). For this star we excluded the spectra taken during flares, computed the nightly average of the Hα flux, and calculated the Lomb-Scargle periodogram, which is

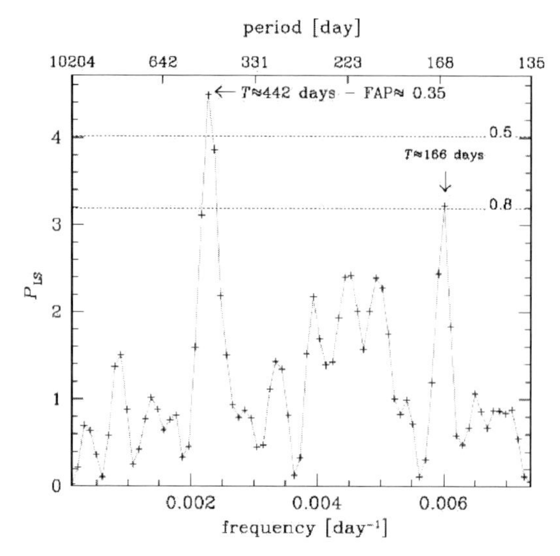

Figure 3. Lomb-Scargle periodogram of our observations. The False Alarm Probability levels of 50 and 80% are shown. (from Cincunegui, Díaz & Mauas 2007b).

shown in Fig. 3. In the periodogram we found strong evidence of a cyclic activity with a period of ∼442 days. Similar values for the period were found using three different techniques in the time domain (see Cincunegui, Díaz & Mauas 2007b for details). We were also able to determine that the activity variations outside of flares amount to 130% in S, three times larger than for the Sun.

Since this star should be fully convective, it cannot support an $\alpha\Omega$ dynamo, and a different mechanism should be found to explain this result. Recently, Chabrier & Küker (2006) showed that these objects can support large-scale magnetic fields by a pure α^2 dynamo process. Moreover, these fields can produce the high levels of activity observed in M stars (see, for example, Mauas & Falchi 1996). This α^2 dynamo does not predict a cyclic activity. However, our observations suggest that this cool star has a clear period.

We also studied the spectroscopic binary system Gl 375 (Díaz *et al.* 2007). We first obtained precise measurements of the orbital period (P = 1.87844 days) and separation (a = 5.665 $R_\odot$), minimum masses and other orbital parameters. We separated the composite spectra into those corresponding to each component, which allows us to confirm that both components are of spectral type dMe 3.5.

To study the variability of Gl 375, besides using the spectra obtained at CASLEO, we also employed photometric observations provided by the All Sky Automated Survey (ASAS Pojmanski 2002). We calculated the Lomb-Scargle periodogram for these data, and obtained a distinct peak corresponding to a period of $P_{\rm phot}$ = 1.876667 days, a period resembling very closely the measured orbital period. We believe this harmonic variability is produced by spots and active regions in the stellar surface carried along with rotation. This would imply that the rotational and orbital periods are synchronized, as is expected for such a close binary.

To verify if this is indeed the case, we phased the data to the obtained period, for two different seasons (Fig. 4). The sinusoidal shape of the variation is evident, although the amplitude of the modulation is different in both cases, probably an indication of different area covered by starspots or active regions, and therefore different activity levels, in each epoch. Therefore, the amplitude of this modulation can be used as an activity proxy, and indicates that the system exhibits a roughly periodic behavior of 2.2 years (or 800 days).

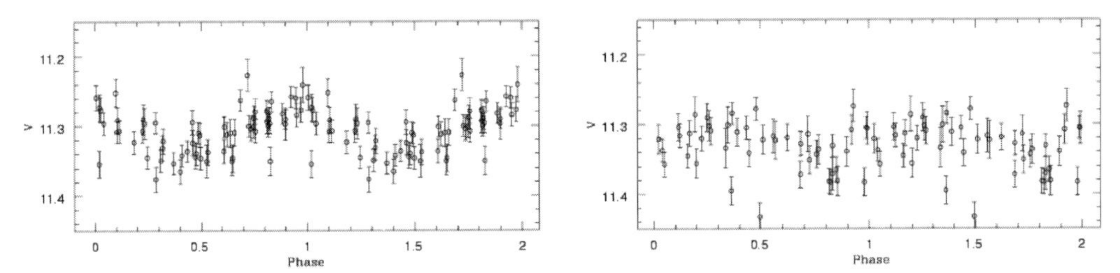

Figure 4. ASAS photometry phased to the orbital period for two different epochs. Left: 2002.5-2003.5 Right: 2006. (from Díaz *et al.* 2007).

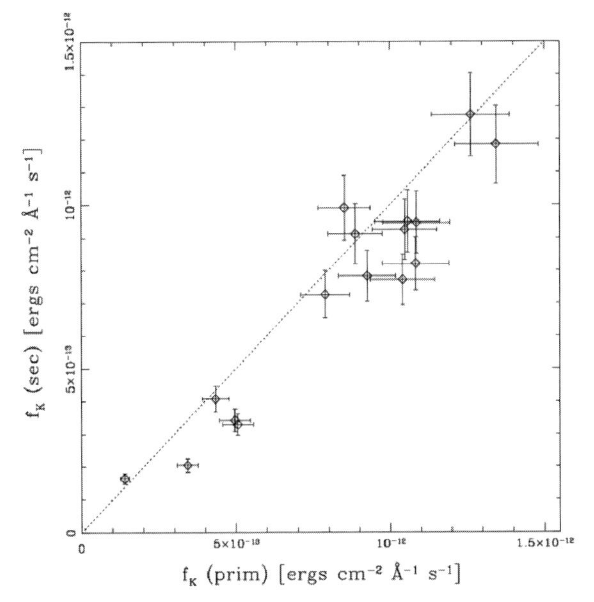

Figure 5. Comparison of the fluxes in the Ca II K line for both components. The error bars correspond to a 10% error in the line fluxes, and the dotted line is the identity relation. (from Díaz *et al.* 2007).

The same period was found in the mean magnitude of the system and in the flux of the Ca II K line, although the Ca flux variations occur 140 days ahead of the photometric ones, a behavior that has been previously observed in other stars (Gray *et al.* 1996). The agreement between the behavior of the three observables is remarkable because of the different nature of the observations and the different instruments and sites where they were obtained.

Another interesting result of this work is that the activity of Gl 375 A and Gl 375 B, as measured in the flux of the Ca II K lines, are in phase, as can be seen in Fig. 5. There is an excellent correlation between the levels of chromospheric emission of both components, implying a magnetic connection between them. Due to its vicinity and relative brightness, this system presents an interesting opportunity to further study this type of interaction.

We also studied the long-term activity of two other M dwarf stars: Gl 229 A and Gl 752 A, using again the Ca II - K line-core fluxes measured on our spectra and the ASAS photometric data. Using the Lomb-Scargle periodogram, we obtained a possible activity cycle of ~4 and ~7 yrs for Gl 229 A and Gl 752 A, respectively (Buccino *et al.* 2011). This work was complemented by other studies, were we investigated the presence of activity cycles using IUE data (Buccino & Mauas 2008 and Buccino & Mauas 2009).

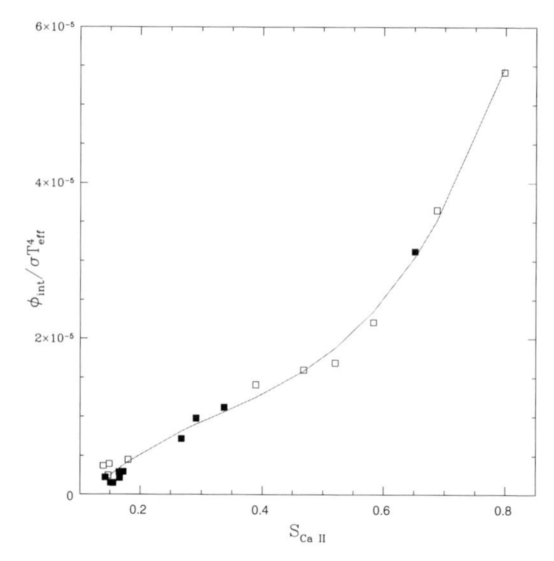

Figure 6. Normalized ϕ_{int} versus the S index. The empty squares represent the K star models from Vieytes, Díaz & Mauas (2009) and the full squares indicate the G star models from Vieytes, Mauas & Cincunegui (2005).

4. Chromospheric modeling

We also used our spectra as observational basis to model the chromosphere of different type of stars ans activity levels, following the path initiated with the model of Ad-Leonis in its quiescent state (Mauas & Falchi 1994), which was completed with a study of other dM 3.5 stars of different activity levels (Mauas *et al.*1997). We first computed models of the Sun as a star and 9 solar analogues (Vieytes, Mauas & Cincunegui 2005) which were followed by models for 6 K stars, "analogs" of epsilon-Eridani (Vieytes, Díaz & Mauas 2009). For most stars, we built two models, to match the observations in its minimum and maximum levels of activity, and found that the differences in the atmospheric structure for a star in its maximum and minimum activity levels are similar to the changes seen between two different stars.

We also found that the integrated chromospheric radiative losses ϕ_{int}, normalized to the surface luminosity, show a unique trend for G and K dwarfs when plotted against the S index, (see Fig. 6). This might indicate that the same physical processes are heating the stellar chromospheres in both cases. We calculated an empirical relationship between S and the energy deposited in the chromosphere, which can be used to estimate the energetic requirements of a given star knowing its chromospheric activity level.

References

Buccino, A. P., Díaz, R. F., Luoni, M. L., Abrevaya, X. C., & Mauas, P. J. D. 2011, *AJ* 141, 34
Buccino, A. P. & Mauas, P. J. D.. 2008, *A&A* 483, 903
Buccino, A. P. & Mauas, P. J. D.. 2009, *A&A* 495, 287
Chabrier, G. & Küker, M. 2006, *A&A* 446, 1027
Cincunegui, C. C. & Mauas, P. J. D.. 2004, *A&A*, 414, 699
Cincunegui, C. C., Díaz, R., & Mauas, P. J. D. 2007a, *A&A*, 469, 309
Cincunegui, C. C., Díaz, R., & Mauas, P. J. D. 2007b, *A&A*, 461, 1107
Díaz, R., Cincunegui, C. C., & Mauas, P. J. D. 2007, *MNRAS*, 378, 1007
Díaz, R., González, F., Cincunegui, C. C., & Mauas, P. J. D. 2007, *A&A*, 474, 345
Gray, D. F., Baliunas, S. L., Lockwood, G. W., & Skiff, B. A. 1996, *ApJ*, 465, 945

Mauas, P. J. D. & Falchi, A. 1994, *A&A* 281, 129
Mauas, P. J. D. & Falchi, A. 1996, *A&A*, 310, 245
Mauas, P. J. D., Falchi, A., Pasquini, L., & Pallavicini, R. 1997, *A&A* 326, 249
Middelkoop, F. 1982, *A&A* 107, 31
Noyes, R. W., Hartmann, L. W., Baliunas, S. L., Duncan, D. K., & Vaughan, A. H. 1984, *ApJ* 279, 763
Pojmanski, G. 2002, *Acta Astron.* 52, 397
Rutten, R. G. M. 1984, *A&A* 130, 353
Vaughan, A. H., Preston, G. W., & Wilson, O. C. 1978, *PASP* 90, 267
Vieytes, M.; Mauas, P. J. D., & Cincunegui, C. C. 2005, *A&A* 441, 701
Vieytes, M., Díaz, R., & Mauas, P. J. D.. 2009, *MNRAS* 398, 1495.

Discussion

Steve Saar: You mentioned ASAS data in a couple of cases, I would encourage you to look at the ASAS data for Proxima Centauri. By an odd coincidence we have been looking for an x-ray cycle in that recently and it has what appears to be a good photometric period but it is more like 8 years rather than the period you quote here. I have been wondering if you see a signal in that range.

Pablo Mauas: I was thinking it was time for an update in that star, since we have four more years of data and we certainly have to use the ASAS data. We will have a look at it.

Jürgen Schmitt: Alpha Cen must be perfectly suited for observations you have – have you looked at Alpha Cen?

Pablo Mauas: Yes, we have, but we had a problem with the gain of the pointing screen at the observatory so we had to end those observations three or four years ago. But we have a hint of a cycle in the K-star, Alpha Cen B.

Mark Giampapa: I think it is great that you are getting cycle periods for M dwarf stars. Just a comment on the H-α and Calcium K. The models that I have constructed and looked at show that they are very segregated in their formation regions and so you could have some, at least in models, disjoint behavior, but I am surprised that you don't see in observations pretty direct correlation. But, in any event I could imagine in the upper atmosphere you could have flare-like behavior that may not propagate into the lower atmosphere and that could lead to some deep correlation between Calcium and H α.

Pablo Mauas: Usually, we can detect flare activity because we take two spectra to eliminate cosmic rays, so before we do anything with our data we check to see if the spectra are more or less the same. So if you had a flare, that wouldn't happen particularly not for M stars integration times are very long. So usually we leave out all the flares. For Proxima for example that's hard work because you can have four or five flares a day. So usually those extremes are eliminated from our data. I think there can be different things for these different correlations between Calcium and H-α. One can be filling factor problems. Perhaps you can explain the same integrated emission with different filling factors and different contrasts but they won't give the same relation between H-α and Calcium. The other one is the orientation - it is not the same if you observe the star pole-on or from the equator. So we are exploring that, and preliminary results are shown in our poster.

Comparative Magnetic Minima:
Characterizing quiet times in the Sun and Stars
Proceedings IAU Symposium No. 286, 2011
C. H. Mandrini & D. F. Webb, eds.

doi:10.1017/S1743921312005030

A statistical analysis of Hα-Ca II relation for solar-type stars of different activity levels

Andrea P. Buccino[1,2], **Mariela C. Vieytes**[1] **and Pablo J. D. Mauas**[1]

[1] Instituto de Astronomía y Física del Espacio (IAFE), CC 67-Suc. 28 (C1428ZAA), Ciudad Autónoma de Bs. As., Argentina
email: abuccino@iafe.uba.ar

[2] Departamento de Física, Facultad de Ciencias Exactas y Naturales, Universidad de Bs. As., Ciudad Autónoma de Bs. As., Argentina

Abstract. Based on our large spectral database obtained at CASLEO Argentinian Observatory, we analyzed the relation between simultaneous measurements of Hα and Ca II H+K fluxes. Although the correlation between both proxies is positive for the solar case, in 2007 our group found that while some stars exhibit correlations between Hα and the Ca II lines, the slopes change from star to star, including cases where no correlation was found. To discern if this flux-flux relation depends on the level of activity of the star and if it is associated with the distribution of active regions in the stellar atmosphere, in this work we analyze the relation between Hα-Ca II fluxes for the whole set of 44 G dwarf stars and individually for a subset of several solar-type stars of different level of activity.

Keywords. stars: activity, late-type, line: profiles, methods: statistical

1. Introduction

It is well known that the correlation between the Ca II K and H line-core fluxes and the Hα emission is positive in the solar case for the whole solar-cycle (Livingston *et al.* 2007).

However, Cincunegui *et al.* (2007) reported that the correlation between Ca II and Hα is not always valid for stars. In that paper they studied this relation for a set of 109 dF to dM stars and found that while some stars exhibit correlations between both proxies, the slopes change from star to star. Furthermore, in several cases Hα and the Ca II fluxes were not correlated, as in the binary system Gl 375 (Díaz *et al.* 2007) and the M dwarfs Gl 229 A and Gl 752 A (Buccino *et al.* 2011), and other stars even exhibit anti-correlations.

Motivated by these results, Meunier & Delfosse (2009) re-analyzed the Ca-Hα relation for the Sun for different solar-cycle phases and they found that this correlation depends on the balance between the emission in plages and absorption in filaments.

2. Observations

Since 1999, we systematically observed more than 140 main-sequence stars, with the Echelle spectrograph on the 2.15 mts telescope of the CASLEO Observatory located in the Argentinean Andes. To date, we have more than 5000 spectra, ranging from 3890 to 6690 Å with R = 13000, which constitute an ideal dataset to study long term activity. Following Cincunegui & Mauas (2004) spectra are calibrated in flux which allow us to simultaneously study different spectral features, from the Ca II lines to Hα.

In this work, we extended the analysis presented in Cincunegui *et al.* (2007) by adding 5 years of simultaneous CASLEO observations of Ca II and Hα chromospheric lines. The 12-year-length of both series allow us to discern if the Ca II-Hα relation depends on the level of activity of each individual star and if it is associated to the distribution of active regions in the stellar atmosphere. In the present work, we specifically summarize our results obtained for the dG stars of our stellar sample.

3. Hα-Ca II H+K for dG stars

For a set of 767 CASLEO spectra of 44 G0-G9 dwarf stars, we obtained simultaneous measurements of Ca II and Hα fluxes. On each spectrum we integrated the Ca II H and K line-core fluxes with a triangular profile of 1.09Å FWHM and we computed the flux in the Hα line as the average surface flux in a 1.5 Å square passband centered on 6562.8 Å. In Fig. 1 we plot both surface fluxes obtained.

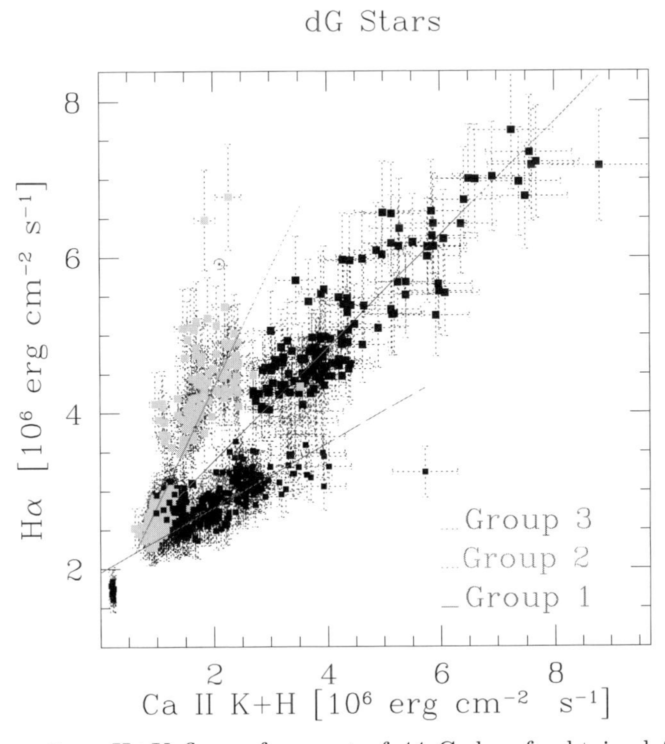

Figure 1. Hα vs. Ca II H+K fluxes for a set of 44 G dwarfs obtained from CASLEO spectra. We also included these values for the Quiet Sun, derived from Kitt Peak spectra (⊙).

In Fig. 1 we observe that there is not a single Hα-Ca II relation for G stars. It seems that we could recognize three groups with different trends, which could be related to the balance of different magnetic structures in the stellar atmosphere. The Quiet Sun (⊙) seems to belong to the group with high Hα emission and low Ca II fluxes, coherent with results presented in Meunier & Delfosse (2009).

Assuming that these three different Hα-Ca II tendencies in dG stars cannot be due to differences in color, we analyzed the relation for each star individually. In Fig. 2 we plot an histogram of the Pearson correlation coefficient and the slope obtained for each group of stars.

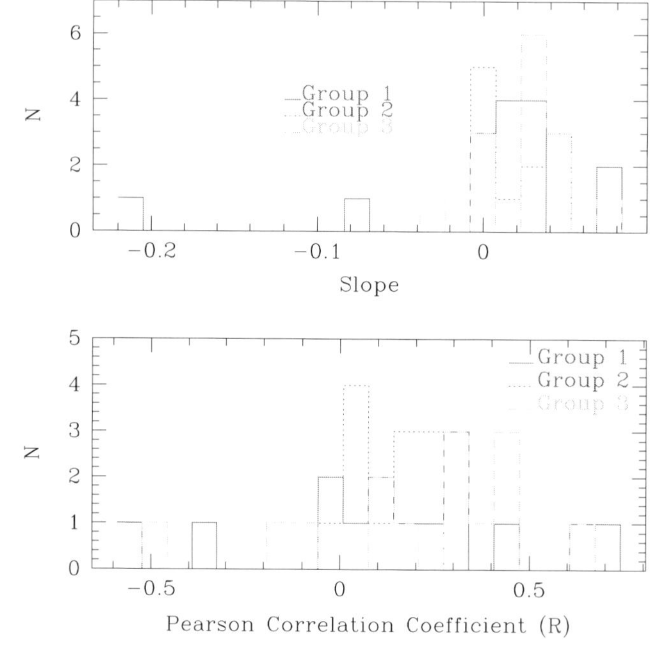

Figure 2. Histogram of Pearson correlation coefficients and slopes of the Hα-Ca II relation for each individual star.

Most of the stars present a positive correlation between Hα and Ca II fluxes as observed in the Sun. Nevertheless we observe a great dispersion in both correlation coefficients and the slopes, confirming Cincunegui *et al.* (2007).

4. Hα-Ca II H+K for solar analogues

Meunier & Delfosse (2009) found that the Hα-Ca II correlation coefficient in the solar case is not the same along the solar cycle, as it is even higher than for the whole solar-cycle at the end of the ascending phase and much lower (R<0.5) at cycle minimum. To analyze this result for stellar cases, in Fig. 3 we plot the Hα and Ca II fluxes for solar analogues of different activity regime (see Vieytes *et al.* 2005), we computed the Hα-Ca II correlation coefficient R for the whole dataset and for the most active phase of the star (R_{act}). We observe that although Hα and Ca II fluxes show a low correlation for the whole series, the correlation is strongly positive during the maximum active phase (indicated with red circled points) .

5. Conclusion

In this work, we extended the analysis between the Hα and Ca II fluxes presented in Cincunegui *et al.* (2007) by adding 5 years of simultaneous CASLEO observations of these chromospheric lines. In particular, we analyzed the fluxes derived from 767 spectra of 44 G0 to G9 stars obtained between 1999 and 2010. As a preliminary result, we found that there is not a single Hα-Ca II relation for G stars. We also analyzed this relation for each G star individually and we obtained a great dispersion in both correlation coefficients and the slopes for the whole stellar sample, confirming Cincunegui *et al.* (2007).

On the other hand, we analyzed the Hα-Ca II relation for several solar-type stars of different activity levels. For HR 6060, the best solar-twin, HD 1835 and HD 172051 we

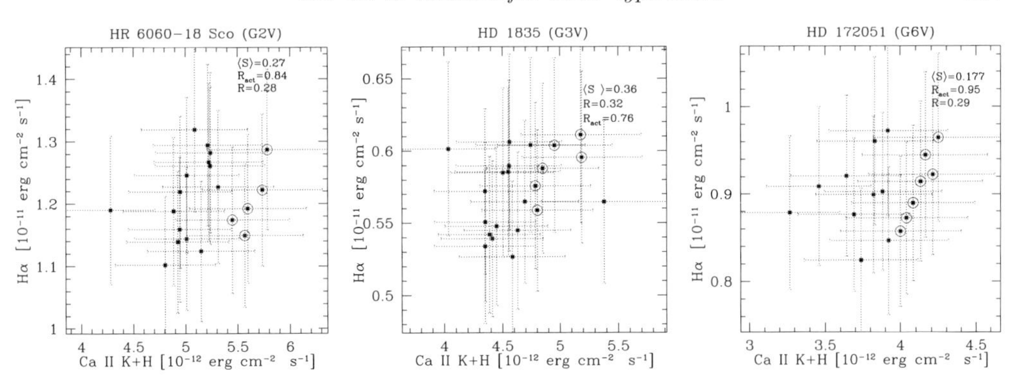

Figure 3. Hα-Ca II relation for three solar-analogues of different active regimes quantified by the mean Mount Wilson index ($\langle S \rangle$). We computed the Pearson correlation coefficient for the whole series (R) and for the red circled points (R_{act}), which indicate those points where the level of activity of the star excceeds in more than 0.5σ the mean.

found that the correlation between both proxies is strongly positive near the maximum active phase. In particular, HR 6060 (about solar-age) and HD 1835 (about Hyades-age, an older Sun) present similar correlation coefficients, probably indicating preservation of this correlation over epochs in which other observables change dramatically.

In future works we will explore these conclusions in detail.

References

Buccino, A. P., Díaz, R. F., Luoni, M. L., Abrevaya, X. C., & Mauas, P. J. D. 2011, AJ, 141, 34
Cincunegui, C., Díaz, R. F., & Mauas, P. J. D. 2007, A&A, 469, 309
Cincunegui, C. & Mauas, P. J. D. 2004, A&A, 414, 699
Díaz, R. F., González, J. F., Cincunegui, C., & Mauas, P. J. D. 2007, A&A, 474, 345
Livingston, W., Wallace, L., White, O. R., & Giampapa, M. S. 2007, ApJ, 657, 1137
Meunier, N. & Delfosse, X. 2009, A&A, 501, 1103
Vieytes, M., Mauas, P., & Cincunegui, C. 2005, A&A, 441, 701

Comparative Magnetic Minima:
Characterizing quiet times in the Sun and Stars
Proceedings IAU Symposium No. 286, 2011
C. H. Mandrini & D. F. Webb, eds.

doi:10.1017/S1743921312005042

Precise effective temperatures of solar analog stars

D. Cornejo-Espinoza[1], **I. Ramírez**[2], **P. S. Barklem**[3] **and W. Guevara-Day**[1]

[1]Departamento de Astrofísica, Agencia Espacial del Perú, CONIDA
[2]McDonald Observatory and Department of Astronomy, University of Texas at Austin, USA
[3]Departament of Physics and Astronomy, Uppsala University, Sweden

Abstract. We perform a study of 62 solar analog stars to compute their effective temperatures ($T_{\rm eff}$) using the Balmer line wing fitting procedure and compare them with $T_{\rm eff}$ values obtained using other commonly employed methods. We use observed Hα spectral lines and a fine grid of theoretical LTE model spectra calculated with the best available atomic data and most recent quantum theory. Our spectroscopic data are of very high quality and have been carefully normalized to recover the proper shape of the Hα line profile. We obtain $T_{\rm eff}$ values with internal errors of about 25 K. Comparison of our results with those from other methods shows reasonably good agreement. Then, combining $T_{\rm eff}$ values obtained from four independent techniques, we are able to determine final $T_{\rm eff}$ values with errors of about 10 K.

Keywords. Lines Profiles, Stellar Atmospheres.

---

1. Introduction

The effective temperature ($T_{\rm eff}$) is one of the most important parameters in the study of stars. For example, precise and accurate $T_{\rm eff}$ values allow us to reliably measure the chemical compositions of stars. Other important stellar parameters such as luminosity, radius, etc., can only be obtained once $T_{\rm eff}$ is known. A number of techniques have been devised to derive $T_{\rm eff}$. In this work, we use the relative flux level in the wings of Hα line profile as an indicator of the star's effective temperature (e.g., Gehren 1981, Barklem *et al.* 2002).

In studies of stars like the Sun, systematic errors can be minimized if the data are carefully treated with a differential analysis. Thus, very precise $T_{\rm eff}$ values can in principle be derived using high quality data of solar analog stars. The aim of this work is to derive $T_{\rm eff}$ values using model fits to the Hα line wings of 62 solar analogs and to compare the results with the $T_{\rm eff}$ values derived using three other methods.

2. Determination of the effective temperature using Hα

The method we use consists of finding the best match to an observed Hα line profile from a theoretical grid (see Fig. 1). Spectroscopic data acquired with the R. G. Tull coudé spectrograph on the 2.7 m Telescope at McDonald Observatory, properly normalized, are employed. The spectral resolution is $R = 60,000$ and the average signal-to-noise ratio is 300. Spectral windows free from telluric lines in our solar spectrum (asteroid reflected sunlight) are identified first and later used for the entire sample. The model grid was calculated as in Barklem *et al.* (2002) and it has a fine spacing of 10 K in $T_{\rm eff}$, 0.05 dex in $\log g$, and 0.05 dex in [Fe/H]. The $T_{\rm eff}$ and its error are derived using least squares minimization. We find $T_{\rm eff} = 5752 \pm 16$ K for the Sun (error bar corresponds

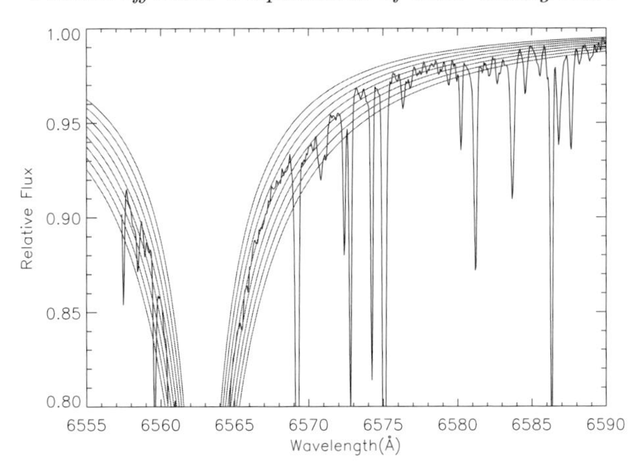

Figure 1. Our observed solar spectrum is superposed on a theoretical grid of Hα line profiles.

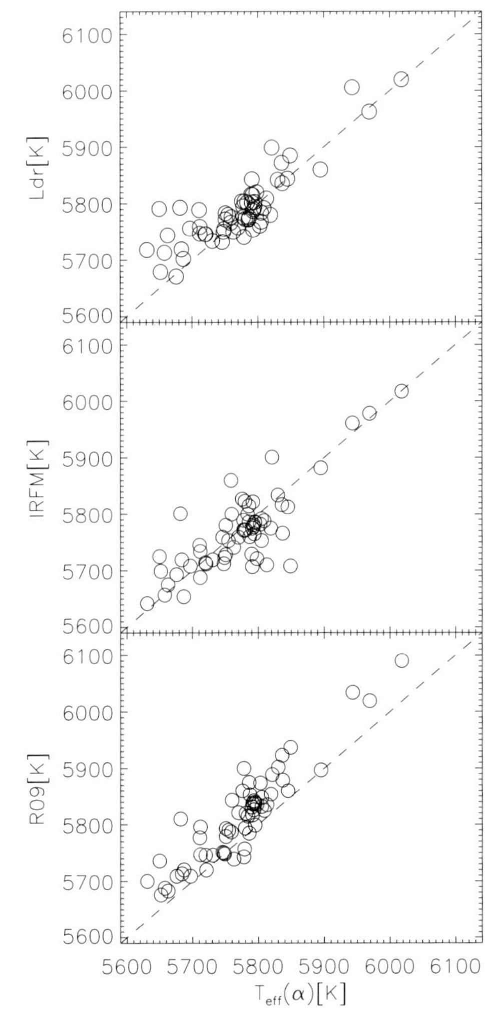

Figure 2. Comparison of our $T_{\rm eff}\,({\rm H}\alpha)$ with $T_{\rm eff}$ values from the Ldr (upper panel), IRFM (middle panel), and R09 (lower panel) methods.

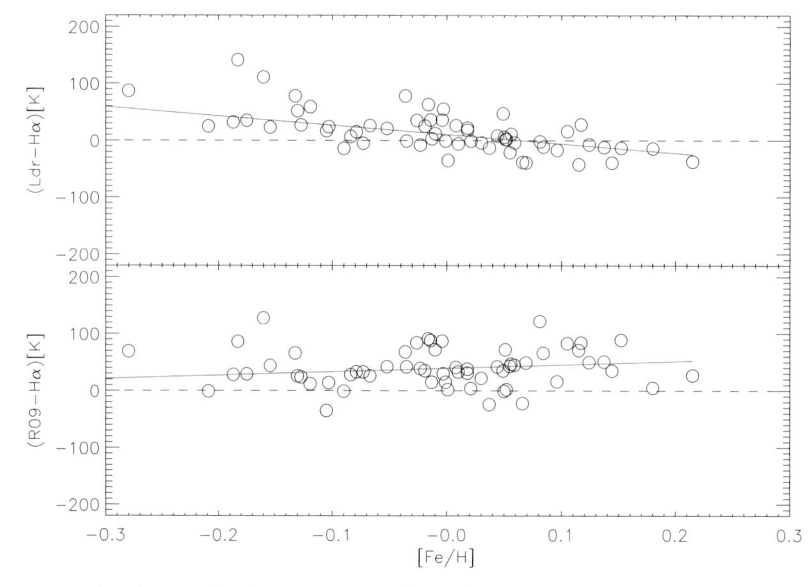

Figure 3. $T_{\rm eff}\,({\rm Ldr})-T_{\rm eff}\,({\rm H}\alpha)$ residuals vs. [Fe/H] (upper panel) and $T_{\rm eff}\,({\rm R09})-T_{\rm eff}\,({\rm H}\alpha)$ residuals vs. [Fe/H] (lower panel). Solid lines are linear fits to the residuals.

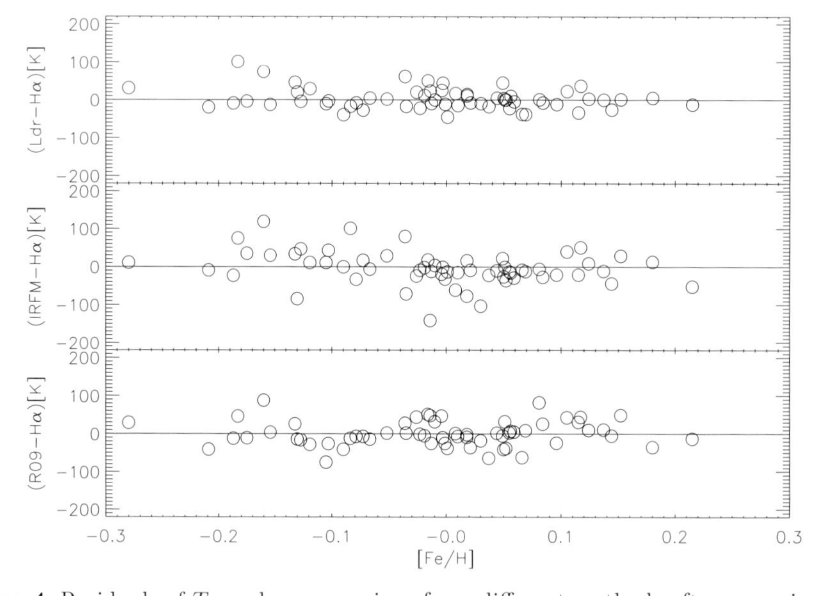

Figure 4. Residuals of $T_{\rm eff}$ value comparison from different methods after removing small trends and offsets.

to observational noise only). We applied zero point corrections to our solar $T_{\rm eff}$'s based on solar spectrum adding the difference in temperature that forces the solar $T_{\rm eff}$ to be equal to 5777 K, adding the same temperature difference on the whole sample. Internal errors in our derived $T_{\rm eff}$ values are about 25 K. Note, however, that Barklem *et al.* (2002) point out that systematic errors can be as large as 80 K. Nevertheless, in our differential analysis of solar analog stars, we expect the systematic errors to have a small impact.

3. Comparison with other methods

Careful inspection of the residuals of the $T_{\rm eff}$ value comparisons revealed small offsets and trends with stellar parameters. For example, Ldr–Hα showed a clear [Fe/H] dependency while R09–Hα revealed an offset of about 40 K, as shown in Fig. 3. The former could be due to the fact that Ldr calibration formulae do not take [Fe/H] into account while the latter may be related to the degeneracy between $T_{\rm eff}$ and $\log g$ derived only from an iron line analysis (i.e., forcing excitation/ionization balance). We re-calculated the values of $T_{\rm eff}$ (R09) and $T_{\rm eff}$ (Ldr), thus eliminating the small trends and offsets with linear corrections. In this way, residuals of the $T_{\rm eff}$ comparisons are dominated by measurement errors (Fig. 4).

We compared our $T_{\rm eff}$ (Hα) with the $T_{\rm eff}$ obtained from the method of the spectral line-depth ratios (Ldr; e.g., Gray & Johansson 1991, Gray 1994), using the calibration formulae by Kovtyukh *et al.* (2003). We also compared our temperatures with those from the infrared flux method (IRFM; e.g., Ramírez *et al.* 2005) $T_{\rm eff}$ scale, using the color calibrations by Casagrande *et al.* (2010). Finally, we also made a comparison with the values of $T_{\rm eff}$ obtained from the excitation equilibrium of Fe I lines, as derived by Ramírez *et al.* (2009, hereafter R09). Our $T_{\rm eff}$ (Hα) values are in reasonably good agreement with those from the Ldr, IRFM, and R09 methods, as shown in Fig. 2.

4. Conclusions

Effective temperatures have been determined using the method of Balmer line fitting for a sample of 62 solar analog stars, with internal errors of about 25 K. The other methods discussed in this work have internal errors of about 35 K. The high precision of our $T_{\rm eff}$ values are useful to find small residual trends in the comparison with other methods. We find reasonably good agreement with the $T_{\rm eff}$'s obtained with the Ldr, IRFM, and R09 methods, but small trends and offsets for the residuals are detected and removed with linear corrections. We argue that high accuracy effective temperatures, with errors of order 10 K, are possible to achieve for solar analog stars if several independent measurement are combined, mainly because the impact of errors is very small and can be understood and removed empirically.

Acknowledgments

I. R.'s work was performed under contract with the California Institute of Technology (Caltech) funded by NASA through the Sagan Fellowship Program. D. C. thanks the Organizing Commitee of the event for the financial support, and J.F. Valle of the direction of Astrophysics of CONIDA - Space Agency of Perú, for his suggestions and CONIDA for its support. P. S. B is a Royal Swedish Academy of Sciences Research Fellow supported by a grant from the Knut and Alice Wallenberg Foundation.

References

Barklem, P. S., Stempels, H. C., Allende Prieto, C. *et al.* 2002, *A&A*, 385, 951
Casagrande, L., Ramírez, I., Meléndez, J. *et al.* 2010, *A&A*, 512, 54
Gehren, T. 1981, *A&A*, 100, 97
Gray, D. F. & Johanson, H. L. 1991, *PASP*, 103, 439
Gray, D. F. 1994, *PASP*, 106, 1248
Kovtyukh, V. V., Soubiran, C., Belik, S. I., & Gorlova, N. I. 2003, *A&A*, 559, 564
Ramírez, I. & Meléndez, J. 2005, *ApJ*, 626, 465
Ramírez, I., Meléndez, J., & Asplund, M. 2009, *A&A*, 508, L17

Session 5

Grand minima and Historical Records

Comparative Magnetic Minima:
Characterizing Quiet Times in the Sun and Stars
Proceedings IAU Symposium No. 286, 2011
C. H. Mandrini & D. F. Webb, eds.

doi:10.1017/S1743921312005066

Stars in magnetic grand minima: where are they and what are they like?

Steven H. Saar and Paola Testa

Smithsonian Astrophysical Observatory,
60 Garden Street, Cambridge,MA 02138, USA
email: saar@cfa.harvard.edu, ptesta@cfa.harvard.edu

Abstract. We explore various ideas of what a star in a Maunder-like magnetic minimum would look like, and ways of finding stars in such a state, and make some estimates of their physical and magnetic activity properties. We discuss new X-ray observations of a small selection of candidates for being in magnetic grand minima. These are then compared with the Sun and other low activity stars.

Keywords. stars: magnetic fields, stars: late-type, stars: evolution, stars: activity, Sun: magnetic fields, Sun: activity

1. Introduction

Solar magnetic grand minima (=MGM), their properties and connection to normal cycle minima are clearly of interest for many reasons (hence, this conference!). Unfortunately, all recent examples of solar MGM fall outside the time horizon of modern instrumentation, so we are left with an imperfect record of sunspot counts and mostly indirect data of other types (e.g., cosmogenic isotopes). What is it about the solar dynamo that causes MGM? What are solar conditions like during these events? What governs the length of MGM and how does the Sun recover its cycle again? The existing data are useful, but more and better information would certainly help.

It has long been realized that solar-like stars may be very useful in better understanding solar activity and cycles (e.g., Wilson 1978). Carefully selected stars could inform not only what solar MGM conditions might have been like, but also how MGM conditions and frequency vary with stellar age, mass, and other properties. The difficulty lies in determining just what a star in an MGM *should* look like. Indeed, it has recently been suggested that there are very few solid MGM star candidates (Wright 2004), implying the Sun may be highly unusual for having these epochs of magnetic somnambulence! Is this view correct? This paper will review the history of various attempts to define MGM star candidates (§2), lay out a new attempt at defining MGM stars (§3), explore their properties (§4), and discuss implications for the solar MGM properties and MGM origins, and future directions (§5).

2. A bit of history

Early in the study of solar-like activity, the possibility of finding solar-like stars in MGM was recognized. Baliunas & Vaughan (1985), in their review of Ca II HK measurements in cool stars, noted, "because all other weak emission-line stars in this range [$0.72 \leqslant$ (B-V) $\leqslant 0.76$] do vary, the implication is that HD 10700 may be in an epoch of a virtual lull in chromospheric activity similar to the Maunder minimum." This star has been repeatedly mentioned as an MGM candidate, and is still often viewed as such (Judge *et al.* 2004).

Baliunas & Jastrow (1990) published an intriguing paper studying a mix of measurements of $S_{\rm HK}$, the Mount Wilson Ca II HK core-to-continuum ratio, over time, including four flat activity stars (with low $\langle S_{\rm HK}\rangle$ and variability σ_S). The sample was restricted to stars of solar-like color ($0.60 \leqslant$ B-V $\leqslant 0.76$). A histogram of the $S_{\rm HK}$ values revealed a bimodal distribution, with ≈30% lying in a low $\langle S_{\rm HK}\rangle$ peak. They suggested this meant that solar-like stars spend ≈30% of the time in MGM. Hall & Lockwood (2004), however, used a larger sample of low activity stars and more even time sampling and found no such bimodality.

Saar & Baliunas (1992) found that about 10-15% of the Mount Wilson survey stars were of the flat activity class, almost all of them concentrated in the F and G stars. These seemed to be the best MGM candidates, and Saar (1998) argued many of them might be in MGM, since their rotation rates were similar to cycling stars, and yet their activity levels were strongly dependent on rotation, suggestive of a turbulent (non cycling) dynamo (Bercik *et al.* 2005). The number of flat stars with known $P_{\rm rot}$, however, was small, making the statistics less than impressive.

Using a large survey of southern hemisphere stars, Henry *et al.* (1996) studied the distribution of calibrated $R'_{\rm HK} = (F_{\rm HK} - F_{\rm phot})/F_{\rm bol}$ values, where $F_{\rm HK}$ is the raw calibrated HK core flux (computed from $S_{\rm HK}$), $F_{\rm phot}$ is the photospheric component of the HK core flux, and $F_{\rm bol}$ is the bolometric flux (Noyes *et al.* 1984). $R'_{\rm HK}$ values in principle should permit better comparison of stars with different temperatures, since different backgrounds $F_{\rm phot}$ and continuum fluxes are accounted for. Henry *et al.* found that ≈5-10% of the their sample formed a distinct low activity group with log $R'_{\rm HK} \leqslant -5.1$, and identified these with MGM candidates.

Wright (2004) studied $R'_{\rm HK}$ derived by Wright *et al.* (2004) from a large database of exoplanet search spectra, comparing $R'_{\rm HK}$ with the star's magnitude separation Δ M_V from an *Hipparcos*-defined main sequence. He found that almost without exception, stars with log $R'_{\rm HK} \leqslant -5.1$ had Δ $M_V > 1$, and were thus not strictly solar-like, as they were either evolved (subgiants) or had strongly different metallicity, or both. He therefore questioned whether there were *any* true MGM candidates, at least with the restriction log $R'_{\rm HK} \leqslant -5.1$. Wright (2004) also presciently noted that there might be problems with the $R'_{\rm HK}$ calibration related to unaccounted-for metallicity and gravity effects. We return to this below.

Giampapa *et al.* (2006) took a different approach and surveyed the roughly solar age (≈ 4 Gyr) cluster M67. They found about 17% of dwarfs in the color range $0.58 \leqslant$ B-V $\leqslant 0.76$ had $\langle S_{\rm HK}\rangle$ below solar minimum levels, and identified these as MGM candidates. For a more restricted range of very solar-like stars ($0.63 \leqslant$ B-V $\leqslant 0.67$), ≈19% (4 of 21) lay below solar minimum in Ca II HK. Jenkins *et al.* (2008) looked at a large sample of southern dwarfs, and again adopting log $R'_{\rm HK} \leqslant -5.1$ for a MGM candidate, found 1–3% might be in such a state.

So, estimates of the fractions of stars in MGM have been generally dropping since the first studies, but is the fraction near zero (Wright 2004) or small but non-negligible (e.g., Giampapa *et al.* 2006)? How unusual *is* the Sun for having MGM episodes?

3. Constructing a new set of MGM candidate criteria

Wright (2004) raised some important issues concerning the most commonly used Ca II HK activity index. He noted that there is both metallicity and gravity effects that are entwined in the $S_{\rm HK}$ index, which complicate defining the activity minimum. Indeed, the traditional calibration to convert $S_{\rm HK}$ to a normalized photosphere-subtracted flux ratio $R'_{\rm HK}$ *does not take gravity or metallicity into account*. Others have complained recently that $R'_{\rm HK}$ does not correlate well with age for older (> 1 Gyr or so) stars (Pace & Pasquini 2004; Saffe *et al.* 2005); it is likely these calibration issues are partly to blame. Inspired by Wright (2004), Saar (2006) tried to remove the effect of gravity and metallicity effects on determining the minimum Ca II HK fluxes (and, hence, where to best search for MGM candidates). By matching stars in the Wright *et al.* (2004) HK database with the detailed spectroscopic modeling of Valenti & Fischer (2005), he compared the $R'_{\rm HK}$ values with accurate $T_{\rm eff}$, metallicities [M/H], gravities g, and $v \sin i$ (Saar 2006; Saar 2011). A fit to the main sequence seen in the log g – $T_{\rm eff}$ plane (Fig. 1) gave > 500 spectroscopically confirmed dwarfs independent of metallicity. Subgiants were then excluded, since their dynamos might well have stopped functioning due to their (evolution-driven) lower rotation rates. When the $R'_{\rm HK}$ of the Wright *et al.* (2004) sample was plotted against the log metallicity [M/H] relative to solar from Valenti & Fischer (2005), a striking dependence of the minimum $R'_{\rm HK}$ values in dwarfs could be seen (Fig 2; Saar 2006;Saar 2011). A rough boundary of the minimum can be sketched as log $R'_{\rm HK}$(min) = -5.125 - 0.213 [M/H]. Since even metal-poor halo dwarfs can have cycles (e.g., HD 103095; Baliunas *et al.* 1995) there seems no reason to exclude stars of arbitrary [M/H] from being candidate MGM stars. Thus, we can expect MGM candidates to reside among the lowest activity stars anywhere along this boundary. The previous criterion of log $R'_{\rm HK} \leqslant -5.1$ for *all* [M/H] would appear to be too restrictive, ruling out all but a few true dwarfs.

Very low average activity is a probably necessary but certainly not sufficient condition to be an MGM candidate. (The caveat "probably" is included because we shouldn't

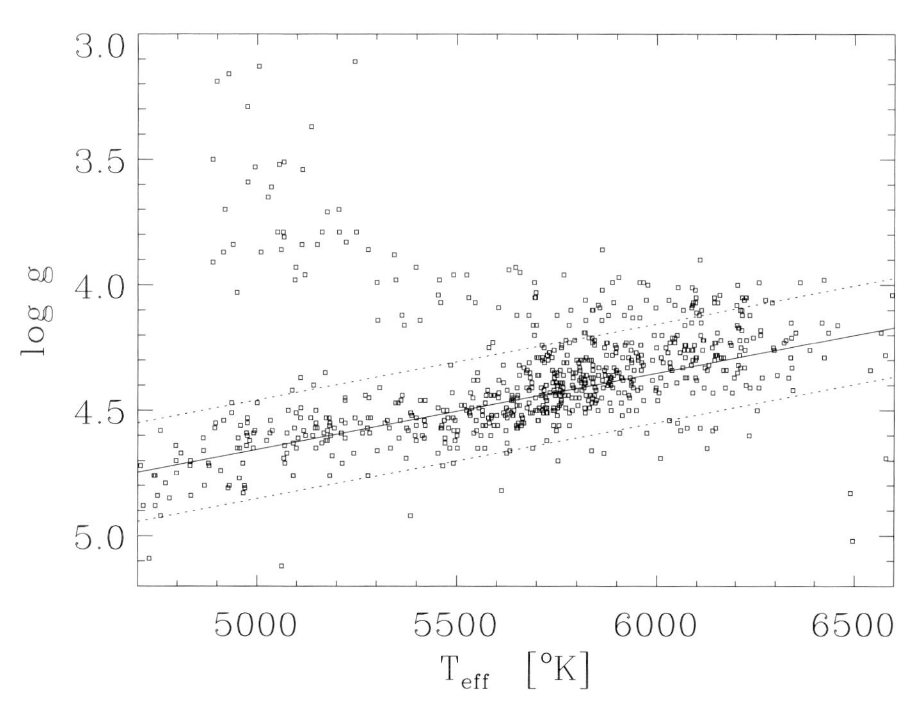

Figure 1. $T_{\rm eff}$ vs. $\log g$ from Valenti & Fischer (2005) for stars also in Wright *et al.* (2004). A linear fit to the main sequence (solid) and $2.3\sigma_{\rm fit}$ boundaries defining dwarfs (dotted) are shown (from Saar 2011).

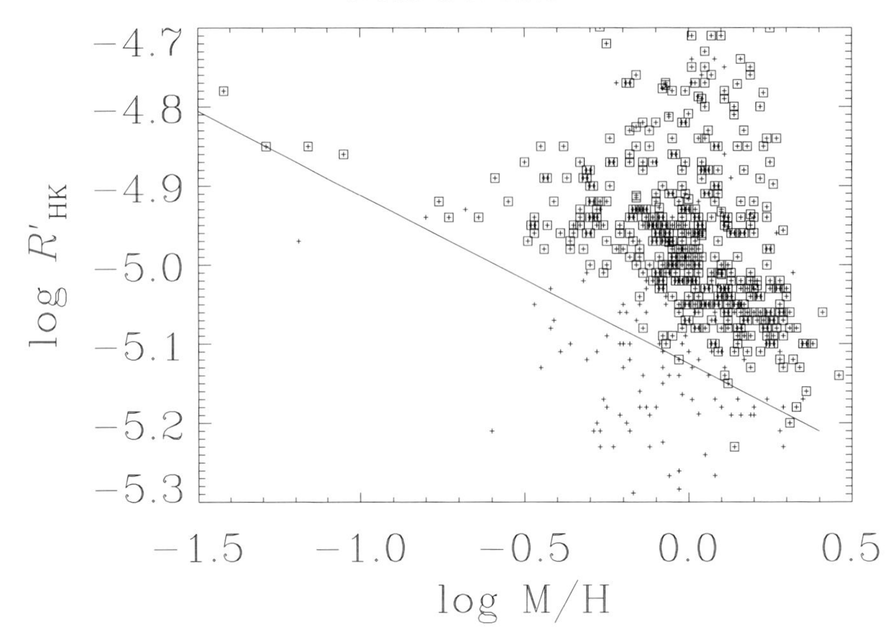

Figure 2. $R'_{\rm HK}$ from Wright *et al.* (2004) vs. log M/H from Valenti & Fischer (2005) for stars in common; dwarfs as defined by Fig. 1 are boxed. A linear lower boundary for $R'_{\rm HK} \approx -0.213$ [M/H] - 5.125, is drawn (solid); only one dwarf falls well below this boundary (from Saar 2011).

completely rule out the possibility that the activity level of an MGM state is related to the the mean activity level and/or rotation rate of the star when not in such a state. Faster rotators *might* have higher $\langle R'_{\rm HK} \rangle$ in their grand minima than slower rotators.) A good MGM candidate should also have low HK *variability* and to have had sustained it for as long a time as possible. At a minimum, σ_S should be derived from data which span an interval longer than the longest normal solar cycle minimum – about 4 years. Interestingly, just above the $R'_{\rm HK}$(min) boundary, for $\Delta \log R'_{\rm HK} = \log R'_{\rm HK} - \log R'_{\rm HK}$(min) $\leqslant$0.054, nearly all stars show $\sigma_S/S_{\rm HK} \leqslant 2\%$ (Fig. 3). This is the variabiliy level Baliunas *et al.* (1995) used to define "flat activity" stars, and we adopt it again here. A 2% HK variation criterion is a factor of ~5 below the Sun's over a typical cycle (as measured from the Sacramento Peak Observatory K line data over the last two cycles) and is ~30% less than the average variation during the last three solar minima.

Our new MGM candidate critera are thus the following: (1) the star is a bona-fide dwarf as determined from spectroscopically determined $T_{\rm eff}$ and gravity (Fig. 1); (2) The star shows log $R'_{\rm HK} < -5.125$ -0.213 [M/H] + δ, where [M/H] is the spectroscopically determined log metal abundance (relative to solar) and $\delta \approx 0.054$; and (3) $\sigma_S/S_{\rm HK} \leqslant 2\%$ spanning at least $t_{\rm obs} \geqslant 4$ years of measurements (i.e., > a solar minimum timescale).

Indirect evidence from cosmogenic isotopes such as ^{10}Be and ^{14}C indicates that cyclic modulation continued through the Maunder minimum (e.g., Usoskin *et al.* 2004), despite minimal spot modulation. It would appear that the magnetic cycle continued, producing weak activity (active network/plage and thus Ca II HK emission) without many spots. Thus, it seems reasonable to expect *some* HK variation during MGM, but this discussion underlines the uncertainties of the problem. Note also that, for example, the Dalton minimum was much shorter and less deep than the Maunder! Clearly, solar grand minima

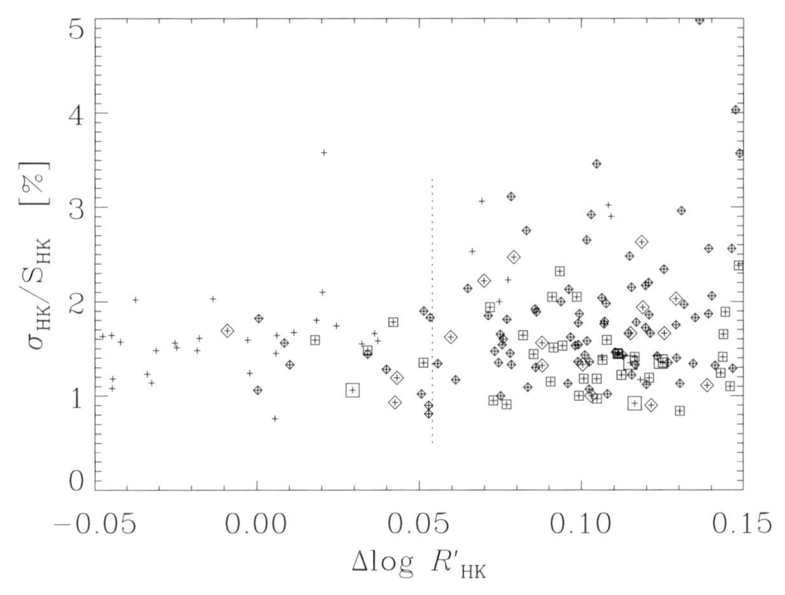

Figure 3. Fractional raw $S_{\rm HK}$ variability from Valenti & Fischer (2005) vs. distance $\Delta \log R'_{\rm HK} = \log R'_{\rm HK} - \log R'_{\rm HK}\,({\rm min})$ from dwarf $R'_{\rm HK}\,({\rm min})$ boundary in Fig. 2 for stars with $t_{\rm obs}$ >4 years; non-dwarfs are +, dwarfs with $t_{\rm obs}$ >4, 5, 6, or 7 years of data are enclosed by small diamonds and small squares, large diamonds and large squares, respectively. Below $\Delta \log R'_{\rm HK}$ = 0.054 (dotted), all dwarfs show $\sigma_S/S_{\rm HK}$ < 2%; above this, maximum variability rapidly increases. We propose dwarfs with $\Delta R'_{\rm HK} \leqslant 0.054$ and data spanning $t_{\rm obs}$ > 4 years are candidate magnetic grand minimum stars (MGM stars; from Saar 2011).

have a range of properties themselves; we can probably expect the same from their stellar analogs.

4. Properties of the new set of MGM candidates

The set of MGM candidates defined above can now be explored to determine their important properties (see also Saar 2011). Overall, ≈7% of the joint Wright *et al.* – Valenti & Fischer sample are MGM candidates. The fraction of candidates peaks near the solar $T_{\rm eff}$ at ≈17%, cuts off sharply for $T_{\rm eff} \geqslant$ 6050 K (~F9), and averages <7% elsewhere (Fig. 4). The distribution with [M/H] is fairly even, with perhaps an increase at [M/H] <-0.65, though statistics are poor. A $T_{\rm eff}$ – [M/H] correlation suggests a relatively small range of convective zone depths are allowed (Fig. 5). The MGM candidate $v \sin i$ distribution is not statistically different from that of stars just above the $\Delta R'_{\rm HK}$ cutoff ($0.054 < \delta \leqslant 0.108$; Saar 2011).

These selection criteria may be overstrict, as they have been set conservatively to avoid false positives. For example, there may be MGM candidates among the low $\sigma_{\rm HK}$ stars with $\Delta R'_{\rm HK} > 0.054$. The criteria certainly exclude some stars which are otherwise good candidates. For example, 51 Peg, a low $\sigma_{\rm HK}$ star (Baliunas *et al.* 1995) with low X-ray emission (Poppenhäger *et al.* 2009) is excluded here only because $t_{\rm obs}$ <1 year in the Wright *et al.* (2004) data set.

The MGM candidates form a fairly distinct group at the bottom of the $\Delta \log R'_{\rm HK}$ distribution (Fig. 6), well separated from the (on average) more variable stars with

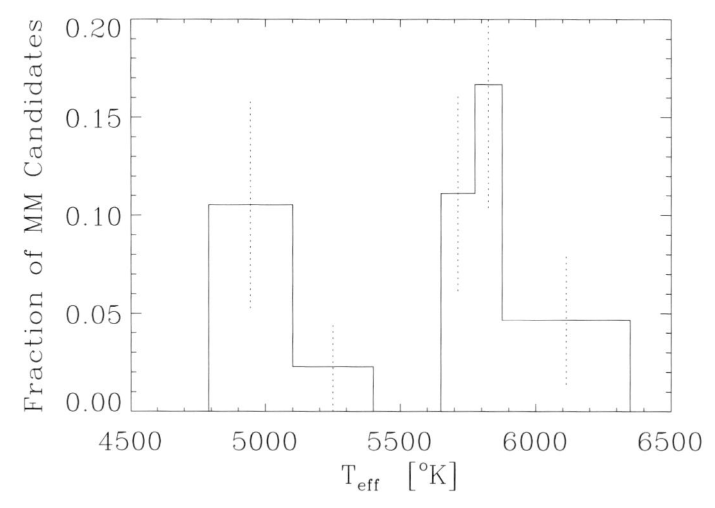

Figure 4. Distribution of MGM candidates observed for a duration $t_{\rm obs} > 4$ years (as a fraction of all dwarfs in the sample with $t_{\rm obs} > 4$ years) as a function of $T_{\rm eff}$ with ≈41 stars per bin. There is a peak at $T_{\rm eff} = 5825 \pm 50$ K, with no candidates showing $T_{\rm eff}$ >6100 (from Saar 2011).

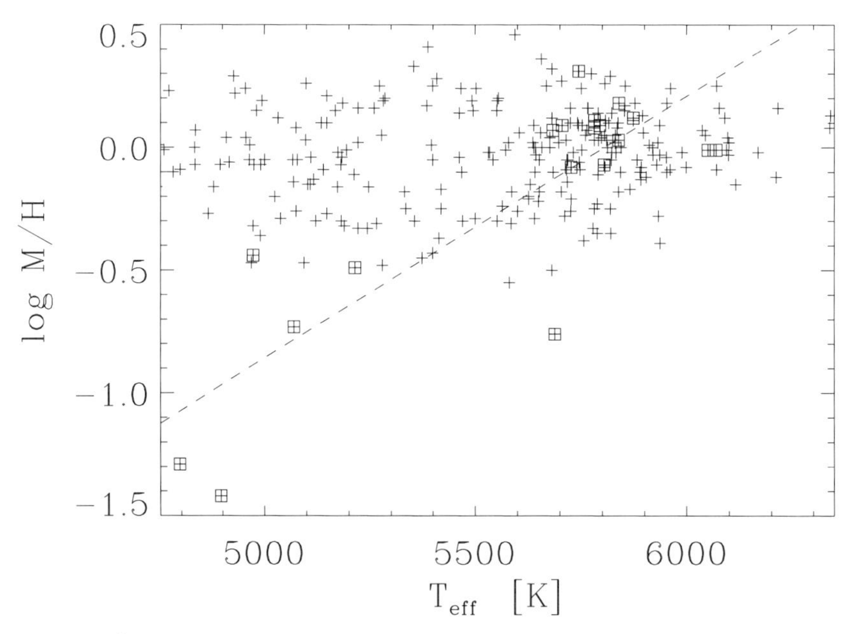

Figure 5. Log M/H vs. $T_{\rm eff}$, for all dwarfs with $t_{\rm obs} > 4$ years and for MGM candidates $t_{\rm obs} > 4$ years (boxed). Unlike the overall sample, the MGM candidates show a distinct trend (dashed) of increasing metallicity with $T_{\rm eff}$ (from Saar 2011).

$\Delta \log R'_{\rm HK}$ values just above. This slight bimodal character brings back the possibility that the MGM is truly a distinct mode of the dynamo, rather than the tail of the distribution of behavior. The metallicity adjustment provided by using $\Delta \log R'_{\rm HK}$ allows this bimodality to be seen; it is obscured without this correction. The gap is quite small though, and requires accurate $R'_{\rm HK}$ data to discern.

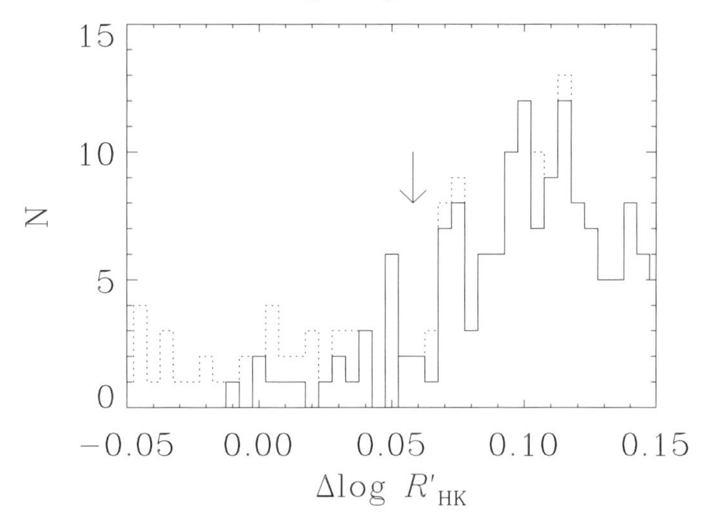

Figure 6. Histogram of $\Delta \log R'_{\rm HK}$ for all stars (dotted), all dwarfs (solid), and MGM candidates defined in Fig. 3 (thick solid); $t_{\rm obs} > 4$ years throughout. The MGM candidates form a fairly distinct population below a gap (arrow; from Saar 2011).

We have recently begun a study of the coronae of selected MGM candidates as defined by the above criteria (§3). X-rays are unambiguous markers of magnetic fields in non-accreting cool stars, since they are impossible to produce in any quantity without magnetic-driven heating. We used *Chandra's* ACIS-S to measure four MGM candidates, and one test case (matching the MGM criteria except for a too-short $t_{\rm obs}$). We are still awaiting the data from one of these targets, but can give a preliminary report on the others here. Counts in the energy range 0.1-5 keV were extracted within a 8-pixel radius circle and compared to a background in an annulus 16 to 30 pixels distant, centered on the apparent source position. A modification of this was needed for one source (HD 179958), whose common proper motion companion (HD 179957) was serendipitously detected 8" away. Here, wedges were removed from the background annuli of each, and we estimated the (small) cross-contamination of the two stars.

An accurate X-ray surface flux F_X requires an good estimate of the mean coronal temperature T_X. Therefore, we also extracted counts for our sources between 0.2-5 keV, and used APEC models to predict how the count ratio $C_{0.1-5}/C_{0.2-5}$ would vary with T_X. We compute the X-ray luminosity L_X using the best estimate T_X (which were generally uncertain), *Hipparcos* distances d, and column density $N_H = 0.07d$ cm^{-2} with d in parsecs. Our preliminary results and comparison stars are gathered in Table 1. In all cases, the coronal temperatures for the MGM candidates were $T_X \leqslant 1$ MK, indicating very cool coronae, similar to 51 Peg. The corresponding F_X (computed using radii from Valenti & Fischer 2005) were in good agreement with Schmitt (1997) and Schmitt & Liefke (2004), with all the new MGM candidates having surface fluxes $F_X \sim 10^4$ ergs cm^{-2} s^{-1} (Fig. 7). The common proper motion (and thus presumably coeval) companion to candidate HD 179958 showed about ~3 times higher F_X and a slightly warmer T_X, demonstrating a notable coronal activity difference at fixed age and similar mass.

The one exceptional star showing $F_X \ll 10^4$ ergs cm^{-2} s^{-1}, HD 157214, was also a test case MGM candidate with $t_{\rm obs} < 1$ year. Indeed, Hall *et al.* (2009) observed the star recently showing notably higher $R'_{\rm HK}$, suggesting Wright *et al.* (2004) observed the star is a *temporary* (cycle?) minimum. In this respect, the star may be similar to α Cen A,

Table 1. Coronae of MGM Candidates and Comparison Stars

HD	V	Spec. type	$T_{\rm eff}^1$ [K]	[M/H][1]	log g^1	log $R'_{\rm HK}$[2]	σ_S^2 [%]	$t_{\rm obs}^2$ [yr]	$L_X/10^{27}$ [ergs s^{-1}]	log T_X	notes
50806	6.04	G5V	5685	0.07	4.36	-5.10	1.28	6.4	2.8	5.8:	new
120066	6.30	G0,5IV-V	5873	0.12	4.23	-5.15	1.82	6.4	1.9	5.9:	new
157214	5.40	G0V	5695	-0.15	4.50	-5.04	0.99	0.2	0.2	5.8:	new
179958	6.57	G4V	5760	0.05	4.39	-5.08	1.34	6.3	2.0	5.8:	new
10700	3.50	G8V	5283	-0.36	4.59	-4.98	1.35	3.0	0.5	6.0:	τ Cet
207014	5.49	G5V	5787	0.15	4.45	-5.08	...	0.2	0.63[3]	<6.0	51 Peg
128620	-0.01	G2V	5801	0.19	4.33	...	...	...	0.20[4]	⩽6.0	α Cen A, in min.
179957	6.75	G6V?	5676	0.00	4.34	-5.05	1.0	6.3	2.4	5.9:	new, non-MGM

Notes: [1]from Valenti & Fischer (2005); [2]from Wright *et al.* (2004); [3]from Poppenhäger *et al.* (2009); [4]from Ayres *et al.* (2008)

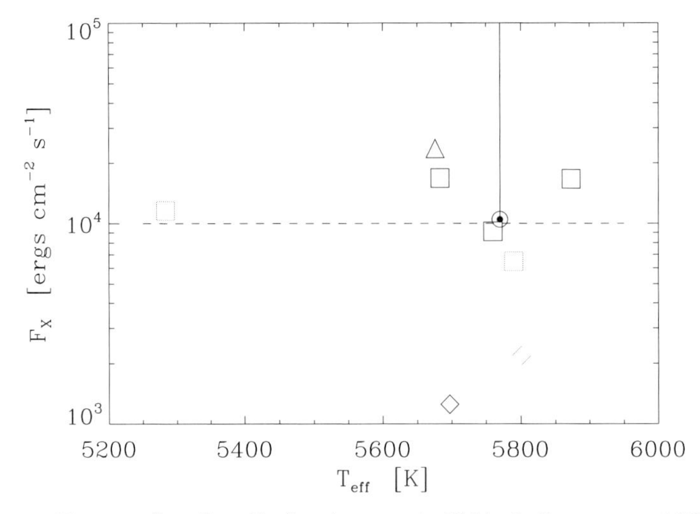

Figure 7. $T_{\rm eff}$ vs. X-ray surface flux F_X for the stars in Table 1. Squares are MGM candidates, diamonds are stars in temporary (cycle?) minima, the triangle is the non-MGM companion of HD 179958, and gray symbols are from the literature. The Sun at cycle minimum is marked ($\odot$) as is the minimum F_X level in dwarfs seen by Schmitt (1997) (dashed).

which recently became extremely faint and cool in X-rays for a period of a few years (Ayres *et al.* 2008).

5. Discussion

The MGM candidate selection method above can be refined to better remove the residual metallicity effects. Recall that $R'_{\rm HK} = (F_{\rm HK} - F_{\rm phot})/F_{\rm bol}$ (Noyes *et al.* 1984). The correction used above (Saar 2011) is a bulk adjustment to $R'_{\rm HK}$, not considering the (possibly different!) individual contributions of $F_{\rm HK}$(M/H), $F_{\rm phot}$(M/H) and $F_{\rm bol}$(M/H). Since $T_{\rm eff}$(M/H) is known for these stars from Valenti & Fischer (2005), $F_{\rm bol}$ can be corrected for separately. If both $F_{\rm HK}$ and $F_{\rm phot}$ scale similarly with M/H, then defining a minimum $R'_{\rm HK}$ as a function M/H *with an M/H corrected* $F_{\rm bol}$ should be a better estimate of the true minimum Ca II HK in dwarfs. An improved analysis along these lines is in progress (Saar 2012, in prep.). Ultimately though, a full recalibration of $R'_{\rm HK}$ might be needed for optimum results.

Radial velocity surveys for exoplanets tend to avoid more active stars, since they have higher levels of velocity "jitter" due to the rotation and evolution of active regions. The stellar sample here, drawn from such a survey, is thus biased towards optically brighter, lower activity stars. Effectively, this means that the sample includes few stars younger than ∼1 Gyr, but should be representative of older nearby dwarfs.

The typical X-ray levels of the MGM candidates are similar to inactive areas of the Sun (Schmitt 1997) and the average Sun at cycle minimum (Judge *et al.* 2003). Dwarf stars can apparently occasionally drop below this level (HD 157214, α Cen A), but this may be only in brief excursions, in which their coronae are very cool ($T_X \ll 10^6$ K). Perhaps these stars become briefly dominated with coronal holes, the coolest solar coronal structures. In other magnetic activity diagnostics, MGM candidates would seem to be only marginally less active ($R'_{\rm HK}$) or similar to (FUV lines; Judge & Saar 2007) the cycle minimum Sun. By implication, solar emission in MGM may not be much different from a deep cycle minimum. Barring actually seeing a star enter or leave an MGM then (e.g., perhaps Donahue *et al.* 1995), low activity *variability* over long timescales may be more indicative of MGM stars than just a low average level.

Comparing with Giampapa *et al.* (2006), the fraction in our sample in the roughly equivalent $T_{\rm eff}$ range ($5700 \leqslant T_{\rm eff} \leqslant 5830$) to the near-solar sample ($0.63 \leqslant$ B-V $\leqslant 0.67$) in M67 is 12.5% (7 out of 56). This would suggest that near-solar stars at 4 Gyr may have a higher rate of MGM behavior (19%) than for our sample of mixed ages. However, the M67 sample includes some binaries. As binaries are not strictly solar-like, and even wide binaries may have their rotational evolution (and, thus, perhaps their activity and dynamos) altered relative to single stars (Meibom *et al.* 2006), it might be safer to exclude these stars from the comparison. If we remove binaries, the MGM fraction of the M67 sample there drops to ≈12% (2 of 17), very similar to our result. This might then be a better estimate of the amount of time the Sun should spend in MGM.

Long term study of older clusters (e.g., M67, NGC 752, Rup 147, Kepler clusters NGC 6819 and 6791) will help better define the dependence of the MGM phenomenon and its properties with age. Also needed are more X-ray and $P_{\rm rot}$ measurements, though these will not be trivial. In particular, if the new MGM candidates are well chosen, they should show a $P_{\rm rot}$ distribution similar to low activity cycling stars of the same mass. Comparing Rossby number distributions may be useful, (e.g., Saar 1998). How does MGM frequency change with stellar age and rotation? How tightly restricted is the MGM mass dependence? How does the spectrum of MGM durations vary with stellar parameters? Some progress has been made, but the present indications need refinement and there is still much left to learn!

Acknowledgments

This work was supported by a Solar Heliospheric Guest Investigator grant NNX10AF29G, and *Chandra* grant GO1-12036X. We are grateful to J. Wright, P. Judge, J. Schmitt, T. Ayres, and K. Poppenhäger for many helpful discussions.

References

Ayres, T. R., Judge, P. G., Saar, S. H., & Schmitt, J. H. M. M. 2008, *ApJ*, 678, L121
Baliunas, S. L., Donahue, R. A., Soon, W. H., *et al.* 1995, *ApJ*, 438, 269
Baliunas, S. L. & Vaughan, A. H. 1985, *ARAA*, 23, 379
Baliunas, S. & Jastrow, R. 1990, *Nature*, 348, 520

Bercik, D. J., Fisher, G. H., Johns-Krull, C. M., & Abbett, W. P. 2005, *ApJ*, 631, 529
Donahue, R. A., Baliunas, S. L., Soon, W. H., & McMillan, F. M. 1995, IAU Symposium, 176, 72P
Giampapa, M. S., Hall, J. C., Radick, R. R., & Baliunas, S. L. 2006, *ApJ*, 651, 444
Hall, J. C., Henry, G. W., Lockwood, G. W., Skiff, B. A., & Saar, S. H. 2009, *AJ*, 138, 312
Hall, J. C. & Lockwood, G. W. 2004, *ApJ*, 614, 942
Henry, T. J., Soderblom, D. R., Donahue, R. A., & Baliunas, S. L. 1996, *AJ*, 111, 439
Jenkins, J. S., Jones, H. R. A., Pavlenko, Y., *et al.* 2008, *A&A*, 485, 571
Judge, P. G. & Saar, S. H. 2007, *ApJ*, 663, 643
Judge, P. G., Saar, S. H., Carlsson, M., & Ayres, T. R. 2004, *ApJ*, 609, 392
Judge, P. G., Solomon, S. C., & Ayres, T. R. 2003, *ApJ*, 593, 534
Meibom, S., Mathieu, R. D., & Stassun, K. G. 2006, *ApJ*, 653, 621
Noyes, R. W., Hartmann, L. W., Baliunas, S. L., Duncan, D. K., & Vaughan, A. H. 1984, *ApJ*, 279, 763
Pace, G. & Pasquini, L. 2004, *A&A*, 426, 1021
Poppenhäger, K., Robrade, J., Schmitt, J. H. M. M., & Hall, J. C. 2009, *A&A*, 508, 1417
Saar, S. H. 1998, Cool Stars, Stellar Systems, and the Sun 10, ASP Conf. Ser. Vol. 154, 211
Saar, S. H. & Baliunas, S. L. 1992, The Solar Cycle, ASP Conf. Ser. Vol. 27, 150
Saar, S. H. 2006, *Bulletin of the American Astronomical Society*, 38, 240
Saar, S. H. 2011, Cool Stars, Stellar Systems, and the Sun 16, online poster volume.
Saffe, C., Gómez, M., & Chavero, C. 2005, *A&A*, 443, 609
Schmitt, J. H. M. M. 1997, *A&A*, 318, 215
Schmitt, J. H. M. M. & Liefke, C. 2004, *A&A*, 417, 651
Usoskin, I. G., Mursula, K., Solanki, S., Schüssler, M., & Alanko, K. 2004, *A&A*, 413, 745
Valenti, J. A. & Fischer, D. A. 2005, *ApJS*, 159, 141
Wilson, O. C. 1978, *ApJ*, 226, 379
Wright, J. T. 2004, *AJ*, 128, 1273
Wright, J. T., Marcy, G. W., Butler, R. P., & Vogt, S. S. 2004, *ApJS*, 152, 261

Discussion

Volker Bothmer: If you had looked at the Sun for the last two or three years, you would have been in the band that you are considering as Maunder Minimum?

Steve Saar: Not quite, at least in Calcium. Its still a little high – in the chromosphere at least, the Maunder Minimum candidates seem to be a little bit lower.

Dibyendu Nandy: It seems to me that your candidates with spots all have different coronal temperatures and different X-ray fluxes and presumably different magnetic fluxes. Given this, is it fair to conclude that there is not a single, absolute for stellar minimum active regions?

Steve Saar: They are sort of scattered, but within the large uncertainties in the temperatures I think we can say that 10^4 is a reasonable estimate within the error bars we currently have. It would be nice to have an X-ray mission that could look nice and soft, and get better values than these, but unfortunately such a mission doesn't exist.

Ramon Lopez: I didn't catch how many stars were actually observed? Given the amount of stars that you are observing , how long would it actually take you to actually catch a "smoking gun" of a star going into or coming out of a Maunder Minimum?

Steve Saar: The number of stars in the survey was about 1200, with about 600 being true dwarfs in the end. The calculation of how long to see a Maunder state is not quite so straightforward: F and K stars are mostly not so good, so only the G stars which are even a smaller subset of the sample that we actually see 13.

JEFF LINSKY: I barely heard the word "solar twin" – a comment: people should look systematically (for many years) at solar twins (stars very nearly like the Sun).

STEVE SAAR: Agreed, but the number of really good "solar twins" is small. We need more of them too!

Comparative Magnetic Minima:
Characterizing quiet times in the Sun and Stars
Proceedings IAU Symposium No. 286, 2011
C. H. Mandrini & D. F. Webb, eds.

doi:10.1017/S1743921312005078

Soft X-ray emission as diagnostics for Maunder minimum stars

Katja Poppenhaeger and Jürgen H. M. M. Schmitt

Hamburger Sternwarte, Gojenbergsweg 112, 21029 Hamburg, Germany
email: katja.poppenhaeger@hs.uni-hamburg.de

Abstract. The identification of stars in a Maunder minimum state purely from their chromospheric emission (for example in Ca II lines) has proven to be difficult. Photospheric contributions, metallicities and possible deviations from the main sequence stage may lead to very low values of the traditional chromospheric activity indicators, while no Maunder minimum state may be present. X-ray observations can be a key tool for identifying possible Maunder minimum stars: We have detected very soft X-ray emission from low-temperature coronal plasma, similar to emission from solar coronal holes, in several stars with very low chromospheric activity indicators. The coronal properties inferred from X-ray observations can therefore yield a crucial piece of information to verify Maunder minimum states in stars.

Keywords. X-rays: stars, stars: activity, stars: coronae, stars: late-type

---

1. Maunder minimum candidates

We know of three phases in the life of the Sun when it displayed very low activity over a long time: the Maunder, Dalton and Spörer Minima, of which the Maunder Minimum was the most extreme. The causes of such Grand minima, however, are still unknown. The recent very low and slightly prolonged minimum of the solar cycle in 2008/2009 may have given us a taste of what a Maunder-minimum state might look like observationally. By observing other stars, we can estimate how frequent such extended minima states are. Initially, a large number of stars (30%, later corrected to 10 – 15%) was assumed to be in a Maunder minumum based on chromospheric activity measured in the Mount Wilson project (Saar & Baliunas 1992). However, it was shown by Wright (2004) that most of these stars were actually evolved stars, making their chromospheric activity indices incomparable to main sequence stars. Additionally, chromospheric activity is always influenced by a small amount of basal heating, making X-ray and EUV measurements necessary to truly assess if a star is in a Maunder minimum (Judge & Saar 2007).

We have started an observational program at X-ray wavelengths to determine the coronal properties of several solar-like stars with very low chromospheric activity, in order to identify possible Maunder minimum stars. We demonstrate here our technique to determine coronal temperatures for inactive stars, and present the derived temperatures for 51 Peg and, as a moderate activity comparison, HD 189733.

2. Inferring coronal temperatures

Stellar (and solar) X-ray emission is a direct consequence of the magnetic heating of coronae, and is therefore an ideal tool to measure magnetic activity. It is well known that both X-ray luminosities and coronal temperatures are correlated with the activity level; for quiet regions on the Sun, coronal temperatures of approximately 1 MK have

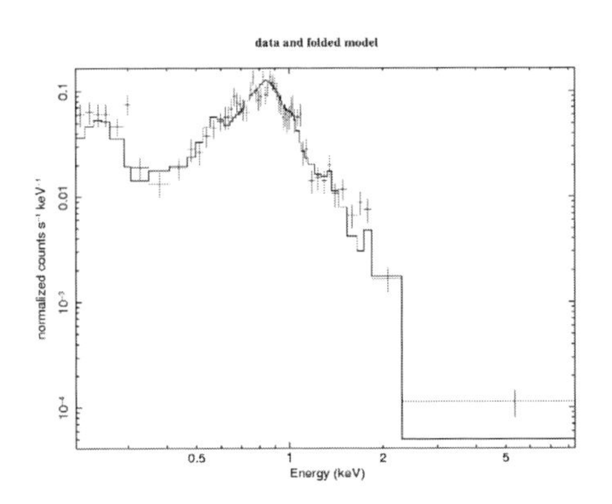

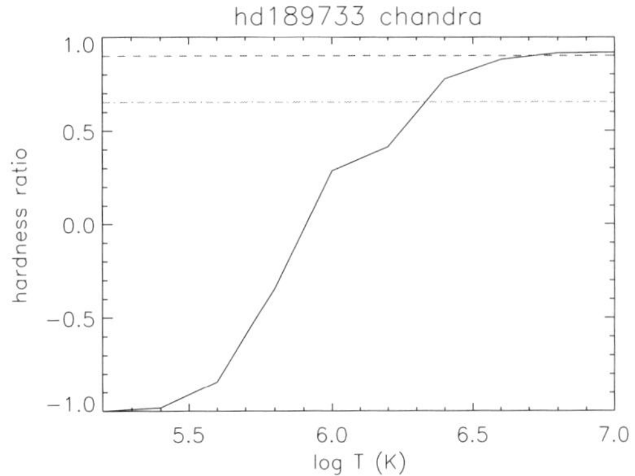

Figure 1. Left panel: Soft X-ray spectrum of the moderately active star HD 189733A. The spectral fit yields a mean coronal temperature of $\log T = 6.59$. **Right panel:** Theoretical temperature-dependent hardness ratio (black solid) together with the observed hardness ratio of HD 189733A, measured from only 20 source counts (green dashed) with 1σ errors (green dash–dotted).

been measured, while more active stars can easily display coronal temperature components above 10 MK. Consequently, stars with a very low activity level such as Maunder minimum candidates will display low X-ray luminosities and coronal temperatures.

Furthermore, an estimate of the coronal temperature is needed to convert a measured X-ray count rate into the stellar X-ray flux. For X-ray bright stars, moderately resolved CCD spectra or even high-resolution grating spectra with sufficient signal to noise ratio can be collected within reasonable exposure times. These spectra can be fitted with an optically-thin thermal plasma model to yield coronal temperatures and emission measures. For low-activity stars, however, one often collects only 20 or fewer source counts, and therefore needs other methods to infer the coronal temperature.

A convenient way to derive an estimate for the coronal temperature of an X-ray dim target is to analyze hardness ratios. A hardness ratio is defined as $HR = \frac{H-S}{H+S}$, with H and S being the number of X-ray photons in a predefined hard and soft energy band. As stellar X-ray spectra shift to lower energies for lower coronal temperatures, hardness ratios are intrinsically temperature-dependent. To demonstrate the accuracy of this method, we compare the coronal temperature estimate for the moderately active K star HD 189733A derived from its X-ray CCD spectrum, consisting of ca. 1200 source counts, as opposed to the temperature derived from the hardness ratio of a short excerpt of the data with only 20 source counts in total.

HD 189733A was observed by the Chandra X-ray telescope with the ACIS-S camera for 20 ks on July 05 2011. The stellar emission is soft, with negligible flux above 5 keV. The X-ray spectrum in the 0.2-5 keV energy band is shown in Fig. 1 (left panel), together with a two-temperature-component fit (details of the fit given in Table 1).

We then analyze a fraction from that data which contains only 20 source counts. We calculate the theoretical hardness ratios of synthetic spectra folded with the ACIS-S response using Xspec v.12 and compare them to the measured hardness ratio, using energy ranges of $200 - 350$ eV and $350 - 1000$ eV as the soft and hard band, respectively. The threshold between the hard and soft band is at a quite low energy; this is because we will need the low-energy diagnostic power when investigating the soft X-ray emission of low-activity stars. For consistency, we use the same energy threshold for HD 189733A as well, even if a higher threshold would reduce the error bars for this moderately active star. The result of our analysis is shown in Fig. 1 (right panel). The observed hardness

Table 1. Coronal properties of HD 189733A, derived from ACIS-S CCD spectra; emission measures given in units of 10^{50} cm^{-3}.

T_1 (K)	EM_1	$\log T_2$ (K)	EM_2	mean $\log T$ (K)	χ^2	L_X (erg/s)
6.14	4.82	6.84	4.03	6.59	1.44	1.25×10^{28}

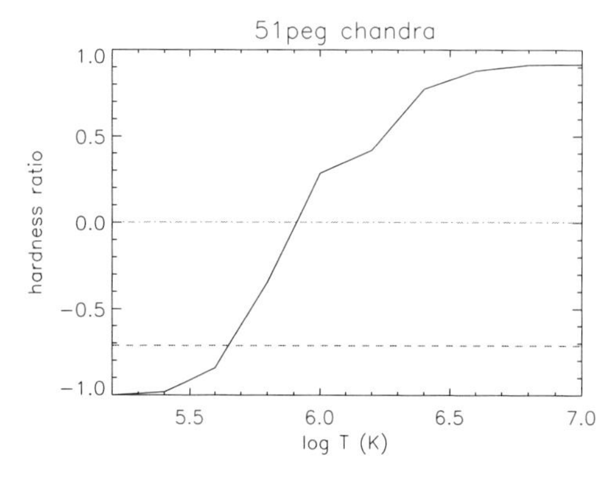

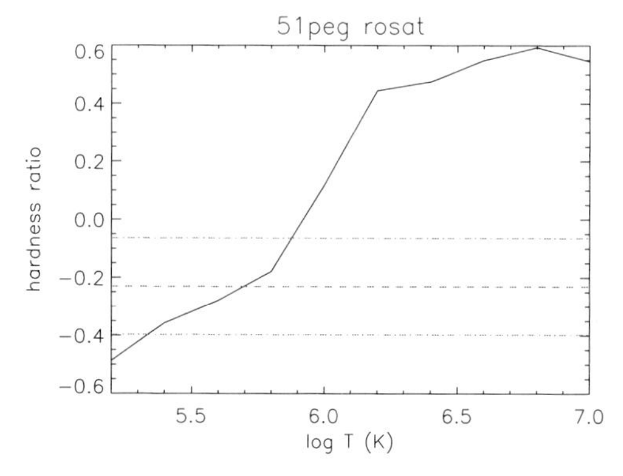

Figure 2. Left panel: As Fig. 1 (right panel), but for the Maunder minimum candidate 51 Peg. The star clearly displays a mean coronal temperature below 1 MK during the Chandra observation in 2008. **Right panel:** Same as left panel, but for archival X-ray data from ROSAT, taken in 1992. The very low coronal temperature seems to be persistent over 16 years.

ratio matches the synthetic spectra at a coronal temperature of $\log T = 6.65$, which is in close agreement with the spectral fit to the full data; however, due to the low energy threshold, it not well constrained towards higher temperatures.

With the validity of the method shown, we can apply it to low activity stars to determine their coronal temperature. The G-type star 51 Peg, host star to a Hot Jupiter, displays a very low magnetic activity level in chromospheric measurements. Its mean Mount-Wilson S index is 0.149 (Baliunas *et al.* 1995); for comparison, the Sun's S index varies between ca. 0.16 and 0.21 during its activity cycle. In a 5 ks observation with Chandra ACIS-S performed in 2008, we collected eight X-ray source photons from 51 Peg. Additionally, an archival ROSAT observations from 1992 exists with an exposure time of 12.5 ks. In Fig. 2 we show the observed and theoretical hardness ratios; we can constrain 51 Peg's mean coronal temperature to be $\log T \leqslant 5.85$ during the Chandra exposure and to $5.35 \leqslant \log T \leqslant 5.9$ during the ROSAT pointing. The X-ray activity indicator $\log L_X/L_{bol}$ has a low value of ≈ -7 (Poppenhäger *et al.* 2009).

Putting these coronal temperatures into the solar context is somewhat complicated by the fact that disk-integrated soft X-ray spectra of the Sun are difficult to construct. However, quiet regions on the Sun seem to be quite well characterized by a coronal temperature of ≈ 1 MK (Orlando *et al.* 2001). This puts 51 Peg's corona at the lowest end of the solar coronal temperature span. The very low coronal temperature and L_X/L_{bol} ratio which are persistent over a long time span suggest that 51 Peg may indeed be in a Maunder minimum state.

The case of 51 Peg indicates that coronae during Maunder minima might observationally not be very different from the solar corona in deep single minima like in 2008/2009, except for the longer duration. Already the compilation by Schmitt (1997) of X-ray fluxes normalized by the stellar surface showed that even in very low-activity stars, the X-ray surface flux does not fall below 10^4 erg s^{-1} cm^{-2} in the 0.1-2.4 energy band. Apparently, coronae experience a low level of heating even during extended minima.

3. Conclusion

With contemporaneous X-ray telescopes, coronal temperature estimates can be inferred even for very inactive cool stars. Our observations show that the planet-hosting star 51 Peg is probably in a Maunder minimum state, characterized by low coronal temperature as well as low X-ray and chromospheric activity indicators over a time span of 16 years. Even in such a long-term minimal activity state, the coronal X-ray emission does not vanish completely; it is similar to the solar coronal emission of quiet regions, indicating that long-term minima may not be fundamentally different from single deep minima of the solar cycle. X-ray observations of a larger sample of inactive stars are under way and will yield more insight into the frequency of extended minimium states in solar-like stars.

References

Baliunas, S. L., Donahue, R. A., Soon, W. H., *et al.* 1995, *ApJ*, 438, 269
Judge, P. G. & Saar, S. H. 2007, *ApJ*, 663, 643
Orlando, S., Peres, G., & Reale, F. 2001, *ApJ*, 560, 499
Poppenhäger, K., Robrade, J., Schmitt, J. H. M. M., & Hall, J. C. 2009, *A&A*, 508, 1417
Saar, S. H. & Baliunas, S. L. 1992, in *Astron. Soc. Pacific CS*, Vol. 27, The Solar Cycle, ed. K. L. Harvey, 150–167
Schmitt, J. H. M. M. 1997, *A&A*, 318, 215
Wright, J. T. 2004, *AJ*, 128, 1273

Discussion

DIBYENDU NANDY: From the point of view of dynamo theory it is important to know whether stars fall in Maunder minima suddenly or gradually. Is there any observation like this from other stars that can tell us about it?

KATJA POPPENHAEGER: It's very difficult to obtain information on transitions from cyclic activity to a Grand Minimum or the other way around from X-ray observations. Usually, our time sampling is rather sparse and we are somewhat limited by the life time of the X-ray satellites (even if cross-calibration with other missions like ROSAT works rather well), so that X-ray data cannot yield strong constraints on such transitions at the moment.

MARK GIAMPAPA: Low activity stars are not just low in X-ray luminosity but also low in coronal temperature. Have you thought about a cause for the structure of the corona of these stars?

KATJA POPPENHAEGER: Also our sample stars follow the canonical picture of low coronal temperature being correlated with low overall X-ray luminosity, My feeling is that the coronae of these stars might consist in a large part of coronal holes, and very small loops in the remaining parts. However, with our few X-ray source counts, we cannot perform the sophisticated analysis needed to really infer the coronal structure.

VOLKER BOTHMER: What definition of coronal temperature do you use?

KATJA POPPENHAEGER: In my analysis, I use the average coronal temperature inferred from the X-ray emission between 0.15 and 2 keV. This is a very strong simplification; when analyzing high signal-to-noise stellar X-ray spectra, one usually needs several temperature components for an adequate fit of the coronal spectra. For the low X-ray signal of our sample stars, however, the mean coronal temperature is the best one can do.

Comparative Magnetic Minima:
Characterizing quiet times in the Sun and Stars
Proceedings IAU Symposium No. 286, 2011
C. H. Mandrini & D. F. Webb, eds.

doi:10.1017/S174392131200508X

Dynamo models of grand minima

Arnab Rai Choudhuri

Department of Physics, Indian Institute of Science, Bangalore-560012
email: arnab@physics.iisc.ernet.in

Abstract. Since a universally accepted dynamo model of grand minima does not exist at the present time, we concentrate on the physical processes which may be behind the grand minima. After summarizing the relevant observational data, we make the point that, while the usual sources of irregularities of solar cycles may be sufficient to cause a grand minimum, the solar dynamo has to operate somewhat differently from the normal to bring the Sun out of the grand minimum. We then consider three possible sources of irregularities in the solar dynamo: (i) nonlinear effects; (ii) fluctuations in the poloidal field generation process; (iii) fluctuations in the meridional circulation. We conclude that (i) is unlikely to be the cause behind grand minima, but a combination of (ii) and (iii) may cause them. If fluctuations make the poloidal field fall much below the average or make the meridional circulation significantly weaker, then the Sun may be pushed into a grand minimum.

Keywords. Sun: dynamo — Sun: activity — sunspots

1. Introduction

At the very outset, I would like to mention that the subject of this invited talk was not chosen by me. The organizers felt that this Symposium should have an invited talk on dynamo models of grand minima and requested me to give it. Only after considerable initial hesitation, I finally agreed. The reason behind my initial hesitation is that at present we have no dynamo model of grand minima which is completely satisfactory or which is generally accepted in the community. No two self-respecting dynamo theorists seem to agree how grand minima are produced! To the best of my knowledge, this is the first time an invited talk on this subject is being given in a major international conference.

Given this situation, I have decided to adopt the following strategy. I shall mainly focus on the various bits of physics which go into making models of grand minima rather than discussing specific models of grand minima in detail. While many of the present-day models of grand minima may eventually fall by the wayside, I believe that the bits of physics that we consider relevant today will still remain relevant after 20 or 30 years when there may be a better understanding of what occasionally pushes the Sun into the grand minima.

2. Observational characteristics of grand minima

Before getting into the theoretical discussion, let us see what we can learn about the characteristics of grand minima from the very limited observational data available to us.

Several authors have studied the archival records of sunspots during the Maunder minimum (Sokoloff & Nesme-Ribes 1994; Hoyt & Schatten 1996). The few sunspots seen during the Maunder minimum mostly appeared in the southern hemisphere. Sokoloff & Nesme-Ribes (1994) have used the archival data to construct a butterfly diagram for a part of the Maunder minimum from 1670, showing a clear trend of hemispheric asymmetry. It is an open question whether hemispheric asymmetry played any crucial role in creating the Maunder minimum (Charbonneau 2005). Usoskin, Mursula & Kovaltsov (2000) argued that the Maunder minimum started abruptly but ended in a gradual manner, indicating that the strength of the dynamo must be building up as the Sun came out of the Maunder minimum. However, some recent evidence suggests that the onset of the Maunder minimum may not be as abrupt as believed earlier (Vaquero *et al.* 2011).

When solar activity is stronger, magnetic fields in the solar wind suppress the cosmic ray flux, reducing the production of ^{10}Be and ^{14}C which can be used as proxies for solar activity. From the analysis of ^{10}Be abundance in a polar ice core, Beer, Tobias & Weiss (1998) concluded that the solar activity cycle continued during the Maunder minimum, although the overall level of the activity was lower than usual. Miyahara *et al.* (2004) drew the same conclusion from their analysis of ^{14}C abundance in tree rings.

This method of using various proxies (like the abundances of ^{10}Be and ^{14}C) for sunspot activity can be extended to study even the earlier grand minima before the Maunder minimum. Usoskin, Solanki & Kovaltsov (2007) estimated that there have been about 27 grand minima in the last 11,000 years. They also identified about 19 grand maxima, i.e. periods during which sunspot activity was unusually high, like what was seen during much of the twentieth century.

3. Are grand minima merely extremes of cycle irregularities?

We know that the solar cycle is only approximately periodic. Both the strength and the period vary from one cycle to another. We begin our theoretical discussion by raising the question whether the grand minima are merely extreme examples of cycle irregularities. Are the theoretical ideas used to model irregularities of solar cycles adequate to explain the occurrences of grand minima, or do we need to invoke some qualitatively different ideas? We do not yet have a definitive answer to this question. Any dynamo theorist is entitled to have his or her own personal opinion. Let me put forth my personal opinion.

Our simulations (to be discussed later) seem to suggest that nothing very extraordinary may be needed to push the Sun into a grand minimum. If the fluctuations which cause the usual cycle irregularities are sufficiently large, they may sometimes cause grand minima. However, even after the Sun is pushed into the grand minimum, a subdued cycle has to continue (as discussed in § 2) and eventually the Sun has to come out of the grand minimum. These are more problematic to explain. There are certain mechanisms of magnetic field generation which crucially depend on the existence of sunspots. Certainly those mechanisms cannot be operative during a grand minimum. So we need to invoke alternative mechanisms.

Let us look at the question how magnetic fields are generated in the dynamo process. The basic idea of solar dynamo is that the toroidal and poloidal components of the solar magnetic field sustain each other through a feedback loop. It is fairly easy to generate the toroidal field by the stretching of the poloidal field due to differential rotation. Since helioseismology has shown that the differential rotation is concentrated in the tachocline, the generation of toroidal field mainly takes place there. To complete the loop, we need to generate the poloidal field from the toroidal field. The historically important idea of Parker (1955) — which was further elaborated by Steenbeck, Krause & Rädler (1966) — is that the cyclonic turbulence in the convection zone twists the toroidal field to produce the poloidal field. This mechanism is often called the α-effect because the crucial parameter describing this process is usually denoted by the symbol α. Within certain approximation schemes, this parameter can be shown to be given by

$$\alpha = -\frac{1}{3}\overline{\mathbf{v}.(\nabla \times \mathbf{v})}\,\tau, \tag{1}$$

where $\mathbf{v}$ is the fluctuating part of the velocity field and τ is the correlation time (see, for example, Choudhuri 1998, § 16.5). This α-effect mechanism can be operative only if the toroidal field is not too strong such that the helical turbulence is able to twist it. However, the flux tube rise simulations by several authors (Choudhuri & Gilman 1987; Choudhuri 1989; D'Silva & Choudhuri 1993; Fan, Fisher & DeLuca 1993; Caligari *et al.* 1995) indicated that the toroidal field at the base of the solar convection zone has to be as strong as 10^5 G. Such a strong field cannot be twisted by helical turbulence and we cannot invoke α-effect to generate the poloidal field from such a strong toroidal field. An alternative mechanism which has been widely used in many recent dynamo simulations is due to Babcock (1961) and Leighton (1969). Bipolar sunspots on the solar surface have a tilt with respect to the solar equator and this tilt increases with latitude. This was discovered by Joy in 1919 and is known as *Joy's law*. D'Silva & Choudhuri (1993) provided the first theoretical explanation of Joy's law by showing that the tilt is produced by the Coriolis force acting on the flux tubes rising through the convection zone due to magnetic buoyancy. When a tilted bipolar sunspot decays, fluxes of opposite polarities diffuse at slightly different latitudes, contributing to the poloidal field. According to this Babcock–Leighton mechanism, a tilted bipolar sunspot pair is a conduit for converting the toroidal field to the poloidal field. The sunspot pair forms due to the buoyant rise of the toroidal field and we get the poloidal field after its decay.

Since we see the Babcock–Leighton mechanism clearly operational at the solar surface, most of the recent flux transport dynamo models take this as the primary generation mechanism of the poloidal field. The α-effect cannot operate on the strong toroidal field at the base of the convection zone. But this strong toroidal field is expected to be highly intermittent (Choudhuri 2003) and the α-effect is likely to be operative in those regions of the convection zone where the toroidal field is weak, although the nature, the spatial distribution and even the algebraic sign of the α parameter remain unclear at the present time. The Babcock–Leighton mechanism presumably cannot work during the grand minimum when there are no sunspots. So we have to fall back upon the α-effect to continue the cycles during the grand minimum and eventually to pull the Sun out of it. Our lack of knowledge about the α-process limits our understanding of these phenomena. In the mean field dynamo equations, the term capturing the Babcock–Leighton process is formally very similar to the term capturing the α-effect — often even using the symbol α. Hence many dynamo models of the grand minima are worked out at the present time by solving the same equations during and outside the grand minima. But it should be kept in mind that the physics behind the symbol α must be very different during and outside the grand minima.

In summary, our view is that the usual sources of irregularities in solar cycles are sufficient for the onset of a grand minimum, but to pull the Sun out of a grand minimum we need some physics different from the physics behind the usual solar cycles. Presumably the situation is somewhat different for grand maxima. Not only are the usual irregularities expected to cause a grand maximum, we also do not require anything unusual to take the Sun out of the grand maximum. The usual poloidal field generation by the Babcock–Leighton mechanism continues during the grand maximum. It is intriguing that Usoskin, Solanki & Kovaltsov (2007) concluded that the lengths of grand maxima correspond to an exponential distribution, but the lengths of grand minima have a more complicated bimodal distribution. Is this connected with the fact that grand maxima do not involve any physical processes different from the normal, but grand minima require the generation process of the poloidal field to be different from the normal situation?

4. The origin of irregularities in the flux transport dynamo

We now discuss the possible sources of irregularities in the flux transport dynamo — the most widely studied model of the solar cycle in recent years. Let us begin by recapitulating some basic facts about the flux transport dynamo. The toroidal field is generated in the tachocline by the strong differential rotation and then rises to the surface due to magnetic buoyancy to form tilted bipolar sunspots. When these sunspots decay, we get the poloidal field by the Babcock–Leighton mechanism. The meridional circulation of the Sun, which is found to be poleward in the upper layers of the convection zone and must have a hitherto unobserved equatorward branch at the bottom of the convection zone in order to conserve mass, plays a very crucial role in the flux transport dynamo. The meridional circulation causes the observed poleward transport of the poloidal field. At the base of the convection zone, it is responsible for making the dynamo wave propagate equatorward, such that sunspots are produced at lower and lower latitudes with the progress of a cycle. In the absence of the meridional circulation, the dynamo wave at the bottom of the convection zone would propagate poleward in accordance with the Parker–Yoshimura sign rule (Parker 1955; Yoshimura 1975) contradicting observations. The flux transport dynamo could become a serious model of the solar cycle only after Choudhuri, Schüssler & Dikpati (1995) demonstrated that a sufficiently strong meridional circulation could overrule the Parker–Yoshimura sign rule and make the dynamo wave propagate in the correct direction.

The original flux transport dynamo model of Choudhuri, Schüssler & Dikpati (1995) led to two offsprings: a high diffusivity model and a low diffusivity model. The diffusion times in these two models are of the order of 5 years and 200 years respectively. The high diffusivity model has been developed by a group working in IISc Bangalore (Choudhuri, Nandy, Chatterjee, Jiang, Karak), whereas the low diffusivity model has been developed by a group working in HAO Boulder (Dikpati, Charbonneau, Gilman, de Toma). The differences between these models have been systematically studied by Jiang, Chatterjee & Choudhuri (2007) and Yeates, Nandy & Mckay (2008). Both these models are capable of giving rise to oscillatory solutions resembling solar cycles. However, when we try to study the irregularities of the cycles, the two models give completely different results. We need to introduce fluctuations to cause irregularities in the cycles. In the high diffusivity model, fluctuations spread all over the convection zone in about 5 years. On the other hand, in the low diffusivity model, fluctuations essentially remain frozen during the cycle period. Thus the behaviours of the two models are totally different on introducing fluctuations. Over the last few years, several independent arguments have been advanced in support of the high diffusivity model (Chatterjee, Nandy & Choudhuri 2004; Chatterjee & Choudhuri 2006; Jiang, Chatterjee & Choudhuri 2007; Goel & Choudhuri 2009; Hotta & Yokoyama 2010). We adopt the point of view here the solar dynamo is most likely a high diffusivity flux transport dynamo.

Three main sources of irregularities in dynamo models have been studied by different authors over the years: (i) chaotic behaviours introduced by nonlinearities of the dynamo process; (ii) fluctuations in the generation of the poloidal field; (iii) fluctuations in the meridional circulations. The three following sections will focus on these three sources of irregularities and discuss the question whether they can cause grand minima. Some of these sources of irregularities have been investigated even before the flux transport dynamo model became popular, by applying them to the earlier solar dynamo models.

5. Effects of nonlinearities

It is well known that nonlinear dynamical systems can show complicated chaotic behaviours. Some of the earliest efforts of modelling solar cycle irregularities invoked the idea of nonlinear chaos. The full dynamo problem is certainly a nonlinear problem in which the magnetic fields produced by the fluid motions react back on the fluid motions.

The simplest way of capturing the effect of this in a kinematic dynamo model (in which the fluid equations are not solved) is to consider a quenching of the α parameter as follows:

$$\alpha = \frac{\alpha_0}{1 + |\overline{B}/B_0|^2}, \tag{2}$$

where $\overline{B}$ is the average of the magnetic field produced by the dynamo and B_0 is the value of magnetic field beyond which nonlinear effects become important. There is a long history of dynamo models studied with such quenching (Stix 1972; Ivanova & Ruzmaikin 1977; Yoshimura 1978; Brandenburg *et al.* 1989; Schmitt & Schüssler 1989). In most of the nonlinear calculations, however, the dynamo eventually settles to a periodic mode with a given amplitude rather than showing sustained irregular behaviour. The reason for this is intuitively obvious. Since a sudden increase in the amplitude of the magnetic field would diminish the dynamo activity by reducing α given by (2) and thereby pull down the amplitude again (a decrease in the amplitude would do the opposite), the α-quenching mechanism tends to lock the system to a stable mode once the system relaxes to it. In fact, Krause & Meinel (1988) and Brandenburg *et al.* (1989) argued that the nonlinear stability may determine the mode in which the dynamo is found. Yoshimura (1978) was able to reproduce some irregular features of the solar cycle by introducing an unrealistic delay time of 29 years between the magnetic field and its effect on the α-coefficient. In some highly truncated models with the suppression of differential rotation, one could find the evidence of chaos in limited parts of the parameter space (Weiss, Cattaneo & Jones 1984). Küker, Arlt & Rüdiger (1999) suggested that the quenching of differential rotation might have caused the Maunder minimum. This seems unlikely now on the ground that torsional oscillations — periodic modulations of differential rotation caused by the dynamo-generated magnetic field — appear like small perturbations.

It does not seem that the irregularities of solar cycles are primarily caused by nonlinearities. But that does not mean that nonlinearities have no important consequences in the currently favoured flux transport dynamo models. In order to explain the even-odd or the Gnevyshev–Ohl effect of solar cycles, Charbonneau, St-Jean & Zacharias (2005) and Charbonneau, Beaubien & St-Jean (2007) made the highly provocative suggestion that the solar dynamo may be sitting in a region of period doubling just beyond the point of nonlinear bifurcation. Recently the effects of nonlinearities introduced by the quenching of turbulent diffusion (Guerrero, Dikpati & de Gouveia Dal Pino 2009) and meridional circulation (Karak & Choudhuri 2012) are being investigated.

6. Fluctuations in poloidal field generation

Since the mean field dynamo equations are derived by averaging over turbulence, we expect fluctuations to be present around the mean. Choudhuri (1992) was the first to suggest that these fluctuations will be particularly important in the poloidal field generation. It is now difficult to believe that this was an unorthodox and radical idea in 1992 when it was proposed, though this idea was explored further by Moss *et al.* (1992), Hoyng (1993) and Ossendrijver, Hoyng & Schmitt (1996). This idea was applied to the flux transport dynamo by Charbonneau & Dikpati (2000).

Let us consider the question how fluctuations in poloidal field generation arise in the flux transport dynamo. The Babcock–Leighton mechanism of poloidal field generation depends on the tilts of bipolar sunspot pairs. While the average tilts are given by Joy's law, one finds a large scatter around this average. Longcope & Choudhuri (2002) provided a theoretical model of this scatter on the basis of the idea that the rising flux tubes are

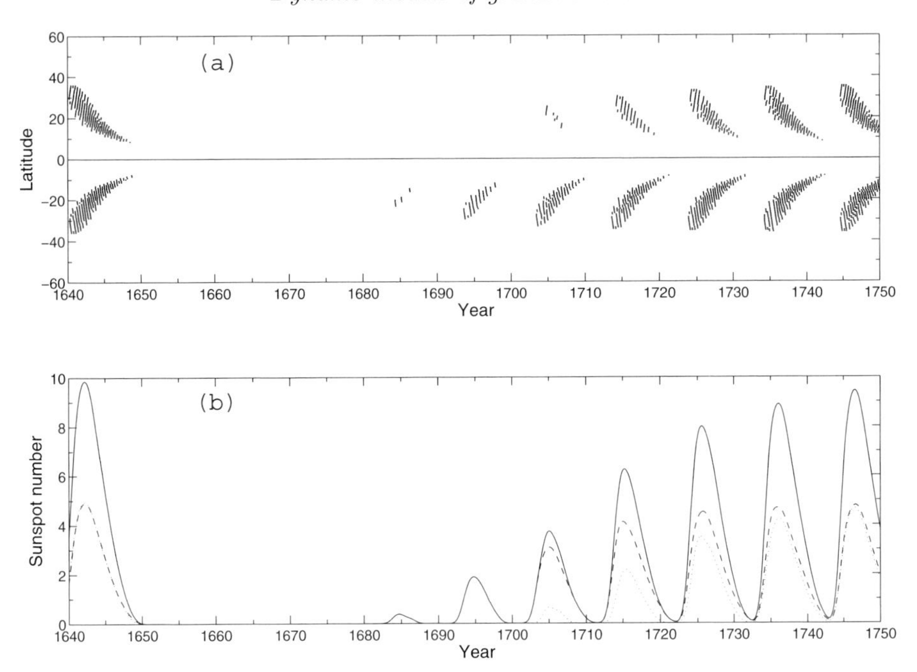

Figure 1. The theoretical model of the Maunder minimum from Choudhuri & Karak (2009). This is based on a simulation in which the poloidal field was reduced to 0.0 and 0.4 of its average value in the two hemispheres. The upper panel shows the butterfly diagram. The dotted and dashed lines in the lower panel are sunspot numbers in the northern and southern hemispheres, whereas the solid line is their sum.

buffeted by turbulence in the convection zone. This scatter around Joy's law produces fluctuations in the poloidal field generation process and we identify this as a primary source of irregularities in the solar cycle. It may be noted that Choudhuri, Chatterjee & Jiang (2007) and Jiang, Chatterjee & Choudhuri (2007) modelled the last few cycles by assuming the fluctuations in poloidal field generation to be the main source of irregularities in solar cycles and predicted that the forthcoming cycle 24 will be weak. This prediction was based on the high diffusivity model. There are enough indications by now that the upcoming cycle is going to be a weak one, providing further support to the high diffusivity model.

We now come to the question whether fluctuations in the poloidal field generation can produce grand minima. Several authors found that intermittencies resembling grand minima can be obtained in simple dynamo models by introducing fluctuations (Schmitt, Schüssler & Ferriz-Mas 1996; Mininni, Gomez & Mindlin 2001; Brandenburg & Spiegel 2008). The effect of such fluctuations on flux transport dynamo models has been investigated only recently. Charbonneau, Blais-Laurier & St-Jean (2004) carried out a simulation by introducing 100% fluctuations in α (the poloidal field generation parameter) in a flux transport dynamo with low diffusivity. They found intermittencies in their simulations resembling grand minima. The low diffusivity of their model ensured that the diffusive decay time or the 'memory' of the dynamo was rather long (of the order of a century) and most probably this long memory played a role in producing intermittencies of similar duration. The important question is whether the high diffusivity dynamo model, which we consider to be the appropriate model for explaining solar cycles and which has a much smaller diffusive decay time, can also produce similar intermittencies on introducing fluctuations in poloidal field generation. This question has been studied by Choudhuri & Karak (2009).

Choudhuri, Chatterjee & Jiang (2007) proposed a simplified procedure for incorporating the cumulative effect of fluctuations in poloidal field generation. Since these fluctuations in poloidal field generation would make the poloidal field at the end of a cycle different from the average poloidal field one would get from the mean field equations without including fluctuations, they suggested that the poloidal field at the end of the cycle may be modified suitably to account for the cumulative effect of the fluctuations. Choudhuri & Karak (2009) found that the dynamo is pushed into a grand minimum if the poloidal field at the end of a cycle falls to 0.2 of its average value. At the time of this work during the depth of a long sunspot minimum, there was considerable speculation whether the Sun was entering another grand minimum. Choudhuri & Karak (2009) concluded that the poloidal field had fallen to only about 0.6 of its average value and hence the Sun should *not* be entering another grand minimum. On making the poloidal field in the northern and southern hemispheres fall to respectively 0.0 and 0.4 of its average value, Choudhuri & Karak (2009) found that many characteristics of the Maunder minimum were reproduced. Fig. 1 shows the theoretical plots of butterfly diagram and sunspot number, which compare favourably with the corresponding observational plots given in Fig. 1(a) of Sokoloff & Nesme-Ribes (1994) and Fig. 1 of Usoskin, Mursula & Kovaltsov (2000). We find that the theoretical model reproduced the fact that the Maunder minimum started abruptly, but ended gradually. It is basically the growth time of the dynamo which determines the duration of the grand minimum during which the magnetic field has to grow up again to return to normalcy. As pointed out in § 3, the operation of the dynamo during the grand minimum presumably depends on the α-effect and the theoretical model shows an ongoing but subdued cycle of magnetic field in the solar wind. We sum up the theoretical results in the following words. If the poloidal field at the end of a cycle turns out to be very weak due to fluctuations in its generation process, then that can push the dynamo into a grand minimum, from which it recovers gradually in the dynamo growth time.

7. Fluctuations in meridional circulation

It is well known that the period of the flux transport dynamo varies roughly as the inverse of the meridional circulation speed. The period of the dynamo is approximately given by the time taken by meridional circulation at the bottom of the convection zone to move from higher latitudes to lower latitudes. In other words, the period of a flux transport dynamo does not depend too much on the details of poloidal field generation mechanism. Probably this is the reason why the period of the dynamo during a grand minimum like the Maunder minimum does not change drastically, even though the poloidal field generation mechanism may be different from normal times as explained § 3.

Since the meridional circulation determines the period of the flux transport dynamo, it is obvious that any fluctuations in meridional circulation would have an effect on the flux transport dynamo. It has been found recently that the meridional circulation has a periodic variation with the solar cycle, becoming weaker at the time of sunspot maximum (Hathaway & Rightmire 2010; Basu & Antia 2010). Presumably the Lorentz force of the dynamo-generated magnetic field slows down the meridional circulation at the time of the sunspot maximum. Karak & Choudhuri (2012) found that this quenching of meridional circulation by the Lorentz force does not produce irregularities in the cycle, provided the diffusivity is high as we believe. We disagree with the model of Nandy, Muñoz-Jaramillo & Martens (2011) which assumes that the meridional circulation changes abruptly at

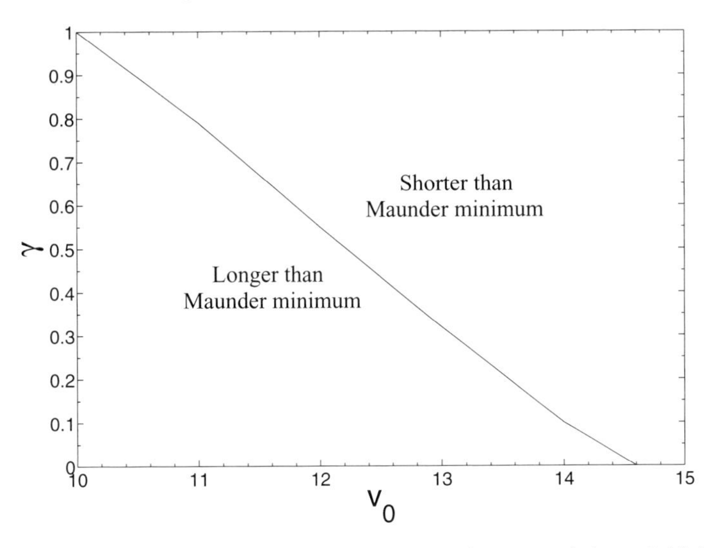

Figure 2. The parameter space indicating the reduction factor γ of the poloidal field and the amplitude of the meridional circulation needed to produce Maunder-like grand minima.

each sunspot maximum. Our point of view is that the periodic variation of meridional circulation due to the Lorentz force cannot be responsible for solar cycle irregularities and we need to consider other kinds of fluctuations in meridional circulation.

We have reliable observational data on the variation of meridional circulation only for a little more than a decade. To draw any conclusions about the variation of meridional circulation at earlier times, we have to rely on indirect arguments. If we assume the cycle period to go inversely as meridional circulation, then we can use periods of different past solar cycles to infer how meridional circulation has varied with time in the last few centuries. On the basis of such considerations, it appears that the meridional circulation had random fluctuations in the last few centuries with correlation time of the order of 30–40 years (Karak & Choudhuri 2011). We now come to the question what effect these random fluctuations of meridional circulation may have on the dynamo. Based on the analysis of Yeates, Nandy & Mckay (2008), we can easily see that dynamos with high and low diffusivity will be affected very differently. Suppose the meridional circulation has suddenly fallen to a low value. This will increase the period of the dynamo and lead to two opposing effects. On the one hand, the differential rotation will have more time to generate the toroidal field and will try to make the cycles stronger. On the other hand, diffusion will also have more time to act on the magnetic fields and will try to make the cycles weaker. Which of these two competing effects wins over will depend on the value of diffusivity. If the diffusivity is high, then the action of diffusivity is more important and the cycles become weaker when the meridional circulation is slower. The opposite happens if the diffusivity is low.

As we pointed out in §4 and §6, there are enough indications that the diffusivity of the solar dynamo is high. If that is the case, then a slowing of the meridional circulation would make the cycles weaker. Karak (2010) found that the flux transport dynamo can be pushed into a grand minimum if the meridional circulation drops to 0.4 of its normal value. This is another possible mechanism for producing a grand minimum.

Miyahara *et al.* (2004) found that cycles during the Maunder minimum became somewhat longer, indicating that the meridional circulation must have slowed down. This supports the theoretical idea that the weakening of meridional circulation might have played an important role in producing the Maunder minimum. It should be noted that the opposite would happen in the low diffusivity model. The weakening of meridional circulation and the lengthening of cycles in a low diffusivity model should be associated with a grand maximum, since longer cycles would allow the differential rotation to generate stronger fields in the low diffusivity model.

8. Concluding remarks

Although there are many uncertainties in our theoretical understanding of grand minima, it appears that fluctuations in poloidal field generation and fluctuations in meridional circulation are the main causes of irregularities in solar cycles and can also produce grand minima. We believe the solar dynamo to be a high diffusivity dynamo in which a fall in the meridional circulation makes cycles weaker. If fluctuations in poloidal field generation alone or fluctuations in meridional circulation alone are to produce grand minima, then the poloidal field at the end of a cycle or the meridional circulation has to fall to rather low values to cause grand minima. The situation is a little less constrained if we consider simultaneous fluctuations in both. Fig. 2 taken from Karak (2010) shows the values to which the poloidal field at the end of a cycle and the meridional circulation have to fall if a grand minimum is to be caused by their simultaneously falling to low values. This seems to be the most likely scenario we have at the present time for explaining grand minima. One important question is whether we can estimate how often this is likely to happen. Can we explain why there were 27 grand minima in the last 11,000 years? We are looking at this question right now.

I end by summarizing what appears to me to be the most plausible theoretical scenario for grand minima based on the flux transport dynamo. Due to fluctuations in poloidal field generation and meridional circulation, if both of them simultaneously happen to become sufficiently weak, that may push the Sun into a grand minimum. Within the grand minimum, the dynamo keeps operating on the basis of the α-effect and ultimately bounces out of the grand minimum in the dynamo growth time.

Acknowledgments: My participation in IAU Symposium 286 was made possible by a JC Bose Fellowship awarded by Department of Science and Technology, Government of India.

References

Babcock, H. W. 1961, *ApJ*, 133, 572
Basu, S. & Antia, H. M. 2010, *ApJ*, 717, 488
Beer, J., Tobias, S., & Weiss, N. 1998, *Solar Phys.*, 181, 237
Brandenburg, A., Krause, F., Meinel, R., & Moss, D., Tuominen I. 1989, *A&A*, 213, 411
Brandenburg, A. & Spiegel, E. A. 2008, *AN*, 329, 351
Charbonneau, P. 2005, *Solar Phys.*, 229, 345
Charbonneau, P., Beaubien, G., & St-Jean, C. 2007, *ApJ*, 658, 657
Charbonneau, P., Blais-Laurier, G., & St-Jean, C. 2004, *ApJ*, 616, L183
Charbonneau, P. & Dikpati, M. 2000, *ApJ*, 543, 1027
Charbonneau, P., St-Jean, C., & Zacharias, P. 2005, *ApJ*, 619, 613
Chatterjee, P. & Choudhuri, A. R. 2006, *Solar Phys.*, 239, 29
Chatterjee, P., Nandy, D., & Choudhuri, A. R. 2004, *A&A*, 427, 1019
Choudhuri, A. R. 1989, *Solar Phys.*, 123, 217
Choudhuri, A. R. 1992, *A&A*, 253, 277
Choudhuri, A. R. 1998, *The Physics of Fluids and Plasmas: An Introduction for Astrophysicists* (Cambridge University Press, Cambridge)
Choudhuri, A. R. 2003, *Solar Phys.*, 215, 31

Choudhuri, A. R., Chatterjee, P., & Jiang, J. 2007, *Phys. Rev. Lett.*, 98, 131103
Choudhuri A. R., Gilman P. A. 1987, *ApJ*, 316, 788
Choudhuri, A. R. & Karak, B. B. 2009, *RAA*, 9, 953
Choudhuri, A. R., Schüssler, M., & Dikpati, M. 1995, *A&A*, 303, L29
D'Silva, S. & Choudhuri, A. R. 1993, *A&A*, 272, 621
Fan, Y., Fisher, G. H., & DeLuca, E. E. 1993, *ApJ*, 405, 390
Goel, A. & Choudhuri, A. R. 2009, *RAA*, 9, 115
Guerrero, G., Dikpati, M., & de Gouveia Dal Pino, E. M. 2009 *ApJ*, 701, 725
Hathaway, D. H. & Rightmire, L. 2010, *Science*, 327, 1350
Hotta, H. & Yokoyama, T. 2010, *ApJ*, 714, L308
Hoyng, P. 1993, *A&A*, 272, 321
Hoyt, D. V. & Schatten, K. H. 1996, *Solar Phys.*, 165, 181
Ivanova, T. S. & Ruzmaikin, A. A. 1977, *SvA*, 21, 479
Jiang, J., Chatterjee, P., & Choudhuri, A. R. 2007, *MNRAS*, 381, 1527
Karak, B. B. 2010, *ApJ*, 724, 1021
Karak, B. B. & Choudhuri, A. R. 2011, *MNRAS*, 410, 1503
Karak, B. B. & Choudhuri, A. R. 2012, *Solar Phys.*, in press
Krause, F. & Meinel, R. 1988, *GAFD*, 43, 95
Küker, M., Arlt, R., & Rüdiger, G. 1999, *A&A* 343, 977
Leighton, R. B. 1969, *ApJ*, 156, 1
Longcope, D. W. & Choudhuri, A. R. 2002, *Solar Phys.*, 205, 63
Mininni, P. D., Gomez, D. O., & Mindlin, G. B. 2001, *Solar Phys.*, 201, 203
Miyahara, H., Masuda, K., Muraki, Y., Furuzawa, H., Menjo, H., & Nakamura, T. 2004, *Solar Phys.*, 224, 317
Moss, D., Brandenburg, A., Tavakol, R., & Tuominen, I. 1992, *A&A*, 265, 843
Nandy, D., Muñoz-Jaramillo, A., & Martens, P. C. H. 2011 *Nature* 471, 80
Ossendrijver, A. J. H., Hoyng, P., & Schmitt, D. 1996, *A&A*, 313, 938
Parker, E. N. 1955, *ApJ*, 122, 293
Schmitt, D. & Schüssler, M. 1989, *A&A*, 223, 343
Schmitt, D., Schüssler, M., & Ferriz-Mas, A. 1996, *A&A*, 311, L1
Sokoloff, D. & Nesme-Ribes, E. 1994, *A&A*, 288, 293
Steenbeck, M., Krause, F., & Rädler, K. H. 1966, *Z. Naturforsch.*, 21, 369
Stix, M. 1972, *A&A*, 20, 9
Usoskin, I. G., Mursula, K., & Kovaltsov, G. A. 2000, *A&A*, 354, L33
Usoskin, I. G., Solanki, S. K., & Kovaltsov, G. A. 2007, *A&A*, 471, 301
Vaquero, J. M., Gallego, M. C., Usoskin, I. G., & Kovaltsov, G. A. 2011, *ApJ*, 731, L24
Weiss, N. O., Cattaaneo, F., & Jones, C. A. 1984, *GAFD*, 30, 305
Yeates, A. R., Nandy, D., & Mackay, D. H. 2008, *ApJ*, 673, 544
Yoshimura, H. 1975, *ApJ*, 201, 740
Yoshimura, H. 1978, *ApJ*, 226, 706

Discussion

Janet Luhmann: What is the role of the hemispheric asymmetric activity?

Arnab Choudhuri: The Maunder Minimum was highly asymmetric in the hemispheres. We do not know whether this was typical for other grand minima. The theoretical implications of hemispheric asymmetry are also very imperfectly understood. So, we are still not in a position to give a good answer to your question.

Mark Giamapapa: Do you get sudden or gradual transitions to the Maunder Minimum?

Arnab Choudhuri: Theoretical models are still rather under-constrained and it is possible to get both sudden and gradual transitions by varying parameters of the models. More observational data on this will help in constraining theoretical models.

Comparative Magnetic Minima:
Characterizing quiet times in the Sun and stars
Proceedings IAU Symposium No. 286, 2011
C. H. Mandrini & D. F. Webb, eds.

doi:10.1017/S1743921312005091

A model for grand minima and geomagnetic reversals

D. D. Sokoloff[1], G. S. Sobko[2], V. I. Trukhin[3] and V. N. Zadkov[4]

Department of Physics, Moscow State University, 119991, Moscow, Russia
email: [1]`d_sokoloff@hotmail.com`
[2]`sobko@physics.msu.ru`
[3]`dean@phys.msu.ru`
[4]`zadkov@phys.msu.ru`

Abstract. We suggest a simple dynamical system which mimics a nonlinear dynamo which is able to provide (in specific domains of its parametric space) the temporal evolution of solar magnetic activity cycles as well as evolution of geomagnetic field including its polarity reversals. A qualitative explanation for the physical nature of both phenomena is presented and discussed.

Keywords. Sun: activity, Earth, magnetic fields, sunspots.

---

1. Introduction

Temporal behaviour of solar magnetic activity is far from just a cycle. Discussions of various features of long-term dynamics of solar activity including Maunder minimum and other Grand minima are presented in many papers included in this volume. Temporal behaviour of geomagnetic field is obviously specific. In contrast to the solar magnetic cycle, geomagnetic dipole field remains usually more or less constant and its secular and more long-term archaeomagnetic variations do not result in its polarity reversals. In geological timescales however geomagnetic field suddenly changed, as geologists believe, its polarity many times (Christensen *et al.* 2010; Hulot *et al.*, 2010). Sequence of geomagnetic reversals as well as that one of solar Grand minima look aperiodic and irregular.

Various solar dynamo models including those discussed in this volume reproduce at least qualitatively solar Grand minima as well as geodynamo model based on direct numerical simulations give hundreds of reversals (Olson *et al.* 2010). Magnetic field reversals similar to some extent to the geomagnetic reversals was reproduced in laboratory dynamo experiments (Berhanu *et al.* 2007). Physical nature of these phenomena remains however not completely clear, cf. Choudhuri (2012).

Solar magnetic activity as well as geomagnetic field are thought to be driven by dynamos in spherical shells based on differential rotation and mirror-asymmetric convection. Particular manifestations of these dynamos are obviously specific. However it looks attractive to present both associated long-term dynamics as manifestations of a unique physical mechanism acting in two separate domains of the parametric space of dynamo governing parameters. Here we present a simple model which demonstrates a mechanism which can be responsible for both long-term dynamics.

We consider fluctuations in regeneration rate of poloidal magnetic field from toroidal one as the physical driver underlying mechanism leading to the magnetic long-term dynamics. To be specific, we discuss this mechanism in the framework of dynamo based on the classical α-effect. However the idea looks applicable as well to the dynamos based on meridional circulation.

α-coefficient being a result of averaging over an ensemble of a moderate number of convective cells (say, $N \approx 10^4$) contains a noisy component which importance for dynamo was stressed e.g. by Hoyng (1993). Hoyng (1993) supposed correlation time and length of α-fluctuations to be comparable with that ones for convective vortexes. Then one need fluctuations amplitude comparable with mean value of α to get an interesting long-term dynamics for dynamo generated magnetic field. Basing on the later results of direct numerical simulations of α-effect, (Brandenburg & Sokoloff 2002; Otmianowska-Mazur, Kowal & Hanasz 2006), determinations of α-coefficient from dynamo shell models (Frick, Stepanov & Sokoloff 2006) as well as laboratory measurements of the α-coefficient (Stepanov *et al.*, 2006) we presume that the correlation time and length of α-fluctuations are comparable for the cycle period and size of solar convective zone; standard $N^{-1/2}$ estimation for the fluctuation size do not contradict to this presumption. Then we need $\delta\alpha/\alpha \approx 10\% - 20\%$ to get a desired behaviour for dynamo generated field (Moss *et al.* 2008; Usoskin, Sokoloff & Moss 2009).

2. Dynamical system

We obtain the desired simple model from general Parker mean-field dynamo models with differential rotation and α-effect decomposing it in the Fourier series and truncating the series keeping in consideration as small number of modes as possible to get generation of magnetic field with nonvanishing magnetic moment (see Nefedov & Sokoloff 2010, for details of calculations). More explicitly, we represent toroidal magnetic field B and toroidal component of magnetic potential A as

$$B = -b_1 \sin 2\theta + b_2 \sin 4\theta, \quad A = a_1 \cos\theta - a_2 \cos 3\theta, \tag{2.1}$$

where θ is co-latitude and truncate Parker (1955) dynamo equation accordingly to get the following dynamical system

$$\frac{da_1}{dt} = \frac{R_\alpha b_1}{2} - a_1 - \frac{3R_\alpha b_1}{8}(b_1^2 + 2b_2^2), \tag{2.2}$$

$$\frac{da_2}{dt} = \frac{R_\alpha}{2}(b_1 + b_2) - 9a_2 - \frac{3R_\alpha(b_1 + b_2)}{8}(b_1^2 + b_1 b_2 + b_2^2), \tag{2.3}$$

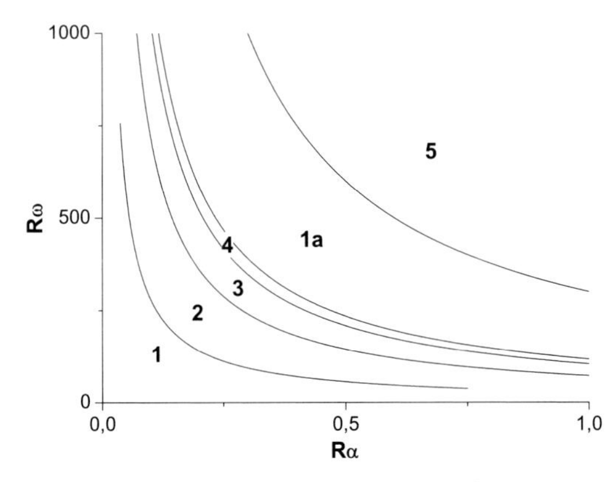

Figure 1. Synoptic map for dynamo regimes: 1, 1a - decay, 2 - stationary field, 3 - vacillations, 4 - dynamo bursts, 5 - oscillations.

$$\frac{db_1}{dt} = \frac{R_\omega}{2}(a_1 - 3a_2) - 4b_1, \tag{2.4}$$

$$\frac{db_2}{dt} = \frac{3R_\omega a_2}{2} - 16b_2. \tag{2.5}$$

Here R_α and R_ω are dimensionless numbers for intensity of α-effect and differential rotation correspondingly. We presume that the radial rotation shear is latitude independent and dominate over the latitudinal one and $\alpha(\theta) \sim \sin\theta$. We presume simple algebraic α-quenching in form $\alpha \sim 1/(1 + B^2/B_{\rm eq}^2)$ and measure magnetic field in units of the equipartition magnetic field, so $B_{\rm eq} = 1$. We neglected regeneration of toroidal field from poloidal one due to the α-effect and connect this regeneration with differential rotation only, i.e. consider so-called $\alpha\omega$-dynamo. Because the dynamo generated toroidal magnetic field is usually much larger then the poloidal one, we neglect in dynamical system nonlinear terms with a_1 and a_2. Replacing variables one can combine dimensionless numbers R_α and R_ω in so-called dynamo number $D = R_\alpha R_\omega$. Note that the coefficient a_1 is proportional to the magnetic moment of dynamo generated magnetic field.

3. Dynamo regimes

We simulate the dynamical system (2.2-2.5) numerically to isolate the following regimes of magnetic field evolution.

Let us start with dynamo regimes which occur for a time-independent α-effect parameterized by dimensionless number R_α. For a very weak dynamo action, i.e. small D

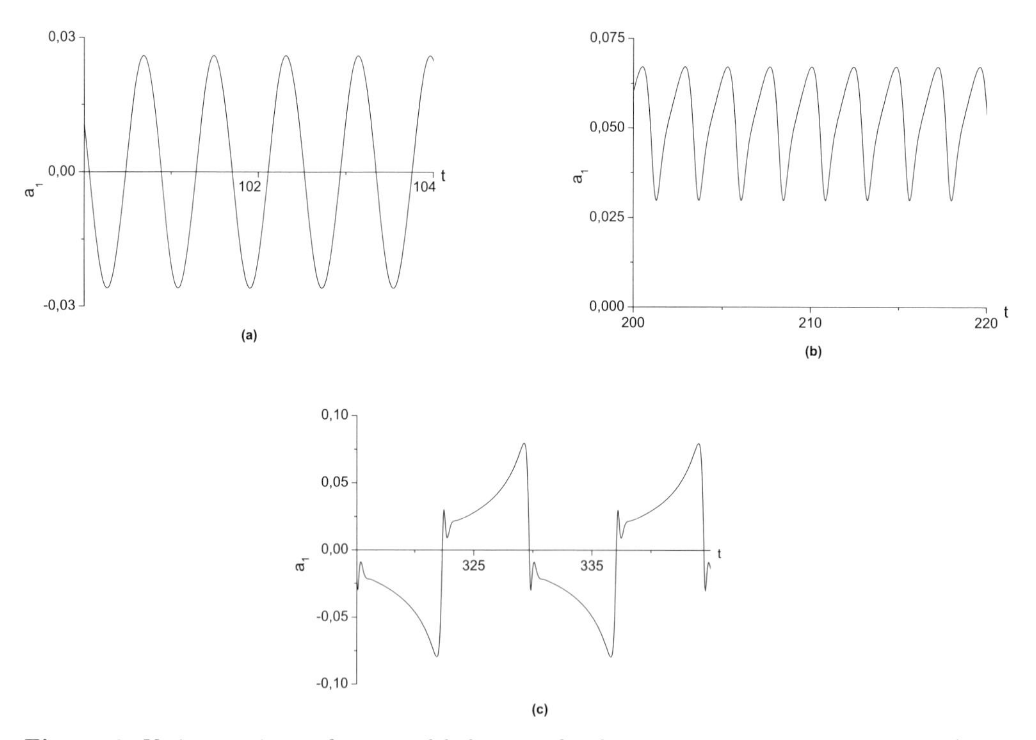

Figure 2. Various regimes of temporal behaviour for dynamo generated magnetic field (coefficient a_1 responsible for magnetic moment versus time): a - oscillations, b - vacillations, c - dynamo bursts.

magnetic field decays (corresponding range in synoptic map Fig. 1 is marked by 1). For slightly larger dynamo numbers dynamo dynamo self-excitation occurs, dynamo generated magnetic field first growths manometrically and then becomes steady due to nonlinear dynamo saturation (domain 2 in Fig. 1). Even larger dynamo number results in a nonlinear regime with so-called vacillations, i.e. periodic almost harmonic magnetic field evolution with nonzero mean value (domain 3 in Fig. 1; form of this solution is shown in Fig. 2a. Further enlargement of D results in a periodic however highly nonharmonic temporal evolution known as dynamo bursts (domain 4 in Fig. 1). Dynamo generated magnetic field grows slowly then suddenly and rapidly decay up to zero, changes its sign and then grows slowly again with the opposite sign until next sign reversal restore initial magnetic configuration (Fig. 2c). Such temporal behaviour of dynamo generated magnetic field is known from dynamo experiments, e.g. Berhanu *et al.* (2007).

If dynamo action becomes even larger it becomes less effective so magnetic field decays again (domain marked by 1a in Fig. 1) however enlarging D further we obtain almost harmonic oscillations with zero mean (domain marked 5 in Fig. 1).

We identify domain 5 in Fig. 1 with the dynamo parameter range responsible for the cyclic solar activity while the parameter range 3 looks similar to the normal behaviour of dipole geomagnetic field (almost steady component and moderate variations). In particular, magnetic moment represented in the model by a_1 changes its sign each oscillation (domain 5) however keeps its sign in course of vacillations (domain 3).

4. Grand minima and reversals

Let us consider now what happens if the averaged value of the fluctuating α-coefficient corresponds to a dynamo regime located in the domain 5 and one need an α-fluctuation as large as several its standard deviations to move the dynamo regime in the domain 1a where magnetic field decays. Fluctuations required to shift the dynamo regime in domain 1a being quite large occur rarely so after many normal activity cycles something like a Grand minima occurs. Moss *et al.* (2008) and Usoskin, Sokoloff & Moss (2009) investi-

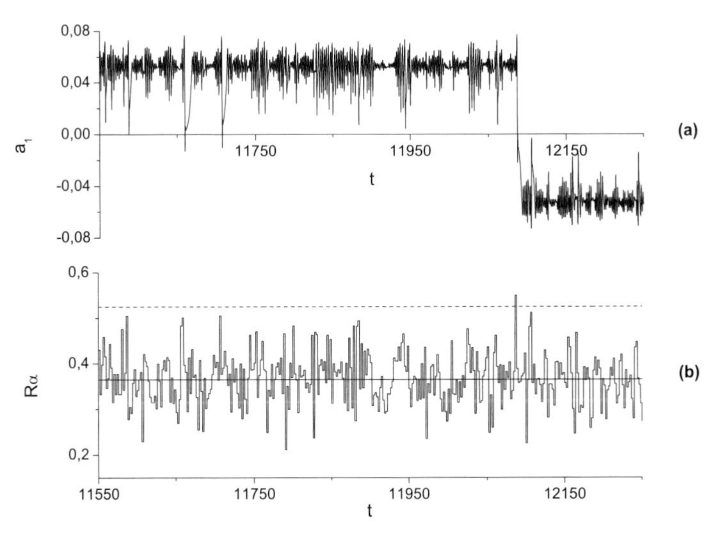

Figure 3. Scenario for magnetic field reversals: a - evolution of the coefficient a_1 which represents magnetic moment; b - evolution of R_α; mean value of α is shown by a horizontal solid line while a horizontal dashed line shows R_α which corresponds to transition from domain 3 to the domain 4 in Fig. 1.

gated (in the framework of Parker migratory dynamo with fluctuating α) corresponding long-term behaviour of such dynamo to demonstrate that when dynamo governing parameters correspond to the domain 1a magnetic field mimics what happens during the Maunder minimum.

A scenario for magnetic field reversals which resembles the geomagnetic field evolution is slightly more complicated. Let the averaged value of α corresponds to a point in domain 3, i.e. magnetic field has vacillations. A moderate positive fluctuations of α shifts this point in domain 4 where dynamo bursts occur (Fig. 3b). Further behaviour (Fig. 3a) depends on the scale of this fluctuations and on the phase of the dynamo burst which develops during the fluctuation. If the fluctuation is strong and the burst has enough time during the fluctuation to change sign of dynamo generated magnetic field, then an inversion occurs. If the fluctuation is weaker and/or the dynamo burst occurred changes

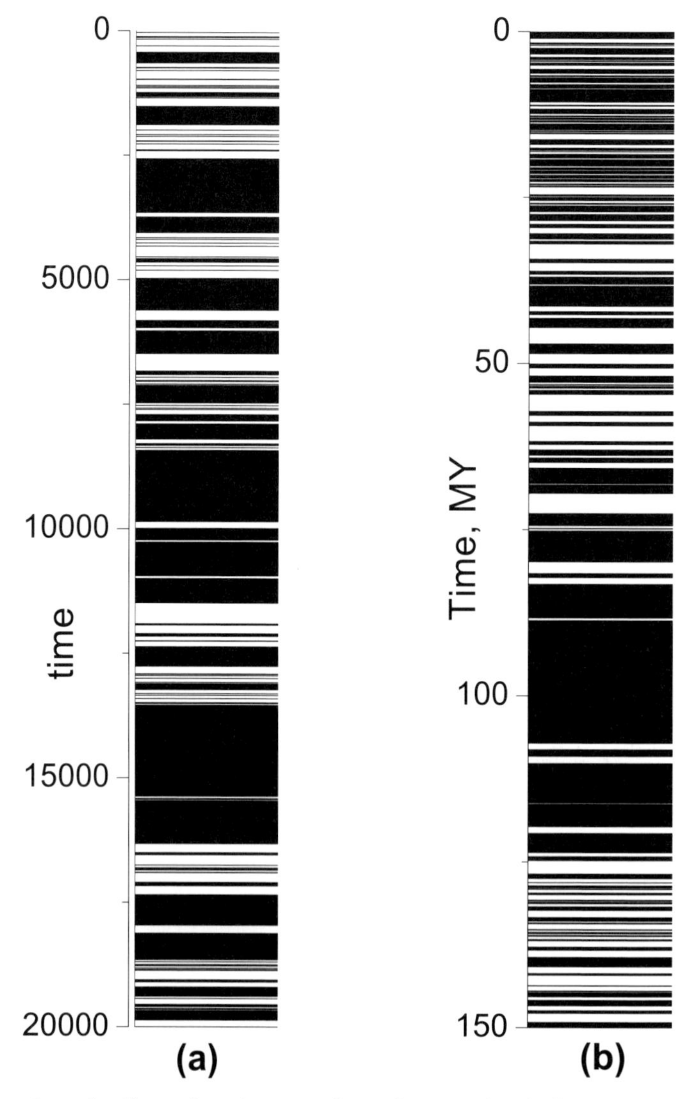

Figure 4. Scenario under discussion gives a geological timescale which is quite similar to that one known from the palaeomagnetic studies: a - simulated geomagnetic timescale; b - geomagnetic timescale for the last 150 million years after Gradstein, Ogg & Smith (2004).

sign too late magnetic moment becomes very weak however keeps its sign and its value recovers after the end of the fluctuation. Such episodes are known in the palaeomagnetic studies as excursions. A substantial negative fluctuation of R_α can lead to a long epoch without reversals as well as without vacillations because the dynamo regime moves in the domain 2. Such epochs look similar to the long epochs without geomagnetic reversals known in geological history as superchrones.

The temporal evolution of the solution for Eqs. (2.2-2.5) with the averaged values of the dynamo governing parameters located nearby the border, between domains 3 and 4, demonstrates a sequence of the dynamo generated field reversals. We illustrate this sequence by a diagram known in the palaeomagnetology as geomagnetic time scale, i.e. a column with time as a vertical coordinate where times of one polarity are shown in black while the times of the opposite polarity are shown in white (Fig. 4a). We compare this simulated diagram with a corresponding timescale known from palaeomagnetic studies (Gradstein, Ogg & Smith 2004) as follows. Note that the dynamical system Eqs. (2.2-2.5) is formulated using a dimensionless time. Of course, palaeomagnetic studies give the instants of the field reversals in dimensional units. We consider, after Gradstein, Ogg & Smith (2004) the geomagnetic timescale for the last 150 million years where it is most elaborated, choose an interval in the simulated timescale which contains the same number of reversals and presume that the lengthes of both timescales are equal. We present both timescales in Fig. 4. Of course, the timescales can not be identical because at least one of them are taken from a realization of a random process. We note however that the general shapes of both scales are quite similar and conclude that the scenario suggested reproduces basic feature of geomagnetic reversals.

In general, we conclude that the simple dynamical system Eqs. (2.2-2.5) reproduces in specific domains of parametric space basic features of cyclic solar activity as well as evolution of geomagnetic field on geological times including sequence of chaotic field reversals.

Acknowledgements

DS is grateful for financial support from RFBR, as well as from the organizing committee to participate in the meeting.

References

Berhanu, M., Monchaux, R., Fauve, S., Mordant, N., Petrelis, F., Chiffaudel, A., Daviaud, F., Dubrulle, B., Marie, L., Ravelet, F., Bourgoin, M., Odier, Ph., Pinton, J.-F., & Volk, R. 2007, *Europhys. Lett.*, 77, 59001

Brandenburg, A. & Sokoloff D. 2002, *Geophys. Astrophs. Fluid Dyn.* 96, 319

Choudhuri, A. R. 2012, *this volume*

Christensen, U. R., Balogh, A., Breuer, D., & Glassmeier, K. H. (Eds.), 2010, *Planetary Magnetism* (Springer)

Frick, P., Stepanov, R., & Sokoloff, D. 2006, *Phys. Rev. E* 74, 066310

Gradstein, F., Ogg, J., & Smith, A. 2004, *A Geological Time Scale-2004*, Cambridge, Univ. Press.

Hoyng, P. 1993, *A&A* 272, 321

Hulot, G., Finlay, C. C., Coustable, C. G., Olsen, N., & Mandea M. 2010, *Space Sci. Rev.* 152, 159

Moss, D., Sokoloff, D., Usoskin, I., & Tutubalin, V. 2008, *Solar Physics*, 250, 221

Nefedov, S. N. & Sokoloff, D. D. 2010, *Astron. Rep.*, 54, 247

Olson, P. L., Coe, R. S., Driscoll, P. E., & Glatzmaier, G. A. 2010, *Phys. Earth Planet. Inter.*, 180, 66

Otmianowska-Mazur, K., Kowal, G., & Hanasz, M. 2006, *A&A* 445, 915
Stepanov, R., Volk, R., Denisov, S., Frick, P., Noskov, V., & Pinton, J.-F. 2006, *Phys. Rev. E* 73, 046310
Usoskin, I. G., Sokoloff, D., & Moss, D. 2009, *Sol. Phys.*, 254, 345

Discussion

ARNAB CHOUDHURI: What kind of diffusivity do you have in your model?

DMITRY SOKOLOFF: We assume that turbulent diffusivity is time-independent. In principle it is no problem to include diffusivity fluctuations in the model.

ARNAB CHOUDHURI: How strong fluctuations of the parameters do you need in your model to get a grand minimum?

DMITRY SOKOLOFF: About 10-20%.

Comparative Magnetic Minima:
Characterizing quiet times in the Sun and Stars
Proceedings IAU Symposium No. 286, 2011
C. H. Mandrini & D. F. Webb, eds.

doi:10.1017/S1743921312005108

Is meridional circulation important in modelling irregularities of the solar cycle?

Bidya Binay Karak and Arnab Rai Choudhuri
Department of Physics, Indian Institute of Science, Bangalore 560012, India
email: bidya_karak@physics.iisc.ernet.in

Abstract. We explore the importance of meridional circulation variations in modelling the irregularities of the solar cycle by using the flux transport dynamo model. We show that a fluctuating meridional circulation can reproduce some features of the solar cycle like the Waldmeier effect and the grand minimum. However, we get all these results only if the value of the turbulent diffusivity in the convection zone is reasonably high.

Keywords. Sun: dynamo — Sun: activity — sunspots

1. Introduction

The solar cycle is not regular. There were several grand minima like the Maunder minimum during which the activity level was extremely low. Although the solar activity has varied approximately cyclically since the Maunder minimum, the amplitudes and the periods of individual cycles varied in an irregular manner. Another important feature of solar cycles is the Waldmeier effect, which is basically the anti-correlation between the rise time and the peak sunspot number. However, we define two different aspects of it, which we call WE1 and WE2. By WE1 we refer to the anti-correlation between the *rise time* and the peak sunspot number, whereas by WE2 we refer to the positive correlation between the *rise rate* and the peak sunspot number.

Our motivation is to model the irregularities of the solar cycle, including features like the Waldmeier effect, by using the flux transport dynamo model (see Choudhuri 2011 and references therein). In this model, the toroidal field is generated near the base of the convection zone by differential rotation and the poloidal field is generated near the solar surface by the decay of tilted bipolar sunspots. These two source regions are connected to each other by several transport agents. One important flux transport agent in this model is the turbulent magnetic diffusivity (η_t). However, its value in the whole convection zone is not properly constrained. This has led to two different classes of models, in which the diffusivity has been taken to be high or low (see Jiang, Chatterjee & Choudhuri 2007; Choudhuri 2011). The values of η_t in these two classes of models are taken in the ranges $\sim 10^{12} - 10^{13}$ cm^2 s^{-1} and $\sim 10^{10} - 10^{11}$ cm^2 s^{-1} respectively. Another important flux transport agent in this model is the meridional circulation. The present understanding of its origin — and, more importantly, its fluctuations — is very primitive. It is believed that the the meridional circulation arises from a slight imbalance between two large terms (the centrifugal force due to the variation of angular velocity with distance from the equatorial plane and the thermal wind due to a temperature variation with latitude). Therefore, we expect that there may be random variations in the meridional circulation due to stochastic fluctuations in any one of these driving forces. Only since 1990s we have some observational data of meridional circulation near the surface. Therefore we do not know whether it had large variations in the past. However, we can get some idea of the variations of the meridional circulation from the observed periods of the past solar cycles.

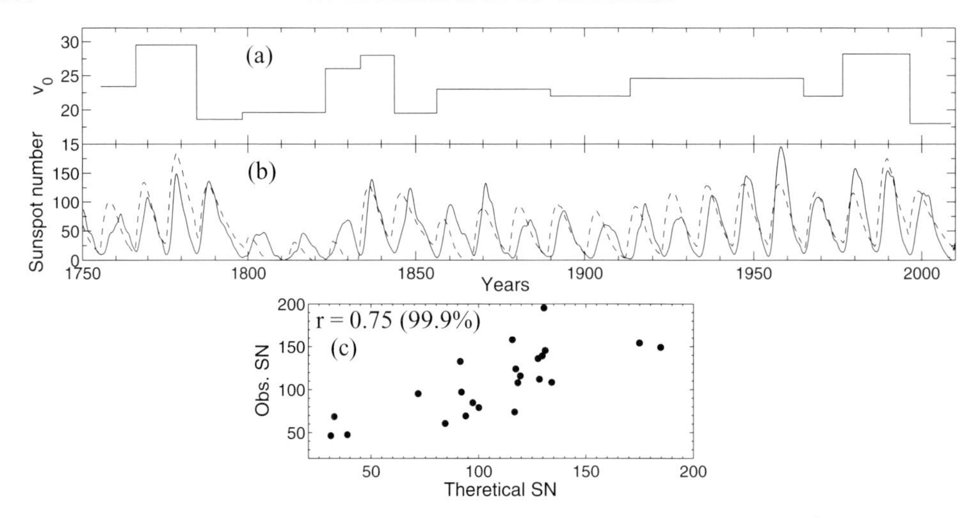

Figure 1. (a) Variation of amplitudes of meridional circulation v_0 (in m s^{-1}) with time (in yr). The solid line is the variation of v_0 used to match the theoretical periods with the observed periods. (b) Variation of theoretical sunspot number (dashed line) and observed sunspot number (solid line) with time. (c) Scatter diagram showing peak theoretical sunspot number and peak observed sunspot number. From Karak (2010).

In the next section, we discuss how this is done by using the flux transport dynamo model and then we model solar cycles using the variable meridional circulation.

2. Methodology and results

We know that the period of the cycle in the flux transport dynamo model is primarily determined by the strength of the meridional circulation (Dikpati & Charbonneau 1999). A stronger meridional circulation makes the dynamo period shorter. Therefore it should be possible to match the observed periods of the last 23 solar cycles by using a variable meridional circulation. Karak (2010) performed this experiment using a high diffusivity model based on the model of Chatterjee, Nandy & Choudhuri (2004). Fig. 1(a) shows the variation of the amplitude of meridional circulation required to match the periods of last 23 solar cycles. From this figure, we see that the meridional circulation varied significantly with time. Therefore, if the flux transport dynamo model is the correct model for the solar cycle, then we have to conclude that the meridional circulation had large variations in the past. Now let us look at the variation of theoretical sunspot number obtained by Karak (2010) shown by the dashed line in Fig. 1(b). For comparison, the observed sunspot number is shown by the solid line. Surprisingly, most of the theoretical sunspot cycle amplitudes are matched with the observed ones. We do get a good correlation between these two as shown in Fig. 1(c). This is a very important result of this analysis because our motivation was only to match the solar cycle periods and to get some idea of the variation of meridional circulation in the past. However, while doing this, we find that most of the solar cycle amplitudes are also matched to some extent. Therefore, this study suggests that a significant amount of fluctuations in the strengths of the cycles is arising from the variations in the meridional circulation, which seems important in modelling not only the solar cycle periods but also the cycle amplitudes. We point out that using a truncated dynamo model Passos (2012) (and references therein) have concluded that a variable meridional circulation is necessary in modeling irregular solar cycle.

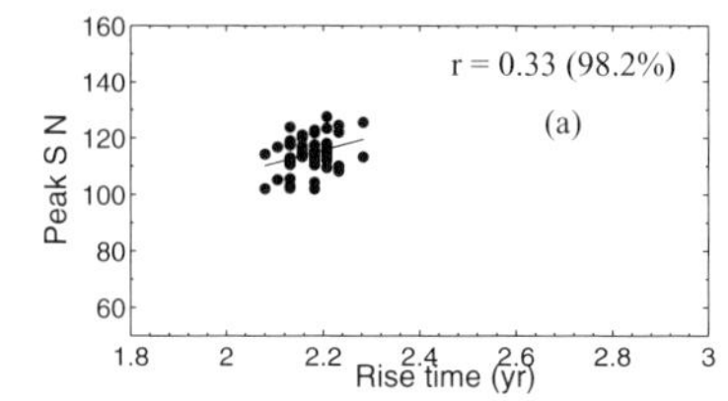

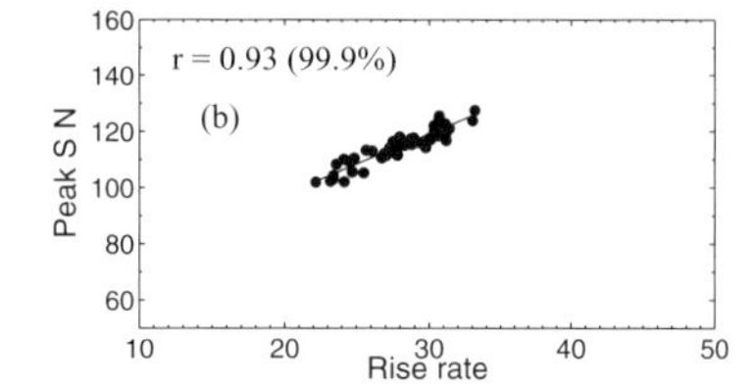

Figure 2. Results for WE1 (left panel) and WE2 (right panel) obtained by introducing fluctuations in the poloidal field at the minima. From Karak & Choudhuri (2011).

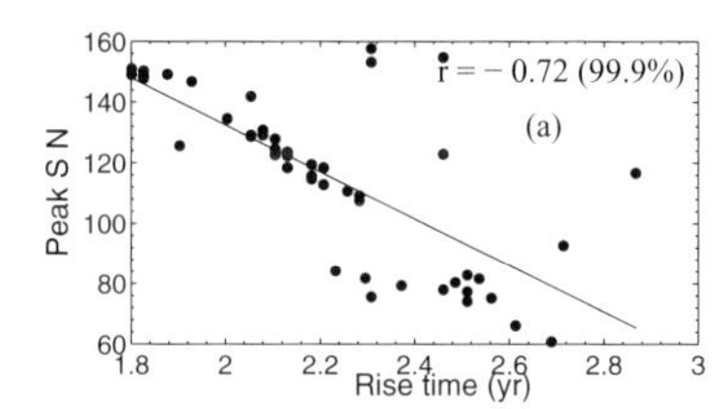

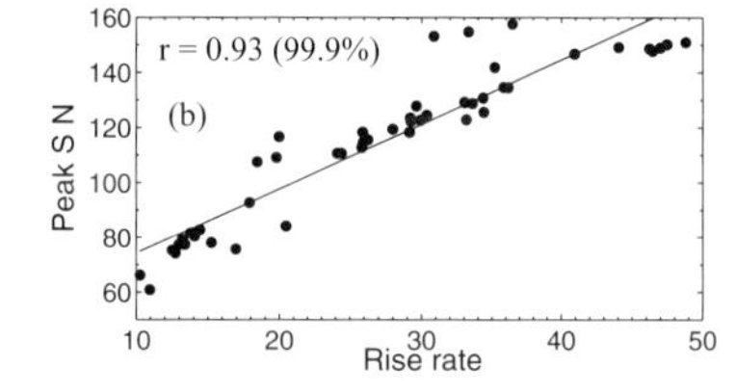

Figure 3. Same as earlier but obtained by introducing fluctuations in the meridional circulation. From Karak & Choudhuri (2011).

Now let us explain the physics of this result based on the arguments given by Yeates, Nandy & Mackay (2008). We know that in the flux transport dynamo, the production of the toroidal field is more if the poloidal field remains in the tachocline for longer time and vice versa. However, the poloidal field also diffuses during its transport through the convection zone. As a result, if the diffusivity is very high, then much of the poloidal field diffuses away and not much of it reaches the tachocline to induct the toroidal field. Therefore, when we decrease v_0 in high diffusivity model to match the period of a longer cycle, the poloidal field gets more time to diffuse during its transport through the convection zone. This ultimately leads to a lesser generation of toroidal field and hence the cycle becomes weaker. On the other hand, when we increase the value of v_0 to match the period of a shorter cycle, the poloidal field does not get much time to diffuse in the convection zone. Hence it produces stronger toroidal field and the cycle becomes stronger. Consequently, we get weaker amplitudes for longer periods and vice versa. However, this is not the case in low diffusivity model because in this model the diffusive decay of the fields is not very important. As a result, the slower v_0 means that the poloidal field remains in the tachocline for longer time and therefore it produces more toroidal field, giving rise to a strong cycle. Therefore, we do not get the correct correlation between the amplitudes of theoretical sunspot number and those of observed sunspot number when we repeat the same analysis in the low diffusivity model.

Next we present some results from Karak & Choudhuri (2011), who studied the Waldmeier effect using the flux transport dynamo model. The stochastic fluctuations in the Babcock–Leighton process of generating poloidal field and the stochastic fluctuations in the meridional circulation are the two main sources of irregularities in this model. Therefore, to study the Waldmeier effect, we first introduce suitable stochastic fluctuations in the poloidal field source term. Fig. 2 shows the result. We see that this study cannot reproduce WE1 (Fig. 2(a)), although it reproduces WE2 (Fig. 2(b)). Next we introduce stochastic fluctuations in the meridional circulation. Fig. 3 shows this result. We see that both WE1 and WE2 are remarkably reproduced in this case. We may mention that very recently Pipin & Kosovichev (2011) are also able to reproduce Waldmeier effect using their mean field dynamo model.

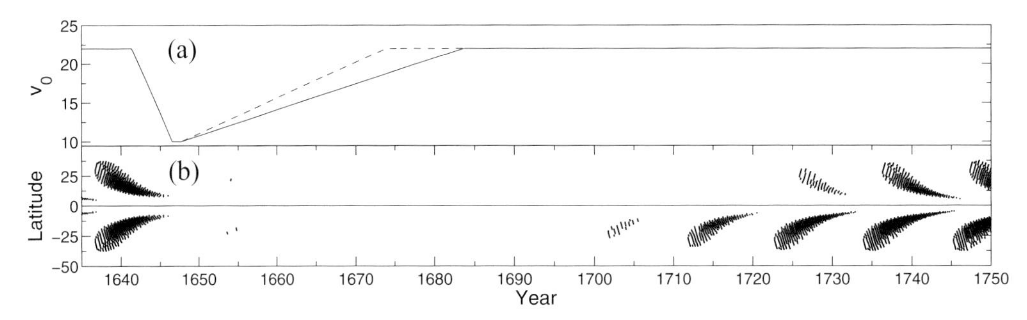

Figure 4. (a) The solid and dashed line show the variations of v_0 (in m s^{-1}) in northern and southern hemispheres with time. (b) The butterfly diagram. From Karak (2010).

Karak (2010) found that a decrease of the meridional circulation to a very low value can reproduce a Maunder-like grand minimum. Fig. 4(a) shows the variation of v_0 required to model a Maunder-like grand minimum. Fig. 4(b) shows the butterfly diagram of sunspot numbers. This result reproduces most of the features of the Maunder minimum e.g., North-South asymmetry of sunspot (Sokoloff & Nesme-Ribes 1994) and the sudden initiation of Maunder minimum (Usoskin, Mursula & Kovaltsov 2000) remarkably well.

We have shown that the temporal variation of the meridional circulation is important in modelling irregularities of the solar cycle. Our studies also suggest that there have been large variations in the meridional circulation in the past. With suitable stochastic fluctuations in the meridional circulation, we are able to reproduce many important irregular features of the solar cycle including the Waldmeier effect and Maunder-like grand minima. However we fail to reproduce these results in a low diffusivity model. Therefore this study along with some other studies (Chatterjee, Nandy & Choudhuri 2004; Chatterjee & Choudhuri 2006; Jiang, Chatterjee & Choudhuri 2007; Choudhuri & Karak 2009; Goel & Choudhuri 2009; Hotta & Yokoyama 2010; Karak & Choudhuri 2012) supports the high diffusivity model for the solar cycle.

Acknowledgments

We thank IAU funding agencies and DST, India, for financial support.

References

Chatterjee, P. & Choudhuri, A. R. 2006, *Solar Phys.*, 239, 29
Chatterjee, P., Nandy, D., & Choudhuri, A. R. 2004, *A&A*, 427, 1019
Choudhuri, A. R. 2011, *Pramana*, 77, 77
Choudhuri, A. R., Chatterjee, P., & Jiang, J. 2007, *Phys. Rev. Lett.*, 98, 1103
Choudhuri, A. R. & Karak, B. B. 2009, *RAA*, 9, 953
Choudhuri, A. R., Schüssler, M., & Dikpati, M. 1995, *A&A*, 303, L29
Dikpati, M. & Charbonneau, P. 1999, *ApJ*, 518, 508
Goel, A. & Choudhuri, A. R. 2009, *RAA*, 9, 115
Hotta, H. & Yokoyama, T. 2010, *ApJ*, 714, L308
Jiang, J., Chatterjee, P., & Choudhuri, A. R. 2007, *MNRAS*, 381, 1527
Karak, B. B. 2010, *ApJ*, 724, 1021
Karak, B. B. & Choudhuri, A. R. 2011, *MNRAS*, 410, 1503
Karak, B. B. & Choudhuri, A. R. 2012, *Solar Phys.*, 278, 137
Passos, D. 2012, *ApJ*, 744, 172
Pipin, V. V. & Kosovichev, A. G. 2011, *ApJ*, 741, 1
Sokoloff, D. & Nesme-Ribes, E. 1994, *A&A*, 288, 293

Usoskin, I. G., Mursula, K., & Kovaltsov, G. A. 2000, *A&A*, 354, L33
Yeates, A. R., Nandy, D., & Mackay, D. H. 2008, *ApJ*, 673, 544

Discussion

ANDREY TLATOV: Solar cycles overlap and changing meridional circulation may affect this overlap. Can you comment on this?

BIDYA KARAK: Yes, there is an overlap between two cycles. This overlap is affected by the meridional circulation speed. If the meridional circulation is weaker, then the overlap is less. Whereas, if the meridional circulation is stronger, then, the overlap is larger. This issue has been discussed by Nandi *et al.* (2011), Nature paper.

JANET LUHMANN: If meridional circulation changes are not stochastic but affected by active regions or sunspots, what are the consequences for the model?

BIDYA KARAK: There is an indication that the meridional circulation has a periodic variation with the solar cycle becoming stronger during solar minimum and weaker during solar maximum. It is not clear whether this variation is coming due to the active regions or due to the Lorentz force of the magnetic field acting on the velocity field (Karak & Choudhuri 2012). As the meridional circulation is coming from a slight imbalance between two large terms, we expect the meridional circulation to vary cycle to cycle stochastically. Indeed from the variation of the observed periods of the solar cycle, we inferred this kind of stochastic variation. Our modeling is based on the stochastic variation of the meridional circulation. However, if the meridional circulation changes are only due to active regions, we cannot model the irregular solar cycle using this model.

Comparative Magnetic Minima:
Characterizing quiet times in the Sun and stars
Proceedings IAU Symposium No. 286, 2011
C. H. Mandrini & D. F. Webb, eds.

doi:10.1017/S174392131200511X

Grand minima of solar activity during the last millennia

Ilya G. Usoskin[1], **Sami K. Solanki** [2,3] **and Gennady A. Kovaltsov**[4]

[1] Sodankyä Geophysical Observatory (Oulu unit), University of Oulu, 90014 Finland
email: ilya.usoskin@oulu.fi

[2] Max-Planck-Institut für Sonnensystemforschung, Max-Planck-Str. 2, 37191 Katlenburg-Lindau, Germany

[3] School of Space Research, Kyung Hee University, Yongin, Gyeonggi, 446-701, Korea

[4] Ioffe Physical-Technical Institute, 194021 St. Petersburg, Russia

Abstract. In this review we discuss the occurrence and statistical properties of Grand minima based on the available data covering the last millennia. In particular, we consider the historical record of sunspot numbers covering the last 400 years as well as records of cosmogenic isotopes in natural terrestrial archives, used to reconstruct solar activity for up to the last 11.5 millennia, i.e. throughout the Holocene. Using a reconstruction of solar activity from cosmogenic isotope data, we analyze statistics of the occurrence of Grand minima. We find that: the Sun spends about most of the time at moderate activity, 1/6 in a Grand minimum and some time also in a Grand maximum state; Occurrence of Grand minima is not a result of long-term cyclic variations but is defined by stochastic/chaotic processes; There is a tendency for Grand minima to cluster with the recurrence rate of roughly 2000-3000 years, with a weak ≈210-yr periodicity existing within the clusters. Grand minima occur of two different types: shorter than 100 years (Maunder-type) and long ≈150 years (Spörer-type). It is also discussed that solar cycles (most possibly not sunspots cycle) could exist during the Grand minima, perhaps with stretched length and asymmetric sunspot latitudinal distribution.

These results set new observational constraints on long-term solar and stellar dynamo models.

Keywords. Sun: activity, (Sun:) solar-terrestrial relations

---

1. Introduction

With the recent decline of overall solar activity, and in particular a very long and quiet minimum between solar cycles Nos. 23 and 24 in 2008–2010, the question of solar activity variability on longer timescales becomes acute (e.g., Schrijver *et al.* 2011). The recent solar cycle minimum appears unusual in many respects, e.g., in solar magnetic features (de Toma *et al.* 2010), in heliospheric modulation of cosmic rays (McDonald *et al.* 2010), in Space weather (Schwadron *et al.* 2010; Barnard *et al.* 2011). The last five solar cycles were very intensive corresponding to the unusual very active state of solar activity, or to a Grand maximum (Solanki *et al.* 2004; Usoskin 2008). Accidentally, the modern Grand maximum coincided with the space era with its numerous, precise and detailed in-situ and remote observation of the Sun, interplanetary medium and geosphere. Accordingly, the decline of solar activity after cycle No. 23 is often considered as an unusual phenomenon, probably leading to a Grand Maunder-like minimum. However, it is fully consistent with the features observed during the period of moderate solar activity in the 19-th century. Although a decline of the activity from the grand maximum state is apparent now, in accord with probabilistic predictions (Solanki *et al.* 2004; Abreu *et al.* 2008), it is unclear wether solar activity can slip into a new Grand minimum in the near future, or the Sun just returns to a moderate activity level, or possibly even returns into a Grand maximum

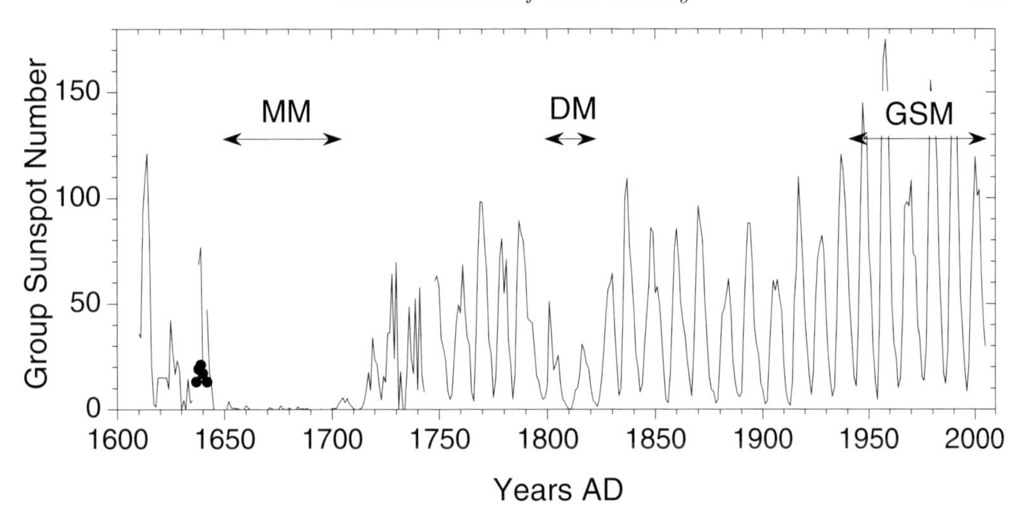

Figure 1. Monthly group sunspot numbers (Hoyt & Schatten 1998) since 1610. International sunspot numbers (Data Centre for the Sunspot Index (SIDC), Royal Observatory of Belgium, http://sidc.oma.be/html/sunspot.html) are used after 1996). Black dots represent newly revised sunspot numbers for years 1637–1641 (Vaquero *et al.* 2011). Maunder minimum (MM), Dalton minimum (DM) and the Grand Solar Maximum (GSM) are denoted.

state (e.g., Solanki & Krivova 2011). Therefore, a question of what is normal and what is unusual in the solar activity needs careful consideration, particularly with regard to minima of activity. This requires using a proxy of solar magnetic activity extending in the past, before the space era of direct measurements.

In this review we discuss the historical record of sunspot numbers for the last 400 years as well as solar activity on a time scale of a dozen millennia reconstructed from cosmogenic isotopes, as well as the method employed to obtain such activity records. We also put together general features of Grand minima of solar activity as deduced from solar activity reconstructed over the Holocene.

2. Solar activity over the last 400 years

For the last 400 years, solar activity is known pretty well in the form of the relative sunspot number (see details, e.g., in Usoskin 2008; Hathaway 2010). The sunspot number is based on original or reproduced drawings and records of sunspots observed by professional or amateur astronomers. Initially introduced by Rudolf Wolf of Zürich in the 19-th century as the relative sunspot number (also known as Wolf sunspot number), it was greatly improved to the group sunspot number series by Hoyt & Schatten (1998) to take its present form (Fig. 1). Still some new records and drawing continue to be found in archives, making further corrections of the sunspot series possible (e.g. Vaquero 2007; Arlt 2008; Vaquero *et al.* 2011), but the main pattern is quite well established.

• The main feature is the 11-year quasi-periodical solar cycle (known as the Schwabe cycle), produced by the solar dynamo (Hathaway 2010).

• The magnitude of the Scwhabe cycle (observed as the envelope) varies greatly on the centennial time scale.

• It includes the Maunder minimum in 1645–1715 (Eddy 1976; Soon & Yaskell 2003) when almost no sunspots were present, and the Dalton minimum at the turn of 18-th to the 19-th centuries.

• Between the 1940s and 2000s, the level of activity was high, exceeding 100 in the peak sunspot number. This period corresponds to the modern Grand solar maximum (Usoskin *et al.* 2003; Solanki *et al.* 2004).

• The length of the Schwabe cycle also varies, being between about 8 and 14 years. It is weakly anti-correlated with the cycle magnitude (Dicke 1978; Hoyng 1993; Solanki *et al.* 2002; Hathaway 2010).

Thus, the sunspot record suggests that solar activity is more often at a moderate level (peak sunspot numbers between 50 and 100), but may display excursions to the very quiet state of a Grand minimum (Maunder minimum) or very active state of a Grand maximum. The Dalton minimum is not considered as a complete Grand minimum but rather as a separate state of the dynamo (Schüssler *et al.* 1997), or an unsuccessful attempt at reaching a Grand minimum (Sokoloff 2004).

The sunspot series is one of the longest regular scientific observations and forms a benchmark for many studies, related, e.g., to solar/stellar physics, solar-terrestrial relations and geophysics. Therefore, the question arises of how representative is the sunspot series for the last 400 years for solar activity on much longer time scales? Are the last 400 years a typical period or special in some way? In order to study solar activity variations in the past, before the start of regular (instrumental) sunspot observations, one has to use more indirect proxies of solar activity. Since such proxies as naked-eye observations of sunspots and aurorae borealis recorded in historical chronicles, cannot be used for quantitative studies (Usoskin 2008), the best way to reconstruct past solar activity is related to naturally archived proxies as discussed in the next section.

3. Solar activity proxy: Cosmogenic isotopes

The magnetic Sun forms the heliosphere – a region of about 200 AU across, totally controlled by the permanently emitted solar wind and frozen-in heliospheric magnetic field (HMF). The dynamic heliosphere is driven by the solar activity (solar wind velocity and density, HMF strength, coronal mass ejections, etc.) and, in turn, modulates the influx of galactic cosmic rays (GCR) observed near Earth (e.g., Scherer *et al.* 2006). Energetic particles of GCR, when entering the Earth's atmosphere, collide with nuclei of atmospheric gases, producing, in particular, radioactive nuclides. Some of them, which have no terrestrial sources except cosmic rays, are called cosmogenic isotopes, and their amount is defined by the GCR flux modulated by solar magnetic activity and an additional geomagnetic shielding. The amount of cosmogenic nuclides can be measured in natural archives and used to reconstruct solar activity in the past, provided the geomagnetic field can be independently reconstructed (see, e.g., Beer 2000; Usoskin 2008). The main advantage of this method is its *off-line* type: natural archival records in independently datable samples (ice cores, stratified sediments, or tree trunks) can be measured nowadays in modern laboratories. As a result, a homogeneous, i.e. of roughly equal quality for different times, data series can obtained for further analysis.

Most important for the long-term reconstruction of solar/heliospheric activity are two nuclides - radiocarbon ^{14}C and ^{10}Be. Details of their production, transport and archiving are given below.

Radiocarbon ^{14}C (half-life 5730 years) is produced as a result of capture of an atmospheric thermal neutron by a ^{14}N nuclei: $^{14}\mathrm{N} + n \rightarrow\ ^{14}\mathrm{C} + p$. It is produced mostly in the upper troposphere – low stratosphere (Masarik & Beer 2009). Once produced it gets oxidized to CO_2 and, in gaseous form, takes part in the global carbon cycle (see Fig. 2), whereby it gets completely mixed in the atmosphere. As the ocean with its huge

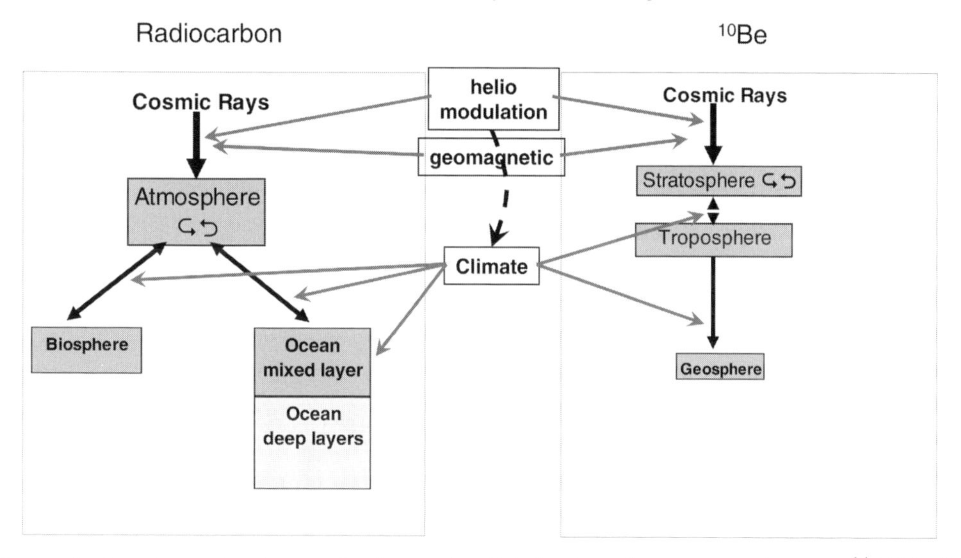

Figure 2. A cartoon showing production and redistribution of cosmogenic isotopes ^{14}C (Radiocarbon, left panel) and ^{10}Be (right panel). The flux of GCR is affected by both the heliospheric modulation and the geomagnetic field. ^{14}C is globally mixed in the atmosphere and within different reservoirs, including the deep and mixed ocean layers, and finally stored in the biosphere (living plants). Climate changes can affect the ^{14}C concentrations via slow changes of the ocean circulation/ventilation, which can play a role on millennial time scales. ^{10}Be is sufficiently mixed in the stratosphere but quickly precipitates from the troposphere. The processes of atmospheric redistribution/precipitation can be severely affected by regional atmospheric dynamics.

capacity and slow response is involved in the carbon cycle, the ^{14}C production changes are damped in magnitude (e.g., by a factor of 100 for the Schwabe cycle and delayed (see, e.g., Siegenthaler *et al.* 1980; Bard *et al.* 1997). However, if the ocean and atmospheric circulation remain roughly constant, as can be validated for the Holocene (Stuiver *et al.* 1991), the carbon cycle can be effectively reduced to a simple Fourier filter (Usoskin & Kromer 2005). Due to the global carbon cycle, ^{14}C is not sensitive to fast regional climate changes, but may be affected by slow trends in the ocean circulation. Production of ^{14}C in the 20-th century is very difficult to study, because of the fossil fuel burning (Suess effect), which inhomogeneously dilutes natural radiocarbon with a large amount of ^{14}C-free CO_2 (Tans *et al.* 1979).

Measurements of the Δ^{14}C (normalized ratio ^{14}C/^{12}C – see, e.g. Damon & Sonett 1991) are done on samples of tree-trunks, where annual tree rings allow absolute dating. As a result, a calibration ^{14}C curve for the last 25 millennia is available (Stuiver *et al.* 1998; Reimer *et al.* 2004) presenting the global ^{14}C signal.

Isotope ^{10}Be (half-life $1.36 \cdot 10^6$ years) is produced in spallation of atmospheric N, O and Ar nuclei by cosmic rays, mainly in the lower stratosphere – upper troposphere (Kovaltsov & Usoskin 2010). It soon gets attached to atmospheric aerosols and thus descends relatively quick. Its residence time in the stratosphere is a few years (Beer 2000) leading to partial mixing. The tropospheric residence time is a few weeks. Concentration of ^{10}Be is usually measured in polar (Greenland or Antarctic) ice cores, allowing for independent dating by glaciological methods. Deposition of ^{10}Be is straightforward, with the dominant precipitation at mid-latitudes and relatively small deposition in polar regions (Field *et al.* 2006; Heikkilä *et al.* 2009). However, it is affectable by the atmospheric circulation and precipitation pattern, and thus the ^{10}Be signal in ice cores can be greatly affected by the regional climate (precipitation), particularly on temporal scales shorter than 100

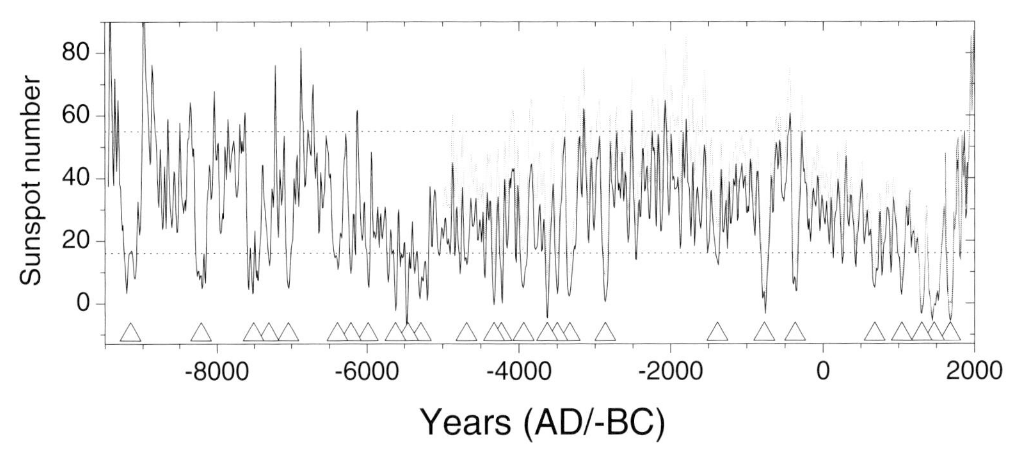

Figure 3. Reconstruction of the decadal (thus, the 11-yr solar cycle is not visible) solar sunspot activity over the Holocene, based on ^{14}C data (Solanki *et al.* 2004; Usoskin *et al.* 2006, 2007). The black and blue curves are built using the paleomagnetic reconstruction by Yang *et al.* (2000) and Korte & Constable (2005), respectively. The thick red curve is the actually measured group sunspot numbers (Hoyt & Schatten 1998). Horizontal dotted lines roughly indicate the levels chosen for Grand minima and maxima. Blue triangles depict centers of the identified Grand minima (Usoskin *et al.* 2007).

years. Presently there is no global ^{10}Be series, and the data are related to individual ice cores which may be prone to local/regional climate variability, whose influence is difficult to estimate.

Because of very different redistributions of the two isotopes in the geosphere, a common signal in their records can be robustly ascribed to the production, viz. solar or geomagnetic, signal. A detailed comparisons between the two isotopes (Bard *et al.* 1997; Usoskin *et al.* 2009a) shows that they agree with each other at time scales between 100 and 1000 year. The ^{14}C data are in good agreement with Antarctic Dom Fuji (Horiuchi *et al.* 2008) and Greenland GISP (Finkel & Nishiizumi 1997) ^{10}Be series, while South Pole (Bard *et al.* 1997) and Greenland Dye-3 (Beer *et al.* 1990) ^{10}Be series yield poor correlation with other data sets. On longer time scales, a systematic discrepancy is observed in the early Holocene (cf. Vonmoos *et al.* 2006), that is probably related to the delayed effect of the deglaciation. The lack of agreement on the short time scale (< 100 years) is likely related to the regional climate (depositional pattern) influence on ^{10}Be content in ice cores and/or to possible dating errors of the ice cores. It is interesting that the pair-wise agreement between ^{14}C and any of the ^{10}Be series is better than between the individual ^{10}Be series, confirming an essential role of the local/regional climate on individual ice core ^{10}Be records. Thus, redistribution of the isotopes in the geosphere, which is to a large extent unknown in the past, may distort the production (viz. solar activity) signal in the record. Polar records of ^{10}Be are prone to short-term regional and long-term global transport variability. Radiocarbon is insensitive to short-term climate changes but can be affected by changes in the large-scale ocean circulation at multi-millennial scales. In order to resolve these uncertainties, a combined result from different proxy records is needed.

4. Solar variability and Grand minima during the past millennia

A reconstruction of solar activity, based on the ^{14}C global INTCAL record, is shown in Fig. 3 for the Holocene, as made using the method by Solanki *et al.* (2004) Usoskin *et al.*

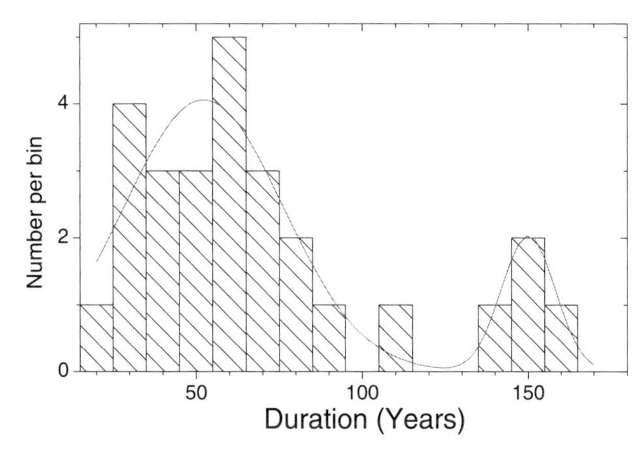

Figure 4. A histogram of the duration of the reconstructed Grand minima (modified after Usoskin *et al.* 2007) along with the best-fit double Gaussian. A clear bimodality, corresponding to Maunder- and Spörer-type minima, can be observed.

(2007) and paleomagnetic reconstructions by Yang *et al.* (2000) and Korte & Constable (2005). Long-term variations of the geomagnetic field, which provides additional shielding for GCR at Earth, are evaluated independently by paleomagnetic methods (Donadini *et al.* 2010). Uncertainties in the paleomagnetic data form the main source of the solar activity reconstructions (Solanki *et al.* 2004; Snowball & Muscheler 2007). One can see from Fig. 3 that using different paleomagnetic data one obtains slightly different overall levels of the reconstructed activity. However, the definition of Grand minima, which form the primary focus of this paper, is quite robust, so that the properties of the identified Grand minima are hardly affected by this uncertainty.

In our recent work (Usoskin *et al.* 2007) we have defined Grand minima of solar activity as periods when the (smoothed) sunspot number is $\leqslant 15$ during at least 20 years or forms a clear dip (the depth $\geqslant 20$ with respect to the surrounding level) with the bottom being $\leqslant 20$ in sunspot numbers. This definition is more robust, as it accounts for possible uncertainties, than a simple threshold definition (e.g., Stuiver *et al.* 1991; Voss *et al.* 1996; Abreu *et al.* 2008). In this way, 27 Grand minima have been identified during the Holocene (see Table 1 in Usoskin *et al.* 2007), with the total duration of about 1900 years, thus about $^1/_6$ of the total time. This list largely agrees with other lists of Grand minima (Eddy 1977; Stuiver & Braziunas 1989; Goslar 2003). We note that Abreu *et al.* (2008) defined Grand minima and maxima differently, as the lowest and uppermost 20% of the reconstructed activity, respectively.

It has been shown that Grand minima present a special state of the dynamo rather than being simply fluctuations of the dynamo parameters (Moss *et al.* 2008). Grand minima tend to appear in clusters with roughly 2400 years separation (the Hallstatt cycle, see e.g., Damon & Sonett 1991). Within the clusters, the Grand minima appear with roughly 210-year quasi-periodicity (de Vries or Suess cycle, see e.g., Suess 1980).

4.1. *Duration and recurrence of Grand minima*

The duration of the Grand minima has a bimodal distribution (see Fig. 4), with shorter Maunder-like (100 years or shorter) and longer Spörer-like (about 150 years) minima (Stuiver & Braziunas 1989; Goslar 2003; Usoskin *et al.* 2007). The two peaks corresponding to these different types of Grand minima are highly significantly distinguishable.

The time intervals (waiting times) between consequent Grand minima are more consistent with a power-law than with an exponential distribution (Usoskin *et al.* 2007).

This feature, observable as clustering of the Grand minima, suggests that the occurrence of Grand minima can be governed by non-Poissonic processes (e.g., self-organized criticality, de Carvalho & Prado 2000) with the presence of an intrinsic long-term memory. However, a thorough statistical analysis (Usoskin *et al.* 2009c) shows that, because of the insufficient number of events (Grand minima), the null hypothesis of the purely Poisson (stochastic) process with an exponential distribution of the waiting times cannot be ruled out. Thus, the result of a non-Poissonic nature of Grand minima occurrence is only barely significant (confidence level 0.93).

4.2. *Solar cycle during Grand minima*

Although the level of the surface magnetic activity drops below the sunspot formation threshold during a Grand minimum, different data sets imply that the global solar dynamo is not completely switched off but keeps operating, although in a special mode. An analysis of sunspot and aurora (Křivský & Pejml 1988) data suggests that the dominant periodicity of the solar cycle during the Maunder minimum was ≈22 years rather than the usual 11-year Schwabe cycle (Silverman 1992; Usoskin *et al.* 2001). An analysis of ^{14}C data also indicates longer cycles during the Maunder minimum (Peristykh & Damon 1998; Miyahara *et al.* 2004). Another Grand minimum not covered by sunspot observations, the Spörer minimum in the turn of 15–16-th centuries, is also characterized by extended cycles according to the ^{14}C data (Miyahara *et al.* 2006).

Results based on ^{10}Be data are less clear. Only Greenland ice cores, e.g. Dye-3 (Beer *et al.* 1990) and NGRIP (Berggren *et al.* 2009), provide sufficient resolution to study solar cycles. Although a band-pass filtered Dye-3 data depicts a weak 10-year cycle during the Maunder minimum (Beer *et al.* 1998), it has incorrect (*in phase*) relation with direct manifestations of solar activity (Usoskin *et al.* 2001). On the other hand, a wavelet or spectral-time analysis of both Dye-3 and NGRIP data sets yields a 15–20-year dominant periodicity (which is however, statistically insignificant) of ^{10}Be data during the Maunder minimum, in agreement with other proxies, and a 4–8-year intermittent variability, likely related to the regional North-Atlantic climate variability mode.

4.3. *North-South asymmetry*

Direct solar observations suggest that sunspot formation was highly asymmetric during the Maunder minimum, with sunspots observed mostly in the south hemisphere (Ribes & Nesme-Ribes 1993; Sokoloff 2004). Newly reconstructed and analyzed data (Arlt 2008) made it possible to reconstruct sunspot positions for the beginning of the Dalton minimum, which also show a significant asymmetry but with the northern hemisphere now being dominant (Usoskin *et al.* 2009b). This does not support the idea of a relic solar magnetic field weakly affecting the sunspot activity (Cowling 1945; Sonett 1983; Mursula *et al.* 2001). The north-south asymmetry (when one hemisphere dominates over the other) seems to be a feature of a Grand minimum, although this observation is based on only two known examples.

4.4. *General scenario of a Grand minimum*

The question on whether the onset of a Grand minimum is sudden or gradual is not fully clear. A widely accepted paradigm, schematically shown in Fig. 5, assumes that a Grand minimum begins suddenly, without apparent precursors, following normal activity cycles (Usoskin 2008). Recovery of the activity level from the deep minimum to normal activity is gradual, via emergence of the 11-year cycle. The transition may take several decades. This scenario puts an important constraint on the solar dynamo theory as, e.g.,

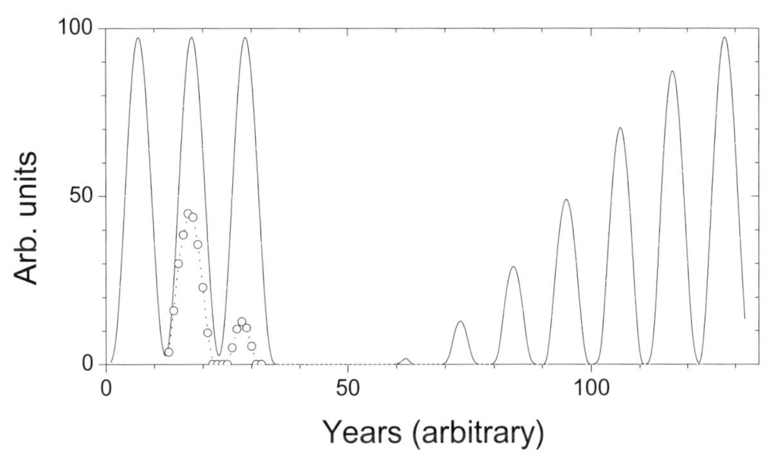

Figure 5. A schematic general scenario of sunspot activity around a Grand minimum (solid curve): sudden termination of active cycles to the very deep minimum followed by a gradual recovery. However, a scenario with a gradual decline of activity (dotted curve) may be more likely on the basis of recent evidence (Vaquero *et al.* 2011). This plot does not represent any particular Grand minimum.

it is consistent with some dynamo models (Charbonneau 2001), but disagree with the others.

However, such a scenario is mostly based on the Maunder minimum and hence suffers from very poor statistical significance. In addition, a recent re-analysis of historical sunspot records by Vaquero *et al.* (2011) yields that a couple of solar cycles before the Maunder minimum were low, suggesting a gradual beginning of the Grand minimum (dotted curve in Fig. 5). Also, there are some indications, based on high resolution ^{10}Be and ^{14}C data, that the length of the solar cycle may be slightly stretched a few cycles before the onset of a Grand minimum (Fligge *et al.* 1999; Miyahara *et al.* 2010). However, this idea is also based on only two examples, Maunder rand Spörer minima.

5. Summary

Solar magnetic activity as reconstructed using direct sunspot observations for the last 400 years and cosmogenic isotopes (^{14}C and ^{10}Be) proxy data for millennia, depicts a great deal of variability on different time scales, from the dynamo-driven 11-year Schwabe solar cycle to the centennial and millennial variability. The level of solar activity varies from a quiet Grand minimum with virtually no sunspots on the solar disc to a Grand maximum characterized by the very active Sun.

In this paper we do not aim at providing an overview of all aspects of long-term solar activity but rather focus on Grand minima. The main features of the Grand minima can be summarized as follows (features, marked with the * sign, are based on one or a few examples only and cannot be considered as statistically grounded):

• The Sun spends about 3/4 of the time at moderate activity, 1/6 in the Grand minimum and about 1/10 in the Grand maximum state (depending on the selection criterion).

• Identification of Grand minima is robust and is not grossly affected by the uncertainties of the paleomagnetic data.

• Occurrence of Grand minima is not a result of long-term cyclic variations but is defined by stochastic or chaotic processes.

• There is a tendency for Grand minima to cluster with the recurrence of roughly

2000-3000 years. Within clusters, a ≈210-yr quasi-periodicity (de Vries or Suess cycle) exists between Grand minima.

• Grand minima occur of two different types: short ($\leqslant$ 100 year) Maunder-type and long (about 150 years) Spörer-type minima.

• Solar cyclic dynamo keeps operating during Grand minima, but at a greatly reduced strength and perhaps with stretched cycle length.

• The presently available data somewhat favors gradual onset of a Grand minima.

• Sunspot activity can be strongly asymmetric between northern and southern hemispheres during a Grand minimum.

Acknowledgements

GAK acknowledges visiting support from the University of Oulu and the Academy of Finland. Work by SSK has been partly supported by WCU grant No. R31-10016 of the Korean Ministry of Education, Science and Technology. IGU thanks organizers of the IAU-286 Symposium for the invitation and partial support.

References

Abreu, J. A., Beer, J., Steinhilber, F., Tobias, S. M., & Weiss, N. O. 2008, *Geophys. Res. Lett.* 352, L20109
Arlt, R. 2008, *Solar Phys.* 247, 399
Bard, E., Raisbeck, G., Yiou, F., & Jouzel, J. 1997, *Earth Planet. Sci. Lett.* 150, 453
Barnard, L., Lockwood, M., Hapgood, M. A., Owens, M. J., Davis, C. J., & Steinhilber, F. 2011, *Geophys. Res. Lett.* 381, L16103
Beer, J. 2000, *Space Sci. Rev.* 94, 53
Beer, J., Tobias, S., & Weiss, N. 1998, *Solar Phys.* 181, 237
Beer, J., Blinov, A., Bonani, G., Hofmann, H., & Finkel, R. 1990, *Nature* 347, 164
Berggren, A.-M., Beer, J., Possnert, G., Aldahan, A., Kubik, P., Christl, M., Johnsen, S. J., Abreu, J., & Vinther, B. M. 2009, *Geophys. Res. Lett.* 36, L11801
Charbonneau, P. 2001, *Solar Phys.* 199, 385
Cowling, T. G. 1945, *MNRAS* 105, 166
Damon, P. & Sonett, C. 1991, in: C. Sonett, M. Giampapa, & M. Matthews (eds.), *The Sun in Time*, (Tucson, U.S.A.: University of Arizona Press), p. 360
de Carvalho, J. & Prado, C. 2000, *Phys. Rev. Lett.* 84
de Toma, G., Gibson, S., Emery, B., & Kozyra, J. 2010, in: M. Maksimovic, K. Issautier, N. Meyer-Vernet, M. Moncuquet, & F. Pantellini (eds.), Proc. AIP Conf. v. 1216 (Melville, New York: AIP) p. 317
Dicke, R. H., 1978, *Nature* 276, 676
Donadini, F., Korte, M., & Constable, C. 2010, *Space Sci. Rev.* 155, 219
Eddy, J. 1976, *Science* 192, 1189
Eddy, J. 1977, *Scientific American* 236, 80
Field, C., Schmidt, G., Koch, D., & Salyk, C. 2006, *J. Geophys. Res.* 111, D15107
Finkel, R. & Nishiizumi, K. 1997, *J. Geophys. Res.* 102, 26699
Fligge, M., Solanki, S. K., & Beer, J. 1999, *A& A* 346, 313
Goslar, T. 2003, *PAGES News* 11, 12
Hathaway, D. H. 2010, *Living Rev. Solar Phys.* 7
Heikkilä, U., Beer, J., & Feichter, J. 2009, *Atmos. Chem. Phys.* 9, 515
Horiuchi, K., Uchida, T., Sakamoto, Y., Ohta, A., Matsuzaki, H., Shibata, Y., & Motoyama, H. 2008, *Quat. Geochronology* 3, 253
Hoyng, P. 1993, *A& A* 272, 321
Hoyt, D. & Schatten, K. 1998, *Solar Phys.* 179, 189
Korte, M. & Constable, C. 2005, *Earth Planet. Sci. Lett.* 236, 348

Kovaltsov, G. A. & Usoskin, I. G. 2010, *Earth Planet. Sci. Lett.* 291, 182
Křivský, L. & Pejml, K. 1988, *Publ. Astron. Inst. Czech Acad. Sci.* 75, 32
Masarik, J. & Beer, J. 2009, *J. Geophys. Res.* 114, D11103
McDonald, F. B., Webber, W. R., & Reames, D. V. 2010, *Geophys. Res. Lett.* 371, L18101
Miyahara, H., Kitazawa, K., Nagaya, K., Yokoyama, Y., Matsuzaki, H., Masuda, K., Nakamura, T., & Muraki, Y. 2010, *J. Cosmol.* 8, 1970
Miyahara, H., Masuda, K., Muraki, Y., Furuzawa, H., Menjo, H., & Nakamura, T. 2004, *Solar Phys.* 224, 317
Miyahara, H., Masuda, K., Muraki, Y., Kitagawa, H., & Nakamura, T. 2006, *J. Geophys. Res.* 111, A03103
Moss, D., Sokoloff, D., Usoskin, I., & Tutubalin, V. 2008, *Solar Phys.* 250, 221
Mursula, K., Usoskin, I., & Kovaltsov, G. 2001, *Solar Phys.* 198, 51
Peristykh, A. & Damon, P. 1998, *Solar Phys.* 177, 343
Reimer, P., Baillie, M., Bard, E., Bayliss, A., Beck, J., Bertrand, C. *et al.* 2004, *Radiocarbon* 46, 1029
Ribes, J. & Nesme-Ribes, E. 1993, *A&A* 276, 549
Scherer, K., Fichtner, H., Borrmann, T., Beer, J., Desorgher, L., Flükiger, E., Fahr, H.-J., Ferreira, S. E. S., Langner, U. W., Potgieter, M. S., Heber, B., Masarik, J., Shaviv, N., & Veizer, J. 2006, *Space Sci. Rev.* 127, 327
Schrijver, C. J., Livingston, W. C., Woods, T. N., & Mewaldt, R. A. 2011, *Geophys. Res. Lett.* 380, L06701
Schüssler, M., Schmitt, D., & Ferriz-Mas, A. 1997, in: B. Schmieder, J. del Toro Iniesta, & M. Vázquez (eds.) *1st Advances in Solar Physics Euroconference: Advances in the Physics of Sunspots*, ASP Conference Series, Vol. 118, (San Francisco, U.S.A.: Astronomical Society of the Pacific), p. 39
Schwadron, N. A., Boyd, A. J., Kozarev, K., Golightly, M., Spence, H., Townsend, L. W., & Owens, M. 2010, *Space Weather* 80, S00E04
Siegenthaler, U., Heimann, M., & Oeschger, H. 1980, *Radiocarbon* 22, 177
Silverman, S. 1992, *Rev. Geophys.* 30, 333
Snowball, I. & Muscheler, R. 2007, *Holocene* 17, 851
Sokoloff, D. 2004, *Solar Phys.* 224, 145
Solanki, S. K. & Krivova, N. 2011, *Science* (in press)
Solanki, S., Krivova, N. A., Schüssler, M., & Fligge, M. 2002, *A& A* 396, 1029
Solanki, S., Usoskin, I., Kromer, B., Schüssler, M., & Beer, J. 2004, *Nature* 431, 1084
Sonett, C. 1983, *J. Geophys. Res.* 88, 3225
Soon, W.-H. & Yaskell, S. 2003, *The Maunder Minimum and the Variable Sun-Earth Connection* (Singapore; River Edge, U.S.A.: World Scientific)
Stuiver, M. & Braziunas, T. 1989, *Nature* 338, 405
Stuiver, M., Braziunas, T., Becker, B., & Kromer, B. 1991, *Quatern. Res.* 35, 1
Stuiver, M., Reimer, P., Bard, E., Burr, G., Hughen, K., Kromer, B., McCormac, G., v.d. Plicht, J., & Spurk, M. 1998, *Radiocarbon* 40, 1041
Suess, H. 1980, *Radiocarbon* 22, 200
Tans, P., de Jong, A., & Mook, W. 1979, *Nature* 280, 826
Usoskin, I. & Kromer, B. 2005, *Radiocarbon* 47, 31
Usoskin, I., Mursula, K., & Kovaltsov, G. 2001, *J. Geophys. Res.* 106, 16039
Usoskin, I., Solanki, S., Schüssler, M., Mursula, K., & Alanko, K. 2003, *Phys. Rev. Lett.* 91
Usoskin, I. G. 2008, *Living Rev. Solar Phys.* 5, 3
Usoskin, I. G., Horiuchi, K., Solanki, S., Kovaltsov, G. A., & Bard, E. 2009a, *J. Geophys. Res.* 114, A03112
Usoskin, I. G., Mursula, K., Arlt, R., & Kovaltsov, G. A. 2009b, *ApJL* 700, L154
Usoskin, I. G., Sokoloff, D., & Moss, D. 2009c, *Solar Phys.* 254, 345
Usoskin, I. G., Solanki, S. K., & Korte, M. 2006, *Geophys. Res. Lett.* 33, 8103
Usoskin, I. G., Solanki, S. K., & Kovaltsov, G. A. 2007, *A&A* 471, 301
Vaquero, J. 2007, *Adv. Space Res.* 40, 929
Vaquero, J. M., Gallego, M. C., Usoskin, I. G., & Kovaltsov, G. A. 2011, *ApJL* 731, L24

Vonmoos, M., Beer, J., & Muscheler, R. 2006, *J. Geophys. Res.* 111, A10105
Voss, H., Kurths, J., & Schwarz, U. 1996, *J. Geophys. Res.* 101, 15637
Yang, S., Odah, H., & Shaw, J. 2000, *Geophys. J. Internat.* 140, 158

Discussion

LEIF SVALGAARD: There is general acceptance that the heliospheric open flux now is similar to 110 years ago. Therefore, shouldn't the modulation of cosmic rays be the same as 110 years ago?

ILYA USOSKIN: Despite some indirect open flux reconstructions, cosmic rays are known to be less modulated 100 years than now, as confirmed by cosmogenic isotopes, ^{10}Be in the polar ice and ^{44}Ti in meteorites.

ARNAB CHOUDHURI: (Comment) You pointed out that the statistics for grand maximum were consistent with fluctuations, but grand minimum statistics were more complicated. I think this is because there is a different mechanism needed for recovery from grand minimum (different form fluctuations), which cloud complicate the statistics.

ILYA USOSKIN: I am glad you can immediately interpret our results.

JEFFREY LINSKY: (Comment) Another physical concept that should be considered with regard to cosmic ray interpretation, the Sun is moving through the in-homogeneous interstellar medium. The size of the heliosphere varies depending on the structure of this medium. We move at 1 parsec/300000 years and the size of the heliosphere is much smaller than 1 parsec. The interstellar medium may change density by significant factors (2, 8, 10).

ILYA USOSKIN: I agree such fluctuations may exist on very long-term scale but they are expected to be small for high-energy cosmic rays because of the diffusion length.

Comparative Magnetic Minima:
Characterizing quiet times in the Sun and stars
Proceedings IAU Symposium No. 286, 2011
C. H. Mandrini & D. F. Webb, eds.

doi:10.1017/S1743921312005121

Historical records of solar grand minima: a review

José M. Vaquero

Departamento de Física, Centro Universitario de Mérida, Universidad de Extremadura,
Avda. Santa Teresa de Jornet, 38, 06800, Mérida, Badajoz, Spain
email: jvaquero@unex.es

Abstract. Knowing solar activity during the past centuries is of great interest for many purposes. Historical documents can help us to know about the behaviour of the Sun during the last centuries. The observation of aurorae and naked-eye sunspots provides us with continuous information through the last few centuries that can be used to improve our knowledge of the long-term solar activity including solar Grand Minima. We have more or less detailed information on only one Grand minimum (the Maunder minimum in the second half of 17th century), which serves as an archetype for Grand minima in general. Telescopic sunspot records and measurements of solar diameter during Maunder minimum are available. In this contribution, I review some recent progress on these issues.

Keywords. Sun: activity, history and philosophy of astronomy.

---

1. Introduction

Historical documents can help us to know about the behaviour of the Sun during the last centuries including Grand Minima (Vaquero & Vázquez 2009). On longer time scales, other procedures can also be used for the reconstruction of solar activity such as cosmogenic radionuclides (Usoskin 2008). There exists a wide range of possibilities to reconstruct solar activity on the basis of documentary sources (Figure 1). On the one hand, one has direct data, i.e. data obtained from observations of the Sun. And on the other hand, one has indirect data based on terrestrial phenomena linked to the behaviour of the Sun. Nonetheless, the indices that have been most used are those related to the observations of sunspots due to their versatility.

Some recent research on solar Grand Minima based on historical documents will be reviewed here. Therefore, the aim of this work is to show that this is a research field that can be fruitfully developed because there is interesting information that is still buried in archives and libraries.

2. Naked-eye sunspot observations

This section is devoted to naked-eye sunspot observations, and especially to their possible use for the reconstruction of solar Grand Minima during the last two millennia. It is occasionally possible to observe a sunspot by eye when it is sufficiently large and there are certain atmospheric conditions (mist, dust, smoke,...) which reduce the intensity of the Sun's light (see Vaquero & Vázquez 2009, specially Chapter 2). The sunspot visibility problem should be approached as a problem of calculating contrast thresholds for the human eye (Schaefer 1991). Schaefer (1993) developed a theoretical model of sunspot visibility that can be applied to naked-eye observation.

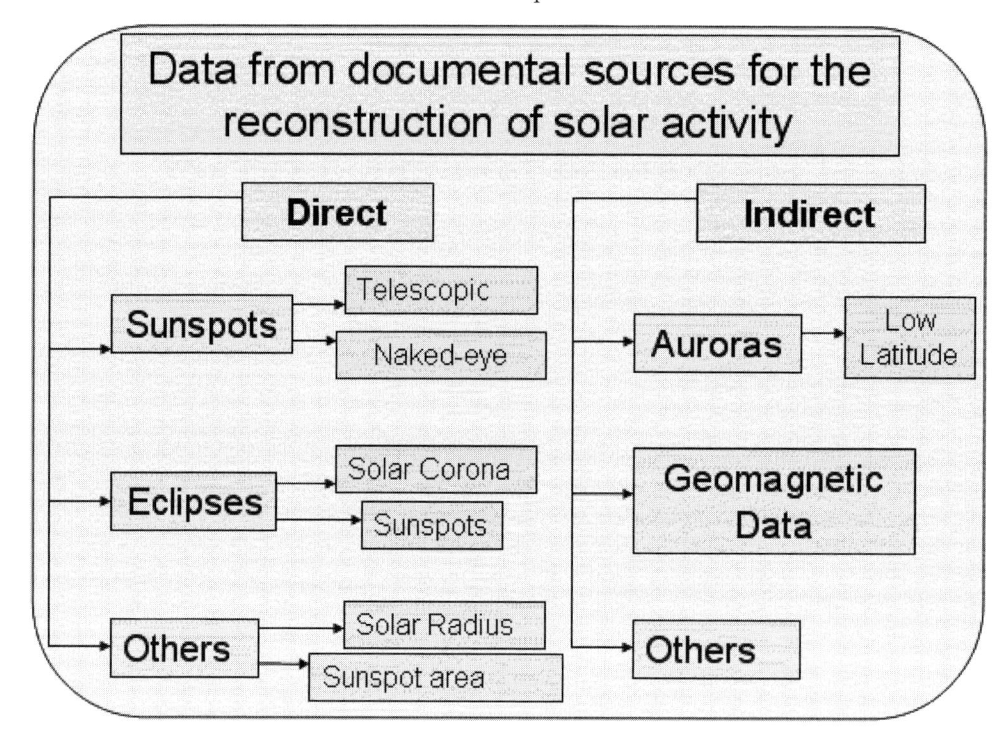

Figure 1. Different types of data from documentary sources for the reconstruction of solar activity.

A great number of historical naked eye sunspots are reported in Oriental historical sources (China, Korea, and Japan). Others are reported in European, Arabic (Vaquero & Gallego 2002), Indian (Malville & Singh 1995), and Maya (Burland 1958) sources. There are various catalogues of naked-eye sunspot observations available in the literature (Wittmann & Xu 1987; Yau & Stephenson 1988). In the last years, some new records appeared (Krivsky 1985; Stephenson & Willis 1999; Vaquero & Gallego 2002; Moore 2003; Lee *et al.* 2004; Vaquero 2004).

The reliability of naked-eye sunspot observations has been assessed by comparing carefully the Oriental sunspot sightings from 1862 onwards with contemporary Occidental white-light images of the Sun (Willis *et al.* 1996). These observations may be related to variations in some meteorological parameters (Willis *et al.* 1980; Eddy *et al.* 1989; Scuderi 1990; Hameed & Gong 1991), such as the number of days of mist or dust storms. Nevertheless, some authors have reconstructed time series of annual numbers of naked-eye sunspots in spite of their evident problems (Nagovitsyn 2001 and Vaquero *et al.* 2002).

Vaquero & Vázquez (2009) made a comparison between the solar activity reconstructed using cosmogenic radionuclides, the most reliable proxy to study the long-term behaviour of the Sun, and the series of naked-eye sunspot observations. Figure 2 shows the reconstruction of sunspot number made by Solanki *et al.* (2004) and a 50-year moving average of the annual number of naked-eye sunspot reconstructed by Vaquero *et al.* (2002) in arbitrary units. The naked-eye record is not useful for the study of long-term solar activity from 200 BC to AD 800 although some peaks of solar activity also appear in the naked-eye record. For the period AD 800-1600 approximately, the naked-eye sunspot record shows the most important periods of the solar activity although the amplitudes are variable. The two curves are quite similar during the period AD 1300-1600. Since the

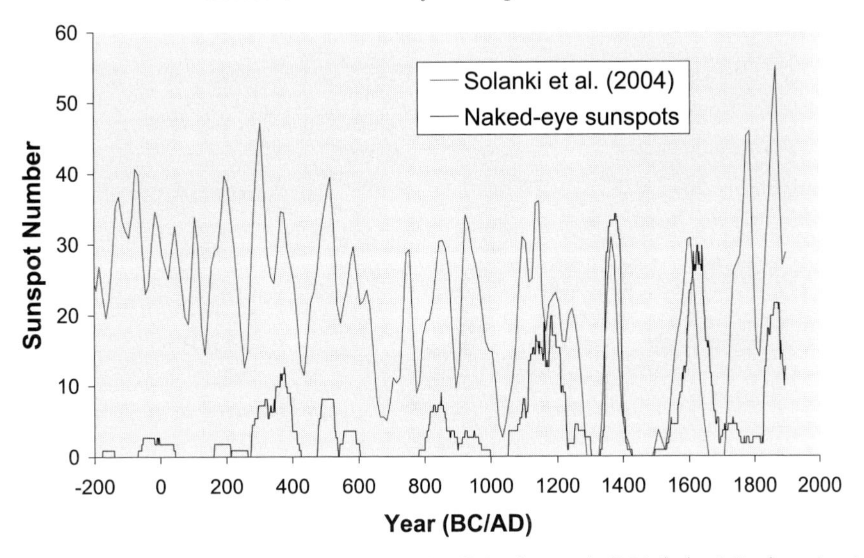

Figure 2. Reconstruction of sunspot number by Solanki *et al.* (2004) (red line) and using the 50-year moving average of the annual number of naked-eye sunspots by Vaquero *et al.* (2002) (black line). Data were provided by I.G. Usoskin.

year 1600, the use of local sources in the Oriental record has made the correlation worse, and major differences can be observed during the period 1700-1825.

3. Auroral observations in History

Few phenomena have made as much of an impression on human beings as the aurora borealis, also known as the northern lights. The aurora is one of the most spectacular and earliest-known manifestations of the links between the Sun and the Earth. The northern and southern lights appear in the night sky with a great variety of colours and forms (Figure 3). The auroral activity is not rare, although few aurorae can be observed from low-latitude sites. Some general works on aurorae are Eather (1980), Brekke & Eggeland (1994), Akasofu (2007), and Bone (2007).

The systematic study of the seasonal and secular variations of the auroral appearances needed data with perfectly dated and correctly compiled records of aurorae. Catalogues of aurora borealis were born of this need. The best-known catalogue of the 19th century was that of Hermann Fritz (1830-1893) who used many sources (Fritz, 1873). A catalogue of southern hemisphere aurorae was written by Boller (1898). Of the 20th century catalogues, we would cite as of special interest that written by Link (1962, 1964). Another useful catalogue was compiled by Krivsky & Pejml (1988)Krivsky & Pejml (1988) for the interval 1000–1900. They give dates of observation only and are restricted to latitudes <55 degrees.

Figure 4 shows the number of days per year (using a 5-year moving average and logarithmic scale) in which aurorae were recorded from the year AD 1000 to 1900 using the catalogue of Krivsky & Pejml (1988). Very few aurorae before the Maunder minimum are recorded there, but the number increases considerably later. The number of auroral days per year during the first part of the total period is very low. During the 16th and 17th centuries, this number begins to increase, although it is still small compared with the number of recorded auroral days during the 18th and 19th centuries. Note that the solar

Figure 3. Sketch of the auroral display observed in Paris (13 May 1869) from Flammarion (1873).

Grand Minima of the last millennium appear clearly in this record in agreement with Solanki *et al.* (2004). Note also a minimum of auroral activity around 1765 (Silverman, 1992).

4. Sunspot observations during the onset of Maunder minimum

During the 400 years of systematic sunspot observations with telescopes (Hoyt & Schatten 1998), several episodes of special interest have been recorded. Probably, the most interesting was the Maunder minimum (Eddy 1976, 1983; Ribes & Nesme–Ribes 1993;

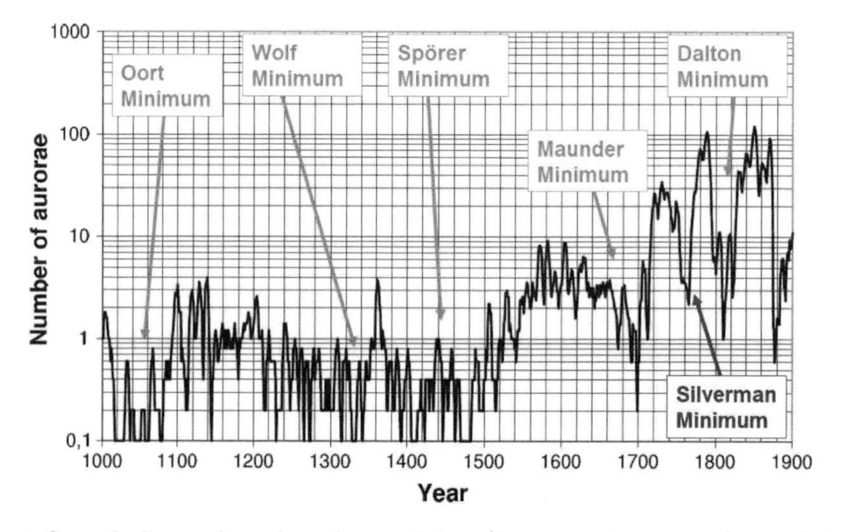

Figure 4. Smoothed annual number of auroral days (5-year moving average) recorded in the catalogue of Krivsky & Pejml (1988).

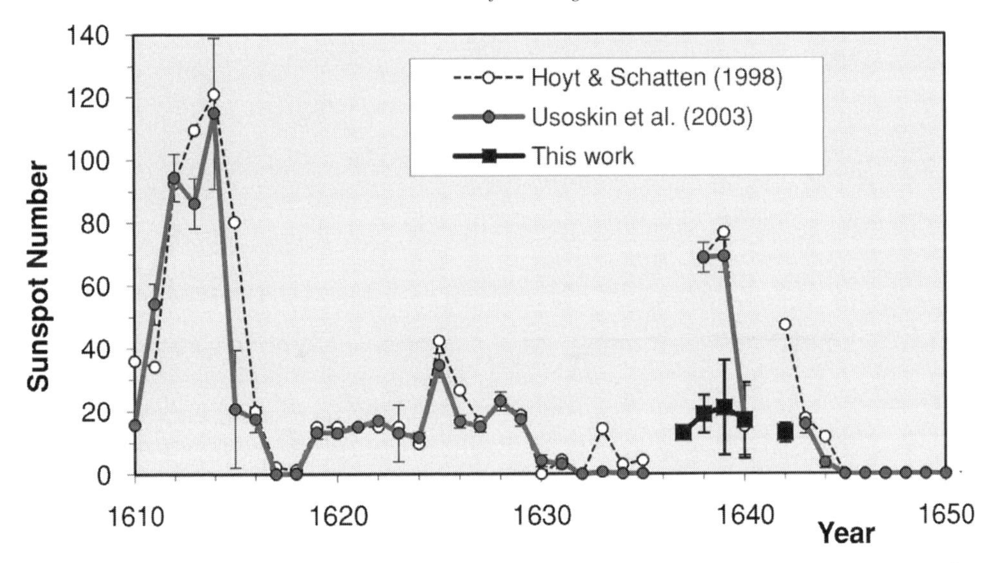

Figure 5. Annual sunspot numbers in the first half of 17th century. Group sunspot numbers (Hoyt & Schatten 1998) are shown by the dotted line, weighted sunspot number, based on the same data set (Usoskin *et al.* 2003) by the red line, and the weighted sunspot number estimated in Vaquero *et al.* (2011) by the black line.

Usoskin *et al.* 2001; Vaquero 2007). In the last few years, numerous studies have dealt with the Maunder minimum, including a review by Soon & Yaskell (2003) and more recently by Miyahara *et al.* (2006). Another recent study considered solar rotation during the 17th century (Casas *et al.* 2006), using observations of a sunspot by Nicholas Bion.

The Maunder minimum forms an archetype for the Grand minima, and detailed knowledge of its temporal development has important consequences for the solar dynamo theory dealing with long-term solar activity evolution. The paradigm of the Grand minimum general scenario (Usoskin 2008) established that (i) transition from the normal activity to the deep minimum was sudden, (ii) a 22-year cycle was dominant in sunspot, and (iii) the recovery of the sunspot activity from the deep minimum to normal activity was gradual. Vaquero *et al.* (2011) reconsidered the paradigm of a Grand minimum general scenario by using newly recovered sunspot observations as well as revising some earlier uncertain data for the period 1636-1642, i.e., one solar cycle before the beginning of the Maunder minimum (Figure 5). The revised and updated data for 1637-1642 have essentially changed the profile of temporal variability of sunspot data before the Maunder minimum. In particular, the last solar cycle before the minimum now appears quite modest, with the peak value of 20 approximately compared with about 70 in the Group Sunspot Number series. This new scenario implies low solar activity roughly two solar cycles before the beginning of the Maunder minimum. Thus, transition from the normal activity to the deep minimum was not as sudden as previously thought. Figure 6 depicts a solar disk drawing by G. Marcgraf showing sunspots from 1637 September 21 to 25 as an example of historical documents used by Vaquero *et al.* (2011).

5. Solar radius measurements during Maunder Minimum

It is a matter of current debate whether the variations of the solar radius are correlated with the 11-year solar cycle. All the observations since 1650 have been compiled by

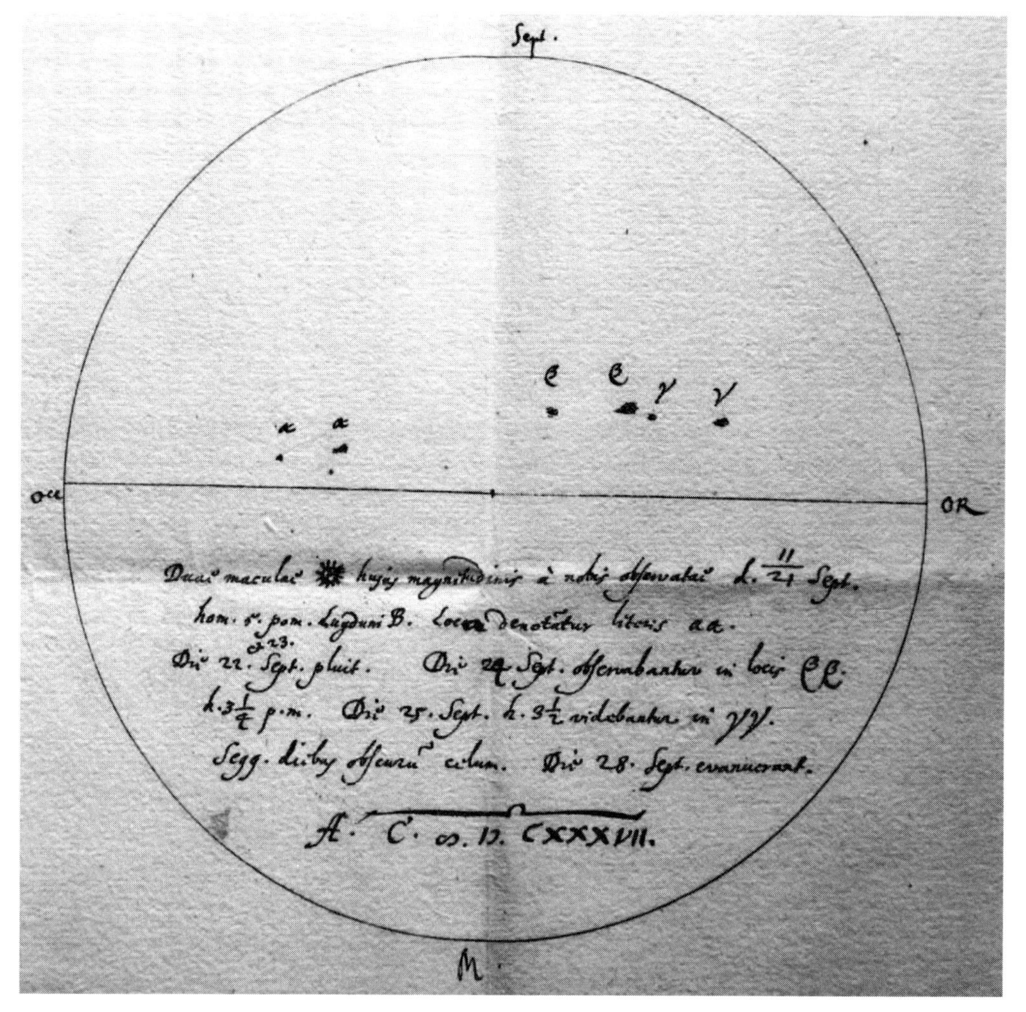

Figure 6. Example of the solar disk drawing by Marcgraft depicting sunspots from 1637 September 21 to 25.

Toulmonde (1997), Pap *et al.* (2001), and Rozelot (2001). Since the 18th century, the values of solar radius rapidly converged towards the current value. Ribes *et al.* (1987) suggested the existence of a larger Sun during Maunder minimum. This claim was criticized by O'Dell & Van Helden (1987), who argued that the error in the measurements made by the Paris observers was larger than Ribes *et al.* (1987) had assumed.

During the 17th and 18th centuries, meridian lines in cathedrals in Bologna, Rome, Florence, and Paris served as solar observatories making careful and continuous study of the Sun possible (Heilbron 1999). The earliest of these observatories was in the cathedral of San Petronio in Bologna. The heliometer of San Petronio consists of two separate pieces. One piece lies on the floor; it is a perfectly horizontal rod running due north for about 67 m from a spot under one of the side chapels to the front door of the church. The other part is a small hole (2.5 cm in diameter) in a horizontal metal plate fixed in the roof of the chapel. The hole is permanently open so as to give free access to solar rays around noon throughout the year. For 80 years (1655-1736), some scientists and

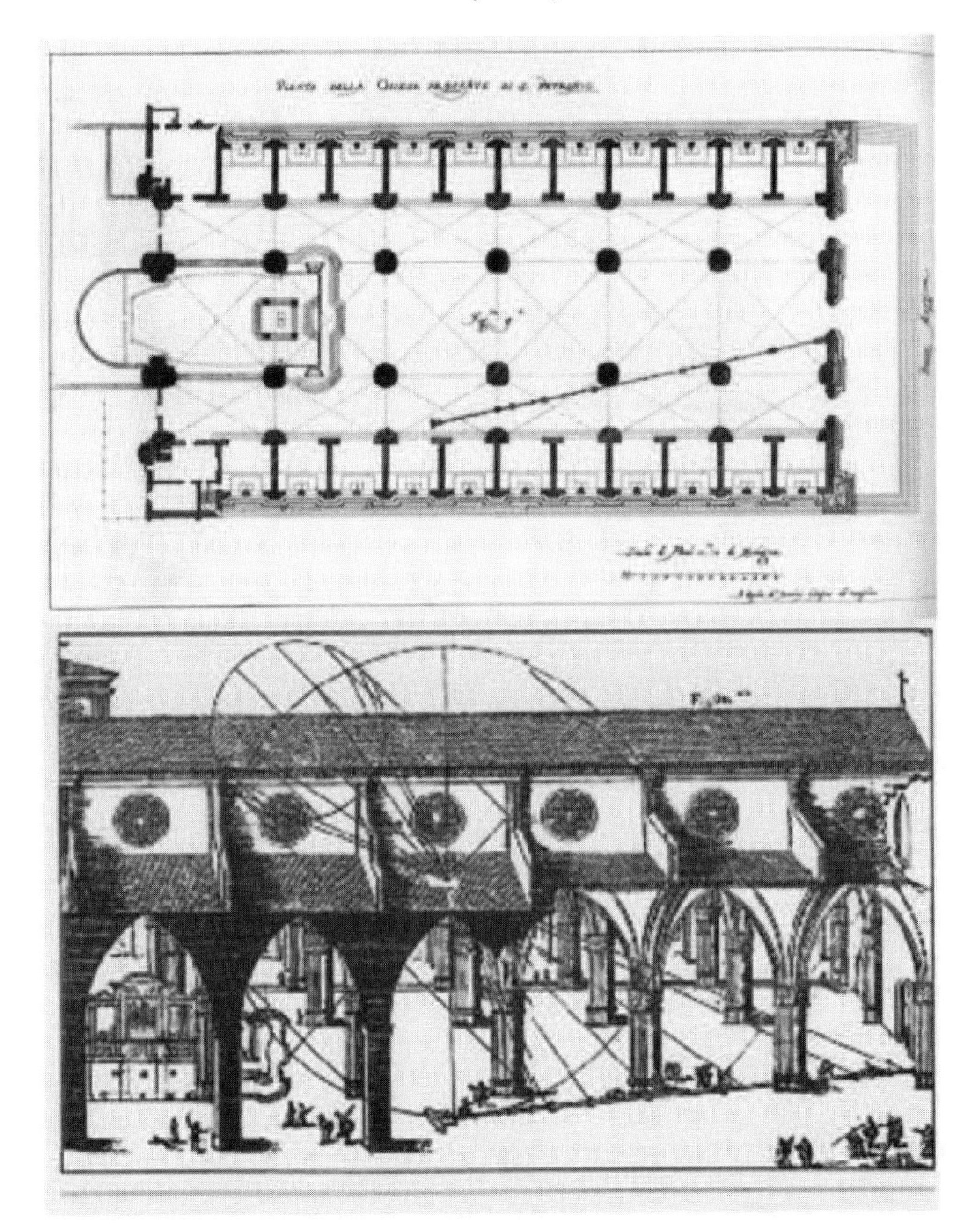

Figure 7. Old schemes of San Petronio Cathedral in Bologna (Heilbron, 1999).

clerics observed the Sun, filling about 300 pages in the register published by Eustachio Manfredi (1736). Each entry includes a description of the weather, the distances of the Sun's limbs from the vertex corrected for the penumbra, and the apparent diameter of the sun, all given to seconds of arc.

Vaquero & Vázquez (2009), see Figure 5.10 in page 227, compiled all measurements of solar radius, in arcminutes, made in San Petronio from 1655 to 1736. This time series is

Figure 8. The *Basilica di San Petronio* in Bologna (picture by Paolo Carboni).

very noisy and it is not easy to obtain a clear conclusion. However, one can see clearly the annual variation of the apparent solar radius in this record in Figure 9, where only the period 1695–1705 is represented. This dataset require further studies.

6. Conclusion

Vaquero & Vázquez (2009) have shown that solar Grand Minima of the last millennium are recorded in documentary solar proxies as naked-eye sunspot and auroral records. The well-known minima of Oort, Wolf, Spoerer, and Maunder stand out sharply in both records. Moreover, revised sunspot data prior to the Maunder minimum lead to the revisited observational scenario of a Grand minimum of solar activity. The new data dramatically change the magnitude of the sunspot cycle just before the Maunder minimum implying a possibly gradual onset of the minimum with reduced activity started two cycles before it (Vaquero *et al.* 2011).

Any information on the state of the Sun in past times is interesting because the Sun is not a laboratory experiment that can be controlled. Of course, there are useful data to study past solar activity preserved in libraries and archives.

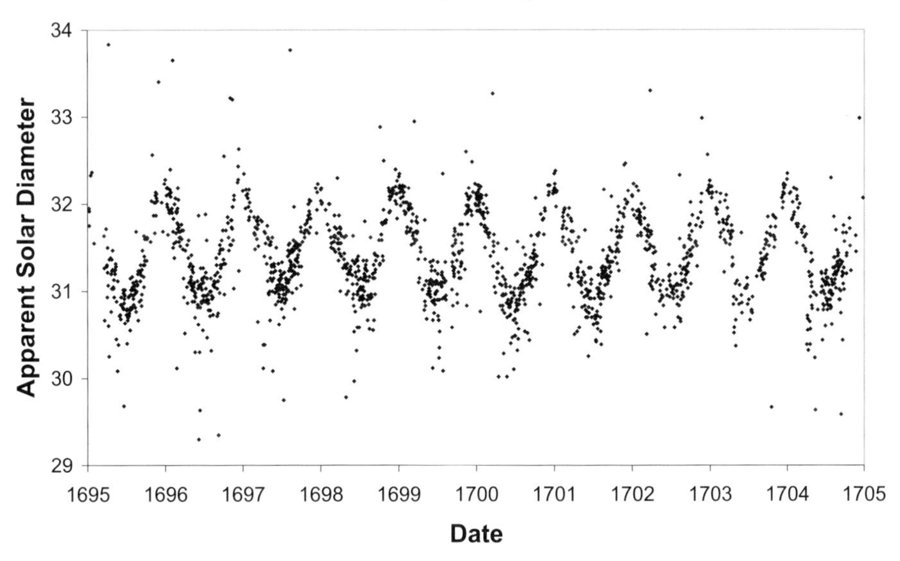

Figure 9. Measurements of solar radius, in arcminutes, made in San Petronio from 1695 to 1705.

Acknowledgements

The author is grateful to I.G. Usoskin for his valuable comments. Support from the Junta de Extremadura and Ministerio de Ciencia e Innovacin of the Spanish Government (AYA2008-04864/AYA and AYA2011-25945) is gratefully acknowledged.

References

Akasofu, S.-I. 2007, *Exploring the Secrets of the Aurora* (Springer, 2nd Edition)

Boller, W. 1898, *Gerlands Beiträge zur Geophysik*, 3, 56, 550

Bone, N. 2007, *Observing and Recording Nature's Spectacular Light Show* (Springer, Patrick Moore's Practical Astronomy Series)

Brekke, A. & Eggeland, A. 1994, *The Northern Lights. Their Heritage and Science* (Reidel)

Burland, C. A. 1958, In: *Proceedings of the 32 International Congress of Americanist (Copenhague)*, 326-330

Casas, R., Vaquero, J. M., & Vázquez, M. 2006, *Solar Phys.* 234, 379

Eather, R. H. 1980, *Majestic Lights* (Washington: American Geophysical Union)

Eddy, J. A. 1976, *Science* 192, 1189

Eddy, J. A. 1983, *Solar Phys.* 89, 195

Eddy, J. A., Stephenson, F. R., & Yau, K. C. 1989, *Q. J. R. Astron. Soc.* 30, 65

Flammarion, C. 1873, *L'Atmosphère*, (Paris: Librairie Hachete et Cia.)

Fritz, H. 1873, *Verzeichniss Beobachteter Polarlichter*, (Wien: C. Gerold's Sohn)

Hameed, S. & Gong, G. 1991, *Solar Phys.* 139, 409

Heilbron, J. L. 2001, *The sun in the church: cathedrals as solar observatories* (Harvard: Harvard University Press)

Hoyt, D. V. & Schatten, K. H. 1998, *Solar Phys.* 179, 189

Krivsky, L. 1985, *Bull. Astron. Inst. Czechosl.*, 36, 60

Krivsky, L. & Pejml, K. 1988, *Solar Activity, Aurorae and Climate in Central Europe in the last 1000 Years* (Publications of the Astronomical Institute of the Czechoslovak Academy of Sciences, Publication No. 75)

Lee, E. H., Ahn, Y. S., Yang, H. J., & Chen, K. Y. 2004, *Solar Phys.*, 224, 373

Link, F. 1962, *Geofysikalni Sbornik*, 173, 297

Link, F. 1964, *Geofysikalni Sbornik*, 212, 501

Malville, J. M. & Singh, R. P. B. 1995, *Vistas in Astronomy*, 39, 431
Manfredi, E. 1736, *De Gnomone Meridiano Bononiensi ad Divi Petronii* (Bononiae: Laeli a Vulpa)
Miyahara, H., Sokoloff, D., & Usoskin, I. G. 2006, *The solar cycle at the Maunder minimum epoch.* In: Ip, W.-H., Duldig, M. (Eds.), Advances in Geosciences (Singapore: World Scientific) , pp. 120
Moore, S. 2003, *J. Br. Astron. Assoc.*, 113(5), 305
Nagovitsyn, Yu. A. 2001, *Geomagn. Aeron.* 41, 711
O'Dell, C. R. & Van Helden, A. 1987, *Nature* 330, 629
Pap, J., Rozelot, J. P., Godier, S., & Varadi, F. 2001, *Astron. Astrophys.* 372, 1005
Ribes, E., Ribes, J. C., & Barthalot, R. 1987, *Nature* 330, 629
Ribes, J. C. & Nesme-Ribes, E. 1993, *Astron. Astrophys.* 276, 549
Rozelot, J. P. 2001, *J. Atmos. Terr. Phys.* 63, 375
Schaefer, B. E. 1991, *Q. J. R. Astron. Soc.*, 32, 35
Schaefer, B. E. 1993, *Astrophys. J.*, 411, 909
Scuderi, L. A. 1990, *Q. J. R. Astron. Soc.* 31, 109
Silverman, S. M. 1992, *Reviews of Geophysics* 30, 333
Solanki, S. K., Usoskin, I. G., Kromer, B., Schüssler, M., & Beer, J. 2004, *Nature*, 431, 1084
Soon, W. W.-H. & Yaskell, S. H. 2003, *Maunder Minimum and the Variable Sun Earth Connections* (Singapore: World Scientific Printers)
Stephenson, F. R. & Willis, D. M. 1999, *Astron. Geophys.*, 40, 21
Toulmonde, M. 1997, *Astron. Astrophys.* 325, 1174
Usoskin, I. G., Mursula, K., & Kovaltsov, G. A. 2001, *J. Geophys. Res.* 106(A8), 16039
Usoskin, I. G. 2008, *Living Rev. Solar Phys.*, 5, 3 (http://www.livingreviews.org/lrsp-2008-3)
Vaquero, J. M. & Gallego, M. C. 2002, *Solar Phys.*, 206, 209
Vaquero, J. M., Gallego, M. C., & García, J. A. 2002, *Geophys. Res. Lett.*, 29, 1997
Vaquero, J. M. 2004, *Solar Phys.*, 223, 283
Vaquero, J. M. 2007, *Adv. Spa. Res.* 40, 929
Vaquero, J. M. & Vázquez, M. 2009, *The Sun Recorded Through History* (New York: Springer)
Vaquero, J. M., Gallego, M. C., Usoskin, I. G., & Kovaltsov, G. A. 2011, *ApJL* 731, L24
Willis, D. M., Easterbrook, M. G., & Stephenson, F. R. 1980, *Nature* 287, 617
Willis, D. M., Davda, V. N., & Stephenson, F. R. 1996, *Q. J. R. Astron. Soc.* 37, 189
Wittmann, A. D. & Xu, Z. T. 1987, *Astron. Astrophys. Suppl. Series*, 70, 83
Yau, K. K. C. & Stephenson, F. R. 1988, *Q. J. R. Astron. Soc.*, 29, 175

Discussion

NADEZHDA ZOLOTOVA: Is it correct that due to uncertainties in the solar images, it is not possible to identify not only the latitude but even the hemisphere where the sunspots are?

JOSÉ VAQUERO: In general, my answer is "yes". However, we have also some detailed and well oriented old sunspot drawings.

Comparative Magnetic Minima:
Characterizing quiet times in the Sun and stars
Proceedings IAU Symposium No. 286, 2011
C. H. Mandrini & D. F. Webb, eds.

doi:10.1017/S1743921312005133

Effects of solar variability on planetary plasma environments and habitability

César Bertucci

Instituto de Astronomía y Física del Espacio, Buenos Aires, Argentina
email: cbertucci@iafe.uba.ar

Abstract. Intrinsic and induced planetary magnetospheres are the result of the transfer of energy and linear momentum between the solar wind and, respectively, the magnetic field and the atmospheres of solar system bodies. This transfer seems to be, however, more critical to the atmospheric evolution of unmagnetized objects such as Mars and Venus, as locally ionized planetary particles are accelerated by solar-wind induced electric fields, leading to atmospheric escape. The nature of the obstacle to the solar wind being different, intrinsic and induced magnetospheres respond differently to solar cycle changes in solar photon flux and solar wind properties. The influence of solar variability on planetary magnetospheres and its implications for atmospheric evolution based upon remote and in situ spacecraft measurements, and numerical simulations are discussed. In particular, the case of unmagnetized objects where non-thermal escape process might have played a role in their habitability conditions is considered.

Keywords. solar activity, solar neighborhood, planets: habitability

---

1. Introduction

The Sun interacts with planets and other solar system bodies through radiation and plasma processes with important consequences on the evolution of their atmospheres. The solar system is permeated by the solar wind, the coronal plasma in supersonic expansion which carries the frozen-in interplanetary magnetic field. Several types of obstacles are formed as a result of the interaction of the solar wind flow with planetary objects, depending on whether these objects are magnetized or not magnetized. Magnetized planets like Earth ($M \approx 7.9 \times 10^{25}$ G cm^3), Saturn ($M \approx 4.6 \times 10^{28}$ G cm^3) or Jupiter ($M \approx 1.53 \times 10^{30}$ G cm^3) generate 'intrinsic magnetospheres' - cavities free from solar wind plasma whose outer boundary, the magnetopause, is located well above the planet's atmosphere. This prevents the direct interaction of the solar wind with their atmospheres.

The absence of strong intrinsic fields at Venus ($M < 3\times10^{21}$ G cm^3, Russell *et al.* 1980) and Mars ($M < 2\times10^{21}$ G cm^3, Acuña *et al.* 1998) results in the direct transfer of energy and momentum from the solar wind and the Interplanetary Magnetic Field (IMF) into the atmospheric ionized particles. This is also the case of Saturn's major moon, Titan ($M < 2 \times 10^{21}$ G cm^3, Ness *et al.* 1982), which interacts most of the time with the co-rotating plasma of Saturn's magnetosphere. In these three cases, a well-defined region of perturbed IMF and populated by solar wind and local plasma is formed. This region is referred to as 'induced magnetosphere' (see, e.g., Bertucci *et al.* 2011).

When the supersonic solar wind approaches the ionized atmosphere of an unmagnetized planet, the transfer of linear momentum and energy from the solar flow to the planetary charged particles occur both abruptly and gradually in plasma boundaries and regions, respectively. According to the magnetohydrodynamic (MHD) theory, two boundaries characterize the interaction: the bow shock (BS), where the flow becomes subsonic, compressed and heated, and the so-called 'ionopause' (IP), a pressure balance boundary that encircles a region inaccessible to the solar wind plasma. Accordingly, MHD theory predicts -in the subsolar direction- a gradual build up of the magnetic pressure in the detriment of the initially dominant (just inside the shock) thermal pressure between the BS and the IP. The difference in the speed of the plasma between the subsolar region and the flanks results in the draping of IMF field lines around the body and the formation of an 'induced' magnetotail consisting of two magnetic lobes with field lines parallel and anti-parallel to the incoming flow. In situ plasma measurements revealed the presence of an additional boundary located between the BS and IP, the induced magnetosphere boundary (IMB). The IMB is the outer boundary of the so called 'induced magnetosphere' and is characterized, among other signatures, by an abrupt increase in the magnetic field draping (Bertucci *et al.* 2003a,b) and the change in the dominant ion population, as solar wind ions start to be outnumbered by planetary ions (e.g., Dubinin *et al.* 2006, Martinecz *et al.* 2008, Bertucci *et al.* 2011). The occurrence of the IMB is associated with the role of the exosphere (in addition to the ionosphere) in the interaction as it is the source of locally ionized planetary ions above the ionosphere which are massively incorporated into the solar wind, reducing its momentum. This process, called massloading (Szego *et al.* 2000) is central in the interaction of the solar wind with active comets and involves the acceleration of local plasma via macroscopic (e.g. convective) and microscopic (e.g. electromagnetic waves) electric fields resulting from the drift between the solar wind and the planetary population in the context of the absence of collisions.

It is precisely inside the IMB that most of the escaping flux of atmospheric ions occurs (e.g., Fedorov *et al.* 2008) as a result of the action of a variety of acceleration processes taking place (Dubinin *et al.* 2011). In some cases, the escape of planetary ions stoichiometrically corresponds to the loss of water (e.g., Lundin *et al.* 2009), suggesting the link between the plasma escape problem and the evidence of dramatic changes in global climate and habitability conditions.

As a result, a better understanding of the nature and efficiency of the processes occurring within induced magnetospheres as well as their structure and variability in response to changes in the interplanetary plasma conditions are important to assess the evolution of the atmospheres of these objects. In this context, current investigations aim at answering the following questions: How much of the atmospheric evolution is caused by solar wind interactions? What is the effect of solar variability on these interactions and resulting atmospheric escape? What lessons can be learned for planets around other stars?

In this work we will not provide answers to them in detail, and we will only cover a few results obtained around planets Venus and Mars. In spite of these limitations, however, the results presented are key elements to build those answers in the future.

The text is organized as follows: An overview of the interaction of the solar wind with Venus and Mars is given in the next section, followed by a discussion on the role of solar cycle in the variability of the interaction region extension, ionosphere and escape rate.

2. Solar wind interaction with Mars and Venus. General morphology

Apart from the moon, Venus and Mars are the most visited solar system objects in the history of space exploration. Since the dawn of the space era, many missions have characterized their plasma environments, providing a basis for comparative studies over different periods of solar activity (see Fig. 2).

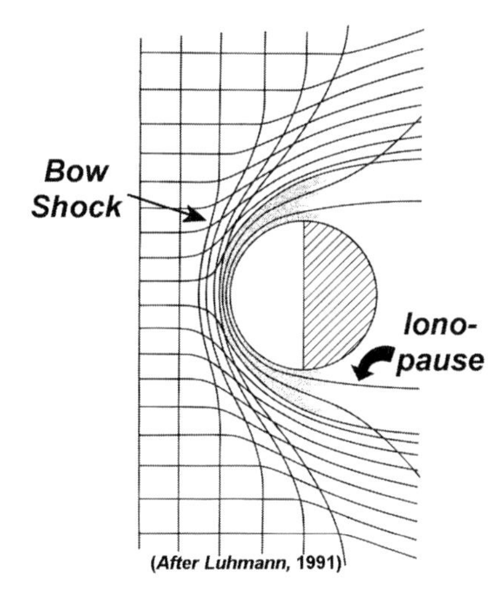

Figure 1. Schematic describing the formation of an induced magnetosphere from the deflection of solar wind stream lines and frozen in magnetic fields (after Luhmann 1991)

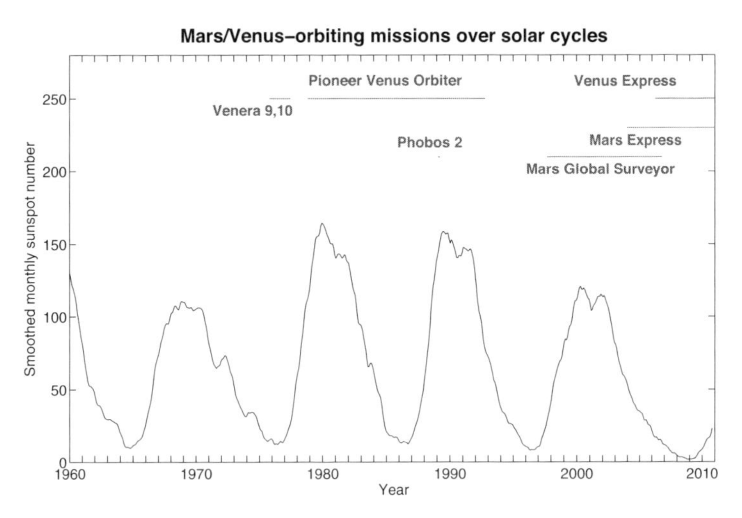

Figure 2. Time coverage of orbiting missions with payload for plasma investigations around Venus (blue, see online version) and Mars (red, see online version) and averaged sunspot number as a function of decimal years since 1960 (source: http://sidc.oma.be/sunspot-data/).

2.1. *Venus*

Current knowledge of the solar wind interaction with Venus comes from the Venera-9 and 10 (e.g., Zeleny and Vaisberg 1985) providing measurements over solar minimum, from the Pioneer Venus Orbiter (PVO) which provided a data set that extended over an entire solar cycle (Russell *et al.* 2006a), and more recently from the Venus Express (VEX) mission, which began its operations in during the last minimum of solar activity.

The plasma environment of Venus has been analyzed using data from the PVO magnetometer OMAG (Russell *et al.* 1980) and plasma analyzers, notably, the Orbiter Retarding Potential Analyzer ORPA (Knudsen *et al.* 1979) because of its higher time resolution. However, compared with VEX's magnetometer MAG (Zhang *et al.* 2006) and the ASPERA-4 plasma analyzer (Barabash *et al.* 2007), PVO instruments had much lower temporal, energy and angular resolution.

Although PVO's temporal coverage includes a whole solar cycle, no direct measurements of the plasma environment during solar minimum were possible due to the high PVO orbital altitude (> 2000 km) at that time. The VEX spacecraft has a constant periapsis altitude of about 250 km and thus can sample this region during solar minimum. Just prior to PVO arrival, the Russian Venera 9 and 10 orbiters (1975, 1976) observed the Venus solar wind interaction, including the bow shock and tail during solar minimum (Verigin *et al.* 1978).

Figure 3 shows VEX ASPERA measurements where the main plasma regions and boundaries around Venus are detected both inbound and outbound from the planet. As the spacecraft approaches Venus, the collisionless bow shock (BS) is clearly defined from the increase in the counts of all measured particles. This corresponds to the increase of the density as the upstream flow is compressed while it becomes subsonic. The heating of the shocked plasma is evident in the outbound leg. Inside the BS, the solar wind plasma becomes increasingly loaded with cold, local plasma and the whole flow continues to decelerate. The IMB (labeled as ICB in the figure) encloses plasma which is mostly cold and of local origin. This is especially noticeable in the oxygen ions. The decrease in all particle counts around the IMB is only apparent. The change in the spacecraft potential prevents electrons with low (a few eVs) energies to be detected. The ionopause (IP) is not clearly observed in the figure, but it is characterized, on the dayside, by the occurrence of local photoelectrons at energies around 20 eV (Coates *et al.* 2011).

These plasma features are correlated with magnetic field signatures. Inside the shock, the magnetosheath is characterized by a high magnetic field variability. At the bottom of the magnetosheath, on the dayside, the magnetic field pileup increases (usually gradually) and a magnetic barrier forms (Zhang *et al.* 1991). The IMB marks the entry into the induced magnetosphere (IM), where piled-up, draped fields coexist with a plasma which is predominantly of planetary origin. The IMB extends to at least 11 planetary radii downstream and encircles the induced tail where planetary plasma escape is concentrated (Saunders and Russell 1986). At the ionopause, the thermal ionospheric pressure is expected to balance the induced magnetosphere's magnetic pressure. When that balance is achieved, the magnetic field is depleted inside the IP. As it will be seen, however, this condition only holds during solar maximum. The lower doses of EUV flux and the increase in periods of high solar wind dynamic pressure during solar minima result in the weakening of the shielding effect of the ionopause and the ionosphere gets magnetized.

2.2. *Mars*

The in situ exploration of the Martian environment dates back to the very beginning of space era. Following Mariner 4, 6 and 7, Mars 2, 3, and Mariner 9 were probably the first spacecraft to enter the Martian induced magnetosphere (Vaisberg and Bogdanov 1974). At the time of those observations, however, it was still uncertain if Mars possessed an intrinsic magnetic field on its own. This issue was resolved with the arrival of Mars Global Surveyor (MGS). MGS observations revealed that the planet has lacked a global intrinsic magnetic field for at least 4 Gyr (Acuña *et al.* 1998). The absence of a global intrinsic magnetic field at Mars led to the re-interpretation of part of previous plasma measurements. A comprehensive review on the comparison between MGS and Phobos-2 plasma observations within the induced magnetosphere of Mars can be found in Nagy *et al.* 2004. Although the magnetic field and the solar wind electron population were efficiently characterized by MGS' magnetometer and electron reflectometer (MAG/ER) (Acuña *et al.* 1992), MGS did not carry ion instruments.

With an overlap of a few years with MGS, Mars Express (MEX) began its observations in 2003, providing a wide range of particle measurements, without a magnetometer. Solar

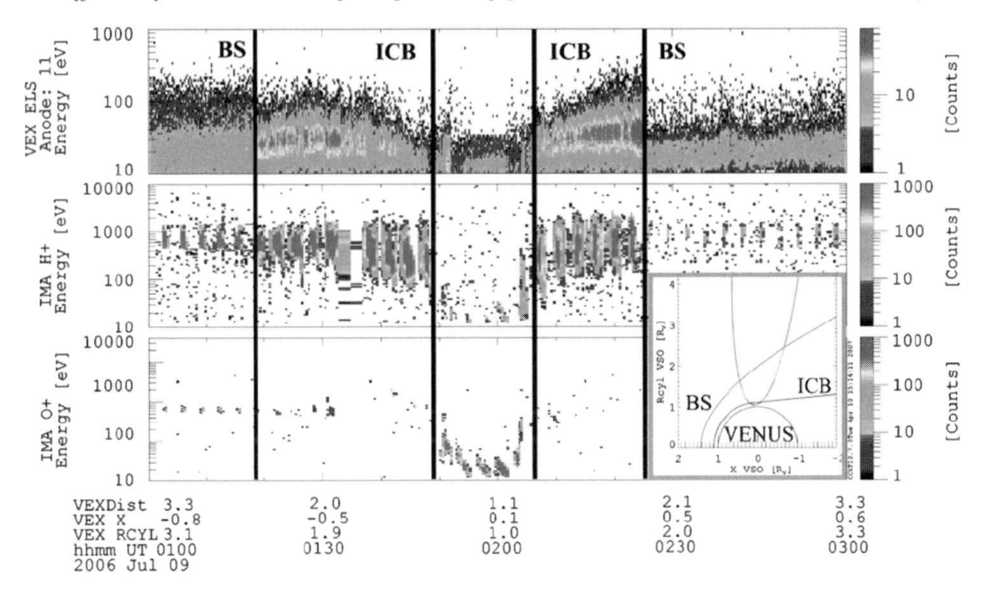

Figure 3. From top to bottom: Electron, H+ and O+ counts as a function of energy and time measured by VEX ASPERA during an orbit around Venus. The inset in the low right corner depicts the trajectory a planet-centred cylindrical coordinates where the X axis points sunward (from Martinecz *et al.* 2008).

wind and planetary particles were measured by the ASPERA-3 instrument, consisting in an electron spectrometer (ELS) and an ion mass analyzer (IMA) (Barabash *et al.* 2006).

Unfortunately, MEX does not carry a magnetometer. Nevertheless, for some orbits, the magnetic field strength and the electron density within the induced magnetosphere can be deduced from the Mars Advanced Radar for Subsurface and Ionospheric Sounding (MARSIS) measurements (Gurnett *et al.* 2005). One capability of the ionospheric sounding mode of MARSIS onboard MEX is the measurement of the local plasma density. Summaries of MEX plasma results can be found in review articles by Franz *et al.* (2006) and Dubinin *et al.* (2006).

Apart from the ULF plasma waves generated by the pick up of exospheric ions, the first perturbation generated by Mars in the supermagnetosonic solar wind flow is the bow shock (Mazelle *et al.* 2004). As for the case of Venus, the Martian bow shock, also decelerates, heats and compresses the solar wind plasma. The end of the turbulent regime of the magnetosheath is marked once again by a well-defined IMB, which precedes the IM and its highly draped magnetic fields and dominating local plasma. The IMB extends out into the downstream sector where it becomes the outer boundary of the magnetic tail. The IMB has several subregions. On the dayside, the region below the IMB and above the ionospheric boundary is referred to as the magnetic pileup region (MPR). At low magnetic latitudes, the interplanetary magnetic field (IMF) within the MPR is connected to tail lobe fields. The ionospheric boundary marks the lower end of the MPR on the dayside. Usually referred to it as photoelectron boundary (PEB), this boundary is usually associated with the upper limit of the collisional ionosphere. The PEB and to a lesser extent the IMB locations are influenced by the magnetic fields from the crustal sources. In the downstream sector, a tail plasma sheet separating both tail lobes is observed. The plasma structure of the plasma sheet close to the planet is still under scrutiny with a strong effect of the crustal magnetic sources in the nightside.

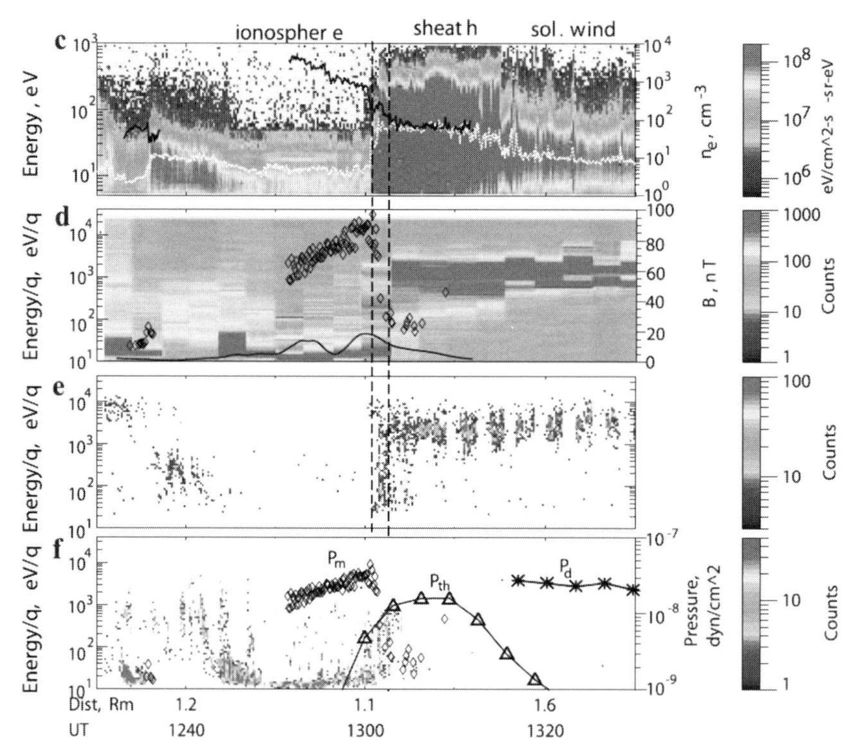

Figure 4. MEX ASPERA/MARSIS measurements (Dubinin *et al.* 2008). The dashed curve corresponds to the position of the IMB. Energy-time spectrograms of the (c) electrons with the imposed curves of pseudo electron density from ASPERA (white) and total electron density from MARSIS (black) and (d) all ion species, (e) He^{++}, and (f) heavy ($m/q > 16$) ions. Imposed curves are: the magnetic field value from the MARSIS observations (diamonds) and Cain *et al.* (2003) crustal field model (solid) (d), the magnetic pressure P_m (diamonds), the thermal proton pressure P_{th} (triangles), and the solar wind ram pressure P_d (asterisks).

Figure 4 shows different plasma boundaries and regions around Mars as seen by ASPERA and MARSIS during the outbound leg of a MEX orbit. Based on similar signatures described for Venus, the IMB and and BS are clearly detected around 13:00 and 13:16. In particular, magnetic field strength calculated from the MARSIS ionospheric sounding experiment clearly shows the increase in the pileup at the IMB as an increase of the magnetic pressure P_m.

3. Solar influence

3.1. *Venus*

Venus' induced magnetosphere has proven to be strongly dependent on the Solar cycle phase. Such an effect is probably linked to the EUV flux variability which influences the extension of the ionosphere and the ion pick up rates.

Alexander and Russell (1985), Russell *et al.* (1988) and Zhang *et al.* (1990) investigated the Venus bow shock location based on nearly 2000 PVO boundary crossings. These works concluded that the shock location depends on the solar cycle and solar EUV flux, the upstream solar wind parameters, and the orientation of the IMF (see also Phillips and McComas 1991).

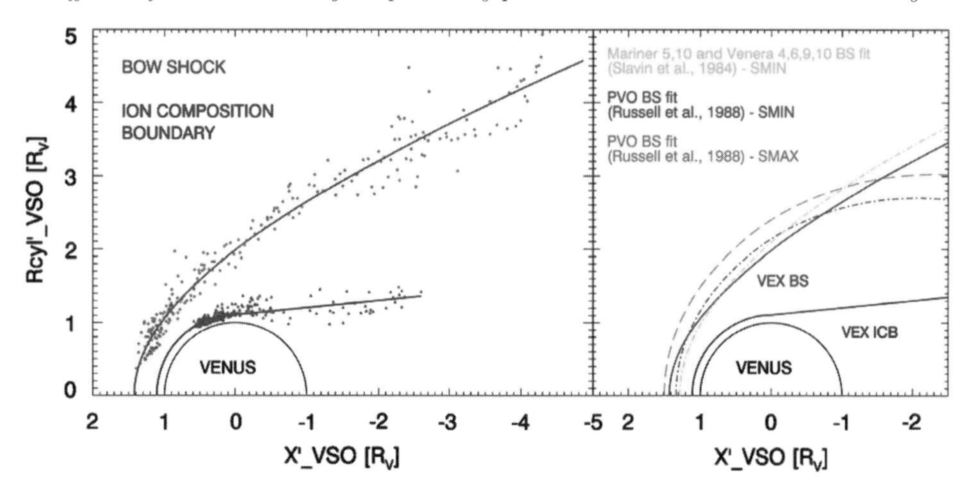

Figure 5. BS (red, see online version) and IMB (blue, see online version) crossings by VEX in aberrated cylindrical VSO coordinates where the X-axis points towards the Sun taking into account the planet's orbital speed (Martinecz, *et al.* 2008)

Figure 5a shows Martinecz *et al.* (2008) VEX BS and IMB fits and in Figure 5b a comparison with other shock models based on other data sets at solar minimum. The VEX BS fit is in good agreement with the model of Slavin *et al.* (1984) based on Mariner 5,10 and Venera 4, 6, 9, 10 observations.

During solar minimum, a comparison between the BS position and solar EUV flux (F50 index: 0.1 - 50 nm integrated photons cm^{-2} s^{-1} and shifted to Venus) derived from SOHO SEM observations shows no apparent dependence (Martinecz *et al.* 2008). This is probably due to the fact that the EUV flux variation is small over the period of observation.

As for the Venusian IMB, VEX provided a fit (Figure 5a) corresponding to the location of the boundary during solar minimum. Unfortunately, there is no comparison with similar fit obtained during solar maximum yet. This is partly due to the low time resolution of the PVO ion plasma measurements.

The ionopause has been shown to be strongly influenced by the solar cycle. During solar minimum, the pressure balance defining the boundary is achieved in general at lower altitudes in comparison with solar maximum (Russell and Vaisberg 1983). Also, the ionosphere is more magnetized during solar minimum, as the thermal pressure in the ionosphere is not enough to withstand the magnetic pressure in the barrier. Also, the ionopause altitude seems to control the ionospheric O^+ flow into the nightside, probably affecting the escape. Figure 6 shows the flow pattern of O^+ ions during solar maximum (Miller and Whitten 1991). During solar minimum, however, this circuit disappears, leading to the idea that the lower location of the pressure balance boundary might disrupt ionospheric flow.

During the VEX era, ASPERA measurements across the magnetotail of Venus have confirmed the fact that planetary ions (H^+ and O^+) escape from Venus due to the solar wind interaction (Barabash *et al.* 2007). In particular, escape rates have been shown to increase by a factor of 2 during the passage of CIR and ICMEs (Edberg *et al.* 2011). This indicates the importance of disturbed solar wind periods in the atmospheric escape at unmagnetized objects.

Figure 5 shows antisunward fluxes of planetary O^+ ions during the passage of CIRs/ICMEs (top panel), during periods of 'quiet' solar wind (middle panel) and the ratio

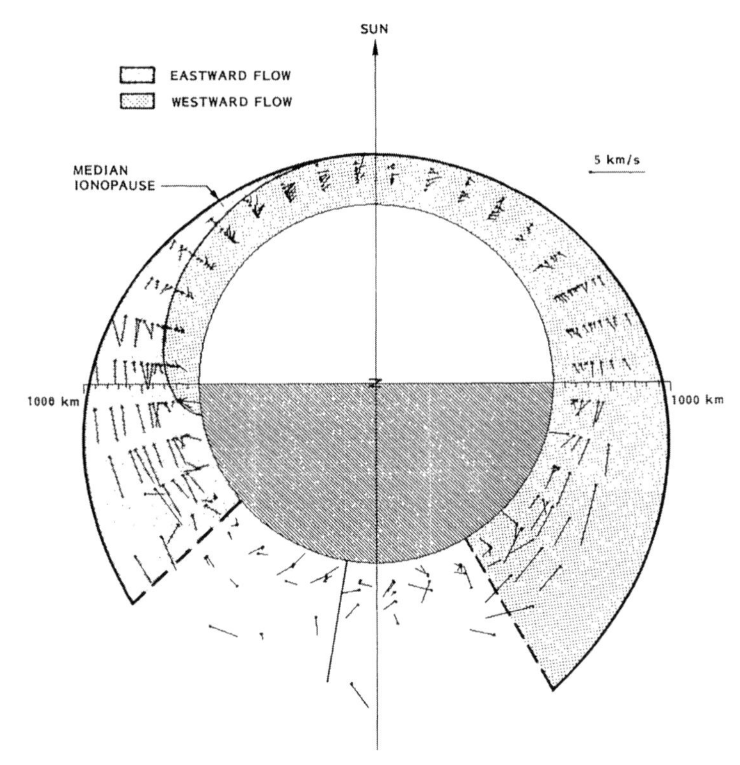

Figure 6. Velocity vectors of O^+ ions within the Venusian ionosphere from PVO observations (Miller and Whitten 1991). The shaded area identifies the westward flow, whereas the white region encloses the area where the flow vectors point eastwards.

between disturbed solar wind times and quiet times in each bin (bottom panel). As shown in the figures (in particular in the bottom panel) the fluxes, concentrated inside the IMB are enhanced during the disturbed periods. Unfortunately, this study has not been done for solar maximum conditions.

McEnulty *et al.* (2010) studied 17 ICME events and demonstrated that the energy (but not the flux) of pickup ions around Venus increases whenever the planet is impacted by an ICME. Earlier, Luhmann *et al.* (2007) showed that atmospheric escape could increase by a factor of 100 during ICMEs, as measured by the PVO spacecraft.

However, case studies of the influence of ICMEs have left ambiguous results, since in 3 out of 4 cases studied by Luhmann *et al.* (2008), the escape rate was not observed to increase. Futaana *et al.* (2008) showed that another single large ICME associated with simultaneous increase in solar energetic particle flux increased the atmospheric escape rate at both Venus and Mars, by a factor of $\approx 5 - 10$. It should be mentioned that McEnulty *et al.* (2010) and Luhmann *et al.* (2007) looked at the escape of high energy ions only while Futaana *et al.* (2008) included all ions.

3.2. *Mars*

The shape and location of the Martian BS and IMB have been studied in the past by e.g. Slavin and Holzer (1981), Vignes *et al.* (2000), Bertucci *et al.* (2005), Trotignon *et al.* (2006) and Edberg *et al.* (2008). During the Phobos 2 mission the number of bow shock and IMB crossings were 127 and 41, whereas during the MGS mission, 573 (Trotignon *et al.* 2006) and 1149 (Bertucci *et al.* 2005), respectively. In the data set of MEX/ASPERA-3 from 2004 until 2008 have 5014 IMB crossings and 3277 BS crossings

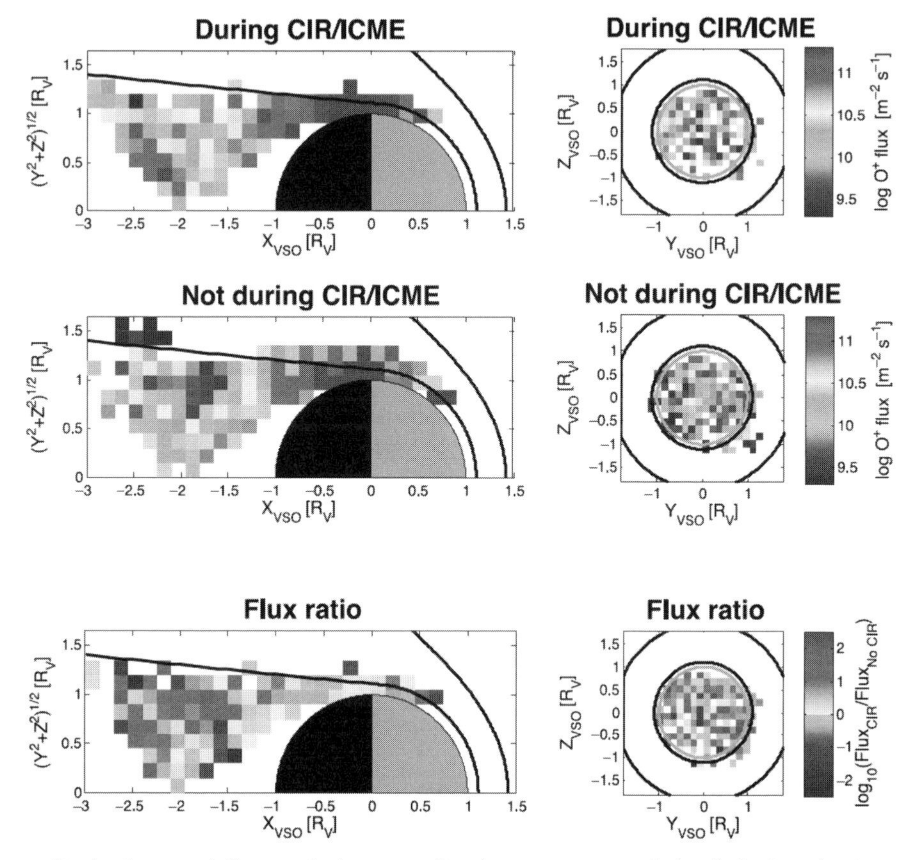

Figure 7. Antisunward fluxes of planetary O+ ions as measured (top) during the impact of CIRs/ICMEs and (middle) during the time of quiet solar wind as well as (bottom) the flux ratio between disturbed solar wind times and quiet times in each bin (Edberg *et al.* 2011). The data is shown (left) in cylindrical VSO coordinates as well as (right) in the VSO y – z plane for -3 RV $< x < -1$ RV. Each bin is 0.15×0.15 RV large and contains at least 4 measurement points. The black lines indicate the average locations of the BS and the IMB from Martinecz *et al.* (2008) and the grey circles indicate the limb of the planet.

been identified (Edberg *et al.* 2009). The number of crossings during the overlapping mission time between MGS and MEX (Feb 2004 - Nov 2006) is 2500 and 1840, respectively.

Figure 8 shows the conic section fits of the BS and IMB from Phobos 2 (solar maximum) and MGS (solar minimum) (Vignes *et al.* 2000). Contrary to the behaviour observed at Venus, there is not appreciable difference in the location of these boundaries with solar cycle.

However, high variability in the location of these boundaries is observed for shorter timescales. The factors that have been studied for possible effects on the location of these boundaries include the IMF direction (Vignes *et al.* 2000, Brain *et al.* 2005), the crustal magnetic fields (Crider *et al.* 2002, Dubinin *et al.* 2006, Fränz *et al.* 2006, Edberg *et al.* 2008, Edberg *et al.* 2009), the solar wind dynamic pressure (Crider *et al.* 2003, Brain *et al.* 2005, Edberg *et al.* 2009), and the magnetosonic Mach number (Edberg *et al.* 2010). In some of these studies, proxies derived from extrapolated in situ measurements around 1 AU have been used (e.g. Edberg *et al.* 2010). Of these studies, it is worth mentioning that of the influence of the crustal magnetic fields (Acuña *et al.* 1999) on boundary locations (Crider *et al.* 2002, Brain *et al.* 2005, Fränz *et al.* 2006, Edberg *et al.* 2008, Edberg *et al.*

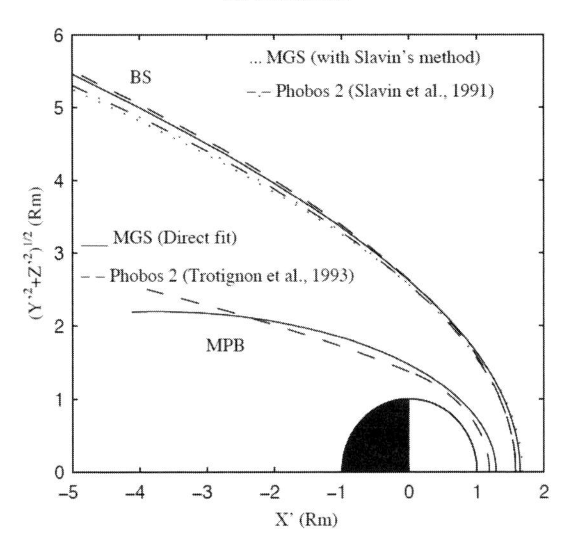

Figure 8. Conic section fits of the Martian BS and IMB obtained from Phobos-2 and MGS spacecraft during Solar maximum and minimum respectively.

2009). All these studies conclude that crustal fields do affect the altitude of the IMB, but that there is ambiguous evidence of the influence on the shock. In spite of the effect on the IMB, however, the role of crustal sources in the rate of plasma escape is thought to be negligible (Dubinin *et al.* 2006).

On the other hand, a clear influence of the solar cycle on the exospheric structure is predicted by current models of exosphere. In particular, Modolo *et al.* (2005) made use of a Chamberlain model of the O and H exospheres for Mars for their global kinetic simulations of the environment of Mars noting an inflation of the exospheres during solar maximum (Figure 9).

Modolo *et al.* (2005) also showed that the oxygen escape, as well as the hydrogen escape, is strongly dependent on the solar EUV flux. Whatever the solar EUV flux is, however, between 85 and 90% of the escaping proton flux comes from charge exchange. The escape flux of H^+ ions is approximately four times higher during solar minimum than during solar maximum, due to the inflation of the neutral hydrogen corona at solar minimum. On the other hand, photoionization is the main process which contributes to the escape of O^+ ions, which maximizes at solar maximum with the extension of oxygen corona.

As for the influence of solar wind energetic transient phenomena, Edberg *et al.* (2010) found that the atmospheric escape rate at Mars is not constant but rather increases by a factor of ~2.5 on average during the pass of CIRs and ICMEs. Dubinin *et al.* (2009) similarly showed that a single large CIR that impacted on Mars increased the scavenging of the ionosphere, with the escape rate again being estimated to increase by a factor of ~10.

4. Summary

In this work, we described how induced magnetospheres are the result of the direct interaction of the atmospheres of unmagnetized objects with the solar wind and how they are more prone to atmospheric evolution than magnetized planets. In particular, the cases of Venus and Mars have been analyzed, by discussing the influence of the solar

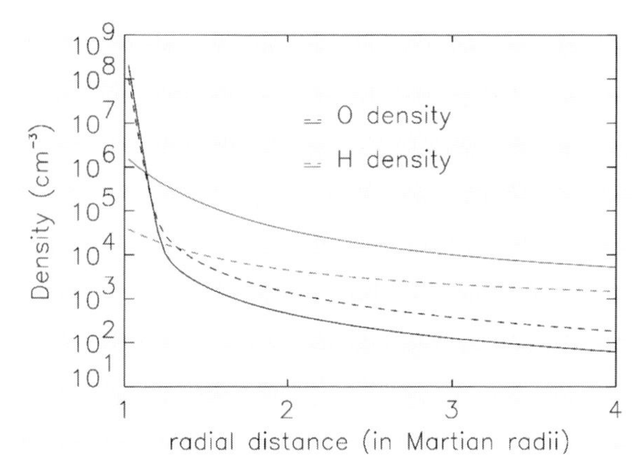

Figure 9. Chamberlain Hydrogen and oxygen vertical density profiles at solar minimum (solid lines) and solar maximum (dashed lines) (Modolo *et al.* 2005)

cycle and solar wind variability with the structure of their induced magnetospheres and resulting escape rate.

Plasma measurements obtained around Venus suggest that plasma regions and escape seem to be strongly influenced by solar cycle. However, most of the studies involving high time resolution plasma measurements are available only for solar minimum. At Mars, on the other hand, the epochal coverage by Phobos-2 and MGS show that plasma boundaries would not be dependent on the solar cycle phase, although simulations suggest that escape is as a result of the change in the EUV flux. Current escape rates of the order of 1025 part/s are estimated (Edberg *et al.* 2011 and references therein) but estimates may double and even increase by an order of magnitude during stormy space weather.

We expect to find similar scenarios at other planetary systems but it is still difficult to assess those interactions in the absence of in situ measurements. Techniques involving remote sensing will have to be used in order to look for observables obtained from models built and validated from local in situ observations.

References

Acuña, M. H. *et al.* 1992, *Jour. of Geophys. Res.*, 97E5, 7799

Acuña, M. H. *et al.* 1998, *Science*, 279, 1676

Acuña, M. H., *et al.* 1999, *Science*, 284, 790

Alexander, C. J. & Russell, C. T. 1985, *Geophys. Res. Lett.*, 12(6), 369

Barabash, S. *et al.* 2006, *Space Sci. Rev.*, 126, 113

Bertucci, C. *et al.* 2003, *Geophys. Res. Lett.*, 30(2), 1099

Bertucci, C. *et al.* 2003, *Geophys. Res. Lett.*, 30(17), 1876

Bertucci, C. *et al.* 2005, *Jour. of Atmos. and Solar-Terres. Phys.*, 67(17-18), 1797

Bertucci, C. *et al.* 2011, *Space Sci. Rev.*, 162(1-4), 113

Brain, D. A. *et al.* 2005, *Geophys. Res. Lett.*, 32, 1820

Cain, J. C. *et al.* 2003, *Jour. of Geophys. Res.*, 108(E2), 5008

Coates, A. J. *et al.* 2011, *Planetary and Space Science*, 59(10), 1019

Crider, D. H. *et al.* 2002, *Geophys. Res. Lett.*, 29, 1170

Crider, D. H. *et al.* 2003, *Jour. of Geophys. Res.*, 108(A12), 12

Dubinin, E. *et al.* 2006, *Space Sci. Rev.*, 126(1-4), 209

Dubinin, E. *et al.* 2008, *Geophys. Res. Lett.*, 35, 11103

Dubinin, E. *et al.* 2009, *Geophys. Res. Lett.*, 36, 1105

Dubinin, E. *et al.* 2011, *Space Sci. Rev.*, 162(1-4), 173

Edberg, N. J. T. *et al.* 2008, *Jour. of Geophys. Res.*, 113(A8), 206
Edberg, N. J. T. *et al.* 2009, *Ann. Geophys.*, 27(9), 3537
Edberg, N. J. T. *et al.* 2010, *Jour. of Geophys. Res.*, 115(A7), 203
Edberg, N. J. T. *et al.* 2011, *Jour. of Geophys. Res.*, 116(A9), 308
Fedorov, A. *et al.* 2008, *Planetary and Space Science*, 56, 812
Fränz, M. *et al.* 2006, *Space Sci. Rev.*, 126(1-4), 165
Futaana, Y. *et al.* 2008, *Planetary and Space Science*, 56(6), 873
Gurnett, D. A. *et al.* 2005, *Science*, 310, 1929
Knudsen *et al.* 1979, *Space Sci. Instrum.*, 4, 351
Luhmann, J. G. 1991, *Rev. Geophys.*, 29, 965
Luhmann, J. G. *et al.* 2007, *Jour. of Geophys. Res.*, 112(A4), 10
Luhmann, J. G. *et al.* 2008, *Jour. of Geophys. Res.*, 113(A5), 2
Martinecz, C. *et al.* 2008, *Planetary and Space Science*, 56(6), 780
Mazelle, C. *et al.* 2004, *Space Sci. Rev.*, 111(1-4), 115
McEnulty, T. R. 2010, *Planetary and Space Science*, 58(14-15), 1784
Miller, K. L. & Whitten, R. C. 1991, *Space Sci. Rev.*, 55, 165
Modolo, G. M. *et al.* 2005, *Ann. Geophys.*, 23), 433
Nagy, A. F. *et al.* 2004, *Space Sci. Rev.*, 111, 33
Ness, N. F. *et al.* 1982, *Jour. of Geophys. Res.*, 87(A3), 1369
Phillips, J. L. & McComas, D. J. 1991, *Space Sci. Rev.*, 55, 1
Russell, C. T. *et al.* 1980, *IEEE trans. Geosci. Electron.*, GE-18, 32
Russell, C. T. & Vaisberg, O. 1983, *Venus, University of Arizona Press, edited by D.M. Hunton, L. Colin, T.M. Donahue, V.I. Moroz*, 873
Russell, C. T. *et al.* 1988, *Jour. of Geophys. Res.*, 93, 5461
Russell, C. T. *et al.* 2006, *Planetary and Space Science*, 54, 1482
Saunders, M. A. & Russell, C. T. 1986, *Jour. of Geophys. Res.*, 91(A5), 589
Slavin, J. A. & Holzer, R. E. 1981, *Jour. of Geophys. Res.*, 86(A11), 401
Slavin, J. A. *et al.* 1984, *Jour. of Geophys. Res.*, 89(A5), 2708
Szego, K., *et al.* 2000, *Space Sci. Rev.*, 94(3-4), 429
Trotignon, J. G. *et al.* 2006, *Planetary and Space Science*, 54, 357
Vaisberg, O. L. & Bogdanov, A. V. 1974, *Cosmic Research*, 12, 253
Verigin, M. I. *et al.* 1978, *Jour. of Geophys. Res.*, 83(A8), 3721
Vignes, D. *et al.* 2000, *Geophys. Res. Lett.*, 27(1), 49
Vignes, D. *et al.* 2002, *Geophys. Res. Lett.*, 29(9), 1329
Zelenyi, L. M. & Vaisberg, O. L. 1985, *Advances of Space Plasma Physics, ed. Buti, World Scientific*, 59
Zhang, T.-L. *et al.* 1990, *Jour. of Geophys. Res.*, 95(A14), 961
Zhang, T.-L. *et al.* 1991, *Jour. of Geophys. Res.*, 96(A11), 153
Zhang, T.-L. *et al.* 1991, *Planetary and Space Science*, 54, 1336

Discussion

Mark Giampapa: This is a source of erosion of planetary magnetospheres. Can you extrapolate this to the past, to times of high solar activity?

César Bertucci: Yes, there are a few works that have addressed the cumulative escape due to solar wind interaction since the Sun reached ZAMS. I suggest you to read the review article by Lammer *et al.* (2006, Space Scien. Rev. 122, 189). However, in my opinion much needs to be further understood before extrapolating current escape rates. Thanks for your interest.

Comparative Magnetic Minima:
Characterizing quiet times in the Sun and Stars
Proceedings IAU Symposium No. 286, 2011
C. H. Mandrini & D. F. Webb, eds.

doi:10.1017/S1743921312005145

Flares and habitability

Ximena C. Abrevaya[1,3], Eduardo Cortón[2,3] and Pablo J. D. Mauas[1,3]

[1]Instituto de Astronomía y Física del Espacio, UBA - CONICET
Pabellón IAFE, Buenos Aires, Argentina
email: abrevaya@iafe.uba.ar

[2]Depto. de Química Biológica, FCEyN, UBA
Pabellón 2, Ciudad Universitaria, Buenos Aires, Argentina

[3]Consejo Nacional de Investigaciones Científicas y Técnicas (CONICET)
Buenos Aires, Argentina

Abstract. At present, dwarf M stars are being considered as potential hosts for habitable planets. However, an important fraction of these stars are flare stars, which among other kind of radiation, emit large amounts of UV radiation during flares, and it is unknown how this events can affect life, since biological systems are particularly vulnerable to UV. In this work we evaluate a well known dMe star, EV Lacertae (GJ 873) as a potential host for the emergence and evolution of life, focusing on the effects of the UV emission associated with flare activity. Since UV-C is particularly harmful for living organisms, we studied the effect of UV-C radiation on halophile archaea cultures. The halophile archaea or haloarchaea are extremophile microorganisms, which inhabit in hypersaline environments and which show several mechanisms to cope with UV radiation since they are naturally exposed to intense solar UV radiation on Earth. To select the irradiance to be tested, we considered a moderate flare on this star. We obtained the mean value for the UV-C irradiance integrating the IUE spectrum in the impulsive phase, and considering a hypothetical planet in the center of the liquid water habitability zone. To select the irradiation times we took the most frequent duration of flares on this star which is from 9 to 27 minutes. Our results show that even after considerable UV damage, the haloarchaeal cells survive at the tested doses, showing that this kind of life could survive in a relatively hostile UV environment.

Keywords. astrobiology, stars: activity, stars: flare, ultraviolet: stars

1. Introduction

Astrobiology is a multidisciplinary area of scientific research. Its main goal is to understand the origin, the evolution and the distribution of life on Earth and elsewhere in the universe. According to the "Mediocrity principle", which proposes that our planetary system, life on Earth, and technological civilizations are an average case in the universe (von Hoerner 1961), life as we know it could emerge in any place with conditions similar to Earth, and would develop following similar selection rules.

Based on this, we wish to study if exoplanets around dMe stars can be suitable places for the emergence and evolution of life. To this end it is necessary to consider the habitability criteria, which defines if a place is suitable for life for a significant period of time. Habitability is based on physical and chemical environmental factors (Cockell 2007) which are necessary for the existence of "life" as we know it on Earth, since it is the unique form of life that we know. Since terrestrial life is dependent on the existence of water usually the most important factor for habitability is the presence of liquid water in the planet's surface which is known as Liquid-Water Habitable Zone (LW-HZ) (Huang 1959; Dole 1964; Hart 1979).

However, other important environmental factors as UV radiation can play a significant role in habitability. On one hand, it is well known that UV radiation can be very damaging and even lethal for life (inducing damage to DNA and other cellular components). In particular, high exposure to UV-C results very damaging and even lethal to most terrestrial biological systems. On the other hand UV could be also a factor necessary for life origin and biological evolution. Therefore, UV radiation should also be considered as an habitability criteria (Buccino *et al.* 2006).

A paradigmatic case are dM stars since it is known that many dMe stars emit large amounts of UV radiation during flares and it is uncertain how these UV events can affect life (Buccino *et al.* 2007). Moreover they are the most common stars in the galaxy and around these stars is easier to detect terrestrial planets which would probably be in the LW-HZ. Therefore we should consider UV and to analyze flaring activity to set an habitability zone on these planets (UV-HZ).

To set this habitability zone we study UV radiation from dMe-stars and we performed biological experiments to see the effects of UV radiation on life. To this end we used extremophile microorganisms, because they are "model" organisms in Astrobiology (Cavicchioli 2002) due to their capacity to survive extreme physicochemical conditions that usually we can find in extraterrestrial environments. In particular the extremophiles the we used in this work are halophilic archaea (haloarchaea), microorganisms hat live at high salt concentrations (3-5 M NaCl; hypersaline environments) such as hypersaline lakes, salterns, etc., because the could be taken as models since they are considered UV resistant microorganisms. Because much halophilic archaea are exposed to intense solar UV radiation in their natural environments, so they are generally regarded as relatively UV tolerant, among other factors (DasSarma 2006; Abrevaya *et al.* 2011).

In this work we examine the effect of UV-C on *Natrialba magadii* and *Haloferax volcanii* two halophilic archaea with different physiological features, initially isolated from Magadi Lake in Kenya, Africa, and the Dead Sea in the Jordan Rift Valley, respectively.

Even though flares emit UV radiation of wide spectra we selected to use UV-C radiation, due it is potentially more damaging that other wavelengths, because DNA has their maximum of absorption at this wavelength range.

2. Methods

To select the irradiance values we considered the UV spectra from different flares on dM stars, from the International Ultraviolet Explorer (IUE) satellite. Using this data we choose as a mean case the dM-star EV Lacertae (EV Lac, M3.5 V) (GJ873). It is a star with frequent periods of high activity in which emits flares. We selected as a sample the flare from September 3rd, 1981. We calculate the mean value for UV-C during impulsive phase, resulting in 3.7 W m^{-2} in the center of the LW-HZ. From literature it is known that most flares last between 9 and 27 minutes Gerschberg, 2005.

Thereafter, cultures of *N. magadii* and *H. volcanii* were grown around an optical density (OD) related to mid-exponential phase, and diluted to an OD=0.5. Samples were divided in five groups: Control (non-irradiated culture), and irradiated for 10 and 30 minutes (doses: 2220 and 6660 J m^{-2}, respectively). Liquid culture was irradiated being exposed to a UV-C source (Phillips 15W Hg lamp $\lambda = 254$ nm, irradiance 3.7 W m^{-2}). Aliquots of the irradiated and control groups were diluted with fresh medium until to reach an OD=0.05 and withdrawn after different irradiation times. The biological effects of the UV treatment were assesed following the changes of the growth kinetics. To obtain growth curves, OD values for each sample were acquired at different times after irradiation and were plotted versus post-irradiation time.

3. Results

Figures 1 and 2 show the growth curves obtained for *H. volcanii* and *N. magadii* after irradiation. As can be observed for the irradiated groups, there is a delay in the growth that can be seen as a displacement of the growth curve, which is significantly different from non-irradiated group.

Additionally it can be observed that this delay is dose dependent for *N. magadii* (between 20 and 30 hours, for 2220 and 6660 J m^{-2} doses, respectively) (Fig. 2) and independent from the dose for *H. volcanii* (around 70 hours for both for 2220 and 6660 J m^{-2} doses) (Fig. 1).

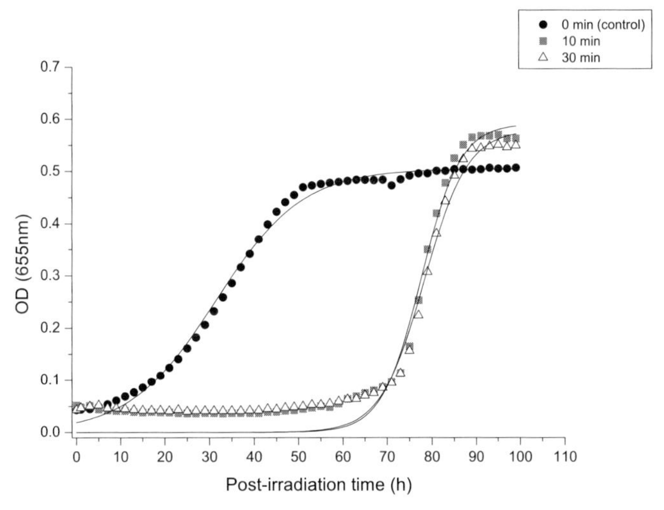

Figure 1. UV survival curves for *H. volcanii*. The black circles corresponds to the control group while the grey squares corresponds to 2220 J m^{-2} dose and the open triangles to the 6660 J m^{-2} dose. Each curve is the average of duplicates for each experimental group

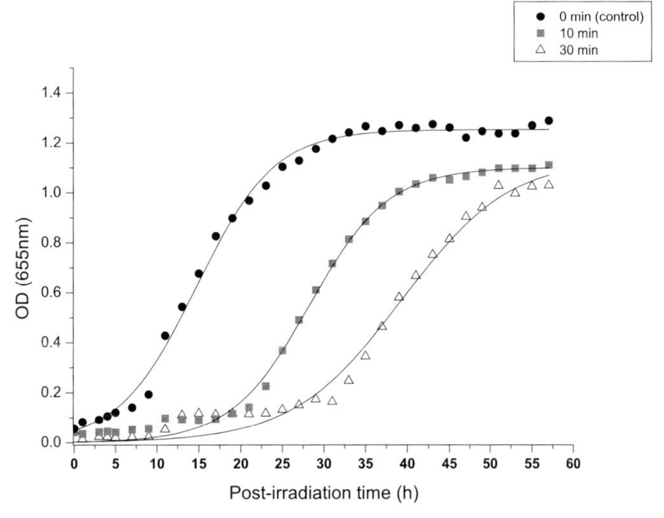

Figure 2. UV survival curves for *N. magadii*. The black circles corresponds to the control group while the grey squares corresponds to 2220 J m^{-2} dose and the open triangles to the 6660 J m^{-2} dose. Each curve is the average of duplicates for each experimental group

4. Conclusions

The tested doses used in this work are equivalent to the doses that life (in this case microorganisms) would receive from a flare of moderate intensity, not very high to sterilize the surface of the planet, considering a terrestrial planet orbiting a dMe star in the center of the LW-HZ.

On the other hand, it is well known that UV induces diverse kind of damage in cells. In particular UV-C is a potent inductor of DNA chemical modifications, mainly formation of cyclobutane pyrimidine dimers and 6-4 photoproducts, which eventually could lead to mutations or cell death if they are not repaired(Mitchell & Nairn 1989; Cadet *et al.* 2005). Growth delay could be explained by processes related to sublethal effects of this kind of cell damage, which can cause blockage of diverse cellular processes. In such cases cells stops to growth and bacterial population remain in lag phase, which is visualized as a displacement of the growth curve.

In our work even the microorganisms seems to be significantly damaged, as was observed by the delay in growth for 20 and 30 hours for *N. magadii* and 70 hours for *H. volcanii*, both species are able to survive at the tested doses. Moreover these results show that both species are capable to survive not only high UV radiation levels, but also UV-C radiation which is not present in their natural environments.

These preliminary results provide evidence that such "kind of life" could possibly survive in a relatively hostile environment from the point of view of UV radiation, in an hypothetical planet around dMe stars with moderate flaring activity.

References

Abrevaya, X. C., Paulino-Lima, I. G., Galante, D., Rodrigues, F., Cortón E., Mauas, P. J. D., & de Alencar Santos Lage, C. 2011, *Astrobiology*, 10, 1034

Buccino, A. P., Lemarchand, G. A., & Mauas, P. J. D. 2006, *Icarus*, 183, 491

Buccino, A. P., Lemarchand, G. A., & Mauas, P. J. D. 2007, *Icarus*, 192, 582

Cadet, J., Sage, E., & Douki, T. 2005, *Mutat. Res.*, 571, 3

Cavicchioli, R. 2005, *Astrobiology*, 2, 281

Cockell, C. S. 2007, in: G., Horneck & P., Rettberg (eds.), *Complete course in Astrobiology* (Wiley-VCH), pp. 151-177

DasSarma, S. 2006, *Microbe*, 1, 120

Dole, S. H. 1964, in: *Habitable Planets for Man* (Blaisdell Pub. Co., 1st ed., New York)

Gershberg, R. E. 2005, in: *Solar-Type Activity in Main-Sequence Stars* (Springer, Heidelberg, The Netherlands)

Huang, S. S. 1959, *Am. Sci.*, 47, 397

Hart, M. H. 1979, *Icarus*, 37, 351

Mitchell, D. L. & Nairn, R. S. 1989, *Photochem. Photobiol.*, 49, 805

Discussion

MARK GIAMPAPA: The food source should have also survived the UV-irradiation. What is you comment on that?

XIMENA ABREVAYA: In our experiment we irradiated the haloarchaea on their own growth medium, but after irradiation we changed this medium putting a fresh one. That is because we suppose that UV could produce some chemical modifications on the "food source". We didn't study that point.

JEFFREY LINSKY: (Comment) Observations of EUV flux are distorted through the interstellar medium. This needs to be corrected.

XIMENA ABREVAYA: It is an interesting observation but our work is focused on UV-C radiation. We should study that to reach a conclusion.

Comparative Magnetic Minima:
Characterizing quiet times in the Sun and Stars
Proceedings IAU Symposium No. 286, 2011
C. H. Mandrini & D. F. Webb, eds.

doi:10.1017/S1743921312005157

Potential energy stored by planets and grand minima events

Rodolfo G. Cionco[1,2]

[1]Universidad Tecnológica Nacional, Facultad Regional San Nicolás,
Colón 332, San Nicolás (2900), Bs. As., Argentina
email: gcionco@frsn.utn.edu.ar

[2]Grupo de Ciencias Planetarias, Universidad Nacional de La Plata,
Paseo del Bosque s/n, La Plata (1900), Bs. As., Argentina
email: cionco@fcaglp.unlp.edu.ar

Abstract. Recently, Wolff & Patrone (2010), have developed a simple but very interesting model by which the movement of the Sun around the barycentre of the Solar system could create potential energy that could be released by flows pre-existing inside the Sun. The authors claim that it is the first mechanism showing how planetary movements can modify internal structure in the Sun that can be related to solar cycle. In this work we point out limitations of mentioned mechanism (which is based on interchange arguments), which could be inapplicable to a real star. Then, we calculate the temporal evolution of potential energy stored in zones of Sun's interior in which the potential energy could be most efficiently stored taking into account detailed barycentric Sun dynamics. We show strong variations of potential energy related to Maunder Minimum, Dalton Minimum and the maximum of Cycle 22, around 1990. We discuss briefly possible implications of this putative mechanism to solar cycle specially Grand Minima events.

Keywords. Sun: activity - Sun: interior, sunspots.

1. Aim and Methods

Wolff & Patrone (2010) presented the first mechanism devoted to explain modifications of stellar interiors by planetary gravity. They claimed that a cell inside a rotating star with orbiting planets (i. e. with measurable barycentric motion), can creates potential energy per unit mass (PE) that can be released by internal processes. The authors used interchange arguments; i. e., the cell is composed by two masses that, by interchanging their positions, can release PE but conserving angular momentum.

There are two points which seem to have not been discussed by these authors, which limit the applicability to a real star (Svalgaard, private communication):

First of all, remaining in an inertial frame, which simplifies the governing equations, but complicates the boundary conditions substantially, because they are now moving (with the star) relative to the static interchange. Wolff & Patrone appear not even to have considered boundary conditions, neither explicitly nor implicitly.

Second, one should also keep in mind that instability can never be proved by interchange arguments, unless one can demonstrate that the interchanges considered can be realized by the fluid. One can, in principle, demonstrate stability; however, by showing that no displacement, realizable or not, can liberate energy to drive the instability. However, when the interchange is carried out in a plausible manner which avoids this complication, as did Rayleigh and Chandrasekhar, the outcome can be usefully suggestive. The interchange considered by Wolff and Patrone leaves the fluid elements (apparently filling the spaces into which they have been displaced, yet) moving with respect to them;

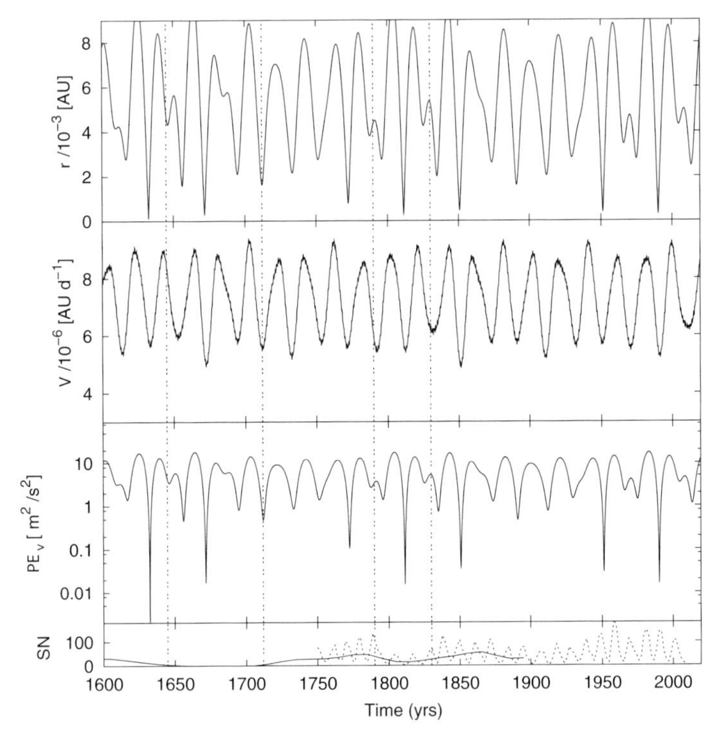

Figure 1. Barycentric solar distance r, barycentric solar velocity V; PE storage at 0.16 solar radius facing the barycentre (PE_v), and the sunspots series (SN) observed/smoothed (dashed line) and reconstructed (solid line). Following to Wolff & Patrone (2010), the particular dynamics around 1632, 1811 and 1990 produce a dramatically decrease of PE in the studied solar zone, related to Maunder Minimum, Dalton Minimum and the maximum of Cycle 22.

therefore it is valid dynamically, for the purposes of energy computation, only for an interval of time of measure zero, which is insufficient to take the temporal derivatives required to determine subsequent evolution, essential, of course, for assessing stability.

In this context, these authors have shown that the strongest case is the vertical one, i.e., the larger storage would occurs when the cell is near in the Sun's centre-barycentre direction (Fig. 3 in that paper). The authors presented graphics showing the values of PE stored in these locations. They mentioned possible effects in PE due to occasional variations of certain dynamical parameters (e. g., barycentric distance r and velocity V of the Sun).

The movement of the Sun around the barycentre of the Solar System (the *solar inertial motion*) has had important irregularities in the past, e.g., angular momentum inversions related to Maunder and Dalton minimums and at the maximum of Cycle 22 in 1990, and related to Gnevyshev-Ohl rule violations (Fairbridge & Shirley 1987; Javaraiah 2005; Perrymann & Schulze-Hartung 2011). Also, the solar inertial motion was related to solar cycle (i.e., the planetary hypothesis of solar cycles, see e.g. Perrymann & Schulze-Hartung 2011). In particular, Fairbridge & Shirley (1988) were able to show that only at times of Grand Minima (GM) events, a substitute of the apsidal axis of the solar barycentric orbit has had strong oscillations, but in their work there is no hypothesis related to any particular forcing mechanism in connection with this phenomenon. They also found a strong oscillation in 1990, so the authors, extrapolating these findings, argued for a new imminent GM event.

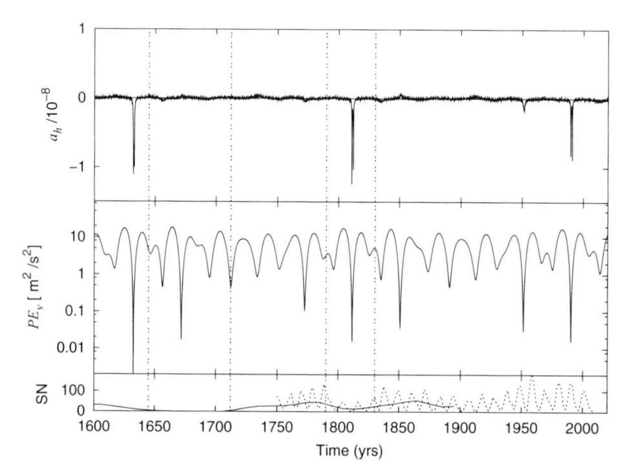

Figure 2. PE storage at 0.16 solar radius facing the barycentre (PE_v), and the sunspots series SN (pointed-line: observed-smoothed; solid line: reconstruction by Solanki *et al.* (2004)). a_h is the angular momentum projection (scaled) in the Sun's acceleration direction. Following to Wolff & Patrone (2010), the particular dynamics around these impulsive events (1632, 1811 and 1990) produce a dramatically decrease of PE in the studied solar zone, related to Maunder Minimum, Dalton Minimum and the maximum of Cycle 22, before the present extended minimum.

Then, in the light of this new hypothesis of Sun-planets interaction, taking into account its limitations and the particularities of solar inertial motion, it is interesting to check what could be the possible PE variations with regards to these last GM events and the mentioned cycle. In addition, Cionco & Compagnucci (this volume) showed that the Sun acceleration has presented impulsive manifestation before the Maunder Minimum (MM), around 1632; in the middle of Dalton Minimum (DM), in 1811; and in the maximum of Cycle 22, around 1990, before the present extended minimum. To show the possible influence of real barycentric sun-dynamics in PE storage, we calculate the specific PE stored in the vertical case (PE_v), supposing a near coplanar solar movement but using the actual velocity and position obtained in our SIM simulations.

2. Results

Fig. 1 shows that the Sun's closest approaches to the barycentre (that also coincide with drops in velocity; keep in mind that the Sun's movement is not keplerian) has associated big drops in PE_v. The dashed vertical lines indicate Maunder Minimum and Dalton Minimum.

Fig. 2 shows the PE_v stored at 0.16 solar radius facing the barycentre and the sunspots series SN (observed-smoothed and reconstructed by Solanki *et al.* 2004). The a_h quantitie is the angular momentum projection on the Sun acceleration direction. The particular dynamics around these impulsive events (1632, 1811 and 1990) produce a dramatically decrease of PE in the studied solar zone, consistent with Maunder Minimum, Dalton Minimum and the maximum of Cycle 22. Clearly PE variations are correlated with these GM events and show that, following this formalism, the Sun barycentric dynamics should be a measurable effect in the Sun's interior.

3. Conclusions and hypothesis

If the results of Wolff & Patrone were valid for a real star and taking into account our results, one can think that Maunder Minimum, Dalton Minimum and the maximum of

Cycle 22, shared certain changes due to planetary accelerations that could have a significant effect in the solar interior, which could be a global support to planetary hypothesis of solar cycles.

Then it is very important to clarify the true scope of the work of Wolff and Patrone due to the chaotic nature of the solar dynamo can amplify the effects of a weak external periodic forcing through resonances, collective synchronization and feedback mechanisms.

Acknowledgments

The author is indebted to Dr. Leif Svalgaard who pointed out the limitations of Wolff and Patrone work. The author acknowledges the support of a grant from IAU to attend the Symposium 286 and PID-UTN 1351 *Forzantes externos al planeta y variabilidad climática* of UTN, Argentina.

References

Fairbridge, R. & Shirley, J., 1987, *Solar Phys.*, 110, 191
Javaraiah, J., 2005, *MNRAS* 362, 1311
Perryman, M. A. C. & Schulze-Hartung, T., 2011, *A&A* 525, A65
Solanki, S. K., Usoskin, I. G., Kromer, B., Schussler, M., & Beer J., 2004, *Nature*, 431, 1084
Wolf, C. & Patrone, E. 2010, *Solar Phys.*, 266, 227

Comparative Magnetic Minima:
Characterizing quiet times in the Sun and Stars
Proceedings IAU Symposium No. 286, 2011
C. H. Mandrini & D. F. Webb, eds.

doi:10.1017/S1743921312005169

A new imminent grand minimum?

Rodolfo G. Cionco[1] **and Rosa H. Compagnucci**[2]

[1]UTN-Facultad Regional San Nicolás, San Nicolás (2900), Bs. As., Argentina
email: gcionco@frsn.utn.edu.ar

[2]Dept. de Ciencias de la Atmósfera y los Océanos, Universidad de Buenos Aires,
Pabellón II, CABA (1428), Argentina
email: rhc@fcen.uba.ar

Abstract. The planetary hypothesis of solar cycle is an old idea by which the planetary gravity acting on the Sun might have a non-negligible effect on the solar magnetic cycle. The advance of this hypothesis is based on phenomenological correlations between dynamical parameters of the Sun's movement around the barycenter of the Solar System and sunspots time series. In addition, several authors have proposed, using different methodologies that the first Grand Minima (GM) event of the new millennium is coming or has already begun. We present new fully three dimensional N-body simulations of the solar inertial motion (SIM) around the barycentre of the solar system in order to perform a phenomenological comparison between relevant SIM dynamical parameters and the occurrences of the last GM events (i.e., Maunder and Dalton). Our fundamental result is that the Sun acceleration decomposed in a co-orbital reference system shows a very particular behaviour that is common to Maunder minimum, Dalton minimum and the maximum of cycle 22 (around 1990), before the present prolonged minimum. We discuss our results in terms of a dynamical characterization of GM with relation to Sun dynamics and possible implications for a new GM event.

Keywords. Sun: activity - Sun: interior, sunspots.

1. Introduction

The planetary hypothesis of the solar cycle is an old idea in which the gravitational influence of the planets has a non-negligible effect on the causes of the solar magnetic cycle (Perrymann & Schulze-Hartung 2011, and references therein). The advance of this hypothesis is based on phenomenological correlations between dynamical parameters of the Sun's movement around the barycenter of the Solar System and sunspots time series. However, at present there is no clear forcing mechanism between these phenomena. In addition, the current exceptionally long minimum of solar activity has attracted the attention with respect to the possibility of developing a new grand minimum in the next decades (see e.g., Feulner & Rahmstorf 2010).

The aim of this work, was to find some distinctive dynamics in planetary forcing related to solar barycentric movement and its possible relationship with Grand Minima (GM) events. Our fundamental result is that the solar acceleration had a unique geometry with respect to the Solar System barycentre before the Maunder Minimum, in the Dalton Minimum and also, at the maximum of cycle 22, around 1990, before the present prolonged minima. These dynamical events are unique at these epochs and never occurred before at least in the past millennium. We discuss our results in terms of a possible dynamical characterization of GM with relation to Sun dynamics. In the light of the planetary hipothesis, these dynamical similarities support the idea of an imminent important minimum.

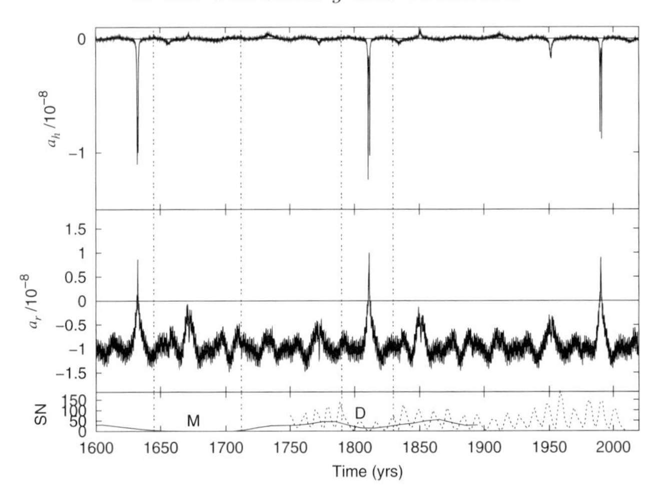

Figure 1. Normal (a_h) and radial (a_r) accelerations; sunspot number SN (solid line:^{14}C reconstruction Solanki *et al.* (2004), dashed: observed-smoothed). The Maunder (M) and Dalton (D) Minima are marked with vertical dashed lines. The impulses are clearly visible starting-during Munder Minimum, Dalton Minimum and at maximum of cycle 22. Physical units: astronomical unit, solar masses and days.

2. Methods

The 3D equations of motion of the solar system planets are integrated using a Bullirsch-Stoer scheme. Then, we obtain position, velocity and acceleration of the Sun in the ecliptical-barycentric system (the 'inertial system'), we also obtain angular momentum L and planetary 'torque' dL/dt. In order to observe the solar acceleration in a reference system more related to solar dynamics, we transformed from the inertial system to a barycentric orbital rotating system. This new system is defined by the *osculating solar orbit* (OSO) and the directions radial ($\breve{r}$), normal ($\breve{h}$) and transversal ($\breve{t}$). Therefore, the Sun's acceleration in this system is (a_r, a_t, a_h).

3. Results

Whereas the Sun acceleration's components show a monotonic harmonic behavior in the inertial system (not showed), their representation in the orbital system (a_r, a_t, a_h), show clearly gravitational impulses (Fig. 1) at the epochs of retrograde solar motion (OSO inclination $i > 90$ deg; $L_z < 0$) around the barycentre. The radial component a_r that is always negative, becomes positive. This is due to the barycentre being left outside the solar orbit, i.e., the Sun fails to loop the barycentre. This is a gravitational impulse due to giant planets quasi-alignments. The normal component a_h also has an exceptional increase at these epochs, which is explained as follows. When the OSO becomes retrograde, the angular momentum is inverted and this inversion aligns the L vector towards the planetary acceleration direction. Therefore, a_h is not an impulsive change in planetary acceleration normal to the solar orbit. This must be seen as the maximum projection of L in the solar acceleration direction, but in a contrary sense. Obviously, changes in this component are due to orbital libration, not to an important increase in z-component of acceleration in the inertial system, because planetary acceleration is always near the ecliptical plane. But these features in a_h are not trivial, they mean that at the times of angular momentum inversions, L is almost anti- parallel to the acceleration vector for a while.

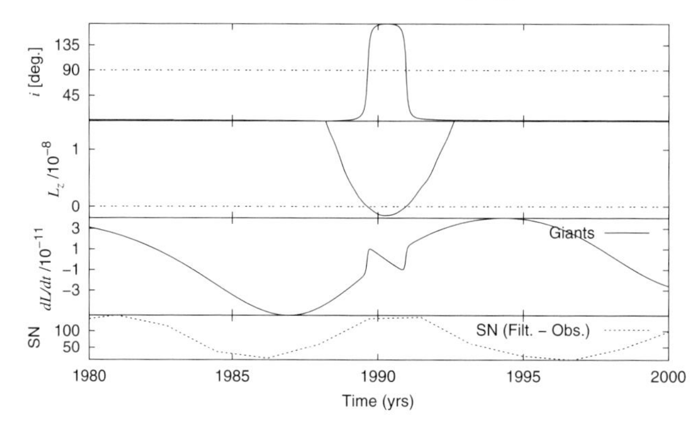

Figure 2. OSO inclination (i); L_z component and dL/dt (only giant planets) of SIM; Sunspot Number (SN) observed-filtered about 1990. When i becomes greater than 90 deg the OSO is retrograde. Note the correlation-anticorrelation between dL/dt and SN and the duration of the maximum of cycle 22. Physical units: astronomical unit, solar masses and days

It is very suggestive that the first detected impulse occurs at the very beginning of the Maunder Minimum, the second one occurs in the middle of the Dalton Minimum; and taking into account these facts, it is straightforward to think about the possibility of a new GM after 1990. This last impulsive event occurred in 1990 and was coincident with the maximum of the cycle 22. We plot in Fig. 2, the OSO inclination, dL/dt, L_z and the monthly observed/smoothed numbered sunspots series (SN). We can see an apparent phase synchronization (correlated, anti-correlated) between SN and dL/dt. The correlation between both quantities taking into account a Schwabe cycler before the radial impulse is 0.76, and after the impulse is -0.52. Notably, the duration of the maximum of the smoothed cycle is coincident with the retrograde motion interval.

It is interesting to note that in 1990 there was another Gnevyshev-Ohl (G-O) rule violation (like during Maunder and Dalton minimum) involving cycle 22 and 23 (Javaraiah 2005; Kane 2008)), i. e., before the radial impulse at the maximum of the cycle 22. Nielsen and Kjeldsen (2011) analysed the spotless days of solar cycle and conclude that the ongoing accumulation of spotless days is comparable to that of cycle 6 near the Dalton Minimum, among other cycles. These facts and our findings strongly suggest that particular epochs with impulsive manifestations in radial acceleration have associated GM events, therefore, support the claims that we are at the onset of a new prolonged minimum.

4. Conclusions and hypothesis

We have shown for the first time the existence of a unique forcing nexus between the Sun and the planets in 1632, 1811 and 1990; that is, at times of retrograde barycentric motion. Although giant planets quasi-alignments repeat every 179 yr, only at epochs of Maunder Minimum, Dalton Minimum and the maximum of cycle 22, is this particular barycentric dynamic shared. Our findings make a global support to planetary hypothesis on a physical basis. Thus, we argue for a classification or dynamical characterization of GM events and also for a possible new imminent prolonged minimum. The general dynamo theory cannot naturally reproduce the cyclicality occurrence of GM-like events; for that, some prescribed changes in the dynamo parameters or external parameterizations are also required (Chouduri & Karak 2011; Chouduri 1992). Then it is important to investigate this forcing in relation to dynamo activity and possibly Sun-planet interactions (see Cionco, this volume).

Acknowledgments

The authors acknowledge the support of PID-UTN 1351 *Forzantes externos al planeta y variabilidad climática* of UTN, Argentina. RGC acknowledges support from IAU to attend the Symposium 286.

References

Choudhuri, A. R., 1992. *A&A*, 253, 277
Feulner, G. & Rahmstorf, S., 2010. *Geophys. Res. Lett.*, 37, L05707
Javaraiah, J., 2005. *MNRAS*, 362, 1311
Kane, R. P., 2008. *Ann. Geophys.*, 26, 3329
Karak, B. D. & Choudhuri, A. R., 2011, in: D. Choudhary & K. Strassmeier (eds.), *Physics of Sun and stars spots* Proc. IAU Symposium No. 273 (Los Angeles), p. 15
Nielsen, M. L. & Kjeldsen, H., 2011 *Sol. Phys.*, 270, 385, 2011.
Perryman, M. A. C. & Schulze-Hartung, T., 2011. *A&A*, 525, A65
Solanki, S. K., Usoskin, I. G., Kromer, B., Schussler, M., & Beer J. 2004. *Nature*, 431, 1084

Comparative Magnetic Minima:
Characterizing quiet times in the Sun and Stars
Proceedings IAU Symposium No. 286, 2011
C. H. Mandrini & D. F. Webb, eds.

doi:10.1017/S1743921312005170

Long term relation between solar activity and surface temperature at different geographical regions

M. P. Souza-Echer[1], W. D. Gonzalez[1], E. Echer[1], D. J. R. Nordemann[1] and N. R. Rigozo[2]

[1]National Institute for Space Research (INPE)
S. J. Campos, SP, Brazil
email: mariza@dge.inpe.br

[2]CRS, Santa Maria, RS, Brazil
email: rigozo@lacesm.ufsm.br

Abstract. Global suface temperature has showed a rise trend in the last 150 years. This has been mainly attributed to the anthropogenic induced grenhouse gases emissions. However, the role of natural processes is not completely understood and should not be underestimated. In this work, we compare the long term variability of solar activity (as quantified by the sunspot number) with several surface temperature series from different geographical regions (global, hemispheric and latitudinal ranges). The interval of analysis is 1880-2005. The data are analyzed with wavelet multiresolution technique. It has been found that the solar activity long term trend has a maximum around 1970, while air surface temperature series showed maximum (still rising) at 2005. There are differences in the long term trend for Northern and Southern hemispheres. These differences and the relation with solar activity are discussed in this work.

Keywords. Solar Activity, Climate Change, Sun Climate relation.

1. Introduction

The air surface temperature is a basic meteorological parameter and its variation is a primary measure of global, regional and local climate changes. During the last 150 years, an upward trend of 0.6^o C in the global surface temperature data has been observed, which has been considered to be the main signature of the so called global climatic warming. The largest part of this climatic warming is usually attributed to the anthropogenic effects due to the enhanced grenhouse gases concentrations. However, there seems to be evidence that natural phenomena can account for part of the climate variability (Souza Echer *et al.* 2009; Souza Echer *et al.* 2012; Scafetta 2010).

Scafetta (2010) found empirical evidences that the climatical oscillations within the secular scales are likely driven by astronomical cycles. It was also found in that work that in all major surface temperature records there are several cycles that are coincident with astronomical cycles.

The sunspot number variability and the associated solar activity cycle are known to have important impacts in the geomagnetic activity and space weather variability (Echer *et al.* 2005). Both the solar irradiance variation and geomagnetic disturbances could have some impact on Earth′s climate, although this is a topic of intense debate and research (Haigh 2007; Souza Echer *et al.* 2009). Furthermore, the influence of these natural solar oscillations on the air surface temperature can be dependent of local conditions (Souza Echer *et al.* 2008; Souza Echer *et al.* 2012).

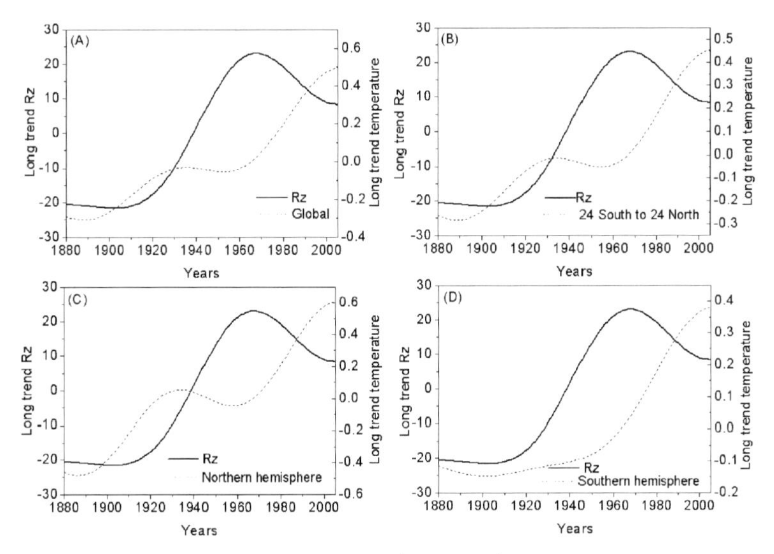

Figure 1. Long term trend for sunspot number (solid line) and for surface temperature (dotted lines). Panels show temperature long term trends for (A),Global, (B) 24° N-24° S, (C) Northern hemisphere and (D) Southern hemisphere averaged temperatures

In this work, we compare the long term variability of the solar activity, quantified by the sunspot number, with several surface temperature series from different geographical regions (global, hemispheric and latitudinal ranges). The period of study is the interval covering NASA/GISS temperature database, 1880-2005. The data are analyzed with wavelet multi-resolution analysis (MRA).

2. Methodology and Data

For studying the terrestrial climate, we have used the compiled NASA/GISS database for global, hemispheric and latitudinal air surface temperature series, available for 1880-2005 (Hansen & Lebedeff 1988, Hansen *et al.* 1996; Hansen *et al.* 1999).

The longest solar activity index is the sunspot number (R_z), which was first compiled by R. Wolf in the XIX century and it is available as annual averages since 1700 (Eddy 1976; Hoyt & Schatten 1997; Echer *et al.* 2005; Hathaway 2010). The annual averages of R_z used in this work were obtained from the Sunspot Index Data Center and the period used is the same as for the surface temperature series, 1880 to 2005.

In order to analyze and determine the long term trend, we used the wavelet multi-resolution technique (Kumar & Foufoula-Georgiou 1997; Torrence & Compo 1998; Percival & Walden 2000; Souza Echer *et al.* 2009). The temperature and sunspot number time series were decomposed in orthonormal frequency levels using the discrete Meyer wavelet transform. The approximation level 5 (period $>$ 64 years) is used in this paper for the study of long term trend in solar activity and Earth's climate.

3. Results

Figure 1 shows the long term trend levels for R_z,Global (A), 24° N-24° S (B), Northern hemisphere (C) and Southern hemisphere (D) averaged surface temperatures.

The Global air surface temperature shows two local maxima, the first one in 1935 and the second one in 2005 (at the end of the series). It also shows two local minima, the first one in 1888, and the second one in 1952 (see Figure 1A).

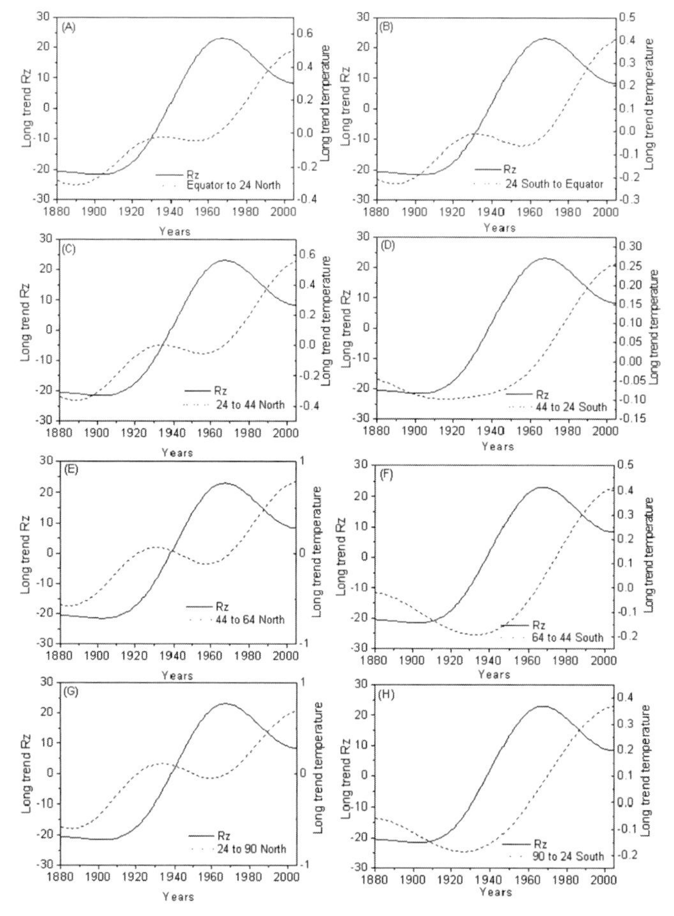

Figure 2. Long term trend for sunspot number (solid line) and for surface temperature (dotted lines). Panels show, on the left, latitudinal trends temperature long term trend for (A) 00°-24° N, (C) 24-44° N, (E) 44-64° N and (G) 24-90° N averaged temperatures; on the right, temperature long term trend for (B) 00°-24° S, (D) 24°-44° S, (F) 44-64° S and (H) 24-90° S averaged temperatures

The latitudinal range of 24° S to 24° N (Figure 1B) shows the same behaviour seen in the global temperature. There are two maxima, the first one in 1934 and the second one in 2005, and two minima, in 1889 and in 1955.

The Northern hemisphere air surface temperature shows a similar long term rising trend. This temperature series has two local maxima, the first one in 1934 and the second one in 2005 (see Figure 1C)

The Southern hemisphere air surface temperature shows that the long term trend is monotonic and upwards from 1880 to 2005. The minimum occurs in 1902 and the maximum in 2005 (see Figure 1D).

The R_z long term curve shows the rising trend from the 1880 until 1970. The plot displays one maximum value in 1968 and two minima. The first minimum occurs in 1903 and the second one in 2005. This long term trend may correspond to the Gleissberg solar cycle (period between 80-100 years).

Figure 2 shows the long term trends for several averaged latitudinal ranges (dotted lines) and again the sunspot number (solid line). See figure caption for identify each latitudinal range.

Following Figures 1 and 2 we can see that, for both hemispheres there is a rising trend in temperature since 1880. But the shape-behavior is different: for the Northern hemisphere regions there are two maxima in all latitudes. For the Southern hemisphere latitudinal ranges, it seems that most regions have a monotonic curve, with exception of the equatorial region.

4. Conclusions

For both hemispheres, there is a rising trend in temperature, since 1880. However, the shape of the long term trend is different. For Northern hemisphere regions, there are two local maxima for all latitudinal ranges. For the Southern hemisphere regions, most latitudinal ranges present a monotonic trend.

The Southern hemisphere temperature series shows a minimum around 1902 and a maximum in 2005. The Northern hemisphere temperature series present two local maxima, in 1934 and in 2005. The temperature series also shows two local minima, the first one in 1886 and the second one in 1958.

The results of this study showed that solar activity and surface temperature have different long term trend over the last 120 years. This means that solar activity effects are not strongly modulating the long term ($>$ 64 years) temperature variations, although they can have some effect on the scales of solar 11 and 22 year cycles (e.g, Souza Echer *et al.* 2009). There are noticeable differences between Northern and Southern hemisphere long term trends, which should be investigated in more detail in future works. Potential factors are regional differences, such as the ocean-land contrast, and anthropogenic effects, as the higher greenhouse gases emissions in Northern hemisphere regions.

Acknowledgements

The authors would like to thanks CNPq (MPSE: project 156109/2009-8 and 170.180/2011-5; EE: project 300211/2008-2; NRR: project 470455/2010), FAPERGS (NRR: project 1013273) and FAPESP (WDG: project 2008/06650-9) agencies for financial support.

References

Echer, E., Gonzalez, W. D., Guarnieri, F. L., Dal Lago, A., & Vieira, L. E. A. 2005, *Adv. Space Sci.*,35, 855

Eddy, J. A. 1976, *Science*, 192, 1189

Haigh, J. D. 2007, *Liv. Rev. Sol. Phys.*, 4

Hansen, J. & Lebedeff, S. 1988, *Geophys. Res. Lett.*, 15, 323

Hansen, J., Ruedy, R., Sato, M., & Reynolds, R. 1996, *Geophys. Res. Lett.*, 23, 1665

Hansen, J., Ruedy, R., Glascoe, J., & Sato, M. 1999, *J. Geophys. Res.*,104, 30997

Hathaway, D. H. 2010, *Liv. Rev. Solar Phys.*, 7

Hoyt, D. V. & Schatten, K. H. 1997, *The Role of the Sun in climate change* (Oxford University Press)

Kumar, P. & Foufoula-Georgiou, E. 1997, *Rev. Geophys.*, 35, 385

Percival, D. B. & Walden, A. T. 2000, *Wavelet methods for time series analysis* (Cambridge University Press)

Scafetta, N. 2010, *J. Atmos. Solar-Terr. Phys.*, 72, 951

Souza-Echer, M. P., Echer, E., Nordemann, D. J. R., Rigozo, N. R., & Prestes, A. 2008, *Climatic Change*, 87, 489

Souza-Echer, M. P., Nordemann, D. J. R., Echer, E., & Rigozo, N. R. 2009, *J. Atmos. Solar-Terr. Phys.*, 71, 41

Souza-Echer, M. P., Echer, E., Rigozo, N. R., Brum, C. G. M., Nordemann, D. J. R., & Gonzalez, W. D 2012, *J. Atmos. Solar-Terr. Phys.*, 74, 87

Torrence, C. & Compo, G. P. 1998, *Bull. American. Meteorol. Soc.*, 79, 61

*Comparative Magnetic Minima:
Characterizing quiet times in the Sun and Stars
Proceedings IAU Symposium No. 286, 2011
C. H. Mandrini & D. F. Webb, eds.*

doi:10.1017/S1743921312005182

Parallels among the "music scores" of solar cycles, space weather and Earth's climate

Zoltán Kolláth[1], **Katalin Oláh**[1] **and Lidia van Driel-Gesztelyi**[1,2,3]

[1]Konkoly Observatory, Budapest, Hungary
email: kollath@konkoly.hu

[2]Observatoire de Paris, LESIA, CNRS, UPMC Univ. Paris 06,
Univ. Paris-Diderot, Meudon, France

[3]University College London, Mullard Space Science Laboratory, UK

Abstract. Solar variability and its effects on the physical variability of our (space) environment produces complex signals. In the indicators of solar activity at least four independent cyclic components can be identified, all of them with temporal variations in their timescales.

Time-frequency distributions (see Kolláth & Oláh 2009) are perfect tools to disclose the "music scores" in these complex time series. Special features in the time-frequency distributions, like frequency splitting, or modulations on different timescales provide clues, which can reveal similar trends among different indices like sunspot numbers, interplanetary magnetic field strength in the Earth's neighborhood and climate data.

On the pseudo-Wigner Distribution (PWD) the frequency splitting of all the three main components (the Gleissberg and Schwabe cycles, and an ≈ 5.5 year signal originating from cycle asymmetry, *i.e.* the Waldmeier effect) can be identified as a "bubble" shaped structure after 1950. The same frequency splitting feature can also be found in the heliospheric magnetic field data and the microwave radio flux.

Keywords. methods: data analysis, Sun: activity, Sun: magnetic fields, solar-terrestrial relations

1. Time-frequency representations

Individual peaks in Fourier spectra have physical meaning only for very specific type of data (e.g. signals with constant periods). In other cases time-frequency distributions (TFD) should be investigated. The most frequently used time-frequency methods are the wavelet and the Short Time Fourier Transform (see e.g. Kolláth & Oláh 2009), however more sophisticated representations exist based on the Wigner distribution (Wigner 1932; Cohen 1995). The generalised time-frequency distribution is given by

$$C(t,f) = \frac{1}{2\pi}\int\int\int exp(-i\xi t - 2\pi i\tau f - i\xi\theta)$$
$$\Phi(\xi,\tau)s^*(\theta-\tau/2)s(\theta+\tau/2)d\theta d\tau d\xi, \tag{1.1}$$

where $s(t)$ is the analysed time series and $\Phi(\xi,\tau)$ is the kernel of the distribution that determines the specific properties of the distribution. The Wigner-Ville transformation is given by $\Phi(\xi,\tau) = 1$. This distribution is heavily contaminated by cross terms of the different components. A bi-Gaussian kernel defines the pseudo-Wigner distribution:

$$\Phi(\xi,\tau) = exp(-\xi^2/\beta^2 - \tau^2/\alpha^2), \tag{1.2}$$

where α and β are parameters that contoll the properties of the distribution.

The figures of time-frequency distributions are colour/grey-scale coded: the darker the shade, the larger the amplitude. "Bubbles" in TFDs are signs of temporary frequency splittings indicating the appearance of additional periodicities. Figure 1 displays the Short-Term Fourier Transform (left) and the pseudo-Wigner Distribution (right) of a test signal with constant frequencies indicated by horizontal lines. The 0.09 c/y (11.1 y) component is always present, while an f=0.115 c/y (8.7 y) component appears at about 1950, which exists only for a limited time interval. This signal results in bubbles in most of the time-frequency representations. Frequency modulation or other non-stationarity effects may further modify these structures.

2. The Waldmeier effect on time-frequency representations

Parallels and differences of the different periodic components within the same dataset, like modulations of the harmonics of the frequency of the 11-year solar cycle, provide additional clues for understanding processes like the Waldmeier effect.

Waldmeier (1935) has shown that an anti-correlation exists between the rise times of sunspot cycles and their strengths (i.e. shorter rise-times are coupled with higher cycle maxima). Shorter rise-times imply faster rise-rates. Thus the Waldmeier effect can also be formulated as a correlation between rise-rates and cycle strengths. The Waldmeier effect is a robust feature of the solar cycle, with consequences on correlations of other cycle measures and even cycle forecast.

In our time-frequency diagram (Fig. 2) the Waldmeier effect appears as a higher than expected amplitude of the half period of the Schwabe cycle, showing clearly the strong asymmetry of the higher amplitude (stronger) cycles.

Petrovay (2009), using harmonic analysis, has shown that the inverse correlation between cycle length and amplitude is the consequence of the Waldmeier effect, i.e. the strong inverse correlation between rise time and cycle amplitude, since the decay times do not correlate with cycle strengths. Cameron & Schüssler (2007) showed an important implication of the Waldmeier effect: since stronger cycles tend to rise faster to their maximum activity, the temporal overlapping of cycles leads to a shift of the minimum epochs that depends on the strength of the following cycle. This information is picked up by precursor methods, widely used for cycle-to-cycle forecasting. However, their finding indicates that the precursors are ultimately a simple consequence of the Waldmeier effect.

3. Solar activity variations

The presence of similar patterns at different frequencies, e.g. the bubbles at 5-, 10- and 80-year cycle lengths during the last 60 years (Fig. 2), provide very strong indication of multiple cycles in solar activity. Data noise cannot possibly produce such coherent multi-frequency patterns. The long-term variation of solar activity is a combination of at least two different cycles with variable lengths and amplitudes. This complex frequency-splitting pattern seems to be an important potential precursor of extreme cycle phases like long, deep minima and grand maxima of solar activity.

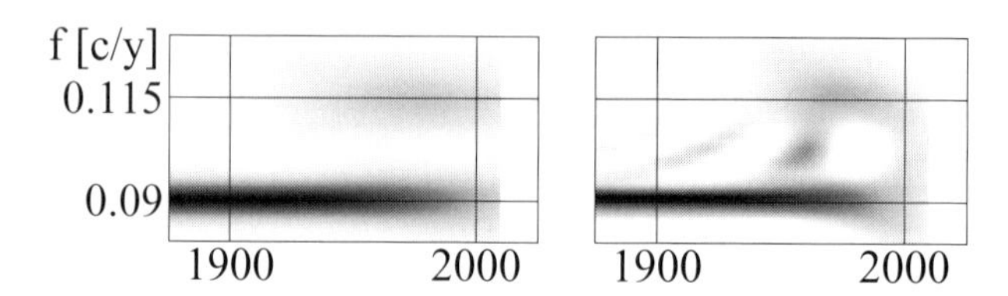

Figure 1. Short-Term Fourier Transform (left) and the pseudo-Wigner Distribution (right) of the test signal. The frequencies are measured in cycles/year (c/y).

Analyzing long-term data with such "music scores" can bring to light recurrent structures hidden in other data representations. These recurrent "tunes", due to their regular nature, can be used for forecasting the phenomena.

Solar activity went through several frequency shifts during the centuries covered by sunspot observations, like the one around 1700 (Fig. 2, left panel). This frequency shift event was strongly related to the termination of the Maunder minimum.

A similar structure can be found just before the Maunder minimum. The period splitting of the Gleissberg cycle in the 16th century, i.e., the appearance of an about 30 year long cycle together with the century-long variation looks the same as the recent period split, which started in 1950. We recall that 400 years ago a similar feature was followed by the Little Ice Age.

Then it is straightforward to compare the "scores" of solar activity to climate data. For this comparison we selected a global dataset, the Monthly Global Ocean Temperature Anomalies (Smith *et al.* (2008), http://www.ncdc.noaa.gov/cmb-faq/anomalies.php) and local temperature observations consistently observed from Armagh (Butler *et al.* 2005). The Pseudo-Wigner distributions of the three data sets are displayed on Fig. 3. Here we used a higher maximum frequency, so even the third harmonic of the Schwabe cycle is visible (fourth bubble on the left diagram). There is no direct (frequency-to-frequency) agreement among the time-frequency distributions of the different data sets – even the different climate proxies show different structures. However, it should be noted that the appearance of the new frequencies in solar activity slightly before 1950 correlates with the changes of the frequency patterns in the global sea temperature data, i.e., a periodicity around 10 years appears with a deceasing trend (Fig. 3, lower middle panel). The result for the Armagh mean temperature (Fig. 3, lower right panel) has a hint of a bubble-structure starting at about the same time as found from the sunspot data.

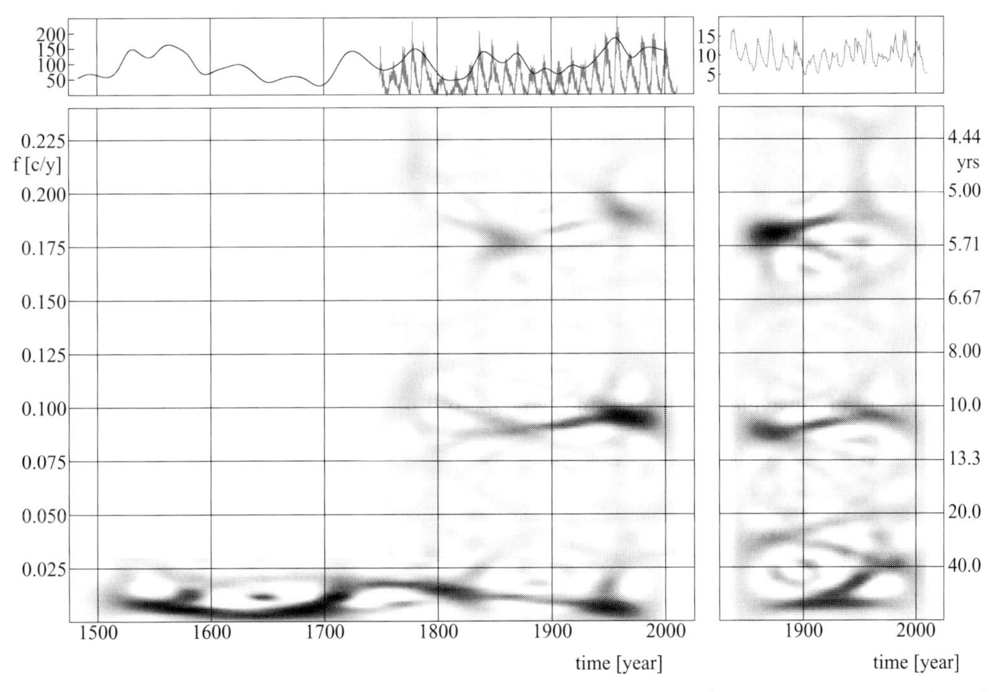

Figure 2. Left panels: Solar activity sunspot number (after 1750). Envelope: Schove (1955) data. Right panels: Heliospheric Magnetic Field – Svalgaard & Cliver (2010). Lower panels: pseudo-Wigner ditributions of the datasets.

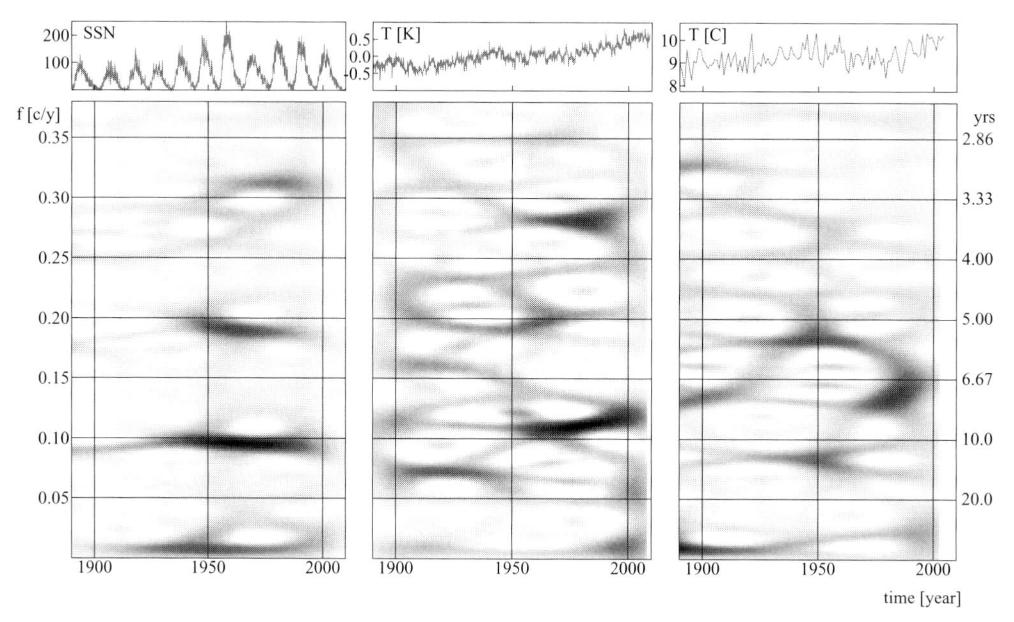

Figure 3. Left panel: Solar activity (sunspot number). Middle panel: The monthly global ocean temperature anomalies. Right panel: Armagh mean temperatures. Upper panels: Datasets. Lower panels: Pseudo-Wigner distributions of the datasets.

Acknowledgements

This work was supported by the Hungarian OTKA grant K-81421.

References

Butler, C. J., García-Suárez, A. M., Coughlin, A. D., & Morrel, C. 2005, *Int. J. Climatol.*, 25, 1055 (see also: Armagh Observatory Climate Series Vol. 2)

Cameron, R. & Schüssler, M. 2007, *ApJ*, 659, 801

Cohen, L. 1995, *Time-Frequency Analysis*, (Prentice-Hall, Englewood Cliffs).

Kolláth, Z. & Oláh, K. 2009, *A&A*, 501, 695

Petrovay, K. 2009, in A. G. Kosovichev, A. H. Andrei & J.-P. Rozelot (Eds) *Solar and Stellar Variability: Impact on Earth and Planets* Proc IAU Symp., No. 264, p. 150.

Schove, D. J. 1955, *J. Geophys Res.*, 60, 127

Smith, T. M., Reynolds, R. W., Peterson, T. C., & Lawrimore, J. 2008, *J. Climate*, 21, 2283

Svalgaard, L. & Cliver, E. W. 2010, *J. Geophys. Res.*, 115, A09111

Waldmeier M. 1935, *Mitt. Eidgen. Sternw. Zurich*, 14, 105

Wigner, J. 1932, *Phys. Rev.*, 40, 749

Comparative Magnetic Minima:
Characterizing quiet times in the Sun and Stars
Proceedings IAU Symposium No. 286, 2011
C. H. Mandrini & D. F. Webb, eds.

doi:10.1017/S1743921312005194

Climate interaction mechanism between solar activity and terrestrial biota

J. Osorio-Rosales[1] **and B. Mendoza**[2]

[1]Instituto de Geofísica, UNAM, México D.F., C.P. 04510
email: jaime@geofisica.unam.mx

[2]Instituto de Geofísica, UNAM, México D.F., C.P. 04510
email: blanca@geofisica.unam.mx

Abstract. The solar activity has been proposed as one of the main factors of Earth's climate variability, however biological processes have been also proposed. Dimethylsulfide (DMS) is the main biogenic sulfur compound in the atmosphere. DMS is mainly produced by the marine biosphere and plays an important role in the atmospheric sulfur cycle. Currently it is accepted that terrestrial biota not only adapts to environmental conditions but influences them through regulations of the chemical composition of the atmosphere. In the present study we used different methods of analysis to investigate the relationship between the DMS, Low Clouds, Ultraviolet Radiation A (UVA) and Sea Surface Temperature (SST) in the Southern Hemisphere. We found that the series analyzed have different periodicities which can be associated with climatic and solar phenomena such as El Niño, the Quasi-Biennial Oscillation (QBO) and the changes in solar activity. Also, we found an anticorrelation between DMS and UVA, the relation between DMS and clouds is mainly non-linear and there is a correlation between DMS and SST. Then, our results suggest a positive feedback interaction among DMS, solar radiation and cloud at time-scales shorter than the solar cycle.

Keywords. Dimethylsulfide, Solar Activity, Sun-Earth Relations, Climate, Wavelet Analysis, Fractals, Vector Autoregressive Analysis.

---

1. Introduction

The solar activity has been proposed as an external factor of Earth's climate change. Solar phenomena such as total and spectral solar irradiance could change the Earth's radiation balance and hence climate (Solanki 2002). However, biological processes have also been proposed as another factor of climate change through its impact on cloud albedo. One of the most important issues regarding the Earth function system is whether the biota in the ocean responds to changes in climate (Charlson *et al.* 1995; Miller *et al.* 2003). According to several authors, the major source of cloud condensation nuclei (CCN) over the oceans is dimethylsulfide (DMS) (Andreae & Crutzen 1997; Vallina *et al.* 2007). Solar radiation is the primary driving mechanism of the geophysical context and is responsible for the growth of the phytoplankton communities. Clouds modify both albedo (short-wave) and long-wave radiation. In particular, for low clouds over oceans, the albedo effect is the most important result of cloud radiation interaction and has a net cooling effect on the climate(Chen *et al.* 1999). The DMS, solar radiation, and cloud albedo are hypothesized to have a negative or positive feedback interaction (Shaw *et al.* 1998; Gunson *et al.* 2006). The purpose of the present study is to examine the relationship between DMS and climatic and solar phenomena, through clouds, sea surface temperature (SST) and the ultraviolet radiation A (UVA) in a selected location and at time-scales shorter than the solar cycle.

2. Region of study and data

The data analysis was performed for the Southern Hemisphere between $40^o - 60^o$S latitude for the entire strip length. We are interested in this area because it is the least polluted in the world, the so-called pristine zone, then solar effects on biota and climate should be more evident. The studied period is 1983-2008, containing almost 25 year of data. The DMS data set was obtained from the site *http://saga.pmel.noaa.gov/dms*. We also use the SST time series, obtained from The Climatic Research Unit *http://www.cru.-uea.ac.uk/cru/data/temperature/hadsst2sh.txt*. Another time series we use is the low cloud cover anomaly data (LCC) from the International Satellite Cloud Climatology Project (ISCCP) *http://isccp.giss.nasa.gov*. Two series of low cloud cover anomaly were obtained: Visible-Infrared (VI-IR)and Infrared (IR). Finally, we work with ultra violet radiation A data (UVA), which comprises from 320 to 400 nm, because the 95% of wavelengths longer than 310 nm reach the surface (Lean *et al.*, 1997) and has a large impact on marine ecosystems (Toole *et al.* 2006; Hefu *et al.* 1997; Slezak *et al.* 2003; Kniventon *et al.* 2003; Häder *et al.* 2011). We use the UVA composite series from the Nimbus7 (1978-1985), NOAA-9 (1985-1989), NOAA-11 (1989-1992)and SUSIM satellites between 1992 and 2008 (DeLand *et al.*, 2008).

3. Results and Discussion

Some of the previous efforts on elucidating a plausible contribution of DMS on the Earth's climate have been mostly based on correlation analysis models. The fact that two series have similar periodicities does not necessarily imply that one is the cause and another is the effect. Here we apply one non-linear analysis to study the time series: *The Wavelet Coherence Analysis*. In Fig. 1 present the coherence analysis between DMS vs SST, DMS vs LCCIR, DMS vs LCCVI-IR and DMS vs UVA respectively along 1992-2008. We choose this time interval because the DMS time series has the largest quantity of data. For each panel, the time series appears at the top, the wavelet coherence spectrum appears at the middle and the global wavelet coherence spectra is at the right. Fig. 1a, shows that the DMS and SST time series have the most persistent and prominent coherence $\sim$4 yrs and tend to be in phase. The DMS and LCCIR time series in Fig. 1b present the most prominent and persistent coherence $\sim$5 yrs, and tend to be in anti-phase. The DMS and LCCVI-IR time series show persistent coherences $\sim$3 and 5 yrs, they tend to be in phase and anti-phase respectively. The DMS and UVA time series show persistent coherence at $\sim$3 yrs but it is not very prominent, in fact the prominent coherences are between $\sim$0.4 and 1.2 yrs, they are very localized in time and tend to be in anti-phase. There is predominance in the periodicity between 3 and 5 years. Peaks shorter than 1 yr may be due to seasonal climatic phenomena. The $\sim$2 yrs period can be associated with the Quasi-Biennial Oscillation (QBO) in the stratosphere(Holton *et al.* 1972; Dunkerton 1997; Baldwin *et al.* 2001; Naujokat 1986; Holton *et al.* 1980) and with the solar activity(Kane, 2005).The periodicities $\sim$3 and 4 yrs could be related to the El Niño-Southern Oscillation (ENSO) (Nuzhdina 2002; Njau 2006: Enfield 1992) that is a large-scale climatic phenomenon. The periodicities $\sim$5 yrs can be a harmonic of the 11-yrs solar cycle (Djurović & Páquet 1996). From Fig. 1, we notice a consistent correlation between DMS and SST and an anticorrelation between DMS and UVA, the relation between DMS and clouds is mainly non-linear. The anticorrelation between UVA and DMS suggest a positive feedback, as discussed in other works (Larsen 2005) or as implied by the findings of other papers (Mendoza & Velasco, 2009; Lockwood 2005; Kristjánsson *et al.* 2002).

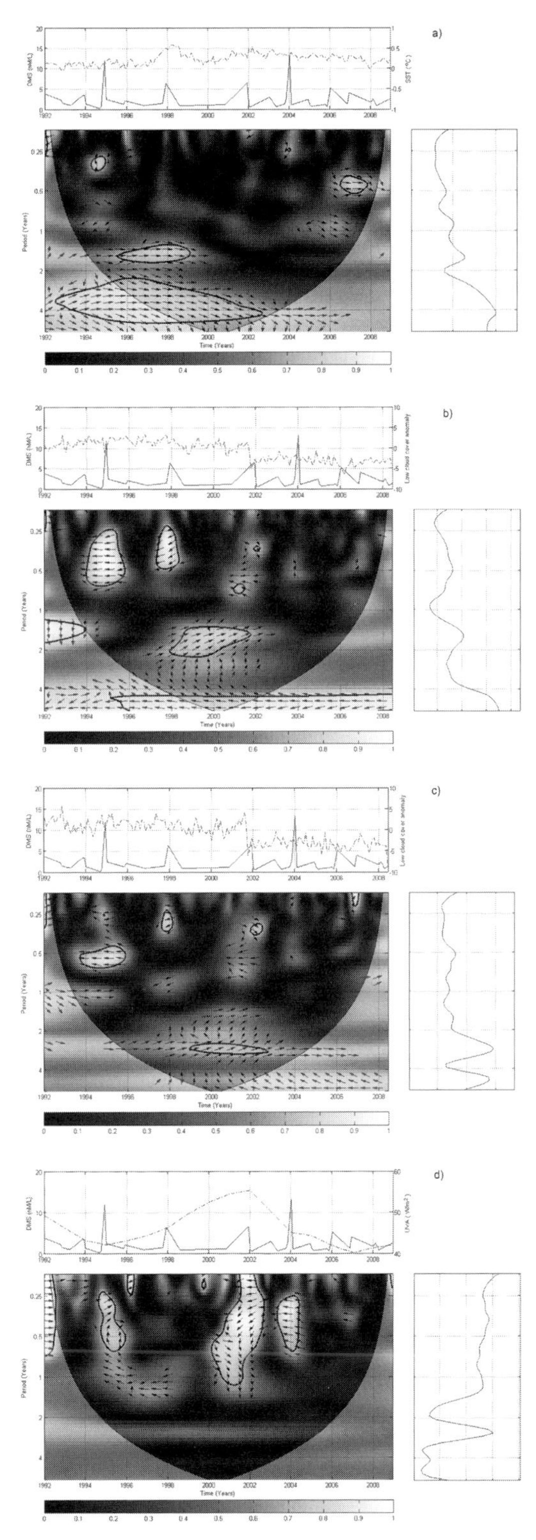

Figure 1. Wavelet coherence analysis. For each panel at the top there is the time series, at the middle the wavelet coherence spectra and at the right the global wavelet coherence spectra. The gray code indicating the statistical significance level for the spectral plots appears at the bottom of the figure; in particular the 95% level is inside the black contours. DMS (pointed line) SST, LCCIR, LCCVI-IR and UVA (dashed line).

4. Conclusions

Here we study relationship between DMS and the SST, LCCIR, LCCVI-IR and UVA using Wavelet Analysis. The DMS, SST, LCCIR and LCCVI-IR series show persistence and therefore have the possibility of a future estimation. For the UVA, the results are not realistic and this is due to the shortness of the series that have prominent periodicities for 11 years. Using the wavelet method of spectral analysis, we found a predominance of periodicity between 3 and 5 yrs. The periodicities ∼3 and 4 yrs could be related to the ENSO. The periodicities ∼5 yrs are associated with solar activity. We found a correlation between DMS and SST and an anticorrelation between DMS and UVA, the relation between DMS and clouds is mainly non-linear. Then, our results suggest a positive feedback interaction among DMS, solar radiation and clouds. The analysis shows the existence of strong relations among clouds, DMS and SST and between the SST and DMS.

References

Andreae, M. & Crutzen, P. 1997, *Science*, 276, 5315
Baldwin, M., Gray, L., Dunkerton, T., & Hamilton, K. 2001, *Rev. of Geophys.*, 39, 2
Charlson, R., Lovelock, J., Andreae, M., & Warren, S. 1995, *Nature*, 326, 655
Chen, T., Rossow, B., & Zhang, Y. 1999, *Journal of Cimate*, 13, 264
DeLand, M. & Cebula, R. 2008, *J. Geophys. Res.*, 113, A11103
Djurović, D. & Páquet, P. 1996, *Solar Physics*, 167, 427
Dunkerton, T. 1997, *J. Geophys. Res.*, 102, 26
Enfield, D. 1992, *Cambridge University Press*, 95
Gunson, J., Spall, S., Anderson, T., & Jones, A. 2006, *Geophys. Res. Lett.*, 33, L07701
Häder, D., Helbling, E., & Williamson, C. 2011, *Photochem. Photobiol. Sci.*, 10, 242
Hefu, Y. & Kirst, G. 1997, *Polar Biol.*, 18, 402
Holton, J. & Lindzen, R. 1972, *J. of the Atmospheric Sciences*, 29, 1076
Holton, J. & Tan, H. 1980, *J. of the Atmospheric Sciences*, 37, 2200
Kane, R. 2005, *J. Geophys. Res.*, 110, 13
Kniventon, D., Todd, M., & Sciare, J. 2003, *Global Biogeochem. Cycles*, 17, 1096
Kristjánsson, J., Staple, A., Kristiansen, J., & Kaas, E. 2002, *Geophys. Res. Lett.*, 29, 2107
Larsen, S. 2005, *Global Biogeochem. Cycles*, 19, 1014
Lean, J., Lee, G., Woods, H., Hickey, T., & Puga, J. 1997, *Geophys. Res. Lett.*, 102, 939
Lockwood, M. 2005, *Saas-Fee Advanced Course*, 34, 109
Mendoza, B. & Velasco, V. 2009, *J. Atm. and Solar-Terr. Phys.*, 71, 33
Miller, A. & Alexander, M. 2003, *Bull. Am. Meteorol. Soc.*, 84, 617
Naujokat, B. 1986, *J. of the Atmospheric Sciences*, 43, 1873
Njau, E. 2006, *Pakistan Journal of Meteorology*, 3, 6
Nuzhdina, M. 2002, *Natural Hazards and Earth System Sciences*, 83
Shaw, G., Benner, R., Cantrell, W., & Veazey, D. 1998, *Climate Change*, 39, 23
Slezak, D. & Herndl, G. 2003, *Mar. Ecol. Prog. Ser.*, 246, 61
Solanki, S. 2002, *Astronomy & Geophysics*, 43, 509
Toole, D., Slezak, D., Kiene, R., & Kieber, D. 2006, *Deep-Sea Res. I*, 53, 136
Vallina, S., Simó, R., & Gassó, S. 2007, *Global Biogeochemical Cycles*, 21,

Session 6

General Topics

Comparative Magnetic Minima:
Characterizing quiet times in the Sun and Stars
Proceedings IAU Symposium No. 286, 2011
C. H. Mandrini & D. F. Webb, eds.

doi:10.1017/S1743921312005212

A cellular automaton model for coronal heating

M. C. López Fuentes[1,2] **and J. A. Klimchuk**[3]

[1]Instituto de Astronomía y Física del Espacio (CONICET-UBA), CC 67, Suc 28,
1428 Buenos Aires, Argentina
email: lopezf@iafe.uba.ar

[2]Facultad de Cs. Exactas y Naturales, Universidad de Buenos Aires, Argentina

[3]NASA Goddard Space Flight Center, Code 671, Greenbelt, MD 20771, USA

Abstract. We present a simple coronal heating model based on a cellular automaton approach. Following Parker's suggestion (1988), we consider the corona to be made up of elemental magnetic strands that accumulate magnetic stress due to the photospheric displacements of their footpoints. Magnetic energy is eventually released in small scale reconnection events. The model consists of a 2D grid in which strand footpoints travel with random displacements simulating convective motions. Each time two strands interact, a critical condition is tested (as in self-organized critical models), and if the condition is fulfilled, the strands reconnect and energy is released. We model the plasma response to the heating events and obtain synthetic observations. We compare the output of the model with real observations from Hinode/XRT and discuss the implications of our results for coronal heating.

Keywords. Sun: corona, Sun: magnetic fields, Sun: activity

1. Introduction

Research works in recent years strongly suggest that the basic constituents of the coronal magnetic structure are unresolved elementary strands rooted in the photosphere (Reale 2010). In the late 80's, Parker (1988) proposed a scenario for coronal heating based on the continual dragging of strand footpoints by photospheric convection. In this scheme, footpoint motions tangle adjacent strands, producing magnetic stress between them and creating favorable conditions for reconnection and energy release. The actual reconnection events occur when neighboring strands reach a critical misalignment. Dahlburg *et al.* (2005) proposed the Secondary instability as a possible a mechanism for such critical release.

The process described above was explored in a wide variety of studies ranging from nanoflare heating models for coronal loop dynamics (see e.g., Cargill & Klimchuk 2004) to self-organized criticality in relation to flare energy power-law distributions (see e.g., Morales & Charbonneau 2008). We recently developed a simple cellular automaton (CA) model based on the above ideas (López Fuentes & Klimchuk 2010) to explain the intensity evolution of Soft X-ray loops. Here, we present a more sophisticated 2D approach. We use our model to construct synthetic light curves and compare them with observations obtained with the X-ray Telescope (XRT) on board Hinode.

2. Description of the model

The model consists of a square grid of sites initially occupied by a uniform distribution of movable points that we associate with magnetic strands (see Figure 1, panel a). At each

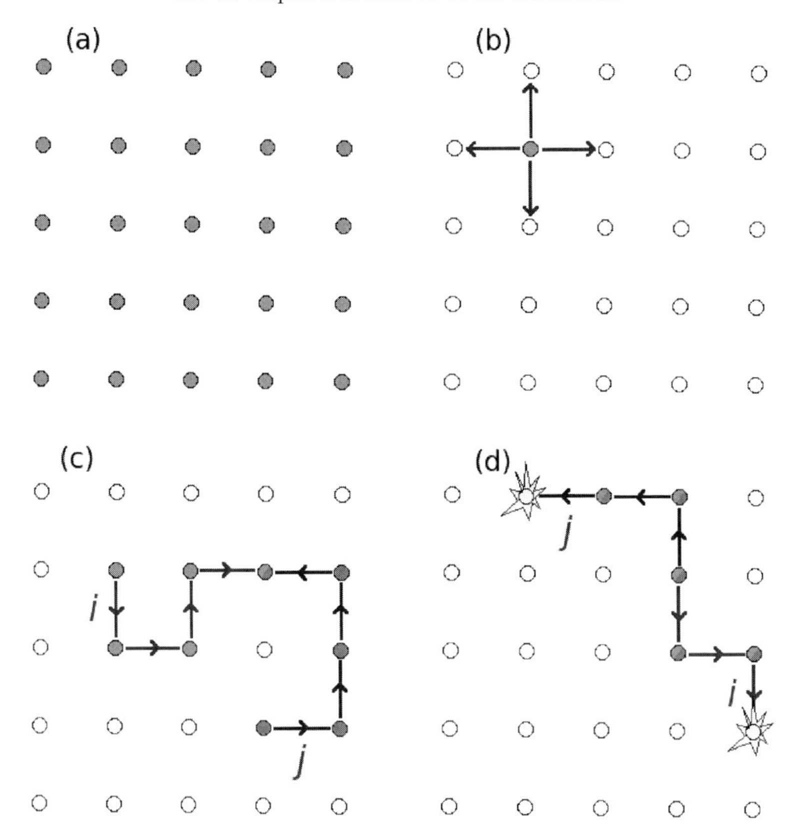

Figure 1. Scheme of the CA model described in Section 2.

time step, the points move to random neighbor positions, one site at a time, simulating photospheric displacements (Figure 1, panel b). Assuming that in the initial distribution all strands are parallel and untangled (panel a), as the system evolves the horizontal magnetic field components of the strands increase. As fully explained in López Fuentes & Klimchuk (2010), the magnitude of the increase in each time step is approximately given by:

$$\delta B_h = B_v d/L, \tag{2.1}$$

where B_v is the vertical component of the magnetic field, L is the strand length and d is a typical length of a photospheric displacement. Here, we consider $d = 1000$ km, corresponding to the approximate distance travelled by a strand footpoint during a convective cell turnover time. Defining S_i as the number of steps travelled by strand i, it is easy to see that, as time goes on, the horizontal component of the field becomes $\approx S_i \delta B_h$. As the system continues evolving, the "strand-points" travel through the grid increasing their paths and their horizontal field components. Whenever two strands (identified with indices i and j in Figure 1, panel c) occupy the same grid position, we consider them to become linked. Strands i and j continue travelling separately but the link is kept (see Figure 1, panel d). We define the critical magnetic field: $B_c = B_v \tan\theta_c$. When the field associated with the mutual displacement of the linked strands surpasses B_c according to:

$$\Delta B = \frac{B_v d}{L}(S_i + S_j) > B_c, \tag{2.2}$$

(i.e., when the misalignment angle exceeds the critical value θ_c), then the strands reconnect and magnetic energy is released. After that, the strands become unlinked and their horizontal field components are diminished in a consistent manner. It can be easily shown that each of these reconnection events (or nanoflares) releases:

$$E_{ij} = \frac{B_v^2 d^2}{8\pi L^2}[(S_i'^2 + S_j'^2) - (S_i^2 + S_j^2)], \tag{2.3}$$

where:

$$S_i' = \alpha(S_i - 1) + (1 - \alpha)(S_j - 1), \tag{2.4}$$

$$S_j' = (1 - \alpha)(S_i - 1) + \alpha(S_j - 1), \tag{2.5}$$

and α is a random parameter ($0 < \alpha < 1$) that accounts for the fact that reconnection between strands is not necessarily symmetrical (the relative lengths of the old and new strands may change).

The output of the model is the set of nanoflares that occurred in each strand after all of the footpoints in the system have been displaced. The nanoflares are modeled as triangular heating functions. To simulate the response of the plasma to the heating we use the EBTEL model (Enthalpy-Based Thermal Evolution of Loops, see Klimchuk *et al.* 2008). Using the known XRT response and the plasma density and temperature output from EBTEL, we obtain the expected emission observed with the XRT instrument. We add the contribution of all strands to the emission and correct for the number of strands covered by a single pixel. We also model the photon noise by adding intensity fluctuations using a Poisson distribution with the amplitude provided by Narukage *et al.* (2011).

3. Comparison with observations

We compare synthetic light curves obtained in this manner with Hinode/XRT observations. The analyzed data were obtained with the Al_poly filter, and correspond to NOAA AR 11147, observed on January 18, 2011. The time span of the data is approximately 8000 sec with a cadence of $\sim$ 10 sec. The images were processed and coaligned using Solar Software routines. In Figure 2, upper panels, we show the light curves of two of the loops selected from the dataset. The lower panels correspond to portions of model light curves with the same durations. For the models we use the following typical solar parameters: $B_v = 100$ G, $L = 100$ Mm, $\tan\theta_c = 0.25$, $N = 121$ (number of strands) and $\tau = 200$ sec (nanoflare duration).

Obviously, given the random nature of the model, it is not reasonable to expect a one to one correspondence between synthetic and observed light curves. For the comparison we rather consider general properties such as mean intensities and standard deviations. This kind of analysis shows that both observed and synthetic light curves in Figure 2 have a mean intensity of $\sim$ 2400 DN/pix and a standard deviation of $\sim$ 300 DN/pix, which is around 12% of the signal.

It is worth noting that part of the observed fluctuations is due to photon noise. However, the photon noise contribution has a smaller amplitude and a shorter characteristic timescale than the longer term fluctuations that produce most of the measured intensity standard deviation. To characterize the short term variation we compute the *rms* of the intensity with respect to the 10-point running average. Both observations and model have a relative *rms* of 0.04. This supports our modeling of the photon noise and suggests that longer term fluctuations are intrinsic, and are due to the variation of the individual strand intensities.

In a recent paper, Terzo *et al.* (2011) found in Hinode/XRT observations a difference between the mean and the median values of intensity fluctuation distributions. They

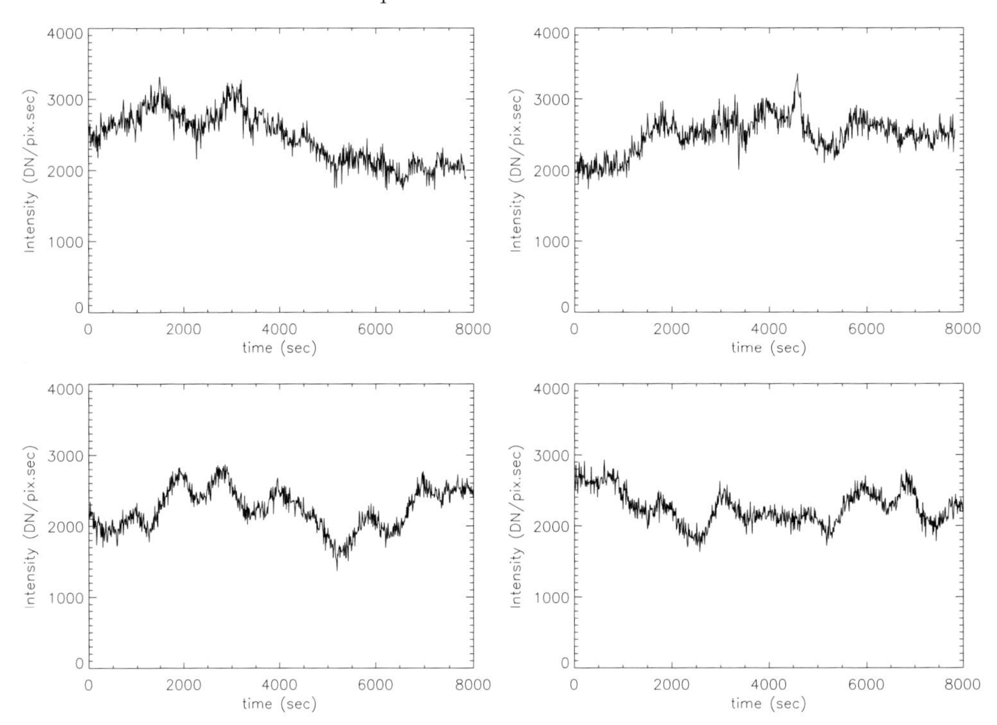

Figure 2. Upper panels: Loop light curves obtained from Hinode/XRT observations. Lower panels: Synthetic light curves obtained with the CA model presented in Section 2.

interpreted this as a signature of nanoflare heating. We not only confirm their findings in the loops studied here, but we also find the same differences between the mean and median values in the model light curves. These are very interesting results and we plan to continue exploring their implications in the near future.

4. Conclusions

We developed a CA model that reproduces the basic characteristics of Hinode/XRT loop light curves. The first results are encouraging. Among future investigations, we plan to explore the full space of parameters of the model to obtain scaling laws of the light curve properties with these parameters. We will also include in our analysis SDO/AIA observations to study how the model compares with the evolution of EUV loops.

References

Cargill, P. J. & Klimchuk, J. A. 2004, *Astrophys. J.*, 605, 911
Dahlburg, R. B., Klimchuk, J. A., & Antiochos, S. K. 2005, *Astrophys. J.*, 622, 1191
Klimchuk, J. A., Patsourakos, S., & Cargill, P. J. 2008, *Astrophys. J.*, 682, 1351
López Fuentes, M. C. & Klimchuk, J. A. 2010, *Astrophys. J.*, 719, 591
Morales, L. & Charbonneau, P. 2008, *Astrophys. J.*, 682, 654
Narukage, N., Sakao, T., Kano, R., *et al.* 2011, *Sol. Phys.*, 269, 169
Parker, E. N. 1988, *Astrophys. J.*, 330, 474
Reale, F. 2010, *Living Reviews in Solar Physics*, 7, 5
Terzo, S., Reale, F., Miceli, M., *et al.* 2011, *Astrophys. J.*, 736, 111

Comparative Magnetic Minima: Characterizing quiet times in the Sun and Stars
Proceedings IAU Symposium No. 286, 2011
C. H. Mandrini & D. F. Webb, eds.

doi:10.1017/S1743921312005224

Magneto-seismology of solar atmospheric loops by means of longitudinal oscillations

M. Luna-Cardozo[1,3], G. Verth[2] and R. Erdélyi[3]

[1]Instituto de Astronomía y Física del Espacio (IAFE), CONICET-UBA, CC. 67, Suc. 28, 1428 Buenos Aires, Argentina. email: mluna@iafe.uba.ar

[2]School of Computing, Engineering and Information Sciences, Northumbria University, Newcastle Upon Tyne, NE1 8ST, UK. email: gary.verth@northumbria.ac.uk

[3]Solar Physics and Space Plasma Research Centre (SP^{2}RC), University of Sheffield, Hicks Building, Hounsfield Road, Sheffield S3 7RH, UK. email: robertus@sheffield.ac.uk

Abstract. There is increasingly strong observational evidence that slow magnetoacoustic modes arise in the solar atmosphere. Solar magneto-seismology is a novel tool to derive otherwise directly un-measurable properties of the solar atmosphere when magnetohydrodynamic (MHD) wave theory is compared to wave observations. Here, MHD wave theory is further developed illustrating how information about the magnetic and density structure along coronal loops can be determined by measuring the frequencies of the slow MHD oscillations. The application to observations of slow magnetoacoustic waves in coronal loops is discussed.

Keywords. (magnetohydrodynamics:) MHD, Sun: corona, Sun: fundamental parameters, Sun: magnetic fields, Sun: oscillations, Waves

1. Introduction

Damped slow MHD oscillations have been observed in the solar atmosphere using high-resolution EUV imager on board space-borne telescopes (see review by Wang 2011). Such oscillations are important because of their potential for the diagnostics of magnetic structures by implementation of the method of magneto-seismology, through matching the MHD wave theory and wave observations in the solar atmosphere to obtain several physical parameters (e.g., magnetic field strength and density scale height).

The theory of MHD wave propagation in solar magnetic structures initially began modelling the magnetic structures as homogeneous cylindrical magnetic flux tubes enclosed within a magnetic environment (Roberts *et al.* 1984). Later on, more advanced equilibrium models to study slow MHD oscillations have also been proposed with e.g., dissipative effects and gravity (Mendoza-Briceño *et al.* 2004, Sigalotti *et al.* 2007), and non-isothermal profiles (Erdélyi *et al.* 2008), while the effect of density and magnetic stratification had been revisited on transversal coronal loop oscillations by Dymova & Ruderman (2006) and Verth & Erdélyi (2008), respectively.

Here, the governing equation of the longitudinal mode is solved numerically for density stratified loops with uniform magnetic field, as well as for expanding magnetic flux tubes with uniform density. The effect of these stratifications on the frequency ratio of the first overtone to the fundamental mode is studied.

2. Governing equation

The ideal MHD equations are linearized by considering small magnetic and velocity perturbations about a plasma in static equilibrium [$\vec{b} = (b_r, 0, b_z)$ and $\vec{v} = (v_r, 0, v_z)$, for r and z the radial and longitudinal coordinates, respectively]. In the derivation, a uniform kinetic plasma pressure is assumed, and the thin flux tube approximation is considered. The second-order ordinary differential equation governing the longitudinal velocity amplitude is (see Luna-Cardozo *et al.* 2012 for a detailed derivation)

$$\frac{d^2 v_z}{dz^2} + \left(\frac{c_s^2 - c_A^2}{c_f^2}\right)\frac{1}{B_z}\frac{\partial B_z}{\partial z}\frac{dv_z}{dz} + \left[\frac{\omega^2}{c_T^2} - \frac{1}{B_z}\frac{\partial^2 B_z}{\partial z^2} - \left(\frac{c_s^2 - c_A^2}{c_f^2}\right)\frac{1}{B_z^2}\left(\frac{\partial B_z}{\partial z}\right)^2\right] v_z = 0, \tag{2.1}$$

where $c_A^2 = (B_z^2/\mu\rho_0)$, $c_s^2 = (\gamma p_0/\rho_0)$, $c_f^2 = c_s^2 + c_A^2$ and $c_T^2 = (c_s^{-2} + c_A^{-2})^{-1}$ are the square of the Alfvén, sound, fast phase and tube speeds, respectively. In this equation, ω is the angular frequency of the oscillations.

Equation (2.1) is numerically solved using the shooting method based on the Runge-Kutta technique, for density stratified loops with uniform magnetic field, as well as for expanding loops with uniform density. Solar waveguides are modelled as axisymmetric cylindrical magnetic tubes with tube ends frozen in a dense photospheric plasma at $z = \pm L$. On average, plasma density and magnetic field strength are expected to decrease with height above the photosphere (Lin *et al.* 2004).

3. Effect of density stratification

The solar coronal loop is modelled by a straight axisymmetric magnetic flux tube with tube length of $2L$ and radius of r_0. The uniform magnetic field is directed along the tube axis, i.e., $\vec{B} = B_z\hat{z}$. In semi-circular coronal loops where the plasma is close to hydrostatic equilibrium, a reasonable assumption for the density profile is

$$\rho_0(z) = \rho_f \exp\left[-\frac{2L}{\pi H}\cos\left(\frac{\pi z}{2L}\right)\right], \tag{3.1}$$

where H is the density scale height and ρ_f the density at the footpoint. To study a standing wave the boundary condition $v_z(\pm L) = 0$ is applied. We solve equation (2.1) using the density profile (3.1). The frequency ratio of the first overtone to the fundamental mode is shown in Figure 1(a) as a function of L/H by the solid line, and it is clearly lower than the canonical value of two. A similar result was obtained for the transversal mode by Dymova & Ruderman (2006) and Verth (2007).

For vertical chromospheric flux tubes the density profile is given by

$$\rho_0(z) = \rho_f \exp\left[-\frac{(z+L)}{H}\right]. \tag{3.2}$$

Longitudinal oscillations in chromospheric flux tubes are studied solving the eigenvalue problem in half of the magnetic bottle, i.e., designating $v_z(-L) = v_z(0) = 0$ as the boundary conditions. The ratio of frequencies against L/H for the density profile (3.2) is shown by the dashed line in Figure 1(a). Now, the frequency ratio is slightly greater than two, indicating that this parameter depends on the functional form chosen of the equilibrium density. This suggests that caution must be used when interpreting the frequency ratio of chromospheric standing modes.

4. Effect of a non-uniform magnetic field

An expanding flux tube with rotational symmetry about the z-axis in cylindrical coordinates (r, θ, z) is used to model a magnetic field equilibrium decreasing in strength with

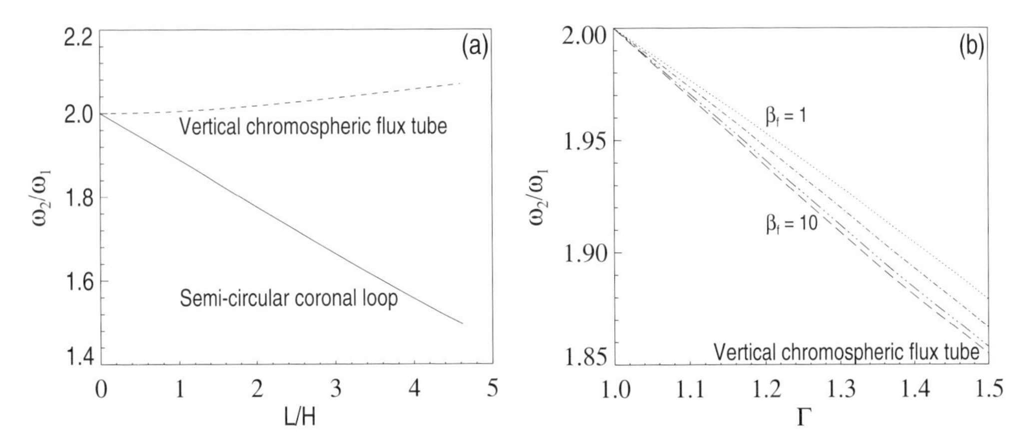

Figure 1. (a) Frequency ratio of the first overtone and fundamental mode against L/H for density stratified coronal (solid line) and chromospheric (dashed line) loops. (b) Frequency ratio against the expansion parameter Γ for different values of $\beta_{\rm f}$ in vertical chromospheric flux tubes. In (b) dotted, dot-dashed, dot-dot-dot-dashed and long-dashed lines correspond to $\beta_{\rm f} = 1$, 2, 5 and 10, respectively.

height above the photosphere. The magnetic field component B_z at the tube boundary can be described explicitly as function of z (see Verth & Erdélyi 2008)

$$B_z(z) \approx B_{z,\rm f}\left\{1 + \frac{(1-\Gamma^2)}{\Gamma^2}\frac{[\cosh(z/L) - \cosh(1)]}{1-\cosh(1)}\right\}, \tag{4.1}$$

where $\Gamma = r_{\rm a}/r_{\rm f}$ is the expansion factor, and $r_{\rm a}$ ($r_{\rm f}$) is the apex (footpoint) radius. The loop expansion has been estimated for various loops, giving mean values of $\Gamma \approx 1.16$ and 1.30 for EUV and soft X-ray loops (Watko & Klimchuk 2000, Klimchuk 2000).

We can compute the numerical solution of equation (2.1) for slow longitudinal oscillations in coronal and chromospheric loops setting the same boundary conditions as in the previous section, and using equation (4.1) for $B_z(z)$. Figure 1(b) shows the frequency ratio as function of the expansion parameter Γ for different values of the footpoint beta plasma $\beta_{\rm f}$ for chromospheric flux tubes. It is found that when the magnetic expansion factor increases the frequency ratio *decreases*, and this effect is more significant for chromospheric flux tubes with higher $\beta_{\rm f}$.

Figure 2 shows the frequency ratio as function of the expansion factor for coronal loops with uniform density in (a) and for typical density stratification (i.e., $L/H = 2$) in (b). It can be seen how these two effects, density stratification and magnetic expansion, contribute to *decrease* the value of ω_2/ω_1. Additionally, the effect of the expansion is stronger in the corona than in the chromosphere.

5. Summary and conclusions

Studying the solutions of the velocity governing equation of the slow standing mode, it is found that density stratification and magnetic expansion cause the *same qualitative effect* on the frequency ratio in coronal loops, giving values of $\omega_2/\omega_1 < 2$. For chromospheric flux tubes density stratification and magnetic expansion cause *opposite effects* on the frequency ratio; however, caution must be taken when studying chromospheric flux tubes since the ratio ω_2/ω_1 depends on the functional form chosen for the density (see Luna-Cardozo *et al.* 2012 for an analytical and numerical detailed study about important issues).

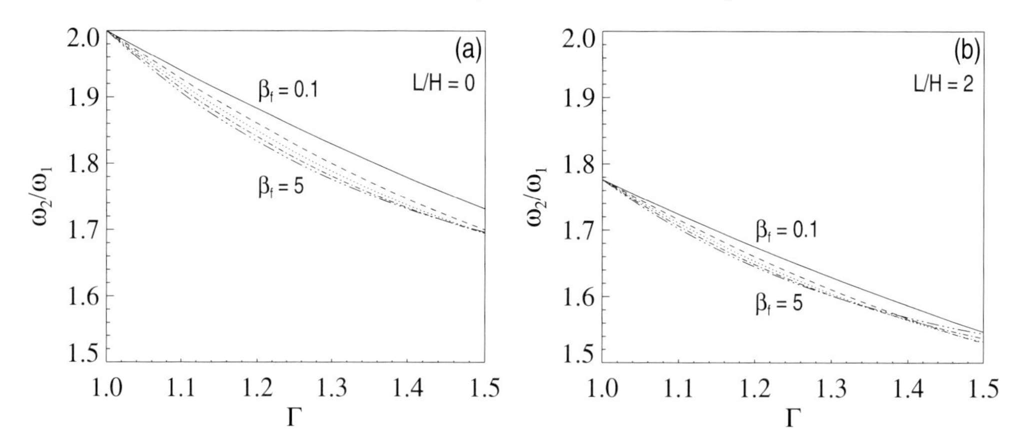

Figure 2. Frequency ratio of coronal loop oscillations against the expansion parameter Γ for different values of β_f. Solid, dashed, dotted, dot-dashed and dot-dot-dot-dashed lines correspond to $\beta_f = 0.1$, 0.5, 1, 2 and 5, respectively. Coronal loops with uniform density ($L/H = 0$) are presented in (a) and with typical density stratification ($L/H = 2$) in (b).

These results are consistent with the values of period ratio of $P_1/P_2 = 1.54$ and 1.84 reported by Srivastava & Dwivedi (2010) while observing slow acoustic oscillations using *Hinode*, in contrast to the theoretical value of $P_1/P_2 = 2$ for a uniform cylindrical flux tube model.

Our results are important for magneto-seismology, where the density scale height of the solar atmosphere can be calculated by using the observed value of the frequency ratio ω_2/ω_1 of longitudinal loop oscillations, to complement both emission measure and magnetic field extrapolation studies. This could provide us with a more complete understanding of the plasma fine structure in the solar atmosphere. These results can be applied in any stage of the solar cycle, including the solar minimum.

Acknowledgements

ML-C thanks the IAU for the travel grant and is grateful for the financial support from PICT 2007-1790 grant (ANPCyT). RE acknowledges M. Kéray for patient encouragement and is also grateful to NSF, Hungary (OTKA, Ref. No. K83133) for financial support received.

References

Dymova, M. V. & Ruderman M. S. 2006, *A&A*, 457, 1059
Erdélyi, R., Luna-Cardozo, M., & Mendoza-Briceño, C. A. 2008, *Sol. Phys.*, 252, 305
Klimchuk, J. A. 2000, *Sol. Phys.*, 193, 53
Lin, H., Khun, J. R., & Coulter, R. 2004, *ApJ*, 613, L177
Luna-Cardozo, M., Verth, G., & Erdélyi, R. 2012, *ApJ*, 748, 110
Mendoza-Briceño, C. A., Erdélyi, R., & Sigalotti, L. Di G. 2004, *ApJ*, 605, 493
Roberts, B., Edwin, P. M., & Benz, A. O. 1984, *ApJ*, 279, 857
Sigalotti, L. Di G., Mendoza-Briceño, C. A., & Luna-Cardozo, M. 2007, *Sol. Phys.*, 246, 187
Srivastava, A. K. & Dwivedi, B. N. 2010, *New Astron.*, 15, 8
Verth, G. 2007, *Astron. Nachr.*, 328, 764
Verth, G. & Erdélyi, R. 2008, *A&A*, 486, 1015
Wang, T. J. 2011, *Space Sci. Rev.*, 158, 397
Watko, J. A. & Klimchuk J. A. 2000, *Sol. Phys.*, 193, 77

Comparative Magnetic Minima:
Characterizing quiet times in the Sun and Stars
Proceedings IAU Symposium No. 286, 2011
C. H. Mandrini & D. F. Webb, eds.

doi:10.1017/S1743921312005236

TTVs study in southern stars

Romina Petrucci[1], Emiliano Jofré[2], Martín Schwartz[1], Andrea Buccino[1] and Pablo Mauas[1]

[1]Instituto de Astronomía y Física del Espacio (IAFE), Buenos Aires, Argentina
email: romina@iafe.uba.ar, mschwartz71@gmail.com, abuccino@iafe.uba.ar, pablo@iafe.uba.ar

[2]Observatorio Astronómico de Córdoba (OAC), Córdoba, Argentina
email: emiliano@mail.oac.uncor.edu

Abstract. In this contribution we present 4 complete planetary transits observed with the 40-cm telescope "Horacio Ghielmetti" located in San Juan(Argentina). These objects correspond to a continuous photometric monitoring program of Southern planet host-stars that we are carrying out since mid-2011. The goal of this project is to detect additional planetary mass objects around stars with known transiting-planets through Transit Timing Variations (TTVs). For all 4 transits the depth and duration are in good agreement with the values published in the discovery papers.

Keywords. extrasolar planets, transiting-planets, TTVs

1. Introduction

So far, over 690 extrasolar planets has been detected by different techniques; radial velocities and transits methods have provided the majority of detections. Period, orbital eccentricity and minimum mass of the planet can be determined by the first technique, meanwhile period, inclination and relative radii of the planet can be obtained by the second one. Therefore, it is possible to calculate the planetary density combining spectroscopic and photometric observations. The disadvantage is that spectroscopic measurements are limited by the star magnitude.

On the other hand, there is another exoplanet detection technique, currently very popular, based on the fact that the time interval between successive transits of an unperturbed planet is always the same. However the presence of another planetary mass body in the system can produce variations of the transiting-planet period due to their mutual gravitational interaction. These transit timing variations (TTVs) depend on the mass of the additional planet, and in some cases terrestrial-mass planets will produce a measurable effect. Furthermore, systems in which two planets transit their star, the masses and radii of each planet can be determined without spectroscopic measurements (Holman & Murray 2005). Therefore, it is possible to compute densities even for faint, low-mass stars. This is one of the aspects that makes TTVs a very important detection technique.

2. The Telescope

The observations of the 4 transits were made remotely with the 40-cm telescope "Horacio Ghielmetti" located on "Cerro Burek" at CASLEO (San Juan, Argentina). This is a MEADE-RCX 400 telescope, $f/8$, with a Ritchey-Chretien advanced optical system, equipped with an APOGEE ALTA U16M camera and UBVRI filters. The image on the CCD covers an area of $49' \times 49'$.

3. First results

Figure 1 (a to d) shows the light-curves corresponding to the exoplanet host-stars: WASP-28, WASP-44, GJ-1214 and WASP-4. All the observations were taken in the clear filter to improve the temporal resolution, and the exposure times were chosen so that the comparison stars and the exoplanet host-stars do not saturate the CCD camera. We corrected the images for bias and dark with standard tasks of the IRAF package program. After calibration, we made aperture photometry for determining instrumental magnitudes using the "DAOPHOT" package. For each star in the stellar field the adopted aperture size was that for which the star magnitude was stable in 0.001 magnitudes (Howell 1989), the annulus was 5 pixels from the aperture, and the dannulus was 5 pixels. Then, we made a first order extinction correction, taking as a reference non variable stars with similar color to the target star. We fitted each resulting light-curve with an iterative method described in Poddany *et al.* 2010.
In the next sections we present more characteristics of each particular case:

3.1. *WASP-28*

It is an object of spectral type G, V=12, with a "hot Jupiter" planet. It has a $M_P = 0.91 M_{jup}$ and P=3.40 days. The transit shown in the figure 1a) was observed on the night of August 27, 2011. The exposure time was 120 seconds and we used 2 reference stars. The data dispersion is $\sigma = 4.3$ mmag and the fit corresponds to a depth of 19.2 ± 1.2 mmag and a duration of 199.4 ± 4.6 min.

3.2. *WASP-44*

This is a V=12.9, G8V star, that posses a planet ($M_P = 0.89 M_{jup}$) with a 2.42 days period (Anderson *et al.* 2011). Figure 1b) shows the transit observed on the night of August 28, 2011. We used an exposure time of 150 seconds and 3 reference stars. The dispersion of the points is $\sigma = 4.7$ mmag and the fit corresponds to a depth of 18.45 ± 1.5 mmag and a duration of 148.3 ± 5.6 min.

3.3. *GJ-1214*

It is a M star with V=14.67 that hosts a super-Earth planet ($M_P = 6.6 M_{earth}$ and $R_P = 2.7 R_{earth}$) (Berta *et al.* 2011). The transit shown in the figure 1c) was observed on the night of July 2, 2011. We adopted an exposure time of 150 seconds and used 7 reference stars. As can be seen, in the points corresponding to the "out-of-transit" after the transit, appears a decrease in magnitudes that could not be removed. This is probably due to an instrumental error or due to the fact that the colors of the comparison stars differ significantly from the color of GJ-1214. The dispersion is σ=3.5 mmag and the fit corresponds to a depth of 16.6 ± 2.9 mmag and a duration of 56.5 ± 10.5 min.

3.4. *WASP-4*

This star has a magnitude of V=12.5 and hosts a planet of $M_P = 1.21 M_{jup}$ (Gillon *et al.* 2009). The transit shown in the figure 1d) was observed on the night of October 17, 2011. The exposure time was 25 seconds and we used 5 reference stars. The data dispersion is $\sigma = 6.6$ mmag and the fit corresponds to a depth of 31.2 ± 1 mmag and a duration of 126.9 ± 2.1 min.

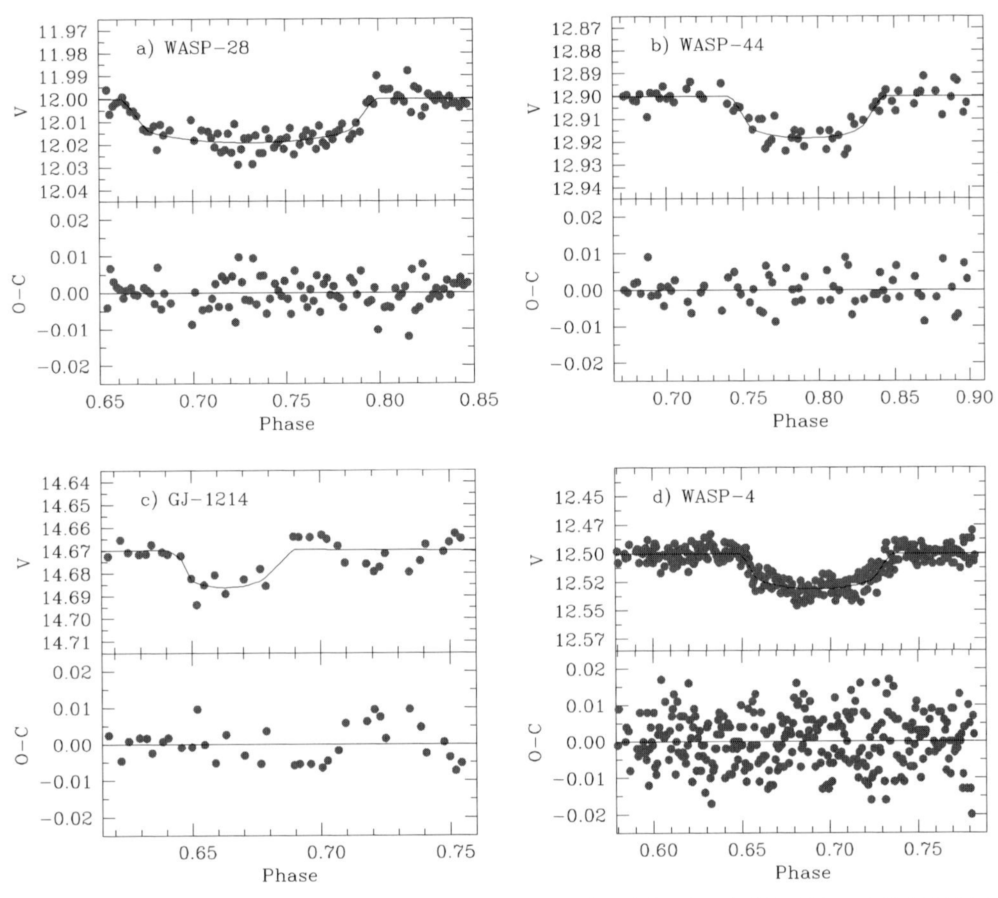

Figure 1. Light-curves and residuals for a) WASP-28, b) WASP-44, c) GJ-1214 and d) WASP-4. In the light-curves the blue dots correspond to the observations and the continuos black line represents the best fit to the data.

4. Conclusion and future steps

In all 4 cases the values of depth and duration of the planetary transits obtained in this work agree with those published in the discovery papers. Our next step is to investigate and test different algorithms to remove systematic effects and decrease data dispersion. In future works, we will use the Mandel and Algol code (Mandel & Algol 2002) for fitting the light-curves and we will improve the precision of minimum time measurements. In the following months we will also continue with the planetary transit observations remotely and will begin with the TTVs analysis.

Acknowledgments

We would like to thank the CASLEO staff for helping to maintain the telescope, to Pablo Perna and Matías Pereyra for providing software support and, finally, Romina Petrucci and Pablo Mauas thank IAU for providing financial support to attend this Symposium.

References

Anderson, D. R., Collier Cameron, A., & Gillon, M., *et al.* 2011, *MNRAS in press.*
Berta, Z. K., Charbonneau, D., & Bean, J., *et al.* 2011, *ApJ*, 736, 12
Gillon, M., Smalley, B., & Hebb, L., *et al.* 2009, *A&A*, 496, 259
Holman, M. J. & Murray, N. W. 2005, *Science*, 307, 1288
Howell, S. B. 1989, *PASP*, 101, 616
Poddany, S., Brat, L., & Pejcha, O. 2010, *New Astron.*, 15, 297

Comparative Magnetic Minima:
Characterizing quiet times in the Sun and Stars
Proceedings IAU Symposium No. 286, 2011
C. H. Mandrini & D. F. Webb, eds.

doi:10.1017/S1743921312005248

The LAGO (Large Aperture GRB Observatory) in Peru

E. Tueros-Cuadros[1], L. Otiniano[1], J. Chirinos[2], C. Soncco[1] and W. Guevara-Day[1]

[1]Departamento de Astrofísica, Agencia Espacial del Perú, CONIDA
[2]Michigan Technology University

Abstract. The Large Aperture GRBs Observatory is a continental-wide observatory devised to detect high energy (around 100 GeV) component of Gamma Ray Bursts (GRBs), by using the single particle technique in arrays of Water Cherenkov Detectors (WCDs) at high mountain sites of Argentina, Bolivia, Colombia, Guatemala, Mexico, Venezuela and Peru. Details of the instalation and operation of the detectors in Marcapomacocha in Peru at 4550 m.a.s.l. are given. The detector calibration method will also be shown.

Keywords. GRB, Gamma Rays.

1. Introduction

The LAGO (Large Aperture GRBs Observatory) international collaboration goal is to observe high energy component (around 100 GeV) of Gamma Ray Bursts (GRBs) through Water Cherenkov Detectors (WCDs) in high mountain sites, (Bertou 2008) using the single particle technique (Aglietta *et al.* 1996). When a GRB occurs, the atmosphere is impacted by a multitude of high energy gamma rays that produce decay showers. Together those showers are detectable at ground level as an instant excess in the flux measured by a detector. Gamma ray showers have a composition of 90 % photons, 9% electrons and $< 1\%$ muons, WCDs are sensitive to all those particles. These showers are absorbed high in the atmosphere, not being able to survive and reach the ground. In order to detect the greatest possible number of particles, the water Cherenkov tank must be placed at high altitude. Peru is in an initial phase. We have built two detectors, located at 4450 m.a.s.l. in Marcapomacocha-Huancayo, and a third one is under construction in the same location.

2. Experimental Setup

The LAGO project is using electronic acquisition and photomultiplier tubes (PMTs) from the engineering array of the Pierre Auger Observatory (PAO), (Pierre Auger Collaboration 2004). As data storage, we use an ARM computer which has a low power consumption (less than 1 watt) and resists extreme temperatures (-40 °C to 85 °C). At high altitude locations in Peru it is dificult to get stable power sources but the low consumption requirements of this system allows the use of solar panels and batteries. The WCD consists of a commercial cylindrical poliethylene tanks of $4m^2$, with a large area PMT located in the central upper section. It has an internal cover of banner to ensure a good reflexivity and diffusivity and it is shielded from external light. The WCDs are filled with clear water up to 1.3m in order to ensure a high probability of photon conversion inside the tank. The water treatment must be done on the site. Filtered water is collected in a poliethylene tank and treated with aluminum sulfate, then is left to sediment

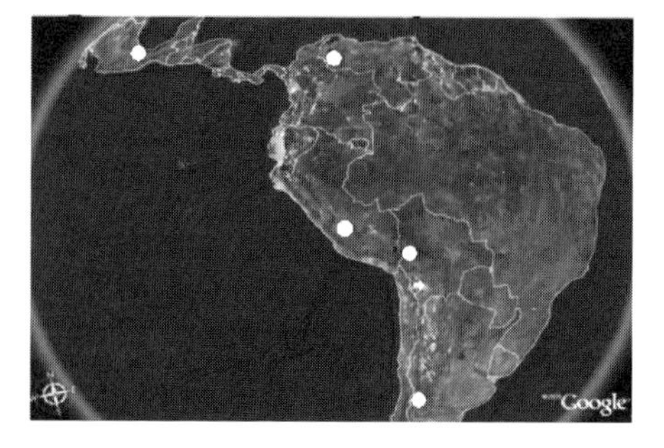

Figure 1. The left pannel hand picture shows the stations of LAGO in Sierra Negra (Mexico), Pico Espejo (Venezuela), Marcapomacocha (Peru), Chacaltaya (Bolivia) and Bariloche (Argentina) . The right pannel hand picture shows a view of the Marcapomacocha site.

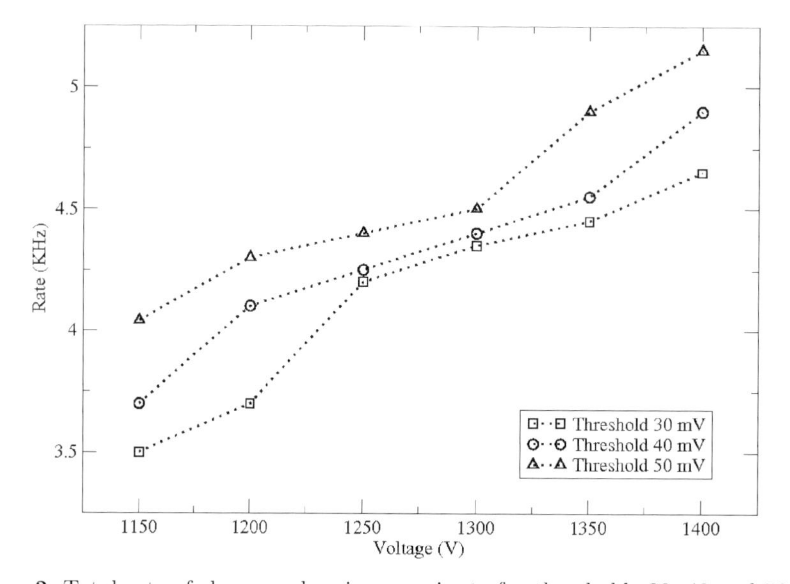

Figure 2. Total rate of charge pulses in one minute for thresholds 30, 40 and 50 mV.

for about a week. After that, the water pass to the WCDs through three filters of 20, 5 and 1 micra and 2 activated carbon filters, leaving behind the sediment, finally chlorination is made inside the WCD.

The signal from the PMT is digitalized by an adquisition board (Local Station, LS) of the engineering array of the PAO. It has six channels that sampled the data at 40 MHz. The content of four scalers per channel is read out every 5 ms. and it is send by serial line to an ARM computer.

3. Calibration of the Detectors

The FPGA(Field Programmable Gate Array) of the LS allows the read out of the distribution of the charge of the pulses generated in the WCD by secondary cosmic rays. We choose the high voltage applied to the PMT at 1250 V by the condition that the rate of cosmic rays is maintained linear in a range of $+/-50V$ (See Fig.2). A secondary

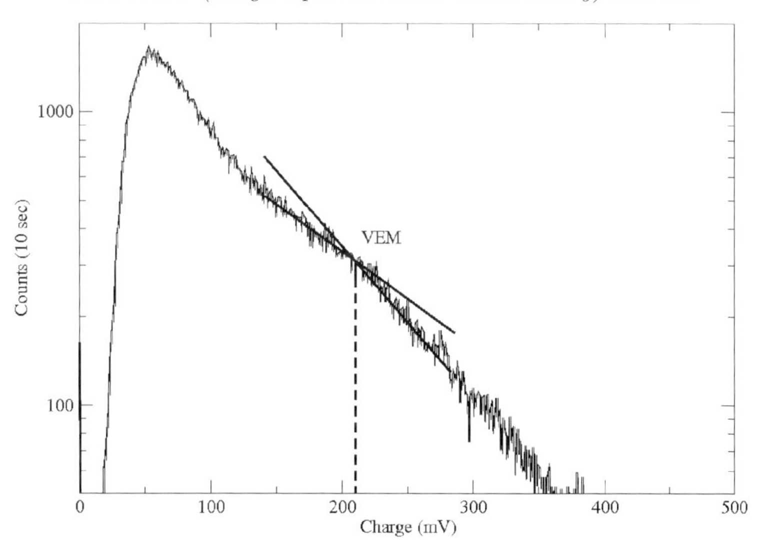

Figure 3. First calibration data: distributions of charge of the signal of the secondary cosmic rays traversing the WCD tank in 10 seconds, the VEM (Vertical Energetic Muon) position is extrapolated from the change in the slope of the charge plot).

peak in the distribution due to atmospheric vertical muons is expected at an energy of 260 MeV (muons deposit 200 MeV/cm when they cross water). Calibration points are obtained by finding a change in the slope of the distribution of charge (see Fig. 3).

4. Conclusions

A first WCD of 4m^2 has been installed in Peru at 4450 m.a.s.l. at Marcapomacocha site. The procedure for water treatment and calibration of the detectors has been developed.

Acknowledgements

We would like to thank the LAGO colaboration, CONIDA (Comisión Nacional de Investigación y Desarrollo Aeroespacial del Perú) for their support to this work and the Pierre Auger Observatory for providing electronic equipment.

References

Aglietta, M. *et al.* 1996, *Astrophys. J.*, 469, 305–310.
Bertou X. for the LAGO Collaboration 2008, *NIM A*, 595, 70–72.
Pierre Auger Collaboration 2004, *NIM A*, 523, 50–95.

Comparative Magnetic Minima:
Characterizing quiet times in the Sun and Stars
Proceedings IAU Symposium No. 286, 2011
C. H. Mandrini & D. F. Webb, eds.

doi:10.1017/S174392131200525X

Seeing measurement on Sasahuine mountain, Moquegua, Perú

C. Ferradas-Alva[1], **G. Ferrero**[1], **M. Huamán**[1], **W. Guevara-Day**[1], **E. Meza**[1], **J. Samanes**[1], **and P. Becerra**[1]

[1]Departamento de Astrofísica, Agencia Espacial del Perú, CONIDA, Perú

Abstract. One of the greatest factors that significantly affect the quality of astronomical images is the atmospheric turbulence causing what we call "seeing". We present results of the reduction and photometry of astronomical images obtained at the Sasahuine mountain astronomical site (4511 m.a.s.l.), located in the Southern Peruvian Andes, in the department of Moquegua, near the town of Cambrune. These data show preliminary seeing measurements for this site. The present work is part of a bigger investigation program called JANAX which seeks to evaluate potential astronomical observation sites in Peruvian territory through a series of observation missions. The program's aim is to gather data to validate the site for the future construction of a National Astronomical Observatory. The observations were made using an SBIG ST-7MX CCD camera and a BVR standard filter set, attached to a MEADE LX200 356mm telescope.

Keywords. site testing, techniques: photometric

1. Introduction

If there were not a terrestrial atmosphere nor interstellar dust between an external source of radiation (eg. a celestial body) and our telescope (on Earth), this radiation would reach the telescope and form a diffraction pattern called Airy disk, due only to the optical effects on the telescope (because of the light diffracting and producing a pattern with concentric dark and bright rings around the image of the object). However, this is not the case. In fact, radiation coming from a star traverses various obstacles, thus affecting its optical path in different ways. If we neglect the effects produced by interstellar material (which is a good approximation) the greatest obstacle between stellar radiation and our telescope is the terrestrial atmosphere. "Seeing" is the astronomical term for the extent of resolution degradation of an image caused by the Earth's atmospheric turbulence. This degradation in image quality results from fluctuations in the refractive index of air as a function of position and time above the site. Seeing is often the limiting factor in the quality of astronomical observations at a given site, and in Astronomy it is quantified using the stellar profile, measuring the full width at half maximum (FWHM) which is the angular size of the image of a star with half the peak intensity level. CONIDA conducted a search for the best potential astronomical sites in the Peruvian territory based on meteorological information from 40 years of data from the International Satellite Land-Surface Climatology Project (ISLSCP), the Surface meteorology and Solar Energy (SSE) database and continously monitoring of several meteorological stations. To accomplish this search, CONIDA created a program denominated JANAX. This program looked for sites over 4000 meters of altitude with the lowest cloud coverage (cf. Barrios 2007), humidity and wind velocity, preferably far from active volcanoes, the lowest geological risk and favourable logistic situations (i. e. roads, water and energy proximity). It was found that the Andes of Moquegua, in Southern Peru, had the best conditions within

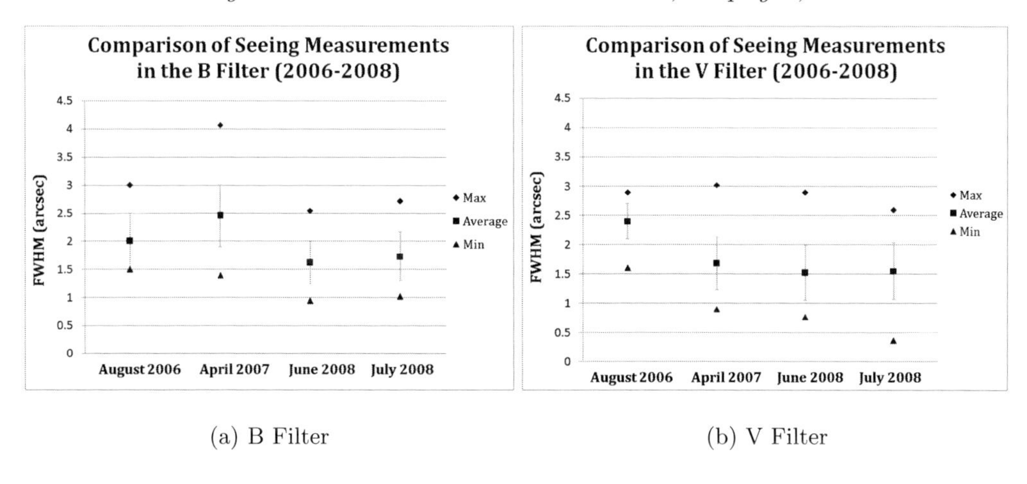

(a) B Filter (b) V Filter

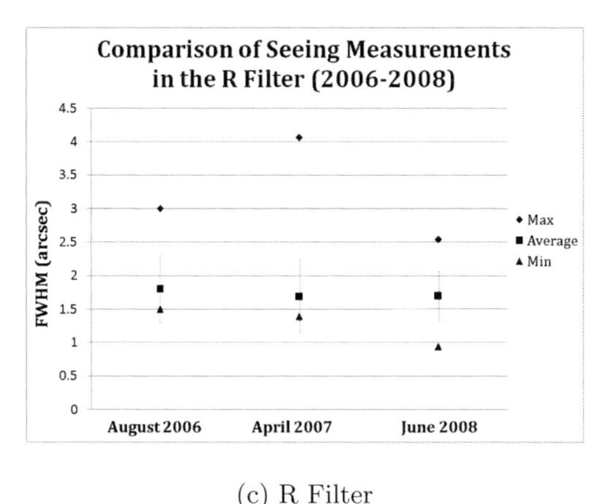

(c) R Filter

Figure 1. Comparison of seeing results for the JANAX missions from 2006 through 2008. Each image shows the average (squares), maximum (diamonds) and minimum (triangles) values.

the country (Ferrero *et al.* 2005; 2006) and, after a visual sites evaluation, measurements of seeing and extinction coefficients were started in Sasahuine mountain.

2. Instruments and Observations

Based on this initial search and as part of the JANAX program, CONIDA decided to send astronomical observation missions to the Sasahuine mountain site, in order to evaluate the quality of the sky there. JANAX, spanned a period of 4 years from 2004 through 2008, was carried out in ten observation missions. These missions were achieved during different times of the year. Digital images were taken for either a single star or a star open cluster during one or two nights of observation per mission. Seeing measurements were made during new moon nights every few minutes during 2 to 3 hours per night, providing that cloud coverage allowed imaging.

An ST7-MX SBIG CCD camera was used, with a BVR standard filter set. This system was attached to a Meade 356 mm/ 14 inch LX200 telescope with Ritchey-Chretien optical design.

3. Data Processing

The processing and reduction of the images was done using the IRAF† (Image Reduction and Analysis Facility) environment in both Unix and Linux operating systems (Coulman *et al.* 1986). This process is necessary because the images need to be corrected for instrumental effects that contaminate the data. The first steps were the mandatory image corrections from bias, dark frames and flat fields. Next, aperture photometry and transformation to the standard magnitude system of Johnson & Morgan were performed (Rolhfs 1949) on each of the stars imaged (single stars such as Achernar (α Eri) or open clusters such as the Jewel box cluster (NGC 4755) and the Butterfly cluster (M6)), using the IRAF `phot` task. The output from this task provided us with instrumental magnitudes and FWHM values for each star, in each image, of each filter. Finally, the airmass was calculated, using the IRAF `astcalc` task. In order to obtain the final seeing measurements in seconds of arc, the airmass for each star in each image in each filter was needed. To find it, the IRAF `astcalc` task was used, with a script whose job was to read an input table containing the necessary parameters (image name, star coordinates, date and time of observation and exposure time), and calculate the airmass based on these parameters. Once the airmass was calculated, the final seeing value for each star was calculated according to the usual formula, which accounts for airmass correction and pixel scale:

$$SEEING = 0.49 \times [FWHM/(airmass^{0.6})] \tag{3.1}$$

Where FWHM is the full width at half maximum value obtained from the `radprof` task, and 0.49 arcsec/pix is the pixel scale measured from the same images.

4. Seeing Results and Conclusions

Figure 1 shows the results of the average seeing values obtained for the visual, blue, and red filters at the Sasahuine mountain astronomical site. We compare the results of the different JANAX missions.

From the first mission in August of 2006 until the last one in July of 2008, we have observed a quantitative improvement in the results, probably due to the placement of a structure to keep the telescope's tracking system warmer so that it could work better. We did this because the extreme cold weather made the tracking system work with difficulty. For the V filter, the highest average seeing value was 2.4 arcsec, corresponding to JANAX 2006, and the minimum value obtained was 0.4 arcsec, corresponding to JANAX July 2008. Even though the highest average value is 2.4 arcsec, a simple comparison of these results with other seeing analysis, like the one done for the Huayao Observatory in Huancayo (Peru) by Pereyra and Baella (2003) demonstrate that Sasahuine mountain has enviable seeing. The average seeing value for all filters is 1.8 arcsec, which is a very good result, considering the type of equipment used. We would expect this result to improve with the use of better equipment. If more observations and seeing calculations confirm the values obtained in the present work, we will be provided with an excellent argument to obtain funding for the construction of a professional astronomical observatory at this site.

† IRAF is distributed by the National Optical Astronomy Observatories, which are operated by the Association of Universities for Research in Astronomy, Inc., under cooperative agreement with the National Science Foundation.

Acknowledgements

The authors thank the entire staff of the Astrophysics department at CONIDA for their support and guidance throughout the whole image reduction process. We also thank the National Meteorology and Hydrology Service for their support in the logistics of the observation missions.

References

Barrios, E., *et al.* 2007, Análisis de Cobertura Nubosa en el Sur del Territorio Peruano Utilizando Imágenes GOES-12 en las Bandas 03 y 04, CONIDA, internal report.

Coulman, C. E., *et al.* 1986, The Observation, Calculation, and Possible Forecasting of Astronomical Seeing. Publications of the Astronomical Society of the Pacific 98, 376-387.

Ferrero, G. *et al.* 2005, Búsqueda de Sitios Astronómicos en el Perú: Resultados de la Expedición Janax I, ECIPERU 2005, ISSN: 1813-0194, vol. 2, n. 1, 17, available at `http://www.cienciaperu.org/images/revista/vol2n1.swf`

Ferrero, G. *et al.* 2006, Análisis de Variables Determinantes en la Búsqueda de Sitios Astronómicos en el Perú, ECIPERU 2006, ISSN: 1813-0194, vol. 3, n. 1, 25, available at `http://www.encuentrocientificointernacional.org/revista/Vol3N1.pdf`

NASA, Goddard Space Flight Center, International Satellite Land-Surface Climatology Project (ISLSCP) available at `http://www.gewex.org/islscp.html`

NASA, Surface Meteorology and Solar Energy available at `http://eosweb.larc.nasa.gov/sse/`

Pereyra, A. & Baella, N. 2003, Medidas de seeing en el Observatorio de Huancayo, REVCIUNI, Vol 7, n. 1, 103.

Rohlfs, K. *et al.* 1949, Lichtelektrische Dreifarben-Photometrie von NGC 6405 (M6), Zeitschrift für Astrophysik, Bd. 47, S. 15-23.

Comparative Magnetic Minima:
Characterizing quiet times in the Sun and stars
Proceedings IAU Symposium No. 286, 2011
C. H. Mandrini & D. F. Webb, eds.

doi:10.1017/S1743921312005261

Creating a sunspot database at the Solar Observatory of Ica National University in Perú

Lurdes Martínez-Meneses

Facultad de Ciencias, Universidad Nacional San Luis Gonzaga de Ica, Perú
email: `lurdesmartinez5@yahoo.es`

Abstract. We describe the database and the method used to analyze the sunspot data recorded at the Solar Observatory of the University of Ica in Peru. The parameters that are measured include the relative sunspot number (R), the sunspot area, their positions on the disk, and an estimate of the constant (k) included in R. Sunspots in the database are classified following the Zurich Classification System. From these observations, the active region area, the sunspot rotation speed, and other active regions properties can be estimated.

Keywords. instrumentation: miscellaneous, methods: data analysis

1. Introduction

The Sun's visible surface is called the photosphere. To observe this first layer of the solar atmosphere, the integrated white light is used. Sunspots are the most relevant features observed at the photospheric level. They are the locations of strong magnetic fields and their lower temperature, when compared to the surrounding photosphere, makes them be seen as dark regions. Their darker center is known as the umbra which is surrounded by the penumbra, with an average diameter 2.5 times that of the umbra. The measured magnetic field in the umbra can be of 2000 to 3000 Gauss, in average, and can reach values larger than 4000 Gauss. The effective temperature of the sunspot umbra is around 3700 K compared to the temperature of the penumbra which is approximately 5600K (Bhatnagar & Livingston 2005, Bhatnagar & Ulmschneider 2010)

The sunspot number is considered a simple measure of solar activity as the solar magnetic field, which is concentrated in sunspots, is the root cause of all solar active phenomena. However, it is not the only activity index. Nowadays, solar activity is measured by several parameters but the oldest, easiest, and best measured is the sunspot number. Houtgast & van Sluiters (1948) have empirically shown that the maximum magnetic flux density B_n in Gauss, at the disk center, is related to the area of the spot, A_i, by the equation $B_n = 3700A_i/(A_i + 60)$ where the sunspot area is measured in millionths of the visible hemisphere of the Sun and can be obtained by projection. This equation applies to stable spots and not to spots which are in developmental phases.

2. Instrument and method

Sunspot data are obtained using a 15 cm Takahashi refractor telescope with a 1050 mm focal length. The database includes observations from June 2003 to January 2006. The Sun's image is projected using an eyepiece which produces a solar disk image with a 15 cm diameter. To sunspots are classified using the Zurich Classification System and the sunspot group evolution is registered.

The relative sunspot number is estimated as defined by Rudolf Wolf in 1848, $R = k(10f + g)$, where k is the observer's correction factor, f is the total number of spots, and g is the total number of sunspot groups. Using this equation we compare or measurements with the values obtained by the other observatories and world data centers and we can estimate the quality of our observations.

In each daily observation, the total area of sunspots is measured and a monthly average is calculated. The area of a sunspot group is calculated as $A_M = A_F N \cos\rho$, where A_M is the area in millionths of the visible hemisphere of the Sun, ρ is the angular distance on the surface of the Sun from the center of the disk to the sunspot, A_F is the area factor that depends on the diameter of the disk and grid size, and N is the number of grid points covering the penumbra and umbra of all spots in a group.

3. Dicussion

The sunspot number is considered as a simple measure of solar activity; however, it is not the only index that can be used but it is the oldest and best measured. We have described the measurements that are obtained at the Solar Observatory of the University San Luis Gonzaga de Ica in Perú. The sunspot number values using the projection method have a 70% reliability. The area, birth, and evolution of sunspot groups (using the Zurich Classification System) is also registered as a contribution to the knowledge of solar activity.

Acknowledgements

My special thanks to Dr. Mutsumi Ishitsuka Komaki, who was the director of Ancon Observatory of the Geophysical Institute of Perú, and Master of Science Trigoso Hugo Avilés for their support and confidence, and to the staff of the Solar Observatory of the University of Ica for the data.

References

Bhatnagar, A. & Ulmschneider, P. 2010, *Lectures on Solar Physics*
Bhatnagar, A. & Livingstone, W. 2005, *Fundamentals of Solar Astronomy*
Houtgast, J. & van Sluiters, A. 2007, *B.A.N.*, 10, 325

*Comparative Magnetic Minima:
Characterizing quiet times in the Sun and stars
Proceedings IAU Symposium No. 286, 2011
C. H. Mandrini & D. F. Webb, eds.*

doi:10.1017/S1743921312005273

A solar station in Ica - Mutsumi Ishitsuka: a research center to improve education at the university and schools

Raúl Terrazas-Ramos

Facultad de Ciencias, Universidad Nacional San Luis Gonzaga de Ica, Perú
email: raulterrazas81@gmail.com

Abstract. The San Luis Gonzaga National University of Ica has built a solar station, in collaboration with the Geophysical Institute of Peru, the National Astronomical Observatory of Japan and the Hida Observatory. The Solar Station has the following equipment: a digital Spectrograph Solar Refractor Telescope Takahashi 15 cm aperture, 60 cm reflector telescope aperture, a magnetometer-MAGDAS / CPNM and a Burst Monitor Telescope Solar-FMT (Project CHAIN). These teams support the development of astronomical science and Ica in Peru, likewise contributing to science worldwide. The development of basic science will be guaranteed when university students, professors and researchers work together. The Solar Station will be useful for studying the different levels of university education and also for the general public. The Solar Station will be a good way to spread science in the region through public disclosure.

Keywords. instrumentation: miscellaneous, methods: data analysis

---

1. Introduction

In 1998, the Faculty of Science at the University of San Luis Gonzaga Ica (UNICA) sought the help of physics lab equipment, the Ancon Observatory Geophysical Institute of Peru (IGP, Itshisuka 2006, Ishitsuka *et al.* 2007). From this date a joint activity was begun between the UNICA and the IGP. In 2002 a Takahashi telescope with electronic drive and an equatorial mount 150 mm aperture and focal length of 1050 mm was installed on the roof of the Faculty of Sciences. Images from the Sun by the projection method have been obtained. A group of students from the Faculty of Sciences was trained to obtain the relative sunspot number, and the first results of the observations were published in a bulletin by the IGP in 2004. In 2003, the UNICA bought a digital camera, Nikon Coolpix 5000 series in order to photograph the solar photosphere. In 2004 the IGP signed a cooperation agreement with the National Astronomical Observatory of Japan (NAOJ), achieving mutual cooperation in the fields of solar physics, astronomy and radio astronomy. Then at the end of 2005, the UNICA decided to give 4.2 acres within the campus of the University for the construction of the Solar Station. In March 2010 Ica Solar Station hosted the following teams: a solar spectrograph, a magnetometer, a 60 cm reflecting telescope and the FMT whose main objective is to establish a network of telescopes, with similar characteristics, installed around the world to monitor the Sun (Project CHAIN-International Network for Continuous Observation of the Sun in H-alpha Images).

2. Solar station

Owing to the good quality of the sky, and the number of hours of sunshine that characterize the city of Ica, it was decided to build the Solar Station. The Solar Station is located within the campus in the city of Ica. This city is located south of Lima. This station is the first observatory of its kind in Perú. The Solar Station construction was funded by UNICA. The Geophysical Institute of Perú, the National Astronomical Observatory of Japan and the Hida Observatory of Kyoto University provide the equipment to be installed at the Solar Station.

Location:
Latitude: 140° 05′ 20.55″ S
Longitude: 75° 44′ 11.28″ W
Altitude: 406 m
300 km south of Lima

Weather: Annual precipitation: 3 mm in average
Average temperature: 23 °C
Average humidity: 20%
Never snows

Equipment:
Digital spectroheliograph
FMT, Flare Monitor Telescope
Reflector Telescope 60 cm
Refractor Telescope Takahashi 15 cm (located on the roof of the Faculty of Science at the University of San Luis Gonzaga Ica)
Magnetometer MAGDAS

3. Research at the Solar Station

Digital spectroheliograph: this instrument can measure the properties of the emission lines and several absorption lines that occur in the solar atmosphere and thus is able to determine the equation of state characterizing the plasma which is part of the chromosphere and solar photosphere. It uses a coelostat to collect sunlight throughout the day.

Study of solar activity through the data captured and analyzed using mathematical models of the relative sunspot number. Currently records of sunspots by the projection method are being made with the Takahashi refractor telescope of 15 cm of aperture on the roof of the Faculty of Sciences. These records are made each day and are sent to the Ancon Observatory where they are processed to determine the Wolf number for each of the observers, and verify the reliability of the sunspot records. The Wolf number (R) of each observer's data are compared with those obtained in the Solar Influence Data Analysis Center (SIDC), Brussels. It also is conducting analysis and creating a database of sunspots, which will allow us to find a systematic methodology for the evolution of sunspot active regions, and also look for a technique to evaluate the consistency of the data quality of sunspots and calculate the area of active regions with sunspots.

FMT - Telescope Flares monitored (solar flares): The FMT is a telescope to observe the Sun in H-alpha and it is part of the international project CHAIN (Continuous H-alpha Imaging Network, Ueno 2007) to continuously monitor the Sun in Earth-based observatories. The FMT consists of 5 solar telescopes that simultaneously observe the entire solar disk at different wavelengths around the H-alpha line. Furthermore, the FMT can measure the velocity field in three-dimensional structures that move over the entire solar disk. The scientific objective of the FMT is to monitor solar flares and eruptive filaments continuously over the whole solar disk and investigate the correlation between the characteristics of these eruptive phenomena and the corresponding efficiency of Coronal Mass Ejections (CMEs).

4. Conclusions

It is important to carry out this joint project between UNICA, Hida Observatory, NAOJ and the IGP, and contribute to the development of solar physics in Ica and Perú.

The development of basic science will be guaranteed when university students, professors and researchers work together.

The Solar Station of Ica is very useful for the study at different levels of university education and also of interest to the general public.

Acknowledgements

Special thanks to Dr. Mutsumi Ishitsuka who was Director of the Ancon-IGP, who initiated this unique and important project to develop the Sslar physics in our region. Thanks to the authorities of the Universidad Nacional San Luis Gonzaga de Ica for promoting the development of the Solar Station of Ica.

References

Ishitsuka, J. K. *et al.* 2007, *Bulletin of the Astronomical Society of India*, 35. 709

Ishitsuka, J. K. 2006, *Proceedings of the* 2nd *UN / NASA Workshop on International Heliophysical Year and Basic Space Science*, 76

Ueno, S. 2007, *Bulletin of the Astronomical Society of India*, 35, 6

Author Index

Subject Index